PERIODIC TABLE OF THE ELEMENTS

	IA	IIA	IIIB	IVB	VB	VIB	VIIB	VIIIB	VIIIB	VIIIB	IB	IIB	IIIA	IVA	VA	VIA	VIIA	0
	s		d										p					
1	1 H 1.00797																	2 He 4.0026
2	3 Li 6.939	4 Be 9.0122											5 B 10.811	6 C 12.01115	7 N 14.0067	8 O 15.9994	9 F 18.9984	10 Ne 20.183
3	11 Na 22.9898	12 Mg 24.312											13 Al 26.9815	14 Si 28.086	15 P 30.9738	16 S 32.064	17 Cl 35.453	18 Ar 39.948
4	19 K 39.102	20 Ca 40.08	21 Sc 44.956	22 Ti 47.90	23 V 50.942	24 Cr 51.996	25 Mn 54.9380	26 Fe 55.847	27 Co 58.9332	28 Ni 58.71	29 Cu 63.54	30 Zn 65.37	31 Ga 69.72	32 Ge 72.59	33 As 74.9216	34 Se 78.96	35 Br 79.909	36 Kr 83.80
5	37 Rb 85.47	38 Sr 87.62	39 Y 88.905	40 Zr 91.22	41 Nb 92.906	42 Mo 95.94	43 Tc (99)	44 Ru 101.07	45 Rh 102.905	46 Pd 106.4	47 Ag 107.870	48 Cd 112.40	49 In 114.82	50 Sn 118.69	51 Sb 121.75	52 Te 127.60	53 I 126.9044	54 Xe 131.30
6	55 Cs 132.905	56 Ba 137.34	57 La 138.91	72 Hf 178.49	73 Ta 180.948	74 W 183.85	75 Re 186.2	76 Os 190.2	77 Ir 192.2	78 Pt 195.09	79 Au 196.967	80 Hg 200.59	81 Tl 204.37	82 Pb 207.19	83 Bi 208.980	84 Po (210)	85 At (210)	86 Rn (222)
7	87 Fr (223)	88 Ra (226)	89 Ac (227)	104 Ku (260)														

f

58 Ce 140.12	59 Pr 140.907	60 Nd 144.24	61 Pm (147)	62 Sm 150.35	63 Eu 151.96	64 Gd 157.25	65 Tb 158.924	66 Dy 162.50	67 Ho 164.930	68 Er 167.26	69 Tm 168.934	70 Yb 173.04	71 Lu 174.97
90 Th 238.03	91 Pa (231)	92 U 238.03	93 Np (237)	94 Pu (242)	95 Am (243)	96 Cm (247)	97 Bk (249)	98 Cf (251)	99 Es (254)	100 Fm (253)	101 Md (256)	102 No (254)	103 Lw (257)

Atomic weight values listed in parentheses are approximate.

GENERAL CHEMISTRY
Revised First Edition

GENERAL CHEMISTRY
Revised First Edition

Interaction of Matter, Energy, and Man

FREDERICK R. LONGO
Professor of Chemistry
Drexel University

McGRAW-HILL BOOK COMPANY

New York St. Louis San Francisco Düsseldorf Johannesburg Kuala Lumpur London Mexico Montreal New Delhi Panama Paris São Paulo Singapore Sydney Tokyo Toronto

GENERAL CHEMISTRY
Revised First Edition
Interaction of Matter, Energy, and Man

34567890MURM798765

Library of Congress Cataloging in Publication Data

Longo, Frederick.
General chemistry.

1. Chemistry. I. Title.
QD31.2.L66 540 73-13814
ISBN 0-07-038685-4

This book was set in Bodoni Book by Progressive Typographers.
The editors were Robert H. Summersgill, Nancy L. Marcus, and Carol First;
the designer was Betty Binns;
and the production supervisor was Sam Ratkewitch.
The drawings were done by Vantage Art, Inc.
The printer was The Murray Printing Company.

To Fred D. Longo, my father,
from whom I inherited my impatience.

To Rose Marie Carioti Longo, my mother,
from whom I inherited my restraint.

To George Stanley Sasin, Professor of Chemistry,
from whom I learned of the excitement of chemical discovery.

CONTENTS

TO THE STUDENT

Most if not all of you who are beginning your study of college science have heard words like *molecule, electron, atom;* in fact, you were probably introduced to these concepts when you were in elementary school. I suppose that there is some advantage in early exposure to modern science but I believe that it is very detrimental in at least two crucial ways: (1) As children you were not very critical of the information that was presented by your teachers and, therefore, you accepted everything you heard. This is absolutely foreign to the spirit of modern science which demands experimental proof to support theories and hypotheses; and (2) by instilling the concepts of science at this early age your chance of looking at something in a truly original, objective manner is greatly diminished. You take on a way of looking at the physical world, a way which is colored by the concepts which were uncritically accepted when you were very young. Just imagine what you would say if you were just introduced to the concepts of electron, atom, and molecule for the first time. Would you accept these crazy ideas without at least saying: How do you know these invisible things really exist?

Well, how *do* we know they exist? The answer is that we really don't know, but, if we assume that these submicroscopic particles do exist, we can explain and predict a lot of natural phenomena fairly accurately.

In this text I have described many of the important experiments which have led to giant scientific, technological, philosophical, and sociological advances. I have tried to show how modern chemists use chemical theory to explain, predict, and invent. You will see that they are quite successful, but please remember their theory is not perfect and they often fail in their research work. Obviously, our present theory, as powerful as it is, needs modification. Perhaps, a radically new theory is needed and will supplant our modern theory, just as the atomic theory supplanted the phlogiston theory.

FREDERICK R. LONGO

TO THE TEACHER

The text has grown from lectures designed for an audience including science, engineering, and business administration students. It would be suitable as an introductory text for students majoring in natural science, premedical science, mathematics, engineering, business administration, and liberal arts. The students should be adept at algebra, but the calculus is not required.

Throughout the text I have emphasized what I believe to be the unique tasks of the chemist: to explain the structure of molecules, to show the relationship between molecular structure and molecular interactions, on the one hand, and the observable properties of matter, on the other; and to build the bridge between the hypothetical microscopic world of atoms and molecules and the macroscopic world which we can see and smell and feel. Frequently, I have used a historical approach but only when I thought I could do so without obscuring the scientific principle involved.

The chapters in this text are arranged in packages. Chapters 1 and 3 through 5 deal with the development of the concept of the atom and of bonding theory. I have tried to emphasize the dependence of bonding theory on atomic theory.

Chapter 2, Basic Concepts, covers chemical nomenclature, the metric system, the unit method of calculation, balancing chemical equations, and stoichiometry. Most students who have had high school chemistry have covered the material in Chap. 2, but Sec. 2-12, Stoichiometry, should be studied carefully since the average student encounters a great deal of difficulty with this topic.

Chapters 6, 7, and 8 deal with the gaseous, solid, and liquid states; Chap. 9, Solutions, may be treated with this group. It introduces the concept of the concentration of solutions and colligative properties.

The laws of thermodynamics are introduced in Chap. 10 and applied in Chaps. 11, 12, and 13. The method used to introduce the internal energy concepts in Chap. 10 is unusual and, hopefully, its importance in chemistry will be more easily understood.

Chapters 15 and 16 deal with two closely related areas, organic chemistry and biochemistry. Chapter 21, Environmental Chemistry, is based on an understanding of the material in these two chapters.

It is not necessary to follow the order of chapters in the text. For instance, one alternative that I would consider myself would be to move Chap. 20, Nuclear Chemistry, into the early part of the course. It can be given after Chap. 3, Atomic Theory, provided that the first order rate equation (considered in Chap. 17) is introduced. After having completed the first eleven chapters almost any order is possible provided that Chap. 15, Organic Chemistry, precedes both Chap. 16, The Chemistry of the Living Cell, and Chap. 21, The Environment.

I feel that the topics in Chaps. 1–10, 11–13, 15, and 17 make up the minimum coverage for an introductory course for engineering and science students. Topics omitted from this list include Biochemistry (Chap. 16), Descriptive and Theoretical Inorganic Chemistry (Chaps. 18, 19), Nuclear Chemistry (Chap. 20), and Environmental Chemistry (Chap. 21). With this minimum coverage the students would understand the basic chemical principles and could read and learn the omitted material.

If I were to teach a chemistry course for liberal arts and business administration majors I would choose Chaps. 1, 3–5, 15, 16, 19, 20, 21. Chapters 1, 3–5, and 19 are concerned with atomic theory and bonding theory, the very heart of chemistry; in addition the material in these chapters is presented within a historical matrix that might be more appealing to nonmajors. Chapters 15, 16, 20, and 21 deal with the biochemical, industrial, and environmental aspects of chemistry with which all educated persons should be familiar.

ACKNOWLEDGMENTS

I began work on this text in 1967 with Dr. John A. O'Malley who at that time was also a member of the Drexel Faculty. When he left to direct a commercial laboratory he was not able to continue with the writing. For his early contributions and stimulating discussions I am very much in his debt. The McGraw-Hill Book Company submitted parts of the manuscript to several reviewers, all of whom remain anonymous to me. Many of these individuals offered good suggestions which I have incorporated. Several of my colleagues at Drexel University have offered constructive criticism to concepts in this text. I am particularly indebted to Drs. Raymond A. Mackay and Erle B. Ayres for reading and criticizing the chapters on atomic theory and valence theory. My own students Drs. Daniel J. Quimby and John J. Leonard read much of the manuscript and checked many of the problems. Dorothy Darrup Longo typed and proofread most of the manuscript. To all these people and to Jim Smith, Nancy Marcus, Bill Orr, and Carol First of McGraw-Hill I am very grateful. But most of all I feel that I have been fortunate to have had the opportunity to teach undergraduates at Drexel University. They have forced me to examine chemical principles more carefully and to reassess the methods by which I teach them.

FREDERICK R. LONGO

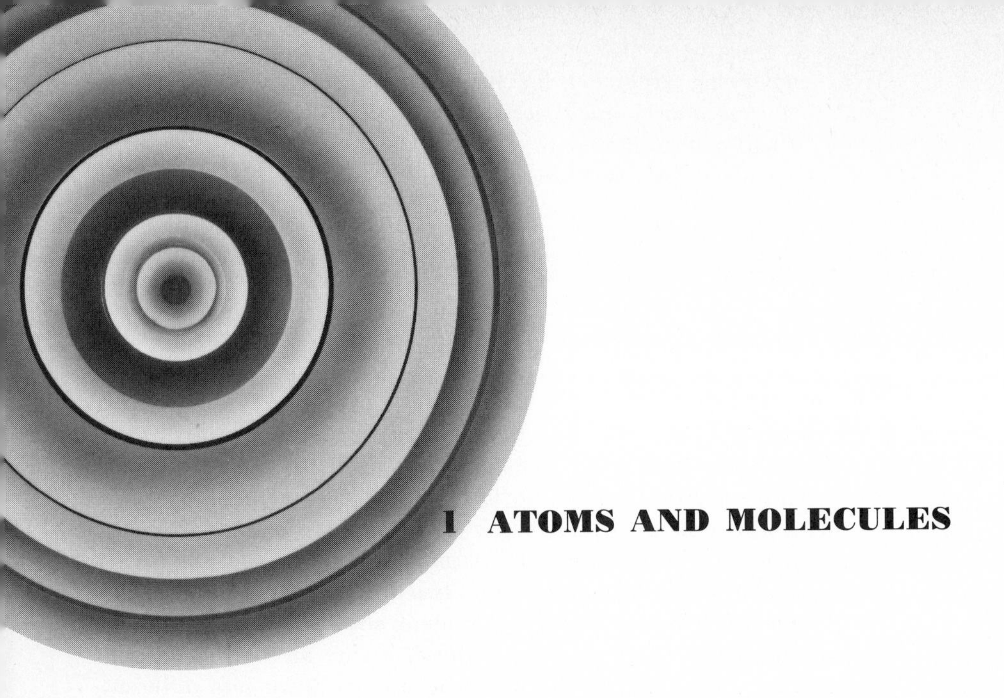

1 ATOMS AND MOLECULES

1.1 THE TASK OF THE CHEMIST

Chemists investigate matter and its properties using a multitude of different techniques and apparatus. Engineers and other scientists are also interested in the properties of matter. What distinguishes the chemist? The answer is his goals, not his techniques. The chemist reviews his observations, results, and measurements in order to find implications concerning the invisible microstructure of matter, the world of atoms, molecules, and electrons. His job is to relate the properties of bulk matter to the behavior of these invisible particles. Let us examine briefly how his strong faith in the existence of atoms and molecules developed.

No one doubts the existence of atoms and molecules today, but is it so obvious that matter is atomic? The atomic theory of the ancient Greeks Leucippus and Democritus was never accepted by Aristotle. It was abandoned and lay dormant for 2300 years, finally being reconsidered by a few scientists, including Newton, in the eighteenth century. In the early nineteenth century Dalton presented a very convincing argument for the existence of atoms, but even Dalton's version of atomic theory was not generally accepted during the first half of the nineteenth century. The atomic nature of matter has never been obvious to the observer.

1.2 THE EXPERIMENTAL BASIS OF THE ATOMIC THEORY

The historical development of the modern atomic theory rests on many observations and measurements, the most important being the decomposition experiments as interpreted by Boyle, the weighing experiments of Lavoisier, and the analytical experiments of Proust.

Robert Boyle (1627–1691) insisted on testing explanations with observations and experiments. Boyle could not accept the old, vague theories which claimed that matter was composed of four *elements* (air, earth, fire, and water). He regarded only those substances which could not be decomposed into two or more different (simpler) substances as elements, and experiment showed that there were many more than three or four of these. Today an *element* is still defined as a material which cannot be broken down into different substances, while a compound of elements, or more simply a *compound*, is composed of more than one element and can undergo decomposition.

A compound differs from a *mixture* of elements. In a mixture, the elements or components retain their individual properties. A compound has properties which differ from those of its elements. A *mixture of nitrogen and oxygen*, air, supports the human respiratory system, thanks to the presence of elemental oxygen in the mixture. One *compound of nitrogen and oxygen*, NO_2 (nitrogen dioxide), is a poisonous gas and causes death if inhaled.

The terms element and compound do not necessarily imply the existence of atoms or molecules because they are terms which describe the *macrostructure* of matter. A lump of pure iron, which we can see and feel, is an element; it cannot be decomposed into simpler substances by ordinary chemical means. A grain of table salt, which we can see, and taste, and decompose into sodium and chlorine, is a compound. The distinction between elements and compounds is based on experimentally observable facts and does not depend on the existence of atoms.

The *phlogiston theory* was one of the first important theories of chemistry. Although it developed into a general theory of chemical reactions, it was originally advanced as an explanation of burning. Substances which burned were believed to give up a fluid called phlogiston and leave behind an ash, which was the original substance minus the phlogiston. This theory leads us to hypothesize that we could regain the original "phlogisticated substance" if we could combine the resulting ash and more phlogiston. For instance, zinc is converted into "ash" when it is heated in air. In terms of the phlogiston theory the zinc gives up its phlogiston:

$$\text{Zinc} \xrightarrow{\text{heat}} \text{zinc ash} + \text{phlogiston}$$

If we now treat the zinc ash with a phlogiston-rich substance, we should be able to recover the zinc. Charcoal is a phlogiston-rich material since it leaves very little ash when it burns. If zinc ash is treated with charcoal, zinc should be regenerated:

$$\text{Zinc ash} + \text{charcoal} \xrightarrow{\text{heat}} \text{zinc}$$

The phlogiston theory predicts the correct result; zinc is obtained when zinc ash is heated in the presence of charcoal. The results of many other experiments were explained and predictable in terms of the phlogiston theory, and it was accepted by most chemists of the eighteenth century.

All good scientific theories lend themselves to predictions which, if correct, strengthen the theory. If the prediction is false, the theory must be modified or discarded.

Antoine Lavoisier. (*Courtesy of the Edgar Fahs Smith Collection, Van Pelt Library, University of Pennsylvania, Philadelphia.*)

In the latter part of the eighteenth century Antoine Lavoisier (1743–1794) performed experiments with oxygen by which he was able to refute the phlogiston theory. Making constant use of the balance, Lavoisier found that sulfur and phosphorus gain weight when burned in the open and, very important, that no change in weight occurs when chemical reactions or other processes, such as distillation, are carried out in sealed vessels. Hence, his experiments implied that the burning process does not involve the *loss* of phlogiston or the loss of anything. In addition, his sealed-vessel experiments supported the principle that there is no net change in weight during a chemical reaction. This was an observation which eventually contributed to the important *law of conservation of mass.* Matter can neither be created nor destroyed although it may be changed in form.

At the beginning of the nineteenth century Joseph Louis Proust (1754–1826) and Claude Louis Berthollet (1748–1822) became involved in an 8-year dispute over the composition of compounds. Berthollet held that the weight fraction of an element in a compound is variable. It was not known then that two elements can form more than one compound. For instance, tin forms two common oxides, stannous oxide, which is 88 percent tin by weight, and stannic oxide, which is 79 percent tin by weight. Analyses of tin oxides, which unknown to Berthollet were mixtures of two distinct compounds, led him to believe that the percentage of tin could vary from 79 to 88 percent. Proust showed that there are two tin oxides and that each is composed of a definite weight fraction of each of its elements. The careful analytical work of Proust finally led all scientists to accept what is now called the *law of definite composition: the weight proportions of the elements in a pure compound are fixed.* This law was extremely important in the development of Dalton's atomic theory.

1.3 DALTON'S ATOMIC HYPOTHESIS

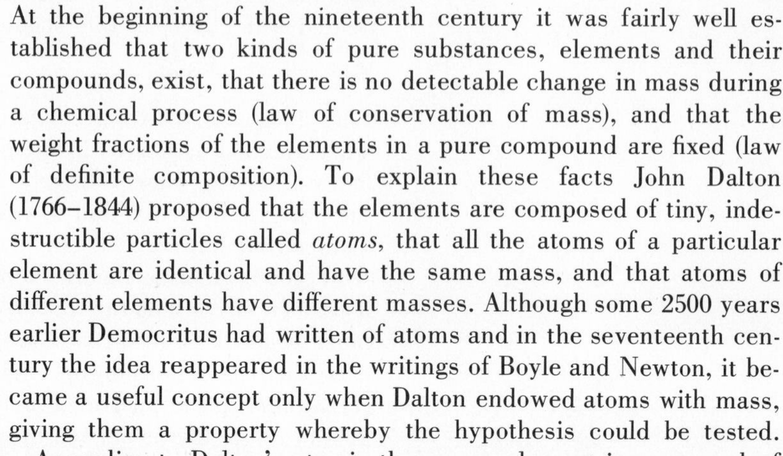

At the beginning of the nineteenth century it was fairly well established that two kinds of pure substances, elements and their compounds, exist, that there is no detectable change in mass during a chemical process (law of conservation of mass), and that the weight fractions of the elements in a pure compound are fixed (law of definite composition). To explain these facts John Dalton (1766–1844) proposed that the elements are composed of tiny, indestructible particles called *atoms,* that all the atoms of a particular element are identical and have the same mass, and that atoms of different elements have different masses. Although some 2500 years earlier Democritus had written of atoms and in the seventeenth century the idea reappeared in the writings of Boyle and Newton, it became a useful concept only when Dalton endowed atoms with mass, giving them a property whereby the hypothesis could be tested.

John Dalton. (*Courtesy of the Edgar Fahs Smith Collection, Van Pelt Library, University of Pennsylvania, Philadelphia.*)

According to Dalton's atomic theory, an element is composed of identical atoms; compounds represent definite combinations of different atoms. When elements react to form a compound, a combination of indestructible atoms (in modern terminology, a molecule) is formed; there is no change in mass; the sum of the mass of the reactants—atoms—is the same as the mass of the product, i.e., molecule. This observation is summarized in the law of conservation of mass. Furthermore, there is a definite ratio of atoms in a molecule, and since atoms of the same element have the same mass, the weight proportion of an element in a pure compound is fixed, in agreement with the law of definite composition.

Proust had discovered that the two elements can form more than one compound, but he failed to see the relationship between the weights of the two elements because he was not analyzing his data from the atomic point of view. He found that in the two tin oxides the percentage of tin is 88 percent in stannous oxide and 79 percent in stannic oxide. If we calculate the weight of tin per gram of oxygen in each compound, we obtain 7.44 grams (g) of tin per gram of oxygen in stannous oxide and 3.72 g of tin per gram of oxygen in stannic oxide. The ratio of the weights of tin to oxygen in the two compounds is 2:1. Dalton discovered that whenever two elements, say A and B, form more than one compound, the ratio of the weight of A per unit weight of B in the different compounds is always a simple whole number ratio. The atomic theory enabled Dalton to un-

TABLE 1.1: DEMONSTRATION OF THE LAW OF MULTIPLE PROPORTIONS

Compounds containing carbon and hydrogen	*Weight percent of carbon*	*Grams of carbon per gram of hydrogen*
Methane	75.0	3.00
Ethylene	85.8	6.00
Compounds containing nitrogen and oxygen	*Weight percent of nitrogen*	*Grams of nitrogen per gram of oxygen*
Nitrous oxide (laughing gas)	63.6	1.75
Nitric oxide	46.7	0.877

derstand this relationship, which we call the law of multiple proportions. Table 1.1 shows two other examples which Dalton himself used to demonstrate the law of multiple proportions. The weight of carbon per unit weight of hydrogen in methane to the weight in ethylene is 3.00:6.00, or 1:2. The weight of nitrogen per unit weight of oxygen in nitrous oxide to the weight of nitrogen per unit weight of oxygen in nitric oxide is 1.75:0.877, or 2:1.

Dalton was able to discover the law of multiple proportions because he was equipped with the atomic theory; given that the same elements can form more than one compound and that all the atoms of the same element have the same mass, the law of multiple proportions follows as a requirement.

Dalton went on to develop an atomic-weight scale which eventually proved to be in error because he assumed that particles of the common gaseous elements are composed of single atoms. For the reaction between hydrogen and oxygen he wrote

1 hydrogen atom + 1 oxygen atom → 1 water molecule
⊙ + ○ → ⊙○

And for the reaction between nitrogen and hydrogen he wrote

1 hydrogen atom + 1 nitrogen atom → 1 ammonia molecule
⊙ + ⦶ → ⊙⦶

Dalton's symbols for the three elements ⊙, ⦶, ○ have been replaced by H, N, and O.

Dalton next arbitrarily assigned a weight of 1 unit to the hydrogen atom and calculated the relative weights of the atoms of the other known elements. He was able to determine that the weight ratio of oxygen to hydrogen is 8:1 in water, and because in his formula for water, HO, there is one oxygen atom and one hydrogen atom, he assigned a relative weight of 8 units to the oxygen atom.

Using the same reasoning, he assigned 4.7 units to each nitrogen atom. In all, he calculated the weights of the atoms of the 20 elements known at that time. These weights disagreed with those proposed by another outstanding chemist, Berzelius, but the discrepancy was not understood until scientists accepted the idea that the molecules of the common gaseous elements consist of two atoms.

1.4 GAY-LUSSAC'S LAW OF COMBINING VOLUMES AND AVOGADRO'S HYPOTHESIS

Amedeo Avogadro. (Courtesy of the Edgar Fahs Smith Collection, Van Pelt Library, University of Pennsylvania, Philadelphia.)

In his investigation of the reactions of gases, Joseph Louis Gay-Lussac (1778–1850) discovered the law of combining volumes: *whenever two or more gases are involved in a chemical reaction, the ratios of the reacting volumes can be expressed as small whole numbers provided that the volumes* (vol) *are measured at the same temperature and pressure*. For example,

2 vol hydrogen (gas) + 1 vol oxygen (gas) $\rightarrow$ 2 vol water vapor (steam)

2 vol carbon (mon)oxide + 1 vol oxygen $\rightarrow$ 2 vol carbon (di)oxide

1 vol hydrogen + 1 vol chlorine $\rightarrow$ 1 vol muriatic acid gas

In order to explain Gay-Lussac's observations in terms of the existence of atoms Amedeo Avogadro (1776–1856) proposed in 1811 that (1) *equal volumes of gases at the same temperature and pressure contain the same number of molecules* and (2) *the molecules of the common gaseous elements are diatomic*.

Gay-Lussac's observation that 2 vol of hydrogen combine with 1 vol of oxygen to produce 2 vol of steam meant to Avogadro that two molecules of hydrogen combine with one molecule of oxygen to produce two molecules of steam. But by the law of definite composition, both molecules of steam contain oxygen atoms (in equal number); the original oxygen molecule must have been composed of more than one atom of oxygen (if atoms are indestructible). Avogadro assumed the oxygen molecule is composed of two oxygen atoms. (He could have chosen any even number, for example, O_2, O_4, O_6.) From the assumption that equal volumes of gases contain equal numbers of molecules he could write

2 hydrogen molecules + 1 oxygen molecule $\rightarrow$ 2 steam molecules

and, assuming hydrogen and oxygen molecules to be diatomic,

$2H_2 + O_2 \rightarrow$ 2 molecules of steam

Since all molecules of steam are identical and atoms are indestructible, the formula for the water molecule must be H_2O and the reaction between hydrogen and oxygen is

$$2H_2 + O_2 \rightarrow 2H_2O$$

From this formula for water, H_2O, and the fact that the weight ratio of hydrogen to oxygen in water is 1:8 Avogadro concluded that oxygen atoms are 16 times heavier than hydrogen atoms. (Dalton had argued that the mass of the oxygen atom is 8 times greater than that of the hydrogen atom since his formula for water was HO.)

The hypotheses that equal volumes of gases contain equal numbers of molecules and that molecules of the common elemental gases are diatomic is Avogadro's interpretation of Gay-Lussac's law of combining volumes in terms of Dalton's atomic theory.

Avogadro's hypothesis, which to Dalton seemed to be incompatible with the atomic theory, finally proved to be the idea which systematized chemistry and allowed it to become a quantitative science. Avogadro's interpretation was not accepted until 50 years after he proposed it. Chemists were then able to develop a consistent set of atomic weights, and the systematic study of the elements, which culminated in the development of the periodic table, became possible.

1.5 SUMMARY

The development of the concepts of atoms and molecules depended on the discovery of rather broad principles:

1. There are two kinds of substances in nature, elements and compounds (Boyle).

2. The law of conservation of mass: Matter can neither be created nor destroyed (Lavoisier).

3. The law of definite composition: The weight fractions of the elements of a pure compound are fixed (Proust).

4. The law of combining volumes: Whenever two or more gases are involved in a chemical reaction, the ratios of their volumes can be expressed as small whole numbers (Gay-Lussac).

These observations, which are summarized in the form of laws, may not be exciting in themselves, but in the hands of Dalton and Avogadro they opened up to view the atoms and molecules of which we now firmly believe matter to consist.

QUESTIONS AND PROBLEMS

1.1 Boyle gave a clear experimental definition of element; the distinction between element and compound was sharpened. Why was this important to the development of the modern atomic theory?

1.2 Distinguish clearly between (*a*) element and atom, (*b*) compound and molecule, and (*c*) atom and molecule. (*d*) Can a molecule be composed of a single atom? More than two atoms? Find examples of each in this book.

1.3 When the metal magnesium is heated in air or oxygen, it burns and forms a white powder. Chemists of the eighteenth century said that magnesium gave up its phlogiston, or had been "dephlogisticated." Devise a simple experiment to test the phlogiston theory in this case.

1.4 Using Table 1.1, calculate (*a*) the grams of hydrogen per gram of carbon in methane and in ethylene and (*b*) the grams of oxygen per gram of nitrogen in nitrous oxide and in nitric oxide. (*c*) Show that these results are consistent with the law of multiple proportions.

1.5 One volume of a gaseous compound called hydrazine forms from 1 vol of nitrogen and 2 vol of hydrogen. Hydrazine consists only of nitrogen and hydrogen and is 12.5 percent hydrogen by weight. (*a*) Write an equation which illustrates the application of Avogadro's hypothesis for this reaction. (*b*) Assuming that nitrogen gas is tetratomic (contains four atoms of nitrogen per molecule) and hydrogen gas is diatomic, how many atoms each of hydrogen and nitrogen are present in a single hydrazine molecule? (*c*) With the assumptions made in part (*b*) calculate the weight of a nitrogen atom relative to a hydrogen atom. (*d*) Considering only the information in this problem, is it possible for a nitrogen molecule to contain an odd number of atoms? Why? (*e*) It is now accepted that both hydrogen and nitrogen are diatomic. Calculate the weight of a nitrogen atom relative to a hydrogen atom on this basis.

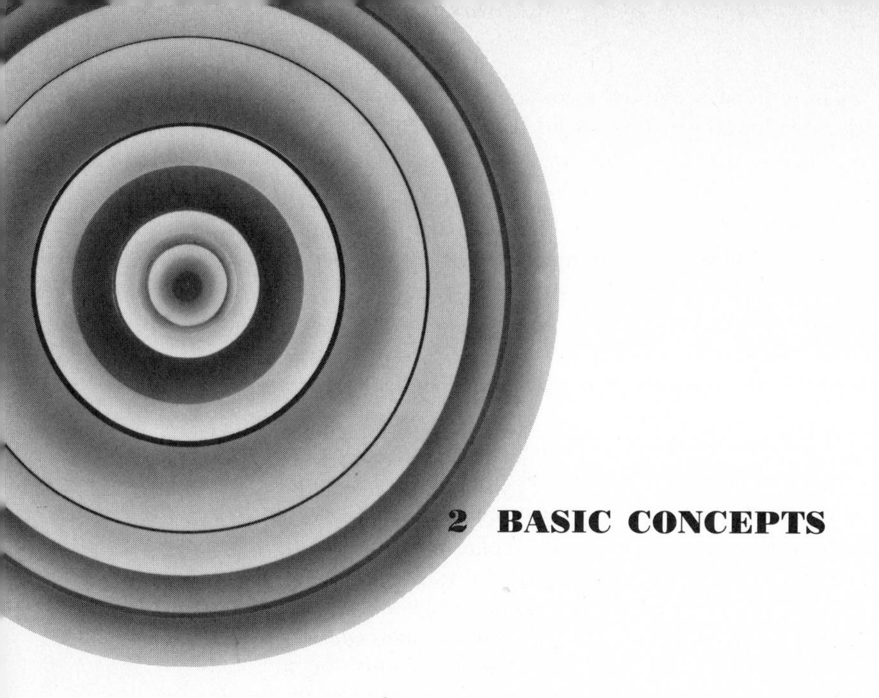

2 BASIC CONCEPTS

2.1 THE METRIC SYSTEM

Chemists have devised a system of nomenclature to help classify and identify all the different substances. In addition, scientists have found it convenient to use the same system of measurement so that a report published by a scientist in Russia, France, or the United States can readily be understood by scientists all over the world. The universally accepted scientific system is the metric system, which is also the legal system for all countries except the United States and Canada. American engineers use the English system, but American scientists have adopted the metric system. The main advantage of the metric system is that it has a decimal base, which simplifies the arithmetic.

The basic measures of any system are length, mass, and time; area, volume, and temperature are derived from the basic measures. Originally the measurements were based on objects readily available, but now more sophisticated and more accurate standards are used.

The unit of length, the meter (m), is actually defined in terms of the wavelength of orange-red light emitted by krypton-86 gas when an electric discharge is passed through it. The wavelength of this light is 1/1,650,763.73 m. More meaningfully, the meter is equivalent to 39.37 in.

The unit of mass is the kilogram (kg); a standard for this mass is a platinum-iridium cylinder at the International Bureau of Weights and Measures in Sèvres, France. A kilogram is equivalent to 2.2046 lb.

The unit of time, the ephemeris second (s), is familiar to all because it is used by scientists and nonscientists. The unit of time in both the metric and English systems, it is exactly defined as a very small fraction, 1/31,556,925.9747, of the tropical year for 1900.

All metric units of length can be defined in terms of the meter (m):

1 *kilo*meter (km) $= 1000$ m $= 10^3$ m
1 *centi*meter (cm) $= \frac{1}{100}$ m $= 10^{-2}$ m
1 *milli*meter (mm) $= \frac{1}{1000}$ m $= 10^{-3}$ m

$$1 \text{ angstrom } (\text{Å}) = \frac{1}{10{,}000{,}000{,}000} \text{ m} = 10^{-10} \text{ m} = 10^{-8} \text{ cm}$$

The meanings of the prefixes used in the metric system are presented in Appendix B, Table B.1.

The important metric-English conversion factors for length are

1 m $= 39.37$ in.
1 in. $= 2.54001$ cm

All metric units of mass can be defined in terms of the kilogram:

1 gram (g) $= \frac{1}{1000}$ kg $= 10^{-3}$ kg

$$1 \text{ milligram (mg)} = \tfrac{1}{1000} \text{ g} = \frac{1}{1{,}000{,}000} \text{ kg} = 10^{-3} \text{ g} = 10^{-6} \text{ kg}$$

One gram is the mass of about 20 drops of water delivered from an ordinary eyedropper. There are two important metric-English conversion factors for mass:

1 kg $= 2.2046$ lb
1 lb $= 453.59$ g

Area and volume are derived from length; area is length squared, and volume is length cubed. The most common units of area in the metric system are square centimeter (cm^2) and square meter (m^2). The important metric-English conversion factors for area are

$$(1 \text{ in.})^2 = (2.54001 \text{ cm})^2 \qquad \text{or} \qquad 1 \text{ in.}^2 = 6.452 \text{ cm}^2$$

and

$$(1\ \text{m})^2 = (39.37\ \text{in.})^2 \qquad \text{or} \qquad 1\ \text{m}^2 = 1550\ \text{in.}^2$$

Note that the relationships between metric and English units of area are derived from the relationships given for length.

Volume in the metric system is often expressed in cubic centimeters (cm^3) and cubic meters (m^3). Again, the metric-English conversion factors are derived from the relationships given for length:

$$(1\ \text{in.})^3 = (2.540\ \text{cm})^3 \qquad \text{or} \qquad 1\ \text{in.}^3 = 16.39\ \text{cm}^3$$
$$(1\ \text{m})^3 = (39.37\ \text{in.})^3 \qquad \text{or} \qquad 1\ \text{m}^3 = 61{,}030\ \text{in.}^3$$

There are special volume units in the metric system. The volume of 1 kg of water at 4°C is called a liter (l), and the volume of 1 g of H_2O at 4°C is called a milliliter (ml). These masses were chosen so that 1 l would be exactly equal to the volume of a cube 10 cm on a side, or 1000 cm^3. It has turned out that this is not exactly true, since 1 l is actually equal to 1000.028 cm^3, but for all except the most accurate work we can equate 1 l with 1000 cm^3 and 1 ml with 1 cm^3. An important metric-English conversion factor for volume is

$$1000\ \text{cm}^3 = 1\ \text{l} = 1.057\ \text{qt}$$

The scientific community uses the Celsius and Kelvin scales for the measurement of temperature (Fig. 2.1). On the Celsius scale the boiling point of water is 100°, and the freezing point of water is 0°. The absolute of zero, theoretically the lowest temperature attainable, is −273.16°C. The magnitude of a degree on the Kelvin scale is equal to the magnitude of a degree on the Celsius scale, but the

FIGURE 2.1

A comparison of temperature scales.

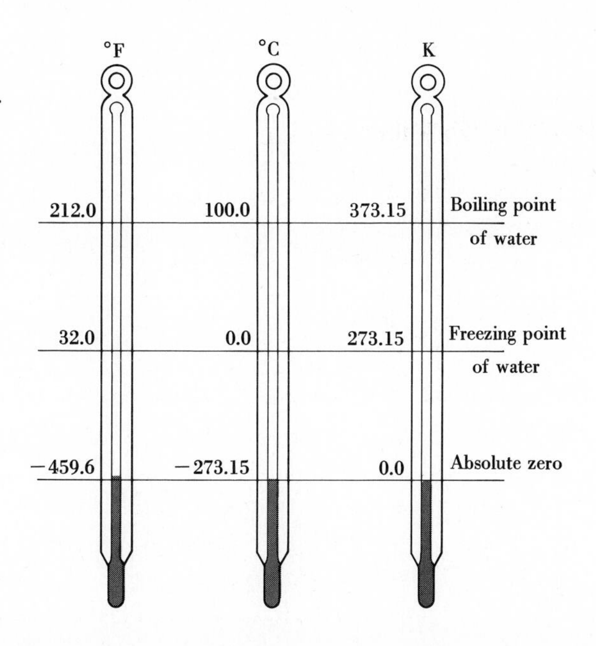

Kelvin scale is displaced by 273.16° so that the absolute of zero is 0 K, the freezing point of water is 273.16 K, and the boiling point is 373.16 K. The degree sign is no longer used in the Kelvin scale, and instead of saying "degrees Kelvin" or "Kelvin degrees" we say *kelvins*.

There are 180° between the freezing point and boiling point of water on the Fahrenheit scale, whereas on the Celsius scale there are 100°. Thus a Celsius degree is $\frac{180}{100}$ or $\frac{9}{5}$ greater than a Fahrenheit degree. Also, the freezing point of water is 32° on the Fahrenheit scale and 0° on the Celsius scale. Therefore, to convert degrees Celsius (°C) to degrees Fahrenheit (°F) multiply the Celsius temperature by $\frac{9}{5}$ and add 32 (see Fig. 2.1).

EXAMPLE 2.1 What is the equivalent of 90.0°F on the Celsius and Kelvin scales?

Solution: 90.0°F is 58.0°F above the freezing point of water (90.0 − 32.0). Since 1 Fahrenheit degree is only $\frac{5}{9}$ of 1 Celsius degree, 58°F represents only ($\frac{5}{9}$ × 58)°C, or 32.2°C.

To convert Fahrenheit temperature to Celsius (1) subtract 32°F from the Fahrenheit temperature (°F − 32) and (2) multiply the result by $\frac{5}{9}$; this is the temperature in Celsius degrees. A formula for this conversion is

$$°C = (°F - 32)\tfrac{5}{9}$$

To obtain the temperature on the Kelvin scale we simply add 273.16° to the temperature on the Celsius scale:

$$K = °C + 273.16 = 32.2° + 273.16 = 305.4$$

EXAMPLE 2.2 Convert 500 K into Celsius and Fahrenheit degrees.

Solution: The conversion from Kelvins to Celsius degrees is

$$°C = K - 273.16 = 500 - 273.16 = 227°$$

To convert 227°C to Fahrenheit degrees we refer to Fig. 2.1 again; 227°C represents the degrees on the Celsius scale *above* the freezing point of water. The freezing point of water on the Fahrenheit scale is 32°F. Therefore, the Fahrenheit equivalent of 227°C must be added to 32°F. Since 1 Celsius degree equals $\frac{9}{5}$ Fahrenheit degree, we multiply 227°C by $\frac{9}{5}$ to find the equivalent of 227°C in degrees Fahrenheit; $\frac{9}{5}$ × 227 is 408°F, which is the number of degrees Fahrenheit above the freezing point of water. To obtain the temperature in degrees Fahrenheit add 32°F to 408°F; 227°C = 408°F + 32°F = 440°F. The formula for the conversion from Celsius to Fahrenheit is

$$°F = \tfrac{9}{5}°C + 32 = \tfrac{9}{5} \times 227 + 32 = 440°$$

Conversions are a frequent necessity in any occupation today, from housewife to engineer. Conversions require a methodical way of thinking that is developed only through practice. It is useful for the smart shopper or the successful scientist or engineer to be able to convert numbers, such as dollars per gram to dollars per pound, and to do it quickly.

2.2 THE UNIT METHOD IN CALCULATIONS

In the physical sciences we almost never deal with pure numbers. All our measurements are associated with units, which, like numbers, obey the laws of algebra.

An experimental quantity, like 5.1 g, can be considered to be a product of 5.1 and g, and this kind of product will obey the laws of algebra. For example,

$$5.1 \text{ g} \times \frac{2 \text{ ml}}{1 \text{ g}} = 10.2 \text{ ml}$$

$$5.1 \text{ g} \times 7.4 \text{ g} = 37.7 \text{ g}^2$$

$$5 \frac{\text{cm}}{\text{s}} \times 10 \text{ s} = 50 \frac{\text{cm-s}}{\text{s}} = 50 \text{ cm}$$

$$5 \frac{\text{g}}{\text{cm}^3} \times 6 \text{ cm}^3 = 30 \frac{\text{g-cm}^3}{\text{cm}^3} = 30 \text{ g}$$

It is often necessary to convert a quantity in one system of units to its equivalent in another system. Many of the *conversion factors* required in such cases were given in Sec. 2.1; additional ones are given in Appendix Table B.2. To refine your technique work through the following examples.

EXAMPLE 2.3 Convert 20 in. into its equivalent in centimeters. The problem is simple enough to solve in your head, but we are more interested here in developing a general technique for solving problems than in obtaining the numerical answer.

Solution: The conversion factor is

$$1 \text{ in.} = 2.54 \text{ cm}$$

To convert 20 in. to centimeters we multiply by 2.54, that is, 2.54 cm/1 in.;

$$20 \text{ in.} \times \frac{2.54 \text{ cm}}{1 \text{ in.}} = 50.8 \text{ cm}$$

Note the canceling of the units. If we had mistakenly divided by

2.54, that is, multiplied by 1 in./2.54 cm, our units would have been $in.^2/cm$, obviously incorrect.

It is important to realize that the factor 2.54 cm/1 in. is equal to unity; the numerator equals the denominator. Therefore, in our calculation we converted 20 in. into its equivalent, 50.8 cm, by multiplying 20 in. by 1.

Likewise, 1 in./2.54 cm equals unity; and in converting length in centimeters to length in inches we would use this form of unity. For instance, 10 cm in inches is given by

$$10 \not{\text{cm}} \times \frac{1 \text{ in.}}{2.54 \not{\text{cm}}} = 3.94 \text{ in.}$$

EXAMPLE 2.4 Convert 5 ft to its equivalent in millimeters.

Solution: We must multiply 5 ft by a unity or sequence of *unities* to convert feet to millimeters. If we multiply by any factor other than unity, we shall make an error because the result will not be equivalent in length to 5 ft. To solve this problem we must use a conversion factor that transforms an English unit into a metric unit. We shall use 1 in. = 2.54 cm. Therefore, we shall convert 5 ft to inches, inches to centimeters and, finally, centimeters to millimeters:

$$\text{Feet to inches: } 5 \not{\text{ft}} \times \frac{12 \text{ in.}}{1 \not{\text{ft}}} = 60 \text{ in.}$$

$$\text{Inches to centimeters: } 60 \not{\text{in.}} \times \frac{2.54 \text{ cm}}{1 \not{\text{in.}}} = 152.4 \text{ cm}$$

$$\text{Centimeters to millimeters: } 152.4 \not{\text{cm}} \times \frac{10 \text{ mm}}{1 \not{\text{cm}}} = 1524 \text{ mm}$$

Each factor, 12 in./ft, 2.54 cm/in., and 10 mm/cm is equal to unity because the numerator equals the denominator. The result, 1524 mm, is not different from 5 ft; it is the equivalent in length expressed in different units.

The whole calculation is more conveniently done in one operation:

$$5 \not{\text{ft}} \times \frac{12 \not{\text{in.}}}{1 \not{\text{ft}}} \times \frac{2.54 \not{\text{cm}}}{1 \not{\text{in.}}} \times \frac{10 \text{ mm}}{1 \not{\text{cm}}} = 1524 \text{ mm}$$

This is a rapid method of solution since no stepwise calculations are necessary. All the arithmetic is done after the problem has been set up and units are checked. In this case the arithmetic amounts to determining the product $5 \times 12 \times 2.54 \times 10$.

Before you proceed, be sure that you are beginning to grasp the units method by trying Examples 2.5 and 2.6. Use a one-operation setup. Be sure that all units are specified and that the units of the answer are correct. If you cannot solve these problems by the units method, try them again when you have finished reading the section.

EXAMPLE 2.5 Convert 0.500 mi to centimeters (1 mi = 5280 ft).

Answer: 80,500 cm or 8.05×10^4 cm.

EXAMPLE 2.6 Convert 5 in. to angstroms.

Answer: 1.27×10^9 Å.

There are two very important advantages of the units method of problem solving: it is rapid and self-testing. A problem must be set up so that the units of the answer are generated. If the units of the setup are not the desired units, the solution is incorrect. The next example illustrates this point.

EXAMPLE 2.7 Convert 20 ft to centimeters.

We shall omit the units in this calculation to show what can happen when a calculation is performed carelessly.

Facts: 1 ft = 12 in.

1 in. = 2.54 cm

Solution: $20 \times \frac{12}{1} \times \frac{1}{2.54} = 94.6 \text{ cm}$

If we now use the units in the calculation, it becomes obvious that the above calculation is incorrect:

$$20 \text{ ft} \times \frac{12 \text{ in.}}{1 \text{ ft}} \times \frac{1 \text{ in.}}{2.54 \text{ cm}} = 94.6 \frac{\text{ft-in.}^2}{\text{ft-cm}} = 94.6 \text{ in.}^2/\text{cm}$$

We have not converted feet to centimeters. We should have used the reciprocal of 1 in./2.54 cm to obtain the desired result:

$$20 \text{ ft} \times \frac{12 \text{ in.}}{1 \text{ ft}} \times \frac{2.54 \text{ cm}}{1 \text{ in.}} = 610 \frac{\text{ft-in.-cm}}{\text{ft-in.}} = 610 \text{ cm}$$

With a little practice the units method becomes extremely rapid. The next example will illustrate another interpretation of this method:

EXAMPLE 2.8 Convert 3500 cm to yards (yd).

Facts: 2.54 cm = 1 in.

36 in. = 1 yd

$$\text{Solution: } 3500 \not{\text{cm}} \times \underset{A}{\frac{1 \not{\text{in.}}}{2.54 \not{\text{cm}}}} \times \underset{B}{\frac{1 \text{ yd}}{36 \not{\text{in.}}}} = 38.3 \text{ yd}$$

What is the meaning of the factor A alone? It is the number of inches per centimeter. If we divide 1 in. by 2.54 cm, we obtain 0.394 in./cm. The multiplication of 3500 cm by A in inches per centimeter gives the length in inches. Now consider factor B; it represents the number of yards per inch: 1 yd/36 in. = 0.0278 yd/in., so that when we multiply the number of inches, $3500 \times 1/2.54$, by factor B, we are converting inches to yards. Let us look at the equivalent solution:

$$3500 \; \cancel{\text{cm}} \times \underbrace{\frac{0.394 \; \cancel{\text{in.}}}{1 \; \cancel{\text{cm}}}}_{A} \times \underbrace{\frac{0.0278 \text{ yd}}{1 \; \cancel{\text{in.}}}}_{B} = 38.3 \text{ yd}$$

EXAMPLE 2.9 Convert 100 Å to inches.

Solution:

$$100 \; \cancel{\text{Å}} \times \frac{10^{-8} \; \cancel{\text{cm}}}{\cancel{\text{Å}}} \times \frac{1 \text{ in.}}{2.54 \; \cancel{\text{cm}}} = \frac{100 \times 10^{-8}}{2.54} \text{ in.} = 39.4 \times 10^{-8} \text{ in.}$$

EXAMPLE 2.10 Convert 15 mi/h to kilometers per second.

Solution: We must change the units in the numerator and denominator. Convert miles to kilometers in the numerator and hours to seconds in the denominator:

$$15 \frac{\cancel{\text{mi}}}{\text{h}} \times \frac{5280 \; \cancel{\text{ft}}}{1 \; \cancel{\text{mi}}} \times \frac{12 \; \cancel{\text{in.}}}{1 \; \cancel{\text{ft}}} \times \frac{1 \; \cancel{\text{m}}}{39.4 \; \cancel{\text{in.}}} \times \frac{1 \text{ km}}{1000 \; \cancel{\text{m}}}$$

In the line above we have converted miles to kilometers. If we stopped to do the computation at this point, we would determine the equivalent of 15 mi/h in kilometers per hour, but since we want to know the number of kilometers per second, we must continue:

$$15 \frac{\cancel{\text{mi}}}{\cancel{\text{h}}} \times \frac{5280 \; \cancel{\text{ft}}}{1 \; \cancel{\text{mi}}} \times \frac{12 \; \cancel{\text{in.}}}{1 \; \cancel{\text{ft}}} \times \frac{1 \; \cancel{\text{m}}}{39.4 \; \cancel{\text{in.}}} \times \frac{1 \text{ km}}{1000 \; \cancel{\text{m}}} \times \frac{1 \; \cancel{\text{h}}}{3600 \text{ s}} = 0.00670 \text{ km/s}$$

EXAMPLE 2.11 Calculate the number of square meters (m^2) in one square mile (1.00 mi^2). We know that

$$1 \text{ mi} = 5280 \text{ ft}$$

Hence

$$(1 \text{ mi})^2 = (5280 \text{ ft})^2$$

Also

$$1 \text{ ft} = 12 \text{ in.}$$
$$1 \text{ ft}^2 = 12^2 \text{ in.}^2$$

$1 \text{ m} = 39.4 \text{ in.}$
$1 \text{ m}^2 = 39.4^2 \text{ in.}^2$

Solution:

$$1 \text{ mi}^2 \times \frac{(5280)^2 \text{ ft}^2}{\text{mi}^2} \times \frac{(12)^2 \text{ in.}^2}{\text{ft}^2} \times \frac{1 \text{ m}^2}{(39.4)^2 \text{ in.}^2}$$

$$= 2.59 \times 10^6 \text{ m}^2 = 2{,}590{,}000 \text{ m}^2$$

EXAMPLE 2.12 The density of water is 62.4 lb/ft^3 (English system). Convert this to the metric units of density, grams per cubic centimeter (or grams per milliliter).

Solution: The numerator in pounds must be converted to grams; the denominator in cubic feet must be converted into cubic centimeters:

$$62.4 \frac{\text{lb}}{\text{ft}^3} \times \frac{454 \text{ g}}{\text{lb}} \times \frac{1 \text{ ft}^3}{(12 \text{ in.})^3} \times \frac{1 \text{ in.}^3}{(2.54 \text{ cm})^3}$$

or

$$62.4 \frac{\text{lb}}{\text{ft}^3} \times \frac{454 \text{ g}}{\text{lb}} \times \frac{1 \text{ ft}^3}{1728 \text{ in.}^3} \times \frac{1 \text{ in.}^3}{16.39 \text{ cm}^3} = 1.00 \text{ g/cm}^3$$

The result, 1.00 g/cm^3, is expected because the mass of 1 cm^3 of water is defined to be 1 g.

The solutions to problems in this text require three significant figures and can be carried out conveniently by slide rule. Appendix A presents elementary discussions of significant figures, quick estimations, error, uncertainty, and calculations including logarithms. This material will assist you in all scientific and engineering computations.

2.3 THE ATOM IS COMPLEX

It is not always as practical as it is interesting to trace the development of chemistry historically and chronologically. Since many important studies were carried out simultaneously, their mutual influence is complicated; attempting to explain them in an introductory course would only obscure the science. In this section, we describe the essential features of our modern concept of the atom and present the nomenclature and ideas associated with this picture.

Dalton's atom was a simple, indestructible particle. The research of the last 75 years has led us to believe that the atom is complex; it has several elementary parts; in fact, approximately 30 subatomic particles have been discovered or postulated. The three most important are the *proton*, the *neutron*, and the *electron* (see Table 2.1). The proton and neutron have almost exactly the same mass and are

TABLE 2.1: PARTICLES IN AN ATOM

Particle	*Symbol*	*Relative mass*†	*Charge* e.u.	*Charge* C
Proton	p	1.0073	+1	$+1.601864 \times 10^{-19}$
Neutron	n	1.0087	0	0
Electron	e	0.00055	−1	$-1.601864 \times 10^{-19}$

† Discussed in Sec. 2.5.

much heavier than the electron. In fact, the mass of an electron is negligible compared to that of a proton or neutron. The proton and the electron are electrically charged; the *neu*tron is *neu*tral. Since the charge on the electron just neutralizes the charge of the proton, we say that the charges of these two particles are equal and opposite.

The charge on the electron is the unit elementary charge, e.u., that is, the smallest unit of electric charge in nature. The total charge on a body must be some whole-number multiple of the charge on the electron (or proton). The unit elementary charge is extremely small:

$$0.000000000000000000160 \text{ coulomb (C)} \qquad \text{or} \qquad 1.60 \times 10^{-19} \text{ C}$$

The nucleus of the atom contains the protons and neutrons; the electrons circulate outside. The nucleus accounts for almost all the mass of an atom but only a small fraction of the volume. This can be seen from the approximate diameters of an atom and a nucleus:

$$\text{Diameter of atom} \approx 0.00000001 \text{ cm} = 10^{-8} \text{ cm} = 1 \text{ Å}$$
$$\text{Diameter of nucleus} = 0.000000000001 \text{ cm} = 10^{-12} \text{ cm} = 10^{-4} \text{ Å}$$

The diameter of an atom is approximately 10,000 times greater than the diameter of the nucleus.

The *atomic number* of an atom is equal to the number of protons in the nucleus, which in turn is equal to the number of electrons which circulate about it. All atoms of the same element have the same atomic number. Dalton postulated that atoms of the same element have the same mass. We now know that atoms of the same element can have different masses; however, atoms of the same atomic number have identical chemical properties even if they have different masses.

Three different atoms have only one proton: protium, deuterium, and tritium. All these species are atoms of the same element, hydrogen, but their nuclear masses are different because they have different numbers of neutrons in their nuclei (Table 2.2).

TABLE 2.2: ISOTOPES OF HYDROGEN

Species	*Symbol*†	*No. of protons*	*No. of neutrons*	*Natural abundance, %*
Protium‡	^1_1H or P	1	0	99.985
Deuterium	^2_1H or D	1	1	0.015
Tritium	^3_1H or T	1	2	0◖

† The symbol for the element hydrogen is H; the superscript indicates the mass number (sum of the number of neutrons and protons); the subscript indicates the atomic number (the number of protons).

‡ Protium, by far the most common isotope of hydrogen, is usually referred to as hydrogen.

◖ Synthetic element.

TABLE 2.3: ISOTOPES OF OXYGEN

Isotope	*Composition of nucleus*		*Natural abundance, %*
	Protons	*Neutrons*	
$^{16}_{8}O$	8	8	99.76
$^{17}_{8}O$	8	9	0.04
$^{18}_{8}O$	8	10	0.20

Protium, deuterium, and tritium are called *isotopes*, i.e., atoms which possess the same number of protons but a different number of neutrons. Isotopes are atoms of the same element with the same atomic number and, of course, identical chemical behavior but different atomic mass.

Except for fluorine and sodium, naturally occurring isotopes of all the common elements have been discovered. Fluorine exhibits only one atomic species, with nine protons and nine neutrons. There are three known isotopes of oxygen which occur naturally (Table 2.3).

2.4 ATOMIC WEIGHT

The weight† of one atom of any element is so small that it would never be detectable by our most sensitive balances. Indirectly, it can be determined that one hydrogen atom has a mass of

$$0.00000000000000000000000167 \text{ g} \qquad \text{or} \qquad 1.67 \times 10^{-24} \text{ g}$$

One carbon atom has a mass of 19.9×10^{-24} g, and one gold atom a mass of 327×10^{-24} g. The smallest mass which can easily be detected in a chemical laboratory is a *microgram*, one-millionth of a gram, written 10^{-6} g or 1 μg. Obviously, we can never hope to determine the mass of one atom by direct weighing.

The atomic weight of an element is defined as the *average mass* of the atoms of that element relative to the mass of the most common isotope of carbon, which has been *assigned* a mass of exactly 12 *atomic mass units* (u). If it were possible, we would make a direct comparison of the mass of an atom with the ^{12}C isotope. We would

† Mass m and weight w can often be used as synonyms, but this is true only because weight is proportional to mass, $w = mg$, where g is the acceleration due to gravity (see Secs. 6.2 and 6.3).

FIGURE 2.2

The mass of the helium atom relative to the carbon-12 isotope.

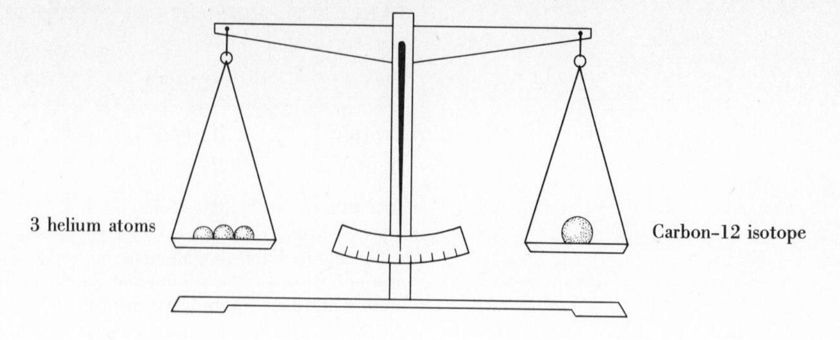

place a ^{12}C isotope on one pan of a scale and a number of atoms of another element on the other pan just sufficient to balance the scale. For helium, He, we would find that three atoms are required to balance the mass of one ^{12}C isotope. Therefore, one helium atom has one-third the mass of a ^{12}C isotope, or a helium atom has a mass of 4 u (see Fig. 2.2). Continuing our imaginary weighing experiments, we would find that three oxygen atoms are required to balance four ^{12}C isotopes, and on the ^{12}C scale oxygen would be assigned 16 u. We could compare the atoms of all the elements to our standard ^{12}C isotope and find the masses of the individual atoms.

Since it is impossible to proceed in this manner because our balances are not sensitive enough to detect the mass of just a few atoms, we have developed a more practical concept, the *gram-atomic weight* or gram atom (g atom).

2.5 GRAM-ATOMIC WEIGHT AND AVOGADRO'S NUMBER

The *gram-atomic weight* is the atomic weight of an element expressed in grams. Since nearly all elements occur as a mixture of isotopes, exact gram-atomic weights generally are not integers; but in many cases they are very close to integers because ordinarily one isotope of an element predominates. For hydrogen, protium, $^{1}_{1}H$, is present to the extent of 99.385 percent, and the gram-atomic weight of hydrogen is almost exactly 1 g (1.00797 g). The gram-atomic weight of oxygen is 16.0 g (15.9994 g) because the isotope $^{16}_{8}O$ is by far the most abundant. However, the gram-atomic weight of chlorine is 35.5 g (35.453 g) because two isotopes of chlorine, $^{35}_{17}Cl$ and $^{37}_{17}Cl$, are abundant in nature, 75 and 25 percent, respectively.

A g atom of one element includes the same number of atoms as a g atom of any other element. Because a carbon atom is 12 times heavier than a hydrogen atom, the number of carbon atoms in 12 g of carbon equals the number of hydrogen atoms in 1 g of hydrogen. The number of atoms in a g atom has been determined by many scientists in a variety of experiments. The number, which is astronomically large, is referred to as *Avogadro's number* N_A, in honor of the great eighteenth-century Italian physicist Amedeo Avogadro. In later chapters we shall describe some of the methods which have been used to determine N_A. The best value obtained so far is 6.02296×10^{23}, or 602,296,000,000,000,000,000,000. If the continental United States were uniformly covered by N_A grains of sand, the depth of the sand would be approximately 10 ft.

Using the relative weights of the atoms (the atomic weights) and Avogadro's number, it is possible to calculate the mass of an atom of any element in grams (or pounds or in any mass unit). For example, the average weight of *one* carbon atom in grams is determined by dividing the gram-atomic weight by the number of atoms in one gram-atomic weight:

$$\frac{12.01115 \text{ g}}{602{,}296{,}000{,}000{,}000{,}000{,}000{,}000 \text{ atoms}}$$

$$0.0000000000000000000000199 \text{ g/atom} = 1.99 \times 10^{-23} \text{ g/atom}$$

One-millionth of a gram (1 μg or 10^{-6} g) is a very small mass. It is barely detectable on our very good balances, but a microgram of any element contains an inconceivably large number of atoms. Let us calculate the number of carbon atoms in 1 μg of carbon:

$$1 \times 10^{-6} \cancel{\text{g C}} \times \frac{6.02 \times 10^{23} \text{ atoms}}{12.0 \cancel{\text{g C}}} = 5 \times 10^{16} \text{ atoms}$$

$$= 50 \text{ thousand million million atoms}$$

In a microgram, 5×10^{16} or 50,000,000,000,000,000, carbon atoms are present, but this huge number will barely cause a deflection on a microbalance. We must admit that even our best instruments are crude in terms of sensing the atoms in which we have such a strong faith.

2.6 MOLECULE, MOLE, AND AVOGADRO'S NUMBER

Just as the ultimate particle of an element is the atom, the ultimate particle of a compound of elements is the *molecule*. The molecule is a collection of atoms bound together in a relatively fixed pattern by a somewhat mysterious force called the *chemical bond*. The distance between two adjacent atoms in a molecule is never greater than 3 to 4 Å (1 Å = 0.00000001 cm = 10^{-8} cm).

The *molecular weight* is the sum of the masses of all of the atoms in the molecule as represented by the molecular formula of the compound. For example, the molecular weight of methane, CH_4, equals the sum of the masses of one carbon atom and four hydrogen atoms:

Mol wt of CH_4 = mass of 1 C atom + 4 × mass of 1 H atom
Mol wt of CH_4 = 12 u + 4 × 1 u = 16 u

A gram-molecular weight (called *gram mole* or simply *mole* and abbreviated mol) is the molecular weight expressed in grams. Thus, 1 mol of CH_4 weighs 16 g. It is very important to understand that in ordinary discussions the chemist uses the term molecular weight to mean gram-molecular weight; this does not ordinarily cause any confusion.

Just as there are N_A atoms in 1 g atom of any element, there are N_A molecules in 1 mol of any compound. In 16 g of CH_4 (1 mol) there are N_A molecules of CH_4, and in 98 g of H_2SO_4 (1 mol) there are N_A molecules of H_2SO_4. The weight of a single molecule in grams is calculated by dividing the gram-molecular weight by N_A:

$$\text{Weight of one CH}_4\text{ molecule} = \frac{16\text{ g}}{\not{\text{mol}}} \times \frac{1\ \not{\text{mol}}}{6.02 \times 10^{23}\text{ molecules}}$$

$$= 2.67 \times 10^{-23}\text{ g/molecule}$$

$$\text{Weight of one H}_2\text{SO}_4\text{ molecule} = \frac{98\text{ g}}{\not{\text{mol}}} \times \frac{1\ \not{\text{mol}}}{6.02 \times 10^{23}\text{ molecules}}$$

$$= 1.63 \times 10^{-22}\text{ g/molecule}$$

The term mole has begun to take on a more general meaning. A mole can now be used as a synonym for Avogadro's number, and we speak of a mole of electrons or a mole of sodium atoms, meaning, respectively, 6.02×10^{23} electrons and 6.02×10^{23} atoms of sodium. We can speak of a mole of sodium chloride, a compound in which there are no molecules (sodium chloride is composed of ions, or charged atoms). When we say "a mole of sodium chloride," we mean an amount of this compound which contains 6.02×10^{23}, or N_A, sodium ions *and* N_A chloride ions. The distinction between compounds composed of ions and compounds composed of molecules will become clear when we discuss the types of chemical compounds and the nature of chemical bonding in Chap. 5. For now, it is important to realize only that 1 mol of a compound of elements is a collection of N_A molecules of that compound.

2.7 SYMBOLS AND FORMULAS

Symbols for the elements are a shorthand used by scientists and should become a part of your vocabulary as you learn chemistry. A complete list of the elements and their symbols is given inside the front cover of this text.

A great deal of information is conveyed by a symbol. A *symbol* identifies an element and may also represent an atom of an element when the symbol is used in the *formula* for a compound. Methanol has the formula CH_4O; this means that the compound is composed of carbon, hydrogen, and oxygen and that the ratio of atoms is 1:4:1. The ratio of iron to chlorine atoms in $FeCl_3$ is 1:3.

The gram formula weight of a compound is the weight calculated by summing the gram-atomic weights of atoms which appear in a formula. For methanol, CH_4O, we have 12 g for 1 g atom of carbon, 4 g for 4 g atoms of hydrogen, and 16 g for 1 g atom of oxygen. The total is 32 g. In this case the gram formula weight equals the *gram-molecular weight*. This is true only when a given formula indicates the actual number of atoms in a molecule of a compound. Such a formula is called a *molecular formula*. An *empirical formula* gives the smallest whole-number ratio of the atoms of the different elements in a compound. The empirical formula for benzene, CH, shows a 1:1 ratio of carbon atoms to hydrogen atoms. The molecular formula for benzene, C_6H_6, shows that there are actually six carbon atoms and six hydrogen atoms in each benzene molecule. For benzene the empirical formula is different from the molecular formula. The gram formula weight equals the gram-molecular weight only when we use the molecular formula for a compound.

The chemical formula for a compound is full of information we must learn to extract. Experimentally we can determine that sodium chloride is 39.4 percent sodium and 60.6 percent chlorine by weight. But since we believe that atoms exist and that the atoms of an element are characterized by an average mass, the weight percentages of the elements permit us to determine the ratio of atoms in a compound. This can best be shown by example.

EXAMPLE 2.13 Calculate the ratio of sodium to chlorine atoms in sodium chloride from the weight percentages given above.

Solution: First we divide the weights (choosing a 100-g sample for convenience) by the respective gram-atomic weights; this will give the ratio of g atoms of the elements, which is the same as the ratio of the atoms of the elements in the sample:

$$\text{Sodium: } 39.4 \cancel{\text{g Na}} \times \frac{1 \text{ g atom}}{23 \cancel{\text{g Na}}} = \frac{39.4}{23} \text{ g atom} = 1.71 \text{ g atom}$$

$$\text{Chlorine: } 60.6 \cancel{\text{g Cl}} \times \frac{1 \text{ g atom}}{35.5 \cancel{\text{g Cl}}} = 1.71 \text{ g atom}$$

We have found the ratio of g atoms and atoms in sodium chloride to be 1.71 Na to 1.71 Cl. Since we believe atoms to be indivisible, we reduce this ratio to the simplest whole numbers and obtain 1:1. We write the formula for sodium chloride as NaCl, which implies that the ratio of the atoms is 1:1. Magnesium chloride would be written as $MgCl_2$ since elemental analysis shows that the ratio of magnesium atoms to chlorine atoms is 1:2. The subscript 1 is never used; it is implied.

We see that the chemical formula indicates the elements of a compound and their atomic ratio; it also implies the weight composition of the compound since the weight of an element in a certain amount of compound is very easily computed from the formula, as shown in Example 2.14.

EXAMPLE 2.14 Calculate the weight of lead in 5.00 g of lead carbonate, $PbCO_3$.

First, calculate the gram formula weight of $PbCO_3$:

1 g atom wt of Pb	207 g
1 g atom wt of C	12 g
3 × g atom wt of O (3 × 16)	48 g
Gram formula weight	267 g

The gram formula weight is 267 g, of which 207 g is Pb. Therefore,

$$267 \text{ g } PbCO_3 = 207 \text{ g Pb}$$

This is our conversion factor for converting a weight of $PbCO_3$ into an equivalent weight of lead:

$$5.00 \cancel{\text{g } PbCO_3} \times \frac{207 \text{ g Pb}}{267 \cancel{\text{g } PbCO_3}} = 3.88 \text{ g Pb}$$

The factor 207 g Pb/267 g $PbCO_3$ equals 0.775 g Pb/g $PbCO_3$. This means there is 0.775 g of lead in *each* gram of $PbCO_3$; 5 times 0.775 gives the weight of Pb in 5 g of $PbCO_3$.

If the elemental analysis and the gram-molecular weight have been determined, it is possible to determine the empirical and molecular formulas for a compound.

EXAMPLE 2.15 A compound containing only nitrogen and hydrogen contains 12.5 percent hydrogen by weight. Its gram-molecular weight is 32. Determine the empirical and molecular formulas.

Solution: First we calculate the ratio of g atoms of each element:

$$\text{Hydrogen: } 12.5 \cancel{\text{g H}} \times \frac{1 \text{ g atom H}}{1 \cancel{\text{g H}}} = 12.5 \text{ g atom}$$

$$\text{Nitrogen: } 87.5 \cancel{\text{g N}} \times \frac{1 \text{ g atom N}}{14 \cancel{\text{g N}}} = 6.25 \text{ g atom}$$

Since the smallest whole-number ratio of H to N is 2:1, the empirical formula is H_2N.

The molecular formula must have hydrogen and nitrogen atoms in a

2:1 ratio; the molecular formula can be H_2N, H_4N_2, H_6N_3, etc. The corresponding gram-molecular weights would be 16 g, 32 g, 46 g, etc. Since the gram-molecular weight is 32 g, the molecular formula must be H_4N_2.

It should be clear that the gram-molecular weight will be some whole-number multiple of the weight given by the empirical formula. You should now try the following unsolved problems; the answers are given.

EXAMPLE 2.16 Teflon is 29.63 percent carbon and 70.37 percent fluorine; determine its empirical formula.

Answer: C_2F_3.

EXAMPLE 2.17 (*a*) Acetic acid, the acid in vinegar, is 40 percent carbon, 6.67 percent hydrogen, and 53.3 percent oxygen by weight. Determine the empirical formula. (*b*) The gram-molecular weight is 60 g. What is the molecular formula?

Answer: (*a*) CH_2O; (*b*) $C_2H_4O_2$ (usually written CH_3COOH).

2.8 OXIDATION NUMBER

The theory of chemical bonding (Chap. 5) describes chemical reactions in terms of the gain, loss, or sharing of electrons of the atoms involved. The number of electrons that an atom loses, gains, or shares when it forms chemical bonds with other atoms is known as the *oxidation number*. Elements which gain electrons have a negative oxidation number; elements which lose electrons have a positive oxidation number.

EXAMPLE 2.18 Consider the formation of table salt from its elements:

$$2Na + Cl_2 \rightarrow 2Na^+Cl^-$$

Here each sodium atom transfers one electron to a chlorine atom, thereby forming charged atoms, or *ions*. The sodium ion, Na^+, and the chloride ion, Cl^-, have respective oxidation numbers of $+1$ and -1.

EXAMPLE 2.19 Consider the combustion of hydrogen, which produces water:

$$2H_2 + O_2 \rightarrow 2H_2O$$

In this case, water molecules form by covalent bonding, i.e., sharing electron pairs. Each hydrogen atom contributes *one* electron to the total bonding, and each oxygen contributes *two* electrons to the total bonding. Since it has been found that oxygen atoms have a greater

TABLE 2.4: POSITIVE OXIDATION NUMBERS OF COMMON ELEMENTS AND RADICALS

1+		*2+*		*3+*		*4+*	
H^+	Hydrogen	Pb^{2+}	Lead	Al^{3+}	Aluminum	Sn^{4+}	Tin
Na^+	Sodium	Zn^{2+}	Zinc	Fe^{3+}	Iron		
K^+	Potassium	Ca^{2+}	Calcium	Cr^{3+}	Chromium		
Ag^+	Silver	Ba^{2+}	Barium				
Cu^+	Copper	Mg^{2+}	Magnesium				
NH_4^+	Ammonium	Cu^{2+}	Copper				
		Hg_2^{2+}	Mercury I				
		Hg^{2+}	Mercury II				
		Fe^{2+}	Iron				
		Sn^{2+}	Tin				
		Cr^{2+}	Chromium				

attraction for electrons than hydrogen atoms have, the oxygen atom is considered negative relative to hydrogen in the water molecule. Therefore, the oxidation number of oxygen is -2, and that of hydrogen is $+1$.

Frequently, a group of atoms behaves like a single atom during chemical combination. A group of atoms which can be transferred as a unit and which exhibits a constant oxidation number is known as a *radical*. The nitrate ion, NO_3^-, and the ammonium ion, NH_4^+,

TABLE 2.5: NEGATIVE OXIDATION NUMBERS OF COMMON ELEMENTS AND RADICALS

1−		*2−*		*3−*	
F^-	Fluoride	O^{2-}	Oxide	PO_4^{3-}	Phosphate
Cl^-	Chloride	S^{2-}	Sulfide		
Br^-	Bromide	SO_3^{2-}	Sulfite		
I^-	Iodide	SO_4^{2-}	Sulfate		
NO_2^-	Nitrite	CO_3^{2-}	Carbonate		
NO_3^-	Nitrate	SiO_3^{2-}	Silicate		
OH^-	Hydroxide	CrO_4^{2-}	Chromate		
ClO^-	Hypochlorite	$Cr_2O_7^{2-}$	Dichromate		
ClO_2^-	Chlorite	O_2^{2-}	Peroxide		
ClO_3^-	Chlorate				
ClO_4^-	Perchlorate				
MnO_4^-	Permanganate				

are radicals and are often transferred without change in oxidation number during a chemical reaction:

$$AgNO_3 + NH_4Cl \rightarrow AgCl + NH_4NO_3$$

The oxidation numbers of the common elements and radicals are given in Tables 2.4 and 2.5; by using them to work problems and examples, you will soon know them well enough to omit reference to the tables. Notice that certain elements, e.g., oxygen, mercury, copper, tin, and iron, exhibit more than one oxidation number.

2.9 WRITING FORMULAS

Since compounds of elements are electrically neutral, the elements must combine so that there is no net charge. We keep this principle in mind when we write formulas for the compounds. For instance, sodium phosphate would be Na_3PO_4 because the sodium ion is singly charged and the phosphate ion triply charged. Three sodium ions just "neutralize" one phosphate ion. Other examples are

Barium chloride, (Ba^{2+}, Cl^-), $BaCl_2$

Ammonium sulfide, (NH_4^+, S^{2-}), $(NH_4)_2S$

Potassium dichromate, (K^+, $Cr_2O_7^{2-}$), $K_2Cr_2O_7$

As we learned from the discussion of Avogadro's hypothesis in Chap. 1, the molecules for many of the common gaseous elements are diatomic. Typical formulas are O_2, N_2, H_2, F_2, and Cl_2.

2.10 NOMENCLATURE OF INORGANIC COMPOUNDS

Because of the great number and variety of chemical compounds, it is necessary to have a systematized nomenclature. The rules of nomenclature for most of the inorganic compounds met in the first year of chemistry are discussed next.

Binary compounds

A binary compound contains only two elements.

Nomenclature: The name of the binary compound is derived from the name of the metal (or element with positive oxidation number) plus the stem name of the nonmetal (element with negative oxidation number) to which is added the suffix *-ide:*

Name of the metal + stem name of the nonmetal + ending *-ide*

NaCl, sodium chlor*ide*

KBr, potassium brom*ide*

CaO, calcium ox*ide*

If a metal exhibits more than one common oxidation number, the number is indicated by a Roman numeral placed in parentheses directly after the name of the element. (This nomenclature replaces the *-ous* and *-ic* endings formerly used.)

Hg_2Br_2, mercury(I) bromide (or mercur*ous* bromide)

$HgBr_2$, mercury(II) bromide (or mercur*ic* bromide)

$SnCl_2$, tin(II) chloride (or stann*ous* chloride)

$SnCl_4$, tin(IV) chloride (or stann*ic* chloride)

Hydroxide compounds

Many *bases* contain the hydroxide, OH^-, radical and an element or radical which has a positive oxidation number.

Nomenclature: Name of the metal (with oxidation number in parentheses) plus the word *hydroxide:*

$NaOH$, sodium hydroxide — $Fe(OH)_2$, iron(II) hydroxide

$Ca(OH)_2$, calcium hydroxide — $Fe(OH)_3$, iron(III) hydroxide

Common acids

Common acids are sour-tasting compounds which cause the corrosion of many metals. Most acids contain hydrogen as well as an element or radical which has a negative oxidation number.

Nonoxygen acids

Nomenclature: For acids (like HCl) which do not contain oxygen the name is formed by attaching the prefix *hydro-* and the suffix *-ic* to the stem name of the key element:

Hydro- + stem name of key element + *-ic* + the word *acid*

For instance, HCl is hydro-chlor-ic acid, hydrochloric acid. HBr is *hydro*brom*ic* acid. H_2S is *hydro*sulfur*ic* acid. When they are not acting as acids, HCl, HBr, and H_2S are called hydrogen chloride, hydrogen bromide, and hydrogen sulfide.

Oxygen acids

Nomenclature: For acids (like HNO_2) which contain oxygen use the stem name of the key element plus the suffix *-ous* or *-ic* (*-ous* indicates fewer oxygen atoms than *-ic*) and add the word *acid.*

HNO_2, nitr*ous* acid — H_2SO_3, sulfur*ous* acid

HNO_3, nitr*ic* acid — H_2SO_4, sulfur*ic* acid

Sometimes an element forms more than two oxygen acids. Chlorine, for example, forms four oxygen acids: $HClO$, $HClO_2$, $HClO_3$, and $HClO_4$. The most common of these acids, $HClO_3$, is called chlo*ric* acid. $HClO_2$, which contains only two oxygen atoms, is called chlor*ous* acid. The name of acid which contains one more oxygen atom than the *-ic* acid is formed by using the prefix *per-*. When there is one atom of oxygen less than the number in the *-ous* acid, the prefix *hypo-* is used.

$HClO$, hypochlorous acid

$HClO_2$, chlorous acid

$HClO_3$, chloric acid

$HClO_4$, perchloric acid

Salts of oxygen acids

Nomenclature: Use the name of the metal (with oxidation number in parentheses) plus the name of the radical. If the radical arises from an *-ous* acid, the name of the salt ends in *-ite*. If the radical arises from an *-ic* acid, the name of the salt ends in *-ate*.

-ous acid → -ite salt
-ic acid → -ate salt

$NaNO_3$, sodium nitrate

The radical NO_3^- arises from HNO_3 (nitric acid), and since *-ic* acids give *-ate* salts, the name is sodium nitr*ate*.

$FeSO_3$, iron(II) sulfite

The oxidation number of Fe is +2. The SO_3^{2-} radical comes from H_2SO_3 (sulfur*ous* acid), and since *-ous* acids give *-ite* salts, the name is iron(II) sulf*ite*.

$Fe(ClO_3)_3$, iron(III) chlorate

The oxidation number of Fe is +3. The radical ClO_3^- comes from $HClO_3$ (chloric acid), and hence the name is iron(III) chlor*ate*.

Hypo- -ous acids give *hypo- -ite* salts, and *per- -ic* acids give *per- -ate* salts. For example, NaClO is named sodium *hypo*chlor*ite*, and $KClO_4$ is named potassium *per*chlor*ate*.

Hypo- -ous acid → hypo- -ite salt
per- -ic acid

2.11 THE CHEMICAL EQUATION

Symbols represent elements, and formulas represent compounds of the elements. Chemical *equations* represent the chemical reactions of the elements and their compounds. The most important rule in writing chemical equations is based on the law of conservation of mass, which holds for all but fission and fusion reactions involving the nucleus. In a chemical equation we write down the formulas for all the reactants on the left side of an arrow

$$H_2SO_4 + NaOH \rightarrow$$

and the formulas for all the products on the right side of the arrow

$$H_2SO_4 + NaOH \rightarrow HOH + Na_2SO_4$$

The products of a reaction are determined by experiment, but we can make good guesses on the basis of the experience of many chemists working for many years.

Since the law of conservation of mass holds, we must *balance* the equation so that the number of atoms of an element is the same on both sides of the arrow. We can balance the above equation by using a coefficient of $\frac{1}{2}$ before both H_2SO_4 and Na_2SO_4 and a coefficient of 1 before NaOH and HOH:

$$\tfrac{1}{2}H_2SO_4 + 1NaOH \rightarrow 1HOH + \tfrac{1}{2}Na_2SO_4$$

It is more conventional to use the smallest whole numbers as coefficients in chemical equations. We can rewrite the above equation as

$$H_2SO_4 + 2NaOH \rightarrow 2HOH + Na_2SO_4 \qquad (2.1)$$

The coefficient of H_2SO_4 (and Na_2SO_4) is 1, which is never written but understood.

Both equations give the same information. The first states that the molar ratio of reactants is $\frac{1}{2}:1$, and the second says that the ratio is 1:2. The equations are equally useful.

Equation (2.1) may be interpreted in terms of molecules, moles, or grams. It tells us that when H_2SO_4 reacts with NaOH, 1 mol of H_2SO_4 will combine with 2 mol of NaOH and produce 2 mol of HOH and 1 mol of Na_2SO_4. If we calculated the gram-molecular weights of all these species, we would conclude that 98 g of H_2SO_4 (1 mol) combines with 80 g of NaOH (2 mol) and produces 36 g of HOH (2 mol) and 142 g of Na_2SO_4 (1 mol). Note that mass is conserved: $98\ g + 80\ g \rightarrow 36\ g + 142\ g$. Also, the number of atoms or g atoms of each element is conserved; the number of molecules or moles is not conserved, except fortuitously.

Chemical equations may also indicate the physical state of the species involved. For instance, in the equation

$$2H_2(g) + O_2(g) \rightarrow H_2O(l)$$

the italic letters mean that H_2 and O_2 are gases and H_2O is a liquid. Other designations include (*s*), (*aq*), and (*c*), which mean solid, dissolved in water (aqueous), and crystalline solid, respectively. The following chemical equations indicate physical states of the species involved:

$$Na(s) + Cl_2(g) \rightarrow NaCl(c)$$
$$NaCl(aq) + AgNO_3(aq) \rightarrow AgCl(s) + NaNO_3(aq)$$

2.12 STOICHIOMETRY

It is extremely important to learn how to obtain quantitative information from chemical equations. A calculation based on a chemical equation is called *stoichiometry*. Go through the following examples of solved stoichiometry problems very carefully.

EXAMPLE 2.20 When iron is heated in the presence of gaseous chlorine, a brown solid is formed. By elemental analysis this compound is found to have the empirical formula $FeCl_3$ [iron(III) chloride, or ferric chloride]. Calculate (*a*) the number of moles of chlorine necessary to produce 10 mol of $FeCl_3$ and (*b*) the weight of chlorine necessary to produce 100 g of $FeCl_3$.

Equation: $Fe + Cl_2 \rightarrow FeCl_3$
Balanced equation: $2Fe + 3Cl_2 \rightarrow 2FeCl_3$

Solution: The balanced equation tells us that 3 mol of Cl_2 produces 2 mol of $FeCl_3$, and we have the conversion factor

$$3 \text{ mol } Cl_2 = 2 \text{ mol } FeCl_3$$

For part (*a*) we must find the number of moles of Cl_2 which will produce 10 mol of $FeCl_3$; that is, we must convert 10 mol of $FeCl_3$ into an equivalent number of moles of Cl_2; this is straightforward:

$$10 \text{ mol } FeCl_3 \times \frac{3 \text{ mol } Cl_2}{2 \text{ mol } FeCl_3} = 15 \text{ mol } Cl_2$$

For part (*b*) we must calculate the number of grams of Cl_2 necessary to produce 100 g of $FeCl_3$. We start by converting 100 g of $FeCl_3$ into moles of $FeCl_3$. The moles of $FeCl_3$ is then converted into its equivalent in moles of Cl_2; from the number of moles of Cl_2 we can easily calculate the grams of Cl_2.

The gram-molecular weights of $FeCl_3$ and Cl_2 are 162.3 g and 71.0 g, respectively, or

$$1 \text{ mol } FeCl_3 = 162.3 \text{ g } FeCl_3$$
$$1 \text{ mol } Cl_2 = 71.0 \text{ g } Cl_2$$

$$100 \cancel{\text{g } FeCl_3} \times \frac{1 \text{ mol } FeCl_3}{162.3 \cancel{\text{g } FeCl_3}} = 0.617 \text{ mol } FeCl_3$$

$$0.617 \cancel{\text{mol } FeCl_3} \times \frac{3 \text{ mol } Cl_2}{2 \cancel{\text{mol } FeCl_3}} = 0.925 \text{ mol } Cl_2$$

$$0.925 \cancel{\text{mol } Cl_2} \times \frac{71.0 \text{ g } Cl_2}{1 \cancel{\text{mol } Cl_2}} = 65.7 \text{ g } Cl_2$$

or, preferably in one equation:

$$100 \text{ g FeCl}_3 \times \frac{1 \text{ mol FeCl}_3}{162.3 \text{ g FeCl}_3} \times \frac{3 \text{ mol Cl}_2}{2 \text{ mol FeCl}_3} \times \frac{71.0 \text{ g Cl}_2}{1 \text{ mol Cl}_2} = 65.7 \text{ g Cl}_2$$

EXAMPLE 2.21 Methane, CH_4, burns in oxygen to produce carbon dioxide gas, CO_2, and water vapor. What weight of water vapor is produced by burning 100 g of CH_4?

Equation: $CH_4(g) + O_2(g) \rightarrow CO_2(g) + H_2O(g)$
Balanced equation: $CH_4(g) + 2O_2(g) \rightarrow CO_2(g) + 2H_2O(g)$

The balanced equation indicates that 1 mol of CH_4 produces 2 mol of H_2O:

1 mol CH_4 = 2 mol H_2O

The gram-molecular weights of CH_4 and H_2O are 16 and 18 g, respectively, or

1 mol CH_4 = 16 g CH_4
1 mol H_2O = 18 g H_2O

We can convert 100 g CH_4 into an equivalent weight of H_2O:

$$100 \text{ g CH}_4 \times \frac{1 \text{ mol CH}_4}{16 \text{ g CH}_4} \times \frac{2 \text{ mol H}_2\text{O}}{1 \text{ mol CH}_4} \times \frac{18 \text{ g H}_2\text{O}}{1 \text{ mol H}_2\text{O}} = 225 \text{ g H}_2\text{O}$$

Grams CH_4 converted to moles CH_4 | Moles CH_4 converted to moles H_2O | Moles H_2O converted to grams H_2O

EXAMPLE 2.22 Magnesium reacts slowly with cold water and produces hydrogen gas and magnesium hydroxide. (*a*) Write a balanced chemical equation. (*b*) How many grams of magnesium hydroxide can be produced from 150 g of magnesium? (*c*) If hydrogen gas has a density of 0.0828 g/l, what volume of gas can be produced from 15.0 g of magnesium? (*d*) How many magnesium atoms (not g atoms) are required to produce 15.0 g of magnesium hydroxide?

Solution: (*a*) The reactants are Mg and H_2O; the products are $Mg(OH)_2$ and H_2.

Unbalanced equation: $Mg + H_2O \rightarrow Mg(OH)_2 + H_2$
Balanced equation: $Mg + 2H_2O \rightarrow Mg(OH)_2 + H_2$

(*b*) First, write the number of moles of each material under the balanced equation. Then calculate the gram-molecular weights and write the number of grams of each material involved under the balanced equation:

	Mg	+	$2H_2O$	$\rightarrow$	$Mg(OH)_2$	+	H_2
Moles:	1		2		1		1
Grams:	24		$2 \times 18 = 36$		58		2

We are interested in the relationship between $Mg(OH)_2$ and Mg. According to the balanced equation, 1 mol of $Mg(OH)_2$ is produced from 1 mol of Mg. Since 1 mol of $Mg(OH)_2$ has a weight of 58 g and 1 mol of Mg has a weight of 24 g, we can say that 58 g $Mg(OH)_2$ is produced from 24 g of Mg. We have a chemical conversion factor: 58 g of $Mg(OH)_2 = 24$ g Mg. We can use this to convert 150 g of Mg into its equivalent weight of $Mg(OH)_2$:

$$150 \cancel{\text{ g Mg}} \times \frac{58 \text{ g Mg(OH)}_2}{24 \cancel{\text{ g Mg}}} = 363 \text{ g Mg(OH)}_2$$

(*c*) From the balanced chemical equation we see that 1 mol of Mg produces 1 mol of H_2 or 24 g of Mg produces 2 g of H_2. Our conversion factor is 24 g Mg = 2 g H_2. From this we can calculate the weight of H_2 produced from 12.0 g of Mg, and with the given density we can calculate the volume of H_2:

$$15.0 \cancel{\text{ g Mg}} \times \frac{2 \text{ g H}_2}{24 \cancel{\text{ g Mg}}} = 1.25 \text{ g H}_2$$

Since each liter of H_2 weighs 0.0828 g, we have the conversion factor, 1.00 l H_2 = 0.0828 g H_2, which can be used to calculate the volume of 1.25 g H_2:

$$1.25 \cancel{\text{ g H}_2} \times \frac{1.00 \text{ l H}_2}{0.0828 \cancel{\text{ g H}_2}} = 15.1 \text{ l H}_2$$

This calculation should be done in one equation:

$$15.0 \cancel{\text{ g Mg}} \times \frac{2 \cancel{\text{ g H}_2}}{24 \cancel{\text{ g Mg}}} \times \frac{1.00 \text{ l H}_2}{0.0828 \cancel{\text{ g H}_2}} = 15.1 \text{ l H}_2$$

(*d*) From the balanced chemical equation we see that 58 g of $Mg(OH)_2$ is produced from 1 mol or g atom of Mg. We also know that there are one 6.02×10^{23} atoms in 1 g atom. First, we convert grams of $Mg(OH)_2$ to g atoms of Mg and then g atoms of Mg to atoms of Mg:

$$15.0 \cancel{\text{ g Mg(OH)}} \times \frac{1 \cancel{\text{ g-atom Mg}}}{58 \cancel{\text{ g Mg(OH)}_2}} \times \frac{6.02 \times 10^{23} \text{ Mg atom}}{\cancel{\text{g-atom Mg}}}$$
$$= 1.56 \times 10^{23} \text{ Mg atom}$$

It is important to become adept at calculations involving the interpretation of a chemical equation. Be sure that you understand all the worked examples and the following unworked examples in this section. Give particular attention to Probs 2.29 to 2.33.

EXAMPLE 2.23 The combustion of octane, C_8H_{18}, a component of gasoline, results in the formation of CO_2 and H_2O. The unbalanced equation is $C_8H_{18} + O_2 \rightarrow CO_2 + H_2O$. What volume of carbon dioxide gas is produced from the combustion of 1 kg of octane? [The density of $CO_2(g)$ is 1.85 g/l.]

Answer: 1669 l.

QUESTIONS AND PROBLEMS

2.1 (*a*) Calculate the radius, diameter, and circumference of the earth in centimeters and angstroms assuming that the earth is a globe having a diameter of about 7918 mi. (*b*) Calculate the surface area of the earth in square centimeters and square meters. (*c*) Calculate the volume of the earth in cubic centimeters, liters, cubic feet, and cubic meters.

2.2 Density is defined as the mass per unit volume of a substance; 1 ft^3 of iron weighs 449 lb. Calculate (*a*) the weight of 1 ft^3 of iron in grams and kilograms and (*b*) the density of iron in grams per milliliter.

2.3 The average velocity of an oxygen molecule under ordinary atmospheric conditions is approximately 400 mi/h. Calculate its velocity in centimeters per second.

2.4 A comfortable room temperature is 70°F; body temperature is about 98.6°F. Convert both these temperatures into their equivalents on the Celsius and Kelvin scales.

2.5 The temperature −273.16°C is the *absolute of zero*, or theoretically lowest attainable temperature. What is the equivalent of the absolute of zero on the Kelvin and Fahrenheit scales?

2.6 Define atomic weight, gram-atomic weight, atomic number, and isotope.

2.7 Six isotopes of sodium, Na, have been identified; their mass numbers are 20, 21, 22, 23, 24, and 25. The atomic number of sodium is 11. How many protons, neutrons, and electrons are present in each isotope?

2.8 From the diameters given in the text for an atom and a nucleus, calculate the ratio of the volume of an atom to the volume of a nucleus. (Volume of a sphere $= \frac{4}{3}\pi r^3$.)

2.9 The mass of one boron, B, atom is 0.45 times smaller than the mass of one carbon-12 atom. (*a*) Calculate the weight in grams of 50 boron atoms. (*b*) How many boron atoms are required to cause a deflection on a balance whose sensitivity is limited to 1 μg?

2.10 Calculate the number of moles in (*a*) 200 g of H_2SO_4, (*b*) 50.0 g of H_2O, (*c*) 85.0 g of chlorine, Cl_2. Consult the inside front cover of the text for a list of the atomic weights of the elements.

2.11 Under ordinary atmospheric conditions the weight of 1 l of oxygen gas is 1.29 g and of 1 l of a gaseous oxide of carbon is 1.14 g. (*a*) Using Avogadro's hypothesis, calculate the ratio of the weight of the oxygen molecule to the carbon oxide molecule. (*b*) Explain the statement: It is impossible to have more than one atom of oxygen in the carbon oxide molecule if oxygen molecules are diatomic. [Consider the ratio of the molecular weights calculated in part (*a*).]

2.12 Calculate the mass in grams of $\frac{1}{2}$ mol of (*a*) neutrons, (*b*) protons, (*c*) electrons.

2.13 Calculate the charge in coulombs on (*a*) 1 million electrons and (*b*) on the protons in 1 g atom of oxygen.

2.14 Calculate the number of gram-atomic weights (g atoms) in (*a*) 50.0 g Al, (*b*) 100 g Ba, (*c*) 155 g Br, and (*d*) 95.5 g S.

2.15 Calculate the number of atoms in (*a*) 9.00 g Al, (*b*) 16.0 g S, (*c*) 24.0 g C, and (*d*) 46.0 g Na.

2.16 Calculate the number of moles in (*a*) 20.0 g H_2SO_4, (*b*) 73.0 g HCl, (*c*) 64.0 g CH_4O (methanol), and (*d*) 5.00 kg $BaCl_2$.

2.17 Calculate the number of molecules *and atoms* in all parts of Prob. 2.16.

2.18 Calculate the weight percentage of each element in the following compounds: (*a*) HI, (*b*) C_2H_6O, (*c*) NO_2, (*d*) $CaCO_3$, (*e*) $K_2Cr_2O_7$.

2.19 A compound of Mg and Br is 13.05 percent by weight Mg. Calculate the empirical formula for the compound.

2.20 A compound of Ca and F is 51.3 percent by weight Ca. Calculate the empirical formula for the compound.

2.21 A compound of Na, O, and Cl is 25.4 percent Na and 35.4 percent O by weight. Calculate the empirical formula for the compound.

2.22 The hydrocarbon ethylene has a gram-molecular weight of 28 g. It is 14.3 percent hydrogen by weight. Calculate its empirical and molecular formulas.

2.23 Isopropanol contains C, H, and O only. It is 60 percent carbon and 13.3 percent hydrogen by weight. Its molecular weight is 60 g. Determine the empirical and molecular formulas.

2.24 Using the mass numbers of the isotopes and the information in Tables 2.2 and 2.3, calculate the average gram-atomic weight of naturally occurring (*a*) hydrogen and (*b*) oxygen.

2.25 Write formulas for (*a*) hydrogen bromide, (*b*) sodium dichromate, (*c*) potassium chromate, (*d*) silver(I) silicate, (*e*) copper(I) sulfate, (*f*) mercury(I) carbonate, (*g*) ammonium sulfite, (*h*) lead(II) sulfide, (*i*) zinc oxide, (*j*) calcium iodide, (*k*) barium hydroxide, (*l*) magnesium nitrate, (*m*) copper(II) nitrite, (*n*) mercury(II) chlorite, (*o*) iron(II) chlorate, (*p*) tin(II) bromide, (*q*) chromium(II) bromide, (*r*) aluminum nitrate, (*s*) iron(III) chloride, (*t*) chromium(III) iodide.

2.26 Name the following: (*a*) $HBrO_2$, (*b*) NaBr, (*c*) KCl, (*d*) $AgClO_3$, (*e*)

$CuClO_2$, (*f*) Hg_2I_2, (*g*) NH_4NO_2, (*h*) $Pb(NO_3)_2$, (*i*) $Zn(OH)_2$, (*j*) CaI_2, (*k*) BaO, (*l*) MgO_2, (*m*) CuS, (*n*) $HgSO_3$, (*o*) $FeCO_3$, (*p*) $SnSO_4$, (*q*) $CrSiO_3$, (*r*) $Al_2(CrO_4)_3$, (*s*) $FePO_4$, (*t*) $Na_2Cr_2O_7$, (*u*) $Cr(NO_3)_3$.

2.27 Balance the following equations; name all reactants and products:
(*a*) $MgCl_2 + AgNO_3 \rightarrow AgCl + Mg(NO_3)_2$
(*b*) $H_2SO_4 + Ba(OH)_2 \rightarrow BaSO_4 + H_2O$
(*c*) $FeCl_3 + NH_4OH \rightarrow Fe(OH)_3 + NH_4Cl$
(*d*) $H_2O_2 \rightarrow H_2O + O_2$
(*e*) $Hg_2(NO_3)_2 + HCl \rightarrow Hg_2Cl_2 + HNO_3$

2.28 The combustion of gasoline can be represented by the unbalanced equation

$$C_8H_{18} + O_2 \rightarrow CO_2 + H_2O$$
Octane

(*a*) Balance the equation. (*b*) How many moles of oxygen are required to burn 10 mol of octane completely? (*c*) What weight of CO_2 is produced when 100 g of octane undergoes complete combustion?

2.29 Silver chloride precipitates as a white solid when an aqueous solution of silver nitrate is treated with hydrogen chloride:

$$AgNO_3(aq) + HCl(g) \rightarrow AgCl(s) + HNO_3(aq)$$

(*a*) Balance the equation. (*b*) How many moles of $AgNO_3$ are required to produce 0.0500 mol of AgCl? (*c*) Calculate the number of grams of AgCl which can be produced from 2.50 kg of $AgNO_3$ and an excess of HCl.

2.30 Phosphorus pentachloride, PCl_5, decomposes into phosphorus trichloride, PCl_3, and chlorine gas, Cl_2. (*a*) Write a balanced chemical equation for this reaction. (*b*) How many grams of Cl_2 can be produced from the decomposition of 100 g of PCl_5? (*c*) If 5 g of Cl_2 is produced during the decomposition of a sample of PCl_5, how many moles of PCl_3 are also produced, and how many grams of PCl_5 are consumed?

2.31 Metallic iron actually burns with a flame when heated in gaseous chlorine. The product is iron(III) chloride. (*a*) Write a balanced equation for the reaction. (*b*) How many grams of iron(III) chloride can be produced from 1.00 lb of iron? (*c*) The density of gaseous chlorine is 2.96 g/l. What volume of chlorine gas is required to burn 1.00 lb of iron?

2.32 Hydrogen peroxide decomposes into oxygen gas and water. (*a*) Write the balanced chemical equation for the decomposition. (*b*) What weight of oxygen gas is produced by the decomposition of 5.00 g of hydrogen peroxide? (*c*) How many oxygen *molecules* are produced by the decomposition of 5.00 g of hydrogen peroxide?

2.33 The combustion of elemental sulfur yields sulfur dioxide gas. (*a*) Keeping in mind that the molecules of elemental sulfur are *oct*atomic, i.e., contain eight atoms per molecule, write a balanced chemical equation for the combustion. (*b*) What weight of sulfur dioxide in grams can be produced from 1.00 ton of sulfur? (*c*) How many oxygen molecules (not moles) are required for the combustion of 1.00 ton of sulfur?

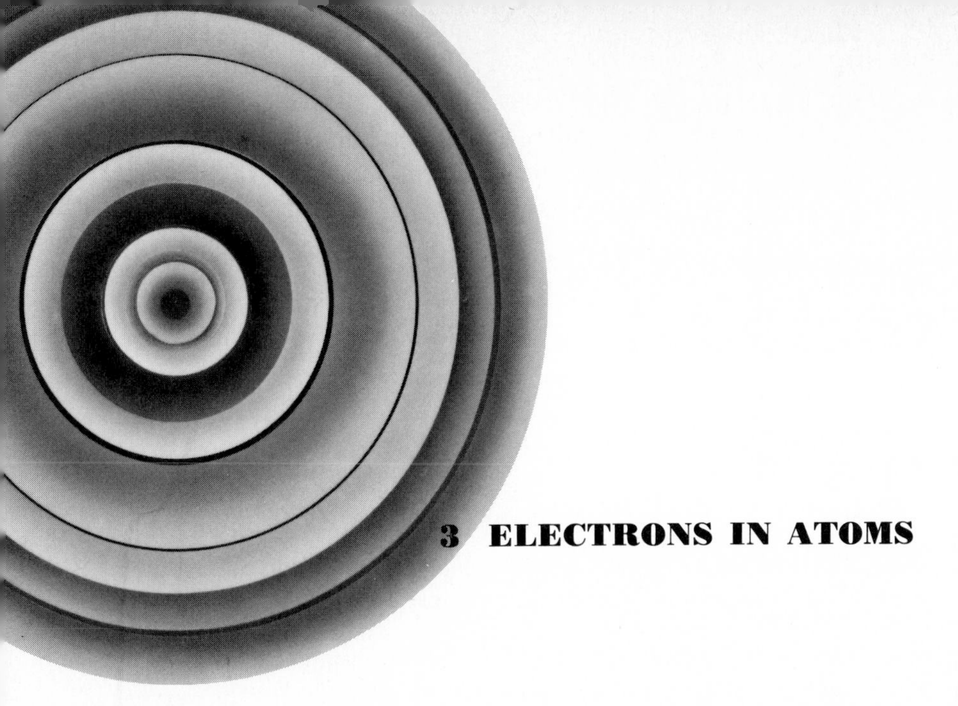

3 ELECTRONS IN ATOMS

The nineteenth century may be characterized as the century of the atom. Much of the chemical activity of the century was expended toward understanding the concept and making it useful to chemistry. The twentieth may be looked upon as the century of the electron, a subatomic particle which related the atom to spectroscopy and led to a new understanding of chemical combination.†

3.1 ELECTRONS AND CHEMICAL BONDS

The basic picture of atomic structure was presented in Chap. 2. In this chapter we describe the important experiments which led to the modern atomic model. It is important to understand *why* we are so concerned with this model and the behavior of electrons in atoms and molecules.

Dalton's model was very simple; his atoms were *micro*spheres,

† Aaron J. Ihde, "The Development of Modern Chemistry," Harper & Row, New York, 1964.

FIGURE 3.1

Structures of the water and methane molecules.

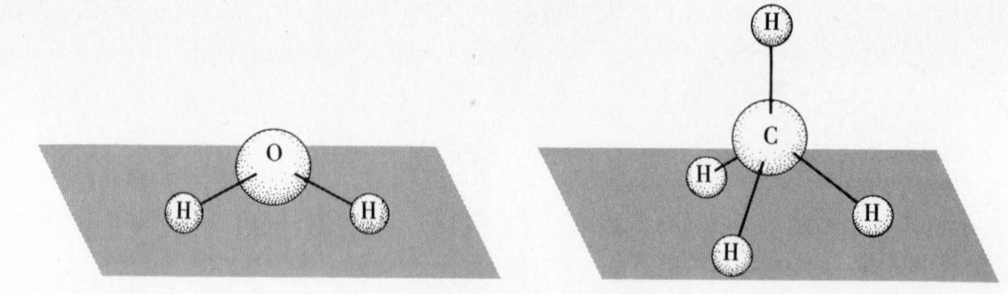

but this simple model cannot account for the chemical bond. In some way, atoms are able to form molecules, i.e., stable aggregates in which the atoms are held together by a strong force. What is the nature of this bonding force?

The geometry of some simple molecules

Consider two relatively simple molecules, water, H_2O, and methane, CH_4. In the water molecule, an oxygen atom is bonded to two hydrogen atoms; all three atoms lie in a plane, and the H—O—H bond angle is 105°. In methane, the carbon atom is bonded to four hydrogen atoms, which may be considered to occupy the vertices of a regular tetrahedron with the carbon atom in the center (Fig. 3.1).

Experimentally, we find that in many different molecules a carbon atom is bonded to four other atoms arranged tetrahedrally about it. An oxygen atom, however, is usually bonded to two other atoms, and the bond angle is always close to 105°. A force acting between atoms causes them to form bonds *in a definite number* and *at definite angles.* This force results in a chemical bond. Because we believe that it is the outer electrons of atoms which are most important in the formation of these bonds, chemists are very much concerned with the electronic structure of atoms.

3.2 IDENTIFICATION OF THE ELECTRON

Cathode rays, which are invisible, can be produced in an apparatus like the one depicted in Fig. 3.2. This apparatus, called a *cathode tube*, is an ancestor of the modern television picture tube. Sealed

FIGURE 3.2

Thomson's cathode tube.

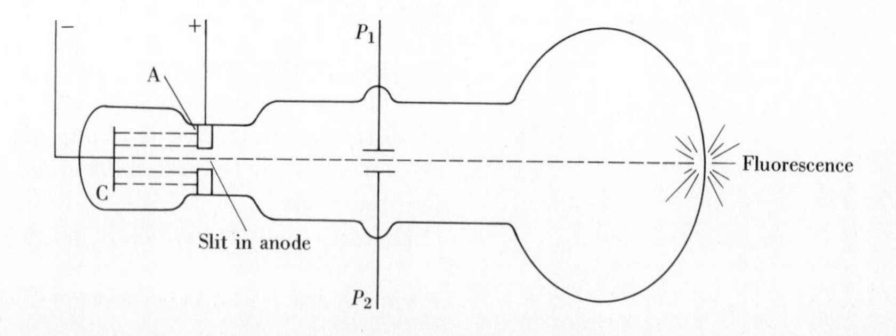

through the left side of the cathode tube is an electrode C, which is connected to the negative side of a high-voltage supply; this is a *cathode*. When the cathode tube is operating the cathode is heated by an electric filament which is not shown in the figure. Another electrode A is sealed through the neck of the cathode tube; it is connected to the positive side of a high-voltage supply and called an *anode*. The anode has a slit in its center. When all the air in the cathode tube is drawn out by a very efficient pump and a high-voltage source is connected across A and C, light is emitted at spot F at the end of the tube. Apparently the voltage generates an invisible ray which emanates from the cathode, passes through the slit in the anode, and strikes the glass at the other end of the cathode tube, causing the glass to light up, or *fluoresce*. In the latter part of the nineteenth century, scientists could not decide whether the cathode ray was a stream of particles or an electromagnetic wave similar to ordinary light. Joseph John Thomson (1856–1940) entered this controversy in the 1890s and showed that the fluorescing spot could be moved up or down by a small electric field produced by connecting the plates (P_1 and P_2 in Fig. 3.2) to a battery. When P_1 was negative, F moved down; when P_2 was negative, F moved up. The deflection indicated that the cathode ray is negative. In addition, Thomson already knew that the rays are deflected by magnetic fields, further indication that the ray is electrically negative. Thomson concluded that the cathode ray is a stream of negatively charged particles. From the extent of the deflection of the rays in electric and magnetic fields he was able to calculate the ratio of the charge to mass, e/m_e. In other experiments, he made a crude determination of the charge e of the particle, which we now call the *electron*. Present values for e/m_e and e are 1.75×10^8 C/g and 1.6021×10^{-19} C, respectively (C = coulomb).

EXAMPLE 3.1 From the ratio of the charge to mass e/m_e and the charge e of an electron calculate the mass of one electron.

Solution: The charge on a single electron is e. If we knew the mass per unit charge, we could multiply by e and determine m_e. The ratio e/m_e is the charge per unit mass, and m_e/e is the mass per unit charge.

Facts: $e = 1.60 \times 10^{-19}$ C/electron

$$\frac{m_e}{e} = \frac{1}{1.75 \times 10^8 \text{ C/g}} = \frac{1}{1.75} \times 10^{-8} \text{ g/C}$$

$$m_e = e \times \frac{m_e}{e} = (1.60 \times 10^{-19} \text{ C/electron}) \left(\frac{1}{1.75} \times 10^{-8} \text{ g/C}\right)$$

$$= 9.16 \times 10^{-28} \text{ g/electron}$$

The properties of cathode rays are independent of the materials from which the cathode tube is constructed. The value of e/m_e is 1.75×10^8 C/g whether the electrodes are made of copper, silver, or other metals; Thomson concluded that the electron is a universal constituent of matter. He proposed that atoms are uniformly dense positive spheres in which electrons are imbedded. Thomson's atomic model, which has been called the "grapes in Jello model," satisfactorily explained the results of his experiments with cathode rays, but it was not useful in explaining the chemical reactions of the elements or the nature of the chemical bond.

3.3 DISCOVERY OF THE NUCLEAR ATOM

Ernest Rutherford (1871–1937), a student of Thomson's, studied the spontaneous emanations, the alpha (α), beta (β), and gamma (γ) rays, produced by radioactive elements, and showed that the α ray is a rapid stream of helium ions, He^{2+}. In 1909 Hans Geiger and Ernest Marsden were continuing the investigation of the properties of α particles in Rutherford's laboratories at Cambridge University (Fig. 3.3). This work and its implications are described in the following excerpt from an address given by Rutherford before the Royal Academy in 1911.

In the early days, I had observed the scattering of α-particles, and Dr. Geiger in my laboratory had examined it in detail. He found in thin pieces of heavy metal that the scattering was usually small, of the order of one degree. One day Geiger came to me and said, "Don't you think that young Marsden, whom I am training in radioactive methods, ought to begin a small research?" Now I had thought that too, so I said, "Why not let him see if any α-particles can be scattered through a large angle?" I may tell you in confidence that I did not believe they would be, since we knew that the α-par-

FIGURE 3.3

α-particle scattering.

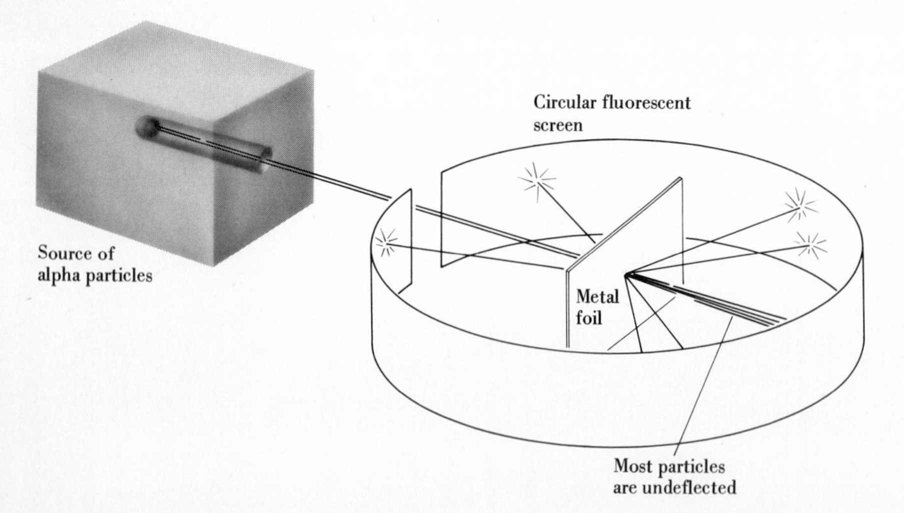

ticle was a very fast massive particle, with a great deal of energy, and you could show that if the scattering was due to the accumulated effect of a number of small scatterings, the chance of an α-particle's being scattered backwards was very small. Then I remember two or three days later, Geiger coming to me in great excitement and saying, "We have been able to get some of the α-particles coming backwards." . . . It was quite the most incredible event that has ever happened to me in my life. It was almost as incredible as if you fired a 15-inch shell at a piece of tissue paper and it came back and hit you. On consideration, I realized that this scattering backwards must be the result of a single collision; and when I made calculations, I saw it was impossible to get anything of that order of magnitude unless you took a system in which the greater part of the mass of the atom was concentrated in a minute nucleus.

Since most of the α particles passed through the metal foil with little or no deflection, Rutherford concluded that the foil was primarily void space. Atomic models like Dalton's or Thomson's, in which there is a uniform density of mass, proved incorrect. Such atoms could not cause α particles to be deflected backward. Rutherford's calculations showed that the part of the atom which could account for the large observed deflections must be extremely dense and positively charged. Eventually Rutherford described a *solar-system atom;* in each atom is a tiny, positively charged nucleus which accounts for almost all the mass of the atom; the electrons are in orbits about the nucleus. Rutherford estimated the nuclear diameter at approximately 10^{-12} cm and the atomic diameter 10,000 times larger, or 10^{-8} cm.

3.4 ELECTROMAGNETIC RADIATION AND VISIBLE LIGHT

Visible light is an electromagnetic radiation to which our eyes are sensitive. The electromagnetic spectrum is shown in Fig. 3.4, where it can be seen that the visible part of this spectrum is only a narrow band. The sun's rays make up a much broader spectrum, but we can only "see" the radiation within narrow limits. Electromagnetic

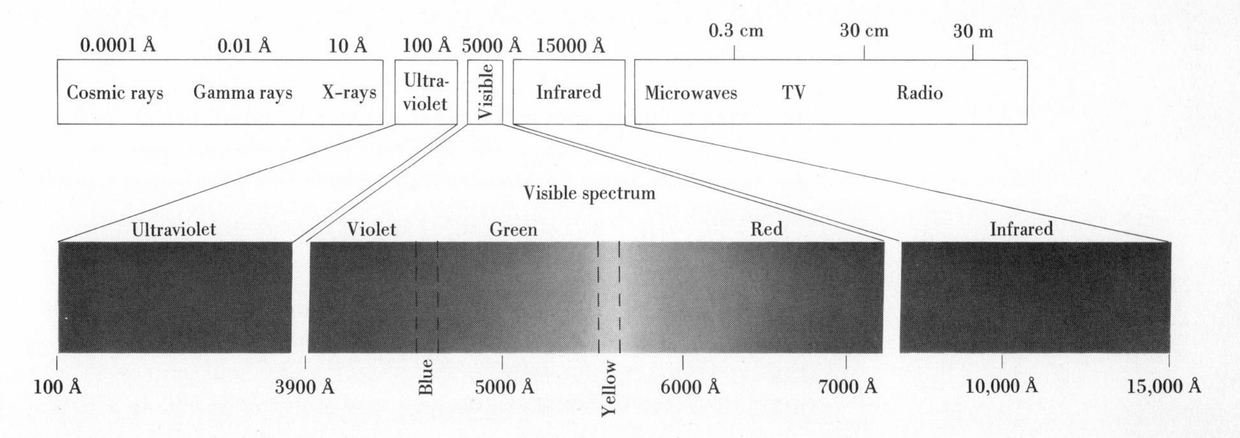

FIGURE 3.4

The electromagnetic spectrum.

radiation produces regularly fluctuating electric and magnetic fields, or waves, as it moves. We can characterize the radiation by its *wavelength* λ and its velocity c. Light is visible electromagnetic radiation with wavelengths between 3900 and 7000 Å.

Experimentally, it has been found that the velocity of light c in a vacuum is 2.9979×10^{10} cm/s no matter what its wavelength. The *frequency* ν of electromagnetic radiation can be calculated from its wavelength and the velocity of light. The frequency is the number of waves passing a point in 1 s.

EXAMPLE 3.2 Calculate the frequency of red radiation having a wavelength of 6500 Å. Use 3.00×10^{10} cm/s for the velocity of light.

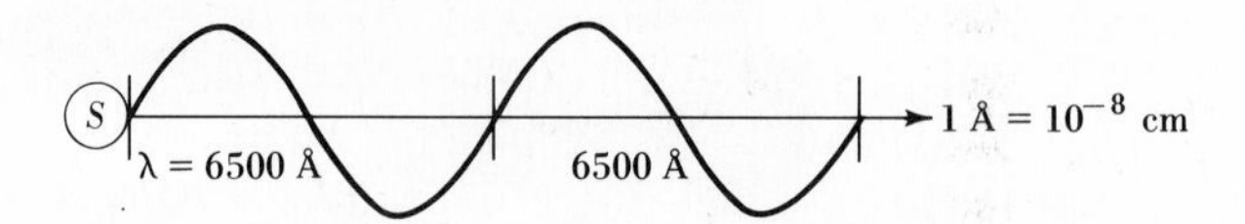

Solution: Each second the radiation travels 3×10^{10} cm. The number of waves having a length of 6.5×10^{-5} cm which pass a point in 1 s is

$$\frac{3 \times 10^{10} \text{ cm/s}}{6.5 \times 10^{-5} \text{ cm}} = 4.6 \times 10^{14} \text{ s}^{-1}$$

Therefore, the frequency of this radiation is 4.6×10^{14} waves per second, or 4.6×10^{14} s^{-1}. The unit "waves per second" is not ordinarily used; s^{-1} is read "per second" or "reciprocal seconds." As can be seen from this simple calculation, the relationship between wavelength and frequency is

$$\frac{c}{\lambda} = \nu \qquad \text{or} \qquad \nu\lambda = c$$

3.5 QUANTUM THEORY

Max Planck (1858–1947) set the stage for the scientific revolution which was to occur in the twentieth century by postulating that electromagnetic radiation can be regarded as a beam of energy packets called *photons* or *quanta*. He deduced that the energy ε of a single photon, or quantum, is directly proportional to the frequency of the radiation:

$$\varepsilon = h\nu$$

The proportionality constant h, called *Planck's constant*, has the units of energy × time. In calorie-seconds $h = 1.58 \times 10^{-34}$, and in erg-seconds $h = 6.63 \times 10^{-27}$. A calorie is the unit of energy required to raise the temperature of 1 g of water at 14.5°C to 15.5°C.

EXAMPLE 3.3 Calculate the energy ε of one photon of red light, $\lambda = 6500$ Å, and the total energy E of 1 mol of such photons.

Solution: From the wavelength in angstrom units we find the wavelength in centimeters:

$$\lambda = 6500\ \text{Å} \times \frac{1\ \text{cm}}{10^8\ \text{Å}} = 6.5 \times 10^{-5}\ \text{cm}$$

since 1 Å $= 10^{-8}$ cm. Then, we calculate the frequency in s^{-1}:

$$\nu = \frac{c}{\lambda} = \frac{3 \times 10^{10}\ \text{cm/s}}{6.5 \times 10^{-5}\ \text{cm}} = 4.6 \times 10^{14}\ \text{s}^{-1}$$

From the frequency, we calculate the energy of one photon, using Planck's equation:

$$\begin{aligned}\varepsilon &= h\nu = (1.58 \times 10^{-34}\ \text{cal-s})\ (4.6 \times 10^{14}\ \text{s}^{-1}) \\ &= 7.4 \times 10^{-20}\ \text{cal per photon}\end{aligned}$$

The energy of 1 mol of photons is calculated by finding the total energy of N_A photons:

$$E = N_A \times \varepsilon = (6.02 \times 10^{23})\ (7.4 \times 10^{-20}\ \text{cal}) = 44{,}500\ \text{cal}$$

3.6 LINE SPECTRA OF THE ELEMENTS

After Robert Bunsen and Gustav Kirchhoff invented the spectroscope (1859), scientists studied the light emitted by the elements when they were heated to incandescence. When gaseous elements are heated or exposed to high voltages, only certain colors or frequencies are emitted; in fact, the emission spectrum is a characteristic of an element and can be used for its identification (Fig. 3.5). Because of this phenomenon, helium was known to exist on the sun 20 years before it was discovered on the earth. During an eclipse in 1868 the spectrum of the solar fringe was obtained by a French astronomer, P. J. C. Janssen. A careful spectroscopic analysis proved that one of the lines could not be attributed to any element known on earth. The line was assigned to a new element, helium, from the Greek *helios*, "sun."

Figure 3.5 shows the visible *emission* spectrum for atomic hydrogen. There are sharp lines (the Balmer series) at 6562 Å (red), 4861 Å (blue), 4340 Å (violet), 4102 Å (violet), and 3970 Å (violet). A sketch of the experimental apparatus used to obtain the emission spectrum of hydrogen and other gases is shown in Fig. 3.6. To obtain an emission spectrum the gas is energized or excited by a high-voltage source; the excited gas emits light, which is separated into its colored components by a prism. The separated components strike a photographic plate and produce five differently colored images of the slit, corresponding to the different wavelengths. When

FIGURE 3.5

Visible emission spectra can be used to identify an element. The characteristic emission lines are superimposed over the continuous spectrum.

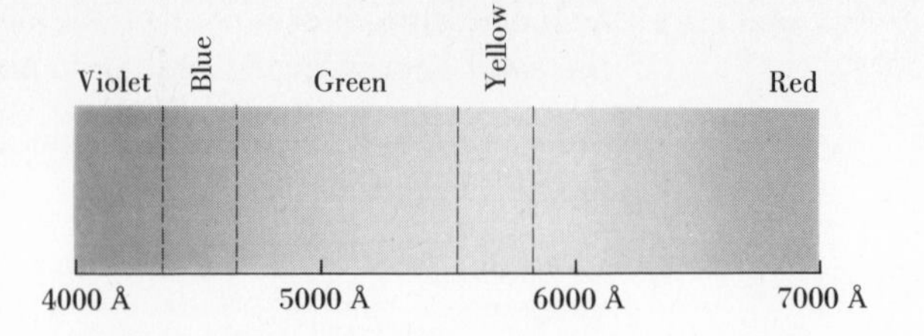

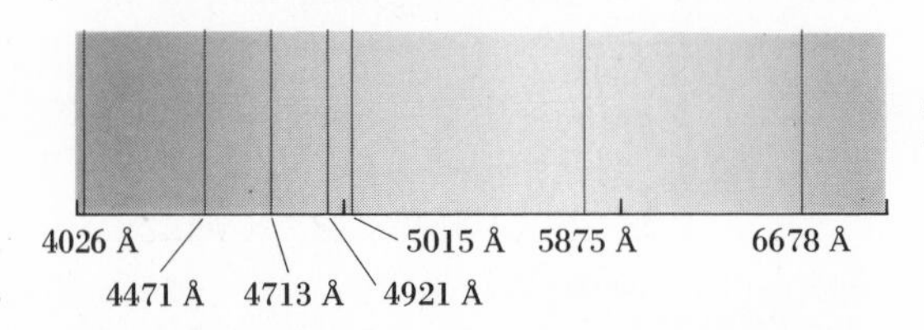

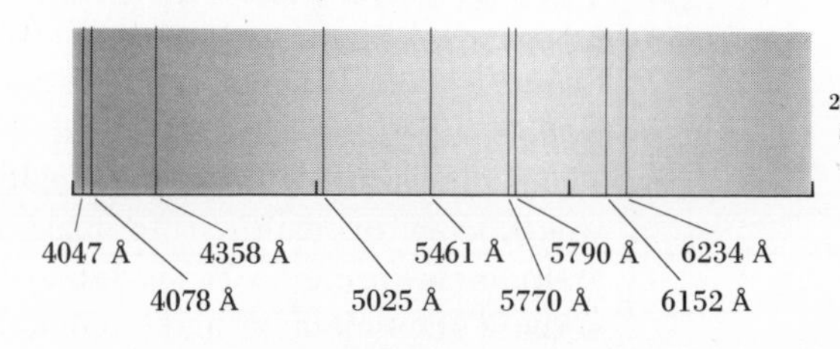

FIGURE 3.6

Apparatus for studying hydrogen emission.

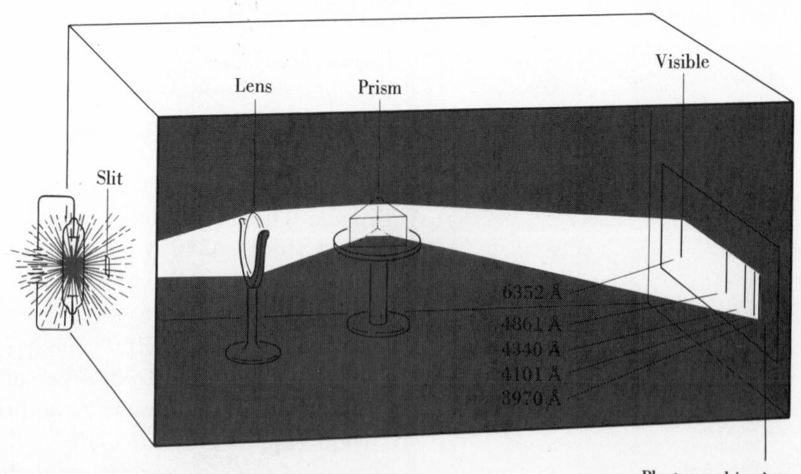

light-absorption experiments are performed, the "opposite" result is obtained. If a beam of "white" light is passed through a sample of atomic hydrogen gas, all the light is transmitted except at the wavelengths 6562, 4861, 4340, 4102, and 3970 Å. The photographic plates from emission and absorption experiments are photographic opposites (Fig. 3.7). During emission experiments energy in the form of light of definite wavelength is emitted; during absorption experiments only light of the same wavelength (or energy) is absorbed.

3.7 THE BOHR MODEL OF THE HYDROGEN ATOM

By 1911 the groundwork for the next giant step in the evolution of the atom concept had been laid.

1. Thomson had identified the electron as a particle of negative charge which is a universal constituent of matter (1898).

2. Planck had presented the quantum theory of electromagnetic radiation (1899).

3. Rutherford had presented evidence for the existence of a nuclear atom (1911).

Niels Bohr (1885–1962) integrated these ideas and produced a model for the atom which permitted him to calculate the line positions in the hydrogen spectrum. His model was very useful in explaining the chemical and physical properties of the elements.

Bohr pictured the hydrogen atom as a system in which a single electron orbits a nucleus having a single positive charge. He started with the Rutherford solar-system model but imposed certain constraints on the energy and motion of the electrons:

1. The single electron moves in circular orbits with definite radii. No other orbits are permitted (Fig. 3.8*a*).

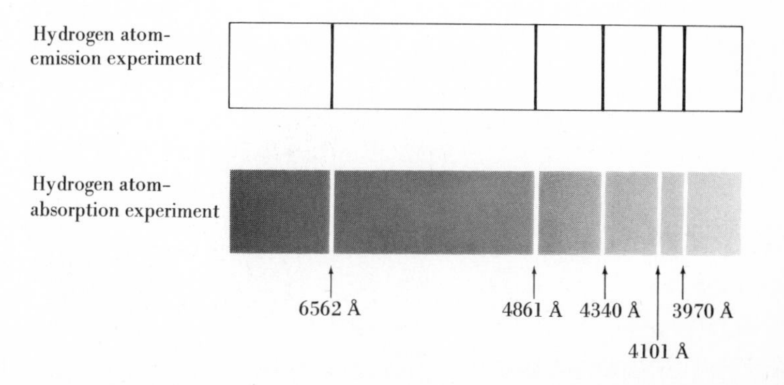

FIGURE 3.7

Plates from emission and absorption experiments.

2. The electrons in such orbits are in *stationary states* of energy; i.e., the electrons possess a definite, fixed energy. While an electron is in a stationary state, it does not emit or absorb energy. When it falls to a lower stationary state, it emits energy in the form of a photon, or quantum of light. When an electron absorbs energy in the form of a photon, or quantum of light, it is elevated to a higher stationary state, i.e., an orbit farther from the nucleus (Fig. 3.8*b*).

3. The energy of a photon emitted by an atom when an electron falls from one stationary state to a lower one is equal to the difference in energy between the two stationary states:

$$\varepsilon_{\text{photon}} = \Delta\varepsilon = \varepsilon_f - \varepsilon_i$$

(Δ, the Greek letter delta, stands for "difference in.") The frequency of the photon can be calculated from Planck's equation:

$$\varepsilon_{\text{photon}} = \varepsilon_f - \varepsilon_i = \Delta\varepsilon = h\nu$$

From this model Bohr was able to obtain mathematical expressions for r_n, the radii of the orbits, and ε_n, their associated energies:

$$r_n = \frac{n^2h^2}{(2\pi)^2me^2} \qquad n = 1, 2, 3, \ldots \tag{3.1}$$

$$\varepsilon_n = -\frac{2\pi^2me^4}{n^2h^2} \qquad n = 1, 2, 3, \ldots \tag{3.2}$$

where m = mass of electron
e = charge of electron
h = Planck's constant
n = number of orbit or stationary state

n is also frequently referred to as the energy level or shell. Since m, e, h, and π are constants, the energy and the radius of an electron in a hydrogen atom depend only on the value of n in Eqs. (3.1) and (3.2). Since n must be an integer, the energy and radius are discon-

FIGURE 3.8

Bohr's hydrogen atom.

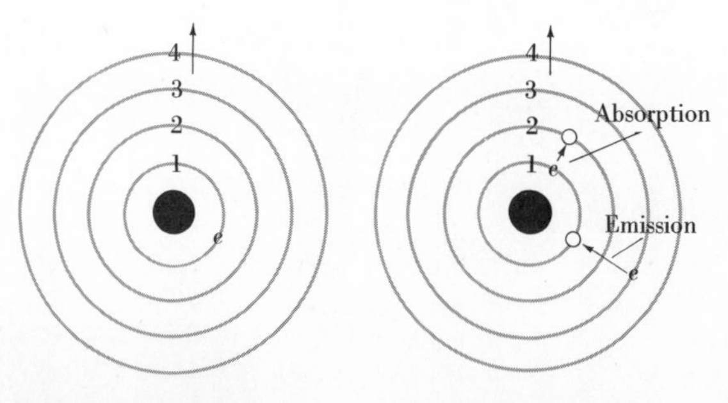

(a) Bohr's hydrogen atom The electron may be in circular orbits of fixed radii or in stationary states

(b) The mechanisms for absorption and emission

TABLE 3.1: ENERGY LEVELS OF THE HYDROGEN ATOM

n	*Shell*	*Radius of electron orbit r* [*Eq.* (*3.1*)], Å	*Energy*† ε *of electron*, cal × 10^{19}
1	*K*	0.529	−5.20928
2	*L*	2.12	−1.30232
3	*M*	4.76	−0.57889
4	*N*	8.46	−0.32581
5	*O*	13.22	−0.20862
6	*P*	19.04	−0.14487
7	*Q*	25.92	−0.10662
∞	—	∞	0

† The energy of the electron when it is at an infinite distance from the nucleus is defined as zero; the energy at any finite distance is lower, and therefore all energy values are negative and approach zero as r approaches infinity.

tinuous, only certain values being allowed. Substitution of the values for the constants in Eq. (3.2) gives

$$\varepsilon_n = -\frac{2.1797 \times 10^{-11}}{n^2} \text{ erg} = -\frac{5.20928 \times 10^{-19}}{n^2} \text{ cal}$$

Table 3.1 shows the energy levels and radii for the electron orbits of the hydrogen atom calculated from Bohr's equations [Eqs. (3.1) and (3.2)]. When the electron is in the first, or lowest, orbit ($n = 1$), its radius is 0.529 Å and its energy is -5.20×10^{-19} cal.

Since the radius of an orbit is some multiple of $h^2/(2\pi)^2me^2$, this quantity is called the *Bohr radius unit* and is symbolized by a_0. For simplification Eq. (3.1) can be rewritten $r = n^2a_0$, where $a_0 =$ 0.529 Å.

We can use the energies associated with the electron orbits to calculate the frequencies of the lines in the hydrogen spectrum. A comparison of the calculated and experimental frequencies is a test of the validity of the Bohr model.

According to Bohr's picture, a photon of light is emitted when an electron falls from an orbit to a lower one. The energy of the emitted photon is equal to the difference between the electron's energy in the higher orbit and in a lower orbit. The Bohr theory permits us to calculate the energy of the photons emitted. From the energy, the frequency and wavelength can be calculated.

Let us calculate the energy of the photon emitted by a hydrogen atom when the electron falls from the third orbit ($n = 3$) to the second orbit ($n = 2$); the difference in energy is

$$\Delta\varepsilon = |\varepsilon_2 - \varepsilon_3|$$

From Table 3.1,

$$\varepsilon_3 = -0.57889 \times 10^{-19} \text{ cal} \qquad \text{and} \qquad \varepsilon_2 = -1.30232 \times 10^{-19} \text{ cal}$$

therefore

$$\Delta\varepsilon = |0.72343 \times 10^{-19} \text{ cal}|$$

The energy of the photon emitted when the electron falls from the third to the second orbit is 0.72×10^{-19} cal. From Planck's equation the frequency of the photon can be calculated:

$$\varepsilon_{\text{photon}} = h\nu$$

or

$$\nu = \frac{\varepsilon}{h} = \frac{0.72343 \times 10^{-19} \text{ cal}}{1.5834 \times 10^{-34} \text{ cal-s}} = 4.5686 \times 10^{14} \text{ s}^{-1}$$

The wavelength of the photon is

$$\lambda = \frac{c}{\nu} = \frac{2.9979 \times 10^{10} \text{ cm/s}}{4.5686 \times 10^{14} \text{ s}^{-1}}$$

$$= 6.562 \times 10^{-5} \text{ cm} = 6562 \times 10^{-8} \text{ cm} = 6562 \text{ Å}$$

This wavelength corresponds exactly to the red line observed in the hydrogen spectrum. It is the line emitted when the electron falls from orbit 3 to orbit 2. In the same way we can calculate the energies of the photons which would be emitted when the electron falls from orbit 4 to orbit 2, from orbit 5 to orbit 2, and so on. The results of these calculations are shown in Table 3.2, which also shows the difference in energy $\Delta\varepsilon$ between these orbits and the frequencies ν and wavelengths λ of the expected radiation. The first five transitions give rise to the visible lines of the hydrogen spectrum which were first observed by Balmer (see Figs. 3.6 and 3.7). The accuracy of Bohr's predictions was incredible and gave strong support to his atomic model.

The Balmer series of lines caused by the electron's falling from orbits above the second into the second orbit is only one such set of lines which can be observed in the hydrogen spectrum. The *Lyman series* arises when the electron falls from the second or higher orbits into the first orbit ($n \geqslant 2 \rightarrow n = 1$). Other series have also been observed. From the Bohr theory the line positions for the Lyman series ($n \geqslant 2 \rightarrow n = 1$), the Paschen series ($n \geqslant 4 \rightarrow n = 3$), and the Brackett series ($n \geqslant 5 \rightarrow n = 4$) can also be predicted with high accuracy. The lines of these series do not fall in the visible region of the spectrum. Figure 3.9 shows the electronic transitions within the hydrogen atom responsible for Lyman, Paschen, and Balmer series.

3.8 THE BOHR THEORY AND OTHER ATOMS

From the physical and chemical properties of the elements, Bohr was able to determine the number of electrons in each orbit for atoms with more than one electron. He concluded that the maximum number of electrons in an orbit, or energy level, n was just equal to $2n^2$. Also, he found that electrons in the second and higher

TABLE 3.2: SUMMARY OF TRANSITIONS FROM $n > 2$ TO $n = 2$ FOR HYDROGEN ATOM (THE BALMER SERIES)

Transition	$\Delta\varepsilon = \varepsilon_n - \varepsilon_2 = \varepsilon_{photon}$, cal $\times 10^{19}$	$\nu = \frac{\varepsilon_{photon}}{h}$, $s^{-1} \times 10^{-14}$	$\lambda \left(= \frac{c}{\nu}\right) \times 10^8$	
$n = 3 \rightarrow n = 2$	0.72343	4.5686	6562 Å, Red	Visible
$n = 4 \rightarrow n = 2$	0.97641	6.1669	4861 Å, Green-blue	Visible
$n = 5 \rightarrow n = 2$	1.09370	6.9069	4340 Å, Blue	Visible
$n = 6 \rightarrow n = 2$	1.15745	7.3090	4102 Å, Violet	Visible
$n = 7 \rightarrow n = 2$	1.19570	7.5513	3970 Å, Violet	Visible
$n = 8 \rightarrow n = 2$	——	——	3889 Å	Invisible Ultraviolet
$n = 9 \rightarrow n = 2$	——	——	3835 Å	Invisible Ultraviolet
$n = \infty \rightarrow n = 2$	——	——	3030 Å	Invisible Ultraviolet

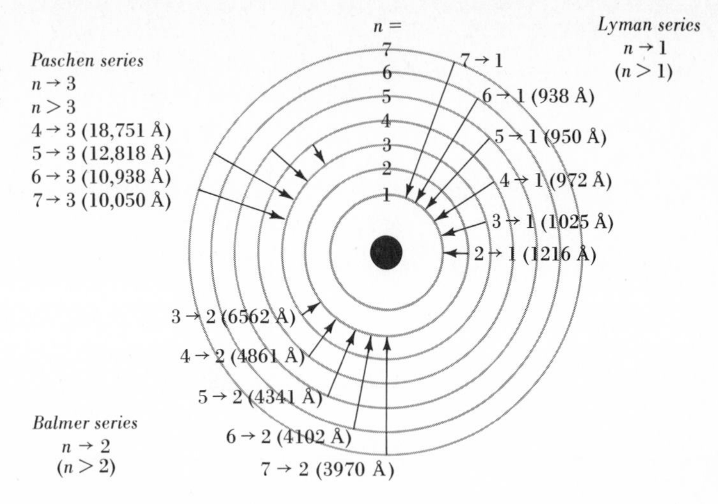

FIGURE 3.9

The transitions which give rise to the Lyman, Balmer, and Paschen series of the hydrogen spectrum.

orbits seemed to form subgroups within an orbit. There are two subgroups within the second orbit; one has a maximum of two electrons, and the other has a maximum of six electrons. There are three subgroups in the third orbit, four subgroups in the fourth orbit, etc. These results are summarized in Table 3.3. Figure 3.10 shows the Bohr models for some of the heavier atoms. Note that the first orbit on the *K* shell never has more than two electrons. The first and second orbits are filled in the neon atom; there are two electrons in the *K* shell and eight electrons in the *L* shell.

The Rutherford-Bohr solar-system atom, which was developed in the first two decades of the twentieth century, led to a unification of principles and a better understanding of many physical and chemical properties of matter. Furthermore, it proved that atoms are extremely complex. However, the Bohr model had its flaws. From

TABLE 3.3: ELECTRONIC STRUCTURE OF THE ATOM ACCORDING TO BOHR

Orbit	*Shell*	*Maximum number of electrons in orbit* ($2n^2$)	*Sublevel grouping*
$n = 1$	*K*	2	2
$n = 2$	*L*	8	2, 6
$n = 3$	*M*	18	2, 6, 10
$n = 4$	*N*	32	2, 6, 10, 14

FIGURE 3.10

Bohr models for the hydrogen, helium, beryllium, oxygen, neon, and sodium.

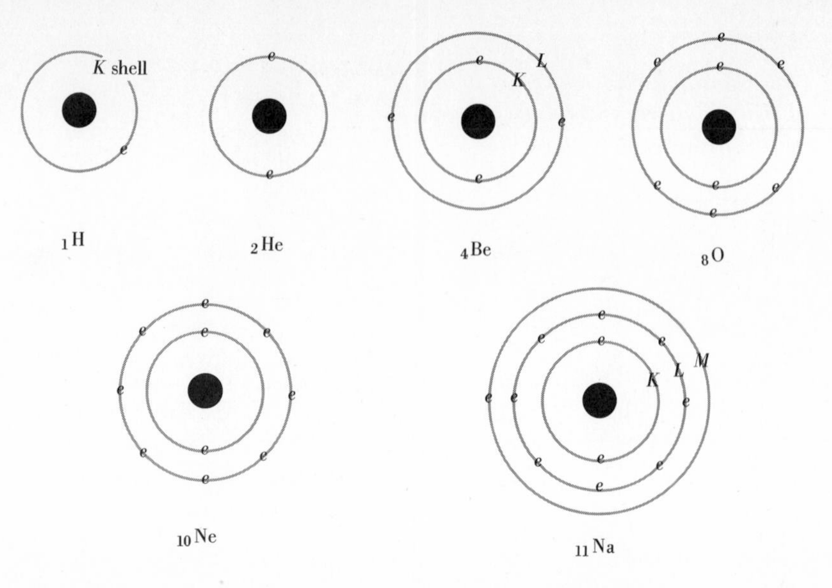

the principles of classical physics it can be shown that an accelerating charge, i.e., a charge moving in a curved path, must radiate energy. Physicists theorized that if an electron moved about the nucleus, it should lose energy by radiation and spiral into the nucleus. Of course, this does not happen, but nevertheless the Bohr atom was not supported by sound theory. It should not be too surprising that the success of the Bohr theory in explaining the hydrogen spectrum was not repeated for any other element. It failed even for the simple helium atom, which has only two electrons and a nucleus. Continued theoretical investigations showed that Bohr's model is accurate only for an atomic system which consists of a nucleus and a single orbiting electron like the hydrogen atom; the helium ion, He^{+}; and the lithium ion, Li^{2+}. Also as spectroscopic techniques became more refined, the single lines (singlets) in the hydrogen spectrum proved to be two closely spaced lines, or *doublets*. The Bohr theory was modified, but it became obvious that another approach would be necessary.

3.9 THE WAVE NATURE OF ELECTRONS

The electrons in the Bohr atom have fixed radii and well-defined energy states. Two discoveries of the 1920s showed that it is impossible to be so certain about the state of affairs in an atom or a molecule. In 1924, Louis de Broglie suggested that electrons may have wave properties; 25 years earlier Max Planck guessed that light waves have some particle nature, and he considered a wave to be a stream of photons, or quanta; de Broglie suggested that the converse may also be true: small particles may have wave properties. He *assigned* a wavelength to the electron, which until then was considered to have only the properties of a particle. He also presented a simple equation relating the wavelength of an electron to its mass m and velocity v:

$$\lambda = \frac{h}{mv} = \frac{h}{p} \qquad mv = \text{momentum} = p$$

where h is Planck's constant. De Broglie's hypothesis was confirmed when it was found that electron beams undergo diffraction, a phenomenon characteristic of waves.

EXAMPLE 3.4 Calculate the wavelength of an electron traveling at one-half the speed of light.

Solution: To solve the problem we need the values of m_e, h, and v. The mass of the electron m_e is 9.108×10^{-28} g. Planck's constant h is 1.58×10^{-34} cal-s or 6.63×10^{-27} erg-s. The speed of light in vacuum is 2.998×10^{10} cm/s. In this calculation it is more convenient to use Planck's constant in erg-seconds than in calorie-seconds. Using de Broglie's equation, we now calculate the wavelength λ of the electron:

$$\lambda = \frac{h}{mv} = \frac{6.63 \times 10^{-27}\ \text{erg-s}}{(9.11 \times 10^{-28}\ \text{g}) \times \frac{1}{2}(3 \times 10^{10}\ \text{cm/s})}$$

$$= \frac{4.83 \times 10^{-10}\ \text{erg-s}}{\text{g-cm/s}}$$

From the fact that

$$1\ \text{erg} = 1\ \text{dyn-cm} = 1\ \frac{\text{g-cm}}{\text{s}^2}\ \text{cm} = 1\ \frac{\text{g-cm}^2}{\text{s}^2}$$

we have

$$\lambda = 4.83 \times 10^{-10}\ \text{cm} = 0.0483 \times 10^{-8}\ \text{cm} = 0.0483\ \text{Å}$$

Such an electron has a wavelength equivalent to gamma rays.

Uncertainty principle

Werner Heisenberg made the next important contribution. His theoretical work on atomic structure led to the *uncertainty principle*, which states that it is impossible to determine simultaneously the exact position and exact velocity of an electron. Actually, the principle applies to all bodies, but it is significant only when the mass is very small. This is not due to any inadequacy of the apparatus employed in an experiment but is a fundamental constraint in nature.

The *uncertainty principle* can best be understood by example. Suppose you were asked to measure the thickness of a sheet of paper with an unmarked meterstick. Obviously, the result would be extremely inaccurate and meaningless. In order to obtain any accuracy you would have to use an instrument graduated in units smaller than the thickness of a sheet of paper. Analogously, in order to measure an electron or determine its position, we must use a "meterstick" calibrated in units smaller than the dimensions of an electron. To sense an electron, we can cause "light" or electromagnetic radiation to interact with it. The "light" must have a wavelength smaller than the dimensions of an electron. But such "light" would possess very high energy, and the photons of such light would change the energy of the electrons by collision. We would be able to calculate the position of the collision, but we would know very little about the energy or velocity of the electron after the collision.

Determining the position of an electron by its interaction with photons is analogous to measuring the length of a room with a bulldozer. By driving the bulldozer across the room we might show that the room *was* $3\frac{1}{2}$ bulldozers long and $2\frac{1}{2}$ bulldozers wide. After the measurement we would be very uncertain of the state (energy) of the room.

In view of de Broglie's hypothesis and Heisenberg's uncertainty principle, Erwin Schrödinger (1927) proposed a *wave equation* to describe the behavior of an electron:

$$\frac{\delta^2\psi}{\delta x^2} + \frac{\delta^2\psi}{\delta y^2} + \frac{\delta\psi}{\delta z^2} + \frac{8\pi^2 m}{h^2}(E - V)\,\psi = 0$$

The Schrödinger equation is a complicated second-order differential equation which is difficult to solve. It is presented here only so that an important qualitative feature can be pointed out: ψ (psi), called the *wave function*, may be thought of as the *amplitude* of the electron wave; m is the mass of the electron. Thus the Schrödinger equation

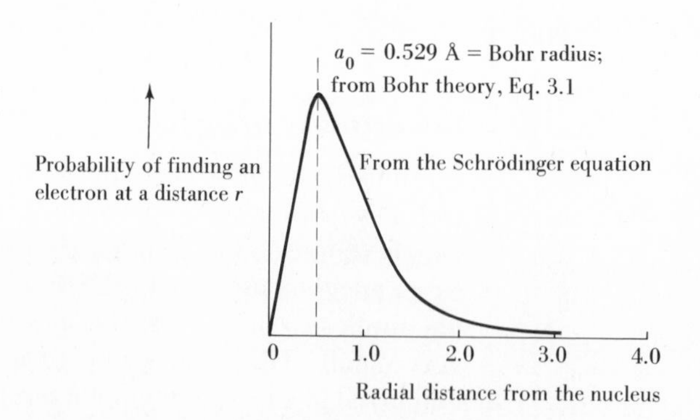

FIGURE 3.11

The electron in the hydrogen atom can be at various distances from the nucleus.

endows the electron with a wave and particle nature simultaneously; it has both a mass and an associated wavelength, in keeping with the experimental facts.

From the Bohr theory, Eq. (3.1), we can determine the *exact* radius of the electron in any orbit of the hydrogen atom. The radii of the first, second, and third orbits in the Bohr model of the hydrogen are a_0, $4a_0$, and $9a_0$, respectively, and no other distances are permitted. From the Schrödinger equation we find that the most probable position of the electron in the unexcited hydrogen atom is at radial distance of a_0, or 0.529 Å, from the nucleus but that there is a probability of finding the electron at other distances (Fig. 3.11).

Because of the uncertainty principle, we must be satisfied with a theory which will permit us to calculate only the probable location of electrons in atoms. Our best atomic models will be abstract, mathematical equations which describe atomic properties and behavior within the theoretical limit of the uncertainty principle. The Bohr model of the atom is too sharp.

To illustrate the wave-mechanical picture of the electron in an atom let us perform an imaginary (impossible) experiment with the ground-state hydrogen atom. We "take a picture" of a single atom thousands of times, using the same photographic plate to obtain a multiple exposure. The hypothetical result is shown in Fig. 3.12*a*, where the dots representing the electron have formed a spherically symmetrical *cloud*, or *density*, about the nucleus. There is no definite radius or boundary for the atom. For convenience, chemists have envisioned imaginary surfaces of constant electron density about the nucleus, and we speak in terms of the shapes of these imaginary surfaces. For the electron in the ground-state hydrogen atom, surfaces of constant electron density would be spherical and symmetrical about the nucleus. In Fig. 3.12*b* we see a constant-density surface within which the probability of finding the electron is 90 percent. The diameter of this imaginary sphere is about 2.8 Å. When the electron in the hydrogen atom absorbs energy, it is elevated to states in which the electron cloud may have a different shape.

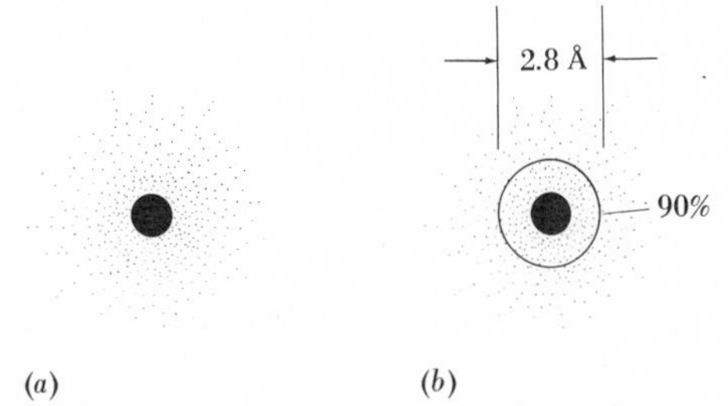

FIGURE 3.12

The electron in the 1s orbital of hydrogen.

2.8 Å

90%

(*a*) (*b*)

3.10 QUANTUM NUMBERS

The mathematical solution of the Schrödinger equation for electrons in atoms cannot be presented at this introductory level, but the results of the solution are meaningful and will be useful when we study the chemical bond and molecular structure. Using the wave-mechanical picture of electrons in atoms, we can begin to understand how electrons are involved in the formation of bonds between atoms. Also, from the shape of the electron cloud, as deduced from the Schrödinger equation, we can predict the structure of a molecule, i.e., the arrangement of atoms in molecules and the bond angles. We shall consider the hydrogen atom in some detail because it is the simplest case and forms the basis for understanding the wave-mechanical treatment of all other atoms.

The Bohr theory involves the use of a *quantum number* or integer n, which we called the *orbit, shell,* or *energy level.* The energy and radius of the orbit of an electron are *precisely determined* in the Bohr theory if the value of the quantum number n is specified. This is obvious from Eqs. (3.1) and (3.2), in which the only variable quantity is n. The solution of the Schrödinger equation for the hydrogen atom requires the introduction of three quantum numbers. These are integers related to the energy and probable location of the electron and the shape of the electron cloud. A fourth quantum number must also be assigned to the electron because the three quantum numbers which arise theoretically from the solution of the Schrödinger equation are not sufficient to explain all the observed properties of the electrons in atoms. The fourth quantum number is introduced to make up for this deficiency. It is called the *spin quantum number* because the electron can be visualized as if it were spinning about its axis as it moves around the nucleus; and it appears to spin in one of two directions. The fourth quantum number indicates spin direction.

3.11 QUANTUM NUMBERS AND ATOMIC ORBITALS

When the Schrödinger equation is solved for the hydrogen atom, we find that the energy of the single orbiting electron is determined solely by the *principal quantum number n*. In fact, the energy as determined by the Schrödinger equation is equal to the energy as determined from the Bohr theory for the hydrogen atom:

$$\varepsilon_{\text{Schrödinger}} = \varepsilon_{\text{Bohr}} = -\frac{2\pi^2 me^4 Z^2}{n^2 h^2} \tag{3.3}$$

where Z is the charge on the nucleus. Since for hydrogen $Z = +1$, Z was not shown in Eq. (3.2). The quantum number n is also called the *energy level* of the electron. The higher the energy level the higher the energy of the electron.

The principal quantum number can have any positive integer value. Its lowest value is 1, and therefore in the first energy level $n = 1$; this is equivalent to the K shell of the Bohr atom. The principal quantum number replaces the capital letters for the shells in the Bohr description. The maximum number of electrons which can be accommodated in any energy level n equals $2n^2$. In the first energy level the maximum number of electrons is 2; in the second the maximum number is 8; and so on, just as Bohr had deduced (see Table 3.3).

The angular-momentum quantum number l specifies the angular momentum of the electron and, more important to us, the "shape" of the electron cloud. Together with the principal quantum number, l determines the *average* distance $\bar{r}$ between the electron and the nucleus:

$$\bar{r} = \frac{n^2}{Z}\left\{1 + \frac{1}{2}\left[1 - \frac{l(l+1)}{n^2}\right]\right\}\frac{h^2}{4\pi^2 me^2} \tag{3.4}$$

or, since $h^2/4\pi^2 me^2 =$ Bohr radius $a_0 = 0.529$ Å,

$$\bar{r} = \frac{n^2}{Z}\left\{1 + \frac{1}{2}\left[1 - \frac{l(l+1)}{n^2}\right]\right\} 0.529 \text{ Å}$$

where n, l = quantum numbers
Z = atomic number = charge on nucleus
m = mass of electron
e = charge of electron

Remember that we cannot know the exact distance separating the electron and the nucleus if at the same time we have an accurate value of the energy. Equation (3.3) gives us an accurate value of the energy. In accordance with uncertainty principle, we cannot also determine the position of the electron exactly. We cannot calculate the exact separation r of the electron and the nucleus, but we can calculate the average separation $\bar{r}$.

The angular momentum quantum number l can have any integral value from 0 to $n - 1$. In the first energy level, l can equal 0 only because the maximum value of l, $n - 1$, is $1 - 1$, or 0. In the second energy level the maximum value of l is 1 because $n - 1$ equals 1, and therefore in the second energy level l can be 0 and 1.

EXAMPLE 3.5 Calculate the average separation between the hydrogen nucleus and the electron in the first energy level.

Solution: In the first energy level $n = 1$, and l can only be 0. The charge in the nucleus is $+1$; $l = 0$, $n = 1$, $Z = +1$. Substitution into Eq. (3.4) gives

$$\bar{r} = \frac{n^2}{Z}\left\{1 + \frac{1}{2}\left[1 - \frac{l(l+1)}{n^2}\right]\right\}\frac{h^2}{4\pi^2 me^2}$$

$$= \frac{1^2}{1}\left\{1 + \frac{1}{2}\left[1 - \frac{0(0+1)}{1^2}\right]\right\}\frac{h^2}{4\pi^2 me^2}$$

But $h^2/4\pi^2 me^2$ = Bohr radius $a_0 = 0.529$ Å, and therefore

$$\bar{r} = \tfrac{3}{2}\frac{h^2}{4\pi^2 me^2} = \tfrac{3}{2}(0.529\ \text{Å}) = 0.793\ \text{Å}$$

The average separation in the first energy level is 1.5 times greater than the Bohr radius. The electron is not to be considered to be at a fixed distance of $1.5a_0$; it may be closer to, or farther from, the nucleus at any instant, but on the average the separation is $1.5a_0$. Do not confuse the *average* separation with the most probable separation. As we learned previously, the most probable separation of the electron and the nucleus is just equal to the Bohr radius.

In the quantum-mechanical description of electrons in atoms, we refer to the *orbital*, in which an electron resides. The term *orbit*, used in the discussion of Bohr theory, connotes a circular or spherical path of motion, but the term orbital does not have this connotation. The orbital is determined by the value of l: when $l = 0$, the electron is in an s orbital. In this case, the *probability* of finding an electron depends only on the distance from the nucleus. Surfaces of constant electron density and the electron cloud of an s orbital are spherically symmetrical about the nucleus (see Fig. 3.13*a*). In the first energy level, where l can only equal 0, the only type of orbital is an s orbital. The s orbital of the first energy level is called the 1s orbital, indicating that $n = 1$ and $l = 0$.

In the second energy level l can be 0 and 1. With $l = 0$ we have a 2s orbital, indicating $n = 2$ and $l = 0$. The 2s orbital is spherically symmetrical about the nucleus. When $l = 1$, the electron is said to be in a p orbital. There are *three* p orbitals in every level above the first. The p orbital of the second energy level is called the 2p orbital, meaning that $n = 2$ and $l = 1$. In such orbitals the surfaces of constant electron density enclose volumes on opposite sides of the nucleus. The electron clouds of the p orbitals resemble a figure eight. Figure 3.13*b* shows the three p orbitals of the second energy level. Note that these clouds are taken to lie along the x, y, and z axes and are therefore mutually perpendicular. In Fig. 3.14*a* cross sections of the electron cloud in the $2p_z$ orbital are shown; the lines of constant electron density represent those surfaces which enclose 50 and 90 percent of the electron cloud.

In the third energy level, $n = 3$, the maximum value of l is 2. Hence, with $n = 3$ we can have $l = 0$, to give the 3s orbital; $l = 1$, to give the 3p orbitals; and $l = 2$. When $l = 2$, the orbital is called a d orbital; there are *five* such orbitals in all energy levels above the sec-

FIGURE 3.13

Surfaces of constant electron density.

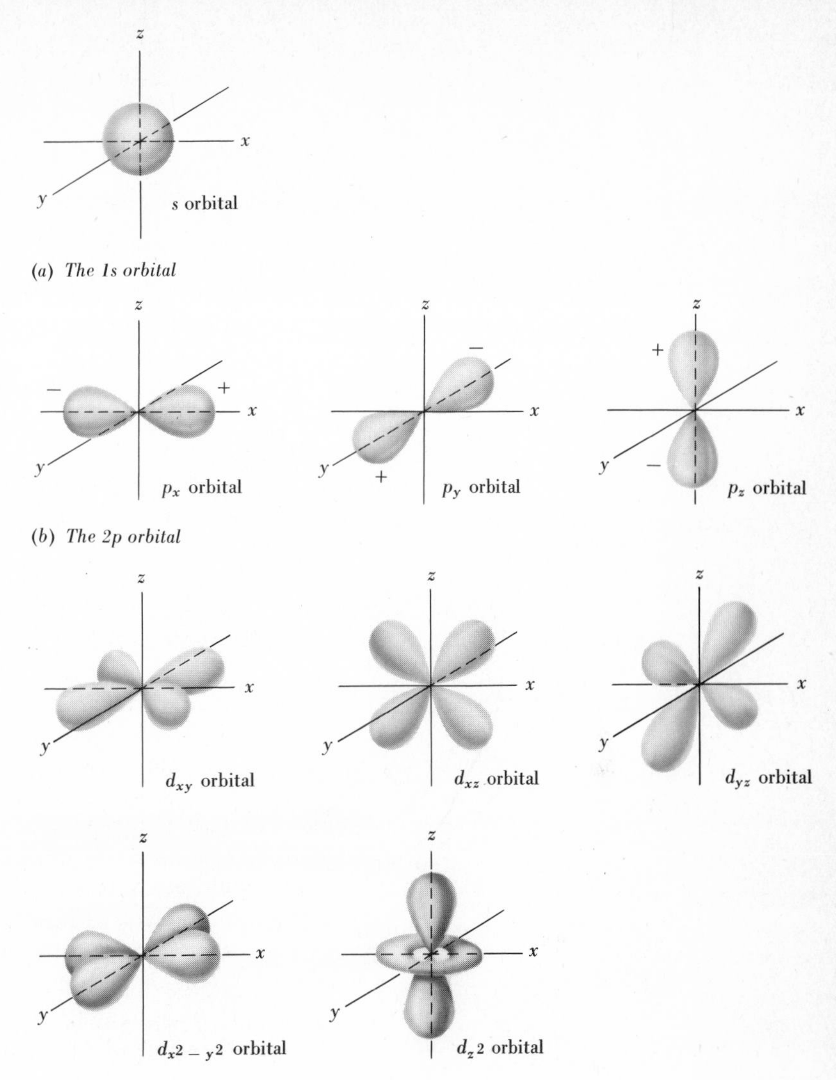

ond. Three-dimensional representations of the electron clouds of the *d* orbitals in the third energy level are shown in Fig. 3.13*c*. Four of the five *d*-orbital clouds resemble a three-dimensional four-leaf clover. The fifth resembles a $2p_z$ cloud with a doughnut about the middle. In Fig. 3.14*b* cross sections of the surfaces of constant electron density for two of these orbitals are shown.

When $l = 3$, we speak of *f* orbitals. Seven *f* orbitals are present in all energy levels above $n = 3$. The electron cloud of a 4*f* orbital has a very complicated geometry and will not be shown here. The *f* orbitals are important for understanding the properties of the rare-earth elements (the lanthanides).

FIGURE 3.14

Cross sections of the electron clouds in the 2p and 3d orbitals.

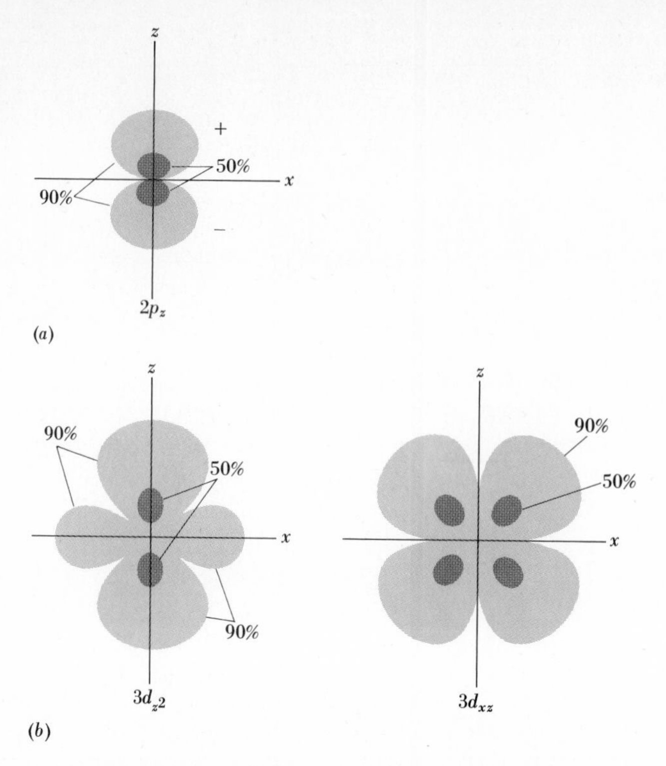

The *magnetic quantum number m* characterizes the orientation of the electron cloud in a magnetic field. To understand why an electron cloud should orient in a magnetic field remember that a moving charge, such as an electron in an atom, sets up a magnetic field perpendicular to the direction of its motion. This is the basis of electromagnets. A coil of wire carrying an electric current sets up a magnetic field. As a result, the coil can cause a deflection in a nearby magnet, or it can be itself deflected by a magnet, as shown in Fig. 3.15.

Because the electron is a moving charge, its motion and therefore the electron cloud can be affected by a magnetic field. We can picture a magnetic field as causing an electron cloud to take up new orientations in space. Because of the nature of the electron in an atom only certain fixed orientations of the cloud are allowed. These correspond to the different permissible values of the magnetic quantum number m, which can have all integral values in the range from $-l$ through 0 to $+l$. When $n = 1$ and $l = 0$, m can only equal 0. The number of different values of m, that is, $2l + 1$, is called the *multiplicity* and equals the number of orbitals of a particular value of l in any energy level. We can now understand why there are only one s orbital, three p orbitals, five d orbitals, and seven f orbitals in a particular energy level. When $l = 0$, m can only equal 0 and $2l + 1$

equals 1; there is only one s orbital. When $l = 1$, m can equal -1, 0, and $+1$; there are three p orbitals. When $l = 2$, m can equal $-2, -1$, 0, $+1$, and $+2$; there are five d orbitals.

Pauli exclusion principle

Before we discuss the fourth quantum number, we must introduce the Pauli exclusion principle, which states that *no two electrons in an atom can have the same set of four quantum numbers*. This principle is not derived from theory. It is an experimentally discovered principle, and it is certainly reasonable, for if there are different electrons in an atom, they must differ in some way capable of description.

Upon closer examination we find that "single" lines of the hydrogen spectrum are really very closely spaced double lines, or *doublets*. In addition, when a beam of electrons is passed through a very strong magnetic field, it is split into two beams. These apparently unrelated phenomena can be explained if we imagine that an electron could rotate about an axis in the same way that the earth rotates about its own axis as it moves in its path around the sun. Since it is known that moving charges set up magnetic fields, the rotation of an electron about an axis should set up a magnetic field. Since a beam of electrons is split into two beams, we must conclude that if electrons are spinning, they can spin in only one of two directions. The concept of a spinning electron is a convenient model which explains our observations, but there is no direct evidence that the electron actually spins.

The *spin quantum number* s specifies the spin direction. It can only be $-\frac{1}{2}$ or $+\frac{1}{2}$.† When $n = 1$, $l = 0$, s can equal $\pm\frac{1}{2}$. Therefore, in

† At this stage it is only important to realize that the spin quantum number can have two values. It is not important to understand why it is $\frac{1}{2}$ or $-\frac{1}{2}$.

FIGURE 3.15

The effect of a magnetic field on the orientation of magnets. Electrons, which because of their motion act like magnets, are also affected by magnetic fields.

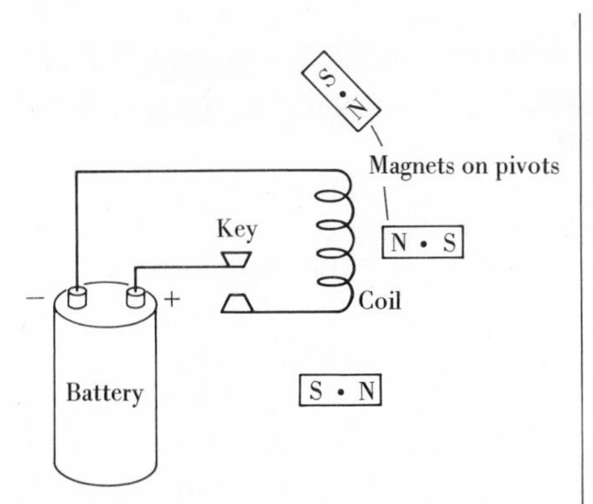

(a) Open circuit
Spatial orientations of magnets is random around the coil when it carries no current

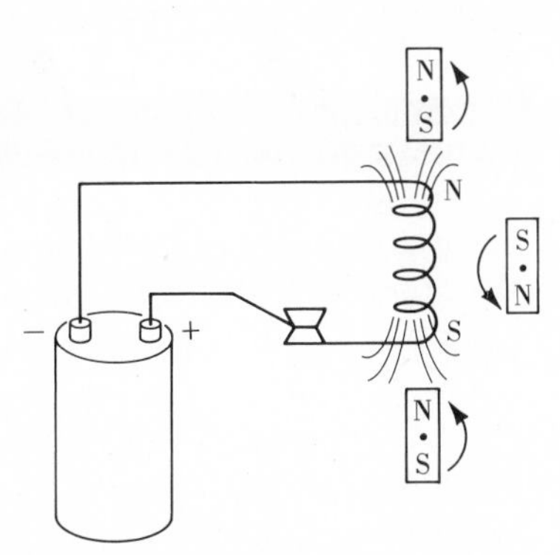

(b) Closed circuit
Magnets align themselves in the magnetic field of the coil when it carries current

the first energy level, i.e., when $n = 1$, there is only one value of l $(=0)$ and one value of m $(=0)$. There are two possible values of s $(\pm\frac{1}{2})$. There is one orbital which is described by the three quantum numbers ($n = 1$, $l = 0$, $m = 0$), and it can be occupied by two electrons which have opposite spins. Therefore, the Pauli exclusion principle can be rephrased: *one orbital can accommodate two electrons provided their spins are opposite, or antiparallel.* The quantum-number description of the two electrons in the helium atom would be

	n	l	m	s
First electron	1	0	0	$-\frac{1}{2}$
Second electron	1	0	0	$+\frac{1}{2}$

or in shorthand notation He $1s^2$, meaning that in the $1s$ orbital there are two electrons having opposite or paired spins. Note that the first energy level is filled when only two electrons are present. This is consistent with the rule that the maximum number of electrons in a given energy level is $2n^2$.

3.12 MULTIELECTRON ATOMS

The electrons in multielectron atoms are assumed to reside in orbitals resembling hydrogen orbitals, and the wave functions for multielectron atoms are modified hydrogen wave functions. Each electron is assigned four quantum numbers, n, l, m, and s. These quantum numbers completely determine the wave function and energy of the electron just as in the hydrogen atom. In multielectron atoms the energy of an electron is determined by the values of n *and* l, in contrast to the one-electron atom, where the electron energy is completely determined by the value of n alone, as indicated by Eq. (3.3). In Fig. 3.16, where the orbital energies for hydrogen, helium, and lithium are compared, it is obvious that the dependence of the energy on the angular-momentum quantum number l increases as the atomic number Z increases.

The specification of the four quantum numbers for each electron in an atom is equivalent to determining the electronic structure of the atom. Before we can learn to assign quantum numbers to the electrons in atoms we must consider the *Aufbau principle,*† which states that in determining the electronic structure of the ground state (unexcited state) of an atom, we can imagine all electrons as being removed from the nucleus and then being fed back one at a time into the available orbitals, filling the lowest-energy orbitals first. Figure 3.17 is a mnemonic diagram of the order in which elec-

† The name comes from the German for "construction" or "building up."

FIGURE 3.16

The energies in multielectron atoms depend on n and l.

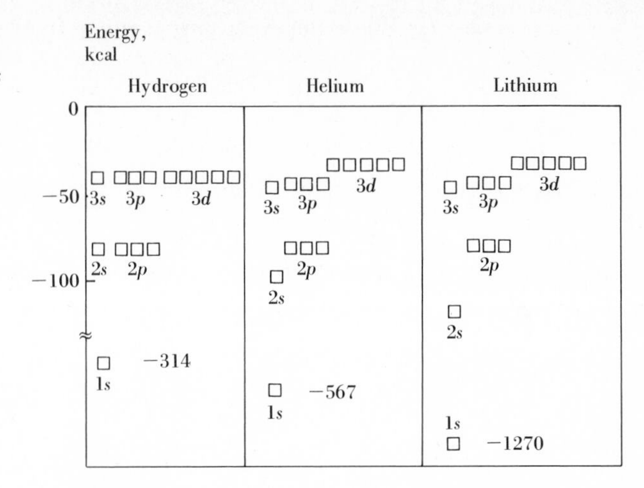

trons fill the available orbitals in atoms other than hydrogen. The order taken from the diagram is 1*s*, 2*s*, 2*p*, 3*s*, 3*p*, 4*s*, 3*d*, 4*p*, 5*s*, 4*d*, 5*p*, 6*s*, etc. This order is not followed unfailingly, yet it is fairly accurate. Note that the 4*s* orbital begins to fill before the 3*d* orbitals.

EXAMPLE 3.6 Write the electronic configuration for lithium; atomic number $Z = 3$.

Solution: The first two electrons have the same quantum numbers as the two electrons in helium, shown on page 60. The third goes into the 2*s* orbital ($n = 2$, $l = 0$, $m = 0$, $s = -\frac{1}{2}$) where we have arbitrarily chosen $s = -\frac{1}{2}$ to represent the lower value of the spin quantum number. The electronic configuration of lithium is

$_3$Li $1s^2$, $2s^1$

The 2*s* orbital is of lower energy than the 2*p*. The order of increasing orbital energy within any energy level is *s*, *p*, *d*, *f*, . . . (or $l = 0$, 1, 2, 3, . . .) for multielectron atoms.

FIGURE 3.17

Order of the orbital energies in multielectron atoms.

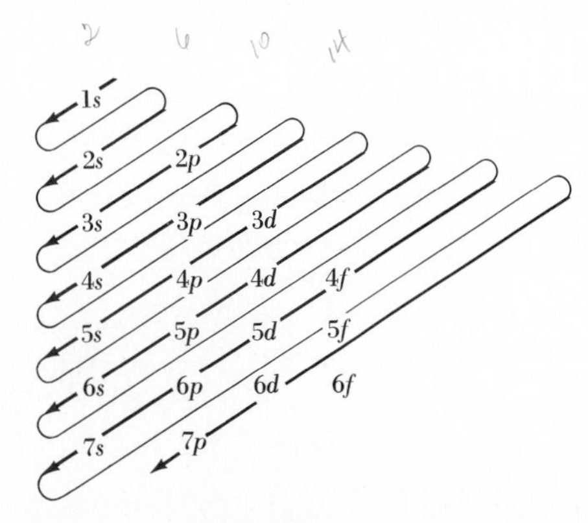

EXAMPLE 3.7 Write the electronic configuration for carbon; $Z = 6$.

Solution: In this case, the first four electrons will fill the 1*s* and 2*s* orbitals. There will be two more electrons available for the 2*p* orbitals which are of the next highest energy. But there are three equivalent *p* orbitals. Do we put the two electrons in the same *p* orbital, or do we put them in separate orbitals? In the ground, or lowest-energy, state, the *p* electrons would be in separate orbitals as explained by *Hund's rule: electrons tend to remain unpaired, or have parallel spins, or have the same spin quantum number.* This is reasonable since electrons have like charges and repel each other. If they were present in the same orbital, the repulsion would be maximized. The electrons tend to occupy different orbitals in order to remain as far apart as possible. Because of repulsion, pairing up, or occupying the same orbital, requires energy, and therefore the electronic description for carbon is

$$_{6}C\ 1s(\uparrow,\downarrow),\ 2s(\uparrow,\downarrow),\ 2p_x(\uparrow,-),\ 2p_y(\uparrow,-),\ 2p_z(-,-)$$

Each arrow represents an electron. When the arrow points up, by convention $s = -\frac{1}{2}$; when it points down, $s = +\frac{1}{2}$; also, we could write

$$_{6}C\ 1s^2,\ 2s^2,\ 2p_x^{\ 1},\ 2p_y^{\ 1},\ 2p_z^{\ 0} \qquad \text{or} \qquad _{6}C\ 1s^2,\ 2s^2,\ 2p^2$$

Two of the *p* orbitals ($2p_x$ and $2p_y$) are half filled; one *p* orbital ($2p_z$) is completely vacant.

The subscripts on the *p*'s indicate that the three *p* orbitals are mutually perpendicular, i.e., directed along the *x*, *y*, and *z* axes. These three electronic descriptions for carbon are equivalent: the first gives the most explicit information, and the second gives more explicit information than the last, but any one description implies the other two.

EXAMPLE 3.8 Write the set of four quantum numbers for each electron in the neon atom, $Z = 10$, in the order in which the electrons would fill the orbitals.

Solution: By Fig. 3.17, the first four electrons will completely fill the 1*s* and 2*s* orbitals. The next three electrons will go into the 2*p* orbitals, but by Hund's rule they will have the same value of the spin quantum number.

Since these three electrons go into the 2*p* orbitals, *n* equals 2 and *l* equals 1 for all of them. Also, since they all have the same spin quantum number, $s = -\frac{1}{2}$ for all of them. Therefore, the *n*, *l*, and s quantum numbers for the fifth, sixth, and seventh electrons are identical:

Electron	*n*	*l*	*m*	s
Fifth	2	1	?	$-\frac{1}{2}$
Sixth	2	1	?	$-\frac{1}{2}$
Seventh	2	1	?	$-\frac{1}{2}$

According to the Pauli exclusion principle, no two electrons in the same atom can have the same set of four quantum numbers. Then the only remaining quantum number m must have different values for these three electrons. The values of m are -1, 0, and $+1$ for the fifth, sixth, and seventh electrons, respectively. The next three electrons, 8, 9, and 10, will then pair up with the fifth, sixth, and seventh electrons by taking an s value of $+\frac{1}{2}$. The complete electronic description of the 10 electrons of neon is tabulated. The fifth and eighth

Orbital	*n*	*l*	*m*	s	*Order in which electrons fill available orbitals*
1*s*	1	0	0	$-\frac{1}{2}$	1
	1	0	0	$+\frac{1}{2}$	2
2*s*	2	0	0	$-\frac{1}{2}$	3
	2	0	0	$+\frac{1}{2}$	4
$2p_x$	2	1	-1	$-\frac{1}{2}$	5
	2	1	-1	$+\frac{1}{2}$	**8**
$2p_y$	2	1	0	$-\frac{1}{2}$	6
	2	1	0	$+\frac{1}{2}$	**9**
$2p_z$	2	1	$+1$	$-\frac{1}{2}$	7
	2	1	$+1$	$+\frac{1}{2}$	**10**

electrons go into the $2p_x$ orbital, the sixth and ninth go into the $2p_y$, and the seventh and tenth go into the $2p_z$. Note that the arbitrary convention is to consider $-\frac{1}{2}$ as the lower energy value. The electronic configuration of neon according to the above assignments is

$_{10}$Ne $1s^2$, $2s^2$, $2p^6$

EXAMPLE 3.9 Write the electronic configuration for scandium; $Z = 21$.

Solution: The first 10 electrons will have the same configuration as neon. The next 8 electrons will fill the 3*s* and 3*p* orbitals completely. From the diagram given in Fig. 3.17, the next orbital to fill is the 4*s*. The next 2 electrons will go into the 4*s*, and the remaining electron will go into one of the 3*d* orbitals. The electronic description of scandium is

$_{21}$Sc $1s^2$	$2s^2$, $2p^6$	$3s^2$, $3p^6$	$4s^2$, $3d^1$
[He]	[Ne]	[Ar]	
2*e*	10*e*	18*e*	

The chemical and physical properties of the elements depend on the electronic configuration of the atoms. The positions of the elements in the families and periods of the periodic table will be discussed in the next chapter. This classification of the elements can be understood only in terms of electronic structure.

3.13 THE MATHEMATICAL FORM OF THE WAVE FUNCTIONS FOR THE HYDROGEN ATOM

The solution of the Schrödinger equation for the hydrogen atom results in a function which contains the quantum numbers n, l, and m as parameters. From this function we can obtain the *orbital wave functions* by substituting particular values of the quantum numbers. For instance, if we substitute $n = 1$, $l = 0$, and $m = 0$, we obtain the wave function for a 1s orbital, symbolized as $\psi(1s)$. This wave function is shown in Table 3.4 along with the wave functions for the 2s and 2p orbitals. The wave functions are in polar coordinates with the nucleus at the origin. The position of the electron is described in terms of its separation from the nucleus r and the two angles θ and ϕ, which are defined by the figure next to the table. The symbol a_0 in the wave functions represents the Bohr radius, 0.529 Å, and Z represents the atomic number. For hydrogen $Z = 1$; these same wave functions are applicable to other one-electron systems like $_2He^+$, $_3Li^{2+}$, etc. In these cases Z would not equal unity.

The probability of finding an electron at a point within an atom is proportional to the square of the orbital wave function at that point. In the ground state the single electron of hydrogen is in the 1s orbital. The probability of finding an electron at point (r,θ,ϕ) depends on the value of $\psi^2(1s)$ at the point (r,θ,ϕ). Examination of the $\psi(1s)$ function in Table 3.4 shows that it has no angular dependence; θ and ϕ do not appear in the function. This means that the square of

$0 < \theta < 180°$
$0 < \phi < 360°$

TABLE 3.4: WAVE FUNCTIONS FOR THE HYDROGEN ATOM

n	l	m	*Orbital*	ψ
1	0	0	1s	$\psi(1s) = \left(\frac{1}{\pi}\right)^{1/2}\left(\frac{Z}{a_0}\right)^{3/2} e^{-Zr/a_0}$
2	0	0	2s	$\psi(2s) = \frac{1}{4}\left(\frac{1}{2\pi}\right)^{1/2}\left(\frac{Z}{a_0}\right)^{3/2}\left(2 - \frac{Zr}{a_0}\right) e^{-Zr/a_0}$
2	1	± 1	$2p_x$ or $2p_y$	$\psi(2p_x) = \frac{1}{4}\left(\frac{1}{2\pi}\right)^{1/2}\left(\frac{Z}{a_0}\right)^{5/2} re^{-Zr/2a_0}\sin\theta\cos\phi$
2	1			$\psi(2p_y) = \frac{1}{4}\left(\frac{1}{2\pi}\right)^{1/2}\left(\frac{Z}{a_0}\right)^{5/2} re^{-Zr/2a_0}\sin\theta\sin\phi$
2	1	0	$2p_z$	$\psi(2p_z) = \frac{1}{4}\left(\frac{1}{2\pi}\right)^{1/2}\left(\frac{Z}{a_0}\right)^{5/2} re^{-Zr/2a_0}\cos\theta$

FIGURE 3.18

Plots of ψ and ψ^2 vs. r.

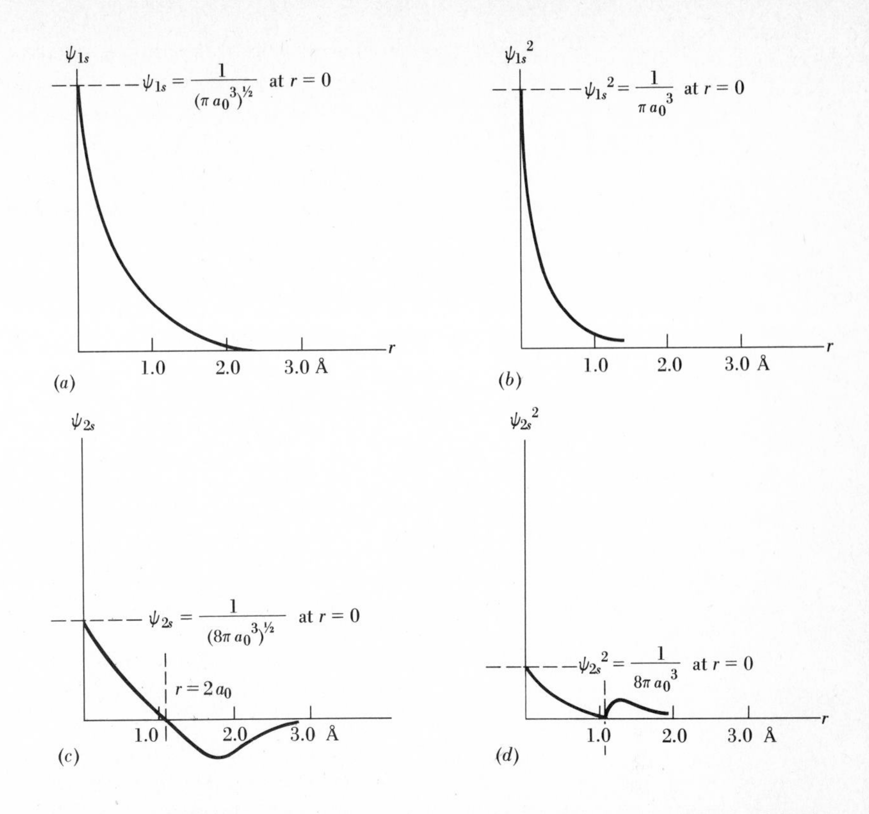

the 1*s* orbital wave function, $\psi^2(1s)$, which is the probability of finding the electron at a point in the 1*s* orbital, depends only on the distance *r* separating the electron from the nucleus. In other words, the 1*s* orbital is spherically symmetrical; the 1*s* electron *cloud* is a sphere with the nucleus at the center. Figure 3.18*a* shows a plot of $\psi(1s)$ versus *r*, and Fig. 3.18*b* shows a plot of $\psi^2(1s)$ versus *r*. As *r* increases, ψ^2 decreases from its maximum value, $1/\pi a_0{}^3$, at $r = 0$ and approaches zero asymptotically. [$\psi^2(1s) \rightarrow 0$ as $r \rightarrow \infty$.] The mathematical equation which describes the 1*s* orbital shows that $\psi(1s)$, and hence $\psi^2(1s)$, will exactly equal zero only when $r = \infty$, but for all practical purposes $\psi(1s)$ and $\psi^2(1s)$ are equal to zero when $r > 4$ Å. The probability of finding an electron in a 1*s* orbital at distances greater than 4 Å is negligible (see Prob. 3.17).

The 2*s* orbital wave function also shows no angular dependence. In fact, no *s* orbital shows angular dependence. Examination of the 2*s* orbital wave function in Table 3.4 shows that it can be positive, zero, or negative, depending on the value of *r*. This is due to the factor $2 - Zr/a_0$. When the factor $2 - Zr/a_0$ is positive, $\psi(2s)$ is positive. This is equivalent to saying that when Zr is less than $2a_0$, $\psi(2s)$ is positive because then the factor $2 - Zr/a_0$ is positive. Since $Z = 1$ for the hydrogen atom, $\psi(2s)$ is positive for all values of *r* less than $2a_0$, or 1.058 Å. By similar reasoning we conclude that $\psi(2s)$

equals zero at $r = 2a_0$ and is negative when r is greater than $2a_0$. To picture a $\psi(2s)$ wave function imagine a spherical boundary surface with a radius of 1.058 Å. Within this surface $\psi(2s)$ is positive, outside the surface $\psi(2s)$ is negative, and exactly at the surface $\psi(2s)$ is zero. The $2s$ orbital is said to have a spherical *nodal* surface at the distance 1.058 Å where $\psi(2s)$ is zero. The probability of finding the electron at some point in a $2s$ orbital is equal to the square of the wave function $\psi^2(2s)$ at that point. The value of $\psi^2(2s)$ will be positive or zero: where $\psi(2s)$ is positive or negative, $\psi^2(2s)$ will be positive; where $\psi(2s)$ is zero, $\psi^2(2s)$ will also be zero. This means that there is a probability of finding an electron at distances closer or further than $2a_0$ but never at a distance exactly equal to $2a_0$.

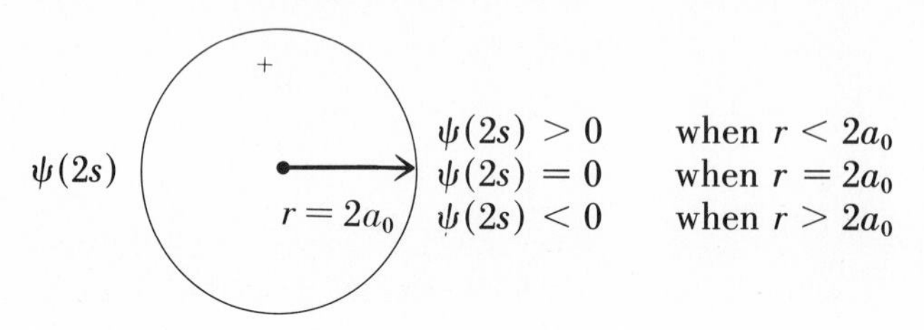

Plots of $\psi(2s)$ versus r and $\psi^2(2s)$ versus r are shown in Fig. 3.18*c* and *d*.

Although the mathematical expression for the $2s$ orbital shows that $\psi(2s)$ can be exactly equal to zero only when $r = 2a_0$ and $r = \infty$, for all practical purposes it reaches a value of zero when r is equal to or greater than 6 Å. This means that the probability of finding an electron in a $2s$ orbital at distances greater than 6 Å from the nucleus is extremely small.

The spherical node in the $\psi(2s)$ function arises from the algebraic term $2 - Zr/a_0$ in the wave function. The three $2p$ orbitals have *planar nodes*, which result from the trigonometric terms in the wave functions for these orbitals. The $2p_x$ orbital has a node in the yz plane because of the cos ϕ factor in this wave function (see Table 3.4). To show that there is a node in the yz plane, we must prove that $\psi(2p_x)$ is zero in this plane. In the yz plane the angle ϕ is either 90 or 270°. In either case, cos ϕ is 0, and since cos ϕ is a multiplier in the wave function, the value of $\psi(2p_x)$ in the yz plane, where ϕ equals 90 or 270°, is 0. By considering the trigonometric factors in $\psi(2p_y)$ and $\psi(2p_z)$ we can show that these orbitals have nodal planes in the xz and xy planes, respectively. The sign of the wave function is different on opposite sides of the nodal planes. For instance, $\psi(2p_x)$ is negative for all negative values of x and positive for all positive values of x (see Prob. 3.18).

Since the probability of finding an electron is proportional to the square of the wave function, we could find an electron in a $2p_x$ orbital on either side of the yz plane but never in the yz plane, which includes

FIGURE 3.19

Plots of the radial probability vs. r.

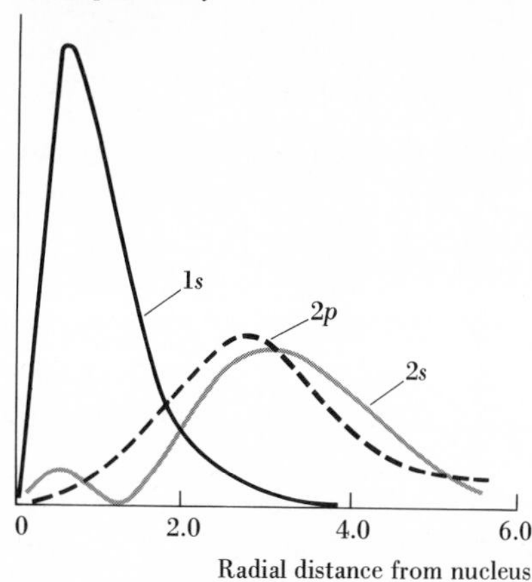

the nucleus. Whenever an electron is in a p orbital, it has a zero probability of being at the nucleus. The electron clouds which represent the *squares* of the $2p$ orbital functions are shown in Fig. 3.14; they resemble figure eights, as mentioned before.

The square of the wave function, ψ^2, gives the probability of finding an electron at a point (r,θ,ϕ) in space. The *radial probability* is the probability of finding an electron in a spherical shell of infinitesimally small thickness Δr at a distance r from the nucleus. The radial probability is the product of the wave function at r, ψ^2, and the volume of the shell, $4\pi r^2\ \Delta r$. For a very thin spherical shell the volume is proportional to the surface area $4\pi r^2$, and the radial probability is defined as $4\pi r^2 \psi^2$. Figure 3.19 shows the radial probabilities for the $1s$, $2s$, and $2p$ wave functions. (Figure 3.11 represents the radial probability for a $1s$ electron alone.)

Even though the wave function ψ may have a finite value at the nucleus where $r = 0$, the radial probability is always zero at the nucleus since the volume element $4\pi r^2\ \Delta r$ is zero when $r = 0$. The maxima in the radial-probability plots of Fig. 3.19 indicate the distances at which the electron is most likely to be found. Note that the electron in the first energy level has a maximum probability closer to the nucleus than an electron in the second energy level. The maximum in the radial probability for a $1s$ electron in the hydrogen atom is just a_0, or the Bohr radius, 0.529 Å (see Prob. 3.19).

QUESTIONS AND PROBLEMS

3.1 It is often said that the mass of the extranuclear electrons in an atom is negligible compared to the mass of the nucleus. Calculate the mass of the electrons and of the nucleus in an atom of (*a*) $^{19}_{9}F$, (*b*) $^{23}_{11}Na$, and (*c*) $^{238}_{92}U$. Use the data given in Table 2.1 for the mass of the proton, neutron, and electron.

3.2 Why did Rutherford expect α particles to pass through a metal foil without undergoing any large deflections?

3.3 Green light has a wavelength of approximately 5500 Å. Calculate the energy in calories of (*a*) one photon of this light and (*b*) 1 mol of photons.

3.4 How did Bohr explain the sharp lines observed in the emission spectra of hydrogen and other gaseous elements?

3.5 (*a*) Using Table 3.1, determine the difference in energy between an electron in the third energy level and one in the fourth energy level. (*b*) What is the wavelength of the radiation emitted when an electron falls from the fourth to the third energy level? (*c*) Is this visible radiation? (*d*) According to Fig. 3.9, to which series of lines would this radiation belong?

3.6 From a study of Eqs. (3.1) and (3.2) and Tables 3.1 and 3.2: (*a*) What happens to the difference between adjacent allowable radii of the electron orbits in the hydrogen atom as the energy level n increases? Compare $r_2 - r_1$ with $r_4 - r_3$ and $r_7 - r_6$. (*b*) What happens to the difference in the allowable energies of adjacent levels as n increases? Compare $\varepsilon_2 - \varepsilon_1$ with $\varepsilon_4 - \varepsilon_3$ and $\varepsilon_7 - \varepsilon_6$. (*c*) What happens to the differences in the wavelengths of the lines emitted as the electrons fall from orbits in which $n > 1$ to the first orbit? (*d*) Make a qualitative sketch of the emission spectrum due to the Balmer series to show what would be expected as n becomes very large.

3.7 Calculate the six missing data in Table 3.2.

3.8 (*a*) In view of de Broglie's correct hypothesis that an electron has wave properties, how must we modify Bohr's model for the hydrogen atom? (*b*) Why is the Schrödinger equation a better description of the behavior of electrons in atoms than the Bohr model?

3.9 (*a*) What does the principal quantum number n signify? (*b*) If an electron is in a very high energy level (high n), will it be possible for the electron to be very near the nucleus? Will it be probable that the electron is very near the nucleus?

3.10 What is the greatest chemical significance of the angular-momentum quantum number l?

3.11 Describe and sketch the "electron clouds" of electrons in 1*s*, 2*p*, and 3*d* orbitals.

3.12 State (*a*) the Pauli exclusion principle, (*b*) the Aufbau principle, and (*c*) Hund's rule.

3.13 Give the four quantum numbers for each electron of a normal (unexcited) atom of (*a*) N ($Z = 7$), (*b*) Cl ($Z = 17$), and (*c*) Mn ($Z = 25$).

3.14 Write the electronic configurations for normal atoms of the following elements: (*a*) $_{37}Rb$, (*b*) $_{15}P$, (*c*) $_{8}O$, (*d*) $_{9}F$, and (*e*) $_{34}Se$.

3.15 Determine the average distance between the nucleus and an electron in the hydrogen atom if the electron is in (*a*) a 2*p* orbital, (*b*) a 3*p* orbital, and (*c*) a 3*d* orbital.

3.16 The grades on a laboratory report for a group of students were 85,

75, 80, 75, 55, 60, 90, 65, 60, 65, 75, 65, 80, 85, 80 and 75. (*a*) Make a plot of the number of students having a particular grade versus that grade. (*b*) What is the most *probable grade;* i.e., what grade was attained most frequently? (*c*) Calculate the average grade. (*d*) Can the most probable grade be very different from the average grade? (*e*) Can the most probable separation of the electron from the nucleus be very different from the average separation?

3.17 (*a*) Calculate the value of ψ^2 (not ψ) for the hydrogen 1*s* orbital at distances of 0.25, 0.5, 1.0, 1.5, 2.0, 3, 4, and 5 Å. (*b*) Make a plot of your values of ψ^2 versus *r*. (Remember that $a_0 = 0.529$ Å and for hydrogen $Z = 1$.)

3.18 The wave function $\psi(2p_x)$ contains the trigonometric factor $\sin\theta\cos\phi$ (see Table 3.4 and the associated figure showing the polar coordinate system). Show that because of the trigonometric term, the wave function $\psi(2p_x)$ is positive for all positive values of x, negative for negative values of x, and zero when $x = 0$. *Hint:* Consider the sign of $\sin\theta$ for positive, negative, and zero values of x.

3.19 (*a*) Calculate the radial probability $4\pi r^2\psi^2$ for an electron in a hydrogen 1*s* orbital at 0.25, 0.50, 0.75, 1.0, 2.0, 3.0, and 5.0 Å. (*b*) Make a plot of these radial probabilities versus *r*. (*c*) At what separation does the maximum value of the radial probabilities occur? (*d*) Use the method of differentiation to determine the separation at which the radial probability is a maximum.

3.20 (*a*) Using $\phi = 45°$ and $r = 1$ Å, calculate the value of $\psi(2p_x)$ and $\psi^2(2p_x)$ at values of $\theta = 0, 30, 60, 90, 120, \ldots$ up to and including 360°. (*b*) Make a table including θ, $\psi(2p_x)$, and $\psi^2(2p_x)$. (*c*) Make a plot of $\psi(2p_x)$ versus θ and $\psi^2(2p_x)$ versus θ.

Stoichiometry practice

3.21 Iron ore, Fe_2O_3, can be reduced to the metal by treatment with hydrogen gas:

$$Fe_2O_3 + H_2 \rightarrow Fe + H_2O$$

(*a*) Balance the equation. (*b*) How much iron ore (in kilograms) would be required to produce 5 tons of iron? (*c*) What volume of hydrogen gas would be required to reduce 5 tons of ore if the density of hydrogen gas is 0.08 g/l?

3.22 When calcium carbonate is heated, it decomposes into the oxides of calcium and carbon:

$$CaCO_3 \rightarrow CaO + CO_2$$

(*a*) Balance the equation. (*b*) How many molecules of CO_2 can be produced from the decomposition of 55.5 g of $CaCO_3$? (*c*) How many grams of CaO would also be produced?

3.23 An aqueous solution contains 15.0 g of $AgNO_3$ per liter. (*a*) How many moles of HCl and (*b*) how many molecules of HCl are required to precipitate all the $AgNO_3$ in 525 ml of this solution? The equation for the reaction is

$$AgNO_3(aq) + HCl(g) \rightarrow AgCl(s) + HNO_3(aq)$$

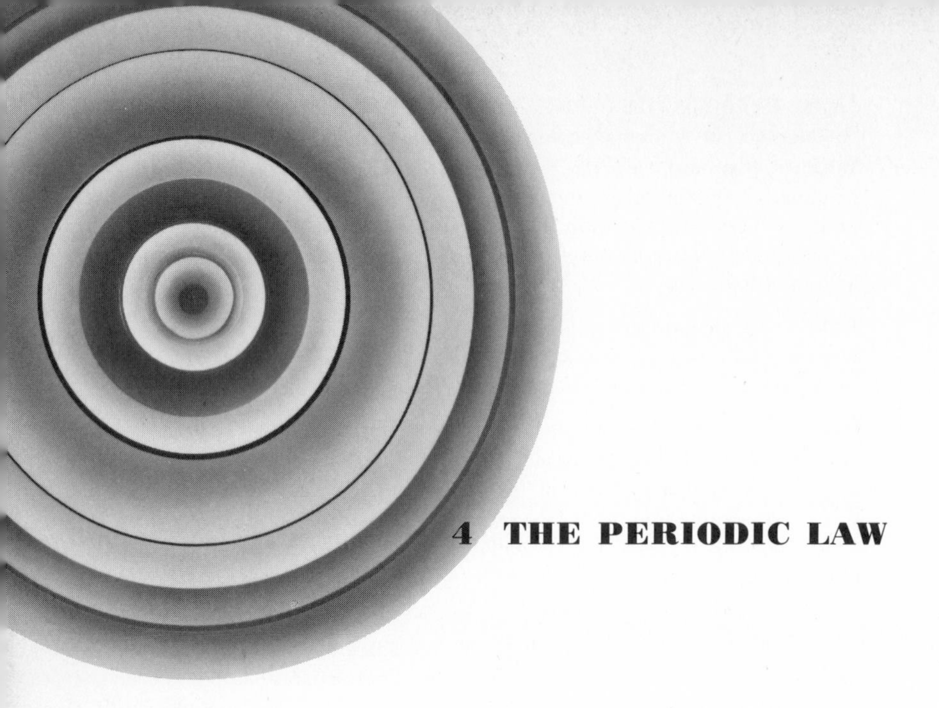

4 THE PERIODIC LAW

4.1 DISCOVERY OF PERIODICITY

In the first 25 years of the nineteenth century, 20 new elements were discovered. As the number of known elements increased, the chemical and physical similarities between certain of them became obvious. Chemists sought some natural principle by which the similar elements could be grouped. In 1817, Döbereiner found that in many cases similar elements can be arranged in *triads* of increasing atomic weight; the middle element of the triad has an atomic weight which is the average of the heaviest and lightest (Table 4.1). After another 100 years, the principle which guides the modern classification of the elements was discovered by H. G.-J. Moseley in 1913: *the properties of the elements are periodic functions of their atomic numbers.* The discovery of this periodicity, or *periodic* law, depended upon two earlier developments: (1) the establishment of a consistent and reliable set of atomic weights and (2) the development of the concept of a nuclear atom with a definite number of protons in the nucleus equal to the number of electrons circulating about it.

TABLE 4.1: DÖBEREINER'S TRIADS

Lithium	7
Sodium	23
Potassium	39
Average	23
Calcium	40
Strontium	88
Barium	137
Average	89
Chlorine	35
Bromine	80
Iodine	129
Average	82
Sulfur	32
Selenium	79
Tellurium	128
Average	80

The first important attempts at a classification of the elements were made after the Karlsruhe Conference (1860), at which the Italian chemist Stanislao Cannizzaro convinced many of those present of the correctness of Avogadro's hypothesis (molecules of common gases are diatomic). The disagreement over the atomic weights of oxygen, carbon, nitrogen, and other elements disappeared (Chap. 1), and a meaningful classification of the elements based on their atomic weights was possible. As we shall see, the order of the elements according to increasing atomic weight is almost identical with order according to atomic number, and classification based on the one is almost equivalent to classification based on the other.

In 1868 the English chemist J. A. R. Newlands read a paper entitled The Law of Octaves and the Causes of Numerical Relations between Atomic Weights before members of the Chemical Society at London. Discussing a table in which he had arranged the elements according to increasing atomic weight, he attempted to demonstrate that every eighth element in the series showed a recurrence of similar physical properties. Unfortunately, at that time there were many undiscovered elements, so that the law of octaves did not hold very well. Newlands' ideas seemed absurd to many of his colleagues, who ridiculed him and discouraged him from continuing his work. This injustice was corrected 19 years later, in 1887, when the Royal Society awarded him the Davy medal, 5 years after a similar award had been made to Mendeleev and Meyer who are considered codiscoverers of the periodic law.

Dmitri Ivanovich Mendeleev (Russian) and Julius Lothar Meyer (German) independently discovered that the elements can be systematized on the basis of their atomic weights. When elements are put into atomic-weight order, the chemical and physical properties are repeated in a regular, or *periodic*, way. Figure 4.1 shows Mendeleev's table representing the periodic regularities of the chemical elements.

Dmitri Ivanovich Mendeleev. (Courtesy of the Edgar Fahs Smith Collection, Van Pelt Library, University of Pennsylvania, Philadelphia.)

Mendeleev did not blindly order the elements according to atomic weight. He was guided by another very important principle. He insisted that only elements with similar physical and chemical properties could be members of the same group, or family. Therefore, since arsenic was similar to phosphorus, he placed it in group V and left vacancies in groups III and IV, guessing correctly that there were two undiscovered elements in series 5 (see Fig. 4.1). Not only did he predict the atomic weights of these undiscovered elements but also their chemical and physical properties. Table 4.2 shows the accuracy of his predictions for the elements now known as gallium and germanium.

Mendeleev's predictions were based on trends in the series or periods (horizontal rows) and groups (vertical columns) of his table. We make predictions and guesses in the same way today. Because

FIGURE 4.1

Mendeleev's table representing the periodic regularities of the chemical elements.

Series	*Group* I — R^2O	*Group* II — RO	*Group* III — R^2O^3	*Group* IV RH^4 RO^2	*Group* V RH^3 R^2O^5	*Group* VI RH^2 RO^3	*Group* VII RH R^2O^7	*Group* VIII — RO^4
1	H = 1							
2	Li = 7	Be = 9.4	B = 11	C = 12	N = 14	O = 16	F = 19	
3	Na = 23	Mg = 24	Al = 27.3	Si = 28	P = 31	S = 32	Cl = 35.5	
4	K = 39	Ca = 40	— = 44	Ti = 48	V = 51	Cr = 52	Mn = 55	Fe = 56, Co = 59, Ni = 59, Cu = 63
5	(Cu = 63)	Zn = 65	— = 68	— = 72	As = 75	Se = 78	Br = 80	
6	Rb = 85	Sr = 87	?Yt = 88	Zr = 90	Nb = 94	Mo = 96	— = 100	Ru = 104, Rh = 104, Pd = 106, Ag = 108
7	(Ag = 108)	Cd = 112	In = 113	Sn = 118	Sb = 122	Te = 125	I = 127	
8	Cs = 133	Ba = 137	?Di = 138	?Ce = 140	—	—	—	— — — —
9	(—)	—	—	—	—	—	—	
10	—	—	?Er = 178	?La = 180	Ta = 182	W = 184	—	Os = 195, Ir = 197, Pt = 198, Au = 199
11	(Au = 199)	Hg = 200	Tl = 204	Pb = 207	Bi = 208	—	—	
12	—	—	—	Th = 231	—	U = 240	—	— — — —

TABLE 4.2: MENDELEEV'S PREDICTIONS

	Ekaaluminum† *(gallium)*		*Ekasilicon (germanium)*	
Property	*Predicted 1871*	*Discovered 1875*	*Predicted 1871*	*Discovered 1886*
Atomic weight	68	69.9	72	72.33
Density, g/ml	5.9	5.93	5.5	5.47
Valence or oxidation number	3	3	4	4
Specific heat, cal/g	0.091	0.089	0.073	0.077
Melting point, °C	Low	30.1	. . .	. . .
Formula for oxide	Ea_2O_3	Ga_2O_3	EsO_2	GeO_2
Formula for chloride	$EaCl_3$	$GaCl_3$	$EsCl_4$	$GeCl_4$
Density of oxide, g/ml	. . .	. . .	4.7	4.703
Boiling point of chloride, °C	. . .	. . .	100	86

† *Eka-*, from Sanskrit for *one*, means "standing before."

Mendeleev insisted that only similar elements should be members of the same group, he not only concluded that certain elements were undiscovered but was forced to make certain transpositions in the order. Thus even though tellurium, Te, has a greater atomic weight than iodine, I, he placed tellurium before iodine. It was obvious to him that tellurium is similar to sulfur and selenium and that iodine is similar to chlorine and bromine.

Mendeleev used his principle of periodicity to correct many errors of judgment which had been made in the interpretation of data. A notable example involves indium, In, to which chemists had assigned an atomic weight of 76.6. By Mendeleev's classification this would have placed indium between arsenic and selenium, but there could be no vacancy between arsenic and selenium; these two elements are members of adjacent groups. Mendeleev demonstrated that the properties of indium are similar to those of aluminum and thallium and intermediate between cadmium and tin. He concluded that indium must be a member of group III, and he added strong evidence for the correctness of this conclusion by the application of the *law of Dulong and Petit.* This empirical law states that *the product of the atomic weight and the specific heat is approximately equal to 6.2.* He measured the specific heat of indium, obtaining 0.055 cal/g, and estimated the atomic weight of indium to be 113; the presently accepted value is 114.82.

Law of Dulong and Petit:

$$\text{Gram-atomic weight} \times \text{specific heat} \approx 6.2 \tag{4.1}$$

FIGURE 4.2

Atomic volume versus atomic weight. (After J. L. Meyer, 1870.)

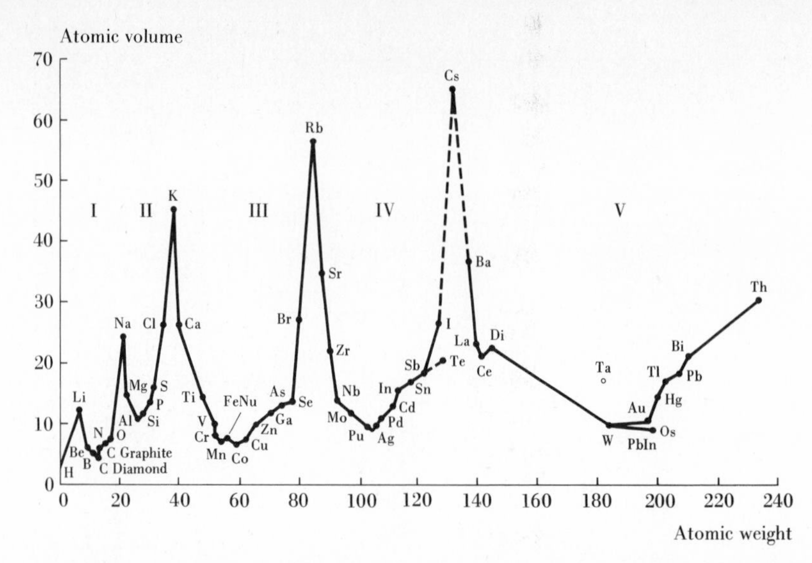

The measured specific heat† for indium is 0.055 cal/g; therefore

$$\text{Gram-atomic weight} = \frac{6.2}{0.055} = 113$$

The accuracy of Mendeleev's predictions attracted attention, and the power of the periodic classification of the elements was well established.

Meyer, who made an independent discovery of the periodic law, demonstrated the principle of periodicity most clearly by plotting the atomic volume of the elements versus their atomic weights (Fig. 4.2).

4.2 ATOMIC NUMBER AND THE MODERN PERIODIC LAW

After arranging the elements in the periodic table, it was natural to give each element a number indicating only its position in the series based on increasing atomic weight. No real physical significance was associated with the concept of atomic number when it first came into use, but after Rutherford proposed the solar-system atom with a tiny nucleus, he estimated that the charge on the nucleus was very close to one-half the atomic weight:

$$\text{Nuclear charge} = Z \approx \tfrac{1}{2}(\text{atomic weight}) \tag{4.2}$$

For many elements, one-half the atomic weight is equal to the atomic number; e.g., helium with an atomic weight of 4 is second in the table; carbon with an atomic weight of 12 is sixth in the table; oxygen, 16, is eighth; and sulfur, 32, is sixteenth. In the second and

† The specific heat is the heat required to raise 1 g of a substance by 1°C.

FIGURE 4.3

A schematic diagram of Moseley's apparatus.

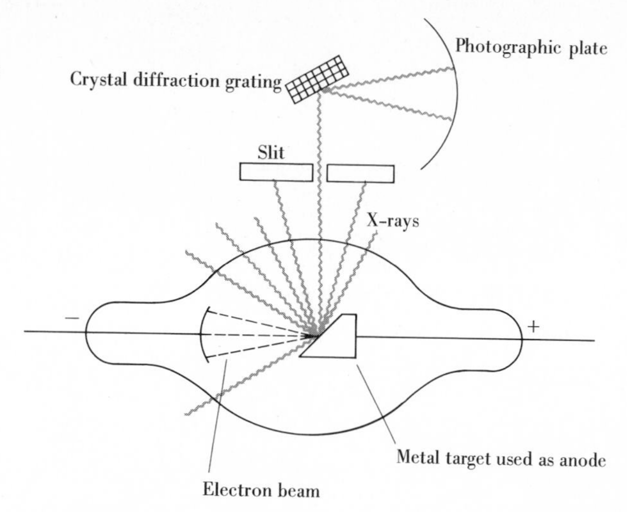

third periods this rule holds well, so that Eq. (4.2) can be extended:

$$\text{Nuclear charge} = Z \simeq \tfrac{1}{2}\,(\text{atomic weight}) \simeq \text{atomic number}$$

Rutherford inferred that the nuclear charge equals the atomic number and further that the magnitude of the nuclear charge is some integral multiple of the electronic charge e

$$Z = \text{nuclear charge} = \text{atomic number} = ne$$

The verification of this hypothesis came from the work of Moseley on the analysis of x-rays produced when cathode rays strike metal targets. When metallic elements are used as targets for cathode rays, which are electron beams (page 40), they emit an x-ray spectrum; there are usually two *lines* in the x-ray spectrum that are very bright. Figure 4.3 shows a diagram of the apparatus used by Moseley, and Fig. 4.4 shows a typical x-ray spectrum produced by a metal target.

FIGURE 4.4

A typical x-ray spectrum.

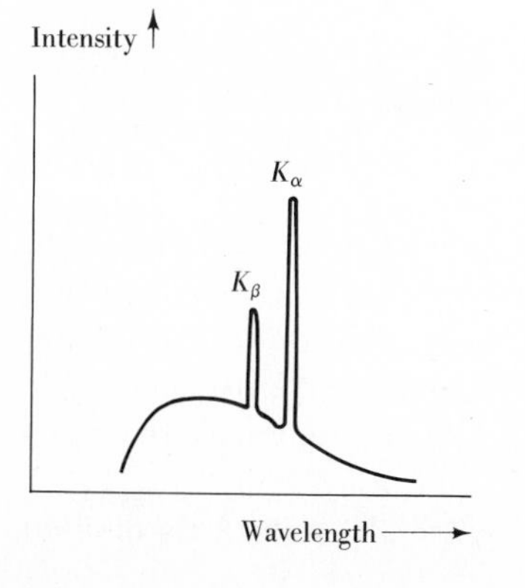

The x-rays are generated when the cathode ray collides with an electron in the K shell, knocking the electron out of the atom. The vacancy in the K shell is filled when electrons in higher energy levels of the atom fall into the K shell. This causes the emission of x-radiation (Fig. 4.5).

As can be seen from Eq. (3.3), the energy of an electron in an atom depends on the nuclear charge Z, which Rutherford identified with the atomic number. As the nuclear charge increases, the force holding the electron increases. Hence, the energy and frequency of the x-rays emitted by metal targets must depend on the atomic number of the target element. Moseley found that the frequency and hence the energy of the bright lines of the x-ray spectrum increases in a regular way as the *atomic number* increases. As shown in Fig.

FIGURE 4.5

The mechanism for the production of x-rays from cathode-ray bombardment.

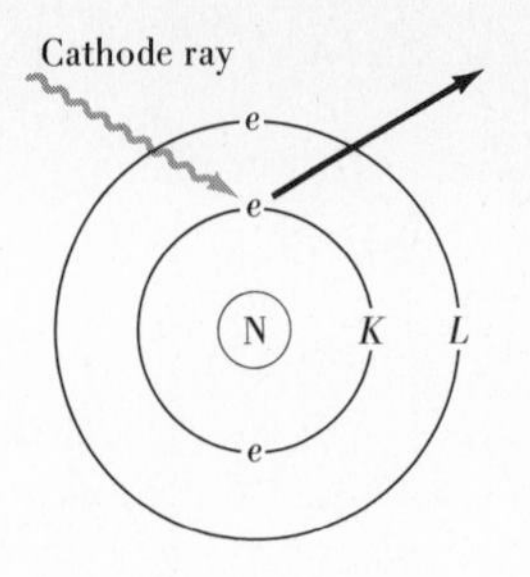

(*a*) *An electron in the K shell is ejected by a cathode ray*

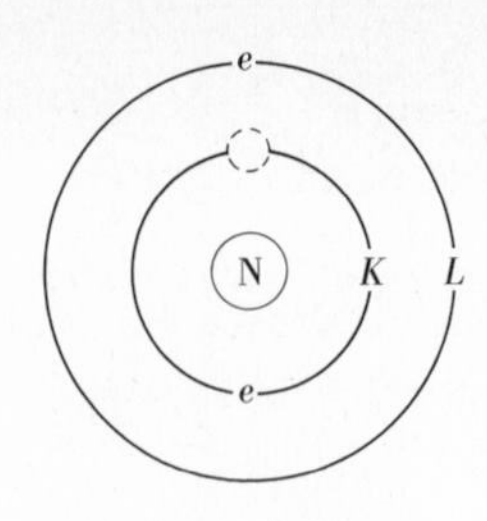

(*b*) *Ejection leaves a vacancy in the K shell*

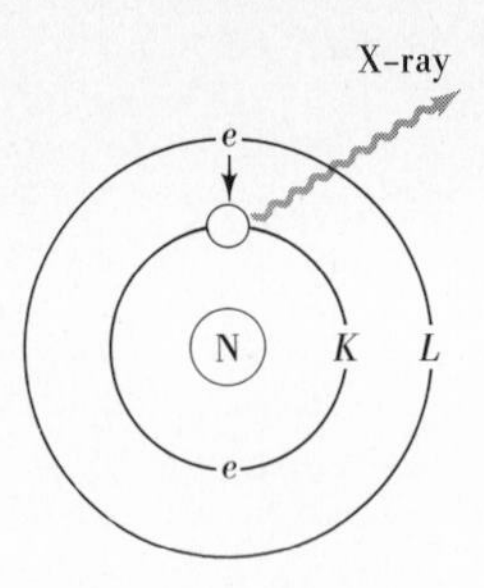

(*c*) *An x-ray is emitted when an electron from a higher orbit falls into the K shell*

4.6, a plot of the atomic numbers versus the square roots of the frequencies of either line is linear for all 38 metallic elements investigated by Moseley. Following the publication of Moseley's work in 1913, it became obvious that the properties of the elements are periodic functions of their atomic *numbers*.

FIGURE 4.6

A plot of the atomic number Z versus the square root of the frequency of the bright lines in the x-ray spectrum.

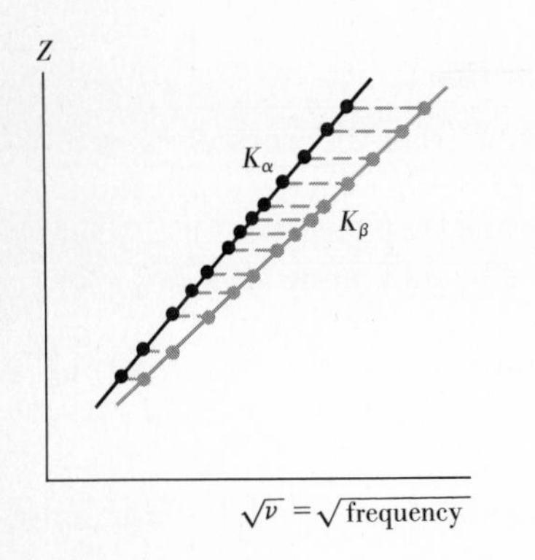

The order of the elements according to atomic number is almost identical with the order according to atomic weight, but there are a few important differences. Strictly followed, the atomic-weight guide would put potassium before argon, nickel before cobalt, and iodine before tellurium:

. . . , K(39.102), Ar(39.948), . . . , Ni(58.71), Co(58.93), . . . , I(126.90), Te(127.60), . . .

The atomic-number order causes inversions of these pairs:

. . . , Ar(18), K(19), . . . , Co(27), Ni(28), . . . , Te(52), I(53) . . .

In order to have similar elements in the same group Mendeleev found it necessary to transpose certain pairs of elements arbitrarily in his periodic table based on atomic weights. When the elements are placed in order of atomic *number*, no transpositions are necessary. Our modern periodic law states that the properties of the elements are periodic functions of their atomic numbers.

4.3 THE ELECTRONIC STRUCTURE OF ATOMS AND THE PERIODIC LAW

Since the atomic number equals the number of electrons in an atom, Moseley's observations added to the growing body of evidence that the electronic structure of the atom governs the physical and chemical properties of the elements and their compounds. Elements of the same family are believed to undergo similar chemical reactions because they have similar electronic configurations.

Figure 4.7 is the modern form of the periodic table. It is divided

FIGURE 4.7

The modern periodic table. (The mass of $^{12}C = 12.000 \cdots u$.)

Periodic Table of the Elements

	IA	IIA	IIIB	IVB	VB	VIB	VIIB	VIIIB			IB	IIB	IIIA	IVA	VA	VIA	VIIA	VIIIA
	s		d										p					
1	1 H 1.00797																	2 He 4.0026
2	3 Li 6.939	4 Be 9.0122											5 B 10.811	6 C 12.01115	7 N 14.0067	8 O 15.9994	9 F 18.9984	10 Ne 20.183
3	11 Na 22.9898	12 Mg 24.312											13 Al 26.9815	14 Si 28.086	15 P 30.9738	16 S 32.064	17 Cl 35.453	18 Ar 39.948
4	19 K 39.102	20 Ca 40.08	21 Sc 44.956	22 Ti 47.90	23 V 50.942	24 Cr 51.996	25 Mn 54.9380	26 Fe 55.847	27 Co 58.9332	28 Ni 58.71	29 Cu 63.54	30 Zn 65.37	31 Ga 69.72	32 Ge 72.59	33 As 74.9216	34 Se 78.96	35 Br 79.909	36 Kr 83.80
5	37 Rb 85.47	38 Sr 87.62	39 Y 88.905	40 Zr 91.22	41 Nb 92.906	42 Mo 95.94	43 Tc (99)	44 Ru 101.07	45 Rh 102.905	46 Pd 106.4	47 Ag 107.870	48 Cd 112.40	49 In 114.82	50 Sn 118.69	51 Sb 121.75	52 Te 127.60	53 I 126.9044	54 Xe 131.30
6	55 Cs 132.905	56 Ba 137.34	57 La 138.91	72 Hf 178.49	73 Ta 180.948	74 W 183.85	75 Re 186.2	76 Os 190.2	77 Ir 192.2	78 Pt 195.09	79 Au 196.967	80 Hg 200.59	81 Tl 204.37	82 Pb 207.19	83 Bi 208.980	84 Po (210)	85 At (210)	86 Rn (222)
7	87 Fr (223)	88 Ra (226)	89 Ac (227)	104 Ku (260)														

f														
58 Ce 140.12	59 Pr 140.907	60 Nd 144.24	61 Pm (147)	62 Sm 150.35	63 Eu 151.96	64 Gd 157.25	65 Tb 158.924	66 Dy 162.50	67 Ho 164.930	68 Er 167.26	69 Tm 168.934	70 Yb 173.04	71 Lu 174.97
90 Th 238.03	91 Pa (231)	92 U 238.03	93 Np (237)	94 Pu (242)	95 Am (243)	96 Cm (247)	97 Bk (249)	98 Cf (251)	99 Es (254)	100 Fm (253)	101 Md (256)	102 No (254)	103 Lw (257)

Atomic weight values listed in parentheses are approximate.

TABLE 4.3: SECOND PERIOD

$_3Li$	$1s^2$	$2s^1$				
$_4Be$	$1s^2$	$2s^2$				
$_5B$	$1s^2$	$2s^2$	$2p_x^1$	$2p_y^0$	$2p_z^0$	
$_6C$	$1s^2$	$2s^2$	$2p_x^1$	$2p_y^1$	$2p_z^0$	
$_7N$	$1s^2$	$2s^2$	$2p_x^1$	$2p_y^1$	$2p_z^1$	
$_8O$	$1s^2$	$2s^2$	$2p_x^2$	$2p_y^1$	$2p_z^1$	
$_9F$	$1s^2$	$2s^2$	$2p_x^2$	$2p_y^2$	$2p_z^1$	
$_{10}Ne$	$1s^2$	$2s^2$	$2p_x^2$	$2p_y^2$	$2p_z^2$	or $1s^2$, $2s^2$, p^6

into four regions labeled *s*, *p*, *d*, and *f*. There are seven periods (rows). The electronic configuration of the first member of each period is characterized by ns^1, where n is the highest occupied energy level of the atom and also the number of the period. For instance, the configuration of hydrogen in the first period is $1s^1$; lithium in the second period is $2s^1$; sodium in the third period is $3s^1$. The columns are called *groups* or *families*. And, as can be seen, there are A families and B families. The members of the A families are called *regular* elements, and the members of the B families are called *transition* elements or *transition-metal* elements. We shall devote most of our discussion in this chapter to the properties of the regular elements.

There are only two elements in the first period, corresponding to the fact that only two electrons can be placed in the first energy level. Hydrogen, with $Z = 1$, must have the electronic configuration $1s^1$ in the ground state. Helium, He, with $Z = 2$, must be in the $1s^2$ state, and the first energy level is filled with two electrons.

Lithium is the next element, with an atomic number of 3; the third electron must go into the second energy level, and the description for Li is $1s^2$, $2s^1$. There are eight members in the second period. They correspond to the atoms in which the highest-energy electrons occupy the $2s$ or $2p$ orbitals. The second energy level is filled in the neon atom, and therefore neon is the last member of the second period. In the third period electrons begin to fill the $3s$ and $3p$ orbitals. Sodium, $_{11}Na$, is the first member of the third period; the eighth and last member is argon, $_{18}Ar$. Their electronic configurations are

$$Na\ 1s^2, 2s^2, 2p^6, 3s^1 \qquad \text{or} \qquad Na = [Ne]\ 3s^1$$

and

$$Ar\ 1s^2, 2s^2, 2p^6, 3s^2, 3p^6 \qquad \text{or} \qquad Ar = [Ne]\ 3s^2, 3p^6$$

The first period is complete when the first energy level is filled; the second period is complete when the second energy level is filled; but the third period is complete *before* the third energy level is

TABLE 4.4: FOURTH PERIOD

s

$_{19}K = [Ar]\ 4s^1$
$_{20}Ca = [Ar]\ 4s^2$

d

$_{21}Sc = [Ar]\ 4s^2, 3d^1$
$_{22}Ti = [Ar]\ 4s^2, 3d^2$
$_{23}V = [Ar]\ 4s^2, 3d^3$
$_{24}Cr^* = [Ar]\ 4s^1, 3d^5$
$_{25}Mn = [Ar]\ 4s^2, 3d^5$
$_{26}Fe = [Ar]\ 4s^2, 3d^6$
$_{27}Co = [Ar]\ 4s^2, 3d^7$
$_{28}Ni = [Ar]\ 4s^2, 3d^8$
$_{29}Cu^* = [Ar]\ 4s^1, 3d^{10}$
$_{30}Zn = [Ar]\ 4s^2, 3d^{10}$

p

$_{31}Ga = [Ar]\ 4s^2, 3d^{10}, 4p^1$
$_{32}Ge = [Ar]\ 4s^2, 3d^{10}, 4p^2$
$_{33}As = [Ar]\ 4s^2, 3d^{10}, 4p^3$
$_{34}Se = [Ar]\ 4s^2, 3d^{10}, 4p^4$
$_{35}Br = [Ar]\ 4s^2, 3d^{10}, 4p^5$
$_{36}Kr = [Ar]\ 4s^2, 3d^{10}, 4p^6$

TABLE 4.5: THE SIXTH PERIOD

s

$_{55}Cs$ = [Xe] $6s^1$

$_{56}Ba$ = [Xe] $6s^2$

d

$_{57}La$ = [Xe] $6s^2$, $5d^1$

$_{72}Hf$ = [Xe] $4f^{14}$, $5d^2$, $6s^2$

$_{73}Ta$ = [Xe] $4f^{14}$, $5d^3$, $6s^2$

$_{74}W$* = [Xe] $4f^{14}$, $5d^5$, $6s^1$

$_{75}Re$ = [Xe] $4f^{14}$, $5d^5$, $6s^2$

$_{76}Os$ = [Xe] $4f^{14}$, $5d^6$, $6s^2$

$_{77}Ir$ = [Xe] $4f^{14}$, $5d^7$, $6s^2$

$_{78}Pt$ = [Xe] $4f^{14}$, $5d^8$, $6s^2$

$_{79}Au$* = [Xe] $4f^{14}$, $5d^{10}$, $6s^1$

$_{80}Hg$ = [Xe] $4f^{14}$, $5d^{10}$, $6s^2$

p

$_{81}Tl$ = [Hg] $6p^1$

$_{82}Pb$ = [Hg] $6p^2$

$_{83}Bi$ = [Hg] $6p^3$

$_{84}Po$ = [Hg] $6p^4$

$_{85}At$ = [Hg] $6p^5$

$_{86}Rn$ = [Hg] $6p^6$

Lanthanides, or inner transition metals

f

$_{58}Ce$ = [Xe] $4f^2$, $6s^2$

$_{59}Pr$ = [Xe] $4f^3$, $6s^2$

$_{60}Na$ = [Xe] $4f^4$, $6s^2$

$_{61}Pm$ = [Xe] $4f^5$, $6s^2$

$_{62}Sm$ = [Xe] $4f^6$, $6s^2$

$_{63}Eu$ = [Xe] $4f^7$, $6s^2$

$_{64}Gd$* = [Xe] $4f^7$, $5d$, $6s^2$

$_{65}Tb$ = [Xe] $4f^9$, $6s^2$

$_{66}Dy$ = [Xe] $4f^{10}$, $6s^2$

$_{67}Ho$ = [Xe] $4f^{11}$, $6s^2$

$_{68}Er$ = [Xe] $4f^{12}$, $6s^2$

$_{69}Tm$ = [Xe] $4f^{13}$, $6s^2$

$_{70}Yb$ = [Xe] $4f^{14}$, $6s^2$

$_{71}Lu$ = [Xe] $4f^{14}$, $5d^1$, $6s^2$

filled. Argon, which has no electrons in $3d$ orbitals, is the last member of the third period. The first element of the fourth period, potassium, has an argon core and an electron in a $4s$ orbital:

$_{19}K$ $1s^2$, $2s^2 2p^6$, $3s^2 3p^6$, $4s^1$ or [Ar] $4s^1$

Since the energy of the $3d$ orbitals is greater than the energy of the $4s$ orbitals, electrons will enter the $4s$ orbitals before entering the $3d$. The $4s$, $3d$, and $4p$ orbitals are of approximately the same energy, and as we inspect the electronic configurations of the atoms from potassium to krypton, the fourth period, we observe that these orbitals are being filled.

Table 4.4 showing the fourth-period elements has been divided into *s*, *d*, and *p* sections. This signifies that the valence electrons occupy the highest *s*, *p*, and *d* orbitals; these are the electrons of the highest energy. The valence electrons determine the physical and chemical properties of the elements.

The 10 elements of the fourth period from scandium, $_{21}Sc$, to zinc, $_{30}Zn$, make up the first transition-metal series. The 10 elements correspond to the 10 electrons which can be accommodated by the five *d* orbitals (see page 58). There are two more series of transition-metal elements; the second series, from ytterbium, $_{39}Y$, to cadmium, $_{48}Cd$, is in the fifth period, and the third series, from lanthanum, $_{57}La$, to mercury, $_{80}Hg$, is in the sixth.

The asterisks beside Cr and Cu in Table 4.4 indicate that their electronic configurations do not strictly follow the rules given in Fig. 3.17. We might expect chromium to have the configuration [Ar] $4s^2$, $3d^4$, but it has been discovered that a combination of half-filled and completely filled *sets* of orbitals are more stable. By going to the configuration [Ar] $4s^1$, $3d^5$ the Cr atom has half-filled $4s$ and $3d$ orbitals. Also, we might have inferred that copper would have the description [Ar] $4s^2$, $3d^9$; but the configuration [Ar] $4s^1$, $3d^{10}$ is more favorable since it involves a half-filled $4s$ orbital and completely filled $3d$ orbitals.

The fifth period also contains 18 elements. The $5s$, $4d$, and $5p$ orbitals are of approximately the same energy. Electrons are present in the $5s$ orbital of rubidium, $_{37}Rb$ = [Kr] $5s^1$. As we inspect the next 17 elements in order of increasing atomic number, we notice that the $5s$, $4d$, and $5p$ orbitals are filled progressively according to the principles given in the last chapter. The last member of the fifth period is the noble gas† xenon, Xe, $_{59}Xe$ = [Kr] $4d^{10}$, $5s^2$, $5p^6$.

The sixth period begins with cesium, $_{55}Cs$, which has one electron in the $6s$ orbital. The energies of the $6s$, $5d$, $4f$, and $6p$ orbitals are approximately equal, and the total number of electrons which can be accommodated by these orbitals is 32. Hence, the sixth period consists of the 32 elements from $_{55}Cs$ to $_{86}Rn$.

† Group 0 in the periodic table consists of six gases, referred to as *noble gases, rare gases,* or *inert gases.* The preferred name for the group is noble gases.

The 14 elements from cerium, $_{58}Ce$, to lutetium, $_{71}Lu$, have properties very much like lanthanum, $_{57}La$. Lanthanum is the third member of the sixth period; it is the first transition metal of this period. The next 14 elements are called *lanthanides* or *inner transition* elements. The similarity among these elements is due to the fact that their valence electrons occupy the 4*f* orbitals and the ionic radii of +3 ions decrease only slightly from the first to the last member. The sixth period is the first one in which electrons occupy *f* orbitals in ground-state atoms.

The electronic configurations for sixth-period elements are presented in Table 4.5, also divided into sections (*s*, *d*, *f*, and *p*) indicating which orbitals are occupied by the highest-energy valence electrons.

The seventh period begins with radioactive francium, $_{87}Fr$, followed by radium, $_{88}Ra$. The next element, actinium, $_{89}Ac$, begins the fourth transition-metal series. The next 14 elements, called *actin*ides, are all similar to actinium. We believe that this is because their valence electrons occupy the 5*f* orbitals and the radii of the most common ions are approximately equal. The actinides are analogs of the lanthanides. The heaviest and last member of the actinide series is the man-made element lawrencium, $_{103}La$.

Actinium, $_{89}Ac$, is considered a transition metal; the next transition metal would have an atomic number of 89 + 14 (for the 14 actinide elements of the seventh period) + 1, or 104. The synthesis of element 104 has been reported by Russian and American workers. It is referred to as khurchatovium, $_{104}Ku$, or ekahafnium, since it would be in the IVB family under hafnium in the periodic table. No known element has a higher atomic number than $_{104}Ku$. Actinium and khurchatovium represent the only known transition metals of the period; francium and radium represent the only known regular elements, or members of the A families, in the seventh period. Therefore the seventh period is as yet incomplete. The atoms of all elements above polonium, $_{84}Po$, in the periodic table have unstable nuclei and decay into other elements by nuclear breakdown. Nuclear breakdown results in the liberation of an incredible amount of energy and, of course, is the basis of uranium-235 atomic bombs and power plants. Nuclear Chemistry, which deals with the transformations of the atomic nucleus, is the subject of Chap. 20.

TABLE 4.6: ELECTRONIC STRUCTURE WITHIN FAMILIES

Halogens (VIIA)

$_{9}F$ = [He] $2s^2$, $2p^5$
$_{17}Cl$ = [Ne] $3s^2$, $3p^5$
$_{35}Br$ = [Ar] $3d^{10}$, $4s^2$, $4p^5$
$_{53}I$ = [Kr] $4d^{10}$, $5s^2$, $5p^5$

Noble gases (0)

$_{2}He$ = $1s^2$
$_{10}Ne$ = [He] $2s^2$, $2p^6$
$_{18}Ar$ = [Ne] $3s^2$, $3p^6$
$_{36}Kr$ = [Ar] $3d^{10}$, $4s^2$, $4p^6$
$_{54}Xe$ = [Kr] $4d^{10}$, $5s^2$, $5p^6$

Alkali metals (IA)

$_{3}Li$ = [He] $2s^1$
$_{11}Na$ = [Ne] $3s^1$
$_{19}K$ = [Ar] $4s^1$
$_{37}Rb$ = [Kr] $5s^1$
$_{55}Cs$ = [Xe] $6s^1$

4.4 PERIODICITY IN ELECTRONIC STRUCTURE

The periodic law based on atomic number is in beautiful harmony with our modern electronic theory of the atom. This is borne out when we compare the electronic configurations of the elements within a family. In Table 4.6 we see the electronic configurations for halogen family (VIIA), the noble-gas family (0), and the alkali-metal family (IA).

There is a characteristic electronic configuration for each family. For the halogens it is ns^2, np^5; for the noble gases ns^2, np^6; and for the alkali metals ns^1, where n represents the period of the element and the highest occupied energy level.

4.5 TRENDS IN VALENCE OR OXIDATION NUMBER

Mendeleev believed that the oxide formula was a group or family characteristic: all group I elements form oxides with the formula R_2O, like Na_2O, K_2O, and Ag_2O; group II elements form oxides with the formula RO, like BeO, MgO, and ZnO. Figure 4.1, the Mendeleev table, emphasizes this since the oxide formula for the group or family is written at the head of each column.

The number of the group or the family is equal to the highest oxidation number which is exhibited by an element in the family. We assign an oxidation number of -2 to oxygen; then, by rules presented in Chap. 2, the group I elements must exhibit $+1$ oxidation numbers, the group II elements must exhibit $+2$ oxidation numbers; the oxidation number is a characteristic of the family. Many elements have more than one oxidation number but the highest value of the oxidation number is a periodic function of the atomic number.

Another trend in oxidation number is obvious if we write the formulas for the hydrides of a period. Table 4.7 shows the second-period hydride formulas. In the metal hydrides, hydrogen is assigned an oxidation number of -1, and the metals have positive oxidation numbers. In the nonmetal hydrides, hydrogen is assigned an oxidation number of $+1$ and the nonmetals have negative oxidation numbers. The sign of the oxidation number is somewhat arbitrary, but it is based on evidence that the metal atoms donate electrons to hydrogen atoms whereas the nonmetals accept electrons from the hydrogen atoms. From a different point of view we say that the nonmetals have a greater affinity for electrons than hydrogen. When they react with hydrogen, the metal atoms become relatively positive. The metals have a smaller affinity for electrons. In methane, CH_4, carbon and hydrogen have almost equal affinities for electrons, and the oxidation number of carbon is not prefixed with a sign.

TABLE 4.7: HYDRIDES AND OXIDATION NUMBERS OF ELEMENTS OF THE SECOND PERIOD

Hydride	*Oxidation number*
LiH	$+1$
BeH_2	$+2$
BH_3†	$(+3)$
CH_4	4
NH_3	-3
OH_2	-2
FH	-1

† BH_3 is hypothetical; the compound has never been isolated.

The oxidation numbers shown in Table 4.7 can be explained in terms of electronic configurations (Table 4.3). When lithium, beryllium, and boron react with hydrogen, they achieve the rare-gas configuration of helium, ${}_2He$ $1s^2$, by losing one, two, and three electrons, respectively. The remaining elements of the second period achieve the rare-gas configuration of neon, ${}_{10}Ne$ $1s^2$, $2s^2$, $2p^6$, when they react with hydrogen. The hydrogen atoms furnish enough electrons to fill the vacancies in the $2p$ orbitals of carbon, nitrogen, oxygen, and fluorine.

In Table 4.3 it can be seen that the carbon atom requires four electrons to achieve the electronic configuration of neon; it bonds to four hydrogen atoms. The nitrogen atom needs three electrons; it bonds to three hydrogen atoms. Further relationships between oxidation number, bonding, and electron configuration will be discussed in the next chapter, but at this point it is important to see the excellent correlation between oxidation number and periodicity.

4.6 METALS AND NONMETALS

One of the most obvious characteristics of the metallic elements is their lustrous sheen. They also have high thermal and electric conductivity. Metals tend to lose electrons when they undergo chemical reactions and are therefore assigned positive oxidation numbers. For example,

$2Na$	$+$	Cl_2	$\rightarrow 2Na^+Cl^-$
Metal; loses electrons		Nonmetal; gains electrons	

Fe	$+$	$\frac{1}{8}S_8$	$\rightarrow Fe^{2+}S^{2-}$
Metal; loses electrons		Nonmetal; gains electrons	

In addition the metal hydroxides like $NaOH$, $Mg(OH)_2$, and $Ba(OH)_2$ are basic. The nonmetals exhibit negative oxidation numbers, and their hydroxides—like sulfuric acid, $SO_2(OH)_2$ or H_2SO_4; phosphoric acid, $PO(OH)_3$ or H_3PO_4; and nitric acid, $NO_2(OH)$ or HNO_3—are acidic.†

As we go across a period, there is a gradual transition from metallic to nonmetallic elements. The broken line in Fig. 4.7 represents the demarcation between metals and nonmetals. Elements bordering on the broken line have properties which are intermediate between metal and nonmetal. For instance, the hydroxide of aluminum can act as an acid or as a base:

$$Al(OH)_3 + 3NaOH \rightarrow Na_3AlO_3 + 3HOH$$
$$Al(OH)_3 + 3HCl \rightarrow AlCl_3 + 3HOH$$

The members of the alkali-metal family (IA) form the strongest bases, and the members of the halogen family (VIIA) form the strongest acids. Because of this we consider the IA elements to be

† As preliminary definitions we might use *acid*, a compound having a sour taste and capable of reddening litmus paper; *base*, a compound having a bitter taste which neutralizes the effects of an acid. Acids and bases are discussed at length in Chap. 13.

the most metallic. These elements are very reactive and are never found in the pure elemental form in nature because they are so easily *oxidized;* i.e., they readily lose their valence electrons (ns^1). Likewise, the halogens are never found in elemental form in nature because they are so easily *reduced;* i.e., they gain electrons to form halide compounds.

To summarize, as we move across a period from left to right there is a steady gradation of properties from very metallic to very nonmetallic (see Fig. 4.7). As we move down in a family, the elements become more metallic. By these trends cesium is the most metallic element and fluorine is the most nonmetallic element. (Francium is probably more metallic than cesium, but it is so rare that its properties have not been studied extensively.)

4.7 TRENDS IN PHYSICAL PROPERTIES

We shall now examine the trends in certain properties of the elements, to demonstrate the usefulness of the periodic classification. The properties we examine will enable us to understand the chemical behavior of the elements and will be helpful in subsequent chapters. At this point we shall restrict our attention to the elements of the regular, or A, families of the periodic table. Trends in the properties of the transition elements will be discussed in Chap. 19.

Atomic radius

As we noted in Chap. 3, we can no longer hope to determine the exact position of an electron in an atom. Therefore, it is incorrect to think of the radius of an atom as the distance from the nucleus to the outermost electron. Instead, we define the radius of an atom as half the distance between two identical atoms in a chemical bond. Table 4.8 shows the bond lengths of the hydrogen and the halogen diatomic molecules and the calculated atomic radii.

TABLE 4.8

Molecule	*Bond length,* Å	*Atomic radius,* Å
H—H	1.10	0.55
F—F	1.44	0.72
Cl—Cl	2.02	1.01
Br—Br	2.30	1.15
I—I	2.66	1.33

Table 4.9 shows the atomic radii of many of the regular elements in their positions in the periodic table. The atomic radius decreases as we go across a period and increases as we go down a group.

The decrease in atomic radius across a period is due to the fact that the valence electrons are experiencing a greater and greater attractive force as the nuclear charge increases resulting in a contraction of the radius. As an illustration, let us consider the third period. All the elements of the third period have a neon core, $1s^2$, $2s^2$, $2p^6$. Sodium can be represented as [Ne] $3s^1$, Mg as [Ne] $3s^2$, and so on. Experimental evidence strongly indicates that the neon core of the third-period elements is not very much disturbed by the addition of electrons to the $3s$ and $3p$ orbitals. As the atomic number increases, the charge on the core increases because the number of protons in

TABLE 4.9: VARIATION IN ATOMIC RADIUS, Å

Period	IA	IIA	IIIA	IV	VA	VI	VIIA	0
1	H 0.55							He 0.93
2	Li 1.33	Be 0.90	B 0.80	C 0.77	N 0.73	O 0.74	F 0.72	Ne 1.12
3	Na 1.54	Mg 1.36	Al 1.25	Si 1.17	P 1.10	S 1.04	Cl 1.01	Ar 1.54
4	K 1.96	Ca	Ga 1.26	Ge 1.22	As	Se 1.16	Br 1.15	Kr 1.69
5	Rb 2.16	Sr 1.92	In 1.44	Sn	Sb 1.38	Ti 1.35	I 1.33	Xe 1.90

the nucleus increases while the number of electrons in the core remains constant at 10. The $3s^1$ valence electron of sodium "sees" a core with a +1 charge, since there are 11 protons in the nucleus and 10 electrons in the first and second energy levels of the core. It is attracted by a +1 charge. In magnesium the core has a +2 charge (12 protons surrounded by 10 electrons of the $1s$, $2s$, and $2p$ orbitals). The two $3s$ electrons of the magnesium atom are each attracted by a core which has a +2 charge. The $3s$ valence electrons of magnesium are pulled closer to the nucleus, and the radius of the magnesium atom is smaller than that of sodium. The core charge increases across the period, and the atoms of the period show a decreasing atomic radius.

When we move down in families, the energy levels of the valence electron of the elements increase but the core charge remains constant. Lithium has its valence electron in the second energy level above a core charge of +1 ($3p + 2e$). The two electrons in the $1s$ orbital *shield* the valence electron from the attractive effect of the nucleus. The next element in the family, sodium, has its valence electron in the third energy level above a core charge of +1 ($11p + 10e$). Each succeeding element has electrons in the next higher energy level, and the valence electrons are further removed from the nucleus.

Now we can understand why the metallic nature of the elements decreases as we move across a period and increases as we move down a group. Across a period the core charge increases, and its attractive force on the valence electrons increases. As we move down a group, the core charge remains constant, but since the valence electron is in a higher energy level, it is more easily removed from the atom.

Ionization potential (ionization energy)

The ionization potential (IP) is the energy required to remove an electron completely from an isolated atom. An atom of a gas can be considered to be an isolated atom if the gas pressure is low. The ionization process can be symbolized as

$$X(g) + IP \rightarrow X^+(g) + e^-$$

where the ionization potential is the minimum energy required to remove one electron from the atom X in the gas phase.

Although there are several exceptions, the ionization potential generally increases across a period and decreases down a group. The trend is exactly opposite to the atomic-radius trend, and this is *not* coincidental. The closer the electron is to the nucleus the higher the energy necessary to remove it. As we go across a period, the atomic radii decrease and the ionization potentials increase; as we go down a group, the atomic radius increases and the ionization potential decreases, as shown in Table 4.10.

The ionization potentials of the elements of the alkali-metal family (IA) decrease continuously as the atomic number increases. This is the general trend within families, but the trend is often violated; e.g., thallium, Tl, the last member of the IIIA family, has a higher ionization potential than any element of the family except boron.

TABLE 4.10: IONIZATION POTENTIALS FOR THE SECOND PERIOD AND THE IA FAMILY

Second period	*Electron configuration in second energy level*	*Ionization potential,* kcal/mol	IA *family*	*State of valence electron*	*Ionization potential,* kcal/mol
Li	$2s^1$	124	$_3Li$	$2s^1$	124
Be	$2s^2$	217†	$_{11}Na$	$3s^1$	120
B	$2s^22p^1$	190	$_{19}K$	$4s^1$	100
C	$2s^22p^2$	261	$_{37}Rb$	$5s^1$	95
N	$2s^22p^3$	340†	$_{55}Cs$	$6s^1$	90
O	$2s^22p^4$	315			
F	$2s^22p^5$	402			
Ne	$2s^22p^6$	500			

† In the second period the ionization potentials violate the predicted trend. The completely filled and half-filled orbitals of $_4Be$ $1s^2$, $2s^2$ and $_7N$ $1s^2$, $2s^2$, $2p^3$, respectively, represent stable electronic configurations; this makes removal of an electron more difficult, and the ionization potentials are high.

Nevertheless, the general ionization-potential trend parallels the nonmetallic character of the elements. As the ionization potential increases (and more energy is required to remove an electron), the elements become more nonmetallic in their chemical behavior. The most metallic elements, $_{55}Cs$ and $_{37}Rb$, have low ionization potentials. The nonmetals oxygen, chlorine, and fluorine have high ionization potentials.

Electron affinity

The electron affinity A is the quantity of energy released when a neutral atom combines with an electron to form a negative ion in the gas phase:

$$X(g) + e \rightarrow X^{-}(g) + \text{energy}$$

As its name suggests, it measures the affinity of an atom for an electron. The nonmetals form stable negative ions. The halogens (VIIA) and some members of the VA and VIA families have highly charged cores which exert high attractive forces on additional electrons. The extra electrons occupy vacancies in the p orbitals. The metals have relatively low core charges, and they have low or even zero values of electron affinity.

It is clear that the electron affinity should increase across a period and decrease down a group. Table 4.11 shows the electron affinities for several elements, and this prediction appears to hold in general but with several exceptions. For instance, the values for fluorine and oxygen seem low, but there is a general increase across period 2. There is no clear trend in the IA family, but there seems to be a general pattern of decreasing electron affinity down a family. Unfor-

TABLE 4.11: ELECTRON AFFINITIES OF SOME OF ELEMENTS, kcal/mol†

Period	IA	IIA	IIIA	IVA	VA	VIA	VIIA
1	H 17						
2	Li 14	Be 14	B 5	C 48	N	O 40	F 80
3	Na 19	Mg 7	Al 12	Si 37	P 18	S 52	Cl 85
4	K 16	Ca	Ga	Ge	As	Se 46	Br 80
5	Rb	Sr	In	Sn	Sb	Te	I 72

† Most of the data come from quantum-mechanical calculations and have not been experimentally verified, but the calculated affinities seem reasonable in terms of normal periodic trends.

tunately, the electron affinity is difficult to measure, and not enough data are available to make a good test. Nevertheless, the concept is very helpful in predicting and explaining the chemical properties of the elements.

4.8 FREONS: AN ASSIST FROM THE PERIODIC TABLE

In the early part of this century ammonia, sulfur dioxide, and propane were used as refrigerant liquids in electric refrigerators for both commercial and home use. Ammonia, NH_3, is poisonous; sulfur dioxide is poisonous and corrosive; and propane is dangerously combustible. What was needed was an odorless, nontoxic, noncorrosive gaseous compound that was easy to liquefy. Of course it would also have to be inexpensive to gain wide acceptance.

How would you search for such a compound? Thomas Midgley, Jr., an American engineer, went to the periodic table. He noticed that only the nonmetals on the right side of the table form compounds that are gases at room temperature and that the flammability of compounds decreases from left to right. In fact, halogen compounds are used as flame retardants. He also noticed that the compounds of heavier elements are generally more toxic. These observations suggested that compounds of fluorine with other lighter nonmetals would be good refrigerants. After 2 years of laboratory work, he had synthesized and tested a group of compounds now called Freons. These are compounds of carbon, fluorine, and chlorine, like CF_4 and CCl_2F_2. They have ideal refrigerant properties and are used almost exclusively in modern refrigerators and air conditioners.

QUESTIONS AND PROBLEMS

4.1 (*a*) Why was Avogadro's hypothesis indispensable for development of the periodic table? (*b*) Why was the periodic law discovered very quickly after Avogadro's hypothesis was finally accepted?

4.2 (*a*) Write the electronic configuration for polonium, $_{84}Po$. (*b*) To which family does $_{84}Po$ belong? How does the electronic configuration bear this out? (*c*) Write the formulas for possible hydrides, oxides, and oxyacids of $_{84}Po$. (*d*) Is the oxidation state of polonium positive or negative in the hydride? Why?

4.3 (*a*) What was the significance of atomic number when it first came into use? (*b*) How was the relationship between atomic number and nuclear charge discovered? (*c*) What relationship did Moseley discover which eventually led to the modern periodic law? (*d*) State the periodic law as given by Mendeleev and Meyer. (*e*) State the modern periodic law.

4.4 The modern periodic table shown in Fig. 4.7 is divided into regions labeled *s*, *p*, *d*, and *f*. What is the significance of this labeling?

4.5 Explain the meaning and give an example of (*a*) a regular nonmetal, (*b*) a regular metal, (*c*) a transition element, (*d*) an inner transition element, (*e*) a lanthanide (*f*) an actinide.

4.6 Assuming that the seventh period has 32 elements, (*a*) what is the atomic number of the first regular element of the eighth period? (*b*) What are the atomic numbers of the first and last transition elements in the eighth period?

4.7 What are the characteristic electronic configurations for (*a*) family IIIA, (*b*) family IVA, (*c*) family VIA, and (*d*) family 0?

4.8 (*a*) What is meant by the term *core?* (*b*) How does the core charge change across a period from left to right? (*c*) Show the core-charge trend in a family by giving the core charges for the elements of family IIA.

4.9 (*a*) Write the electronic configuration of francium, $_{87}Fr$. (*b*) To which family does it belong? How does the electronic configuration bear this out? (*c*) Write the formulas for possible hydrides, sulfides, and bromides. (*d*) Would the hydroxide of francium be acidic? Why?

4.10 What family was undiscovered when Mendeleev proposed the periodic table of the elements?

4.11 Arrange the following groups of elements in order of increasing metallic behavior: (*a*) S, Cl, Na, Si, Al; (*b*) H, F, Se, I, Pb.

4.12 Relate (*a*) the trends in atomic radius to the trend in electron affinity; (*b*) the trend in ionization potential to the trend in metallic properties; (*c*) the trend in atomic radius to the trend in metallic properties.

4.13 Let the unknown element 113 be symbolized as eTl (e for eka). One oxide of thallium is 10.5 percent by weight oxygen. Determine the empirical formula for the analogous oxide of eTl.

4.14 Element 113 has not been synthesized or discovered. Using the principle of periodicity, predict its density, melting point, oxidation state, and the formulas for its oxide and chloride. (You will need data from a handbook to predict its density and melting point.)

4.15 Using Table 4.9, predict or give a range for the atomic radii of calcium, arsenic, and tin.

Stoichiometry practice

4.16 From the following series of reactions determine the number of pounds of sulfuric acid which can be produced from 1 kg of Fe_2S_3:

$$Fe_2S_3 + \tfrac{9}{2}O_2 \rightarrow Fe_2O_3 + 3SO_2$$
$$SO_2 + \tfrac{1}{2}O_2 \rightarrow SO_3$$
$$SO_3 + H_2O \rightarrow H_2SO_4$$

4.17 Bromine, Br_2, and iron react to form a compound which is 81.1 percent bromine. (*a*) Determine the empirical formula for this iron bromide and (*b*) write the balanced chemical equation for the reaction.

4.18 When carbon is burned in an excess of oxygen, it produces about

94.1 kcal of heat per mole, while elemental sulfur produces 70.9 kcal/mol (32 g). A coal sample contains 81.2 percent elemental carbon and 3.3 percent elemental sulfur; the rest is noncombustible. (*a*) How much heat will 1 ton of this coal produce? What weight (in grams) of (*b*) SO_2 and (*c*) CO_2 will be produced by burning 1 ton of coal?

4.19 The density of hydrogen gas and of the vapor of an organic compound are 0.0893 and 2.05 g/l, respectively, under the same conditions of temperature and pressure. The organic compound is composed of carbon, hydrogen, and oxygen only. The percentages of carbon and hydrogen by weight are 52.2 and 13.03 percent. (*a*) Recalling Avogadro's hypothesis, determine the molecular weight of the compound. (*b*) Determine the empirical formula of the compound. (*c*) What is the molecular formula?

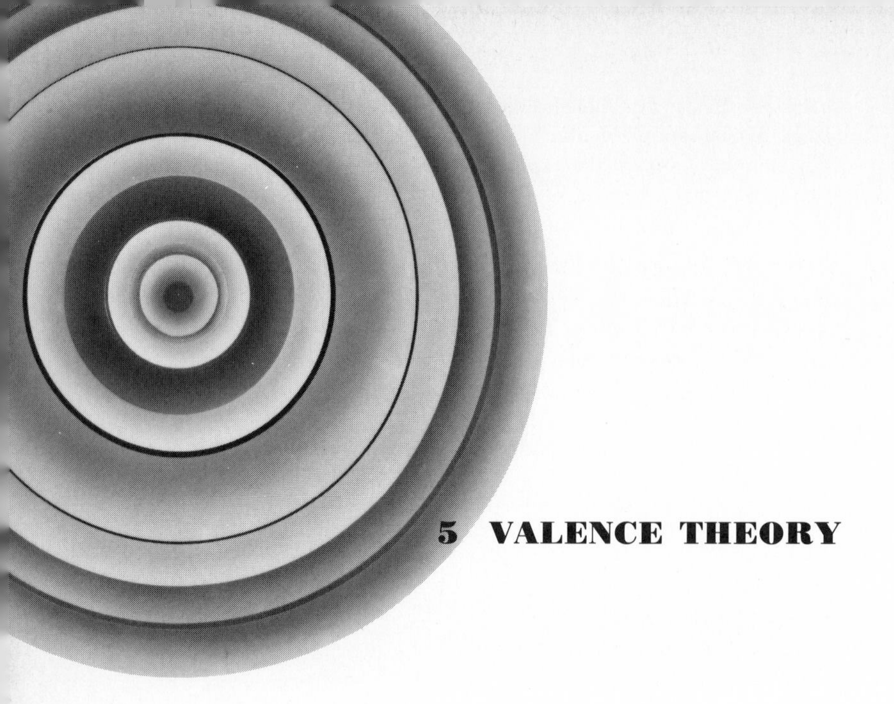

5 VALENCE THEORY

5.1 THE ARCHITECTURE OF MOLECULES

Modern analytical tools and techniques have enabled chemists to conclude that molecules have a definite architecture. The kinds of atoms, their relative positions, and the separations between them are identical in each molecule of the same compound. For example, in the ammonia molecule, NH_3, one nitrogen atom is attached to three hydrogen atoms; the molecule is pyramidal, three hydrogen atoms forming the corners of the base and the nitrogen atom forming the apex of the pyramid. Each H—N—H angle is 107°, the distance from the nitrogen to any hydrogen atom being 1.10 Å. There are many other molecules in which nitrogen forms three bonds and confers a pyramidal shape on the molecule.

In the water molecule two hydrogen atoms are attached to one oxygen atom, and the molecule is bent so that the H—O—H angle is 105°. The distance from the oxygen atom to either hydrogen atom is 1.11 Å. In a host of molecules the oxygen atom forms two bonds and causes bending of about 105° around itself.

Obviously, there is some explanation for the way particular atoms cause definite patterns. There must be a directing force acting between atoms within a molecule causing the atoms to take up a definite configuration. This force results in a linkage between atoms called a *chemical bond*. When a chemist says two atoms are bonded, he means that the bonded atoms are very close together and are mutually held by strong forces. The center-to-center separation between two atoms in a chemical bond is usually less than 3 Å and almost never greater than 4 Å. The chemical bond represents a stable configuration, and consequently energy is required to pull the bonded atoms apart. The energy necessary to break a chemical bond ranges from 20,000 to 200,000 cal per mole of bonds (N_A bonds).

In order to explain how chemical bonds are formed many theories, called *valence* theories, have been proposed. A good valence theory must answer three basic questions: (1) Why does a chemical bond form? (2) Why do different atoms form different numbers of bonds? (3) Why is there a definite arrangement of atoms within a molecule?

The term *valence* is somewhat ambiguous. When we say the valence of element X is 3, we *may* mean that element X is bonded to three other atoms *or* that element X forms three bonds. Very often the number of bonded atoms may equal the number of bonds, but there can be a difference. For example, in the ammonia molecule the nitrogen atom forms three bonds, *one* to each of the three hydrogen atoms. Hence, the number of bonds equals the number of bonded atoms. Now consider the carbon atom in ethylene, C_2H_4 (Fig. 5.1). The molecule is planar. Each carbon atom is bonded to two hydrogen atoms and to another carbon atom, or each carbon is bonded to a total of three atoms. By our first definition, the valence

FIGURE 5.1

The bonding in ethylene, C_2H_4.

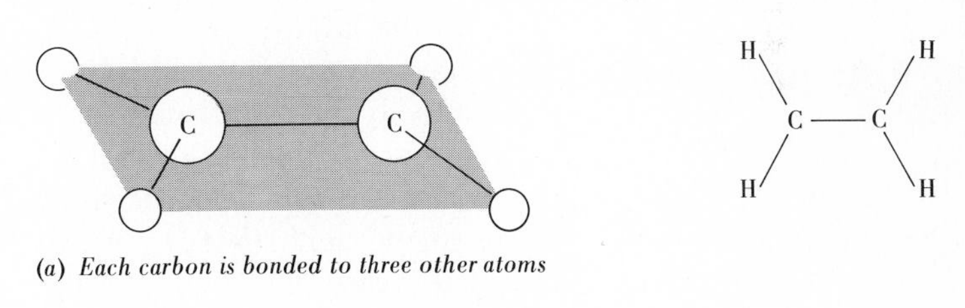

(a) Each carbon is bonded to three other atoms

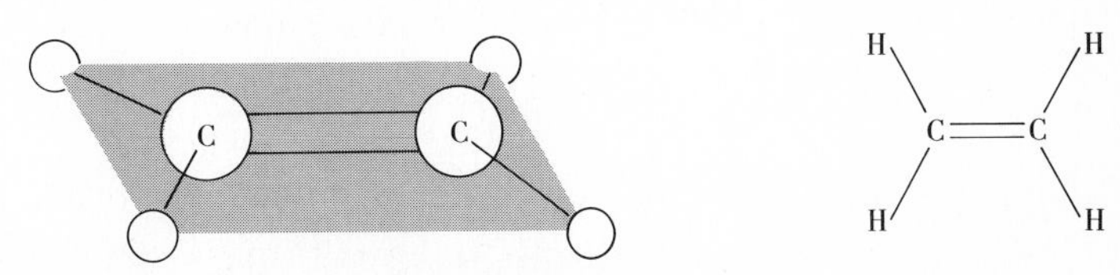

(b) There is a double bond between the two carbon atoms of ethylene

of carbon is 3. However, theories of valence indicate that a *double* bond exists between the two carbon atoms, and, by our second definition, the valence of carbon is 4. We ordinarily use the term valence to mean the number of bonds an atom forms, which is not necessarily the number of other atoms to which it is bonded. Hence, in ordinary usage the valence of carbon is 4.

5.2 THE DEVELOPMENT OF VALENCE THEORY IN NINETEENTH CENTURY

By elemental analysis, propane gas contains 81.8 percent carbon and 18.2 percent hydrogen by weight. Gaseous propane is 22 times denser than gaseous hydrogen under the same condition; therefore, by Avogadro's hypothesis, the molecular weight of propane is 44 if the molecular weight of hydrogen is taken as 2 mass units. These data allow us to determine that the molecular formula for propane is C_3H_8. (See page 23 for the method of determining chemical formulas.)

Nothing we have said so far gives a clue to the arrangement of atoms within the propane molecule. We might consider a molecule as a sort of bag in which there is a definite number and kind of atoms but with an indefinite arrangement. But today when a chemist says that molecules of the same compound are identical, he means that there are equal numbers of atoms of the same elements and that these atoms have a definite pattern within the molecule.

In the mid-nineteenth century, the concept of internal molecular structure was beginning to take form, but many leading scientists felt that the determination of the positions of the atoms within a molecule was beyond the range of any conceivable experiment.

Organic chemists had developed the habit of representing a compound by a *rational* formula, i.e., one which explains (or rationalizes) the largest number of the reactions of the compound. These formulas were not devised to represent molecular structure, but as the body of knowledge of chemical reactions grew, it became increasingly difficult to hold that no relationship existed between the rational formulas and the structure of a molecule. Consider the case of acetic acid, $C_2H_4O_2$, which had been well studied as early as 1850. Acetic acid was known to undergo several reactions; it reacted with sodium hydroxide, to produce a compound in which one of the four hydrogen atoms was replaced by a sodium atom:

$$\underset{\text{Acetic acid}}{C_2H_4O_2} + \underset{\text{Sodium hydroxide}}{NaOH} \rightarrow \underset{\text{Sodium acetate}}{C_2H_3O_2Na} + H_2O \qquad (5.1)$$

The rational formula based only on this reaction would be

$C_2H_3O_2 \cdot H$, the dot in the formula indicating that one of the hydrogen atoms is chemically different from all the others; i.e., one of the hydrogens reacted. Heating the product, sodium acetate, with more NaOH produces sodium carbonate and methane:

$$\underset{}{C_2H_3O_2Na} + NaOH \rightarrow \underset{\text{Sodium carbonate}}{Na_2CO_3} + \underset{\text{Methane}}{CH_4} \tag{5.2}$$

Considering methane, CH_4, to be derived from $C_2H_3O_2Na$, it appears that one carbon atom and two oxygen atoms are removed as a group and are replaced by a single hydrogen atom. On this basis a rational formula for $C_2H_3O_2Na$ would be $CH_3 \cdot CO_2 \cdot Na$, which, coupled with Eq. (5.1), implies a rational formula for acetic acid of $CH_3 \cdot CO_2 \cdot H$. Acetic acid reacts with phosphorus pentachloride; one atom of oxygen and one atom of hydrogen, or a hydroxyl group, is replaced by a single atom of chlorine:

$$C_2H_4O_2 + PCl_5 \rightarrow C_2H_3OCl + POCl_3 + HCl \tag{5.3}$$

Based on this reaction only, the rational formula for acetic acid would be $C_2H_3O \cdot OH$, which indicates that the hydroxyl group is severed as a unit. Considering the three reactions together, the best rational formula would be $CH_3 \cdot CO \cdot O \cdot H$ since it explains all the reactions. The comparison between this formula, which was first proposed in the mid-nineteenth century on the basis of chemical evidence alone, and the accepted modern structural formula for acetic acid is striking:

$$CH_3 \cdot CO \cdot O \cdot H \qquad \begin{array}{c} H\;\;O \\ |\;\;\;\| \\ H-C-C-O \\ |\diagdown \\ HH \end{array} \qquad \text{or} \qquad CH_3-CO-O-H$$

Rational formula — Structural formula

A structural interpretation of rational formulas was presented in 1858 by the German chemist Freidrich August Kekulé, who believed that "Atomicity [valence] is a fundamental property of the atom, which must be constant and invariable like the weight of the atom itself. . . ." Some of the rational formulas in use in the mid-nineteenth century are shown in Table 5.1. A study of this table shows, as Kekulé suggested, that the atoms of certain elements can attach themselves to a fixed number of other atoms. Hydrogen attaches itself to only one other atom, and oxygen usually attaches itself to two other atoms. We say that hydrogen forms only one bond; oxygen forms two bonds. Also, chlorine forms one bond and carbon forms four bonds. The use of rational formulas led to the concept of valence, which immediately implies a definite structure within the molecule.

TABLE 5.1: MOLECULAR AND RATIONAL FORMULAS

Compound	*Molecular formula*	*Rational formula*
Marsh gas (methane)	CH_4	CH_4
Chloroform	$CHCl_3$	$CH \cdot Cl_3$
Wood alcohol	CH_4O	$CH_3 \cdot O \cdot H$
Grain alcohol	C_2H_6O	$CH_3 \cdot CH_2 \cdot O \cdot H$
Methyl ether	C_2H_6O	$CH_3 \cdot O \cdot CH_3$
Ethyl ether	$C_4H_{10}O$	$C_2H_5 \cdot O \cdot C_2H_5$
Acetic acid	$C_2H_4O_2$	$CH_3 \cdot CO \cdot O \cdot H$
Ethylene	C_2H_4	$CH_2 \cdot CH_2$

The structures of marsh gas, chloroform, wood alcohol, and methyl ether as deduced from the rational formulas in Table 5.1 are

```
   H             Cl              H               H       H
   |             |               |               |       |
H—C—H        H—C—Cl        H—C—O—H        H—C—O—C—H
   |             |               |               |       |
   H             Cl              H               H       H
```

Marsh gas   Chloroform   Wood alcohol   Methyl ether

To write structural formulas for many of the carbon compounds like grain alcohol and ethyl ether, another hypothesis was required: the carbon atom can form bonds with other carbon atoms. A pair of separate carbon atoms can form a total of eight bonds, but if the pair is bonded, only six bonds to other atoms are required:

```
  |      |         |  |
—C—    —C—      —C—C—
  |      |         |  |
```

Separate pair, eight bonds unsatisfied   Bonded pair, six bonds unsatisfied

```
   H  H                H  H     H  H
   |  |                |  |     |  |
H—C—C—O—H          H—C—C—O—C—C—H
   |  |                |  |     |  |
   H  H                H  H     H  H
```

Ethyl alcohol   Ethyl ether

Table 5.1 shows that the molecular formulas for grain alcohol and methyl ether are identical, but the difference in the rational formulas accounts for the fact that the compounds have very different chemical and physical properties. Substances with the same molecular formulas but with different rational and structural formulas are called *isomers*. The rational or structural formulas account for the different properties of the isomers. In addition, structural formulas explain why substances with different formulas are similar. Methyl and ethyl alcohol, CH_4O and C_2H_6O, have similar properties because of the presence of a C—O—H group in both structures. Methyl and ethyl ether are similar; both formulas show the presence of the C—O—C group.

To support his contention that the valence of an atom is constant, Kekulé was forced to invent the *multiple-bond* concept. The last two compounds in Table 5.1 can be given structural formulas commensurate with the principle of constant valence if multiple bonds are used:

```
H         H       H       O
 \       /         \     //
  C === C       H — C — C
 /       \         /     \
H         H       H       O — H

   Ethylene           Acetic acid
```

In acetic acid the double bond between carbon and oxygen maintains the valence of 2 for oxygen and 4 for carbon; in ethylene we maintain a valence of 4 for carbon by connecting the two carbon atoms with a *double* bond. However, too many examples of violations of the principle of constant valence, especially for inorganic compounds, were uncovered, and the principle became untenable, but the structural ideas and multiple-bond concept to which it gave rise are extremely important in modern chemistry.

5.3 ELECTRONS AND VALENCE

In order to move from nineteenth-century valence theory to the modern theory two important relationships had to merge. In Mendeleev's periodic table (Fig. 4.1) the oxide and hydride formulas are shown to be a characteristic property of a chemical family. This must mean that valence, or bonding capacity, is a characteristic property of a chemical family. From 1898 to 1914, thanks largely to the labors of J. J. Thomson, Ernest Rutherford, Niels Bohr, and Henry Moseley, it became obvious that the electronic configuration of an atom is also a characteristic property of a family of elements. The merger of these two periodic relationships led to the conclusion that the valence of an atom is determined by its electronic configuration. The outer electrons, which are believed to be involved in bonding, are therefore called *valence electrons*.

5.4 THE OCTET RULE

In 1916, after the Bohr-Rutherford solar-system atom was developed, Walter Kossel proposed that atoms tend to acquire the electronic configuration of a rare gas by their chemical reactions. This proposal came to be called *the octet rule* or *the rule of eight* because achieving the rare-gas configuration is equivalent to having eight electrons in the valence shell. Kossel proposed that when a chemical reaction occurs, electrons are transferred from one atom to another, and the net result is that both atoms attain a rare-gas configuration. For example, consider the transformation from potassium and chlorine into potassium chloride:

$$:\overset{\cdot\cdot}{\underset{\cdot\cdot}{\mathrm{K}}}:\cdot + \,{}^{\circ}\overset{\circ\circ}{\underset{\circ\circ}{\mathrm{Cl}}}{}^{\circ}_{\circ} \rightarrow :\overset{\cdot\cdot}{\underset{\cdot\cdot}{\mathrm{K}}}:^{+} + :\overset{\circ\circ}{\underset{\circ\circ}{\mathrm{Cl}}}{}^{\circ}_{\circ}{}^{-}$$

The potassium ion and the chloride ion have achieved the configuration of argon, with eight electrons in the outer, or valence, shell. The resulting oppositely charged ions are held together by a strong electrostatic attraction. Kossel's explanation of bond formation was successful for salts or electrolytes only. These substances are composed of charged atoms, or *ions*. When salts melt or are dissolved, the ions are free to move and conduct electricity.

5.5 THE LEWIS-LANGMUIR VALENCE THEORY

The American chemists G. N. Lewis and Irving Langmuir developed the octet rule further and were able to use it to account for the formation of ionic and nonionic compounds. Lewis and Langmuir independently proposed that the octet rule does not require a complete transfer of electrons from one atom to another. Lewis (1923) stated that "Two atoms may conform to the rule of eight, or octet rule, not only by the transfer of electrons from one atom to another, but also by *sharing* one or more pairs of electrons. These electrons which are held in common by two atoms may be considered to belong to the outer shells of both atoms." Lewis showed that a chemical bond between two atoms results from sharing two electrons. In most cases the sharing of the bonding electron pairs results in the octet configuration for each atom in the bond, but not always; hydrogen forms molecules with only 2 electrons in its outer shell; boron forms molecules in which only 6 electrons are present in the outer shell; phosphorus can have 10 electrons in the outer shell; many of the transition-metal atoms show the tendency to share 6 electron pairs with other atoms, giving 12 electrons in the outer shell. But all bonds result from the sharing of an electron pair.

Lewis held that the "rule of two is even more fundamental than the rule of eight."

5.6 APPLICATIONS OF THE LEWIS-LANGMUIR VALENCE THEORY

Although the Lewis-Langmuir theory, which is based on the Bohr-Rutherford model of the atom, has been superseded by the more realistic quantum-mechanical theory. *Lewis structures* for molecules are still widely used today. The hydrogen molecule is represented very simply. Each hydrogen atom has only one electron; when a bond forms, two atoms share an electron pair *equally:*

$$H\cdot + H\cdot \rightarrow H:H$$

The formation of the H_2 molecule is attributed to the tendency of the atoms to achieve a rare-gas configuration; in this case each hydrogen atom, by sharing an electron pair, has achieved the configuration of helium.

The Bohr-Rutherford models for the atoms of the elements of the first period are

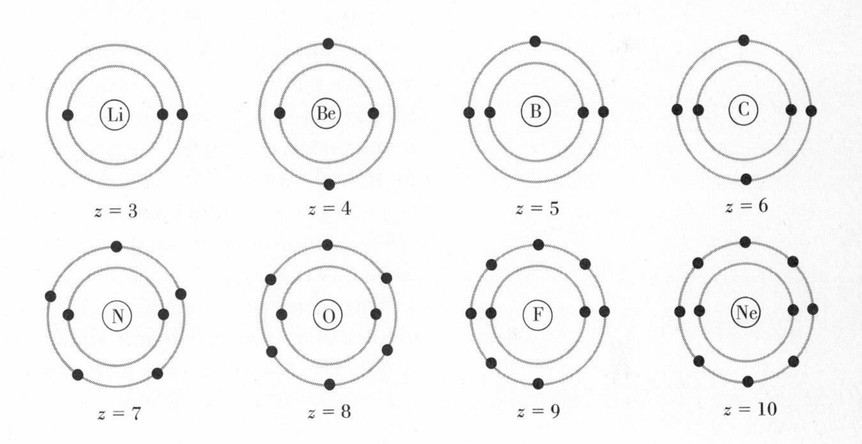

Note that each atom of the period has an inner shell containing two electrons. This is a complete shell, and the electrons in it are considered chemically inert. Only the electrons in the next shell are available for bonding. The second shell is the valence shell for elements of the second period. The part of the atom which consists of the two inner electrons and the nucleus, called the *kernel* or *core* of the atom, remains intact during chemical transformations. We can represent the atoms of the second period more simply by omitting the inner shell.

In order for carbon, $_6C$, to conform to the octet rule it would have to gain four electrons to complete its valence shell. This can be accomplished by sharing electrons with four other atoms as in methane, CH_4; ethane, C_2H_6; and many other carbon compounds:

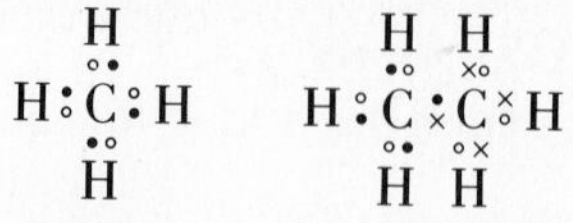

```
   H            H  H
   ∘∘           •∘  ×∘
H∘C∘H       H:C×C×H
   ••           ∘•  ∘×
   H            H  H
```

Methane   Ethane

We note that the electronic structures for methane and ethane allow the carbon atoms and the hydrogen atoms to achieve the rare-gas configuration. For simplicity we use a line between the symbols for the elements to represent an electron-pair bond between two atoms:

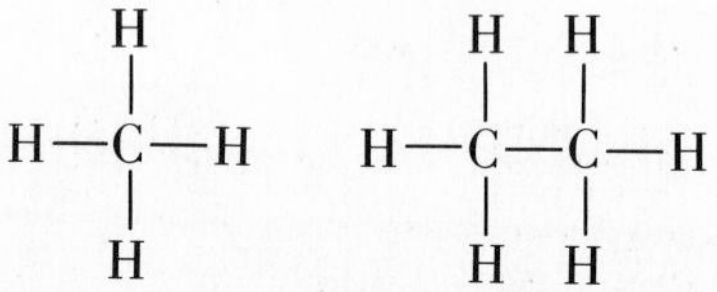

```
     H              H   H
     |              |   |
  H—C—H        H—C—C—H
     |              |   |
     H              H   H
```

This is exactly how the organic chemists of the nineteenth century represented molecules, but now the line representing a bond implies that two electrons are present and that somehow they intimately bind the atoms together.

The formation of electron-pair bonds in nitrogen, oxygen, and fluorine compounds can also be attributed to a tendency to conform to the octet rule or to achieve the rare-gas configuration. This is demonstrated by the Lewis structures for the hydrides of these elements, shown in Fig. 5.2. Each atom involved contributes one electron to each bonding pair of electrons. The resulting pair of electrons is shared by both atoms. This type of bond is called a *covalent bond.*

The nitrogen, oxygen, and fluorine atoms have more than four electrons in their valence shells before bonding. For example, nitrogen has five electrons about its kernel and needs only three elec-

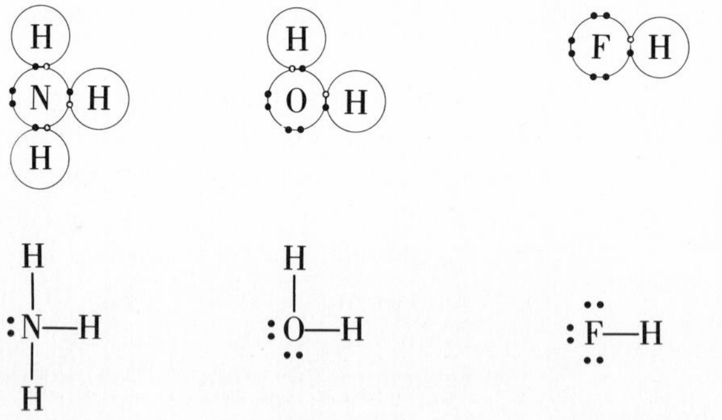

FIGURE 5.2

Lewis structures for ammonia, water, and hydrogen fluoride.

```
  H            H
  |            |
:N—H         :O—H          :F̤̈—H
  |            ••
  H
```

Ammonia   *Water*   *Hydrogen fluoride*

trons to complete its valence shell. In ammonia (Fig. 5.2) nitrogen shares three electron pairs with three hydrogen atoms; in addition, there is another pair which is unshared. This is often referred to as a *nonbonding pair*. In the water molecule the oxygen shows two nonbonding pairs, and in HF the fluorine atom shows three nonbonding pairs.

We see that four elements of the second period, carbon, nitrogen, oxygen, and fluorine, can form covalent bonds. The last element of the period, neon, is a noble gas and it forms no compounds. The noble gases, helium, neon, argon, xenon, and radon, have eight electrons in their outer shells and are in a chemically inert configuration. Only a few noble-gas compounds are known.

5.7 THE IONIC, OR ELECTROVALENT, BOND

When a metallic element reacts with a nonmetallic element, the resulting compound is generally ionic. The first two members of the second period, $_3Li$ and $_4Be$, are metals; these elements tend to achieve the rare-gas configuration of helium, $_2He$, rather than that of $_{10}Ne$. Lithium loses one electron in its reactions with nonmetals and forms ionic compounds; e.g., in reaction with fluorine, which readily accepts a single electron to complete its valence shell, the lithium atom donates one electron to a fluorine atom; lithium achieves the helium configuration, and fluorine achieves the neon configuration. Each atom has become charged, and the compound lithium fluoride is ionic:

$$\mathrm{Li}\cdot + \circ\underset{\circ\circ}{\overset{\circ\circ}{\mathrm{F}}}\!:\; \rightarrow \mathrm{Li}^{+}\; :\underset{\circ\circ}{\overset{\circ\circ}{\mathrm{F}}}\!:^{-}$$

Beryllium reacts in an analogous manner, but one beryllium atom donates two electrons in forming ionic compounds:

$$\mathrm{Be}: + 2\,\circ\underset{\circ\circ}{\overset{\circ\circ}{\mathrm{F}}}\!:\; \rightarrow :\underset{\circ\circ}{\overset{\circ\circ}{\mathrm{F}}}\!:^{-}\; \mathrm{Be}^{2+}\; :\underset{\circ\circ}{\overset{\circ\circ}{\mathrm{F}}}\!:^{-} \qquad \text{or} \qquad \mathrm{BeF_2}$$

According to Lewis theory, the bonding in the ionic compounds is essentially due to the simple attractive electrostatic forces which result when the atoms become oppositely charged. Ionic, or electrovalent, bonds are stronger than covalent bonds. By this we mean that the work that must be done to separate the two ions of an ionic bond is greater than the work required to separate two atoms of a covalent bond. The work or energy required to sever a chemical bond is called the *bond-dissociation energy, D*. The bond-dissociation energy for simple ionic compounds like NaCl is in the range of 200,000 cal/mol, whereas the bond-dissociation energy of most covalent compounds lies between 50,000 and 100,000 cal/mol.

The third element of the second period, boron, is anomalous. With three electrons in its valence shell, the atom is in a quandary; achieving the rare-gas configuration of $_2He$ would require removal of three electrons, but the removal of three electrons from a very small atom such as boron requires a large amount of energy. Achieving the rare-gas configuration of $_{10}Ne$ would require the addition of five electrons. Therefore, in many of its compounds boron compromises. It forms covalent molecules with only six electrons, three pairs, in the valence shell. The electronic structure of BF_3 gives the fluorine atoms completed octets, but the boron atom has a sextet:

$$B + \tfrac{3}{2}F_2 \rightarrow BF_3$$

The structure of boron compounds supports Lewis's realization that the rule of two is more fundamental than the rule of eight. As we shall see, boron often forms compounds in which it completes the octet, but it never forms B^{3+} ionic compounds.

5.8 FAMILY RELATIONSHIPS AND LEWIS STRUCTURES

Although we have chosen compounds of the elements of the second period to demonstrate the application of Lewis-Langmuir theory, the ordinary periodic trends and family relationships hold and analogous Lewis structures can be drawn for many compounds of other elements (see Fig. 5.3).

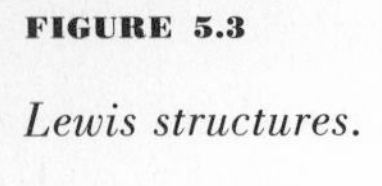

FIGURE 5.3

Lewis structures.

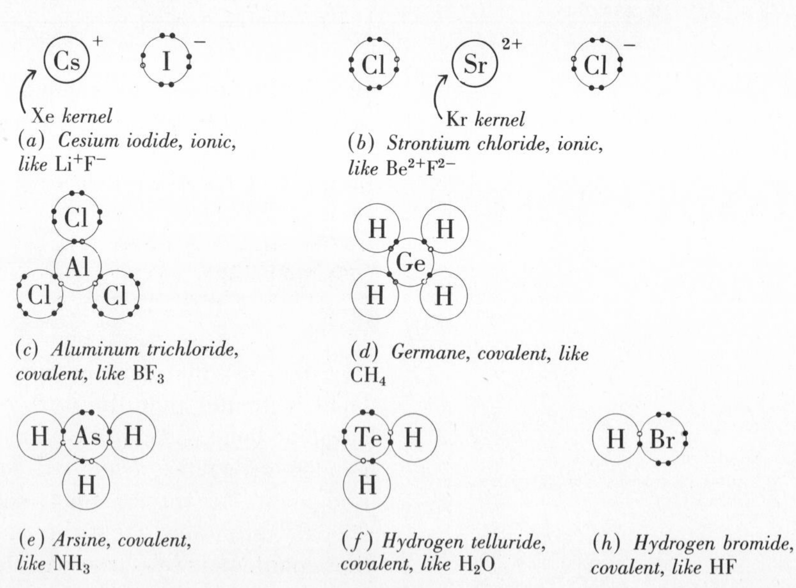

(*a*) *Cesium iodide, ionic, like* Li^+F^-

(*b*) *Strontium chloride, ionic, like* $Be^{2+}F^{2-}$

(*c*) *Aluminum trichloride, covalent, like* BF_3

(*d*) *Germane, covalent, like* CH_4

(*e*) *Arsine, covalent, like* NH_3

(*f*) *Hydrogen telluride, covalent, like* H_2O

(*h*) *Hydrogen bromide, covalent, like* HF

EXAMPLE 5.1 Show the Lewis structures for (*a*) H_2S and (*b*) PCl_3.

Solution: (*a*) The sulfur atom has six valence electrons surrounding its kernel. It needs two more electrons to complete the octet. The hydrogen atom has one valence electron; it must gain one more electron to achieve the helium rare-gas configuration:

:S: (with dots above and below) and H°

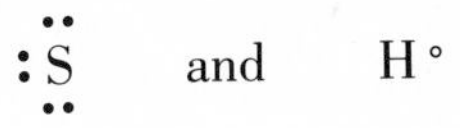

Both sulfur and hydrogen can be satisfied if sulfur forms two covalent bonds with two hydrogen atoms:

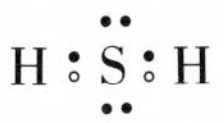

H : S : H (with electron pairs above and below S)

(*b*) Draw electron-dot structures depicting the valence electrons of phosphorus and chlorine:

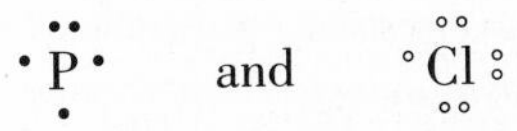

·P· (with dots) and °Cl (with circles)

Since phosphorus has five electrons in its valence shell and chlorine has seven, both atoms can complete the octet if phosphorus forms three covalent bonds with three chlorine atoms:

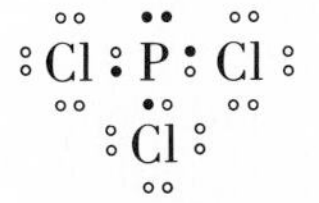

:Cl : P : Cl:
:Cl:

To draw Lewis structures the Bohr models of each atom showing the number of electrons in the valence shell must be drawn. The valence electrons are then arranged so that each achieves the rare-gas configuration.

5.9 MULTIPLE COVALENT BONDS

When two atoms are joined by more than one bond, we say that a multiple bond exists. The second-period elements, carbon, nitrogen, and oxygen (and to some extent phosphorus and sulfur of the third period) can form multiple bonds. In the Lewis-Langmuir theory the multiple bond is a bond formed by two atoms when they share two or more electron pairs. The use of the multiple bond allows us to draw Lewis structures in conformity with the octet rule. The presence of a multiple bond is verified, as we shall see, by the physical properties of such a bond. Examples of molecules with multiple bonding are shown in Fig. 5.4.

Table 5.2 shows the bond distances and dissociation energies for single, double, and triple bonds. The data can be explained if we understand how an electron pair causes two atoms to bond. The pres-

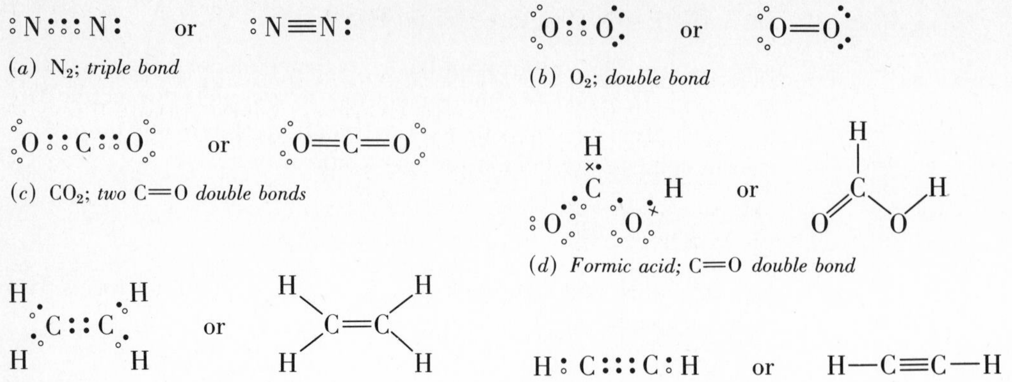

:N:::N: or :N≡N:

(a) N_2; *triple bond*

:O::O: or :O=O:

(b) O_2; *double bond*

:O::C::O: or :O=C=O:

(c) CO_2; *two* C=O *double bonds*

H:C(:O:)O:H or H—C(=O)—O—H

(d) *Formic acid;* C=O *double bond*

H₂C::CH₂ or H₂C=CH₂

(e) *Ethylene;* C=C *double bond*

H:C:::C:H or H—C≡C—H

(f) *Acetylene;* C≡C *triple bond*

FIGURE 5.4

Multiple bonding.

ence of an electron pair between two atoms reduces the repulsion between the positively charged nuclei; in addition, the electron pair is attracted by both nuclei. The net effect of the presence of an electron pair is attraction between the two atoms. The presence of more than one electron pair tends to increase the attraction; i.e., it increases the bond strength and shortens the separation between the bonded atoms. This is borne out by the data in Table 5.2, which show that, compared to a single bond, atoms of a double bond are closer together and more energy is required to dissociate them. For the same reasons the length of a triple bond is even shorter and its dissociation energy greater.

TABLE 5.2: EFFECT OF THE NUMBER OF BONDING ELECTRON PAIRS ON BOND LENGTH AND DISSOCIATION ENERGY

Bond	*Type*	*Example*	*Bond length,* Å	*Dissociation energy in kilocalories per mole of bonds*
—C—C—	Single	Ethane, C_2H_6	1.54	80
C=C	Double	Ethylene, C_2H_4	1.34	145
—C≡C—	Triple	Acetylene, C_2H_2	1.20	198
—C—O—	Single	Ethyl alcohol, C_2H_5—OH	1.43	79
C=O	Double	Acetone, CH_3—C(=O)CH_3	1.22	173
—C—N	Single	Glycine, H_2N—CH_2COOH	1.47	66
C=N	Double	Dimethylimine, $(CH_3)_2C$=N—H	1.28	147
—C≡N	Triple	Methyl cyanide, CH_3C≡N	1.16	209

5.10 THE COORDINATE COVALENT BOND

In order to draw acceptable structures for many of the common inorganic compounds the Lewis-Langmuir theory employs the coordinate covalent, or dative, bond. In the coordinate covalent bond one atom furnishes both electrons for the formation of an electron-pair bond. Once it has formed, the coordinate covalent bond is in all respects identical with an ordinary covalent bond.

For H_2SO_4, H_3PO_4, $HClO_4$, and many other molecules it is impossible to draw Lewis structures which conform to the octet rule without the use of the coordinate covalent bond. In the H_2SO_4 molecule the sulfur atom is bonded to four oxygen atoms. Two of the oxygens are bonded by ordinary covalent bonds and two by coordinate covalent bonds. The Lewis structure for H_2SO_4 and other molecules exhibiting coordinate covalence are shown in Fig. 5.5. Dots represent electrons "from oxygen" and small open circles represent electrons "from sulfur." We cannot actually distinguish the source of electrons; we use the different symbols solely for the purpose of accounting. The coordinate covalent bond is designated by an arrow directed away from the atom that furnishes the electron pair. Note that each atom of the structures has achieved the rare-gas configuration.

FIGURE 5.5

Lewis structures with coordinate covalent bonds.

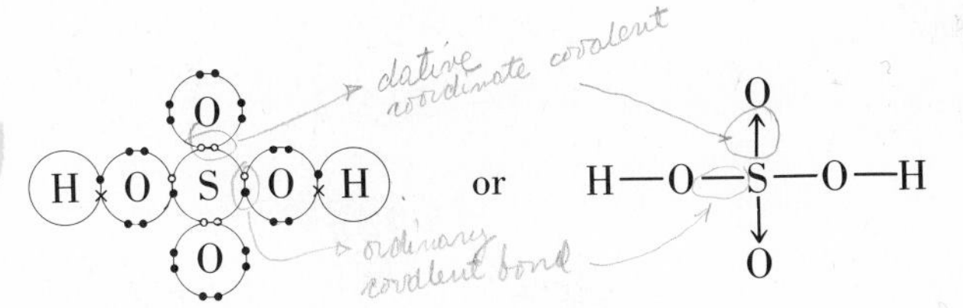

(*a*) H_2SO_4

(*b*) H_3PO_4

(*c*) *Ammonia boron trifluoride*

5.11 GEOMETRY OF MOLECULES BASED ON ELECTROSTATIC-REPULSION FACTORS

We can predict the shapes of molecules with fair accuracy by using a simple concept from electrostatics: like charges repel each other. If the electron pairs in the valence shell of an atom are free to move in an imaginary sphere, they will move as far apart as possible because they are negative charges. The repulsion causes them to take up particular orientations, and this affects the shape of a molecule, as we shall show by example.

Family IIIA

In many of their compounds boron and aluminum, family IIIA, are trivalent, for example, BF_3 and $AlCl_3$. The valence shells of boron and aluminum contain only three electron pairs. If three electron pairs were confined to the surface of a sphere, they would repel each other and move to points representing the vertices of an equi-

FIGURE 5.6

Electrostatic considerations suggest that trivalent boron and aluminum molecules are planar.

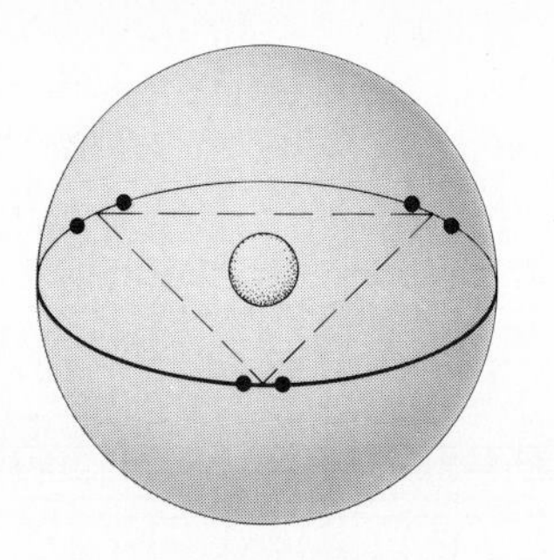

(a) An equilateral triangle inscribed in a sphere with the center of the triangle coincident with the center of the sphere

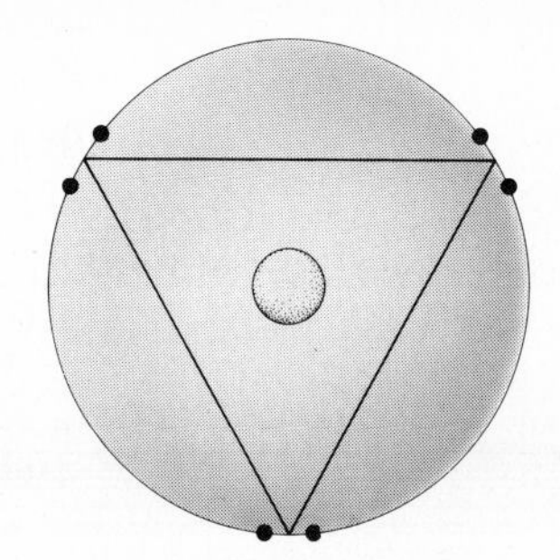

(b) Top view

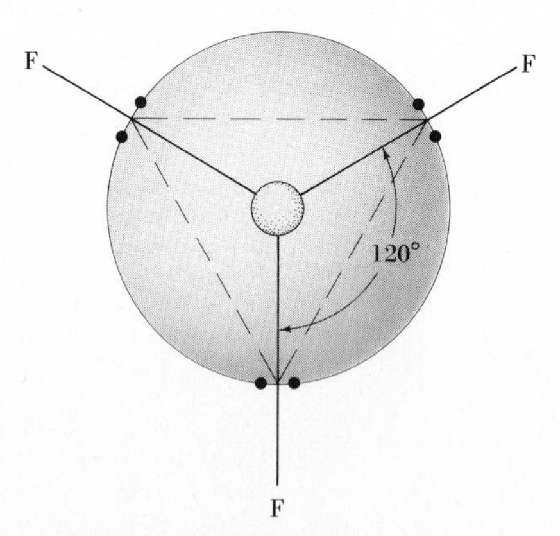

(c) The BF_3 structure is planar

lateral triangle, i.e., points of maximum separation. This means that the nucleus at the center of the equilateral triangle and the three electron pairs are coplanar. Since bonds are formed by the sharing of electron pairs, all atoms bonded to aluminum or boron should be coplanar. The electrostatic-repulsion theory predicts correctly that trivalent boron and aluminum molecules are planar and all bond angles are 120°, as shown in Fig. 5.6 for BF_3.

Family IVA

In most molecules the valence shells of the atoms contain four electron pairs. Electrostatic repulsion means that the optimum arrangement of the four pairs in a spherical shell is tetrahedral, the four electron pairs occupying the corners of a regular tetrahedron inscribed in the sphere (see Fig. 5.7*a*). The shape of a molecule depends on the number of electron pairs used for bonding. For elements in family IVA, which often form four covalent bonds, the molecules have a tetrahedral geometry. For example, CH_4, CCl_4, and $SiCl_4$ are molecules in which the group IVA atom is at the center of a regular tetrahedron and four other atoms are at the vertices (see Fig. 5.7*b*). The Cl—Si—Cl bond angle is 109°, which is the value expected for a regular tetrahedron.

Family VA

In some compounds of the group VA elements, like NH_3 and PCl_3, only three of the four electron pairs are used for bonding. Since the electron pairs are in a tetrahedral orientation, NH_3 would have a pyramidal shape with three hydrogen atoms forming the base and the nitrogen atom at the apex, as shown in Fig. 5.7*c*. Since the ammonia structure is based on the tetrahedral arrangement of the four electron pairs about the nitrogen atom, the H—N—H bond angle is expected to be 109°. The experimental value, 107°, is very close.

Because of the unshared electron pair in the valence shell of nitrogen, the ammonia molecule, NH_3, can react with a proton, H^+, and form the ammonium ion, NH_4^+, which is tetrahedral, like CH_4 and $SiCl_4$. The H—N—H bond angle in NH_4^+ is 109°, as predicted.

Family VIA

The hydrides of the VIA family have the molecular formulas H_2O, H_2S, H_2Te, etc. Since in these cases the atoms have completed valence shells of four electron pairs, there must be *two* nonbonding pairs around each of the VIA atoms. We therefore predict that H_2O, H_2S, and H_2Te are angular, or bent, molecules with bond angles of approximately 109°, as shown in Fig. 5.7*e* for the water molecule. The H—O—H angle is about 104.5°, which is fairly close to the prediction. The H—S—H and H—Te—H molecules are bent, but the bond angles are close to 90°.

FIGURE 5.7

Structures with four electron pairs in the valence shell.

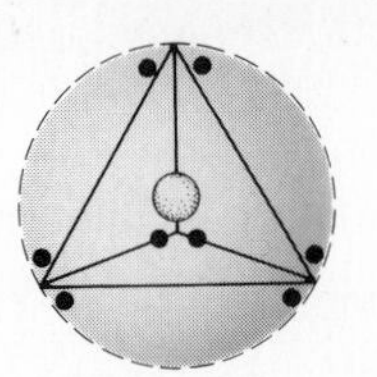

(a) A regular tetrahedron inscribed in a sphere. The center of the tetrahedron and the sphere are coincident.

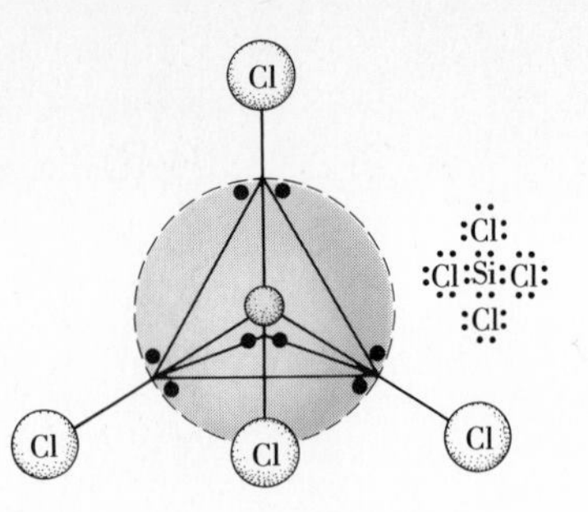

(b) A structure representing the tetrahedral $SiCl_4$ *molecule.*

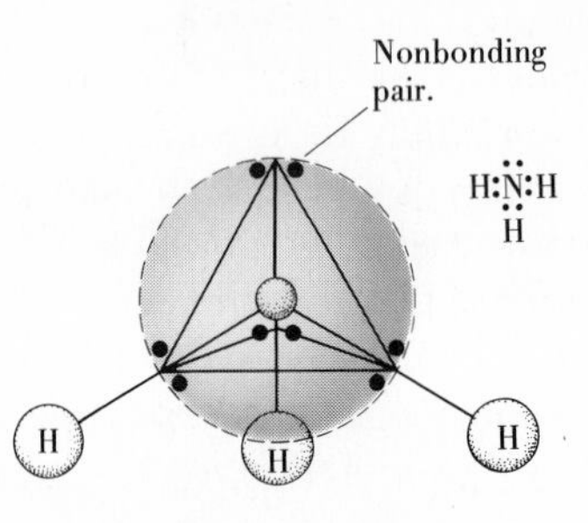

(c) A structure representing the pyramidal ammonia, or PCl_3*, molecule.*

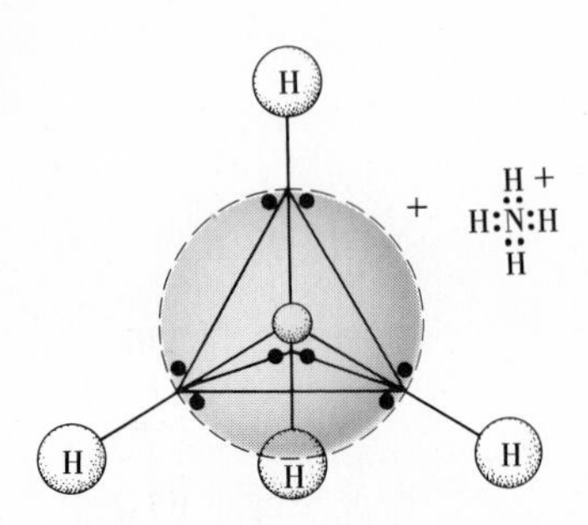

(d) The ammonium ion is tetrahedral.

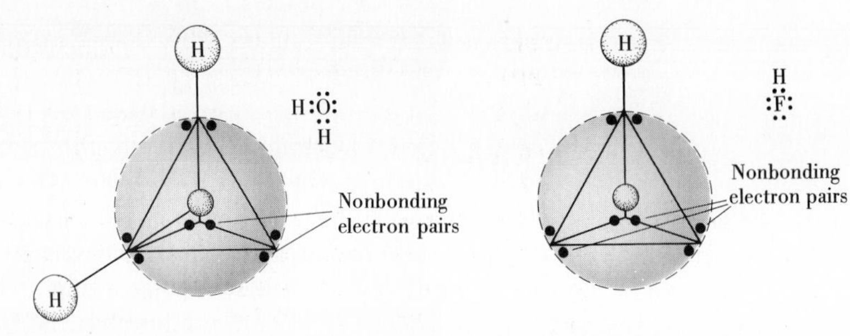

(e) The H_2O *molecule is angular.*

(f) The diatomic hydrogen halide molecule; e.g., HF, HCl, *etc.*

Family VIIA

The hydrogen halides are diatomic, and therefore they must be linear molecules. Figure 5.7*f* shows the electronic structure of HF, which emphasizes the presence of three nonbonding or unshared electron pairs in the valence shell of the fluorine atom. As we shall see, the presence of the unshared electron pairs in a molecule affects its physical and chemical properties.

Molecular geometry and multiple bonds

When multiple bonds are present, we can usually predict the structure of the molecule if we ignore one of the electron pairs of the double bond. For instance, the Lewis structure for SO_3 is

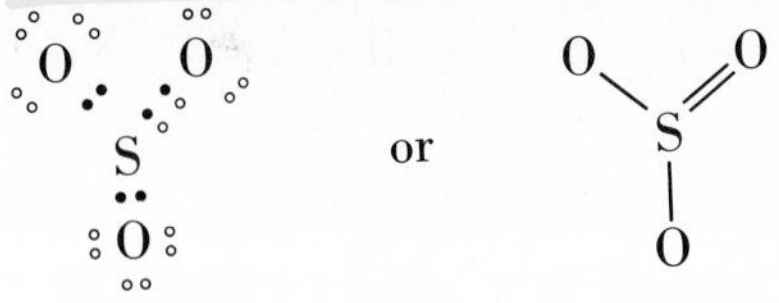

$$\begin{array}{c} :\ddot{O}\!:\;\;:\ddot{O}\!: \\ \ddot{S} \\ :\ddot{O}: \end{array} \qquad \text{or} \qquad \begin{array}{c} O\diagdown\;\;\diagup\!\!\!\diagup O \\ S \\ | \\ O \end{array}$$

If we ignore one of the electron pairs of the S═O double bond, we count only three electron pairs in the valence shell of the sulfur atom. This means that the structure of SO_3 would be similar to planar BF_3. The SO_3 molecule is in fact planar and has the shape of an equilateral triangle with the sulfur atom at the center and the oxygen atoms at the corners.

We see that the Lewis-Langmuir theory coupled with a simple idea from electrostatics allows us to predict the geometry of molecules fairly well. The same kind of reasoning can be applied in predicting the shapes of transition-metal ions like $CoCl_6^{3-}$ and $Fe(OH_2)_6^{2+}$, where the number of electron pairs in the valence shell is often six instead of four.

5.12 BOND POLARITY

For the sake of simplicity chemists have categorized bonds as *covalent* or *ionic*. A purely covalent bond is one in which the electron pair is equally shared by the two atoms of the bond. A purely ionic, or electrovalent, bond is one in which the electron pair is held entirely by one of the atoms of the bond. Actually, these are extreme situations, and most chemical bonds are of an intermediate nature.

The closest approach to a purely covalent bond exists in homonuclear diatomic molecules, i.e., those in which both atoms (nuclei) are of the same element, like H_2, F_2, and Na_2. In these cases the atoms have identical affinities† for the bonding electrons, and there is equal sharing. In a compound like potassium chloride we might suppose that an intermediate bond forms when the single valence electron of a potassium atom pairs with one of the seven valence electrons of a chlorine atom:

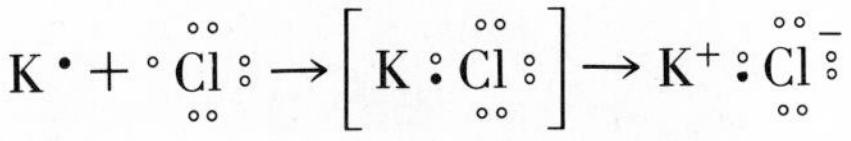

$$K\cdot + \cdot\ddot{\underset{..}{Cl}}: \rightarrow \underset{\text{Intermediate}}{\left[K:\ddot{\underset{..}{Cl}}:\right]} \rightarrow K^+ :\ddot{\underset{..}{Cl}}:^-$$

† See p. 86 for discussion of electron affinity.

But since chlorine has a much greater affinity for electrons than potassium, the electron pair cannot be equally shared; in fact, in this case it is best to consider that the electron pair is controlled by the chlorine atom and the KCl bond is ionic. Although chemists usually consider K^+Cl^- to be a typical ionic compound, our reasoning tells us that Cs^+F^- would be "more ionic"; cesium has a lower electron affinity than potassium, and fluorine has a higher affinity than chlorine. The result is that the F^- ion in Cs^+F^- has greater control over the electron pair than the Cl^- ion does in K^+Cl^-.

The bonds of the diatomic molecules and compounds like KCl and CsF represent the extremes. Ordinarily when two atoms of different elements share an electron pair, the result is a *polar bond*. One of the atoms will have a greater affinity for electrons and the electron pair will be shifted toward it, but not completely controlled as in an ionic bond. The center of charge due to the electron pair does not coincide with the midpoint of the bond. The polar bond is considered to be covalent, with some ionic character.

In the H_2 molecule, the bonding electron pair, on the average, lies midway between the bonded atoms:

H ⁚ H

Nonpolar H_2 molecule

The H_2 molecule and all other homonuclear diatomic molecules have *nonpolar* bonds. In the HCl molecule the bonding electron pair is shifted toward chlorine. The hydrogen end of the HCl molecule is electrically positive, and the chlorine end is negative. We often represent bond polarity by writing δ^+ over the positive end and δ^- over the negative end:

$$\mathrm{H} \overset{\circ}{\rightarrow} \mathrm{\ddot{\underset{..}{Cl}}:} \qquad \overset{\delta^+}{\mathrm{H}}-\overset{\delta^-}{\mathrm{Cl}}$$

Polar HCl molecule

Since the HCl molecule has an electrically positive end, or pole, and an electrically negative pole, we say that HCl is an *electric dipole*.

5.13 THE ELECTRONEGATIVITY SCALE

By analyzing data for many compounds Linus Pauling was able to assign a number, called the *electronegativity*, to each element. The electronegativity indicates the degree of attraction an atom has for bonding electrons. In general, the electronegativity follows the same trend as the electron affinity. When elements with very different electronegativities react, they form highly polar or ionic compounds; elements with similar electronegativity values form nonpolar compounds. As a rule of thumb, whenever the electronegativity difference between elements is greater than 1.7, the resulting compound is ionic.

The trends in the electronegativities shown in Table 5.3 are those we would expect on the basis of what we learned about the periodic trends in the properties of the elements. The metals have the lowest electronegativities, so that electronegativity decreases as we descend a family and increases as we go across a period. The most electronegative element is fluorine and the least electronegative (or most electropositive) is cesium.

All bonds between atoms of different elements must be polar because atoms of different elements have different electron affinities and electronegativities. The atom of the more electronegative element will exert a greater attractive force on a bonding electron pair, and the pair will be shifted away from the midpoint of the bond. When the electronegativities are accidentally equal (as is probably true for the important case of carbon and hydrogen and also for iodine and hydrogen), the resulting bonds are almost purely covalent. The C—H bonds of methane and other hydrocarbons and the H—I bond of hydrogen iodide are said to be nonpolar; they are actually polar, but the polarity is so slight that it is negligible. When bonded atoms have extremely different electron affinities or electronegativities, as when a metal of either group IA or IIA combines with a nonmetal from group VIA or VIIA, the nonmetal atom almost completely controls the electron pair. The result is an ionic bond.

EXAMPLE 5.2 Using the electronegativity data in Table 5.3, (*a*) arrange the following bonds in order of increasing polarity, indicate (*b*) any ionic bonds and (*c*) any nonpolar bonds, and (*d*) label the positive and negative ends of the polar *covalent* bonds: (1) As—F, (2) As—I, (3) Li—I, (4) B—As, (5) Na—Cl.

TABLE 5.3: ELECTRONEGATIVITY VALUES FOR SOME COMMON ELEMENTS

Period	IA	IIA	IIIA	IVA	VA	VIA	VIIA
1	H 2.1						
2	Li 1.0	Be 1.5	B 2.0	C 2.5	N 3.0	O 3.5	F 4.0
3	Na 0.9	Mg 1.2	Al 1.5	Si 1.8	P 2.1	S 2.5	Cl 3.0
4	K 0.8	Ca 1.0	Ga 1.6	Ge 1.8	As 2.0		Br 2.8
5		Sr 1.0		Sn 1.8			I 2.5
6	Cs 0.7	Ba 0.9		Pb 1.8			

Solution: Find the absolute difference in electronegativities between the elements in each bond: As—F (2); As—I (0.5); Li—I (1.5); B—As (0); Na—Cl (2.1). (*a*) The order of increasing polarity is in the order of increasing electronegativity difference: B—As, As—I, Li—I, As—F, and Na—Cl. (*b*) As—I and Na—Cl are ionic bonds because the electronegativity difference is greater than 1.7. (*c*) B—As is a nonpolar bond because the electronegativity difference is zero. (*d*) Only As—I and Li—I are polar covalent bonds (B—As is nonpolar; As—F and Na—Cl are ionic).

$$\overset{\delta^+}{\text{As}}-\overset{\delta^-}{\text{I}} \qquad \overset{\delta^+}{\text{Li}}-\overset{\delta^-}{\text{I}}$$

5.14 THE HYDROGEN BOND

Water is the most common liquid on earth, and we usually do not consider it to be abnormal in any way. However, a careful study of thousands of liquids shows that water, the hydride of oxygen, is very unusual. In fact, the hydrides of the three most electronegative elements, NH_3, H_2O, and HF, all exhibit abnormal properties. The melting points and boiling points of these substances are much higher than would be predicted from family trends. Figure 5.8 shows a plot of the boiling and melting points of the hydrides of family VIA (O, S, Se, and Te) versus the atomic number of the family VIA element. Excluding H_2O, we see that both melting points and boiling points increase with atomic number. We would expect the melting point of H_2O to be below the melting point of H_2S (−83°C) and the boiling point of H_2O to be less than the boiling point of H_2S (−62°C). Many of the other properties of H_2O are anomalous as we shall see in later chapters. In addition, NH_3 is anomalous among the hydrides of family VA and HF among the hydrides of family VIIA. In many respects NH_3 and HF exhibit waterlike properties.

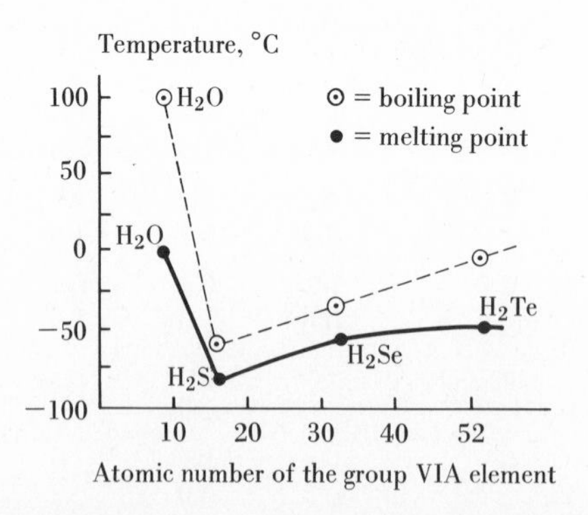

FIGURE 5.8

The anomalous behavior of water.

The electronegativities of nitrogen, oxygen, and fluorine are very high, and the bonds in the hydrides are very polar:

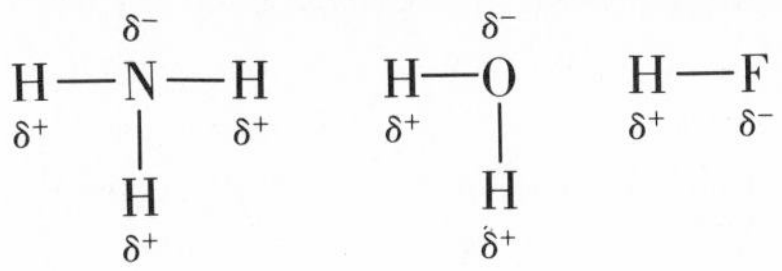

$$\underset{\delta^+}{H}-\overset{\delta^-}{N}-\underset{\delta^+}{H} \qquad \underset{\delta^+}{H}-\overset{\delta^-}{O} \qquad \underset{\delta^+}{H}-\underset{\delta^-}{F}$$
$$\text{(with } \underset{\delta^+}{H} \text{ bonded below N and below O)}$$

In addition, these molecules possess nonbonding electron pairs:

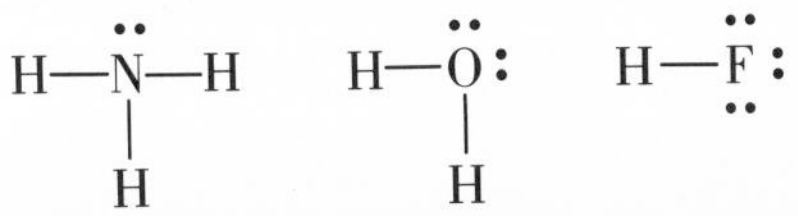

$$H-\ddot{N}-H \qquad H-\ddot{O}: \qquad H-\ddot{\underset{..}{F}}:$$
$$\text{(with H bonded below N and below O)}$$

Since the molecules are polar, they can attract each other: the negative end of one molecule attracts the positive end of another molecule. As the molecules come close together, a nonbonding pair on one molecule can be loosely shared with a hydrogen atom on an approaching molecule. The result is a *hydrogen bond:*

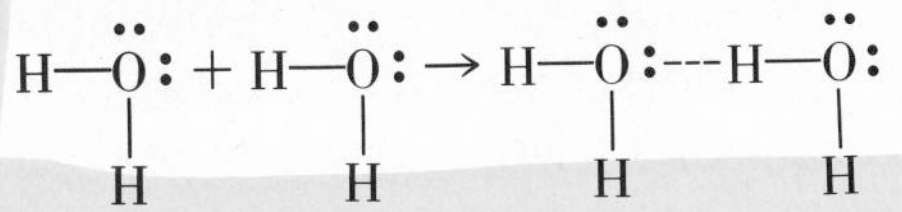

$$H-\ddot{O}: + H-\ddot{O}: \rightarrow H-\ddot{O}:\text{---}H-\ddot{O}:$$
$$\text{(each O bonded to an H below)}$$

In water we believe that aggregates of five to six molecules of H_2O are held together by hydrogen bonds.

The bond energy of a covalent bond between hydrogen and nitrogen, oxygen, or fluorine is around 100,000 cal/mol, and the bond length is 1.0 Å. The energy of a hydrogen bond involving nitrogen, oxygen, or fluorine is approximately 5000 cal/mol, and the hydrogen-bond length is close to 3 Å. The hydrogen bond is weaker and longer than a covalent bond, but its effect is easily detected in almost every property that can be measured.

5.15 RESONANCE AND DELOCALIZED ELECTRON PAIRS

Sulfur trioxide, SO_3, is a planar molecule with a geometry like BF_3. The sulfur atom can be considered to be at the center of an equilateral triangle with oxygen atoms at the corners. All the S—O bonds are of equal length, and each O—S—O bond angle is 120°. When we attempt to write a Lewis structure for SO_3, we find that in order to conform to the octet rule one of the S—O linkages must be a double bond (see Fig. 5.9*a*). This should mean that one of the S—O bonds is shorter and stronger than the other two, but all our evidence shows that the three S—O bonds in SO_3 are identical. Therefore, a simple Lewis structure is not a good description for SO_3. In order to represent the SO_3 molecule more accurately we say that the true electronic structure of SO_3 is a composite of the structures repre-

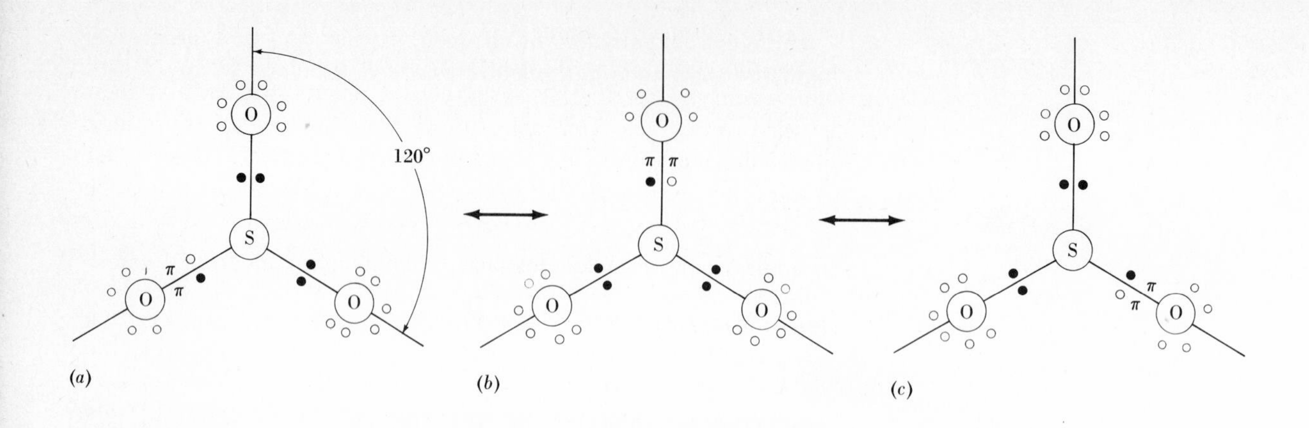

FIGURE 5.9

Resonance structures of SO_3.

sented by Fig. 5.9*a* to *c*. The S—O bonds are neither single nor double bonds; they are intermediate between single and double bonds. Each S—O combination is bonded by a pair of electrons which are *localized* between a particular S—O pair. There is an extra bonding character in the SO_3 molecule due to the presence of an electron pair which is *delocalized;* it is not limited to the space between one of the oxygen atoms and the central sulfur atom but is free to move through the entire molecule and adds strength to each of the S—O bonds. In Fig. 5.9 we indicate the localized electron pairs by small dots and circles; the presence of the delocalized electron pairs is indicated by π. The π electrons are shown in different positions in the structures representing SO_3 to emphasize that the π electrons are *de*localized and contribute to the strength of each S—O bond. The existence of extra bonding in the SO_3 molecule is verified by the fact that S—O bond in this molecule is shorter than an ordinary single bond but longer than a double bond.

When we must draw more than one Lewis structure to represent a molecule more accurately, we say that the molecule exhibits *resonance.* This term does not imply that molecules are rapidly alternating from one Lewis structure to another. Resonance means that a single Lewis structure does not suffice to represent a molecule because some of the bonding electron pairs are delocalized and contribute to the bonding of more than two atoms in the molecule. Resonance is simply a device for representing this. No one Lewis structure can be isolated.

The double-headed arrows in Fig. 5.9 are used to group together the Lewis structures, which, taken as a simultaneous composite, give an acceptable representation of the electronic structure of a molecule. The double-headed arrow signifies resonance, and the individual structures *a*, *b*, and *c* are called *resonance structures.*

There are many examples of resonance in chemical structures. The nitrate ion, NO_3^-, and the carbonate ion, CO_3^{2-}, which have the same geometry as SO_3, are examples of common materials. There are many examples of organic substances which exhibit resonance; one of the most important examples, benzene, C_6H_6, is known to possess the symmetry of a perfect planar hexagon: the carbon atoms occupy the six vertices. The six hydrogen atoms are located on rays emanating from the center of the hexagon and passing through the vertices (see Fig. 5.10*a*). Each carbon is bonded to one hydrogen atom and two carbon atoms; in order to achieve the rare-gas configuration and have eight electrons in the valence shell, each carbon atom must form one double bond. This can be achieved by representing benzene with alternating double and single bonds between the ring carbon atoms. Unfortunately, this would imply that three of the C—C bonds are single bonds and three are double bonds. As a matter of fact all the C—C bonds of benzene are identical in every respect; all have a length of 1.39 Å, which is intermediate between a C—C single bond (1.54 Å) and C═C double bond (1.34 Å). A single Lewis structure is not a good representation, and therefore benzene exhibits resonance. The chief resonance structures of benzene are shown in Fig. 5.10*b* and *c*. As suggested by the composite resonance structures, each C—C bond is an intermediate between a single and double bond; and there are three delocalized electron pairs which contribute to the *extra* bonding of the carbon atoms of the benzene ring.

5.16 ODD MOLECULES AND PARAMAGNETISM

Nitrogen and oxygen form many compounds. The simplest is nitric oxide, NO, a diatomic molecule. Since nitrogen has five valence electrons and oxygen has six, it is impossible for all the bonding electrons in the NO molecule to be paired. Two Lewis structures, one with a double bond and one with a triple bond, can be drawn.

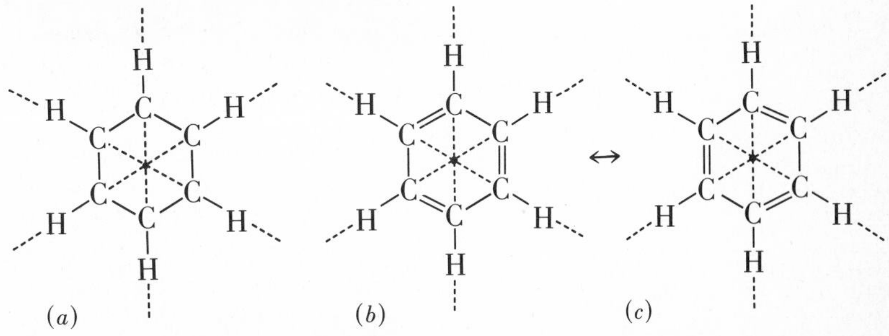

FIGURE 5.10

Resonance structure for benzene.

(*a*) (*b*) ↔ (*c*)

FIGURE 5.11

Lewis structures for NO *and* NO^+.

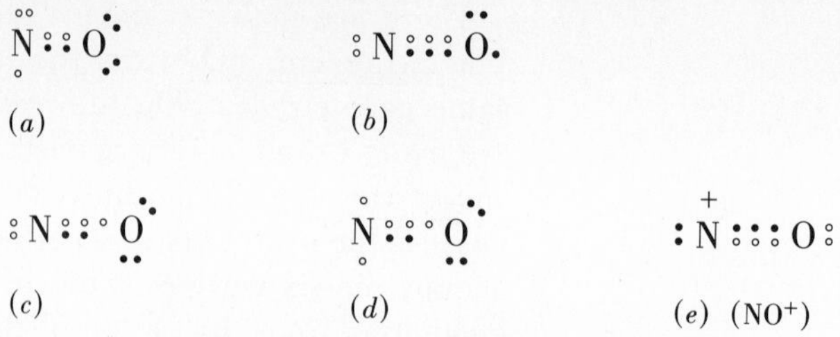

N :: O (*a*)

:N ::: O. (*b*)

:N :: O (*c*)

N :: O (*d*)

:N ::: O: (*e*) (NO^+)

Neither conforms to the rule of eight since in one case nitrogen has a deficiency of electrons and in the other oxygen has an excess (see Fig. 5.11*a* and *b*). The available evidence suggests that the bond in the NO molecule is intermediate in character between a double and a triple bond. Figure 5.11*c* is a representation of NO with five bonding electrons which give "$2\frac{1}{2}$ bonds." This is a better structure than either *a* or *b* but it still does not conform to the rule of eight.

Many molecules have an *odd* number of bonding electrons. Lewis named them *odd molecules;* and, of course, they can never conform to the octet rule. It is always true that substances composed of odd molecules are *paramagnetic*. Due to the presence of unpaired electrons such substances are drawn into a magnetic field. *Diamagnetic* substances are composed of molecules in which all electrons are paired and are always forced out of a magnetic field. The force with which a paramagnetic substance is drawn into a magnetic field is a measure of the density of the unpaired electrons in the paramagnetic sample. Therefore, it is possible to determine the number of unpaired electrons in a molecule by weighing a paramagnetic material in and out of a magnetic field. The instrument used, a Gouy balance, is shown in Fig. 5.12. (Gouy is pronounced "gwee.")

Gouy balance experiments indicate that nitric oxide possesses *one* unpaired electron, so that Fig. 5.11*c* seems to be a good structure for NO in that it shows a bond intermediate between a double and a triple bond and has a single unpaired electron. We can draw other structures, like Fig. 5.11*d*, with more than one unpaired electron; these structures do not fit our observations and must be rejected.

In its reactions NO has a tendency to lose an electron to form the NO^+ ion, which is diamagnetic. The bond length and bond-dissociation energy show that the bonding in the NO^+ ion is stronger than in the NO molecule:

Species	*Bond length,* Å	*Bond dissociation energy,* kcal
NO	1.15	162
NO^+	1.06	240

FIGURE 5.12

Gouy balance. A paramagnetic substance is pulled into a magnetic field. The weight in and out of the field permits us to calculate the number of unpaired electrons present.

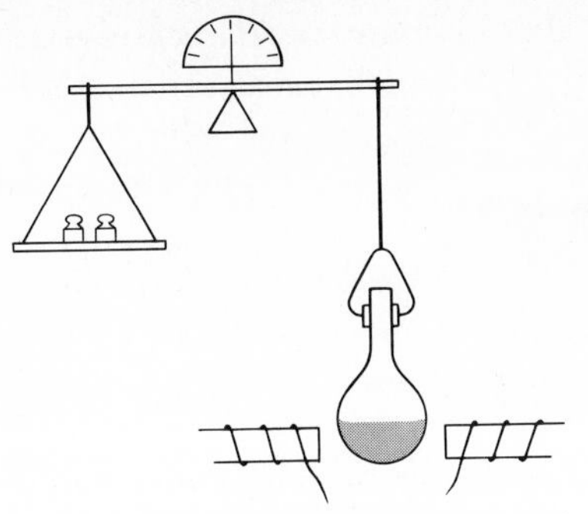

(*a*) *Electromagnet is off; sample is balanced*

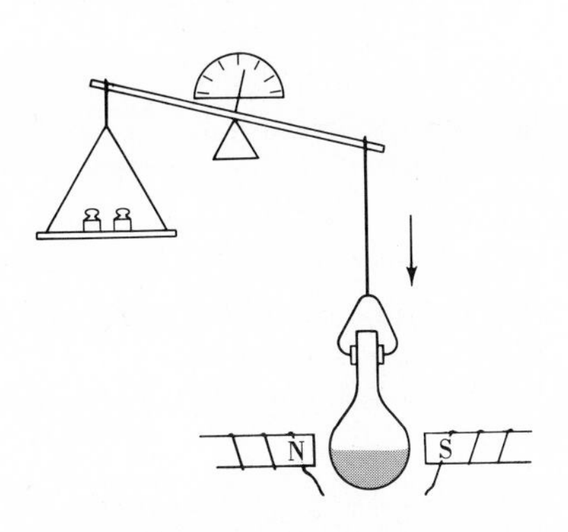

(*b*) *Electromagnet is on; sample is unbalanced*

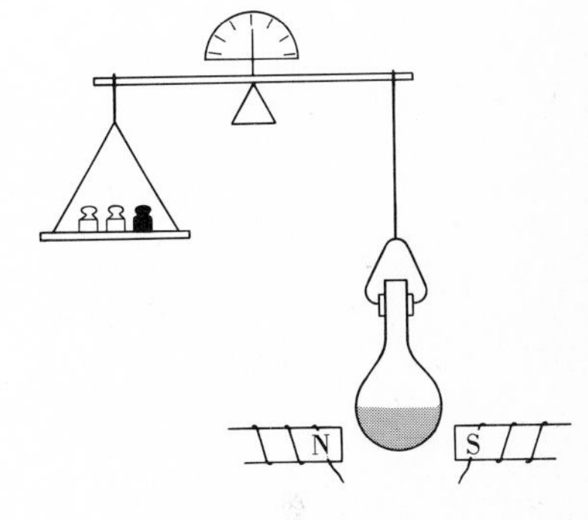

(*c*) *Electromagnet is on; sample is balanced by adding an extra weight*

This evidence suggests that the generation of the NO^+ ion from the NO molecule is accompanied by the formation of a triple bond, as shown in Fig. 5.11*e*. In this Lewis model for NO^+ each atom has achieved the rare-gas configuration, and it is not surprising that the ion is more stable than the molecule. Since all electrons are paired, the Lewis model also is consistent with the fact that NO^+ is diamagnetic.

Lewis structures for the O_2 molecule which conform to the octet rule are shown in Fig. 5.4*b*. The Lewis structure predicts that O_2 should have a double bond, and the measured bond length and dissociation energy of O_2 support the presence of a double bond in this molecule. Unfortunately, O_2 is paramagnetic, and our experiments indicate the presence of two unpaired electrons per molecule. We could draw Lewis structures of O_2 with a double bond and two unpaired electrons but the after-the-fact manner by which we do this is not very satisfactory:

$$\overset{\circ\circ}{\underset{\circ\circ}{}}\mathrm{O} \mathbin{\vdots\vdots} \overset{\circ\circ}{\underset{\bullet}{\mathrm{O}}}\bullet \leftrightarrow \bullet\overset{\circ\circ}{\underset{\bullet}{\mathrm{O}}} \mathbin{\vdots\vdots} \mathrm{O}\overset{\circ\circ}{\underset{\circ\circ}{}}$$

The quantum-mechanical description of the oxygen molecule (Sec. 5.23) is much more satisfactory than the Lewis structure. We see that odd molecules *must* be paramagnetic, but molecules like O_2, which contain an even number of electrons, may also be paramagnetic.

There are several examples of stable, odd molecules, but there are many more examples of odd molecules which are too reactive to exist for more than a fraction of a second. These species are often present as intermediates during chemical reactions. Intermediates, i.e., species which form during a reaction and then change into the final product, can be detected by various methods. Examples of very reactive odd molecules which have been detected include $\cdot OH$, $^{\cdot}CH_3$, and $O{=}^{\cdot}CH$. The single dot indicates the presence of an unpaired electron. These species and the more stable odd molecules are often referred to as *free radicals* or *radicals.*

5.17 THE LEWIS-LANGMUIR THEORY AND QUANTUM MECHANICS

The Lewis-Langmuir theory of valence is very useful in correlating molecular structure and the observable properties of matter. There are many problems which the Lewis-Langmuir theory cannot solve, but the most serious *weakness* lies in its use of the Bohr model of the atom to derive the electronic structure of molecules. In the Bohr model electrons are considered to be particles moving in definite orbits around a nucleus. Experiments show that electrons in atoms have wave and particle properties; a realistic valence theory must consider the wave nature of electrons, which means that we must use our quantum-mechanical model of the atom to develop a valence theory. Modern chemists still use Lewis structures to represent molecules, but these symbols have taken on a new meaning since the advent of quantum mechanics.

Two theories based on the quantum-mechanical atom have developed, the molecular-orbital (MO) theory and the valence-bond (VB) theory. The remainder of this chapter will be devoted to a discussion and comparison of these theories.

The MO and VB theories differ from the Lewis-Langmuir theory in that a thorough understanding requires mathematical skill much above the level of the ordinary undergraduate, but a good appreciation of the methods provided by these theories can be achieved without mathematics. To simplify our discussions we shall consider relatively small molecules which exist in the gaseous state. In such molecules the internal bonding between atoms is not affected by neighbor molecules. In the solid and liquid states, where the separation between molecules is small, the bonding within a molecule can be greatly affected by neighbors—a complication we shall avoid.

5.18 THE MOLECULAR ORBITAL THEORY AND THE H_2 MOLECULE

If we consider the wave nature of electrons in atoms, we can understand how *molecular orbitals* are formed from the atomic orbitals as two hydrogen atoms approach. When two hydrogen atoms are gradually brought together, there is a continuous change in the electron density about the nuclei. In the separate hydrogen atoms the single electrons are in 1*s* orbitals. As the atoms approach, the individual 1*s* wave functions may be in phase so that they reinforce each other, or they may be out of phase so that they destructively interfere with each other. This is precisely the way waves behave, as shown in Fig. 5.13. The interaction of the two 1*s* atomic orbitals in two different ways gives rise to two different molecular orbitals. One is a *bonding* molecular orbital, which results from the in-phase reinforcement of the two atomic orbitals. When electrons occupy the bonding molecular orbital, there is a high probability of finding electrons between the bonded atoms; there is a piling up of charge which reduces the repulsion between the nuclei, and they are allowed to come close together. There is net attraction between the two atoms, and a stable molecule is formed. The energy of the molecule is lower than the energy of the separate atoms. This is a quantum-mechanical description of a chemical bond.

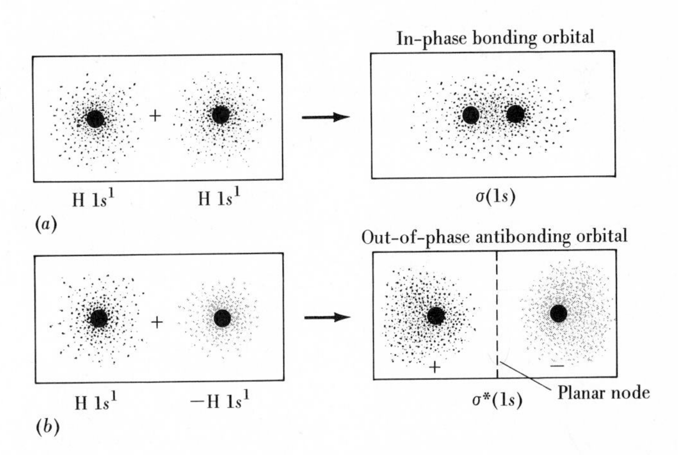

FIGURE 5.13

Formation of the molecular orbitals for H_2.

When two hydrogen atoms approach so that the 1*s* orbitals are out of phase, destructive interference occurs and an *antibonding* molecular orbital is formed. When electrons occupy an antibonding orbital, the electron density between the two atoms is very low. Midway between the two atoms is a plane perpendicular to the bond axis in which the probability of finding an electron is zero. The plane represents a *node* in the molecular orbital. The electron density at the node is zero; the orbital has a different sign on either side. This is shown in Fig. 5.13*b*. When electrons occupy the antibonding orbital, the resulting "molecule" is not stable and decomposes into the separate atoms. The energy of the separate atoms is lower than the energy of the molecule.

In Fig. 5.13 we have used the symbols $\sigma(1s)$ and $\sigma^*(1s)$ for the bonding and antibonding orbitals, respectively. A σ (sigma) molecular orbital, or bond, is one in which the electron density would be symmetrical about the bond axis, so that there would be no apparent change if the molecule were rotated about the bond axis. Both the bonding and antibonding orbitals are of the σ type when they are formed from 1*s* orbitals. The fact that a molecular orbital is antibonding is indicated by an asterisk as σ^*, read "sigma star."

Another fruitful way of picturing the formation of molecular orbitals from atomic orbitals is shown in Fig. 5.14. When the atomic wave functions ψ are plotted versus the distance r from the nucleus, we see how the atomic wave functions change as the nuclei approach. When the atomic wave functions or orbitals are in phase, they reinforce each other and the σ bonding orbital, represented by the dark line in Fig. 5.14*a*, has a high value between the atoms. The dashed lines represent the undisturbed atomic orbitals of the individual atoms. When the atomic orbitals of the individual atoms are out of phase, as shown in Fig. 5.14*b*, an antibonding σ^* molecular orbital results and the electron density between the atoms is very low. A node results midway between the nuclei in the formation of an antibonding orbital. With a low density between the nuclei the molecule is not stable when electrons occupy the σ^* orbital, and such a molecule decomposes into its atoms.

The Pauli exclusion principle, Hund's rule, and the Aufbau principle apply to the electron occupation of molecular orbitals just as they do to atomic orbitals. As we consider more complicated molecules, we shall make use of the principles learned when we studied the quantum-mechanical description of electrons in atoms. For the H_2 molecule we must consider placing two electrons in the available molecular orbitals. The lowest-energy molecular orbital is the $\sigma(1s)$; the first electron is placed in this orbital. The second electron also goes into this orbital since the $\sigma^*(1s)$ is at a higher energy; the second electron must have a different spin in accordance with the Pauli

FIGURE 5.14

Formation of bonding and antibonding orbitals.

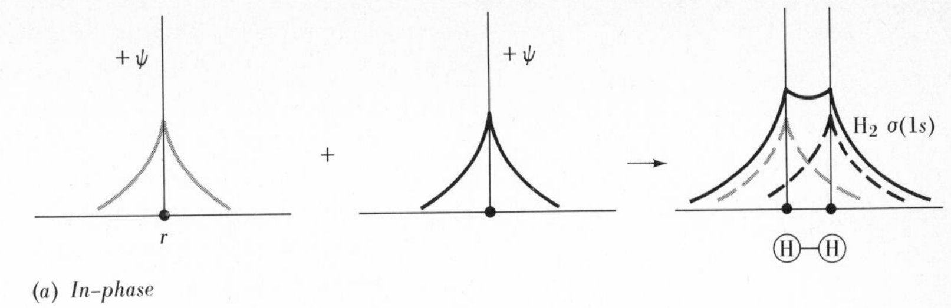

(*a*) *In-phase*

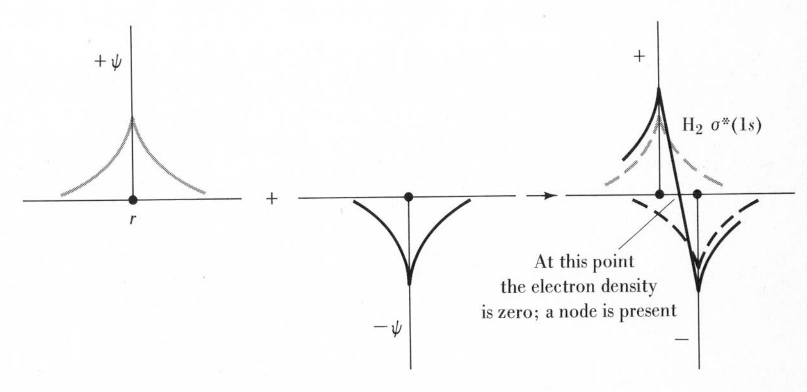

(*b*) *Out of phase*

exclusion principle. The electronic description of the H_2 molecule is shown in Fig. 5.15. This figure indicates that two molecular orbitals are formed from two atomic orbitals and that the $\sigma(1s)$ molecular orbital is of lower energy; it is lower in energy than the separate atoms. The $\sigma^*(1s)$ molecular orbital is of higher energy than the separate atoms. Since the $\sigma(1s)$ orbital is occupied by two electrons, it is filled; the single bond between the two hydrogen atoms is due to the fact that the bonding $\sigma(1s)$ molecular orbital is occupied by two electrons.

As we said above, the presence of electrons in the bonding orbital permits a piling up of electron density between the atoms, reduces

FIGURE 5.15

The molecular orbitals for H_2.

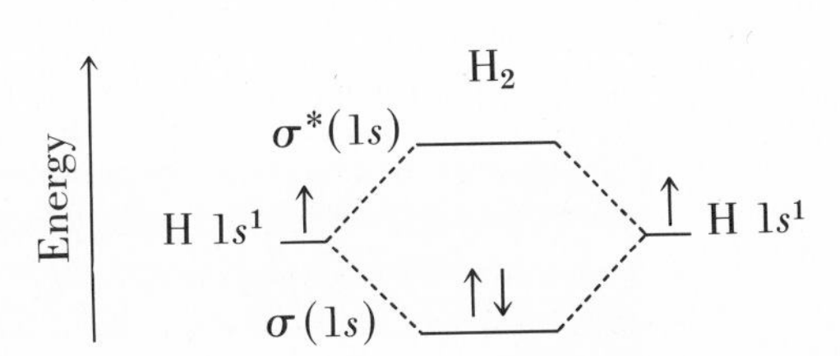

nuclear repulsions, and allows the molecule to be stable when the separation between nuclei is small. It is not necessary for a bonding orbital to contain two electrons to cause bonding. It should be understood that even one electron in a bonding orbital would cause the formation of a stable molecule. The ion-molecule H_2^+ is stable. It has a single electron in the $\sigma(1s)$ molecular orbital. As we might guess, the bond between the two hydrogen atoms in H_2^+ is weaker than the bond in H_2. This is indicated by the longer bond length of H_2^+ (1.06 Å) and the lower bond energy (61 kcal). The presence of one electron in a bonding orbital reduces nuclear repulsion sufficiently, and a chemical bond forms. If we say that a single bond exists between the two atoms in H_2, we must admit that a half bond exists in H_2^+. We shall find other molecules in which half bonds exist. The concept of the one-electron bond is at variance with Lewis's rule of two; it will be of use in understanding the bonding in other simple molecules.

5.19 THE He_2 MOLECULE IS UNSTABLE

FIGURE 5.16

The molecular orbitals for "He_2."

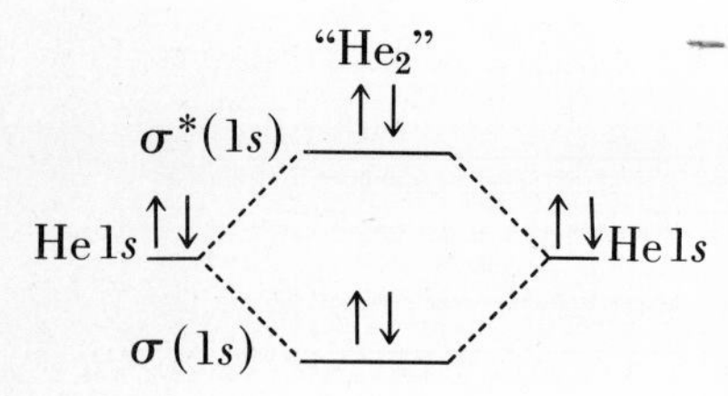

It is well known that helium does not form diatomic molecules. We can understand why by considering how the molecular orbitals of "He_2" are formed and occupied (see Fig. 5.16). Since helium has an atomic number of 2, its electronic configuration is He $1s^2$. As two atoms approach and their atomic orbitals begin to interact, $\sigma(1s)$ and $\sigma^*(1s)$ molecular orbitals form, just as for H_2. However, in this instance, we have four electrons to place in molecular orbitals. The first two go into the bonding molecular orbitals, $\sigma(1s)$, and the second two go into the antibonding molecular orbitals, $\sigma^*(1s)$. Since both orbitals are filled, the effect of the antibonding orbital cancels the effect of the bonding orbital: there is no net bond between the two atoms, and the He_2 molecule does not form.

5.20 HOMONUCLEAR DIATOMIC MOLECULES OF THE SECOND PERIOD

All the elements of the second period except beryllium and neon form diatomic molecules in the gas phase. The diatomic molecule Li_2 may be unfamiliar, but it does exist in the vapor above metallic lithium when it is heated to 500 to 600°C. The species N_2, O_2, and F_2 are familiar, but the unfamiliar B_2, C_2, and Li_2 are also stable molecules.

FIGURE 5.17

The molecular orbitals for Li_2.

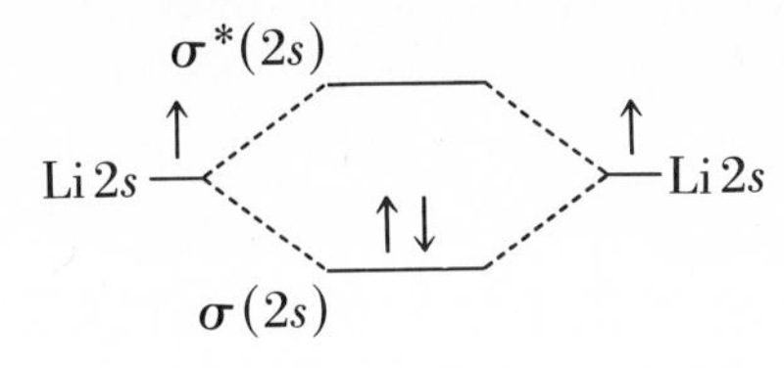

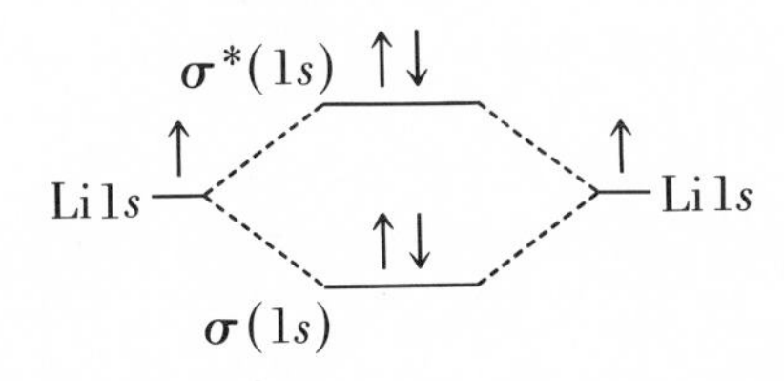

The bonding in Li_2 is weak, as can be guessed from the fact that the bond length is very high (2.68 Å) and the bond energy low (25 kcal/mol). The average distance of a $1s$ electron from the nucleus is ~0.34 Å in the lithium atom. Hence, if the bond length of Li_2 is 2.68 Å, it is obvious that the $1s$ electrons of the individual lithium atoms do not interact very strongly. The average radius of an electron in a $2s$ orbital of the lithium atom is close to 2.5 Å. These facts imply that the bonding in Li_2 is chiefly due to the interaction of the individual $2s$ orbitals. It seems as though each atom of the molecule is present at the point of the highest electron density offered by the $2s$ electron of its partner.

We can picture the formation of the Li_2 molecule by allowing two lithium atoms to approach until they reach the bond separation 2.68 Å. The $1s$ orbitals will again form the $\sigma(1s)$ and $\sigma^*(1s)$ molecular orbitals; but in addition we must consider the creation of higher-energy molecular orbitals from the atomic orbitals of the second energy level. In Li $1s^2, 2s^1$ the $2s$ orbitals may approach in phase and form a bonding molecular orbital, $\sigma(2s)$, or they may approach out of phase and form an antibonding molecular orbital, $\sigma^*(2s)$. The occupied molecular orbitals of the Li_2 molecule are shown in Fig. 5.17. The $\sigma(1s)$ and $\sigma^*(1s)$ molecular orbitals are filled, and their effects cancel. There is a single bond in the Li_2 molecule, due to the bonding by the $\sigma(2s)$ molecular orbital. The antibonding orbital, $\sigma^*(2s)$, is unoccupied. Because the interaction between the $1s$ orbitals of the bonded lithium atoms is very slight, another picture of a chemical bond is suggested: the bond formation is due to the interaction of the individual $2s$ orbitals; the driving force for bond formation is the tendency to have two electrons in each atomic orbital. This is a restatement of the Lewis idea that bond formation is due to the tendency to share electron pairs and that it is the outer, or valence, electrons that are involved in bonding. We shall say more about this idea in Sec. 5.29.

5.22 The Be_2 MOLECULE DOES NOT FORM

The element beryllium, $_4$Be $1s^2, 2s^2$, does not form diatomic molecules. The beryllium atom has one more electron than lithium, and reference to the molecular orbital scheme for Li_2 (Fig. 5.17) makes the reason for the nonexistence of Be_2 clear. The additional electrons of Be_2 would just fill the $\sigma^*(2s)$ molecular orbital. There would be two filled bonding orbitals, $\sigma(1s)$ and $\sigma(2s)$, and two filled antibonding orbitals $\sigma^*(1s)$ and $\sigma^*(2s)$, and thus there is no net bonding.

The next three elements of the second period form diatomic molecules; B_2 exhibits a single bond, C_2 a double bond, and N_2 a triple bond. We shall first look at the bonding in O_2, whereupon the bonding in B_2, C_2, and N_2 will also be understood.

5.23 THE O_2 MOLECULE

FIGURE 5.18

The orbital occupancy of atomic oxygen.

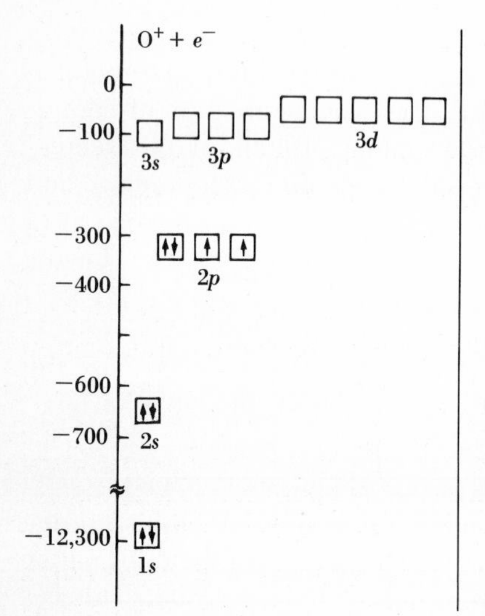

The bonding in the O_2 molecule involves 16 electrons. The four molecular orbitals formed from the interaction of the $1s$ and $2s$ atomic orbitals are of lower energy than the molecular orbitals formed from the $2p$ atomic orbitals. Hence, by the Aufbau principle, these will be filled first. Eight of the sixteen electrons of O_2 are accommodated by the four molecular orbitals which form from the $1s$ and $2s$ atomic orbitals, as shown in Fig. 5.17. We must consider what kinds of molecular orbitals arise from the interaction of the $2p$ orbitals in order to understand the complete bonding in O_2.

The three p orbitals in the separate atoms are of equal energy. This is shown in Fig. 5.18, where the atomic orbitals of oxygen are plotted according to increasing energy. In the ground state, the oxygen atom has the electronic configuration $1s^2$, $2s^2$, $2p_x{}^2$, $2p_y{}^1$, $2p_z{}^1$, indicated in Fig. 5.18 by squares representing the orbitals. The filled $1s$, $2s$, and $2p_x$ orbitals are shown as $\uparrow\downarrow$; the half-filled $2p_y$ and $2p_z$ orbitals are shown as $\uparrow$. All the unfilled orbitals are shown as empty squares.

When two oxygen atoms approach, we can imagine, as with lithium, that the $1s$ and $2s$ orbitals give rise to four molecular orbitals, $\sigma(1s)$, $\sigma^*(1s)$, $\sigma(2s)$, and $\sigma^*(2s)$. Let us further imagine that the axis of approach of the two atoms is the x axis of a three-dimensional coordinate system. The individual $2p_x$ orbitals may be in phase or out of phase as the two atoms approach; if they are in phase, a bonding molecular orbital will result and there will be increased probability of finding electrons between the two bonded atoms. If the orbitals are out of phase, an antibonding orbital will result. In this case there is a decrease of electron density between the two atoms, and a planar node midway between the two atoms perpendicular to the bond axis results. In order to understand the in-phase and out-of-phase approach of two $2p_x$ orbitals we must remember that all three $2p$ orbitals have the dumbbell-shaped electron cloud, which is separated by a nodal plane through the nucleus. The orbital is negative on one side of the plane and positive on the other side. The in-phase approach of two $2p_x$ orbitals along the x axis requires that they come at each other with the same sign: positive end toward positive end or negative end toward negative end. In this

FIGURE 5.19

Formation of molecular orbitals from two $2p_x$ atomic orbitals.

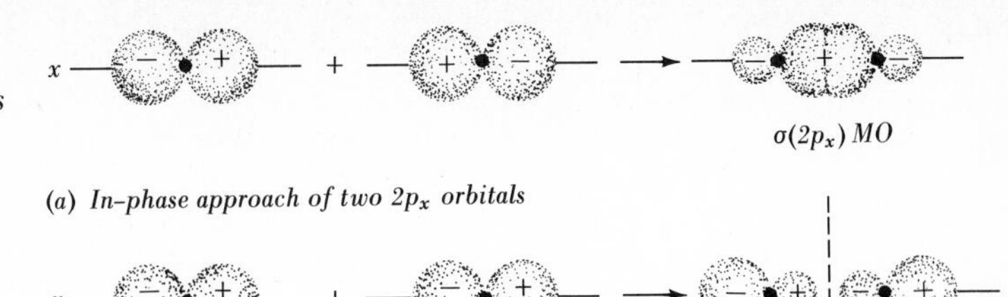

(a) *In-phase approach of two $2p_x$ orbitals*

(b) *Out-of-phase approach of two $2p_x$ orbitals*

way a bonding orbital results. If a positive end approaches a negative end, we have an out-of-phase approach and an antibonding orbital. All these details are depicted in Fig. 5.19. The molecular orbitals resulting from the interaction of the $2p_x$ atomic orbitals are σ orbitals because their appearance does not change when the molecule is rotated about the bond axis, i.e., the x axis. In the language of the chemist "a σ orbital is symmetric with respect to rotation about the bond axis."

We see that the $1s$, $2s$, and $2p_x$ orbitals generate six molecular orbitals which can accommodate only 12 electrons. Since the oxygen molecule contains 16 extranuclear electrons, we must find orbital space for at least 4 more electrons. Let us then consider the interactions of the $2p_y$ and $2p_z$ orbitals as two atoms approach along the x axis. These atomic orbitals are perpendicular to the x axis and to each other. As we shall see, each pair, the two $2p_y$ orbitals and the two $2p_z$ orbitals, gives rise to identical bonding and antibonding molecular orbitals. Let us consider the $2p_y$ orbital first. The atoms approach along the x axis, both $2p_y$ orbitals are perpendicular to this axis, and the $2p_y$ orbitals are parallel to each other. They can approach so that the positive parts and the negative parts of the two orbitals interact. This would be the in-phase approach of the two atomic orbitals and the result is a *bonding* molecular orbital (Fig. 5.20*a*). Note that in this case there is an increase in electron density above and below the bond axis but *not* along the bond axis which lies in a nodal plane. The nodal plane represents a plane in which there is no probability of finding an electron. The sign of the orbital changes in passing through the nodal plane.

The molecular orbital formed from the interaction of two $2p_y$ orbitals is not symmetric with respect to rotation about the bond axis. A rotation of 180° will place the positive part of the orbital in the region occupied by the negative part (see Fig. 5.20*a*). Hence, the orbital we have just discussed is not a σ orbital but a π orbital. Since it is also a bonding orbital formed from two $2p_y$ atomic orbitals, we sym-

FIGURE 5.20

Formation of molecular orbitals from two $2p_y$ atomic orbitals.

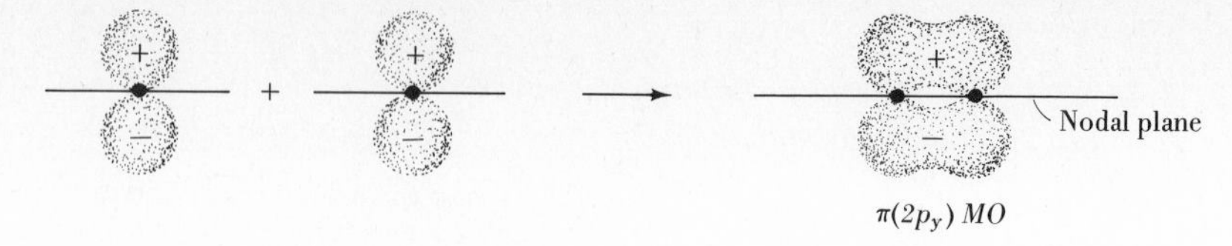

(*a*) *In-phase approach of two $2p_y$ orbitals*

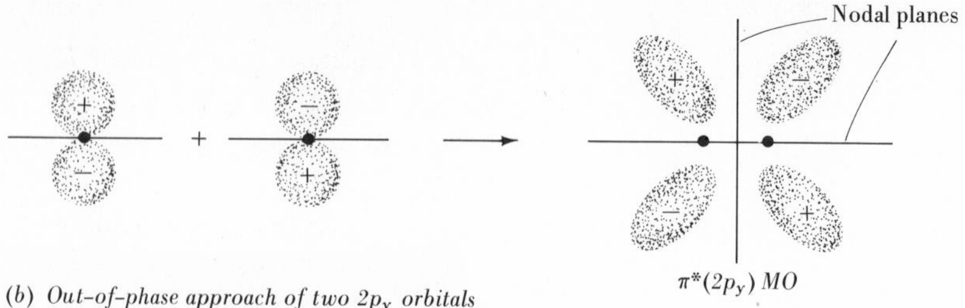

(*b*) *Out-of-phase approach of two $2p_y$ orbitals*

bolize it as $\pi(2p_y)$. The π molecular orbital is unsymmetrical with respect to rotation about the bond axis.

The out-of-phase approach of two $2p_y$ atomic orbitals is shown in Fig. 5.20*b*. There is a decrease in electron density between the approaching atoms, and a nodal plane perpendicular to the bond axis is formed. This interaction of the two $2p_y$ atomic orbitals results in the formation of an antibonding orbital which is unsymmetrical with respect to rotation about the bond axis; it is designated $\pi^*(2p_y)$, read "pi star $2p_y$."

The $2p_z$ orbitals interact with each other in the same way that the $2p_y$ orbitals do, as described above. They approach so that they are parallel and give rise to the bonding orbital $\pi(2p_z)$ and the antibonding orbital $\pi^*(2p_z)$. The $\pi(2p_z)$ is identical to the $\pi(2p_y)$. The two are of equal energy and are said to be *degenerate*. The $\pi^*(2p_y)$ and $\pi^*(2p_z)$ are also degenerate, but since they are antibonding orbitals they are at higher energy than the corresponding π bonding orbitals.

We have now considered how all the atomic orbitals of the first and second energy levels can interact to form molecular orbitals. The pairs of 1*s*, 2*s*, $2p_x$, $2p_y$, and $2p_z$ atomic orbitals gave rise to 10 molecular orbitals, half of which are antibonding. With 10 molecular orbitals we can accommodate 20 electrons; we have enough orbitals to handle the 16 electrons of molecular oxygen, O_2. Before we can place the electrons in the proper molecular orbitals we must know the exact order of increasing energy because we have made the assumption that the Aufbau principle holds for molecules as well as for atoms. Figure 5.21 shows the molecular orbitals of oxygen, their order in terms of increasing energy, and their parentage.

The order of the molecular orbitals in terms of increasing energy is $\sigma(1s)$, $\sigma^*(1s)$, $\sigma(2s)$, $\sigma^*(2s)$, $\sigma(2p_x)$, $\pi(2p_y) = \pi(2p_z)$, $\pi^*(2p_y) = \pi^*(2p_z)$, $\sigma^*(2p_x)$. Note that we place the two bonding π orbitals at the same level and the two antibonding π^* orbitals at the same level in Fig. 5.21 since they are at the same energy. This has very important consequences in the MO description of O_2. It is also important to realize that the σ^* antibonding orbital formed from the out-of-phase interaction of the two $2p_x$ orbitals is at higher energy than the π^* antibonding orbitals formed from the out-of-phase interactions of the other p orbitals.

Now let us feed the 16 electrons of O_2 into the molecular orbitals we have just described. The first 8 electrons go into the σ orbitals, $\sigma(1s)$, $\sigma^*(1s)$, $\sigma(2s)$, $\sigma^*(2s)$, just as in "Be_2." There is no net bonding. The next 2 electrons go into the $\sigma(2p_x)$ and this brings the total number of electrons to ten. Now we come to the degenerate pair. $\pi(2p_y)$ and $\pi(2p_z)$, which can accommodate a total of 4 electrons. Since we have 6 more electrons, this pair of orbitals will be filled, but we should give some attention to the order in which they fill. If the first electron goes into the $\pi(2p_y)$ orbital, the second will go into the equal-energy orbital $\pi(2p_z)$ because electrons repel each other. By Hund's rule we say that electrons tend to have the same spin. With 2 more electrons the $\pi(2p_y)$ and $\pi(2p_z)$ orbitals are filled. This takes

FIGURE 5.21

The energy-level diagram for O_2.

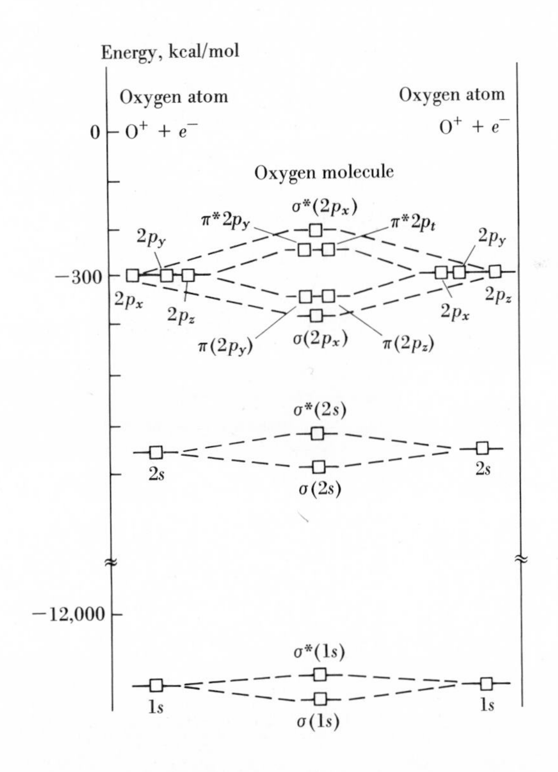

care of the MO occupancy of 14 electrons. The remaining 2 electrons must go into the next highest-energy orbitals. Again we have a degenerate pair, $\pi^*(2p_y)$ and $\pi^*(2p_z)$. In view of Hund's rule, we place 1 electron in each orbital.

Now let us examine the bonding we are predicting from the MO description of O_2:

Orbital	Occupancy		
$\sigma^*(2p_x)$	—		
$\pi^*(2p_y) = \pi^*(2p_z)$	↑	↑	
$\pi(2p_y) = \pi(2p_z)$	↑↓	↑↓	
$\sigma(2p_x)$	↑↓		
$\sigma^*(2s)$	↑↓		"Be_2"
$\sigma(2s)$	↑↓		"Be_2"
$\sigma^*(1s)$	↑↓		"Be_2"
$\sigma(1s)$	↑↓		"Be_2"

The bonding effect of the first four molecular orbitals is self-canceling; it is the same as in the "Be_2" molecule. The next three orbitals are bonding orbitals, and they are completely filled. This would give oxygen three bonds, but this bonding is partially canceled by the occupancy of π^* orbitals. Since there are two electrons in these antibonding orbitals, the effect of one of the bonding orbitals is canceled and according to the MO theory the O_2 molecule should have two bonds or a *bond order*† of 2. In addition, because the two π^* orbitals are only half filled, the MO theory explains the fact that O_2 is paramagnetic because there are two unpaired electrons in each molecule. This is in agreement with experiments on molecular oxygen discussed in Sec. 5.16. In addition, consider some facts which have been learned from spectroscopic studies on oxygen. When one electron is removed from O_2 to produce O_2^+, the bond energy increases from 118 kcal/mol in O_2 to 149 kcal/mol in O_2^+ and the bond length decreases from 1.21 to 1.12 Å in O_2^+ (see Table 5.4). This is consistent with our MO picture, which shows that the highest occupied orbitals are antibonding; removal of an electron should increase the bond order from 2 in O_2 to $2\frac{1}{2}$ in O_2^+. It is also known that when O_2 gains electrons to form O_2^- and O_2^{2-}, the bond length increases, and this is also explained by our MO diagram. The two additional electrons would go into the incompletely filled π^* an-

† The bond order equals the number of electrons in bonding orbitals, B, minus the number of electrons in antibonding orbitals, A, all divided by 2: $BO = (B - A)/2$.

tibonding orbitals and reduce the bonding strength in the molecule.

Whereas it was awkward to account for the paramagnetism of O_2 by the Lewis theory, the MO theory takes care of this problem naturally; in addition, it nicely accounts for many other properties of the O_2 molecule. The success of the MO theory in this well-known case was very important in winning adherents away from older bonding theories.

The MO system we used to explain the bonding in O_2 can be used to explain the bonding in F_2 and the nonexistence of Ne_2; and with a slight modification we can use these molecular orbitals for B_2, C_2, and N_2, also. In the latter three molecules the $\sigma(2p_x)$ molecular orbital is moved to a higher energy, so that the degenerate bonding orbitals $\pi(2p_y)$ and $\pi(2p_z)$ are filled first. In the boron, carbon, and nitrogen atoms the $2s$ and $2p$ orbitals are close in energy. As a result the molecular orbitals formed from these atomic orbitals are also fairly close in energy, and the $\sigma(2s)$ and the $\sigma(2p_x)$ have high electron density in the same region of space. Due to the repulsive energy between the electrons in the $\sigma(2s)$ and the $\sigma(2p_x)$, the $\sigma(2p_x)$ orbital is at a higher energy level in B_2, C_2, and N_2 than in O_2.

5.24 THE B_2 MOLECULE

Since the boron atom has the configuration ${}_5B\ 1s^2, 2s^2, 2p_x^1$, in B_2 there will be 10 electrons to accommodate. The first 8 will fill the four molecular orbitals up to $\sigma^*(2s)$. According to Fig. 5.22a, the next highest unfilled orbitals are the degenerate bonding pair $\pi(2p_y)$ and $\pi(2p_z)$. By Hund's rule the 2 remaining electrons must occupy these orbitals singly. The configuration for B_2 is

$$[\sigma(1s)]^2, [\sigma^*(1s)]^2, [\sigma(2s)]^2, [\sigma^*(2s)]^2, [\pi(2p_y)]^1, [\pi(2p_z)]^1$$

The net result is that there are two electrons in bonding orbitals, and they are unpaired. Hence, the bond order in B_2 is 1, and there are two unpaired electrons in the molecule; B_2 should be a singly bonded paramagnetic species.

The bond between the two boron atoms is a π bond. Comparison of Figs. 5.13, 5.19, and 5.20 shows that the electron density between

TABLE 5.4: SOME PROPERTIES OF THE O_2 SPECIES

Species	*Bond length,* Å	*Bond energy,* kcal/mol	*Bond order*	*Magnetic properties*
O_2	1.21	118	2	Paramagnetic
O_2^+	1.12	149	$2\frac{1}{2}$	Paramagnetic
O_2^-	1.28	. . .	$1\frac{1}{2}$	Paramagnetic
O_2^{2-}	1.49	. . .	1	Diamagnetic

two atoms joined by a bonding π orbital will not be as great as the electron density between two atoms joined by a bonding σ orbital. In the π orbital there is a node containing the bond axis, whereas in the σ orbital the highest electron density exists along the bond axis. This means that the nuclear repulsions between two atoms of a π bond will not be well shielded. The net effect is that the π-bond length will be greater than the σ-bond length.

Our experiments are in agreement with the prediction of MO theory: B_2 is a paramagnetic species with two unpaired electrons per molecule; the bond length is high, at 1.59 Å, and the bond energy is relatively low, 69 kcal/mol.

5.25 THE C_2 MOLECULE

Carbon has an atomic number of 6, and its electronic configuration is $1s^2$, $2s^2$, $2p_x{}^1$, $2p_y{}^1$. In C_2 we have 12 electrons to account for. Using the molecular orbitals shown in Fig. 5.22*a*, we find the molecular electronic configuration of C_2 to be

$$[\sigma(1s)]^2, [\sigma^*(1s)]^2, [\sigma(2s)]^2, [\sigma^*(2s)]^2, [\pi(2p_y)]^2, [\pi(2p_z)]^2$$

All occupied orbitals are filled, and there are no unpaired electrons. Accordingly, we should expect the C_2 molecule to be diamagnetic and its bond order to be 2; the two filled bonding π orbitals are responsible for the stability of the C_2 molecule. By experiment we

FIGURE 5.22

Inversion of the $\sigma(2p_x)$ and the $\pi(2p)$'s.

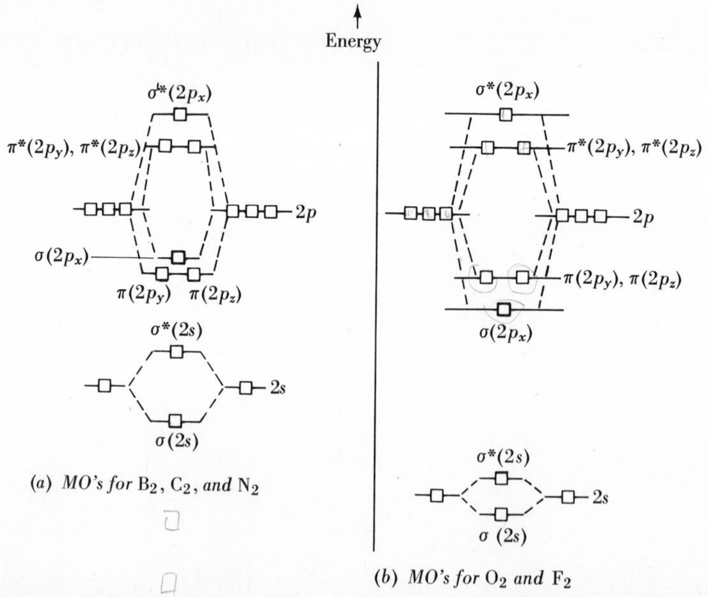

(*a*) *MO's for* B_2, C_2, *and* N_2

(*b*) *MO's for* O_2 *and* F_2

find that C_2 is diamagnetic, the bond length is 1.13 Å, and the bond energy is 150 kcal. The bond energy of this C_2 species with a double π bond is twice that of B_2, in which the atoms are held by a single π bond. Again, the MO theory seems to be in good agreement with the known facts.

5.26 THE N_2 AND F_2 MOLECULES

Observation and experiment show that N_2 is a very stable, inert species under ordinary circumstances. Its bond length is very short, and its bond energy is high; it is diamagnetic. This is consistent with the MO predictions for this molecule, which show that N_2 has a triple bond. In contrast to N_2, F_2 is very reactive. The bonding in the F_2 molecule is rather weak, as indicated by the high bond length and low bond energy. The bonding characteristics of the diatomic molecules of the second-period elements are summarized in Table 5.5. The student is asked to show that the MO theory is consistent with the data for N_2 and F_2 and with the observation that the Ne_2 molecule does not form (see Prob. 5.13).

5.27 HETERONUCLEAR DIATOMIC MOLECULES OF THE SECOND PERIOD

The simple MO theory used to explain the bonding in homonuclear diatomic molecules, like Li_2 and O_2, can be used for *hetero*nuclear diatomic molecules, like CN, CO, and NO, where the atoms of the

TABLE 5.5: SUMMARY OF DATA FOR HOMONUCLEAR DIATOMIC MOLECULES OF THE SECOND PERIOD

Species	*Bond energy,* kcal/mol	*Bond length,* Å	*Bond order*	*Magnetic?*
Li_2	25	2.68	1	No
"Be_2"	†	†	†	†
B_2	69	1.59	1	Yes
C_2	150	1.24	2	No
N_2	225	1.10	3	No
O_2	118	1.21	2	Yes
F_2	36	1.44	1	No
"Ne_2"	†	†	†	†

† Molecule does not form.

molecule are not identical. The MO energies in CN and CO go in the same order as C_2 and N_2, as shown in Fig. 5.22*a*; the $\sigma(2p_x)$ molecular orbital is at a higher energy than the two $\pi(2p)$ orbitals. The energy levels in NO resemble O_2 and F_2, where the $\pi(2p)$ orbitals are higher than the $\sigma(2p_x)$ (Fig. 5.22*b*).

The bonding characteristics of CN, CO, and NO are summarized in Table 5.6. Note that carbon monoxide, CO, is *isoelectronic* with N_2; that is, it has the same number of electrons, 14, and its bonding characteristics are very similar. Like N_2, it has a very high bond energy and a short bond distance. The effects of the first four molecular orbitals, which are filled, cancel each other, and the bonding is due to the presence of six electrons in the three bonding orbitals $\pi(2p_y)$, $\pi(2p_z)$, and $\sigma(2p_x)$. Because these orbitals are filled, CO possesses a triple bond and is diamagnetic. CN has one electron less than CO, and since this gives the CN molecule an odd number (13) of electrons, CN must be paramagnetic. Its MO description is similar to that of CO except that the highest occupied orbital, the bonding $\sigma(2p_x)$, is only half filled. Therefore, the bond order is $2\frac{1}{2}$ and the bond energy of CN is lower than that of triply bonded CO. When we compare CN with CN^-, we find that the ion is more stable; experiments show that the CN^- is diamagnetic. The bond length is 1.14 Å, which is shorter than the bond length in the uncharged CN molecule. From MO theory we would predict that the addition of an electron to the CN molecule would cause the bond order to increase from $2\frac{1}{2}$ to 3 since the $\sigma(2p_x)$ bonding orbital would then be filled; this would cause a decrease in bond length and eliminate the paramagnetism of the CN molecule.

Nitric oxide is one of the species for which no good Lewis structure can be drawn. With 15 electrons, of which 11 are considered valence electrons, it is impossible to draw a structure which conforms to the octet rule (see page 114). The MO theory solves the bonding problem of nitric oxide easily and accurately. Table 5.6 shows that there are 6 electrons in the bonding orbitals $\sigma(2p_x)$, $\pi(2p_y)$, and $\pi(2p_z)$ and 1 electron in an antibonding orbital $\pi^*(2p)$. This means that there are a net of 5 bonding electrons in the NO species. The bond order of NO is $2\frac{1}{2}$, and the bond energy and length in NO are similar to those of the CN species, also with a bond order of $2\frac{1}{2}$.

A real test of the MO theory is afforded by consideration of the ion NO^+, with 14 electrons. It forms by removal of the electron from a $\pi^*(2p)$ orbital of NO. Removal of an electron from this antibonding orbital should cause an increase in bonding strength. This remarkable prediction is borne out by the fact that the bond length of NO^+ is shorter than that of NO. The NO^+ ion has a triple bond $[\sigma(2p_x)]^2$, $[\pi(2p_y)]^2$, $[\pi(2p_z)]^2$, and since all its electrons are paired, it should be diamagnetic, as is observed. Compare the bonding characteristics of NO and NO^+ in Table 5.6.

TABLE 5.6: MOLECULAR ORBITALS FOR SOME HETERONUCLEAR DIATOMICS

	13 electrons	14 electrons	
	CN	CN^-	CO
$\sigma^*(2p_x)$	—	—	—
$\pi^*(2p)$	— —	— —	— —
$\sigma(2p_x)$	↑	⇅	⇅
$\pi(2p)$	⇅ ⇅	⇅ ⇅	⇅ ⇅
$\sigma^*(2s)$	⇅	⇅	⇅
$\sigma(2s)$	⇅	⇅	⇅
$\sigma^*(1s)$	⇅	⇅	⇅
$\sigma(1s)$	⇅	⇅	⇅

	15 electrons	14 electrons
	NO	NO^+
$\sigma^*(2p_x)$	—	—
$\pi^*(2p)$	↑ —	— —
$\pi(2p)$	⇅ ⇅	⇅ ⇅
$\sigma(2p_x)$	⇅	⇅
$\sigma^*(2s)$	⇅	⇅
$\sigma(2s)$	⇅	⇅
$\sigma^*(1s)$	⇅	⇅
$\sigma(1s)$	⇅	⇅

	CN	CN^-	CO	NO	NO^+
Bond energy, kcal/mol	188	?	256	162	?
Bond length, Å	1.18	1.07	1.13	1.15	1.06
Bond order	$2\frac{1}{2}$	3	3	$2\frac{1}{2}$	3
Magnetic?	Yes	No	No	Yes	No

5.28 MOLECULAR ORBITALS FOR POLYATOMIC MOLECULES

The MO theory is particularly easy to apply to the diatomic molecules of the second period. As we might guess from the periodicity of the behavior of the elements, the bonding in analogous heavier molecules is similar; i.e., the molecular orbitals of Na_2 and Cs_2 are similar to those of Li_2, and the molecular orbitals of Cl_2 and Br_2 are similar to those of F_2. In addition, the MO theory has been successful in explaining the bonding in molecules with many atoms. In this section we briefly consider the case of the triatomic molecule BeH_2, beryllium hydride. BeH_2 has not yet been detected in the gaseous phase probably because such a molecule would be very reactive. It would be similar in structure to BeF_2 and other halides of the alkaline-earth family (III). Our knowledge of these known molecules provides a guide to our discussion on BeH_2. When and if this molecule is prepared, we can test our predictions.

For BeH_2 the MO theory does not use the Be 1*s* orbital in the consideration of the bonding. The average radius of the 1*s* orbital of the Be atom is very small (0.24 Å) compared to the estimated Be—H bond length (1.2 Å), and therefore it cannot really be involved in any strong interactions with the 1*s* orbitals of the H atoms. Because of this fact, we are using an important assumption: *the filled atomic orbitals in energy levels below the valence energy level have no signifi-*

FIGURE 5.23

Molecular orbitals for BeH_2.

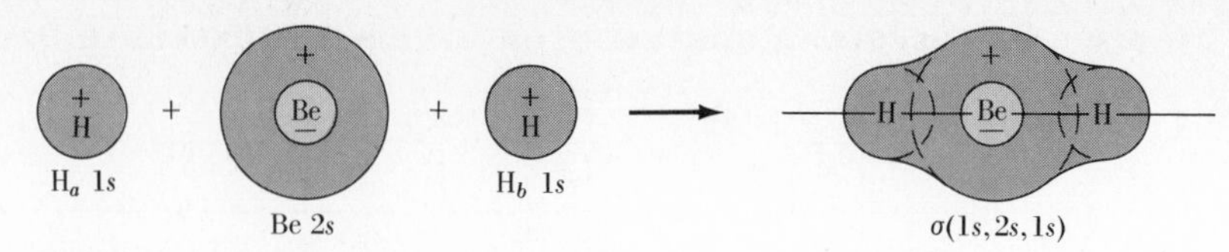

(*a*) *In-phase interactions of* H 1s, Be 2s, *and* H 1s

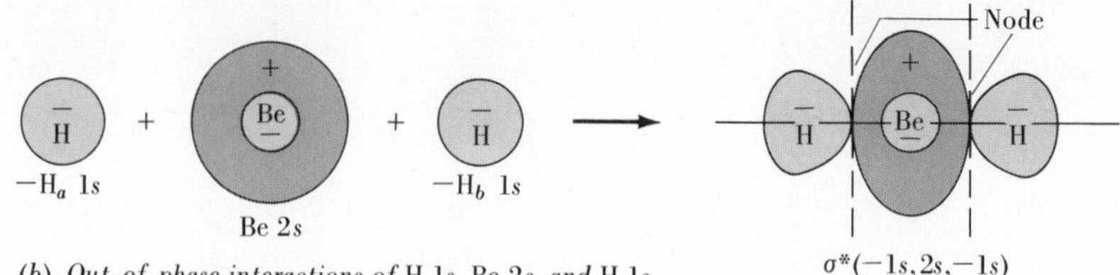

(*b*) *Out-of-phase interactions of* H 1s, Be 2s, *and* H 1s

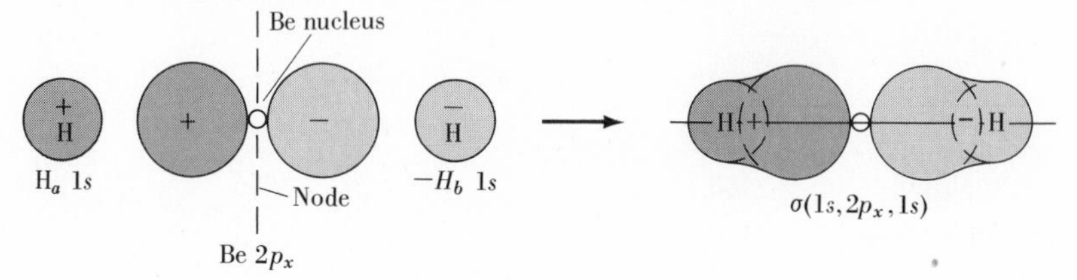

(*c*) *In-phase interactions of* H 1s, Be $2p_x$, *and* H 1s

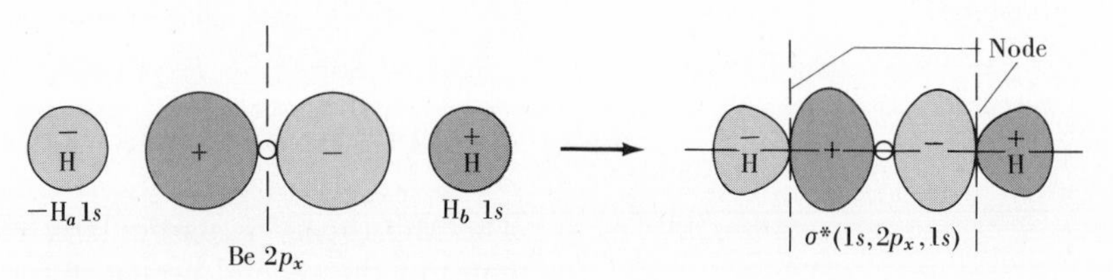

(*d*) *Out-of-phase interactions of* H 1s, Be $2p_x$, *and* H 1s

cant effect on the chemical bonding. The valence energy level of beryllium is 2; it is in the second period. The 2*s* and 2*p* orbitals may take part in the bonding, but we shall assume that the filled 1*s* orbital cannot.

When we study the bonding in a triatomic molecule, we again try to imagine how electrons in atomic orbitals would interact if the separated atoms approached each other to take up their regular positions in the molecule. The in-phase interactions again lead to an increase in electron density between the atoms of the molecule and result in the formation of bonding orbitals. In BeH_2 the interaction of H_a 1*s*, Be 2*s*, and H_b 1*s* can result in such a molecular orbital (see Fig. 5.23*a*). The out-of-phase interaction of these orbitals can lead to antibonding orbitals with nodal planes perpendicular to the two Be—H bond axes, as shown in Fig. 5.23*b*.

The H 1*s* orbitals can also interact with one of the Be 2*p* orbitals. The in-phase and out-of-phase interactions give rise to σ orbitals, one bonding and the other antibonding. This is shown in Fig. 5.23*c* and *d*. All the BeH_2 molecular orbitals we have discussed are σ molecular orbitals because they are symmetrical with respect to rotation about the bond axis. The in-phase interactions cause a buildup of electron charge between the beryllium atom and its two hydrogen atom partners; the result is a bonding molecular orbital. The out-of-phase interactions result in nodes between adjacent atoms and a decrease of electron charge between the beryllium atom and its hydrogen-atom partners; the result is an antibonding molecular orbital.

There are four valence electrons to accommodate in the BeH_2 molecule; two come from the single beryllium atom and two from the two hydrogen atoms. The four electrons occupy the two lowest molecular orbitals. The orbital of lowest energy is the $\sigma(1s,2s,1s)$, and the next lowest is the $\sigma(1s,2p_x,1s)$ (Fig. 5.23*a* and *c*). Both orbitals are filled and extend over the entire molecule.

The MO description of the bonding in BeH_2 is significantly different from the Lewis picture, which would show electron pairs localized between two bonded atoms: H:Be:H. Each electron pair is separately involved in bonding one hydrogen atom to the beryllium atom. The two Lewis bonds are identical.

The lowest-energy molecular orbital in BeH_2 is the $\sigma(1s,2s,2s)$; it is a bonding orbital which extends over the entire molecule, as shown in Fig. 5.23*a*. This differs from the Lewis picture, which would show electron pairs *localized* between two bonded atoms: H:Be:H. The Lewis model for BeH_2 shows that we have two electron pairs involved in the bonding and two identical bonds. The MO theory also indicates that there are two bonds but the two bonds are not identical. Most important, since each molecular orbital extends over the entire molecule, each bonding molecular orbital of BeH_2 involves all three atoms; i.e., each electron pair contributes to the bonding of the entire molecule. In this sense the MO theory shows that the bonding electrons in BeH_2 are *delocalized;* they are not constrained to exist between a pair of bonded atoms but have some freedom to move throughout the molecule and add to the overall bonding. The MO picture is consistent with the wave nature of electrons.

We should make one more point about the MO theory that may be obvious but has not been explicitly mentioned. The number of molecular orbitals developed will equal the total number of atomic orbitals considered. In BeH_2 we used two H 1*s* orbitals, the Be 2*s* and the Be $2p_x$, a total of four. These resulted in four molecular orbitals, two σ and two σ^*. As we might imagine, the MO theory becomes very complicated when we consider molecules which have a large number of atoms or which contain heavy atoms; in each case there is a large number of valence electrons, and a large number of atomic orbitals must be considered in deriving the proper molecular orbitals. Nevertheless, using chemical intuition (as we did when we ignored the Be 1*s* orbital) and high-speed computers, theoretical chemists are applying the MO theory to larger molecules with a good deal of success.

5.29 THE VALENCE-BOND THEORY

The valence-bond (VB) theory, developed before the MO theory, is the quantum-mechanical analog of the Lewis-Langmuir theory. Lewis held that when two electrons on different atoms form a pair which is shared, a chemical bond is formed. The simplest and most important principle of the VB theory holds that when two half-filled orbitals on different atoms "overlap," a chemical bond is formed; atomic orbitals which already contain paired electrons are not available for bonding. In Fig. 5.24 three representations of the bonding in H_2 are contrasted. The Lewis structure of H_2 involves sharing an electron pair. In the VB picture two half-filled 1*s* orbitals overlap as the hydrogen atoms approach, and a molecule is formed. The Lewis-Langmuir theory is based on the Bohr atom, in which electrons are considered as minute charged *particles* moving in shells of fixed radii about the nucleus. The VB theory is based on the quantum-mechanical model of the atom in which the electrons have both a particle and a wave nature. It should be obvious that VB theory is a recasting of the Lewis-Langmuir theory of chemical bonding in terms of the quantum-mechanical model for the bonded atoms. In the MO theory, new molecular orbitals encompassing the entire molecule are formed from the atomic orbitals as the atoms approach each other. The H_2 bond is due to the fact that a bonding molecular orbital is occupied by a pair of electrons while the antibonding molecular orbital is vacant.

FIGURE 5.24

Three representations of the bonding in H_2.

H : H

(*a*) *Lewis electron-pair structure*

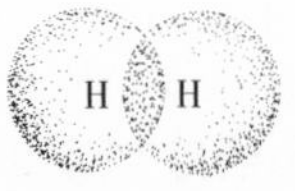

H H

H $1s^1$ + H $1s^1$

(*b*) *Valence-bond orbital overlap structure*

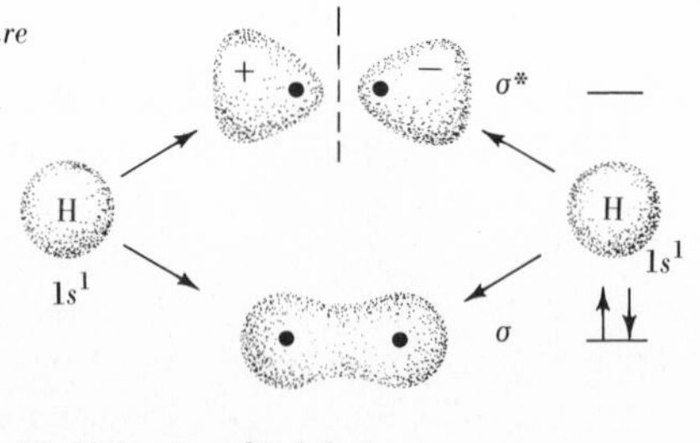

H $1s^1$ H $1s^1$ + − σ^* σ

(*c*) *Molecular orbital theory*
Molecular orbitals form from atomic orbitals

5.30 THE N_2 AND O_2 MOLECULES (VALENCE-BOND THEORY)

Let us consider the bonding in N_2 and O_2 according to the VB picture. The electronic structure of the nitrogen atom is $_7N\ 1s^2,\ 2s^2,\ 2p_x^1,\ 2p_y^1,\ 2p_z^1$; it has three half-filled orbitals which can be used for bonding. If the atoms approach along the x axis, a bond will form by the overlap of the individual $2p_x$ orbitals:

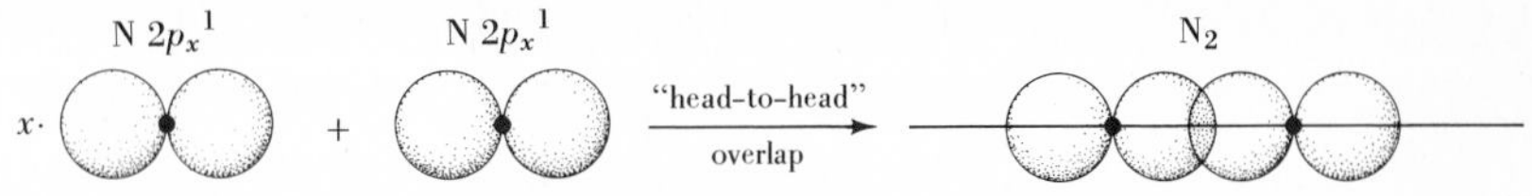

At the same time two π bonds will form from the overlap of a pair of $2p_y$ orbitals and a pair of $2p_z$ orbitals:

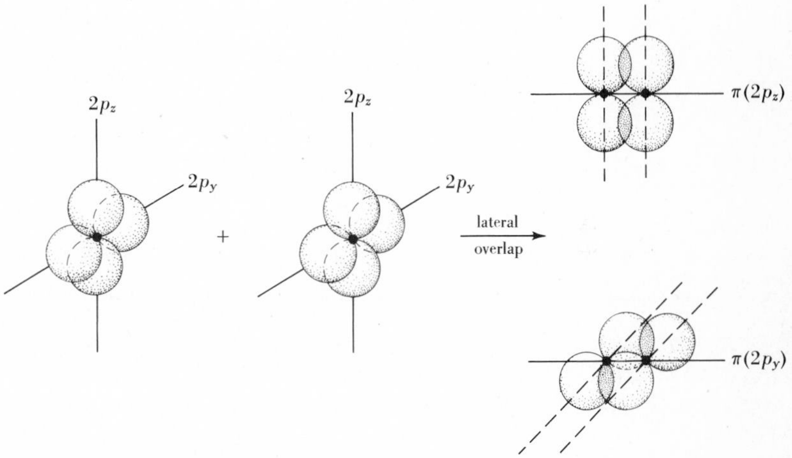

The head-to-head overlap of the two $2p_x$ atomic orbitals results in a bonding orbital which is symmetrical with respect to rotation about the bond axis, and therefore $2p_x$-$2p_x$ overlap gives rise to a σ bond. The overlap of the other $2p$ atomic orbitals occurs laterally instead of in a head-to-head manner. This results in two more bonds, but they are unsymmetrical with respect to rotation about the bond axis. Hence, the lateral overlap of $2p$ orbitals results in a π bond.

The bonding suggested by the theory is conveniently represented by a diagram which shows *atomic* orbitals as boxes. Arrows in the boxes indicate the occupancy of the orbitals, and the direction of the arrow indicates the value of the electron spin: $\uparrow$ means $s = -\frac{1}{2}$ and $\downarrow$ means $s = +\frac{1}{2}$. In the nitrogen atom the $1s$ and $2s$ orbitals are completely filled. They are not involved in bonding in the N_2 molecule. The three pairs of $2p$ orbitals overlap to form three bonds;

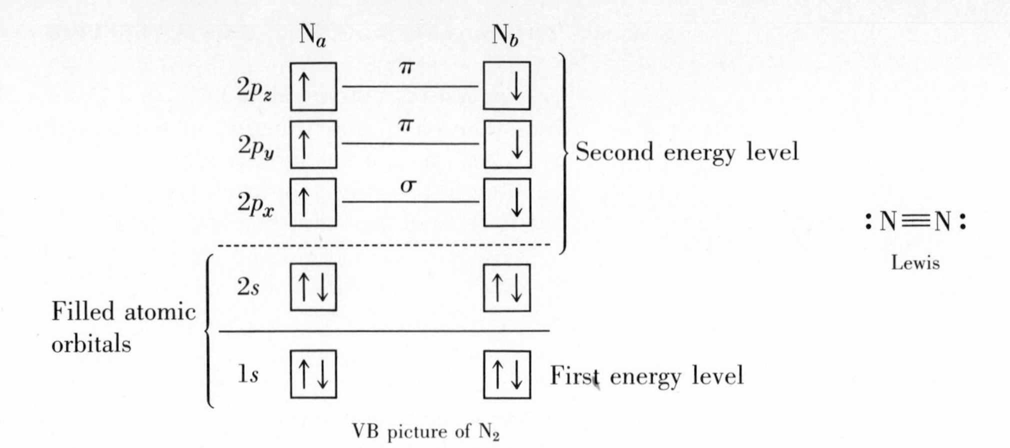

VB picture of N_2

this is indicated by the horizontal lines connecting the boxes. Note the similarity of the VB structure to the Lewis structure for N_2. In the Lewis structure there are three shared pairs of electrons; in addition, on each atom there is one unshared or nonbonding pair of electrons. These correspond to the nonbonding electrons of the filled 2*s* orbitals in the VB description of N_2.

The VB description of O_2 is easy to set up by using the orbital-box description:

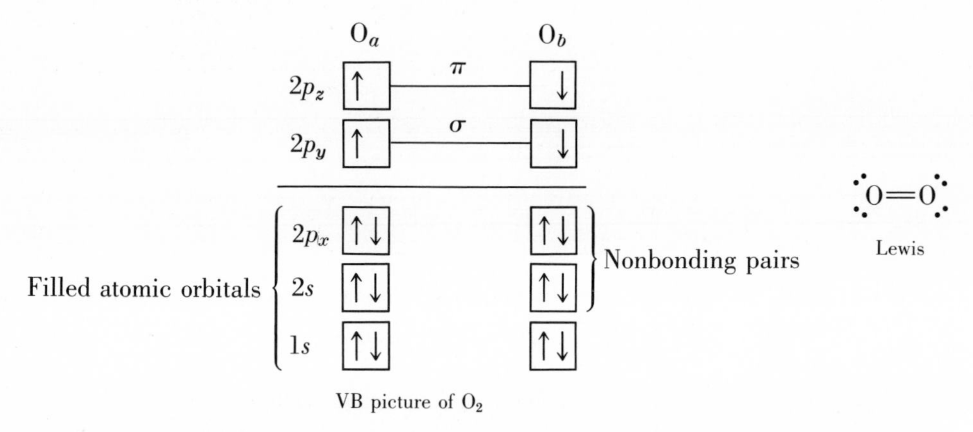

VB picture of O_2

Again we see the similarity to the Lewis electron-dot structure. In each case there are two bonds due to two pairs of bonding electrons, and on each separate atom there are two pairs of nonbonding electrons. Unfortunately the similarity between the VB and Lewis structure does not end here. Neither structure for O_2 explains the paramagnetism of this molecule. This deficiency in the VB theory should not be magnified, however, for, as we shall see, the theory is useful in predicting the bonding and structure in molecules and is often simpler to apply than the MO theory.

Let us now turn our attention to some of the hydrides of the second period. In ammonia, NH_3, we have three hydrogen atoms bonded to a central nitrogen atom. The orbital-box description for NH_3 is easy to write out. Each of the three half-filled p orbitals of the nitrogen atom will overlap with a half-filled 1s orbital of a hydrogen atom. This will result in three σ bonds between nitrogen and the three hydrogen atoms. The details of the bonding are diagrammed in Fig. 5.25. All the N—H bonds are identical, and they are σ bonds because they are symmetrical with respect to rotation about the bond axis. Again, we see the similarities with the Lewis structure, which shows three identical bonds formed when Bohr atoms share electron pairs. Also, the VB and Lewis structures show a pair of nonbonded electrons. One important implication in the VB theory finds no analogy in the Lewis theory. We can predict the shape of molecules when we use the VB theory, but there is no natural feature of the Lewis-Langmuir theory which permits this. Since the VB theory employs the 2p orbitals of the nitrogen atom in the bonding, and since these orbitals are mutually perpendicular, the resulting NH_3 molecule must have pyramidal shape. Each N—H bond should be perpendicular to the other two, and all the H—N—H bond angles should be 90°. Experiments show that the NH_3 molecule is pyramidal but the H—N—H bond angles are close to 107°. A simple modification of the VB theory, called *hybridization*, permits us to explain

FIGURE 5.25

The bonding in NH_3.

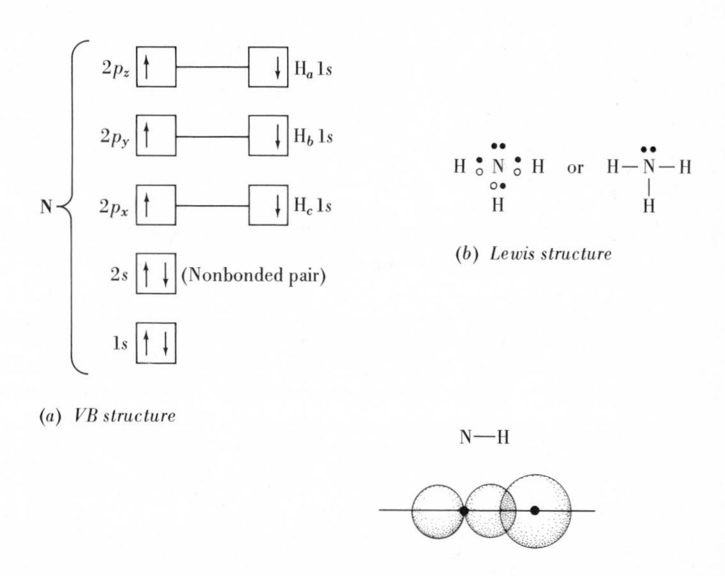

(*a*) *VB structure*

(*b*) *Lewis structure*

(*c*) *Overlap of* N 2p *and* H 1s *leads to* σ bond

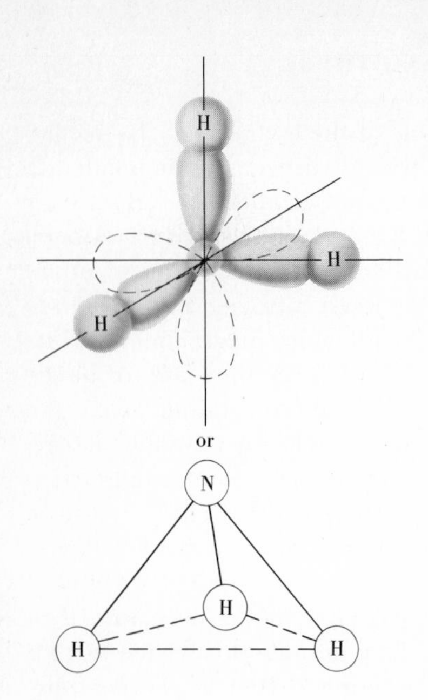

why the bond angles are greater than 90°. Hybridization will be the topic of the next section.

The orbital-box descriptions for H_2O and HF are given in Fig. 5.26. Again, the similarity between the VB and Lewis structures is obvious. In H_2O each theory shows two identical O—H bonds and two nonbonded electron pairs on the oxygen atom. The VB structure, however, suggests that the H_2O molecule is bent: two of the mutually perpendicular p orbitals of oxygen are used to form the O—H bonds. Therefore, the H—O—H bond angle should be 90°. Experiment does show that the H_2O molecule is bent, but the H—O—H bond angle is 104.5°.

Since HF is diatomic, it must be linear. The VB and Lewis structures show a single bond between the two atoms and three nonbonded pairs. The VB theory shows that the bond formed is due to the overlap of the H 1s with an F 2p and is therefore a σ bond.

We see that the VB theory, by resembling the Lewis-Langmuir theory, enables a chemist to obtain a quick insight into the bonding within a molecule. In order to apply the VB theory the chemist needs

FIGURE 5.26

Bonding in H_2O *and* HF.

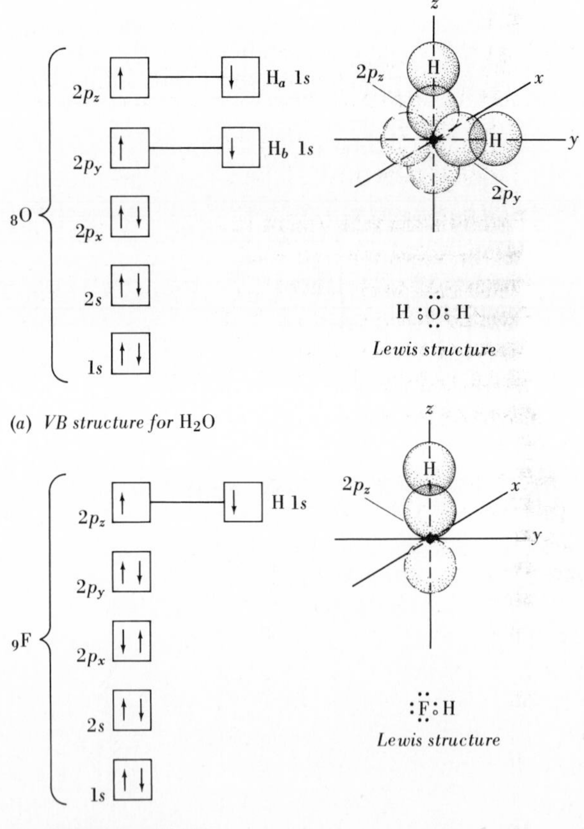

only a good quantum-mechanical description of the individual atoms where the electrons are given a wave and a particle nature. To apply Lewis-Langmuir theory the chemist needs a good Bohr description of the individual atoms, where the electrons are considered as simple particles moving in shells of fixed radii. The chief practical advantage of the VB theory over the Lewis-Langmuir theory (for our purposes) is that it permits us to guess the shapes of molecules; in addition, it is philosophically more satisfying in that it recognizes that electrons in atoms possess a wave and particle nature. Both the VB and Lewis-Langmuir theories differ from the MO theory in one very important respect, namely, that the bonding electrons are *localized* between two bonded atoms. In the MO picture the atomic orbitals give rise to molecular orbitals which range over the entire molecule, and electrons are delocalized over all atoms of the molecule. The MO picture appears to be closest to the true description of electrons in molecules.

5.32 EXTENSION OF THE VALENCE-BOND THEORY BY HYBRIDIZATION

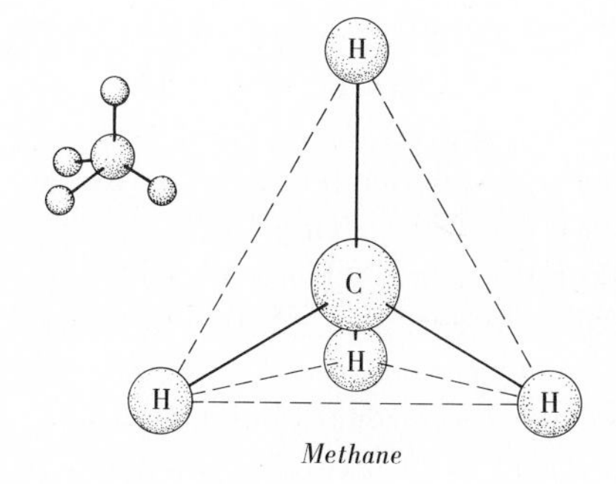

Methane

We have already learned that the methane molecule, CH_4, has a tetrahedral shape, and we have already explained it in terms of the electron-repulsion theory. The four C—H bonds of methane are identical and all oriented toward the vertices of a regular tetrahedron with the carbon atom at the center. If we attempt to explain the bonding in CH_4 in terms of simple VB theory, we run into trouble. The ground-state configuration of the carbon atom is $_6C\ 1s^2$, $2s^2$, $2p_x^1$, $2p_y^1$, $2p_z^0$; two of the p orbitals are half filled and can be involved in bonding by overlap with half-filled orbitals of two other atoms. We might predict that carbon would be divalent and would form molecules like CH_2,† with an H—C—H bond angle of 90°. One way to explain the bonding in CH_4 is to assume that during reaction one of the electrons in the $2s$ orbital is elevated to the vacant $2p_z$. This would give carbon the configuration $_6C\ 1s^2$, $2s^1$, $2p_x^1$, $2p_y^1$, $2p_z^1$, with four half-filled orbitals, and it could then bond four hydrogen atoms to form CH_4. The promotion of the electron from the $2s$ orbital to the $2p_z$ explains the *tetra*valency of carbon, but it does not explain the shape of the CH_4 molecule. In this case one of the C—H bonds would result from H $1s$–C $2s$ overlap; the other three would result from H $1s$–C $2p$ overlap, and, contrary to observation, there would be two different kinds of C—H bonds in the CH_4 molecule. Furthermore, the H—C—H bond angle between any pair of the three bonds formed from C $2p$ orbitals would be 90°, whereas all H—C—H bond angles in CH_4 are known to be 109°. We can completely account for the bonding in CH_4 if we *mix* the C $2s$ with the

† CH_2 does exist under special conditions; it is a very reactive, *linear* molecule.

three C $2p$ orbitals and produce four identical hybrid orbitals. *Hybridization* is a mathematical procedure whereby the wave functions of different atomic orbitals are combined to produce identical hybridized orbitals. Hybridization does not increase or decrease the number of orbitals; it changes their character. To explain the bonding in methane we take the one C $2s$ and the three C $2p$ orbitals and produce four identical orbitals. Since each resultant orbital has a character which is three-fourths $2p$ and one-fourth $2s$, we designate the hybrid orbitals as sp^3 (read "*s p* three"). A comparison of the ground state of the carbon atom and the hybridized reaction state is given:

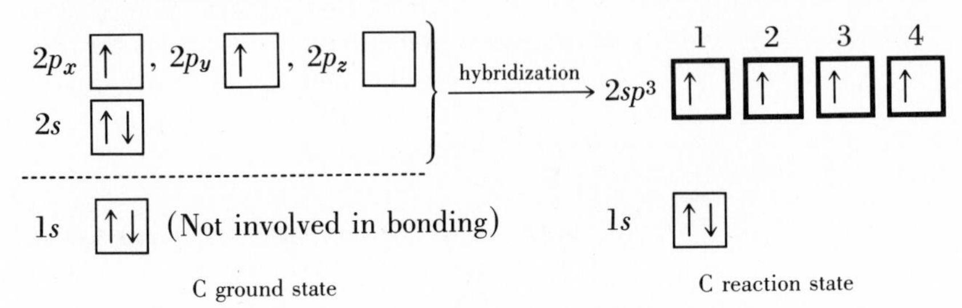

Hybridization produces four identical orbitals from the $2s$ and the three $2p$ orbitals. By Hund's rule we would expect each hybrid orbital to be half filled: there are four electrons to be placed in four orbitals of equal energy. Comparing the ground-state carbon atom with the hybridized carbon atom, we see two driving forces which would favor hybridization: (1) the four valence electrons occupy separate orbitals, reducing the repulsive forces, and (2) the bonding capacity of carbon is increased.

The four sp^3-hybridized orbitals have directional character since they are derived from the p orbitals. The most favorable orientation results when each orbital is directed toward the vertex of a regular tetrahedron, as shown in Fig. 5.27. In this way the electrons, which are mutually repulsive, can more successfully avoid each other.

FIGURE 5.27

The sp^3-hybrid orbitals.

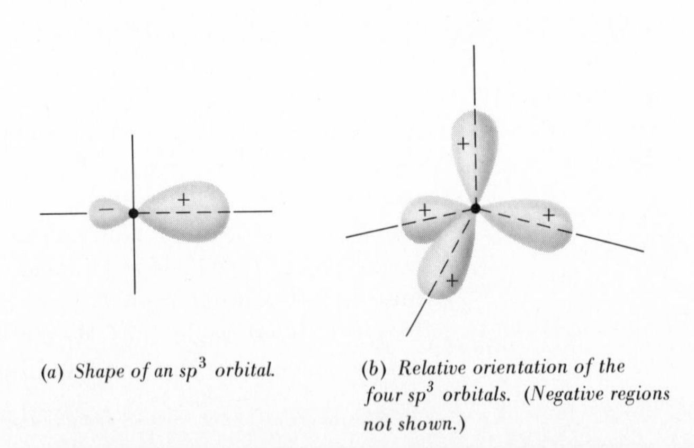

(*a*) *Shape of an* sp^3 *orbital.*

(*b*) *Relative orientation of the four* sp^3 *orbitals. (Negative regions not shown.)*

FIGURE 5.28

The bonding in CH_4.

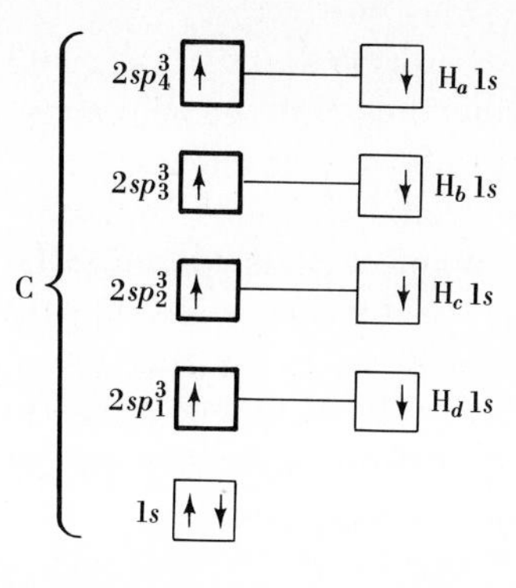

(*a*) *VB structure*

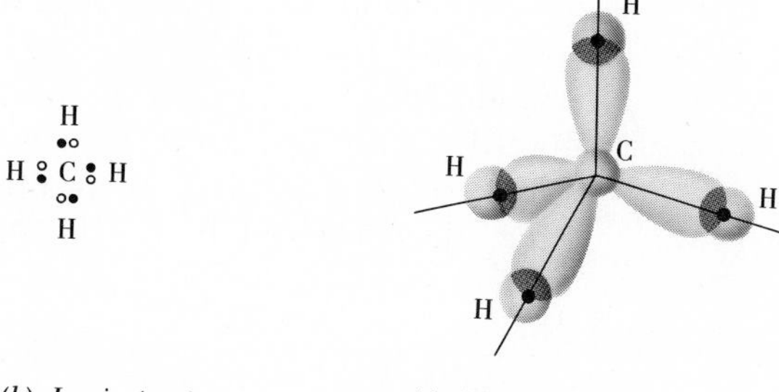

(*b*) *Lewis structure* (*c*) *VB model*

The hybridized carbon atom can now be used in the VB system to account for the bonding in CH_4. The orbital-box description of the bonding is shown in Fig. 5.28.

The bonding in CH_4, according to the VB theory, involves the overlap of the carbon $2sp^3$ orbitals with the hydrogen $1s$ orbitals. Again, we see the similarity between the VB and Lewis structures.

Although we have treated the CH_4 problem as if the hybridization of the carbon atom and the formation of the C—H bonds were separate steps, it must be stressed that in the chemical reaction to form methane the two processes are not sequential. There is no evidence that the carbon atom has an independent existence in the hybridized state C $1s^2$, $(sp_1{}^3)^1$, $(sp_2{}^3)^1$, $(sp_3{}^3)^1$, $(sp_4{}^3)^1$; but given this state, we can explain the structures of many carbon molecules. The bond angles for halogenated methane derivatives and other methane derivatives show that the substitution does not appreciably distort the methane structure and the sp^3 hybridization is preserved:

Compound	*Bond*	*Angle*
CH_4	H—C—H	109°28′
CF_4	F—C—F	109°28′
CH_2F_2	F—C—H	109°11′
	F—C—F	108°13′

In our discussions of NH_3 and H_2O we used simple VB ideas to describe the bonding. We assumed that the bonding was due to overlap of the hydrogen 1*s* and the nitrogen or oxygen 2*p* orbitals; we neglected the filled 2*s* orbitals in each case (see Figs. 5.25 and 5.26). When we considered CH_4, the tetravalency of carbon forced us to consider the possibility that the 2*s* orbitals were involved in bonding. If we guess that the 2*s* orbitals are also involved in the bonding of the H_2O and NH_3 structures, and if in addition we guess that sp^3 hybridization occurs, we obtain better agreement between theory and experiment.

In the ground state the nitrogen atom has three half-filled *p* orbitals, and from simple VB ideas we expect nitrogen to be trivalent with 90° angles between bonds. Nitrogen does form three bonds, but we know the bond angles are 107°. Using sp^3 hybridization, we produce four equivalent orbitals on the nitrogen atom oriented toward the vertices of a regular tetrahedron. Since we have five electrons to place in the four sp^3 orbitals, one orbital will be filled and the other three will be half filled:

$$\underset{\text{Ground state}}{{}_7\text{N}\ 1s^2, 2s^2, 2p_x{}^1, 2p_y{}^1, 2p_z{}^1} \xrightarrow[\text{hybridization}]{sp^3} \underset{\text{Reaction state}}{{}_7\text{N}\ 1s^2,(2sp^3)^2,(2sp^3)^1,(2sp^3)^1,(2sp^3)^1}$$

With the hybridized model we still have the trivalent nitrogen and a nonbonding pair of electrons, but the predicted H—N—H bond angle for NH_3 is 109°, which is much closer to the experimental value.

When ammonia reacts with H^+, it forms the ammonium ion, NH_4^+. The added proton carries no electrons. It is bonded by the pair of electrons which are in the $2sp^3$ nonbonding orbital of the nitrogen atom in the NH_3 molecule. The ammonium ion is perfectly tetrahedral like the CH_4 molecule. The VB models for NH_3 and NH_4^+ are shown in Fig. 5.29.

Now let us make our model for the H_2O molecule using the sp^3-hybridized oxygen atom. Upon hybridization there will be four equivalent sp^3 orbitals. Since there are six electrons in the valence level of oxygen, two of the orbitals will be completely filled and two will be half filled. The half-filled orbitals will be involved in bonding by overlap with the half-filled 1*s* orbitals of two hydrogen atoms, as shown in Fig. 5.30. The predicted H—O—H bond angle using the hybridized oxygen atom is 109°28′; we find by experiment that for molecules in the gaseous state it is closer to 105°; the agreement of theory and observation is fairly good. Furthermore, when water freezes to form crystalline ice, the H—O—H bond angle is exactly 109°28′.

Hybridization involving sp^3 orbitals has been very useful in explaining the structure of many molecules, especially hydrocarbons

FIGURE 5.29

The VB picture of NH_3 *and* NH_4^+ *using* sp^3 *hybridization on the nitrogen atom*

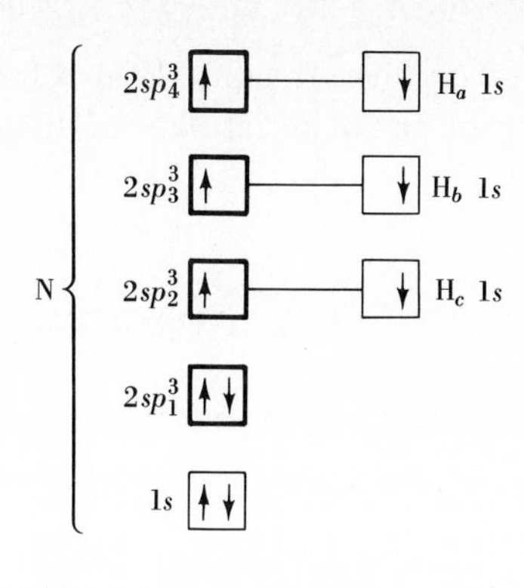

(a)

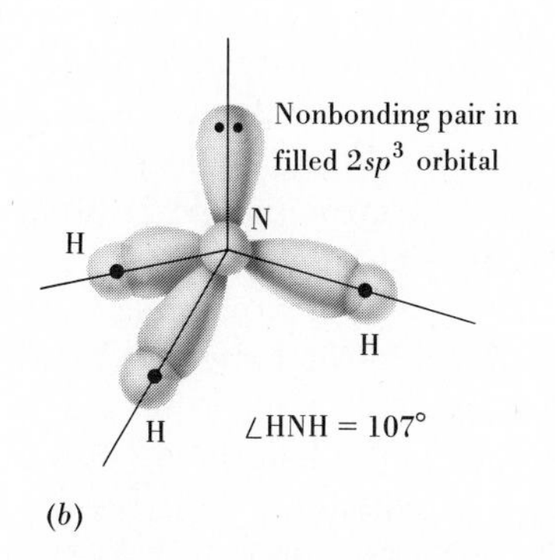

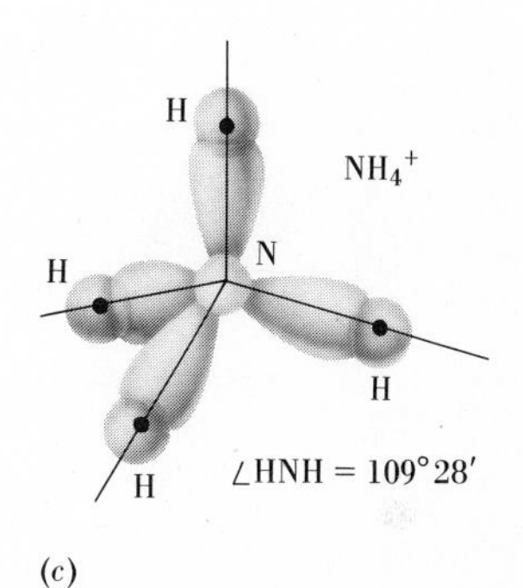

(b) (c)

FIGURE 5.30

The VB picture of H_2O *using* sp^3 *hybridization on the oxygen atom.*

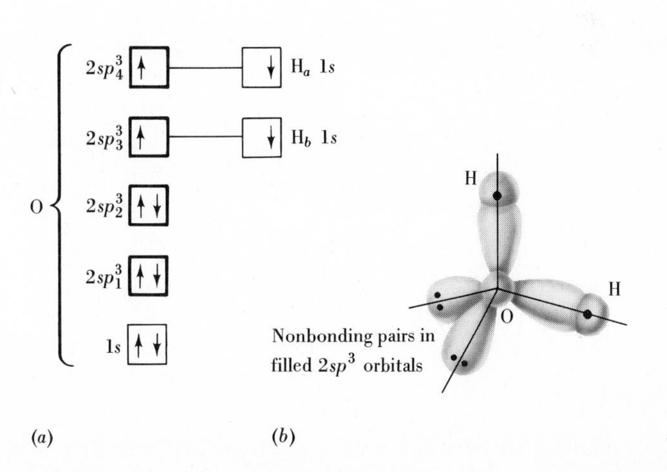

(a) (b)

and their derivatives. It is also helpful in understanding the structure and behavior of the complex ions formed by the transition metals.

5.33 TRIGONAL, OR sp^2, HYBRIDIZATION

The simplest molecule formed from the elements boron and fluorine is BF_3. It is a planar structure of high symmetry. The boron atom can be considered to be present at the center of an equilateral triangle with a fluorine atom placed at each vertex. All B—F bond lengths are 1.30 Å, and all F—B—F angles are 120°. Without this experimental knowledge we might predict that boron and fluorine would form a diatomic molecule BF. The boron atom has a ground-state electronic description of ${}_5B\ 1s^2, 2s^2, 2p_x{}^1$, and fluorine has the description ${}_9F\ 1s^2, 2s^2, 2p_x{}^2, 2p_y{}^2, 2p_z{}^1$. Overlap of the half-filled B $2p_x$ with the F $2p_z$ would satisfy the VB condition for bond formation. The electron-box description of the hypothetical BF species is shown in Fig. 5.31.

The problem of explaining why boron forms three bonds does not arise in the Lewis-Langmuir theory because it is based on the Bohr model for atoms. Since the atomic number of boron is 5, the Bohr model for boron has three equivalent electrons in the valence shell and boron may form three bonds.

From the quantum-mechanical point of view, the fact that boron is trivalent indicates that the 2s electrons must be involved in the bonding. Furthermore, since the three bonds of BF_3 are identical, hybridization of s and p orbitals is suggested. In this case we consider the 2s orbital and only two of the 2p orbitals for hybridization

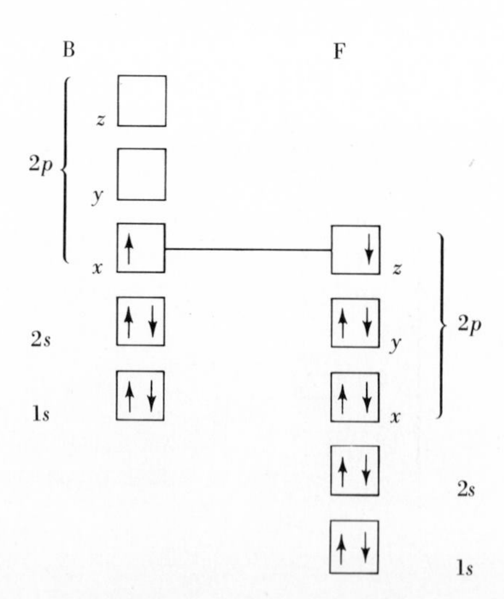

FIGURE 5.31

The bonding in hypothetical BF.

because we require a total of three equivalent bonding orbitals. (When we were discussing CH_4, we chose the $2s$ orbital and *three* $2p$ orbitals because we required a total of four equivalent bonding orbitals.) Since each resultant orbital is one-third s and two-thirds p in character we represent the hybrids as sp^2 (read "s p two"). Because of the form of the resulting molecule the sp^2 orbitals are also called *trigonal* orbitals. A comparison of the ground state and the hybridized reaction state of boron is given:

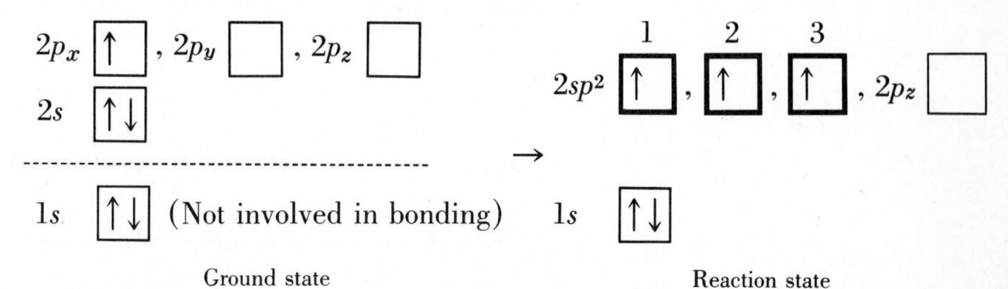

Again we see that the electron repulsion is lower and the bonding capacity is higher in the hybridized state. There are now three equivalent half-filled orbitals. The three sp^2 orbitals will be coplanar, and the angle between the axes of any pair of orbitals is 120°, as shown in Fig. 5.32*b*. This is exactly the kind of atomic orbital neces-

FIGURE 5.32

The sp^2 hybridization of boron and the VB picture of BF_3.

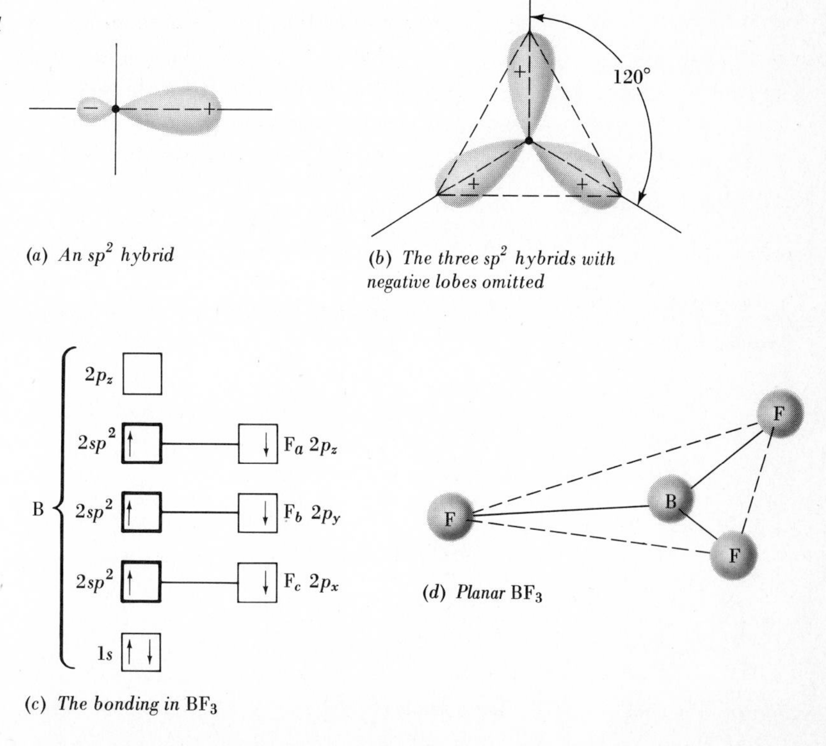

sary to explain the bonding in BF_3 and molecules of other boron compounds. The half-filled $2p_z$ orbitals of three fluorine atoms can overlap with each of the half-filled $2sp^2$ orbitals on boron. The orbital-box description of the bonding in BF_3 is shown in Fig. 5.32*b*.

The sp^2 hybridization which explains the bonding in BF_3 is useful in many other molecules and ions, such as ethylene, $CH_2{=}CH_2$; benzene, C_6H_6; sulfur trioxide, SO_3; and nitrate ion, NO_3^-. All these species are planar with bond angles of 120°, as shown in Fig. 5.33. The details of VB theory for ethylene and benzene are considered in Chap. 15.

5.34 LINEAR, OR *sp*, HYBRIDIZATION

Members of the alkaline-earth family (IIA) exhibit a valence of 2. Dihalide molecules, for example, MgF_2, $CaCl_2$, $SrBr_2$, and BaI_2, have been studied in the gaseous state, and the experimental evidence indicates that these triatomic molecules are linear. No dihalide of beryllium, the first member of family IIA, has ever been detected. But we suppose that when it is finally found it will also be linear.

All members of the IIA family have the electronic configuration of a filled *s* orbital above a rare-gas core (see page 78). Since there are no half-filled orbitals in the ground state, the application of VB theory is not immediately clear. However, the fact that these elements form divalent compounds with two identical bonds suggests that hybridization is occurring. For $_4Be$ we would choose to form hybrid orbitals from the single 2*s* orbital and one of the 2*p* orbitals since we require two orbitals for bonding. Figure 5.34*a* compares the ground state and the hybridized reaction state of the beryllium atom. The hybridized state has lower electron repulsion and increased bonding capacity. The two *sp* orbitals are directed along the same line but in opposite directions so that their axes form a 180° angle, as shown in Fig. 5.34*b*. The hypothetical electron-box descriptions of BeH_2 and BeF_2 are given in Fig. 5.35*a* and *b*. The two beryllium *sp* hybrid orbitals overlap with the 1*s* orbitals of two hydrogen atoms to form linear BeH_2 and with the $2p_z$ orbitals of two fluorine atoms to form linear BeF_2.

FIGURE 5.33

Structures exhibiting sp^2 hybridization.

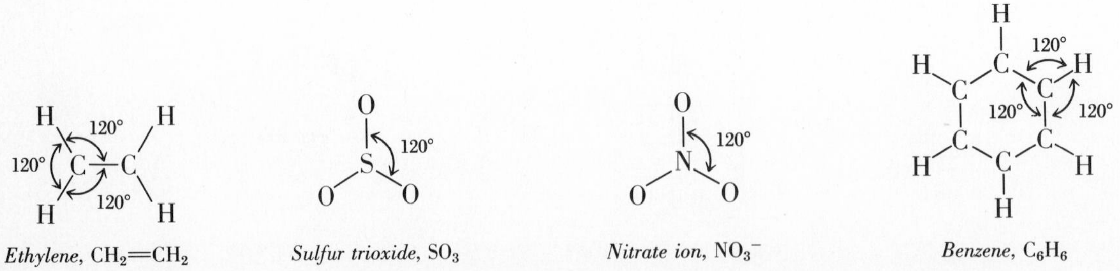

Ethylene, $CH_2{=}CH_2$ *Sulfur trioxide*, SO_3 *Nitrate ion*, NO_3^- *Benzene*, C_6H_6

FIGURE 5.34

The sp hybridization of beryllium.

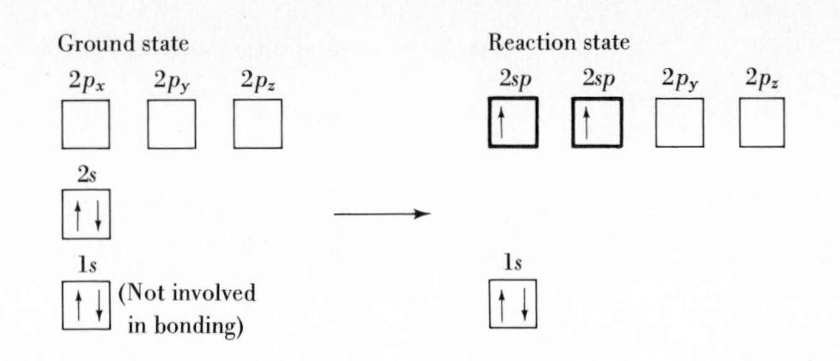

(*a*) *Comparison of the* Be *ground and reaction states*

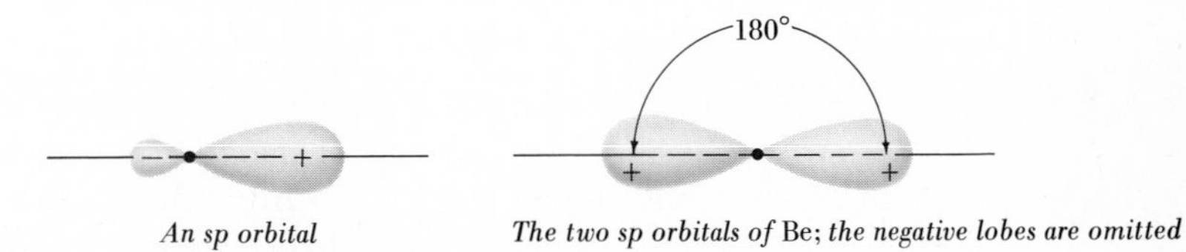

An sp orbital *The two sp orbitals of* Be; *the negative lobes are omitted*

(*b*) *The sp orbitals are colinear*

FIGURE 5.35

Electron-box description and VB structures of BeH_2 *and* BeF_2.

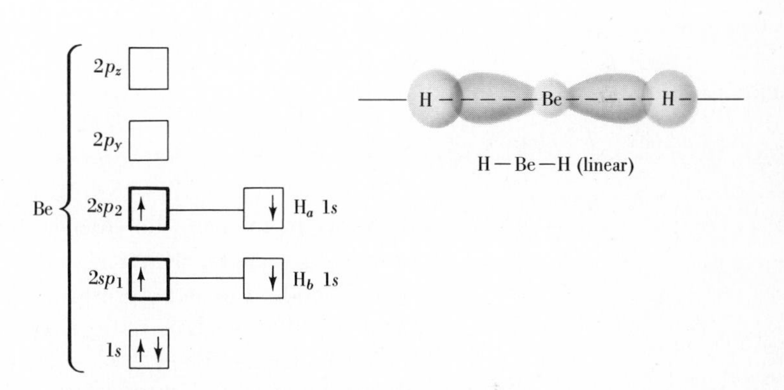

(*a*) *The VB models of* BeH_2 *based on sp hybridization*

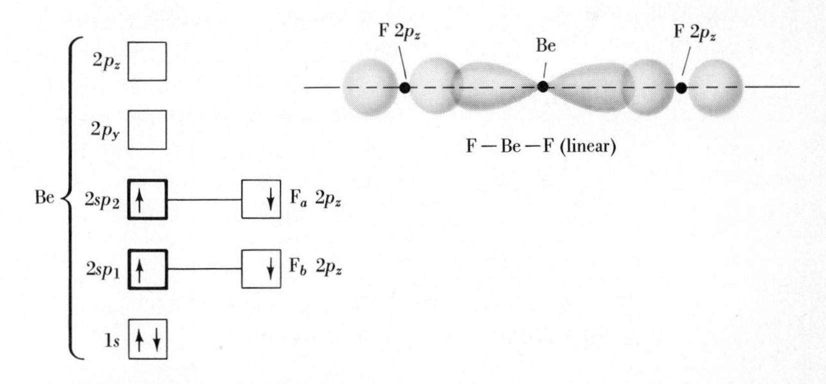

(*b*) *The VB models of* BeF_2 *based on sp hybridization*

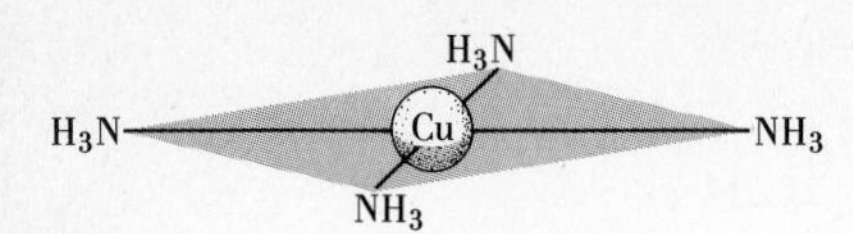

Square planar $Cu(NH_3)_4{}^{2+}$ *dsp^2 hybrid*

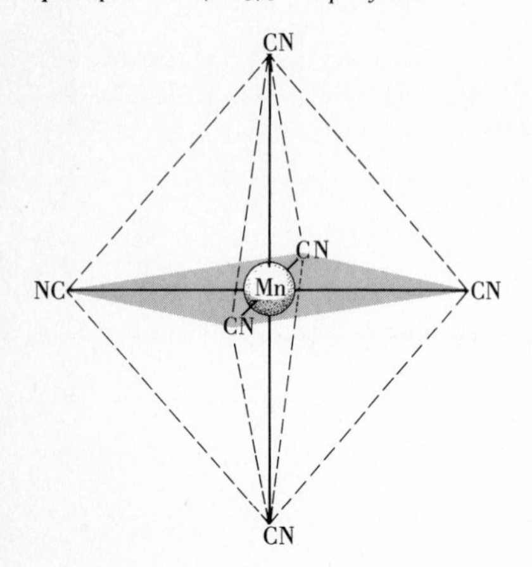

Octahedral $Mn(CN)_6{}^{3-}$ *d^2sp^3 hybrid*

TABLE 5.7: SUMMARY OR HYBRID ORBITALS

Hybrid	*Number of bonds*	*Molecular shape*	*Bond angle*	*Example*
sp^3	4	Tetrahedral	109°28′	CCl_4
sp^2	3	Trigonal (plane)	120°	BF_3
sp	2	Linear	180°	$MgCl_2(g)$
dsp^2	4	Square planar	90°	$Cu(NH_3)_4{}^{2+}$
d^2sp^3	6	Octahedral	90°	$Mn(CN)_6{}^{3-}$

Both BeH_2 and BeF_2 are predicted to be linear because of *sp* hybrid orbitals of the central beryllium atom. The VB structures for other alkaline-earth hydrides and halides are similar. For example, $MgCl_2$ involves hybridization of a 3*s* and a 3*p* orbital on the magnesium atom to form two half-filled 3*sp* hybrid orbitals. These orbitals overlap with the half-filled $3p_z$ orbitals of two chlorine atoms, and the linear Cl—Mg—Cl molecule is formed as shown in Fig. 5.36.

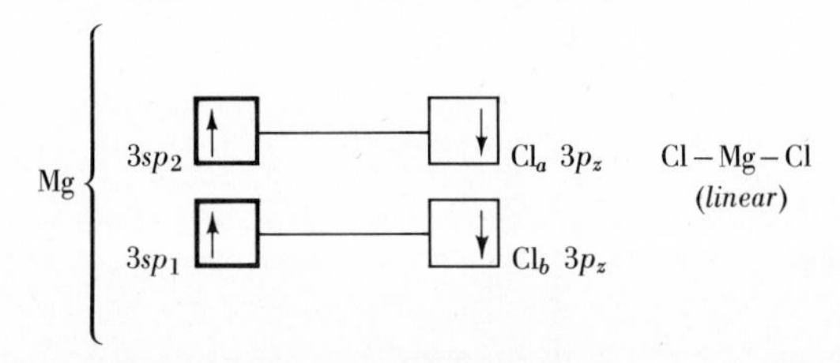

FIGURE 5.36

The electron-box description of $MgCl_2$.

As we have seen, the carbon atom exhibits sp^3 and sp^2 hybridization. In acetylene, HC≡CH, and other carbon compounds involving *triple* bonds, the structure of the molecules can be explained if we suppose that carbon undergoes *sp* hybridization. Acetylene is a linear molecule, H—C≡C—H, and whenever we have linear polyatomic molecules, we should suspect *sp* hybridization. The bonding details of acetylene are dealt with in Chap. 15.

We have discussed hybridization involving the *s* and *p* orbitals. This kind of hybridization is important in considering molecules of the lighter elements. With heavier elements, like the transition metals, the *d* orbitals are involved in the bonding. The VB theory as it applies to transition-metal ions and molecules is discussed in Chap. 19. A summary of some common hybrid types and the bonding and structural characteristics they confer on molecules is given in Table 5.7.

QUESTIONS AND PROBLEMS

5.1 What are the chief questions which must be answered by a good valence theory?

5.2 Organic chemists in the mid-nineteenth century began to write rational formulas to explain the chemical reactions of compounds. (*a*) How did these formulas suggest a definite atomic arrangement within a molecule? (*b*) Why did the early valence theory developed by the organic chemists fail for inorganic compounds?

5.3 When and why did scientists begin to suspect that the outer electrons of atoms are related to valence or bonding properties?

5.4 (*a*) What are the most important tenets of the Lewis-Langmuir valence theory? (*b*) What did Lewis mean by "the rule of two is even more fundamental than the rule of eight"? (*c*) In view of the MO theory (see MO discussion of O_2), is the rule of two as fundamental as Lewis believed?

5.5 Show Lewis structures for $MgCl_2$, RbI, H_2S, PCl_3, NH_2OH [H—N(—H)—OH], CH_3Cl, H_2SO_4 [$(HO)_2SO_2$], SO_4^{2-}, $S_2O_3^{2-}$ [$S—SO_3^{2-}$], $Al(OH)_3$, HNO_3 [$HONO_2$], NO_3^-, $GeCl_4$, $AsBr_3$, and NO_2. (*b*) Which structures require coordinate covalent bonds? (*c*) Which structures are paramagnetic? (*d*) Why is resonance required with the Lewis structures to explain that all the bonds in CO_3^{2-} are identical?

5.6 Use the electron-repulsion factor to predict the shape of (*a*) PCl_3, (*b*) H_2S, (*c*) $GeCl_4$, (*d*) SO_4^{2-}, (*e*) PO_4^{3-}, (*f*) ClO_4^-, and (*g*) ClO_3^-.

5.7 Using the electronegativities listed in Table 5.3, (*a*) arrange the following bonds in order of increasing polarity: Be—Cl, C—I, Ba—F, Al—Br, S—O, P—Cl, and C—O. (*b*) Are any of these bonds ionic?

5.8 Using the data in Table 5.3, (*a*) arrange the following bonds in order of increasing polarity. (*b*) Label the positive and negative ends as δ^+ and δ^-. (*c*) Which bonds are ionic according to the electronegativity data? (*d*) Are any of the bonds nonpolar? (1) P—S, (2) P—Cl, (3) Si—Cl, (4) Si—F, (5) K—F, (6) Na—Br, (7) Ge—O, (8) Li—S, (9) C—I.

5.9 Using Fig. 5.8, estimate what the boiling point and the melting point of water would be if hydrogen bonding were absent.

5.10 (*a*) What is the basic difference between the Lewis-Langmuir theory on one hand and the MO and VB theories on the other? (*b*) What is the difference between the MO theory and the VB theory? (*c*) Which of the newer theories is more similar to the Lewis-Langmuir theory? Why?

5.11 Show the Lewis structure, VB structure, and the MO structure of O_2 and NO. Discuss the merits of each in terms of the bond lengths and magnetic properties of O_2 and NO as described in this chapter.

5.12 Draw the Lewis electronic structure of the reactive free radicals (*a*) O—H, (*b*) CH_3, and (*c*) O=C—H, all of which have one unpaired electron.

5.13 The molecule N_2 is an extraordinarily stable one with a short bond length; the bonding in F_2 is relatively weak; and the Ne_2 molecule does not form. Using the diagrams in Fig. 5.22, show that the MO theory is in agreement with the properties of N_2, F_2, and "Ne_2" given in Table 5.5.

5.14 (*a*) Show the Lewis and VB structures for NH_3, PH_3, H_2O, and H_2S. (Do not use orbital hybridization.) (*b*) NH_3 and H_2O have bond angles of 107 and 105° respectively, whereas both PH_3 and H_2S have bond angles of 93°. Do you believe that the 3*s* orbitals have an important role in the bonding in PH_3 and H_2S? (*c*) Using Lewis structures and the electron-repulsion theory, predict the structure of $GeCl_4$. (*d*) Using VB theory, predict the structure of $GeCl_4$.

5.15 Using VB theory, explain that $AlCl_3$ is a planar structure possessing three equivalent Al—Cl bonds and 120° bond angles.

5.16 Give the Lewis and VB structures for MgF_2, a linear molecule.

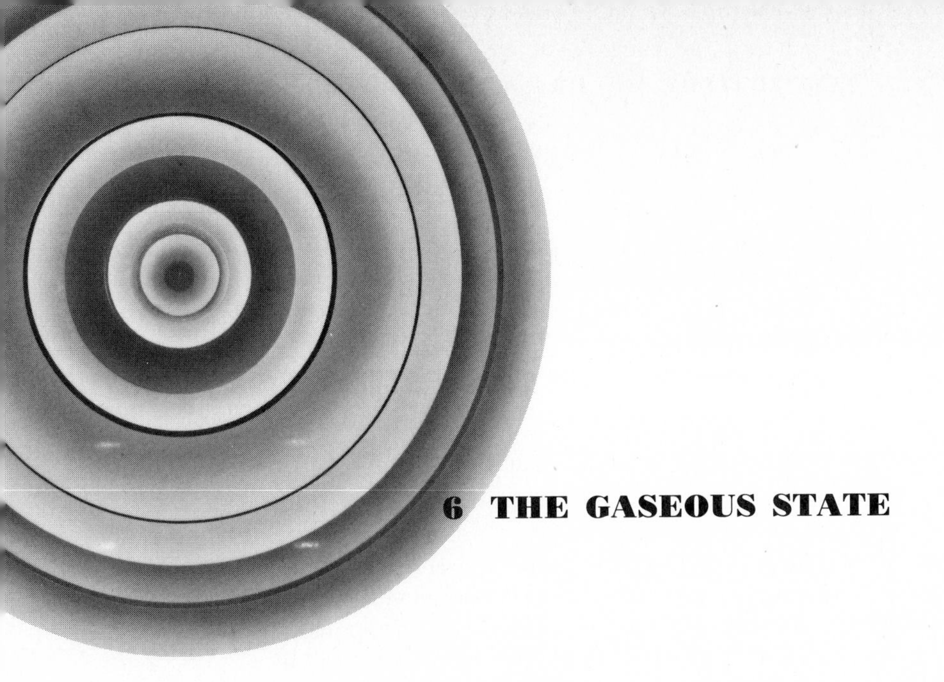

6 THE GASEOUS STATE

6.1 INTRODUCTION

At first it might seem impossible to account for the properties of matter in terms of atoms and molecules because any sample large enough to study contains an astronomical number of molecules. It is even less realistic to attempt to enumerate the number of ways the individual molecules can affect each other by their motions or by electrical and magnetic properties. Fortunately, we can resort to the theory of statistics to relate many of the properties of solids, liquids, and gases to the nature of the molecules and atoms. This has been a particularly fruitful approach in the development of a theory for gases because even the smallest sample of a gas represents an ideal statistical sample.

Since the gaseous state has been much more intensively studied and is better understood than the other two, we can make a very thorough study of gases even at this introductory level. A gas is a fluid which completely fills the vessel which contains it; it exerts an equal pressure on the entire surface of the container walls. Liquids and solids exhibit a very different behavior.

6.2 TEMPERATURE AND PRESSURE

The temperature and pressure of a gas are two of its most important properties. Not only are temperature and pressure easily measured, but they are directly related to other macroscopic and microscopic properties. For example, if oxygen gas is at 25°C and 760 Torr, we can calculate the density, a macroscopic property (1.31 g/l), and the average velocity of the molecules, a microscopic property (about 50,000 cm/s or 30 mi/s).

The Kelvin temperature scale is the most convenient for the gaseous state. The reason will become obvious when we discuss Charles' law.

The pressure of a gas may be expressed in several equivalent ways: pounds per square inch, inches of mercury, centimeters of mercury, Torr, and others. The Torr is a unit meaning millimeters of mercury. Before we can really appreciate the equivalence of these different pressure units, we must examine the definition of pressure. Pressure equals the *force* which acts on a surface divided by the area of the surface. It is the force per unit area, or

$$P = \frac{\text{force}}{\text{area}} = \frac{f}{A} \tag{6.1}$$

The *force* exerted on a body is the product of the mass of the body and its acceleration. This is simply a statement of Newton's second law of motion:

$$f = ma$$

The weight of a body is a *force* due to its acceleration in a gravitational field. The weight (or gravitational force) is calculated by multiplying the mass of a body by g, the acceleration due to gravity.

$$\begin{aligned}\text{Weight} &= \text{gravitational force}\\ &= \text{mass} \times \text{acceleration due to gravity} = mg \qquad (6.2)\end{aligned}$$

The last equation shows that weight and mass are not equal but are directly proportional. Hence, we often use the terms interchangeably. Combining Figs. 6.1 and 6.2, we find that the pressure exerted by a mass m on a surface of area A is mg/A:

$$P = \frac{mg}{A}$$

6.3 THE MEASUREMENT OF PRESSURE

The mercury barometer is a type of balance which permits us to measure the pressure of the air in the atmosphere. Suppose we clamp a piece of glass tube vertically in a beaker of mercury (Fig.

FIGURE 6.1

The barometer is a balance.

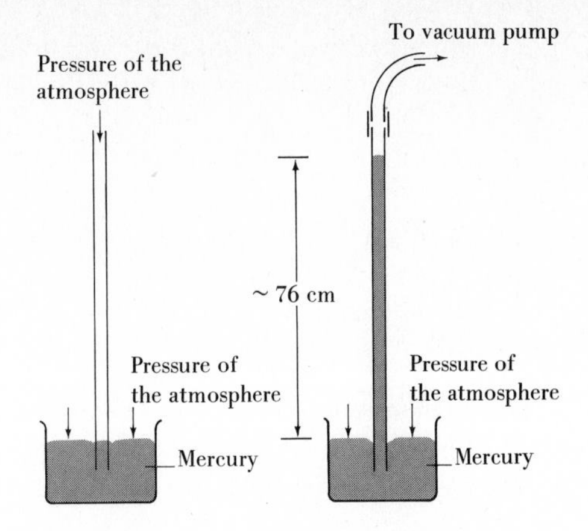

6.1*a*). The atmospheric gases act on the entire surface of the mercury, inside and outside the tube. The mercury levels inside and outside the tube are at the same height, and the system is *in balance*. Next, we connect the tube to a high-vacuum pump so that the atmospheric gases inside the tube can be exhausted. As we remove the air, the mercury rises inside the tube. The constant force of the atmosphere is acting on the surface of the mercury outside the tube as the force inside is being reduced. Finally, when all the air has been removed, the level of the mercury inside the tube rises to about 76 cm. At this point the pressure of the atmospheric gases on the surface of the mercury just equals the pressure caused by the column of mercury on the same surface. The two pressures are in balance, and the column can neither rise nor fall. Hence, by calculating the pressure caused by the column of mercury we also calculate the pressure of the atmosphere.

The pressure is the force of the mercury divided by the area over which the force acts. To calculate the force, we must determine the mass of the mercury and multiply by the acceleration due to gravity g:

$$f = ma = mg = \text{weight}$$

The mass can be calculated from the volume and density of the mercury:

$$m = vd$$

The volume of the mercury is just the volume of a cylindrical tube $\pi r^2 h$, and therefore the mass is

$$m = (\pi r^2 h)d$$

Hence, the force is

$$f = mg = [(\pi r^2 h)d]g$$

The area over which the force acts is just the cross-sectional area of the cylindrical tube πr^2. By Eq. (6.1) the pressure is

$$P = \frac{f}{A} = \frac{mg}{A} \tag{6.3a}$$

or

$$P = \frac{[(\pi r^2 h)d]g}{\pi r^2} \tag{6.3b}$$

Inspection of Eq. (6.3*a*) shows that the pressure is directly proportional to the mass per unit area since g, the acceleration due to gravity, is a constant:

$$P \propto \frac{m}{A} \qquad g = \text{const}$$

Therefore, we can report pressures in pounds per square inch or in kilograms per square meter. A mercury column 76 cm or 29.9 in. high and 1 in.^2 in cross-sectional area has a mass of 14.7 lb, so that the average pressure of the atmosphere is given as 14.7 lb/in.^2.

Equation (6.3*b*) is factorable since πr^2 appears in both numerator and denominator, and we can write

$$P = hdg \tag{6.4}$$

But the density of mercury d and the acceleration due to gravity g are constants; and pressure is directly proportional to the height of the column:

$$P \propto h$$

Therefore, we can report the pressure in inches of mercury, centimeters of mercury, Torr (mm Hg), or feet of water. It is conventional in reporting pressure in terms of the height of a mercury column to give the height which the column would reach if the mercury were at 0°C. This is necessary because the density of mercury, and hence the height of the column, depends on the temperature of the mercury. When necessary, the pressure in force per unit area can be calculated from the height of a mercury column or from a report in pounds per square inch.

The true units of pressure must be in terms of force and area. In the metric system force is expressed in dynes (or g-cm/s^2) and area in square centimeters. The pressure, in force per unit area, can be calculated from Eq. (6.4), $P = hdg$. For the experiment described above, $h = 76.00$ cm, the density of mercury at 0°C is 13.5955 g/cm^3 and $g = 980.2\ \text{cm/s}^2$. Hence,

$$P = (76.00\ \text{cm})(13.60\ \text{g/cm}^3)(980.2\ \text{cm/s}^2)$$

$$P = 1.013 \times 10^6 \frac{\text{cm}^2\ \text{g}}{\text{cm}^3\ \text{s}^2} = 1.013 \times 10^6 \frac{\text{g-cm}}{\text{s}^2} \frac{1}{\text{cm}^2}$$

$$P = 1.013 \times 10^6\ \text{dyn/cm}^2$$

At sea level, the average pressure of the atmosphere can support a mercury column 76 cm, or 760 mm, high. When the pressure of a gas is equivalent to 760 mm Hg, or 760 Torr, we say that the pressure is 1 atmosphere (atm); 380 Torr would be $\frac{1}{2}$ atm, 1520 Torr would be 2 atm, and so on. The unit of atmospheres for pressure is very convenient.

EXAMPLE 6.1 The gas pressure inside a reaction vessel is 200 Torr. Calculate the pressure in (*a*) dynes per square centimeter, (*b*) pounds per square foot, and (*c*) atmospheres.

Solution: (*a*) Using Eq. (6.4) we can calculate the pressure in dynes per square centimeter:

$$P = hdg = (20.0\ \text{cm})(13.60\ \text{g/cm}^3)(980.2\ \text{cm/s}^2)$$
$$P = 2.66 \times 10^5\ \text{dyn/cm}^2$$

We can also solve this problem by converting 200 Torr to dynes per square centimeter using the fact that $1.013 \times 10^6\ \text{dyn/cm}^2 = 760$ Torr:

$$200\ \text{Torr} \times \frac{1.013 \times 10^6\ \text{dyn/cm}^2}{760\ \text{Torr}} = 2.66 \times 10^5\ \text{dyn/cm}^2 = P$$

(*b*) A pressure of 760 Torr is equivalent to 14.7 lb/in^2:

$$200\ \text{Torr} \times \frac{14.7\ \text{lb/ft}^2}{760\ \text{Torr}} = 3.87\ \text{lb/in}^2 = P$$

(*c*) A pressure of 760 Torr is equivalent to 1.00 atm:

$$200\ \text{Torr} \times \frac{1\ \text{atm}}{760\ \text{Torr}} = 0.263\ \text{atm} = P$$

6.4 THE BEHAVIOR OF GASES UNDER ORDINARY CONDITIONS

Several experimental laws involving the temperature, pressure, and volume of a gas can be succinctly stated in simple mathematical form. These laws involve only macroscopic properties. Later in the chapter we shall find a microscopic or molecular basis for these experimental laws.

Boyle's law

In the latter part of the seventeenth century an English chemist, Robert Boyle (1627–1691), and his friend Robert Hooke (1635–1703) carried out experiments with an air pump which led them to dis-

FIGURE 6.2

Boyle's law behavior.

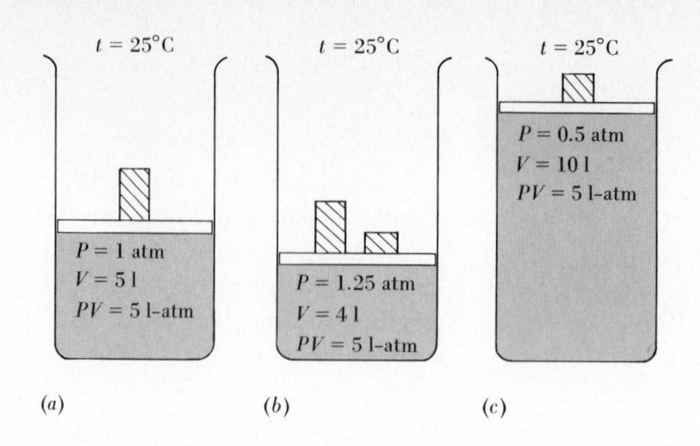

cover an important, simple relationship: *the product of the pressure and volume of a fixed amount of gas is a constant provided that the temperature of the gas is not permitted to change*. This observation, called *Boyle's law*, can be stated mathematically:

$$PV = k_B \qquad T = \text{const} \tag{6.5}$$

where k_B is a constant. Equation (6.5) holds when the temperature and the mass of the gas are fixed.

Figure 6.2 depicts Boyle's law behavior. A cylinder containing a fixed mass of gas is thermostated at 25°C. The volume and pressure of the gas are initially 5 l and 1 atm. When the pressure is increased to 1.25 atm, the volume decreases to 4 l; when the pressure is decreased to 0.5 atm, the volume increases to 10 l. The PV product remains constant at 5 l-atm. At relatively low pressures ($P \leqslant 10$ atm) and high temperatures ($T \geqslant 200$ K), the PV product remains constant if the temperature is constant. The PV product depends only on the gas temperature; in fact, the PV product is directly proportional to the temperature. Figure 6.3*a* shows that PV remains constant as long as the temperature does not change, and Fig. 6.3*b* shows that PV is a linear function of the temperature. This observation is very important in the molecular theory of gases dealt with in detail later.

Charles' law

Air and all other gases undergo large volume changes when they are heated at constant pressure. Charles and Gay-Lussac made quantitative studies of this behavior and discovered what is now generally called *Charles' law*. Solids and liquids undergo such small changes in volume that thermal expansion in these cases is often neglected. Table 6.1 compares the volume expansions of 1.00-ml samples of air, water, and iron when they are heated from 0 to 100°C at 1 atm pressure.

FIGURE 6.3

The behavior of the PV product for 1 mol of gas.

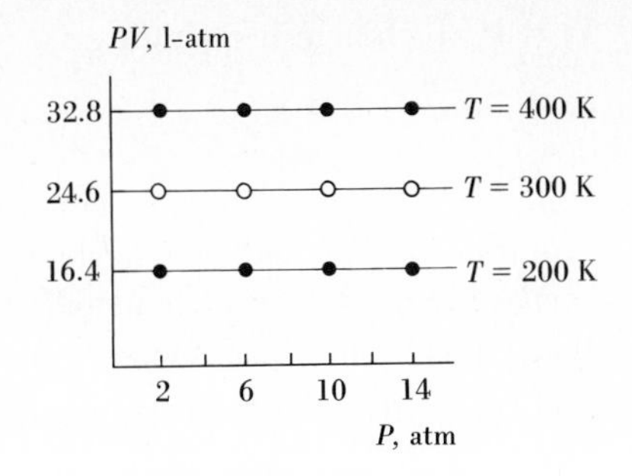

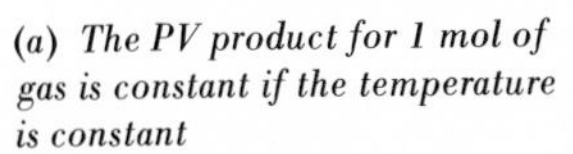

(a) The PV product for 1 mol of gas is constant if the temperature is constant

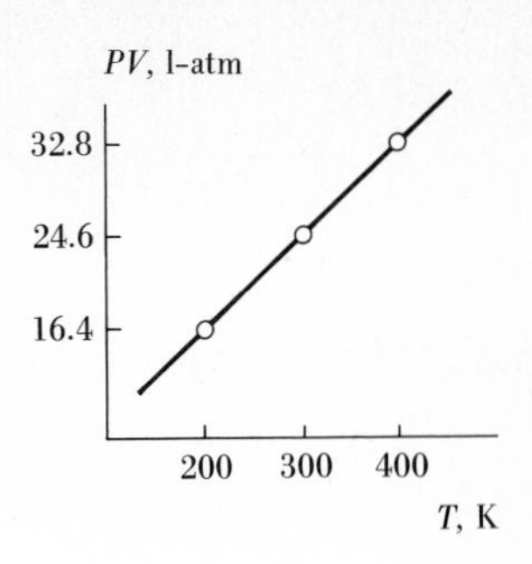

(b) The PV product for 1 mol of gas is a linear function of the temperature

TABLE 6.1: THE EFFECT OF TEMPERATURE ON THE VOLUME OF DIFFERENT SUBSTANCES

	Volume, ml		*Volume change*	
	At 0°C	*At 100°C*	ml	%
Air	1.00	1.37	0.37	37
Water	1.00	1.04	0.04	4
Iron	1.00	1.03	0.03	3

Let us examine the effect of temperature on the volume of a gas, like nitrogen, enclosed in a cylinder by means of a free-riding piston which allows us to maintain a constant pressure. If we start at 0°C with an original volume of gas designated as V_0, we find that this volume increases linearly with temperature, or

$$V \propto t$$

For each 1°C rise in temperature the volume of the gas increases by the amount $\frac{1}{273}V_0$, provided the pressure remains constant (see Fig. 6.4). Then the volume V_t at any temperature t is just the volume V_0 at 0°C plus the change in volume which occurs when the gas goes from 0°C to the temperature t:

FIGURE 6.4

Effect of temperature on gas volume.

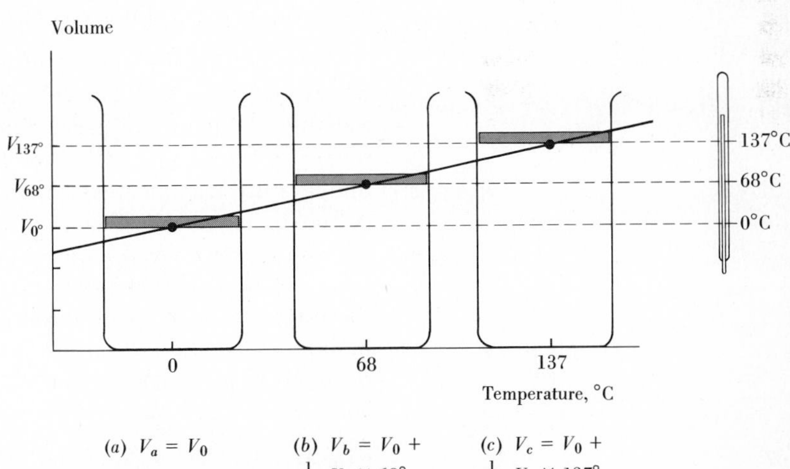

(a) $V_a = V_0$ (b) $V_b = V_0 + \frac{1}{273} V_0 \times 68°$ (c) $V_c = V_0 + \frac{1}{273} V_0 \times 137°$

$$V_t = V_0 + \text{change in volume}$$

or

$$V_t = V_0 + \tfrac{1}{273} V_0 t \tag{6.6}$$

Equation (6.6) is called Charles' law. The fraction $\frac{1}{273}$, the *cubical coefficient of expansion,* is usually represented by the Greek letter α. We can rewrite Eq. (6.6) as

$$V_t = V_0 + \alpha V_0 t$$

or

$$V_t = V_0(1 + \alpha t) \tag{6.7}$$

Many different gases, for example, O_2, N_2, He, NH_3, CO_2, SO_2, Ar, and others, have approximately the same value of α ($\sim \frac{1}{273}$).

EXAMPLE 6.2 A 150-ml sample of a gas is heated at constant pressure of from 0 to 25°C. What is the volume at 25°C?

Solution: $V_0 = 150$ ml, and $t = 25$°C. Substitution into Eq. (6.6) gives the volume at 25°C:

$$V_{25^\circ} = V_0 + \tfrac{1}{273} V_0 t = 150 + (\tfrac{1}{273} \times 150 \times 25) = 164 \text{ ml}$$

Since the volume of a gas is directly proportional to the temperature, as the temperature is decreased the volume decreases linearly. The absolute zero of temperature, the lowest conceivable temperature, is defined as the temperature at which the volume of a gas would reach zero. This temperature can be calculated by extrapolation of a plot of V versus t to $V = 0$. Alternatively, we can obtain this temperature from Eq. (6.7) by substituting $V_t = 0$ and solving for t:

$$0 = V_0(1 + \tfrac{1}{273} t)$$
$$t = -273^\circ\text{C} = \text{absolute zero of temperature}$$

Theoretically, the volume of an ordinary gas would reach a value of zero at −273°C,† but this temperature is experimentally unattainable. Actually, all gases liquefy before reaching this temperature, and furthermore Charles' law [Eq. (6.4)] is not obeyed as the temperature becomes extremely low.

Looking at Charles' law from a different point of view, we see that the volume of a fixed amount of gas is a measure of the temperature; this is indicated by Fig. 6.4: the volume of a gas V_t determines the temperature t. In fact, the scientists at the National Bureau of Standards use the volume of a fixed amount of gas as a primary thermometer to establish a standard temperature scale. The readings of

† The exact value of the absolute of zero is −273.16°C.

other thermometers are corrected by comparison with the National Bureau of Standards primary thermometer.

Equation (6.6) is simplified by introducing the Kelvin, or absolute temperature, scale. The Kelvin temperature is represented by T whereas the Celsius temperature is represented by t. The relationship between the two temperature scales is

$$T = t + 273 \qquad \text{or} \qquad t = T - 273 \tag{6.8}$$

When Eq. (6.8) is substituted into Eq. (6.6), we have

$$V = V_0 + \tfrac{1}{273} V_0 (T - 273)$$

or

$$V = V_0 + \tfrac{1}{273} V_0 T - \tfrac{273}{273} V_0$$

or

$$V = \frac{V_0}{273} T \tag{6.9}$$

Equation (6.9) shows that the *volume of a fixed mass of gas at constant pressure is directly proportional to the absolute temperature* and that the volume of a gas at the absolute zero should be zero. This is the conventional statement of Charles' law. It is usually written simply as

$$V = k_c T \qquad P = \text{const} \tag{6.10}$$

where $k_c = V_0/273$ in Eq. (6.9).

Dalton's law of partial pressures

One of John Dalton's chief interests was climate and the nature of the atmosphere. For 57 years he kept a notebook, "Observations on the Weather," in which he made daily entries. The notebook included descriptions of meteorological instruments: thermometers, barometers, and dew-point apparatus. His consideration of the composition of the atmosphere led him to his version of the atomic theory, which we have already discussed. In addition, his study of the variation of the pressure of water vapor in the atmosphere permitted him to discover a simple *empirical law* involving the total pressure of a gaseous mixture. He found that the water-vapor saturation point (dew point) of air depends only on temperature. At 25°C the highest pressure that water vapor can reach is approximately 24 Torr. This means that if water is allowed to evaporate in a closed vessel, the pressure of water vapor will reach a maximum pressure of 24 Torr at 25°C. If air at a pressure of 760 Torr were initially present in the vessel, the final pressure would be 784 Torr (760 + 24 Torr). If the vessel had been evacuated by means of a pump so that the air pressure was virtually 0 Torr, the final pressure would equal the pressure due to the water vapor only (24 Torr). Dalton substituted other gases and found the same result. If a closed vessel contained nitrogen gas at 500 Torr, after water was placed in it, the pressure rose to 524 Torr at 25°C. On the basis of these studies he concluded that *the total pressure of a gaseous mixture is equal to the sum of the pressures that the individual gases would exert.*

If we have three gas-filled 1-l vessels all at the same temperature, one containing N_2 gas at 0.5 atm, one containing O_2 gas at 0.75 atm, and the third containing H_2 gas at 0.1 atm, and we then force the three gases into a single 1-l vessel, the total pressure is 1.35 atm. This is just the sum of the pressures of the individual isolated gases at the same temperature and volume. The pressure of an individual gas in a mixture of gases is called the *partial pressure. Dalton's law of partial pressures* can be stated succinctly: *the total pressure P_T of a gaseous mixture is equal to the sum of the partial pressures,* or mathematically, if there are n gases in the mixture:

$$P_T = P_1 + P_2 \cdot \cdot \cdot + P_n = \sum_{i=1}^{i=n} P_i \tag{6.11}$$

We have discussed three simple experimental laws which gases obey under ordinary conditions. In addition, we have given each law a mathematical formulation. This is extremely important in the development of any science. We shall now show how the gas laws are used in calculations.

EXAMPLE 6.3 A 5.00-l sample of oxygen at 25°C is enclosed in a cylinder by a free-riding piston. The pressure is *isothermally* decreased from 10.0 to 3.00 atm. What is the final volume?

Solution 1: Since the process occurs *isothermally,* i.e., at constant temperature, Boyle's law holds:

$$PV = k_B \qquad T = \text{const}$$

Therefore, the initial value of PV equals the final value of PV, or

$$P_1V_1 = k_B = P_2V_2$$
$$(10.0 \text{ atm})(5.00 \text{ l}) = (3.00 \text{ atm})(V_2)$$
$$V_2 = 16.7 \text{ l} = \text{final volume}$$

Solution 2: Another way of considering this problem is to keep in mind that the volume of a gas is inversely proportional to its pressure, $V = k_B/P$; a decrease in pressure must be accompanied by an increase in volume if the temperature is constant. *We can simply multiply the initial volume by a ratio of the pressures; the ratio must be greater than 1* since the final volume must be greater than the initial volume:

$$V_2 = 5.00 \text{ l} \times \frac{10.0 \text{ atm}}{3.00 \text{ atm}} = 16.7 \text{ l}$$

EXAMPLE 6.4 A 2.00-l sample of air initially at 1 atm and 25°C is heated *isobarically* until its temperature is 100°C. Calculate the final volume.

Solution 1: Since a fixed mass of gas is heated *isobarically*, i.e., at constant pressure, Charles' law holds:

$$V = k_c T \quad \text{when} \quad P = \text{const}$$

where T is absolute temperature, or

$$\frac{V}{T} = k_c$$

Therefore, the initial value of V/T equals the final value of V/T, or

$$\frac{V_1}{T_1} = k_c = \frac{V_2}{T_2}$$

$$\frac{2.00 \text{ l}}{273 + 25 \text{ K}} = \frac{V_2}{273 + 100 \text{ K}}$$

$$V_2 = 2.50 \text{ l}$$

Solution 2: This problem is solved in another way by considering the effect of increasing temperature on the volume of a gas: the volume increases in direct proportion to the increase in temperature. Since the temperature of the gas is increased, we need only multiply the initial volume by a ratio of temperatures which is greater than unity:

$$V_2 = 2.00 \text{ l} \times \frac{373 \text{ K}}{298 \text{ K}} = 2.50 \text{ l}$$

EXAMPLE 6.5 A 10.0-l sample of H_2 is initially at 100°C and 1.00-atm pressure. After cooling and compression, it is at 0°C and 5.00 atm. Calculate the final volume. (Charles' and Boyle's laws may be applied separately.)

Solution: The temperature decrease will cause a volume decrease in direct proportion. The pressure increase will cause a volume decrease in inverse proportion. We must multiply the initial volume by a ratio of temperatures less than unity and by a ratio of pressures less than unity:

$$V_2 = 10.0 \text{ l} \times \frac{273 \text{ K}}{373 \text{ K}} \times \frac{1.00 \text{ atm}}{5.00 \text{ atm}} = 1.46 \text{ l}$$

EXAMPLE 6.6 A 1.00-l vessel containing argon at 25°C and 2.50 atm is connected to a 2.00-l vessel containing helium at 3.00 atm and 25°C. If the two gases are allowed to mix isothermally, what are the final partial pressures of each gas and what is the total pressure of the mixture?

Solution: The final volume for each gas will be 3.00 l since each gas will have access to the 1.00-l volume originally occupied only by argon and to the 2.00-l volume originally occupied only by helium.

The gases will expand isothermally until each has a volume of 3.00 l. Therefore, Boyle's law holds for each.

$$PV = k_B$$

Since the volume of each gas increases, the pressure of each gas must decrease. We must multiply the initial pressures by volume ratios less than unity.

For argon the volume increases from 1.00 to 3.00 l:

$$\text{Partial pressure of Ar} = P_{\text{Ar}} = 2.50 \text{ atm} \times \frac{1 \text{ l}}{3 \text{ l}} = 0.83 \text{ atm}$$

For helium, the volume increases from 2.00 to 3.00 l:

$$P_{\text{He}} = 3.00 \text{ atm} \times \frac{2 \text{ l}}{3 \text{ l}} = 2.00 \text{ atm}$$

By Dalton's law the total pressure is the sum of the partial pressures:

$$P_{\text{total}} = P_{\text{Ar}} + P_{\text{He}} = 0.83 \text{ atm} + 2.00 \text{ atm} = 2.83 \text{ atm}$$

EXAMPLE 6.7 Hydrogen gas is generated by the action of HCl on Zn:

$$Zn^0 + 2HCl \rightarrow ZnCl_2 + H_2(g)$$

As the gas is evolved, it is trapped in a bottle inverted over water (see Fig. 6.5). At the end of the experiment the volume of the gas in the bottle is 523 ml, the pressure is 752 Torr, and the temperature is 20°C. The H_2 gas is saturated with water vapor from the trough. What volume would the *dry hydrogen* occupy at 0°C and 1-atm pressure?

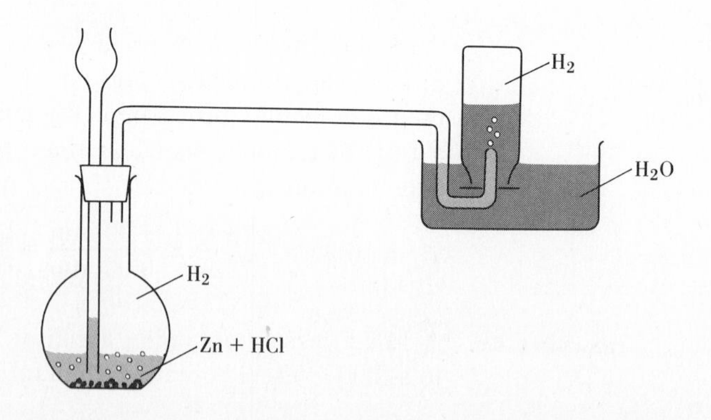

FIGURE 6.5

Collection of hydrogen.

Solution: The gas in the bottle is a mixture of H_2 and water vapor. The maximum pressure of water vapor depends only on the temperature. At 20°C the saturation pressure of water vapor is 17.4 Torr (see Table B.4 in Appendix B). Using Dalton's law, we can calculate the partial pressure of H_2:

$$P_{\text{total}} = P_{H_2} + P_{H_2O}$$
$$P_{H_2} = P_{\text{total}} - P_{H_2O} = 752 \text{ Torr} - 17.4 \text{ Torr} = 735 \text{ Torr}$$

Hence, we have a 523-ml sample of H_2 gas at a pressure of 735 Torr and 293 K. We are interested in the volume of H_2 at 760 Torr and 273 K. Taking the gas from 735 to 760 Torr increases its pressure and will cause V to decrease. We must multiply the volume, 523 ml, by a ratio of pressures less than unity. Cooling the gas from 293 to 273 K will cause V to decrease, and we must multiply the volume by a ratio of temperatures less than unity.

$$V_2 = 523 \text{ ml} \times \frac{735 \text{ Torr}}{760 \text{ Torr}} \times \frac{273 \text{ K}}{293 \text{ K}}$$

$= 472$ ml = volume of dry hydrogen at 0°C and 1 atm

6.5 THE IDEAL-GAS LAW

Charles' and Boyle's laws can be combined to give an equation which describes the interdependence of the pressure, the volume, and the temperature of a given mass of gas. This equation, called the *ideal-gas law*, shows that the quotient PV/T is a constant:

$$\frac{PV}{T} = R \qquad R = \text{const} \tag{6.12}$$

Avogadro's hypothesis suggests a way of evaluating the constant R. According to Avogadro's hypothesis, equal volumes of gases at the same temperature and pressure contain the same number of molecules or moles. The temperature 0°C and the pressure 1 atm have been designated as *standard temperature and pressure* (STP) for gases. Under these conditions a 22.414-l sample of a gas contains exactly 1 mol. From this observation R can be calculated. With $T = 273.16$ K, $P = 1$ atm, and $V = 22.414$ l, R will be 0.08205 l-atm/K:

$$\frac{PV}{T} = R = 1 \text{ atm} \times \frac{22.414 \text{ l}}{273.16 \text{ K}} = 0.08205 \text{ l-atm/K}$$

Obviously, the value of R will depend on the mass or number of moles of the gas; 44.828 l of gas at STP would contain 2 mol, and PV/T would equal 2×0.08205; 11.207 l at STP would contain $\frac{1}{2}$ mol, and PV/T would equal $\frac{1}{2} \times 0.08205$. But *$PV/T$ divided by the number*

TABLE 6.2: VALUE OF PV/nT FOR DRY AIR†

n, mol	P, atm	T, K	V, l	$\frac{PV}{nT}$
0.10	1.0	273	0.224	0.082
0.10	1.0	303	0.246	0.081
0.10	1.0	373	0.310	0.083
0.22	1.0	273	0.510	0.085
0.22	1.5	303	0.360	0.081
0.22	2.0	373	0.336	0.082
			Average	0.082

† Data taken from a freshman chemistry laboratory report at Drexel University, December 1968.

of moles is a constant and equal to 0.08205 l-atm/mol-K, as demonstrated by the data in Table 6.2.

We can incorporate n, the number of moles, into Eq. (6.12), and we have the most general form of the ideal-gas law:

$$\frac{PV}{nT} = 0.08205 \text{ l-atm/mol-K}$$

or

$$PV = 0.08205nT$$

or

$$PV = nRT \tag{6.13}$$

R equals 0.08205 l-atm/mol-K and is called the *gas constant*. In most of our calculations we shall use 0.082 for R.

Very precise and accurate measurements show that no gas obeys the ideal-gas law exactly, but the deviations from it under ordinary atmospheric conditions are usually negligible. As the temperature of a gas increases and its pressure decreases, gas behavior becomes more accurately described by the ideal-gas law. The hypothetical gas which would obey this law under all conditions is called the *ideal gas*. It is a fruitful concept; deviations from ideal-gas behavior can be interpreted in terms of the interactions between molecules, as we shall see later in this chapter.

We shall now derive Eq. (6.12) from Boyle's law and Charles' law, demonstrating that the ideal-gas law is based on the experiments of Charles and Boyle.

Suppose that we have a fixed amount of gas in an initial state A, where the pressure, volume, and temperature are P_A, V_A, and T_A.

Then the gas undergoes a change of conditions so that it is in a final state C, where the pressure, volume, and temperature are P_C, V_C, and T_C. If Eq. (6.12) holds, PV/T is a constant for a fixed amount of gas. No matter what the conditions are, the value of PV/T cannot change. It should have the same values in states A and C, and we should be able to show that

$$\frac{P_A V_A}{T_A} = \frac{P_C V_C}{T_C} \tag{6.14}$$

To derive Eq. (6.12) imagine that the process which takes the gas from state A to state C occurs in two steps (see Fig. 6.6). The first step is isothermal, stopping at an intermediate state B. In this first step we change the pressure until it reaches the final value P_C. In the second step we keep the pressure constant and adjust the temperature until it reaches the final value T_C. Boyle's law applies to step 1:

$$P_A V_A = P_B V_B \tag{6.15}$$

Charles' law applies to step 2:

$$\frac{V_B}{T_B} = \frac{V_C}{T_C}$$

or (6.16)

$$V_B = \frac{T_B V_C}{T_C}$$

In Eq. (6.15), P_B is the pressure of the gas in the intermediate state B; but this is also equal to the final pressure since the second step is isobaric

$$P_B = P_C \tag{6.17}$$

In Eq. (6.16), T_B can be replaced by T_A since the first step, connecting states A and B, is isothermal: $T_A = T_B$. Rewriting Eq. (6.15) with $P_B = P_C$, we have

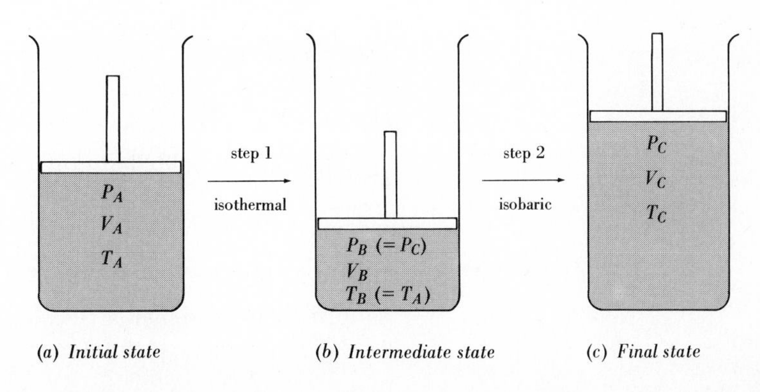

FIGURE 6.6

The quotient PV/T is a constant when a gas obeys Boyle's and Charles' laws.

$$P_A V_A = P_C V_B \tag{6.18}$$

and rewriting (6.16) with $T_A = T_B$, we have

$$V_B = \frac{T_A V_C}{T_C} \tag{6.19}$$

Substituting the value of V_B from Eq. (6.19) into (6.18) gives

$$P_A V_A = P_C \frac{T_A V_C}{T_C}$$

or

$$\frac{P_A V_A}{T_A} = \frac{P_C V_C}{T_C} \tag{6.14}$$

The last equation is the result we sought. It substantiates the fact that for gases which obey Boyle's and Charles' laws PV/T is a constant. Such gases are *ideal gases*.

Now, we shall analyze the ideal-gas law, Eq. (6.13), and show that it reduces to the experimental equations from which it was derived. According to Eq. (6.13) when the mass or number of moles n of gas and the temperature T are fixed, PV must be a constant in agreement with Boyle's law. Also, if P and n are fixed, V is proportional to T; $V = (nR/P)T = (\text{const})T$, in agreement with Charles' law. In addition, the ideal-gas law indicates that the pressure should be directly proportional to the temperature at constant volume. We did not use this condition in proving Eq. (6.14) so that we might look upon this as a theoretical prediction of the ideal-gas law; this prediction is verified by experiment.

The ideal-gas law can be used to solve any problem for which we previously used the separate gas laws, and in most cases this is the recommended method for gas-law problems.

EXAMPLE 6.8 What is the pressure in a 5.00-l vessel which contains 3.00 mol of helium at $-30°C$?

Solution: According to the ideal-gas law [Eq. (6.13)],

$$P = \frac{nRT}{V}$$

Hence,

$$P = \frac{3.00 \text{ mol} \times 0.082 \text{ l-atm/mol-K} \times 243 \text{ K}}{5.00 \text{ l}} = 11.9 \text{ atm}$$

EXAMPLE 6.9 Solve Example 6.5 using the ideal-gas law.

Solution: The problem is to calculate a final volume from the ideal-gas law:

$$V = \frac{nRT}{P}$$

We know that the final temperature is 0°C and the final pressure is 5 atm. To calculate the final volume we must also know n, the number of moles. Since n does not change, we can calculate it from the initial conditions, $P_i = 1.00$ atm, $V_i = 10.0$ l, and $T_i = 373$ K.

$$n = \frac{PV}{RT}$$

$$= \frac{1 \text{ atm} \times 10 \text{ l}}{0.082 \text{ l-atm/mol-K} \times 273 \text{ deg}} = \frac{10}{0.082 \times 373} \text{ mol}$$

The final volume is therefore

$$V_f = n \frac{RT_f}{P_f}$$

$$= \frac{10 \text{ mol}}{0.082 \times 373} \times \frac{0.082 \text{ l-atm/mol-K} \times 273 \text{ K}}{5 \text{ atm}} = 1.46 \text{ l}$$

(Compare with Example 6.5.)

Note that problem solution with the ideal-gas law is straightforward and usually simpler. Furthermore, it keeps us aware of the interdependence of P, V, T, and n.

EXAMPLE 6.10 Calculate the number of molecules (not moles) of carbon dioxide present in a 2.00-l sample at 5°C and 25 Torr.

Solution 1: If we can calculate the number of moles, we can calculate the number of molecules because each mole consists of 6.023×10^{23} molecules. By the ideal-gas law

$$n = \frac{PV}{RT} = \frac{25/760 \text{ atm} \times 2.00 \text{ l}}{0.082 \text{ l-atm/mol-K} \times 278 \text{ K}}$$

$$= 0.00289 \text{ mol}$$

Note that the pressure in atmospheres must be used in the ideal-gas law because R was determined by allowing P to equal 1 atm and R has the units of liter-atmospheres per mole-kelvin. The number of molecules n' is calculated by multiplying the number of moles n by Avogadro's number N_A:

$$n' = N_A n = 6.02 \times 10^{23} \text{ molecules/mol} \times 0.00289 \text{ mol}$$

$$= 1.74 \times 10^{21} \text{ molecules}$$

Solution 2: The problem can be solved in a single step by introducing $n' = N_A n$ directly into the ideal-gas law:

$$n' = N_A n = \frac{N_A PV}{RT}$$

6.6 USING THE GAS LAWS TO DETERMINE MOLECULAR WEIGHT

One of the most convenient methods of determining the molecular weights of substances which can be studied as gases involves the use of the gas laws. We have learned that 1 mol of a gas occupies 22.414 l at STP. Hence, if we can measure or calculate the weight of 22.4 l of a gas at STP, we have determined the molecular weight of the gas.

EXAMPLE 6.11 A sample of a hydrocarbon vapor weighing 7.53 g occupies 2.31 l at STP. Calculate the molecular weight of the hydrocarbon.

Solution: We must determine the weight of 22.4 l at STP. We know the weight of 2.31 l. The problem involves a simple conversion:

$$\frac{7.53 \text{ g}}{2.31 \text{ l}} \times 22.4 \text{ l/mol} = 73.0 \text{ g/mol}$$

Gram-molecular weight = 73.0 g/mol

EXAMPLE 6.12 A gas sample having a volume of 2.55 l at 25°C and 873 Torr weighs 6.05 g. Calculate the molecular weight of the gas.

Solution: First determine the volume at STP. Since we must cool the gas to take it to standard temperature, its volume will decrease and we must multiply the given volume by a temperature ratio less than unity; and since we must reduce the pressure to take the gas to standard pressure, its volume will increase and we must multiply by a pressure ratio greater than unity:

$$\text{Volume at STP} = 2.55 \text{ l} \times \frac{273 \text{ K}}{298 \text{ K}} \times \frac{873 \text{ Torr}}{760 \text{ Torr}} = 2.66 \text{ l}$$

Now we know that 6.05 g of the gas would occupy 2.66 l at STP. We must calculate the weight of 22.4 l just as we did in Example 6.11:

$$\frac{6.05 \text{ g}}{2.66 \text{ l}} \times \frac{22.4 \text{ l}}{1 \text{ mol}} = 50.5 \text{ g/mol}$$

Gram-molecular weight = 50.5 g/mol

Examples 6.11 and 6.12 could have been solved using the ideal-gas law. We know the mass of the gaseous sample; with the ideal-gas law we can calculate the number of moles, since P, V, and T are given. Then the molecular weight is obtained by dividing the mass of the gas by the number of moles.

EXAMPLE 6.13 Solve Example 6.12 using the ideal-gas law.

Solution: Calculate the number of moles n.

$$n = \frac{PV}{RT} = \frac{873/760 \text{ atm} \times 2.55 \text{ l}}{0.082 \text{ l-atm/mol-K} \times 298 \text{ K}} = 0.120 \text{ mol}$$

(Note that the pressure was converted to atmospheres.)

Now we know that 6.05 g is equal to 0.120 mol. The molecular weight is just equal to the mass divided by the number of moles:

$$\text{Molecular weight} = \frac{\text{mass}}{\text{number of moles}}$$

$$M = \frac{g}{n} = \frac{6.05 \text{ g}}{0.120 \text{ mol}} = 50.5 \text{ g/mol}$$

EXAMPLE 6.14 The density of chlorine gas at 20°C and 1 atm is 2.96 g/l. Calculate the molecular weight of chlorine.

Solution 1: The density is the mass of gas in 1 l. From the ideal-gas law we can calculate the number of moles in 1 l since we know T and P:

$$n = \frac{PV}{RT} = \frac{1 \text{ atm} \times 1 \text{ l}}{0.082 \text{ l-atm/mol-K} \times 293 \text{ K}} = 0.0414 \text{ mol}$$

The molecular weight is obtained by dividing the mass of 1 l, 2.96 g, by the number of moles in 1 l:

$$M = \frac{g}{n} = \frac{2.96 \text{ g/l}}{0.0414 \text{ mol/l}} = 71 \text{ g/mol}$$

Solution 2: We can substitute the definition for mole, $n = g/M$, into the ideal-gas law and solve directly for M:

$$PV = nRT = \frac{g}{M} RT$$

Rearrangement gives

$$M = \frac{g}{V} \frac{RT}{P}$$

where g/V equals the density d in grams per liter. Finally,

$$M = d\frac{RT}{P} \tag{6.20}$$

The information in Example 6.12 can be substituted into Eq. (6.20):

$$M = 2.96 \text{ g/l} \times \frac{0.082 \text{ l-atm/mol-K} \times 293 \text{ K}}{1 \text{ atm}} = 71 \text{ g/mol}$$

Equation (6.20) is very convenient for calculating the molecular weight or density of a gas.

6.7 A MOLECULAR MODEL OF THE IDEAL GAS

The smallest sample of a gas we can study experimentally contains an astronomical number of molecules. For instance, 1 ml of air, which is a small volume, contains about 2×10^{19} molecules. We believe that the observable behavior of a gas depends on the way these invisible molecules interact. Our next task will be to interpret the behavior of real gases and thereby obtain a molecular description or model for an ideal gas.

Molecular implications from gas pressure

Experiments show that the pressure is the same at every point within a gaseous sample.† If we believe that gases are ultimately composed of molecules, we must conclude that the molecules are in rapid, random motion and that they continually collide with each other and with the container walls. The collisions with the container walls cause the observed pressure of a gas. If we place a gas in an insulated vessel, so that the temperature remains constant, it will be found that the pressure will also remain constant. This very fact, coupled with our hypothesis that the molecules are continually colliding, leads us to an important conclusion about the average velocity of gaseous molecules. An individual molecule may experience a change in velocity upon collision, but the average velocity of all the molecules must remain constant. If, after a two-molecule collision, one molecule has experienced an increase in velocity, the other molecule must experience a decrease in velocity. If this were not true, the average velocity of the molecules would change; the force and frequency of the molecular collisions would change, and this in turn would cause a change in pressure, which is not observed.

Average velocity and kinetic energy

When we say that the average velocity of the molecules of a gas remains constant, we do not mean that all molecules have the same velocity. This would be inconsistent with the randomness of our molecular model. There is a wide range of velocities in a gaseous system. In principle, we would calculate the average velocity of a group of molecules in the same way that we would calculate the average test grade of a group of students. We simply sum up the velocities of all the molecules and divide by the number of molecules:

$$\text{Average velocity} = \bar{v} = \frac{v_1 + v_2 + v_3 + \cdots + v_N}{N} = \frac{1}{N}\sum_{i=1}^{i=N} v_i$$

† This would not be true in a tall column of gas, where there would be an appreciable gravitational effect and the pressure would be greatest at the bottom of the column.

where $\bar{v}$ = average velocity
$v_1, v_2, \ldots$ = velocity of molecule 1,2, . . .
N = number of molecules in sample

We shall also be concerned with the average of the square velocity $\overline{v^2}$. To obtain the average of the square velocity we find the squares of the velocities of all the molecules, sum them, and divide by the total number of molecules:

Average square velocity = $\overline{v^2}$

$$\overline{v^2} = \frac{v_1^2 + v_2^2 + v_3^2 + \cdots + v_N^2}{N}$$

$$\overline{v^2} = \frac{1}{N} \sum_{i=1}^{i=N} v_i^2$$

The average of the square velocity v^2 is *not* identical with the square of the average velocity $\bar{v}^2$, as can be seen by comparing $\overline{v^2}$ and $\bar{v}^2$:

Square of the average velocity = $\bar{v}^2$

$$\bar{v}^2 = \left(\frac{v_1 + v_2 + v_3 + \cdots + v_N}{N}\right)^2$$

$$\bar{v}^2 = \left(\frac{1}{N} \sum_{i=1}^{i=N} v_i\right)^2$$

However, for our purposes in the study of gases $\overline{v^2}$ and $\bar{v}^2$ are almost of equal magnitude, and at times we shall substitute one for the other. (Note that the use of the bar over a symbol means the average value of the quantity under the bar: v represents the velocity, and $\bar{v}$ represents the average velocity; v^2 represents the square of the velocity, and $\overline{v^2}$ represents the average of the square of the velocity.)

Since the average velocity of the gaseous molecules remains constant, we must also conclude that the average *translational kinetic* energy remains constant. *Kinetic energy* is energy which a body possesses by virtue of its motion. If the center of gravity of a body is moving with respect to some arbitrary observation point, the body is undergoing *translational motion* and possesses *translational kinetic energy*. The velocity and mass of a body determine its translational kinetic energy, as can be seen from the following defining equation:

Translational kinetic energy = $\varepsilon_k = \frac{1}{2}mv^2$

where m = mass of body
v = velocity

Therefore, if the average velocity of the molecules of a gas remain constant, the average kinetic energy remains constant. When bodies undergo collisions in such a way that the average kinetic energy does not change, we say that the bodies undergo *elastic* collisions. As shown in Fig. 6.7, gaseous molecules undergo elastic collisions;

FIGURE 6.7

Elastic and inelastic collisions.

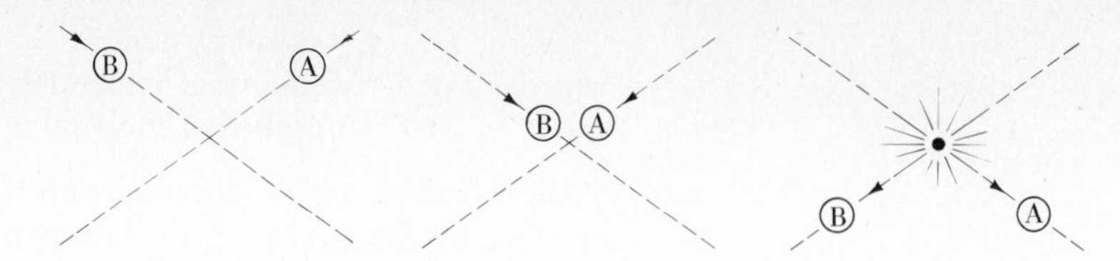

(a) Molecules undergo elastic collisions. The total kinetic energy is the same before and after collision.

(b) Automobiles do not undergo elastic collisions. Kinetic energy is used up in bending, cracking, etc., and in formation of heat.

automobiles do not. When two automobiles collide, the translational kinetic energy is used to do work (bending, cracking, breaking) and to create heat. After the collision both automobiles possess less energy than they did before. If the collisions of gaseous molecules are elastic, the observed pressure of gas in an insulated vessel will be constant, and this prediction is correct.

Molecular implications from gas temperature

As we raise the temperature of an enclosed sample of gas which obeys the ideal-gas law, the pressure increases proportionally. Since we interpret pressure as the effect of the molecular bombardment on the walls, the increase of pressure with temperature must mean that the molecules collide more *frequently* and more *forcefully* as the temperature of the gas increases. The molecules must have a higher average velocity and therefore a higher average kinetic energy. The effect of temperature on pressure leads us to conclude that the average kinetic energy of the molecules is directly proportional to the temperature.

Molecular implications of the condensation of gases

Our model becomes more detailed as we add the implications of other observations. For a long time the gases N_2, O_2, Ar, H_2, etc., were thought to be incondensable because compression does not lead to liquefaction. By considering the molecular mechanism of condensation we shall conclude that the attractive forces which act among the molecules of the "incondensable" gases must be very weak.

When a gas under ordinary conditions is converted into a liquid, a dramatic decrease in volume occurs. For instance, when 1000 ml of steam condenses, only 0.6 ml of liquid water is formed (at 100°C and 1 atm pressure):

	Steam $\xrightarrow[1\ \text{atm}]{100°\text{C}}$	water
Vol, ml:	1000	0.6
Wt, g:	0.6	0.6

The volume of the steam is approximately 1700 times greater than the volume of an equal mass of water. If we believe in the existence of molecules, we must conclude that the average separation of molecules in steam must be much greater than in liquid water. Even though their average separation is high, the H_2O molecules in steam must possess attractive forces, otherwise the liquid would never form; but as long as the kinetic energy of the molecules is high, the attractive forces will not be strong enough to permit the molecules to form stable aggregates, capable of survival under molecular bombardment. If small aggregates do form in the vapor state, they are torn asunder by collisions with other energetic molecules. By reducing the temperature, i.e., removing heat, the average kinetic energy of the molecules in the steam is reduced. The attractive forces which operate between the molecules overcome the kinetic, collisional effects; large aggregates form and continue to grow; finally a liquid phase appears (see Fig. 6.8).

FIGURE 6.8

Attractive forces may cause formation of molecular aggregates.

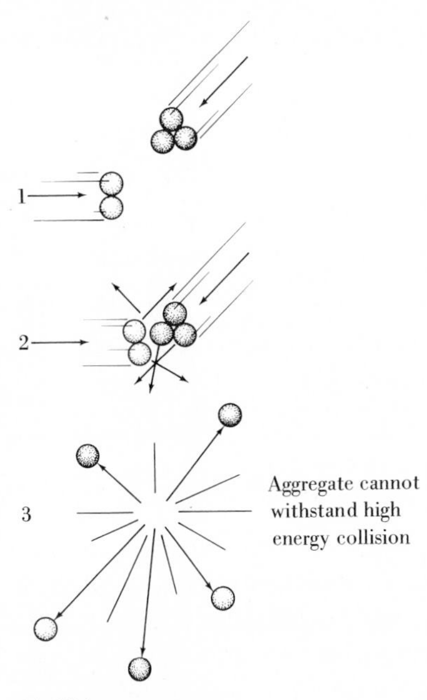

(*a*) *High temperature: Aggregates do not survive.*

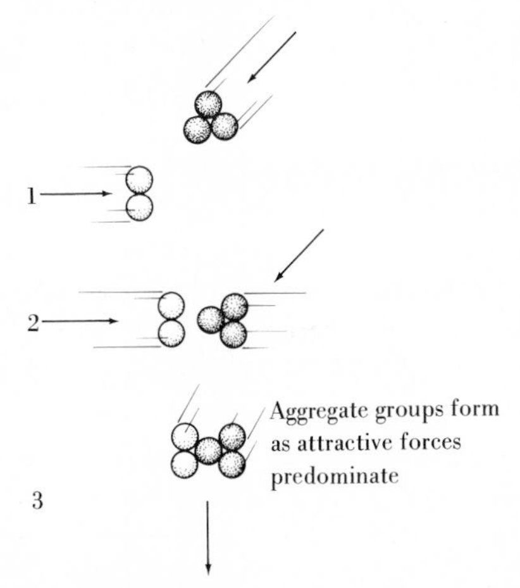

(*b*) *Low temperature: Aggregates are stable. Liquid droplets may form.*

We see that condensation of a gas to form a liquid requires the operation of attractive forces between molecules (intermolecular forces). If a gas cannot be condensed, it is natural to conclude that attractive forces do not exist or are very weak. Since it is very difficult to condense many of the ordinary gases, we may conclude that the molecules of the ordinary gases which obey the ideal-gas law exert no attractive forces on each other.

Molecular implications from gas compressibility

The fact that gases are highly compressible leads us to conclude that the volume available to gaseous molecules must be much greater than the volume of the molecules themselves. When gases are liquefied, the volume of the resulting liquid (which contains the same number of molecules) is a very small fraction of the original volume of the gas. All our experience with ordinary gases leads us to conclude that the volume of the molecules is negligible with respect to the volume occupied by the gas. Therefore, in our model of an ideal gas the molecules are considered to be *point* masses; they have mass but no volume. To say that the molecules of an ideal gas are point masses is another way of saying that the molecules exhibit no repulsive forces. The distance between hypothetical point-mass molecules can be reduced to zero so that the points coincide; the center points of real molecules can never be made to coincide. When the molecular centers are less than 3 Å apart, as in liquids and solids, the electron clouds come very close and the repulsion between the molecules becomes quite high. A great deal of force is required to cause the molecules to come any closer. But in a gas under ordinary conditions, when the average separation is near 40 Å, the repulsive forces between real molecules are extremely weak, and therefore we say that the molecules of a gas which obeys the ideal-gas law are point masses which exhibit no repulsive forces.

A summary of the important features of the ideal gas

The ideal-gas model was developed by considering and idealizing the behavior of real gases under ordinary conditions. The molecular properties of ideal gases are summarized below:

1. Macroscopic property: The ideal gas obeys the equation $PV = nRT$.
2. Microscopic properties:
   a. Gas molecules are in rapid, random motion, and their collisions with the walls of the vessel are the cause of gas pressure.
   b. All collisions of gaseous molecules are elastic.
   c. The average velocity and the average kinetic energy of gas molecules are constant if the temperature is constant.
   d. The average translational kinetic energy is directly proportional to the absolute temperature; $\overline{\varepsilon}_k \propto T$.
   e. The molecules of an ideal gas are point masses which do not exert intermolecular attractive or repulsive forces.

6.8 INTRODUCTION TO KINETIC THEORY OF GASES

Many physicists and chemists have derived equations which relate the physical properties of a gas to the behavior of its molecules. One of the first relationships of this kind was derived in 1738 by Daniel Bernoulli. He developed a theoretical equation relating pressure to molecular velocities by using a model for an ideal gas like the one which we have just described. If n point-mass molecules, all of mass m and of average square velocity $\overline{v^2}$, are in a cubical container of side s, the pressure according to Bernoulli is

$$P = \frac{1}{3} \frac{n' m \overline{v^2}}{s^3} \tag{6.21}$$

Note that the volume V of the gas in a cubical container of side s must be s^3. Therefore, Eq. (6.18) can be written

$$Ps^3 = PV = \tfrac{1}{3}n'm\overline{v^2} \tag{6.22}$$

We can rewrite Eq. (6.22) to show that the PV product is proportional to the average kinetic energy, $\tfrac{1}{2}m\overline{v^2}$:

$$PV = \tfrac{2}{2}\,(\tfrac{1}{3}n'm\overline{v^2})$$

or

$$PV = \tfrac{2}{3}n'\,(\tfrac{1}{2}m\overline{v^2}) = \tfrac{2}{3}n'\bar{\varepsilon}_k \tag{6.23}$$

To show a more practical relationship between the velocity of the molecules and the temperature and between the kinetic energy and the temperature, we must convert the molecular equations (6.22) and (6.23) into *molar* equations. This can be done by letting n' equal Avogadro's number. Then PV in Eqs. (6.22) and (6.23) will be equal to PV in the ideal-gas equation for 1 mol, $PV = RT$. From Eq. (6.22):

$$PV = \tfrac{1}{3}n'm\overline{v^2} = \tfrac{1}{3}N_A m\overline{v^2}$$

From Eq. (6.23):

$$PV = \tfrac{2}{3}n'\bar{\varepsilon}_k = \tfrac{2}{3}N_A\bar{\varepsilon}_k$$

Note that the quantity $N_A m$ is just the weight of the N_A molecules, which is, by definition, the gram-molecular weight M; and $N_A\bar{\varepsilon}_k$ is just the average kinetic energy of 1 mol, $\bar{E}_k$. Hence,

$$PV = \tfrac{1}{3}M\overline{v^2} \qquad \text{and} \qquad PV = \tfrac{2}{3}\bar{E}_k$$

From the ideal-gas law, $PV = RT$, and from the last two equations we obtain two important relationships:

$$\tfrac{1}{3}M\overline{v^2} = RT \qquad \text{or} \qquad \overline{v^2} = \frac{3RT}{M} \tag{6.24}$$

$$\tfrac{2}{3}\bar{E}_k = RT \qquad \text{or} \qquad \bar{E}_k = \tfrac{3}{2}RT \tag{6.25}$$

From the molecular weight and temperature of a gas we can determine the average of the square velocity of the molecules. From the temperature alone we can calculate the average translational kinetic energy.

The square root of the average of the square velocity, called the root-mean-square (rms) velocity is just $\sqrt{\overline{v^2}}$ or

$$\text{rms velocity} = \sqrt{\overline{v^2}} = \sqrt{\frac{3RT}{M}} = v_{\text{rms}} \tag{6.26}$$

The root-mean-square velocity is very nearly equal to the average velocity $\bar{v}$. By methods which are beyond the level of general chemistry it can be shown that

$$\bar{v} = \sqrt{\frac{8}{\pi} \times \frac{RT}{M}} = \sqrt{\frac{2.55RT}{M}} \tag{6.27}$$

Comparison of Eqs. (6.26) and (6.27) shows that $v_{rms} = 1.086\bar{v}$.

EXAMPLE 6.15 Calculate (*a*) the root-mean-square velocity of H_2 at 300 K and (*b*) the average kinetic energy of the molecules.

Solution: (*a*) We begin from Eq. (6.26):

$$v_{rms} = \sqrt{\frac{3RT}{M}}$$

$T = 300$ K, $M = 2$ g/mol, *but* if $R = 0.082$ l-atm/mol-K the result appears ridiculous, since velocity should be in units of length per time, e.g., centimeters per second:

$$v_{rms} = \sqrt{\frac{3 \times 0.082 \text{ l-atm/mol-K}}{2 \text{ g/mol}}}$$

The problem of units is solved by considering our evaluation of R on page 164. We obtain R in liter-atmospheres per mole-kelvin because we used $P = 1$ atm and $V = 22.414$ l. In order to solve the present problem we must reevaluate R using P in dynes per square centimeter and V in cubic centimeters; i.e., we must use the cgs system of units to reevaluate R. At STP 1 mol of an ideal gas occupies 22.414 cm³. Standard pressure in the cgs system is 1.013×10^6 dyn/cm²; with $V = 22.414$ cm³, $T = 273.16$ K, and $n = 1$ mol:

$$R = \frac{PV}{nT} = \frac{1.013 \times 10^6 \text{ dyn/cm}^2 \times 22.414 \times 10^3 \text{ cm}^3}{1 \text{ mol} \times 273.16 \text{ K}}$$

$$= 8.314 \times 10^7 \text{ dyn-cm/mol-K}$$

In the cgs system the dyne is the unit of force. When a force of 1 dyn acts on a body through a distance of 1 cm, an amount of energy equal to 1 *erg* has been expended. Therefore, 1 erg = 1 dyn-cm, and

$$R = 8.314 \times 10^7 \frac{\text{dyn-cm}}{\text{mol-K}} = 8.314 \times 10^7 \text{ erg/mol-K}$$

The evaluation of R in cgs units shows that R is in units of *energy* per mole-kelvin. This is implicit in Eq. (6.25), $\frac{2}{3}\bar{E}_k = RT$. The only quantity with units on the left side is $\bar{E}_k$, which is the *energy per mole*. In order to have a consistent equation the right side must also be in energy per mole. Since T is in kelvins, R must be in energy per mole-kelvin, so that the product, RT, will be in energy per mole:

$$RT = \frac{\text{energy}}{\text{mol-K}} \times \text{K} = \text{energy/mol}$$

R can be expressed in any energy unit. It is important to learn the

TABLE 6.3: VALUES OF R

0.08205 l-atm/mol-K
8.314×10^7 ergs/mol-K
8.314 J/mol-K
1.987 cal/mol-K

value of R in the systems shown in Table 6.3. These values of R serve as convenient conversion factors in going from one system of energy to another:

$$0.08205 \text{ l-atm} = 8.314 \times 10^7 \text{ ergs} = 8.314 \text{ J} = 1.987 \text{ cal}$$

Now back to Example 6.15. To obtain root-mean-square velocity from Eq. (6.26) we must use R in ergs (or g-cm^2/s^2), and the root-mean-square velocity of H_2 at 300 K comes out in units of centimeters per second:

$$v_{\text{rms}} = \sqrt{\frac{3RT}{M}} = \left[\frac{3(8.314 \times 10^7 (\text{g-cm}^2/\text{s}^2)/\text{mol-K})(300 \text{ K})}{2 \text{ g/mol}}\right]^{1/2}$$

$$= (3.84 \times 10^{10} \text{ cm}^2/\text{s}^2)^{1/2} = 1.96 \times 10^5 \text{ cm/s}$$

(*b*) The average kinetic energy of 1 mol of H_2 gas at 300 K can be calculated from Eq. (6.25):

$$\overline{E}_k = \tfrac{3}{2}RT$$

Since chemists ordinarily use the calorie as an energy unit, we shall use a value of 1.987 cal/mol-K for R in this calculation:

$$\overline{E}_k = \frac{3}{2}\,\frac{1.987 \text{ cal}}{1 \text{ mol-K}} \times 300 \text{ K} = 894 \text{ cal/mol}$$

The kinetic-molecular theory of gases is highly developed. Many molecular properties have been quantitatively related to observable, measurable properties. For instance, we can calculate the average distance molecules travel between collisions, the number of collisions occurring between molecules, and the number of collisions molecules make on a container wall in a given time period. Many other molecular properties can also be calculated. One of the main reasons the molecular theory of gases is so well developed is that a gas is virtually a perfect statistical sample. There is an astronomical number of identical species in completely random motion. Statistical predictions which are nothing but good guesses for the behavior of small groups become exact laws for a system of gaseous molecules. We shall develop the kinetic-molecular theory for gases further in Chap. 17.

6.9 GRAHAM'S LAW: A TEST OF THE KINETIC THEORY

The studies of Thomas Graham (1805–1869) on the diffusion of gases through porous membranes form an experimental test of Eq. (6.26), which states that the root-mean-square velocity of a gas is inversely proportional to the square root of its molecular weight:

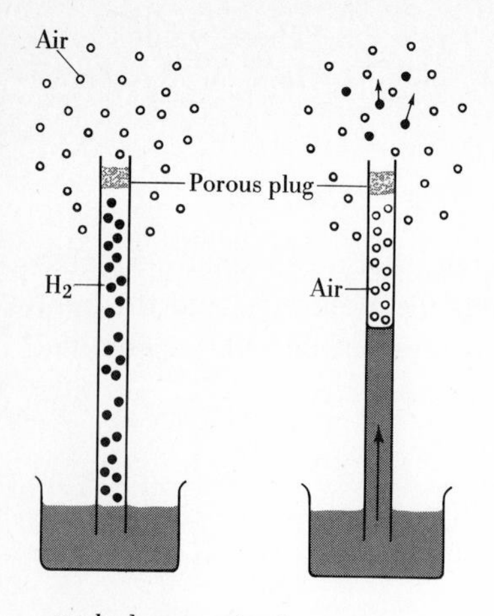

FIGURE 6.9

The water level rises because hydrogen escapes more rapidly than it is replaced by air.

$$v_{\text{rms}} = \sqrt{\frac{3RT}{M}} = \sqrt{\frac{k_g}{M}}$$

The quantity $3RT$ is a constant if T is constant, a condition which prevailed in Graham's experiments.

If hydrogen gas is trapped over water in a tube which has a porous plug at its upper end (Fig. 6.9), the water level rises inside the tube. The hydrogen escapes by diffusing out through the porous plug and is replaced by air which diffuses in. But the hydrogen diffuses more rapidly than air, and therefore the pressure inside the tube is lower than atmospheric pressure. As a result the water is forced up inside the tube. When Graham maintained a constant pressure during the experiment, he found that the *ratio* of the initial volume of hydrogen and the final volume of air was equal to the inverse ratio of the square roots of the densities of the gases:

$$\frac{V_{H_2}}{V_{\text{air}}} = \frac{\sqrt{d_{\text{air}}}}{\sqrt{d_{H_2}}}$$

Graham investigated many gases and found that this result is quite general. When two gases interdiffuse at uniform pressure, the ratio of the diffusing volumes is equal to the inverse ratio of the square roots of their densities (the densities of the gases measured under identical conditions). Table 6.4 shows some of the results obtained by Graham. The second column shows the volume ratio of the diffusing gases and the third column shows the square root of the inverse density ratio, which was determined independently.

TABLE 6.4: GAS-AIR DIFFUSION DATA

Gas	$\frac{V_{\text{gas}}}{V_{\text{air}}}$	$\left(\frac{d_{\text{air}}}{d_{\text{gas}}}\right)^{1/2}$
CH_4	1.34	1.34
CO	1.01	1.02
H_2S	0.95	0.92
SO_2	0.68	0.67

The volume ratios equal the square root of inverse density ratios. The results can be stated mathematically:

$$\frac{V_A}{V_B} = \left(\frac{d_B}{d_A}\right)^{1/2} \tag{6.28}$$

The volume terms in Eq. (6.28) are actually volume flow rates through the porous plug; that is, V_A and V_B represent the volumes per unit time of flow. In Graham's experiments both gases diffuse in the same time period, and the volume ratio equals the volume-flow-rate ratio. We can rewrite Eq. (6.28) using *rates* in place of volumes:

$$\frac{V_A}{V_B} = \frac{\text{rate}_A}{\text{rate}_B} = \left(\frac{d_B}{d_A}\right)^{1/2} \tag{6.29}$$

Also, from Eq. (6.20), $M = d\,RT/P$, we know that the molecular weight is directly proportional to the density of a gas. If two gases are compared under the same conditions of temperature and pressure, the ratio of their densities equals the ratio of their molecular weights:

$$\frac{M_A}{M_B} = \frac{d_A RT/P}{d_B RT/P} = \frac{d_A}{d_B}$$

When the last result is substituted in Eq. (6.29), we find that the *rates of diffusion of gases must be inversely proportional to the square roots of their molecular weights* (Graham's law); or when two gases interdiffuse, the ratio of the rates of diffusion is equal to the inverse ratio of the square roots of their molecular weights:

$$\frac{\text{Rate}_A}{\text{Rate}_B} = \left(\frac{M_B}{M_A}\right)^{1/2} \qquad \text{Graham's law} \tag{6.30}$$

It would seem reasonable for the rate of diffusion of a gas to be directly proportional to the average or the root-mean-square velocity of its molecules. Therefore, we can substitute the root-mean-square velocity into Eq. (6.30):

$$\frac{v_{\text{rms},A}}{v_{\text{rms},B}} = \left(\frac{M_B}{M_A}\right)^{1/2} \qquad \text{experimental law}$$

The last equation means that the root-mean-square velocity of a gas is inversely proportional to its molecular weight. This is a result derived from Graham's experiments. The kinetic theory for gases predicts the same result. If we apply Eq. (6.26) of the kinetic theory to two interdiffusing gases, A and B, we obtain

$$v_{\text{rms},A} = \sqrt{\frac{3RT}{M_A}} \qquad \text{for } A$$

$$v_{\text{rms},B} = \sqrt{\frac{3RT}{M_B}} \qquad \text{for } B$$

By division we have

$$\frac{v_{\text{rms},A}}{v_{\text{rms},B}} = \sqrt{\frac{3RT/M_A}{3RT/M_B}} = \left(\frac{M_B}{M_A}\right)^{1/2} \qquad \text{(theoretical law)}$$

which is Graham's law derived from kinetic theory. The theory is in agreement with experiment, and therefore the theory is supported.

EXAMPLE 6.16 Calculate the ratio of the diffusion rates of HCl and Cl_2.

Solution: We must use Eq. (6.30). Since $M_{Cl_2} = 71$ g/mol and $M_{HCl} = 36.5$ g/mol, we have

$$\frac{\text{Rate}_{HCl}}{\text{Rate}_{Cl_2}} = \left(\frac{M_{Cl_2}}{M_{HCl}}\right)^{1/2} = \left(\frac{71}{36.5}\right)^{1/2} = 1.40$$

Hydrogen chloride gas diffuses 1.40 times faster than Cl_2.

6.10 A CRITIQUE OF THE IDEAL-GAS MODEL

The macroscopic behavior of an *ideal gas* is exactly described by the equation $PV = nRT$. On a microscopic level we picture the molecules as rapidly, randomly moving point masses without intermolecular attractive or repulsive forces. Actually, there are no such gases, but the behavior of many ordinary gases is fairly well predicted by the ideal-gas law. This probably means that our microscopic picture is reasonably accurate; the gaseous molecules are not point masses, but their volumes must be very small compared to the volume of their container; the intermolecular forces must exist (there must at least be gravitational forces), but these forces are extremely weak. A simple calculation will show that at STP the average separation between gaseous molecules is 34 Å (1 Å = 10^{-8} cm). Since the length of a simple gaseous molecule is approximately 2 Å, the average separation between molecules is 17 molecule lengths at STP. If people were separated by 17 *people lengths*, or about 100 ft, they would not interact very strongly either. (People who interact like to be closer.) The low intermolecular attractive force in gases is consistent with the high average separation between molecules.

That all gases can be liquefied shows that all molecules do exhibit intermolecular attractive forces. If there were no attractive forces operating between molecules, the formation of condensed phases (liquids and solids) would be impossible.

6.11 DEVIATIONS FROM IDEAL-GAS BEHAVIOR

No real gas is accurately described for all conditions by our model of the ideal gas. At very low temperatures and high pressures the deviation of a real gas from ideal behavior becomes appreciable. The molecular interpretation for these observations is quite simple. At high pressures the molecules are forced closer together. The volumes of the molecules may now be an appreciable fraction of the volume available. We can no longer consider the molecules as point masses (see Fig. 6.10). Also, since the average separation is much smaller at high pressure, the attractive forces are no longer neg-

FIGURE 6.10

At high pressure the volume of the individual molecules is an appreciable fraction of the volume of the container.

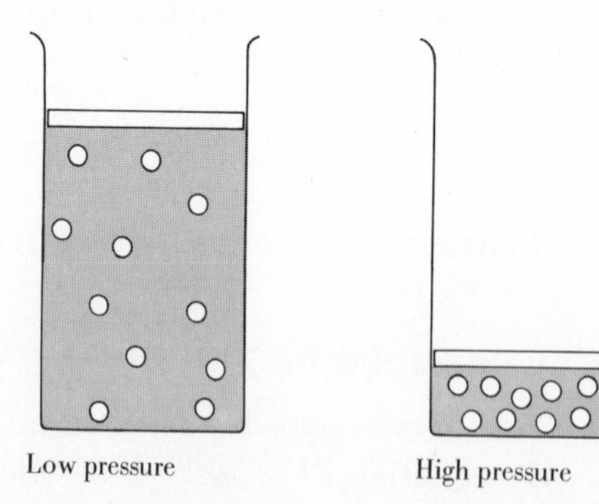

FIGURE 6.11

Attractive forces have their greatest effect when the velocity or kinetic energy is very low.

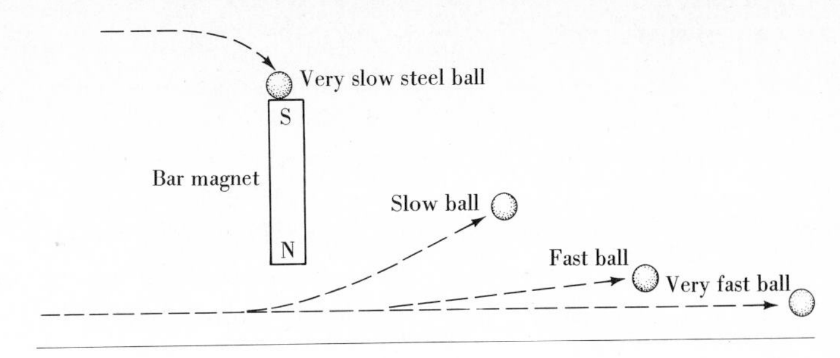

ligible. (The far-off planet Pluto has a negligible effect on the earth, but the nearby moon has many noticeable effects.) Two postulates of the ideal-gas model do not hold: (1) real-gas molecules have volume and cannot be treated as point masses, and (2) real-gas molecules exhibit intermolecular attractive forces. These facts must be considered when the pressure is high.

At low temperature the average velocity and kinetic energy of the molecules are low, and the attractive forces are appreciable. The situation may be compared to a steel ball rolling past the pole of a magnet, as shown in Fig. 6.11. If the ball is moving very fast, it is hardly deflected by the magnet; if it is moving slowly, its path may be appreciably altered; and a very slow ball may even be drawn to the pole and stopped there. At low temperature, when the velocities of the molecules are lower, the attractive forces cause the pressure to be lower than the pressure predicted by the ideal-gas law.

J. C. van der Waals proposed a gas law to account for the fact that real-gas molecules exhibit intermolecular attractive and repulsive forces. For 1 mol of gas the van der Waals equation is

$$\left(P + \frac{a}{V^2}\right)(V - b) = RT \tag{6.31}$$

P, V, R, and T have the usual meanings. The constant a accounts for the intermolecular attractive forces between molecules. Since the forces between molecules must depend on their unique structures, a has a specific value for a given gas. The a values for the noble gases are low because the intermolecular forces are low; the a values for condensable gases like H_2O, NH_3, and SO_2, are relatively high because the intermolecular forces are high. The constant b accounts for the fact that the real molecules exhibit repulsive forces at small separations; in other words, they have a volume which must be reckoned with when the volume of the container is small. b generally increases as the number of atoms in a molecule increases. The volume of 1 mol of a liquid is often a good approximation of the

TABLE 6.5: VAN DER WAALS CONSTANTS

Gas	a, l^2-atm/mol^2	b, l/mol
He	0.0341	0.0237
Ar	1.35	0.0322
O_2	1.32	0.0312
N_2	1.38	0.0394
H_2O	5.46	0.0305
NH_3	4.17	0.0371
SO_2	6.71	0.0564
CCl_4	20.39	0.1383
CO_2	3.60	0.0428

van der Waals b, and this is reasonable since molecules of a liquid are in close contact; the volume of the liquid should be a good approximation of the volume of the molecules. Values of a and b for several gases are shown in Table 6.5.

We may look at the van der Waals equation as a "corrected" ideal-gas law. The quantity a/V^2 "corrects" the pressure, and b "corrects" the volume. It should not be surprising that this equation gives a more accurate description of gas behavior than the ideal-gas law does. Table 6.6 shows the observed value of the PV product for 1 mol of N_2 gas at 40° and at various pressures, along with the PV values predicted by the ideal-gas law and by the van der Waals equation.

Although, the van der Waals equation is not in perfect agreement with observations, it is obviously much better than the ideal-gas result. The real value of the van der Waals equation lies not in its accuracy but in what it suggests about the nature of molecular interactions.

To use the ideal-gas law we need only the value of R, which is the same for all gases. We must know more to use the van der Waals equation: we must have values of a and of b for each gas. Many other equations have been proposed to describe real-gas behavior, and some are extremely accurate. The accuracy of a gas equation increases as the number of constants increases. The ideal-gas law contains one constant, the van der Waals equation contains three, the virial equations, which are the most accurate, contain several.

EXAMPLE 6.17 1.00 mol of O_2 is contained in a 0.100-l vessel at 25°C. Calculate the pressure using (*a*) the ideal-gas law and (*b*) the van der Waals equation.

Solution: (*a*) $PV = RT$

$$P = \frac{RT}{V} = \frac{0.082 \text{ l-atm/mol-K} \times 298 \text{ K}}{0.100 \text{ l}} = 244 \text{ atm}$$

For (*b*)

TABLE 6.6: ***PV*** **PRODUCT FOR 1 MOL OF N_2 AT 40°C**

	P, atm			
PV	*1*	*10*	*100*	*1000*
Observed	25.57	24.49	6.93	40.00
Ideal†	25.60	25.60	25.60	25.60
Van der Waals	25.60	24.71	8.89	54.20

† *PV* for an ideal gas is a function of temperature only.

$$\left(P + \frac{a}{V^2}\right)(V - b) = RT$$

$$P = \frac{RT}{V - b} - \frac{a}{V^2}$$

Find a and b in Table 6.5

$$P = \frac{0.082 \text{ l-atm/mol-K} \times 298 \text{ K}}{0.100 \text{ l/mol} - 0.0312 \text{ l/mol}} - \frac{1.32 \text{ l}^2\text{-atm/mol}^2}{(0.100 \text{ l/mol})^2} = 223 \text{ atm}$$

The measured pressure under these conditions is 221 atm; the van der Waals equation is in fair agreement, but the ideal-gas law is very poor.

The calculation of the volume of a gas from the van der Waals equation is more difficult because the van der Waals equation will give three values for the volume for each pressure-temperature pair; i.e., this equation is cubic in the volume. This can be seen if the left side is multiplied out:

$$V^3 - \left(\frac{RT}{P} + b\right) V^2 + \frac{a}{P} V - \frac{ab}{P} = 0$$

Two of the three values at the volume which satisfy the equation will be physically meaningless (see Prob. 6.33).

QUESTIONS AND PROBLEMS

6.1 What characteristics distinguish gases from the condensed phases (solids and liquids)?

6.2 A column of mercury 75.5 cm high and at 0°C is supported by atmospheric pressure. (*a*) Using Eq. (6.4), calculate the pressure in dynes per square centimeter. Calculate the pressure in (*b*) grams per square centimeter, (*c*) pounds per square foot, (*d*) inches of mercury, (*e*) Torr, and (*f*) atmospheres.

6.3 Calculate the pressure in atmospheres and in Torr exerted by a column of water ($d = 1.00$ g/ml) 30 ft high.

6.4 Calculate the pressure in atmospheres and in grams per centimeter exerted by a column of mercury 60.0 cm high and at 50°C. The density of mercury at 50°C is 13.4 g/ml.

6.5 A 500-ml sample of argon gas at 2.00 atm is expanded *isothermally* to a volume of 1500 ml. What is the final pressure? (Use Boyle's law.)

6.6 If 1 mol of oxygen gas at STP is heated *isobarically* until the temperature is doubled, what are the final values for the temperature, pressure, and volume? (Use Charles' law.)

6.7 A sample of H_2 at STP has a volume of 1300 ml. Calculate its volume after it is compressed isothermally to 2500 Torr.

6.8 A 2.00-l gas sample at 5°C and 1.50 atm is heated to 50°C at constant pressure. Calculate the new volume.

6.9 A 10-l sample of helium at 10.0 atm and 500°C is brought to STP. What is the new volume? (Use Boyle's and Charles' laws.)

6.10 A sample of CO_2 has a volume of 150 ml at 20°C and 760 Torr. Determine the volume of this sample of STP.

6.11 1.00 l of helium at 2.00 atm and 2.00 l of nitrogen at 2.00 atm are allowed to mix isothermally so that the final volume is 3.00 l (*a*) Calculate the partial pressures of He and N_2 and (*b*) the total pressure of the mixture.

6.12 If 1.00 l of hydrogen at 5 atm and 9.00 l of argon at 3 atm are forced into a vessel which has a volume of 2.50 l with no change in temperature, what are (*a*) the final partial pressures of helium and argon and (*b*) the total pressure?

6.13 Oxygen gas is collected in a vessel inverted over water at 21°C. The pressure in the vessel at the end of the experiment is 575 Torr, and the volume of the gas in the vessel is 176 ml. (*a*) What is the partial pressure of oxygen in the vessel? Calculate (*b*) the volume that the dry oxygen would occupy at STP and (*c*) the number of moles of oxygen collected.

6.14 Calculate the pressure exerted by 5.00 g of Cl_2 at 400 K in a 25.0-l vessel. (Use the ideal-gas law.)

6.15 Nitrogen gas is trapped in a bottle inverted over benzene at 26°. The pressure inside the bottle is 800 Torr, and the gas volume is 225 ml. (*a*) Calculate the partial pressure of N_2. (*b*) Calculate the number of moles of N_2 gas and benzene vapor in the bottle. (The vapor pressure of benzene at 26°C is 100 Torr.)

6.16 Oxygen is formed from the thermal decomposition of $KClO_3$:

$$2KClO_3 \xrightarrow{\Delta} 2KCl + 3O_2(g)$$

(*a*) What number of moles of O_2 can be formed from 100 g of $KClO_3$? (*b*) What pressure would this amount of oxygen exert at a temperature of 25°C in a vessel of 5.50 l?

6.17 Hydrogen is formed by the action of sulfuric acid on zinc:

$$H_2SO_4 + Zn^0 \rightarrow ZnSO_4 + H_2(g)$$

(*a*) How many moles of H_2 can be formed from 50.0 g of Zn and excess acid? (*b*) If the generated H_2 is to be stored at 5.00 atm and −50°C, what must the volume of the container be?

6.18 The pressure of N_2 in a 50.0-l vessel is 1500 lb/in.^{2}. Its temperature is 22.0°C. Calculate (*a*) the number of moles, (*b*) the weight, and (*c*) the number of molecules of N_2 in the vessel.

6.19 A mixture of CO_2 and N_2 gas are present in a 10.0-l vessel at 20°C. The total pressure is 1500 Torr. The mixture is 25.0 percent by weight N_2.

Calculate the (*a*) total number of moles of gas, (*b*) the partial pressure of each gas, and (*c*) the number of moles of each gas.

6.20 On page 174 five important microscopic properties of gases were listed. Present a short argument for the reasonableness of each property.

6.21 (*a*) Give a verbal and a mathematical definition of translational kinetic energy. (*b*) What is root-mean-square velocity, v_{rms}? (*c*) Is v_{rms} identical with the average velocity (by definition)? (*d*) Is v_{rms} identical with the square root of the square of average velocity $\sqrt{\bar{v}^2}$ (by definition)?

6.22 (*a*) Calculate $\bar{v}$, $\overline{v^2}$, $\bar{v}^2$, and v_{rms} for a group of 14 molecules having the following velocities in kilometers per second: 1.5, 1.9, 2.0, 2.0, 2.5, 8.6, 9.3, 27.1, 30.3, 46.1, 50.0, 50.0, 54.3, and 59.6. (*b*) Is the average velocity equal to the root-mean-square velocity? (*c*) Why do you suppose that $\bar{v} \approx v_{rms}$ for an ordinary gas sample?

6.23 (*a*) Does each molecule in a sample of gas have a velocity equal to the root-mean-square velocity? (*b*) Is it possible that no molecule in a sample has a velocity equal to the root-mean-square velocity?

6.24 (*a*) Calculate the root-mean-square velocity in centimeters per second for He and Cl_2 molecules when these gases are at STP. (*b*) Calculate the average kinetic energy in ergs and calories for a molecule and for a mole of each of these gases at STP. (*c*) Calculate the ratio of the diffusion rates of He to Cl_2 (under the same conditions). Use only the molecular weights of the gases.

6.25 Uranium 235 and uranium 238 are separated by a process which depends on the slight difference in the rate of diffusion of their hexafluorides. Calculate the ratio of the diffusion rates of $^{235}UF_6$ and $^{238}UF_6$.

6.26 The density of a gas at STP is 1.74 g/l. Calculate its molecular weight.

6.27 Calculate the density of ethylene gas, C_2H_4, at STP.

6.28 A 5.00-g sample of a hydrocarbon gas has a volume of 2.30 l at 20°C and 755 Torr. Calculate the molecular weight of the hydrocarbon.

6.29 Calculate (*a*) the number of molecules in 1.00 l of an ideal gas at 25°C and 1.00 atm and (*b*) the average separation between the molecules. State any assumptions you must make to perform this calculation.

6.30 Why is the van der Waals equation more realistic than the ideal-gas law?

6.31 Calculate the pressure of 1.00 mol of CO_2 in a 0.500-l cylinder at 600 K from (*a*) the ideal-gas law and (*b*) the van der Waals equation (see Table 6.6).

6.32 The van der Waals equation for *n* moles of gas is

$$\left(P + \frac{n^2a}{V^2}\right)(V - nb) = nRT$$

Using this equation, calculate the temperature of 5 mol of argon in a 100-l cylinder at 150 atm.

6.33 From the van der Waals equation calculate the volume of 1 mol of CCl_4 at 50.0 atm and 1200°C. (This problem will require the solution of a cubic equation in V; it is often faster to solve this kind of equation by trial. Use the volume as calculated from the ideal-gas law as a first guess.)

6.34 (*a*) Use the fact that 1 mol of hydrogen under 1.00 atm pressure occupies 22.4 l at 0°C, 26.4 l at 50°C, and 30.4 l at 100°C to determine the value of the absolute of zero in Celsius degrees by graphical means. (*b*) Calculate the coefficient of cubical expansion α for H_2 gas.

6.35 Assuming that CO_2 obeys the ideal-gas law, calculate the density of this gas (*a*) at 780 Torr and 50°C and (*b*) at 1000 Torr and 75°C.

6.36 (*a*) The density of air is very nearly 1.20 g/l under ordinary atmospheric conditions (70°F and 30 in. Hg); calculate the average molecular weight of air. (*b*) To a good approximation air may be considered to be a mixture containing nitrogen, oxygen, and argon in molar percentages of 78, 21, and 1 percent, respectively; calculate the average molecular weight of air from these percentages and compare with the result in part (*a*).

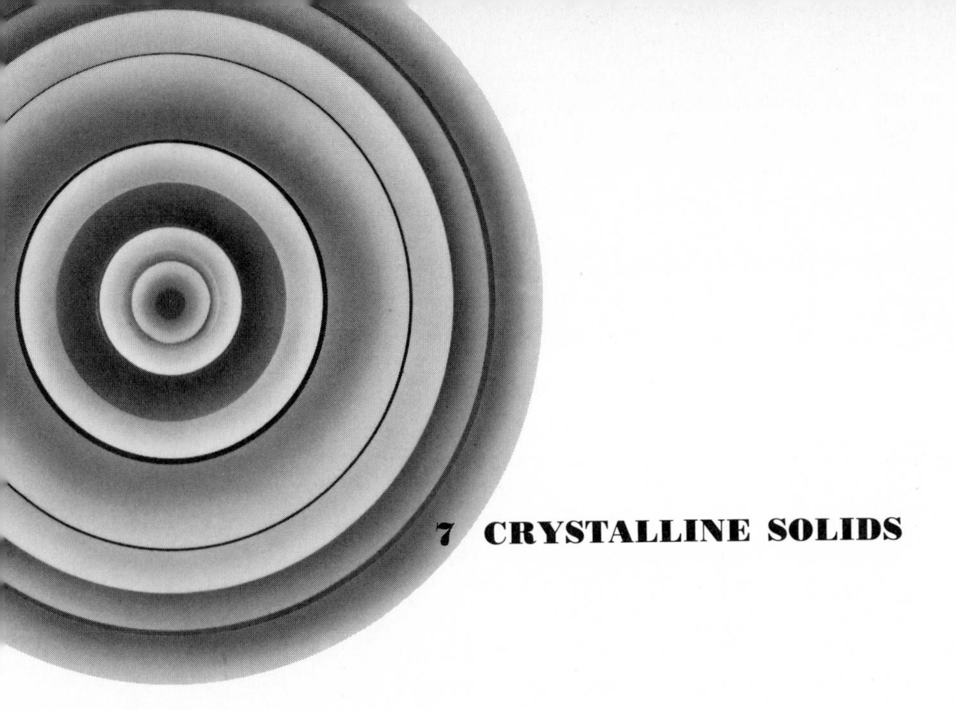

7 CRYSTALLINE SOLIDS

7.1 INTRODUCTION

Some of the most familiar materials of everyday living are solids. The most notable physical properties of solids are rigidity and virtual incompressibility. A solid can support a stress or a strain and undergo only a very slight and temporary change in shape. As soon as the stress is removed, provided that it is not too great, the solid returns to its original dimensions. Many solids are *crystalline;* i.e., they form regular three-dimensional structures, or *polyhedra,* which often are large enough to be seen with the unaided eye. Table salt, Na^+Cl^-, forms cubes; sulfur forms regular rhombic crystals; emerald forms long, six-sided prisms; and metallic copper forms octahedra. The crystalline form of a substance is an identifying characteristic since all crystals of the same substance have the same angles between crystal faces. By measurement of the angles it is possible to identify the chemical composition of the substance.

Beautiful crystalline materials have been treasured as gems since the beginning of history. It is not surprising that from very early

FIGURE 7.1

Photograph of quartz crystals.

times there was conjecture about the cause of their external regularity. In the seventeenth century Robert Hooke suggested that the crystallinity was due to the regular packing of small spherical, cylindrical, or other regularly shaped bodies. He claimed "that there was not any regular Figure" that he could not imitate "with the composition of bullets or globules, and one or two other bodies."

In the nineteenth century, as scientists began to accept the atomic theory, they naturally began to hypothesize that the observable patterns of crystalline materials are a result of microscopic regularity. In the early 1800s René Haüy, of the University of Paris, inferred correctly that the regular external pattern of crystals reflects the orderly arrangement of the atoms and molecules in the crystal. His model of a crystal structure is shown in Fig. 7.2. The experimental proof that crystals possess microscopic regularity had to wait until the development of the wave theory of light and discovery of x-radiation.

7.2 THE USE OF X-RAYS IN THE STUDY OF CRYSTAL STRUCTURE

FIGURE 7.2

Haüy's model of a crystal (about 1800).

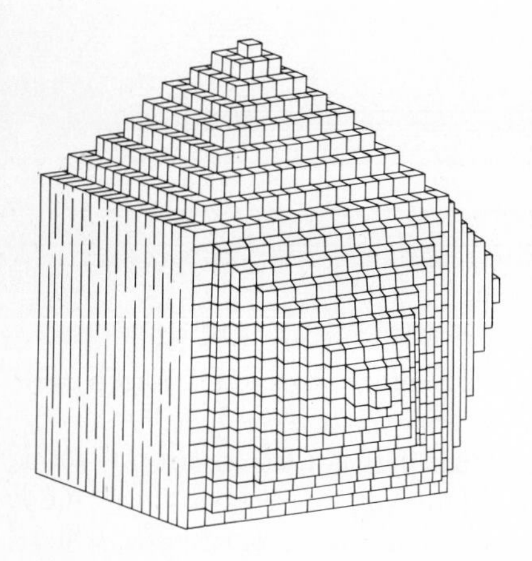

X-radiation was discovered by Wilhelm Konrad Röntgen in 1895. Within 20 years it had become the most important tool in the study of crystal structure at the atomic level. In order to understand how x-ray analysis enables us to determine the positions of atoms in a crystal we must realize that x-rays are electromagnetic waves like light. As a ray of light or an x-ray moves through space, it sets up electric and magnetic fields in the form of moving waves. The electric field E is always perpendicular to the direction of the ray. The magnetic field H is perpendicular to both the direction of the ray and the plane of the electric field, as shown in Fig. 7.3. The maximum value of the electric and magnetic field caused by the wave is called the amplitude A, and the direction between successive maxima is the *wavelength* λ.

Light waves which are perfectly in phase, i.e., whose maxima coincide, add to (reinforce) each other; light waves which are completely out of phase, i.e., whose maxima and minima coincide, subtract from (destroy) each other. This is similar to the mathematical addition of two sine or cosine functions. When two sine functions are in phase, they add to, or reinforce, each other; when they are 180° out of phase, they subtract from, or destroy, each other (Fig. 7.4*a*). The reinforcement or destruction of electromagnetic waves is called *interference*. When x-rays impinge on a crystalline surface, they have such short wavelengths and high energy that individual atoms scatter the x-rays in all directions. Because of interference effects the scattered rays can be detected only at certain

FIGURE 7.3

Electromagnetic radiation.

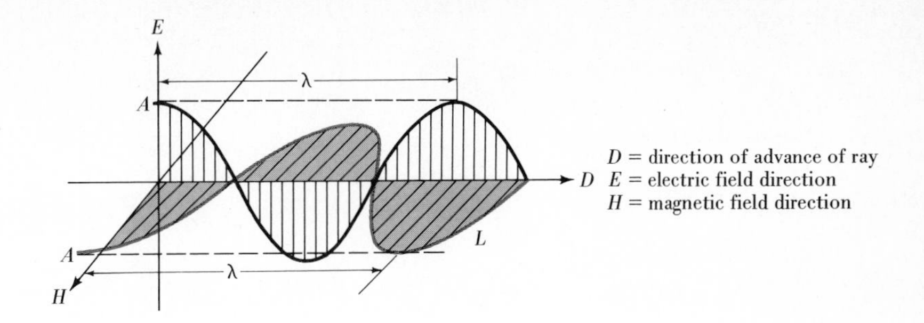

angles. The emitted x-rays give rise to a *diffraction pattern* which depends on the structure of the crystal and wavelength of the x-rays. If the wavelength of the x-radiation is known, the atomic structure of the crystal can be deduced from the diffraction pattern. A photograph of the x-ray diffraction pattern of a crystal of beryl, $Be_3Al_2Si_6O_{18}$, is shown in Fig. 7.5.

The diffraction of x-rays by crystal surfaces can be treated like a reflection with certain constraints. When ordinary light strikes the smooth surface of a crystal, the rays can be reflected at every angle, as shown in Fig. 7.6. The angle of reflection equals the angle of incidence. Also, when x-rays impinge on a crystalline surface the *angle of reflection*† equals the *angle of incidence*, but because of interference effects there are only a few, definite angles at which reflection can occur. The angle at which an x-ray is reflected depends on the wavelength λ of the x-radiation and the spacing d between the atomic layers in the crystal. The equation which relates the angle to λ and d was derived by W. H. Bragg and his son, W. L. Bragg, in the early part of the twentieth century:

† X-radiation undergoes diffraction at a crystal surface, and *reflection* in this discussion actually means reinforcement.

FIGURE 7.4

The addition of sine functions is a basic for understanding the diffraction phenomenon.

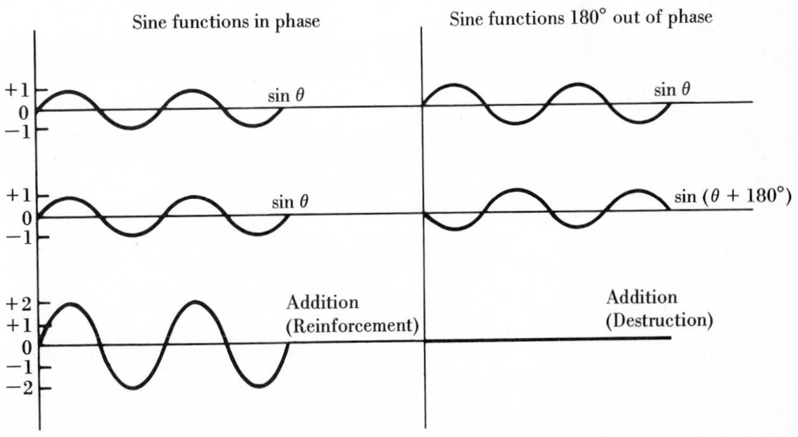

FIGURE 7.5

Laue x-ray diffraction pattern produced by a crystal of beryl. (Eastman Kodak Research Laboratories.)

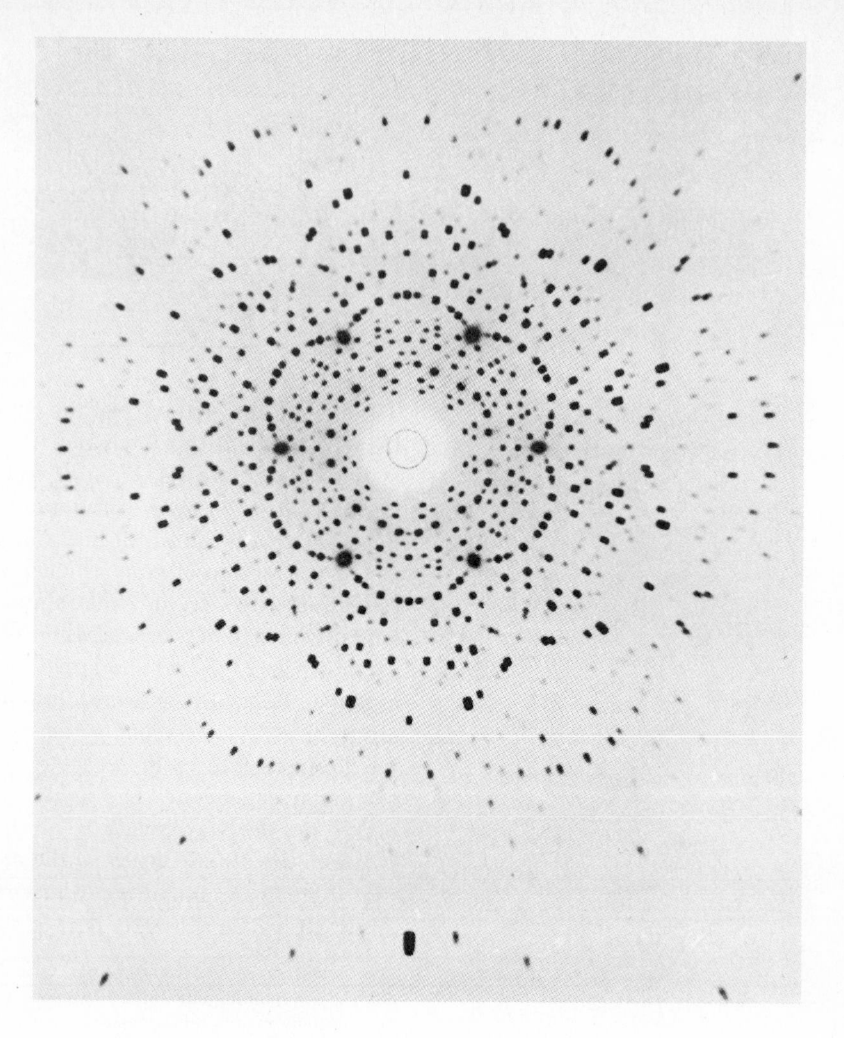

$$\sin \theta = \frac{n\lambda}{2d} \qquad \text{Bragg equation}$$

n is an integer and is called the *order;* its meaning will become clear in the derivation of the Bragg equation given below. The Bragg equation is a simple, concise statement of the condition which must prevail if reflection of x-rays is to be observed; it is, therefore, often referred to as the *Bragg condition.*

To derive the Bragg equation consider Fig. 7.7, which represents a side view of a set of parallel atomic planes in a crystalline lattice. A, B, and C represent incident x-rays which are in phase. A', B', and C' represent emergent x-rays. In order to observe reflection the emergent rays must be in phase. In general, this will not be the case because the paths which the rays take through the crystal will not be of equal length. The path BRB' is longer than path AOA' by the

FIGURE 7.6

Reflection of ordinary light.

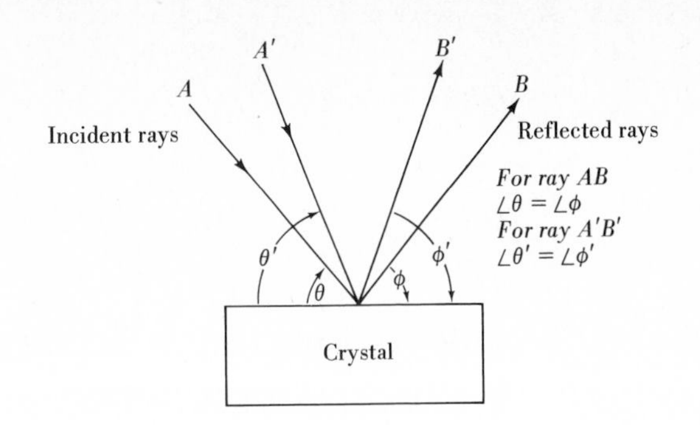

distance QRS, and rays A' and B' will probably be out of phase. If, however, the distance QRS is equal to a whole number of wavelengths, say $n\lambda$, where n is an integer, the rays A' and B' will be in phase. Reinforcement will occur, and x-ray reflection will be observed. Thus, the condition for reinforcement is

$$QRS = n\lambda \tag{7.1}$$

As the angle θ is varied from 0 to 90° the condition for reinforcement may be met several times; i.e., there are several angles at which QRS may equal a whole number of wavelengths.

The distance QRS can easily be related to the perpendicular distance of separation d between the atomic planes and the angle of incidence θ. The ray A is incident upon the atomic plane α; let θ equal the angle between ray A and plane α. Line OQ is perpendicular to ray A, and line OR is perpendicular to plane α; since θ is the angle enclosed by the intersection of ray A and plane α, it will also be the angle enclosed by the intersection of their perpendiculars. Therefore, angle QOR (and angle ROS) is equal to θ. From Fig. 7.6,

$$\sin \theta = \frac{QR}{OR} = \frac{QR}{d}$$

FIGURE 7.7

X-rays striking a crystal.

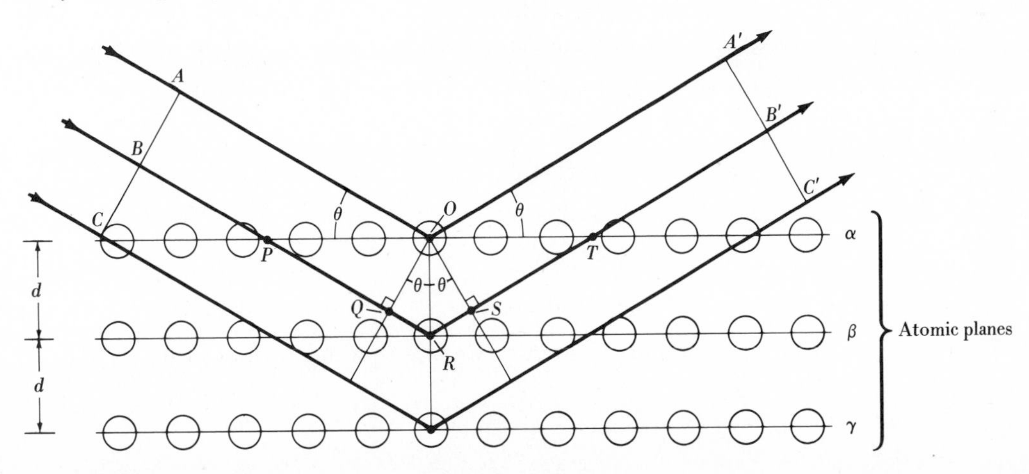

or

$$QR = d \sin \theta$$

The distance RS is also equal to $d \sin \theta$, so that

$$QR + RS = 2(d \sin \theta)$$

or

$$QRS = 2d \sin \theta \tag{7.2}$$

According to Eq. (7.1), reinforcement occurs only when QRS is equal to a whole number of wavelengths. Combining Eqs. (7.1) and (7.2) permits us to determine the conditions under which reflection of x-rays from crystal surfaces can be observed:

$$n\lambda = 2d \sin \theta \tag{7.3a}$$

or

$$\sin \theta = \frac{n\lambda}{2d} \qquad \text{Bragg equation} \tag{7.3b}$$

Equation (7.3*b*) means that x-ray reflection from a crystal can occur only at angles having sine values of $n\lambda/2d$. Since n is an integer, there are only a limited number of angles at which reflection will be observed.

The significance of the Bragg equation can be illustrated by an example.

EXAMPLE 7.1 One of the crystalline modifications of metallic chromium has a lattice which is cubic. The whole crystal is built up from cubes, in which there is an atom at each corner and at each center, as shown in Fig. 7.8. The length of the cube side is 2.878 Å. Calculate the angle θ at which reflection of x-rays ($\lambda = 1.54$ Å) will be caused (*a*) by planes like A, B, and C, in Fig. 7.8*a*, which are parallel to the cube faces, and (*b*) by diagonal planes like D, E, and F in Fig. 7.8*b*. Do the calculation for first-order reflection, i.e., for $n = 1$.

Solution: (*a*) The perpendicular distance between planes like A, B, C, etc., is just equal to one-half the cube side.

$$d = \tfrac{1}{2}(2.878 \text{ Å}) = 1.44 \text{ Å}$$

By the Bragg equation

$$\sin \theta = \frac{n\lambda}{2d} = \frac{1 \times 1.54 \text{ Å}}{2 \times 1.44 \text{ Å}} = 0.535$$

From mathematical tables we find that the angle whose sine is 0.535 is 32.3°.

$$\theta = 32.3°$$

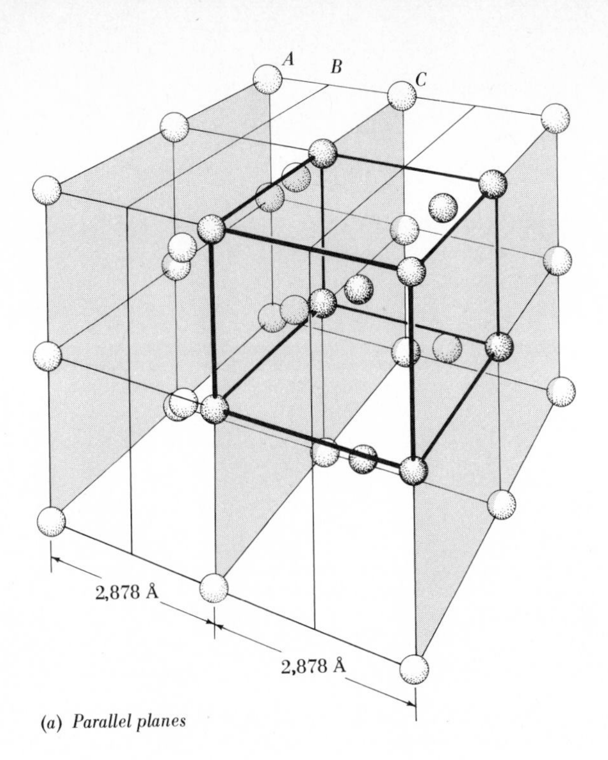

(a) *Parallel planes*

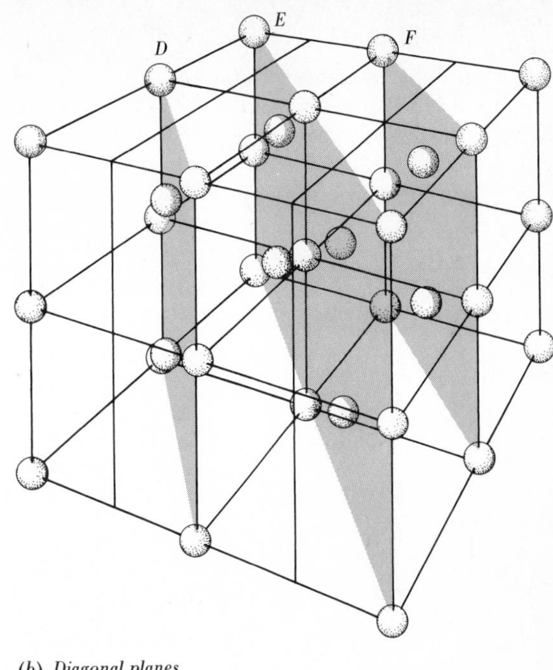

(b) *Diagonal planes*

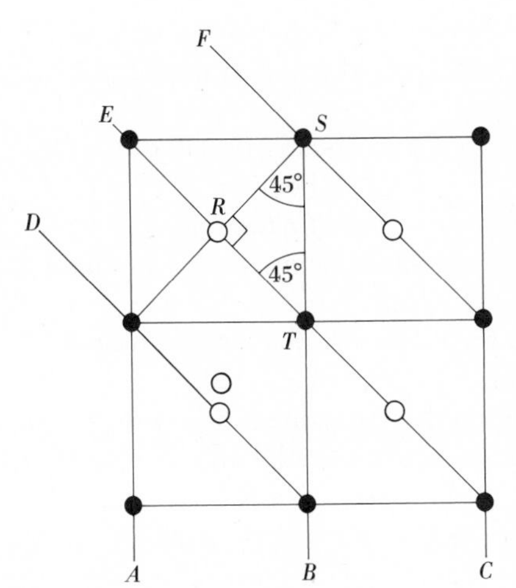

(c) *Top view of body-centered cubic lattice (chromium)*

FIGURE 7.8

Structure of chromium.

Atomic planes with a spacing of 1.44 Å will reflect x-rays with a wavelength of 1.54 Å at 32.3°.

(*b*) The perpendicular distance between planes of the type *D*, *E*, and *F*, can be calculated from simple trigonometric considerations. Figure 7.8*c* represents top view of the crystal showing four unit lattices. *RS* is the shortest distance between the planes of interest; *RS* therefore equals *d* in the Bragg equation. *RST* is an isosceles right triangle whose base length is 2.878 Å. From the pythagorean theorem

$$(RS)^2 + (RT)^2 = (ST)^2$$

but

$$RS = RT$$

and

$$2(RS)^2 = (ST)^2 = (2.878\ \text{Å})^2$$

$$RS = 2.03\ \text{Å}$$

The value of *d* for parallel planes *D*, *E*, *F*, etc., is 2.03 Å. We can now calculate the angle at which reflection due to these planes is permitted:

FIGURE 7.9

Apparatus for x-ray analysis. When an electron beam strikes a metal target x-rays are generated. See page 75.

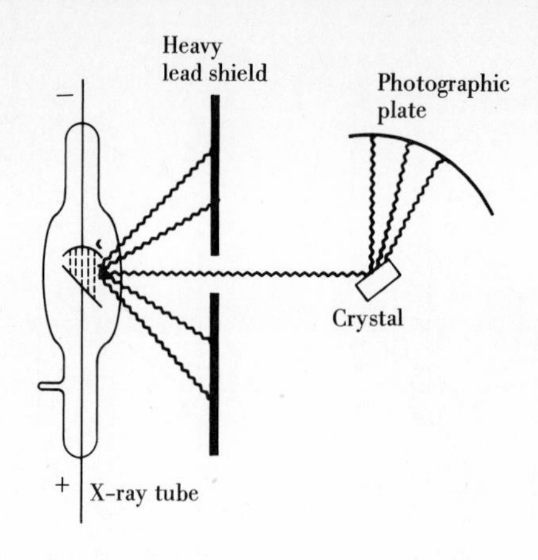

$$\sin\theta = \frac{n\lambda}{2d} = \frac{1 \times 1.54\ \text{Å}}{2 \times 2.02\ \text{Å}} = 0.379$$

$$\theta = 22.3°$$

Figure 7.9 shows the basic principles of x-ray crystallography. A crystal is mounted on a rotating table, and a narrow x-ray beam from a stationary source is aimed at it. As the crystal is rotated, the angle of incidence is varied from 0 to 90°. When the Bragg condition is satisfied, reinforcement of x-rays can be detected using a photographic plate or electronic devices like a Geiger counter.

If a crystal of chromium is rotated around an axis parallel to both planes A and D (Fig. 7.8a and b), a reflected beam will flash out when the incident x-ray beam makes an angle of 32.3° with the parallel planes A, B, and C and when the incident beam makes an angle of 22.3° with the parallel planes D, E, and F.

The Bragg equation is the basic relationship for determining crystal structure by x-ray diffraction. The angles at which reflection of x-rays occur are measured, enabling the crystallographer to calculate the distances between the various sets of parallel atomic planes. The intensity of the reflection depends upon the *atomic density* of the plane; the greater the number of atoms per unit area of the reflecting plane the more intense the reflected ray. Hence, the measured intensity allows the crystallographer to determine the atomic density of the planes in a crystal. Figure 7.10 represents the arrangement of particles in a crystal face. The most intense reflections would be observed for the set of parallel planes marked A. All other sets have a lower atomic density. The intensity also depends on the atomic number, i.e., the number of electrons surrounding the nucleus. The lighter elements give relatively weak reflections; hydrogen atoms are almost undetectable by x-ray analysis. Potassium

FIGURE 7.10

Various sets of planes in a crystal.

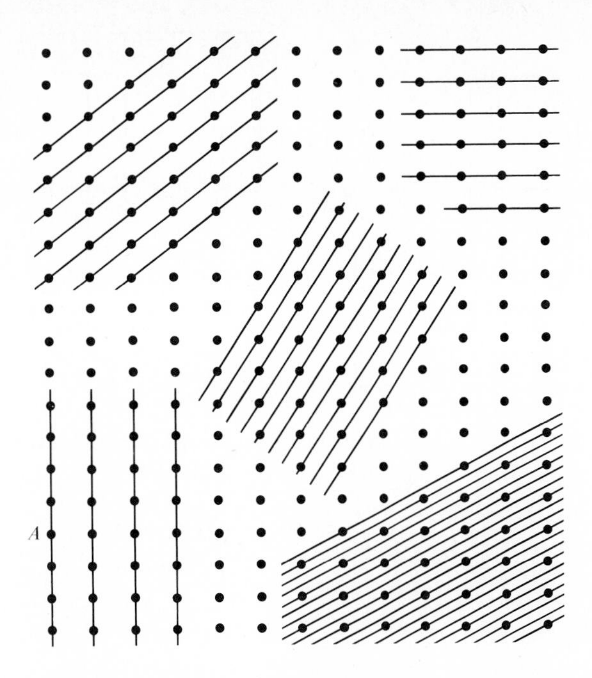

and chloride *ions*, K^+ and Cl^-, are isoelectronic, so that the intensities of the reflections from a KCl crystal depend only on the atomic density of the reflecting planes and not on the nature of the particles in the planes. Very heavy atoms like mercury, lead, and gold give rise to intense spots in the x-ray diffraction pattern.

A study of the angles and the relative intensities of the reflections leads to the determination of the structural pattern of a crystal and the location of each atom within it, so that if a crystal is made up of molecules, an x-ray analysis is equivalent to determining the structure of the molecules. For if the location of each atom in the crystal can be determined, the relative positions of the atoms within the molecules of the crystal are also known. Figure 7.11*a* shows the relative positions of the atoms in a hypothetical crystal structure. Since the separation between bonded adjacent atoms in a molecule is from 1 to 2 Å, it is easy to decide which atoms are bonded. The same atomic arrangement minus the lattice lines is pictured in Fig. 7.11*b*, but the bonds between the atoms within a molecule are shown.

7.3 THE CRYSTAL LATTICE

As a result of x-ray studies we have experimental evidence that the atoms and molecules of a crystal make up a repeating, or *periodic*, three-dimensional pattern. A pattern *unit* in each crystal is repeated indefinitely in three directions to generate the crystal structure. This is similar to wallpaper, on which a two-dimensional pattern unit is

FIGURE 7.11

X-ray analysis permits the determination of molecular structure.

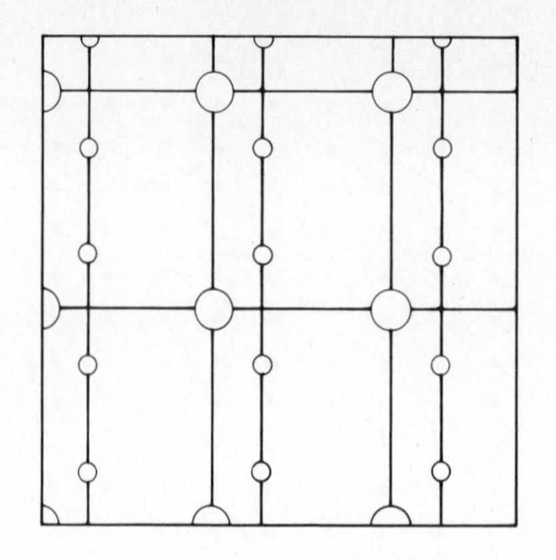

(a) A side view of the atomic arrangement in a hypothetical crystal.

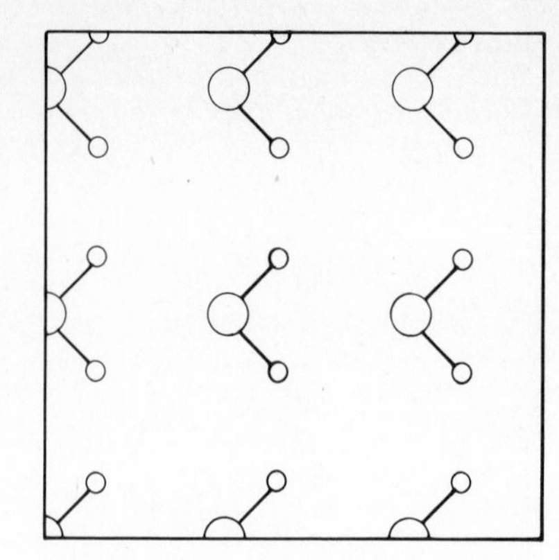

(b) A side view of the atomic arrangement in a hypothetical crystal. The chemical bonds are shown.

FIGURE 7.12

The crystal lattice.

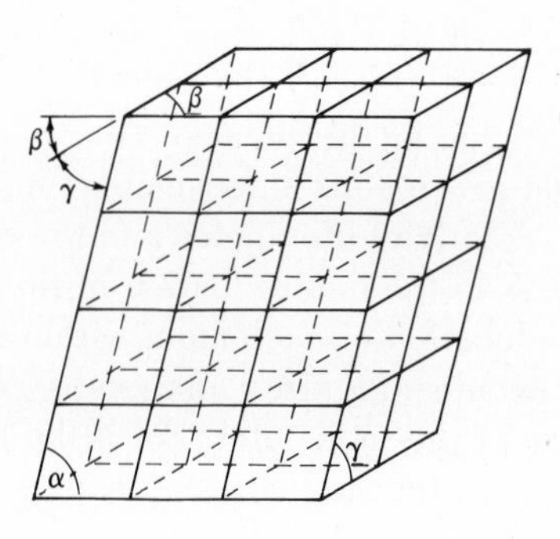

(a) Unoccupied lattice

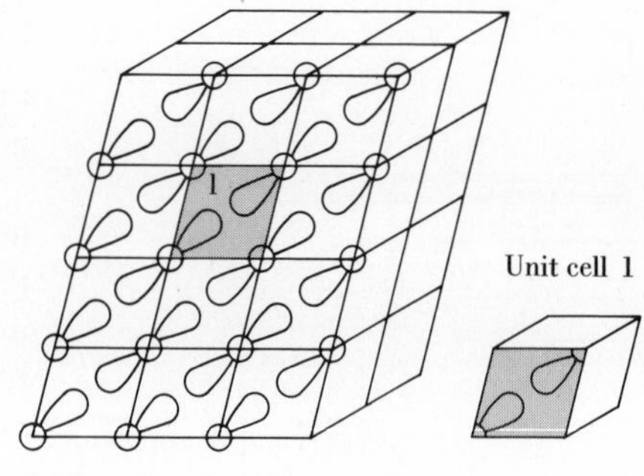

(b) Particles at lattice intersections
A unit lattice is shaded

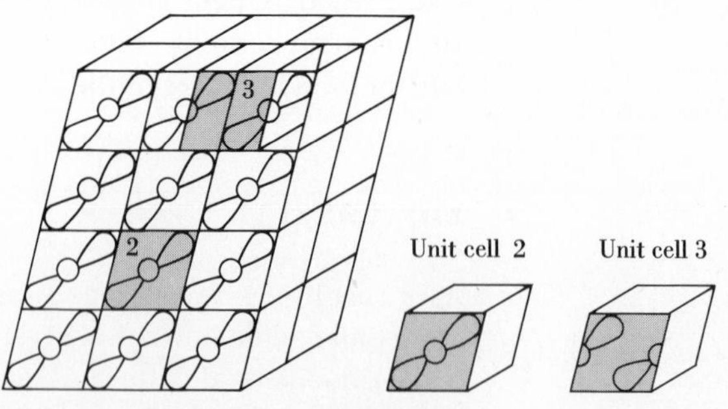

(c) Particles away from lattice intersections
Two possible unit lattices are shaded

repeated indefinitely in two directions and generates a design. The basic pattern of a crystal is called the *unit lattice* or *unit cell*. There are many equivalent ways to choose the unit cell in a crystal, but often one choice is easier to visualize and remember. Figure 7.12*b* and *c* represent the same crystal with different choices for the unit cell in each case. Of the three choices unit cell 2 of Fig. 7.12*c* is simpler and easier to visualize, but the three really are equivalent.

There are three axes along which the unit cell can be moved to generate the entire crystal. These axes need not be mutually perpendicular. If replicas of unit cells are stacked along each of the three directions indicated (Fig. 7.12*a*), the entire crystal is generated.

The task of an x-ray crystallographer is to determine the dimensions of a unit lattice and the positions of the atoms within it. The types of lattices which crystals form range from the very unsymmetrical *triclinic* to the highly symmetrical *cubic* form. The lengths of the three sides of a triclinic unit lattice are unequal, the angles enclosed by its faces are also unequal, and none is 90°. In a cubic unit lattice the lengths of all the sides are equal, and the angles enclosed by the faces are all 90°. We shall consider materials which crystallize in a cubic form in some detail.

7.4 CUBIC CRYSTALS

There are three possible cubic unit lattices, shown in Fig. 7.13. In the *simple*, or primitive, cubic system only the corners of the cube are occupied; in the *face-centered* cube the center of each face and the corners are occupied; and in the *body-centered* cube the corners and the center are occupied.

The rare gases and many metals crystallize to give a face-centered-cubic lattice. The unit-lattice dimensions for some of these elements are given in Table 7.1, which also gives the atomic radius, defined as one-half the distance of the closest approach between two atoms in the crystal. Figure 7.13*b* shows that the distance of closest

FIGURE 7.13

The cubic unit lattices.

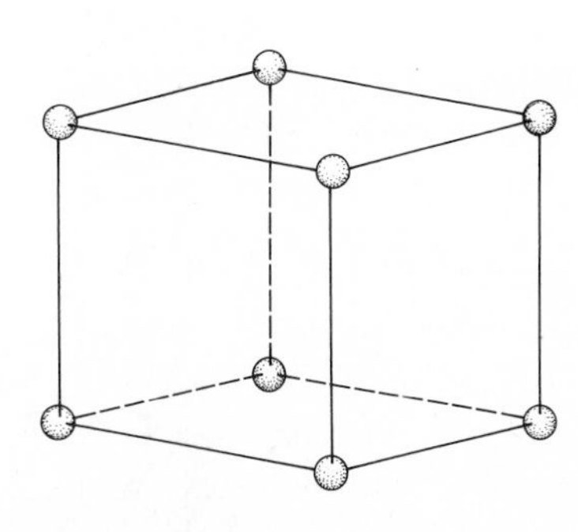

(*a*) *Simple cubic*

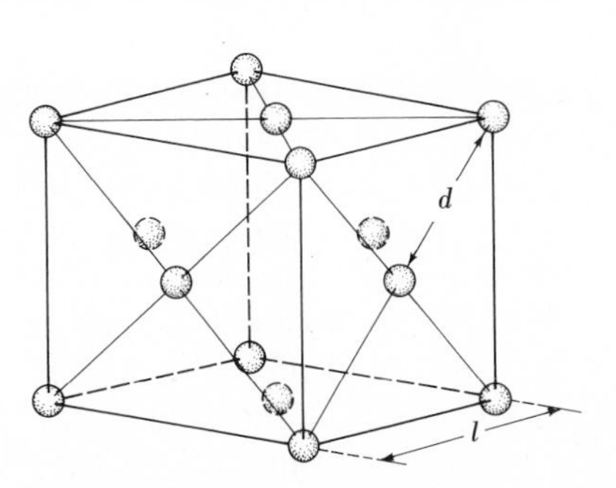

(*b*) *Face-centered cubic* $d = l/\sqrt{2}$

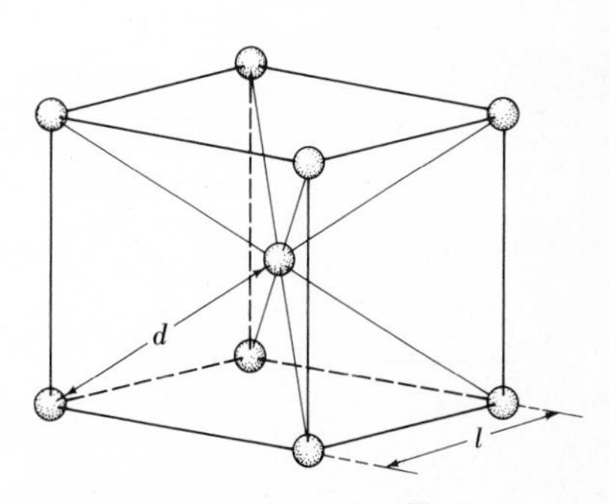

(*c*) *Body-centered cubic* $d = \sqrt{3}\,l/2$

TABLE 7.1: LATTICE DIMENSIONS IN ANGSTROMS FOR SOME CUBIC CRYSTALS

Element	*Length of unit lattice l*	*Atomic radius r*
Face-centered unit cell, $r = \frac{l}{\sqrt{8}}$		
Ne	4.52	1.60
Ar	5.43	1.92
Cu	3.60	1.27
Au	4.07	1.44
Body-centered unit cell, $r = \sqrt{3}\,\frac{l}{4}$		
Li	3.50	1.52
Na	4.30	1.86
K	5.20	2.25
Rb	5.62	2.43
Cs	6.05	2.62

approach d will be the distance from the center of a square face of side l to a corner. The atomic radius r will be $\frac{1}{2}d$. From simple geometric considerations it can be shown that $r = \frac{1}{2}d = l/\sqrt{8}$. The proof of this is left as an exercise (see Prob. 7.5).

The alkali metals crystallize in a body-centered-cubic lattice. Table 7.1 gives the unit cell lengths and atomic radii for these elements. The atomic radius is equal to one-half the distance from the center to a corner of the unit cell, or $r = \frac{1}{2}d = \frac{1}{2}(\sqrt{3}l/2)$ (see Prob. 7.5).

The halides of the alkali metals, except for CsCl, CsBr, and CsI, crystallize in a cubic lattice like the one shown for NaCl in Fig. 7.14*a*. Each sodium ion is surrounded by six equidistant chloride ions and vice versa. The length of the NaCl unit cell is 5.628 Å, or twice the distance between the nearest Na^+ and Cl^- ions. If for a moment we concentrate only on the sodium ions in Fig. 7.14*a* and ignore the chloride ions, we can see that the sodium ions form a face-centered cube. Although it is not quite so evident, the chloride ions also form a face-centered-cubic network. The NaCl crystal is made up of two interpenetrating face-centered-cubic lattices; one contains only Na^+ and the other only Cl^-. There is no unique pair of Na^+ and Cl^- ions. Each Cl^- ion is equidistant from the six nearest Na^+ ions and vice versa. There are no molecules in crystalline NaCl; the formula NaCl represents only the ratio of atoms present in the pure material, not the number of atoms in a molecule.

From the dimensions and mass of the unit cell we can determine the density of the crystal. The mass of the unit cell depends on the number of atoms or ions within it. Let us count the total number of ions in the NaCl unit lattice shown in Fig. 7.14*a*. Each of the sodium ions at the corners is shared by eight unit cells since eight cubes meet at a corner in the lattice. Each corner sodium ion contributes only one-eighth of an ion to the unit cell. There is a sodium ion at the center of each of the six faces. Since faces are shared by two cells, each face ion contributes one-half of an ion to the unit cell. With eight corners contributing one-eighth of an ion and six faces contributing one-half of an ion, there are four sodium ions in the unit cell. The chloride ions are located on the edges and at the center of the cube. Since each edge is shared by four cells, an edge ion contributes one-fourth of an ion to the unit cell. The chloride ion at the center is unshared and contributes one ion to the unit cell. Since there are twelve edges, the total number of chloride ions in the unit lattice is also four, as we would expect from the formula, NaCl, which indicates that the atomic ratio is 1 : 1.

With the number of ions and the dimensions of the unit lattice it is possible to calculate the density of crystalline NaCl. The mass of the unit lattice is the sum of the masses of four sodium and four chloride ions; the mass of a single ion is just the gram-atomic weight divided by Avogadro's number. The volume is just the cube of the

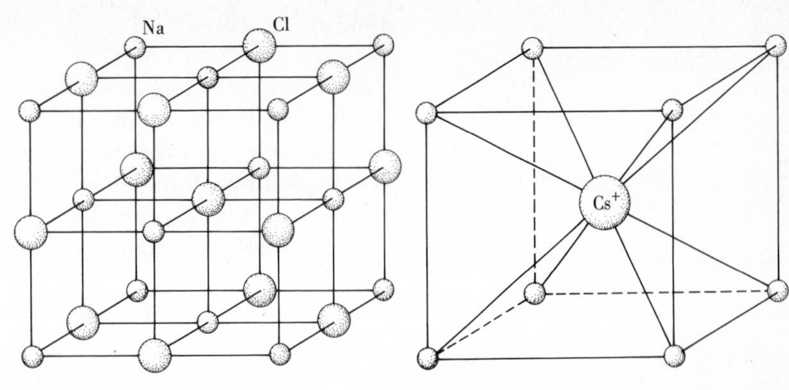

(a) *The sodium chloride lattice*

(b) *The cesium halide unit lattice* (e.g., CsBr)

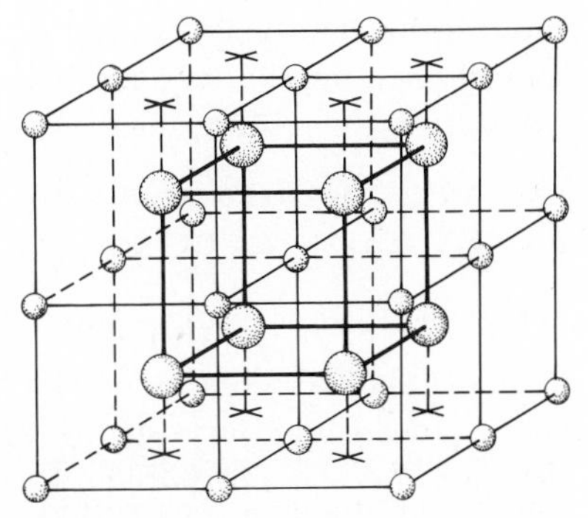

(c) *The cesium halide structure is an interpenetrating simple cubic lattice*

FIGURE 7.14

The alkali halide lattices.

side, 5.628 Å:

$$\text{Mass of unit cell} = 4(\text{mass of one Na ion}) + 4(\text{mass of one Cl ion})$$

$$= 4\,\frac{22.99 \text{ g Na}}{6.023 \times 10^{23} \text{ ions}} + 4\,\frac{35.45 \text{ g Cl}}{6.023 \times 10^{23} \text{ ions}}$$

$$= 3.89 \times 10^{-22} \text{ g}$$

$$\text{Volume of unit cell} = l^3 = (5.628 \times 10^{-8} \text{ cm})^3$$

$$= 1.783 \times 10^{-22} \text{ cm}^3$$

$$\text{Density of NaCl} = \frac{\text{mass}}{\text{volume}} = \frac{3.89 \times 10^{-22} \text{ g}}{1.783 \times 10^{-22} \text{ cm}^3} = 2.177 \text{ g/cm}^3$$

The density of NaCl as determined by conventional methods is 2.165 g/cm^3. If we start with an accurate value of the density of a crystalline material and the unit-lattice dimensions, we can calculate the Avogadro's number N_A. In fact, this is one of the most accurate methods of determining N_A.

Cesium Halides

Cesium fluoride has a lattice structure like NaCl, but the other cesium halides have a different geometry. The CsBr lattice is shown in Fig. 7.14*b*; this lattice is cubic but not face-centered. Each corner is occupied by a bromide ion, whereas the center of each cube is occupied by a Cs^+ ion. Separately the anions and cations form simple cubic lattices. This becomes apparent from a study of Fig. 7.14*c*, where a simple cube formed from Cs^+ ions is prominent. The number of Cs^+ and Br^- ions in the unit lattice (Fig. 7.14*b*) is 1 each since there are eight Br^- ions at each corner and a single unshared Cs^+ ion at the center. Note that each Cs^+ ion is surrounded by eight equidistant Br^- ions. These are referred to as the *nearest neighbors* of the central ion. In addition, the number of nearest neighbors is termed the *coordination number* of the central ion. In CsBr, CsCl, and CsI the coordination number is 8. In the NaCl type of lattice the coordination number is 6 (see Fig. 7.14*a*).

7.5 THE IONIC RADIUS AND ITS EFFECT ON CRYSTAL STRUCTURE

In a crystal of an element all the atoms are identical, and the atomic radius is simply taken as one-half the distance between the closest atoms. In an ionic crystal the anion and cation are different, and although there is no problem in determining the distance between adjacent ions from x-ray experiments, we cannot unambiguously determine the contribution each ion makes to this distance. By comparing the crystallographic data for many ionic substances like NaF, KCl, RbBr, CsI, and Li_2O, Linus Pauling derived a set of self-consistent ionic radii which are presented in Table 7.2. Pauling's ionic radii can be used to estimate the distance between adjacent ions in a crystal. A comparison of Table 7.2 with Table 4.9 shows that the cations are generally smaller than the atoms from which they are derived. Furthermore, considering elements of the same period, the cations are much smaller and more variable in size than the anions.

The cation-anion radius ratio, r^+/r^-, has an important relationship to the crystal structure of ionic materials. When the cation is small compared to the anion, as in NaCl or NaI, the coordination number of the metal ion is 6. When the radii are comparable, as in CsCl, the coordination number is 8. It seems that the larger cations can accommodate a greater number of anion neighbors. In fact, it is possible to predict the number of neighbors, or the coordination number, fairly accurately from a knowledge of the value of r^+/r^- for salts in which the ions have charges of equal magnitude, like Na^+Cl^- and $Zn^{2+}O^{2-}$. When r^+/r^- is equal to or greater than 0.732, the coordination number is 8; when the ratio lies between 0.414 and

TABLE 7.2: IONIC RADII OF SOME OF THE ELEMENTS, Å

Period	IA	IIA	VIA	VIIA
2	Li^+ 0.60	Be^{2+} 0.31	O^{2-} 1.40	F^- 1.36
3	Na^+ 0.95	Mg^{2+} 0.65	S^{2-} 1.84	Cl^- 1.81
4	K^+ 1.33	Ca^{2+} 0.99	Se^{2-} 1.98	Br^- 1.95
5	Rb^+ 1.48	Sr^{2+} 1.13	Te^{2-} 2.21	I^- 2.16
6	Cs^+ 1.69	Ba^{2+} 1.35		
	(NH_4^+, 1.42)	(Zn^{2+}, 0.74)		

0.732, the coordination number is 6; below 0.414 the coordination number is 4. For example, r^+/r^- is 0.402 for ZnS, and the coordination number is 4.

7.6 DIAMOND AND GRAPHITE

The element carbon occurs naturally in two crystalline forms, the precious diamond and the more common graphite. The Braggs determined the structure of diamond in 1913 and found that the unit cell is a modified face-centered cube (Fig. 7.15*a*). To understand this, view the unit cell in the figure as a three-dimensional coordinate system with the ordinary eight octants. It will be seen that, in addition to the atoms at the corners and the centers of each face, there are four interior atoms at the centers of four nonadjacent octants. As can be seen from Fig. 7.15*b*, every interior atom resides at the center of a regular tetrahedron with four other carbon atoms at the vertices. The tetrahedral bonding of the diamond structure is obvious in Fig. 7.15*c*.

The diamond structure is strongly reminiscent of the structure of molecules in which carbon exhibits sp^3 hybridization with four valence bonds directed toward the vertices of a regular tetrahedron. The shortest carbon-carbon distance in diamond is 1.54 Å, which is the same distance exhibited in ethane, $H_3C—CH_3$, and many other carbon compounds. The carbon atoms of diamond are covalently bonded throughout the crystal and form a giant covalent molecule. When atoms are held in a giant network by extensive covalent bonding, the resulting structure is referred to as a *covalent-network lattice*.

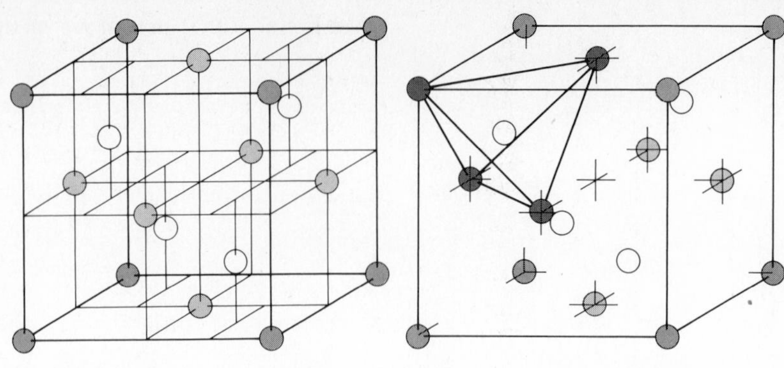

(a) The diamond unit cell is a modified face-centered cube

(b) Every interior atom resides at the center of a regular tetrahedron

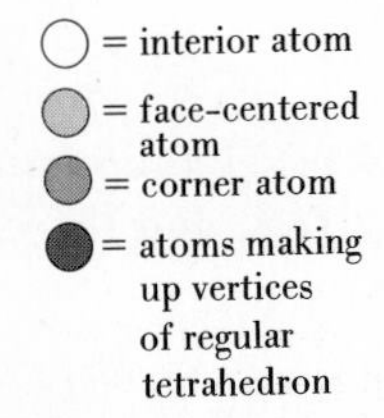

FIGURE 7.15

The unit cell of the diamond structure.

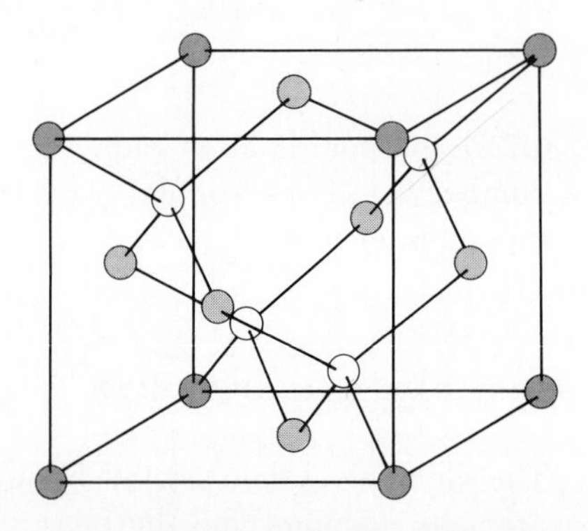

(c) The tetrahedral bonding in the diamond structure

Diamond is a very hard material used for cutting, drilling, and scratching. Other materials which have the covalent tetrahedral structure of diamond are also very hard, e.g., carborundum, SiC, an important abrasive used in sanding wood, polishing metals, and sharpening tools.

When an element occurs in more than one crystalline form, we say it exhibits *polymorphism*. Carbon is polymorphic because it exhibits two crystalline forms, diamond and graphite. The carbon atoms in graphite are arranged in planes; within the planes the atoms occupy the vertices of a network of regular hexagons. As shown in Fig. 7.16, the planar lattice resembles a chicken-wire fence. Each carbon atom is bonded to three others, and all bond angles are 120°. It is obvious that in the graphite lattice the carbon exhibits sp^2, or trigonal, hybridization as in benzene. The carbon-carbon bond length in graphite is 1.42 Å, which is very close to the covalent-bond length in benzene and its derivatives (page 146). The atoms within the planes are covalently bonded in a two-dimensional network and held rigidly in place. The planes them-

FIGURE 7.16

The crystal structure of graphite.

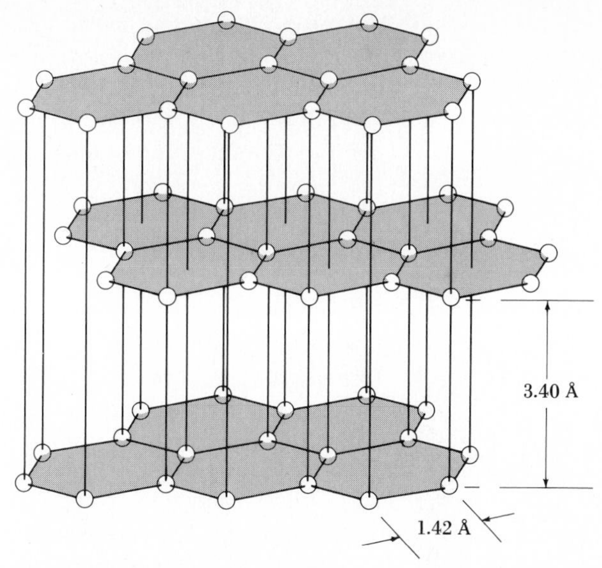

(a) The hexagonal structure of graphite

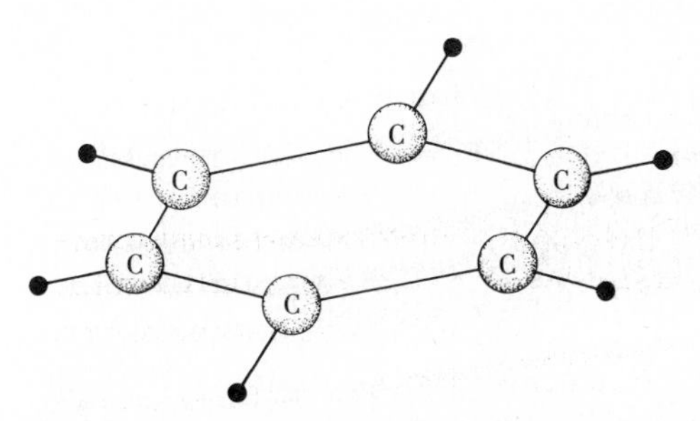

(b) Benzene, C_6H_6, a planar molecule. Graphite and benzene have similar structures. Carbon exhibits sp^2, or trigonal hybridization, in both cases.

selves are relatively far apart (3.40 Å) and are held together by the much weaker forces. When pressure is exerted on graphite, the planes are able to glide over one another, making graphite a fairly good lubricant.

When the electric resistivity of graphite is measured parallel to the hexagonal planes, it is 10^{-4} ohm-cm (or Ω-cm), but measured perpendicular to these planes it is 25,000 times greater. Other properties of graphite, such as mechanical strength and thermal conductivity, depend on direction, and graphite is not unique in this aspect. Many crystalline substances have properties which depend on the direction along which the properties are measured. This is due to the atomic regularity of these substances. The properties of gases and liquids are independent of the direction of observation because the random motion of the atoms and molecules causes all directions through these substances to be equivalent. When the properties of a substance do *not* depend on direction, the substance is said to be *isotropic*. Many crystalline substances are *anisotropic* because their properties depend on the direction of observation through the crystal.

7.7 SNOW AND ICE

Snowflakes are special forms of ice; they are tiny crystals which form when water vapor is converted directly to the solid without forming the liquid first (Fig. 7.17). Snowflakes form high in the atmosphere, where the temperature and the water-vapor pressure are low. Many high clouds are composed of crystalline snowflakes. Ordinary ice forms when the liquid, water, is cooled to 0°C and solidified.

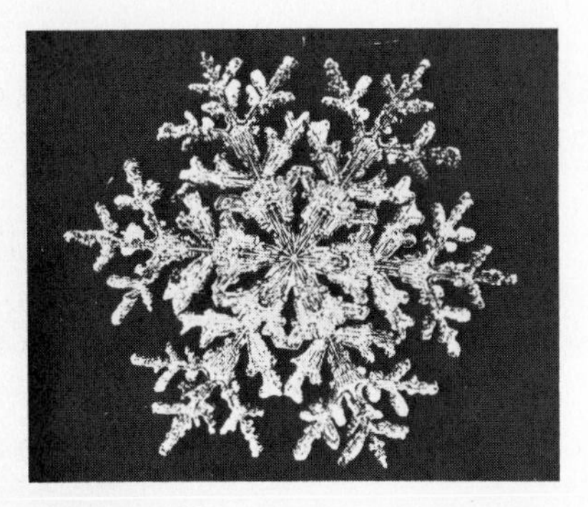

FIGURE 7.17

Photograph of a snowflake. (Courtesy of the American Museum of Natural History.)

The formation of snowflakes has attracted the attention of many scientists because of the importance of snow to precipitation in general. It is believed that much of the rain, even in summer, begins with the formation of snowflakes in the cold upper atmosphere and that melting occurs when the heavier flakes fall toward the earth. Understanding the formation of snowflakes would be a step toward controlling the weather.

Although ice and snowflakes are quite different in appearance, on a molecular level their crystal structures are identical. X-ray studies indicate that the oxygen atoms of the H_2O molecules of ice and snow form a diamondlike structure: each oxygen atom can be considered to be present at the center of a regular tetrahedron with four other oxygen atoms at the vertices, as shown in Fig. 7.18. The distance from the center of the tetrahedron to any vertex is 2.76 Å. If we travel from the center out to the four corners, we find two hydrogen atoms at a distance of 1 Å and two others at a distance of 1.76 Å. The covalent O—H bond length in water is about 1 Å. The 1.76 Å distance is the length of the hydrogen bond, discussed on page 110.

Actually, we can determine the positions of the oxygen atoms from x-ray experiments on the ice structure, but we cannot determine the positions of the hydrogen atoms. The scattering of x-rays depends on the electron density about a nucleus. Since the atomic number of hydrogen is 1, the electron density about the hydrogen nucleus is very low and no scattering is detectable from such a center. The positions of hydrogen atoms must be determined from other experiments.

The difference between the appearance of snow crystals and ordinary ice is related to how the two forms of crystalline H_2O are produced. A snow crystal may form on the surface of a tiny particle

FIGURE 7.18

The structure of ice. (Reprinted from Linus Pauling, "The Nature of the Chemical Bond." Copyright 1939 and 1940 by Cornell University; 3d ed. © 1960 by Cornell University. Used by permission of Cornell University Press.)

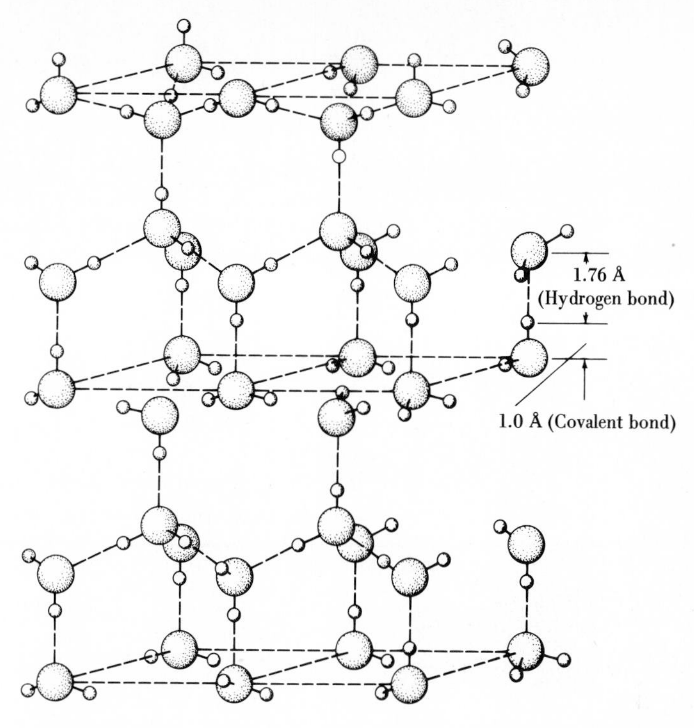

of dust (or ice). Since H_2O molecules are very polar, when they are adsorbed on a surface of a dust particle, they begin to control the orientation of additional molecules as they approach the growing crystal from the gas phase. A pattern is set up, and the crystal grows along preferred axes. This control is absent in the formation of ordinary ice because ice forms from water, not water vapor. Since the molecules of a liquid are very close together, when a tiny crystal begins to form, it is surrounded by many H_2O molecules and the possibility of preferential growth in one direction is limited.

7.8 THE STRUCTURE AND BONDING IN METALS

Table 7.1 shows that metals form relatively simple crystalline lattices. The bonding between the metal atoms in these lattices must explain the unique properties of metals. One of the earliest attempts at explaining the bonding and properties of metals, offered almost immediately after the identification of the electron, is called the *free-electron-gas model*. In this model the valence electrons are detached from their parent atoms and are free to move throughout the entire metal lattice. The valence electrons move randomly, colliding with the ions and the metal surfaces, which are analogous to the walls of a vessel containing a gas. Hence, the metal lattice is permeated by a "gas" made up of the free valence electrons. Bonding results because the negative electron gas reduces the repulsions between positive metal ions of the lattice. Each electron, since it moves through the entire crystal, contributes to the bonding of every pair of ions in the metal lattice.

Figure 7.19 compares the metallic lattice with covalent-network and ionic lattices. In a covalent-network lattice, like diamond, the valence electrons are involved in bonds between the atoms; two atoms share a pair of electrons. In the ionic lattice electrons have been transferred from metal to nonmetal, and the resulting lattice is held together by strong electrostatic forces between adjacent, oppositely charged ions. In the metallic lattice a regular array of positive ions is permeated by a negative electron gas, which bonds the ions together.

The free-electron-gas model explains why the electric conductance of metals is high, why carrying an electric current does not

FIGURE 7.19

A comparison of covalent-network, ionic, and metallic crystal structures.

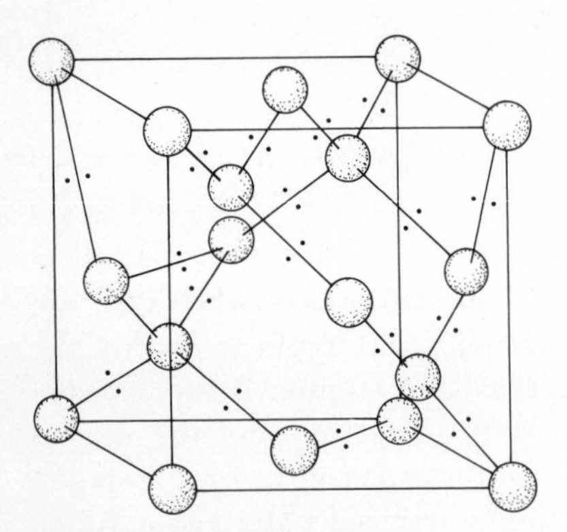

(a) Covalent-network lattice of diamond. Electrons are localized in the covalent bonds of the covalent network of the diamond structure.

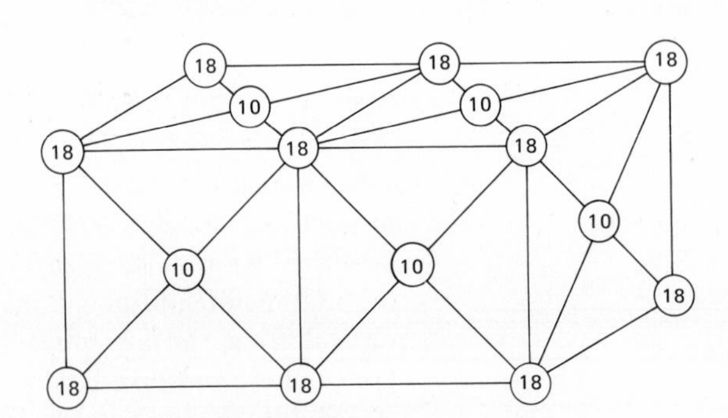

(b) Ionic lattice of Na^+Cl^-. Electrons are localized at the ions in the ionic lattice. There are 18 electrons at the Cl nucleus and 10 at the Na nucleus.

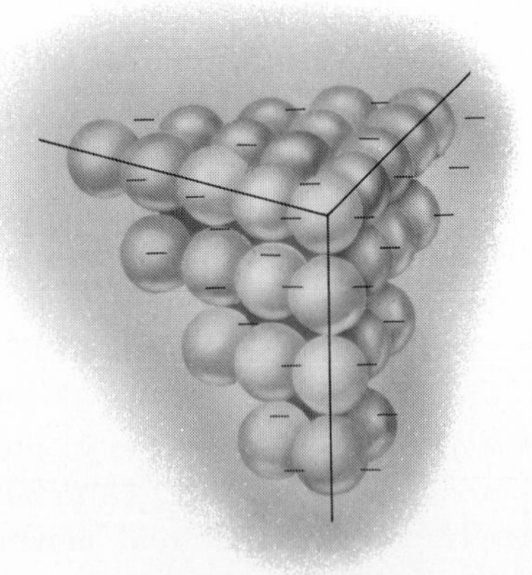

(c) Metallic lattice. Electrons move randomly through the regular array of positive ions.

cause any change in a metal, and why the electric conductance decreases as the temperature increases. When a metal sample is connected to a battery, an electric field is set up in the sample; the electrons are repelled by the negative side of the field and attracted by the positive side. The free-electron gas moves in a preferred direction, and an electric current is generated. If the metal sample is disconnected from the battery, there is no noticeable change in any of its properties because the battery injects electrons at the negative pole and abstracts electrons at the positive pole so that there is no net change in the sample. As the temperature is raised, the ions of the lattice begin to vibrate with higher frequencies and amplitudes. As the electrons move through the lattice, they collide more frequently with the ions. The flow of electrons is impeded, and the conductance is decreased.

Anyone who has stirred hot coffee with a silver spoon knows that the thermal conductance of metals is also high. In the free-electron-gas model the average translational kinetic energy of the electrons is proportional to the temperature, as in an ideal gas. As the temperature of part of a metal sample is raised, the translational kinetic energy of the electrons in that part is raised. This excess thermal energy is quickly transmitted to other parts by electrons which travel through metals with an average speed of $\sim 10^6$ cm/s at room temperature.

The free-electron-gas model explains many of the properties of a metal qualitatively, but since this model does not take into account the wave properties of electrons, we should not be surprised to find that it does not always give good quantitative results.

We get a better understanding of the metallic state from the quantum-mechanical approach because it recognizes that electrons have wave properties. To understand the quantum-mechanical picture of a metal we must recall how the atomic orbitals of individual atoms give rise to molecular orbitals. Lithium forms diatomic molecules in the gas phase. In the molecular-orbital picture of Li_2, a bonding and an antibonding molecular orbital are formed from the 2*s* atomic orbitals on the individual atoms, as shown in Figs. 5.17 and 7.20*a*. Only the bonding orbital is filled, since there are only two valence electrons, one each from the contributing 2*s* atomic orbitals of the lithium atoms. If we consider the bonding of a hypothetical triatomic molecule, Li_3, we have three molecular orbitals formed from the three individual 2*s* atomic orbitals. The lowest would be a bonding orbital, the highest an antibonding orbital, and the intermediate orbital would be nonbonding. The three molecular orbitals could accommodate a total of six electrons, but there would only be three electrons available. As shown in Fig. 7.20*b*, two electrons would go into the bonding molecular orbital and one electron into the *non*bonding molecular orbital. As we add more lithium atoms to our

FIGURE 7.20

The MO models for metals and insulators.

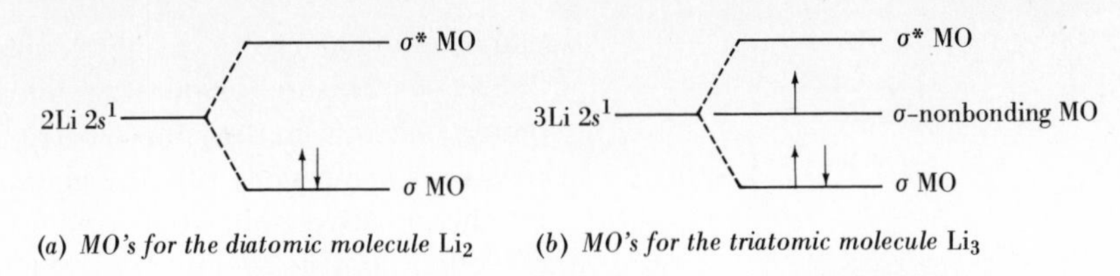

(*a*) *MO's for the diatomic molecule* Li_2 (*b*) *MO's for the triatomic molecule* Li_3

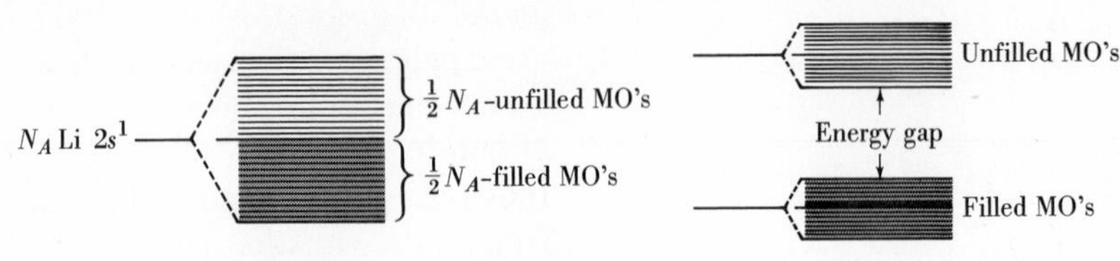

(*c*) *The band model; MO's for metallic lithium* Li_{N_A} (*d*) *The band model for an insulator*

hypothetical molecule, the energy differences between the resulting molecular orbitals diminish; and in the limit of a large number of atoms, i.e., in the limit of an ordinary metallic lattice, the energy difference is very small, so that not very much energy is required to promote an electron from one molecular orbital to the next higher one. Let us picture a sample of metallic lithium containing N_A atoms, as shown in Fig. 7.20*c*. In this giant molecule there will be N_A molecular orbitals, half of which are doubly occupied by electrons. Each molecular orbital extends over the entire array of the lithium atoms and therefore contributes to the bonding of the entire array. An electron in one of these orbitals has a probability of being at any point in the array. (In this sense the quantum-mechanical picture is similar to the free-electron-gas model, in which electrons are free to move through the entire metal lattice.) Since the number of orbitals is extremely large and the energy difference very small, an energy diagram (Fig. 7.20*c*) gives rise to a *band* of closely spaced lines; the quantum-mechanical model is therefore called the *band model*. As we shall see, only the electrons in the highest filled orbitals of the band can give rise to electric and thermal conductance.

In order to observe electric conductance it must be possible to have a net flow of electrons in one direction. The two electrons of a filled molecular orbital tend to stay as far apart as possible. If one electron moves west, the other will move east, and vice versa. Hence, the electrons of a filled molecular orbital will not contribute to the net flow, even in an electric field. Only the single electrons of a half-filled molecular orbital will contribute to the net flow or electric conductance of a metal.

In our picture of the lithium molecule we have a band of closely spaced molecular orbitals, half of which are doubly occupied and half of which are empty. This is a very good picture, but it must be slightly inaccurate, otherwise with all electrons in doubly occupied orbitals lithium would be a nonconductor and a nonmetal. In order to explain metallic conduction single electrons must be present in some of the molecular orbitals. The electrons in the highest filled molecular orbitals of the band can be easily promoted into the lowest unfilled molecular orbitals since only a small amount of energy is required to do this. The electrons in the lower filled molecular orbitals cannot be promoted because too much energy is required to move them into the unoccupied molecular orbitals. Hence, although the valence electrons have access to the entire array, just as in the free-electron-gas model, in the quantum-mechanical picture only a small fraction ($\sim$0.01) is available for electric or thermal conductance.

The conduction in lithium and other metals is due to the small energy gap which separates the highest filled and lowest unfilled molecular orbitals of the band. As shown in Fig. 7.20*d*, the energy gap between the filled and the unfilled orbitals of an insulator is very large, no single electrons are present, and no conduction can occur. The large energy gap in the insulator arises from the fact that the molecular orbitals of an insulator result from different kinds of atomic orbitals in the atoms of the lattice. Since the atomic orbitals are of different energy in the isolated atom, they give rise to bands of different energy. The electrons present in a completely filled lower band cannot be elevated to the unfilled upper band, and the material cannot conduct electricity.

7.9 SEMICONDUCTORS

A vacuum tube and an equivalent semiconductor device.

Vacuum tubes are electron valves which permit the flow of current in only one direction. They are relatively large, especially when they are designed to carry high currents, and they require a lot of energy, since they can operate only when they are very hot. A *transistor*, which is made from a material called a *semiconductor*, is also a one-way valve, but it operates without heat activation, is much simpler and smaller, and has a very low energy requirement. The application of semiconductor materials has revolutionized the electronics industry.

Typical semiconductors are made from extremely pure silicon to which traces of aluminum or phosphorus have been added. Note that silicon is in family IVA whereas aluminum is in family IIIA and phosphorus is in family VA. Pure silicon is an insulator, having low electric conductance. When it is *doped* with a trace of aluminum to the extent of 1 part in 10 million, its electric resistivity drops from 60,000 to 2 Ω.

Silicon has four valence electrons and forms a diamondlike crystal; i.e., each silicon atom is bonded to four others through

FIGURE 7.21

Positive and negative semiconductors.

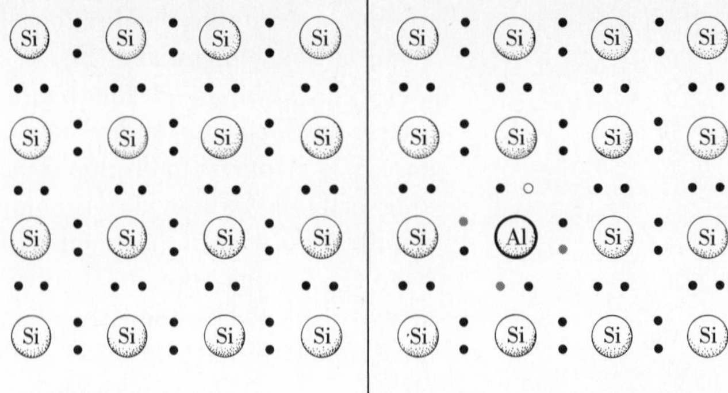

(*a*) *Pure silicon*

(*b*) *Silicon doped with aluminum*
○ = vacancy; ● = electrons "from Al atom"

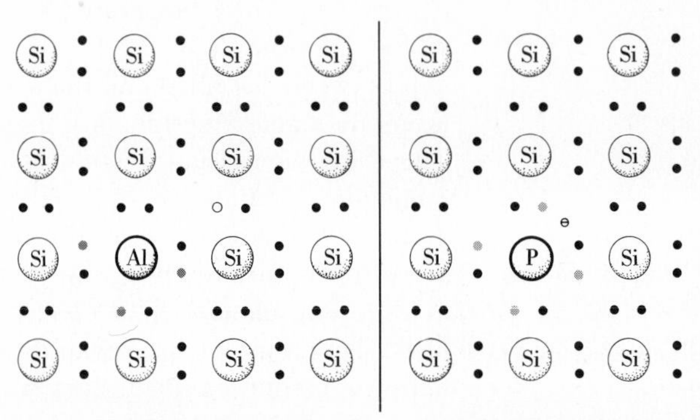

(*c*) *The vacancy which was originally on* Al *migrates to an adjacent* Si *atom*

(*d*) *A phosphorus atom causes an excess electron in the silicon lattice*
⊖ = excess electron;
● = electrons "from P atom"

covalent, two-electron bonds, as shown schematically in Fig. 7.21. Let us examine the mechanism by which the addition of traces of group IIIA or VA elements to pure silicon causes the electric conductivity to increase disproportionately.

When an aluminum atom takes the place of silicon in the lattice, it can contribute only three valence electrons to the bonding. The aluminum atom is pictured in the silicon lattice in Fig. 7.21*b* with three regular two-electron bonds and one one-electron bond. The one-electron bond consists of an electron and a *vacancy*. Nearby electrons can fill this vacancy, simultaneously causing a new vacancy. The net effect is that a vacancy can migrate through the lattice, as shown in Fig. 7.21*c*. The migration of a vacancy in one direction is equivalent to the migration of an electron in the opposite direction. A semiconductor which has a deficiency of valence electrons, i.e., in which vacancies are present, is called a *positive* or *p*-type semiconductor.

When silicon is doped with phosphorus, the electric conductance is again dramatically increased. When a phosphorus atom takes the place of silicon, it contributes five valence electrons instead of four. This excess electron is only loosely bonded to the atom and can migrate through the doped silicon lattice. Such a semiconductor, called a *negative* or *n*-type semiconductor, is depicted schematically in Fig. 7.21*d*.

When a *p*-type and an *n*-type semiconductor are joined together, the junction permits the flow of electricity in only one direction. At the junction the excess electrons of the *n*-type semiconductors may jump into the neighboring vacancies presented by the *p*-type semiconductors. As a result some of the excess electrons and vacancies are depleted near the junction, as shown in Fig. 7.22*a* and *b*. This leaves the *n*-type semiconductor positively charged and the *p*-type negatively charged. If we now connect the negative pole of battery to the *n*-type and the positive pole to the *p*-type, as shown in Fig. 7.22*c*, electrons will be pumped into the *n*-type and vacancies will be pumped into the *p*-type semiconductors. The excess electrons and vacancies migrate toward each other under the electric field produced by the battery; at the junction, they annihilate each

FIGURE 7.22

The p-n junction permits electron flow in one direction only.

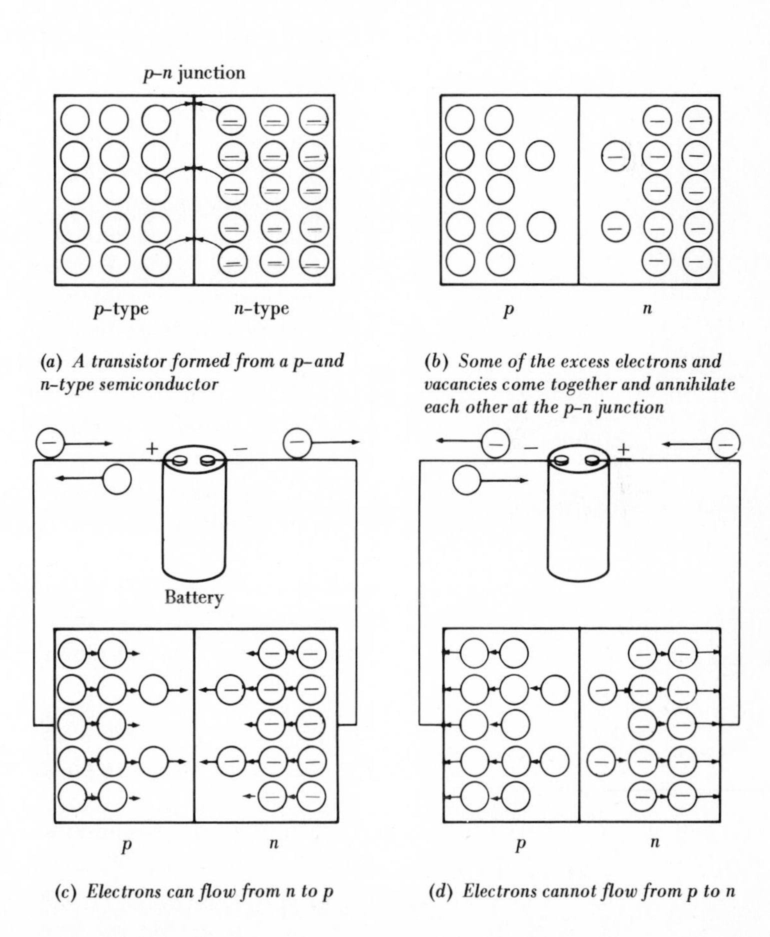

(*a*) *A transistor formed from a p– and n–type semiconductor*

(*b*) *Some of the excess electrons and vacancies come together and annihilate each other at the p–n junction*

(*c*) *Electrons can flow from n to p*

(*d*) *Electrons cannot flow from p to n*

FIGURE 7.23

Band model for doped semiconductors.

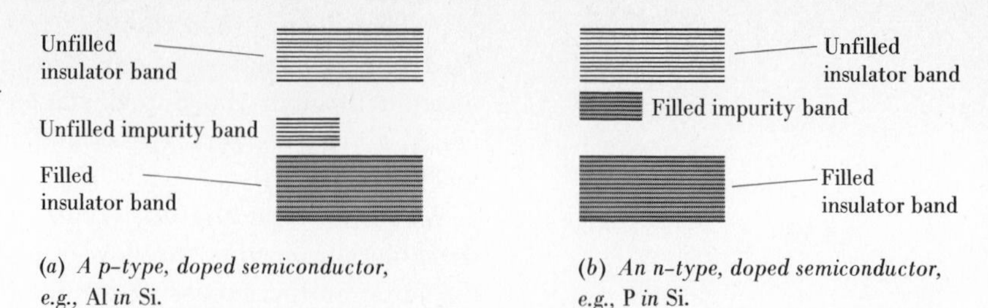

(*a*) *A p–type, doped semiconductor,* e.g., Al *in* Si.

(*b*) *An n–type, doped semiconductor,* e.g., P *in* Si.

other so that an electric balance is maintained and current flows. If we reverse polarity, we pump electrons into the p-type and vacancies into the n-type semiconductor. This would quickly deplete the two types of *carriers* (electrons and vacancies). The semiconductor would then resemble an insulator, and no current would flow. In summary, electrons can pass across the p-n junction in the direction n to p but not vice versa, and, like a vacuum tube, the p-n junction is a one-way valve for electron flow.

Semiconductor devices called *transistors* can be made wafer-thin and extremely small. Since no heating is required for their operation, the solid-state circuit of a transistor radio responds almost immediately when it is turned on. More important, such a circuit consumes much less energy than a vacuum-tube circuit.

Silicon is the most commonly used material for fabrication of semiconductors, but in general any very pure family IVA element, like silicon or germanium, can be doped with family IIIA elements, such as aluminum, gallium, and indium to make p-type semiconductors, or with family VA elements, such as phosphorus, arsenic, and antimony, to make n-type semiconductors.

The behavior of doped semiconductors is nicely explained by the band model. The doped semiconductor is an insulator to which an impurity has been added. When an electron-deficient material like aluminum is added, a low-energy band of unfilled molecular orbitals becomes accessible to the paired electrons of the insulator band, as shown in Fig. 7.23*a*. The set of molecular orbitals introduced by doping with aluminum is called an *impurity band*. When an electron-rich material like phosphorus is added, a high-energy band of *filled* molecular orbitals is introduced, as shown in Fig. 7.23*b*. The electrons in the impurity band can be elevated to the unfilled molecular orbitals of the insulator, and conduction results.

7.10 SUBLIMATION ENERGY AND THE BONDING IN CRYSTALLINE SOLIDS

Sublimation is the process whereby a solid is converted into a vapor. Assuming that the average separation between particles in the vapor is very high, so that no appreciable forces exist among

TABLE 7.3: SUBLIMATION ENERGIES, kcal/mol

Ionic crystals	
NaCl	181.6
KCl	165.4
LiBr	184.4
NaI	161.9
Metallic crystals	
K	20.9
Cu	81.52
Ag	69.12
Au	82.29
Fe	96.68
W	201.6
Covalent-network crystals	
C, graphite	171.698
diamond	171.245
Si, diamond structure	88.04
Molecular crystals	
Ar	1.58
I_2	14.876
P_4	13.12
H_2O	12.15†
CO_2	8.17

† Hydrogen-bonded.

them, the energy required for sublimation must just equal the work which must be done to overcome the forces holding the particles in place in the crystalline lattice. Hence, the energy required for sublimation of a crystalline material is a good indication of the attractive forces which operate between the particles.

The sublimation energy for ionic crystalline materials is very high, requiring approximately 200 kcal/mol in many cases (see Table 7.3). Quite strong forces are also operative in covalent-network crystals like diamond and graphite, so that when ionic or covalent bonds form a giant network, the energy required to separate the ionic or atomic units of the solid crystal is quite high.

The sublimation energy for metals, also quite high, is roughly proportional to the number of valence electrons. Both the free-electron-gas model and the molecular orbital bond model for metals have electrons completely delocalized. Each electron contributes to the bonding of every possible pair of atoms in the structure. This multiple bonding results in relatively high sublimation energies.

Crystals composed of relatively small molecules, like CO_2, H_2O, or argon, have low sublimation energies. This is indicative of the fact that only relatively weak *inter*molecular forces hold the molecules in their lattice positions. Molecular crystals are conveniently divided into two categories, *van der Waals* crystals and hydrogen-bonded crystals. In the van der Waals crystals the attractive forces at work are the same as those which required the introduction of the van der Waals constant a to account for the attractive forces between gaseous molecules. The van der Waals forces are weak electric forces, exclusive of hydrogen bonding, which cause real gas molecules to deviate from ideal-gas behavior and are responsible for liquefaction of gases and solidification. The attraction between polar molecules is an example of the action of van der Waals forces.

Ice is an example of a molecular crystal which involves hydrogen bonding. There are many organic and biological crystals which involve hydrogen bonding. Since van der Waals forces also operate in these cases, the sublimation energies of the hydrogen-bonded crystals are generally higher than the sublimation energies of other molecular crystals.

7.11 THE FUTURE FOR X-RAY DIFFRACTION ANALYSIS

The power of x-ray crystallographic methods is now being used to elucidate complicated biological molecules, on the natural assumption that we shall have a clearer understanding of life processes if we can determine the structure of the molecules involved. The importance of such studies can be gauged by the fact that the 1962 Nobel prize for chemistry was shared by Max F. Perutz, for his determination of the structure of horse hemoglobin (molecular weight = 65,000), and by John C. Kendrew, for his determination of the structure of the muscle protein myoglobin (molecular weight = 17,000). In the same year Francis H. C. Crick, James D. Watson, and Maurice H. F. Wilkins received the Nobel prize for medicine and physiology for the brilliant inference, based on x-ray data, that deoxyribonucleic acid (DNA), a material which is of utmost importance in the transfer of genetic information, has the form of a double helix (see Figs. 16.4 and 16.8 for the structures of these important biological materials).

QUESTIONS AND PROBLEMS

7.1 (*a*) What is the basic similarity of light waves and x-rays? (*b*) What is the chief difference?

7.2 In what way do electromagnetic waves behave like the trigonometric sine or cosine functions?

7.3 (*a*) In the derivation of the Bragg equation, Eq. (7.3), we spoke of the reflection of x-rays by crystal planes. Why is this incorrect? (*b*) What is diffraction? (Refer to an introductory physics text.)

7.4 Using the information in Table 7.1, make sketches of the argon and potassium unit cells, including dimensions.

7.5 (*a*) Show that the atomic radius r for an element crystallizing in a face-centered unit cube is just $r = l/\sqrt{8}$, where l is the length of the cube (see Fig. 7.13). (*b*) Show that the atomic radius for an element crystallizing in a body-centered unit cube is just $r = (\sqrt{3}/4)l$. (*c*) Calculate the atomic radius for copper and rubidium from the length of the unit lattice (Table 7.1).

7.6 (*a*) From Table 7.1 determine the number of atoms present in the unit cell of gold. (*b*) Calculate the density of gold. (The accepted value is 19.3 g/cm^3.)

7.7 The density of lithium is 0.534 g/cm^3. Using the information in Table 7.1, determine (*a*) the number of atoms in the lithium unit cell and (*b*) Avogadro's number.

7.8 Using Table 7.2 and the criteria given in Sec. 7.5, what are the coordination numbers in the following compounds: (*a*) LiI, (*b*) LiF, (*c*) CsF, (*d*) CsI, (*e*) CaO, (*f*) SrSe, and (*g*) MgO?

7.9 The unit cell of diamond is a modified face-centered unit cube. (*a*) What is the difference between a truly face-centered unit cube and the diamond unit cube? (*b*) How many more atoms are present in the diamond unit cube?

7.10 On the basis of the crystal structure of graphite, (*a*) why is it a good lubricant and (*b*) why is it anisotropic?

7.11 Ice and snow crystals are said to have a *diamondlike* structure.

What does this mean? (Compare the positions of the oxygen atoms in ice with the carbon atoms of diamond.)

7.12 How does the free-electron-gas model explain (*a*) how the ions of a metal lattice are bonded together and (*b*) the high electric and thermal conductivity of metals?

7.13 How does the band model explain (*a*) the bonding in metals and (*b*) why only a small fraction of the valence electrons is available for electric or thermal conduction?

7.14 (*a*) According to the band model, what is the difference between metals and insulators? (*b*) How does doping convert an insulator into a semiconductor?

7.15 State whether the following materials would be *n*-type or *p*-type semiconductors: (*a*) Si doped with Ga, (*b*) Si doped with As, (*c*) Ge doped with In, (*d*) Ge doped with Al.

7.16 Make a sketch of (*a*) a germanium lattice (diamondlike) doped with In and (*b*) a silicon lattice doped with As.

7.17 Explain why electrons can flow in only one direction across a *p-n* junction.

7.18 How does the band model explain electric conduction in (*a*) *p*-type semiconductors and (*b*) *n*-type semiconductors?

7.19 (*a*) What is meant by van der Waals forces? (*b*) How do they affect the properties of gases? (*c*) Explain the statement: all substances exhibit van der Waals forces because all substances, including the noble gases (He, Ne, etc.), can be condensed.

7.20 Describe and give an example of (*a*) a covalent-network crystal, (*b*) an ionic crystal, (*c*) a van der Waals molecular crystal, and (*d*) a hydrogen-bonded molecular crystal.

7.21 On the basis of the nature of the bonding and the data in Table 7.3 arrange the following substances in order of increasing sublimation energy: (*a*) sodium metal, (*b*) $Mg^{2+}O^{2-}$, (*c*) silicon, (*d*) solid carbon monoxide (CO is a polar molecule), (*e*) Dry Ice (CO_2 is a linear molecule, O=C=O), (*f*) ammonia, NH_3.

Stoichiometry practice

7.22 Many precious elements are present in seawater in the form of compounds. The weight percentages of copper, silver, and gold are 5×10^{-7}, 2×10^{-8}, and 6×10^{-10}, respectively. How many g atoms of each are present in 1 ton of seawater?

7.23 Zinc reacts with dilute sulfuric acid to produce hydrogen gas and zinc sulfate. (*a*) Write a balanced equation for the reaction. (*b*) How many moles of H_2 can be produced from 100 g of Zn and excess sulfuric acid? (*c*) What is the volume of the H_2 at 25°C and 750 Torr?

7.24 Gaseous acetylene, C_2H_2, can be produced by the hydrolysis of calcium carbide: $CaC_2 + H_2O \rightarrow C_2H_2 + Ca(OH)_2$. Calcium carbide can be manufactured by heating a mixture of coal and CaO: $CaO + C(coal) \rightarrow CaC_2 + CO$. (*a*) Balance both equations. How many grams of (*b*) coal (80 percent carbon by weight) and (*c*) CaO are necessary to produce 150 l of acetylene at 25°C and 100 atm?

7.25 When zinc is treated with hydrochloric acid, hydrogen gas and zinc chloride are formed. (*a*) Write a balanced equation for the reaction. (*b*) The density of hydrogen gas is 0.0828 g/l. What volume of gas would be generated from 10.0 g of zinc? (*c*) If 20.0 g of zinc chloride was formed in a reaction how many moles of zinc were decomposed and how many molecules of hydrogen were formed?

8 LIQUIDS AND A COMPARISON OF THE THREE STATES OF MATTER

8.1 INTRODUCTION

Before discussing the liquid state and comparing it with the two other states of matter we should examine some of the simple, distinguishing features of gases, solids, and liquids; features that almost everyone knows. Gases have no definite shape or volume and flow readily under an applied stress. Liquids also flow readily, but like solids, they have a definite volume; 18 g of water has a volume of 18 ml, and 18 g of ice has a volume of 19.7 ml. On the other hand 18 g of water vapor has an indefinite volume; its volume is exactly equal to the volume of its container, no matter how large the container may be. A solid is rigid and has a tendency to maintain its shape even under high stress.

There is much evidence to indicate that the liquid state is a natural intermediate between the solid and gaseous states. Many materials can be observed to pass from solid, to liquid, to gas as their temperatures are raised; the liquid state appears as an intermediate in these transformations. The density of a substance almost invariably decreases as it passes from solid, to liquid, to gas. As we shall see, there are many other properties which demonstrate the intermediate position of liquids. A good molecular model for the liquid state must reflect its intermediacy.

8.2 EVAPORATION AND EQUILIBRIUM VAPOR PRESSURE

It is common knowledge that when water and many other liquids are allowed to stand indefinitely in an open vessel, they eventually evaporate. This must mean that some molecules in the liquid reach a sufficiently high kinetic energy to escape the attractive forces exerted by neighboring molecules. We might imagine a molecule near the surface being ejected from the body of the liquid as a result of a collision with a very fast molecule, as shown in Fig. 8.1.

We learned in Chap. 6 that the average velocity and kinetic energy of the molecules increase as the temperature increases. It should not be surprising, therefore, that a liquid evaporates faster as its temperature increases. The rate at which the molecules are knocked out of the liquid surface and into the vapor phase should increase as the average velocity and kinetic energy of the molecules increase. At higher temperatures a greater fraction of the molecules can attain sufficient energy to overcome the intermolecular forces of the liquid phase.

Since the molecules must reach the surface of a liquid before they can escape, it is obvious that evaporation should occur more rapidly as the surface area is increased. Common observations support this view: whereas days may be required for the complete evaporation of a glass of water, the same amount of water spilled on the floor over a wide area will evaporate in a few hours. Although the rate at which a liquid evaporates depends on temperature and surface area, there is another important consideration. Benzene evaporates faster than

FIGURE 8.1

Ejection of a molecule from liquid into vapor by collision with a high-energy molecule.

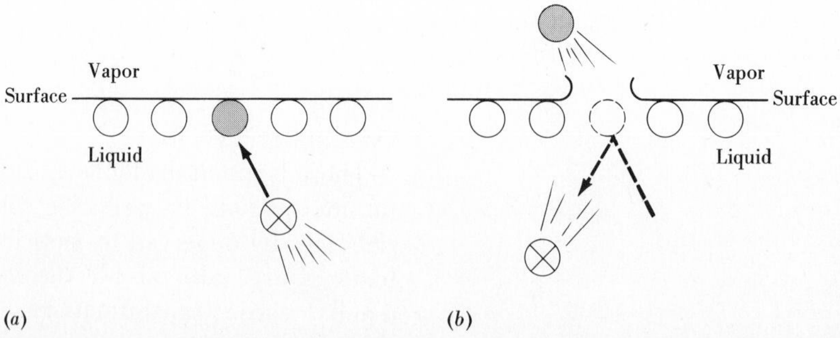

Vapor
Surface
Liquid
(a)

Vapor
Surface
Liquid
(b)

water even at the same temperature and with the same surface area, because benzene has a higher *equilibrium vapor pressure*. The equilibrium vapor pressure of a liquid is the pressure that the vapor above the liquid reaches when the liquid is in a sealed, thermostatted vessel.

The equilibrium vapor pressure of a pure liquid depends only on the temperature. When a pure liquid is thermostated in a closed vessel, its vapor pressure reaches a constant value. In terms of a molecular picture we imagine that molecules escape from the body of the liquid and speed about randomly in the vapor space above it. As the number of molecules in the vapor phase increases, they begin to collide with each other. In their random motions some molecules attain trajectories which cause them to return to the liquid. Eventually the population in the vapor reaches such a value that the number of molecules returning to the liquid phase equals the number escaping. At this point the number of molecules or moles in the vapor phase becomes constant, and since the temperature and volume of the vapor are fixed in a thermostated, rigid vessel, the pressure also is fixed. This can be seen by analysis of the ideal-gas law which many vapors obey: $P = nRT/V$; if n, T, and V are fixed, the pressure is fixed.

Let us reexamine the above explanation of the fact that the equilibrium vapor pressure is constant if the temperature is fixed. At the macroscopic, observable level we do not see the individual molecules escaping from and returning to the liquid; we can only determine that the vapor pressure above the thermostated liquid becomes constant or *static*. We could say that the liquid has stopped evaporating and the liquid and vapor are in *static* equilibrium. But since we believe that liquids and vapors are made up of molecules which are in incessant, random motion, we are forced to construct a *dynamic*, microscopic picture. In this case we choose a *dynamic equilibrium:* Two opposing molecular processes (escape from the return to the liquid) are occurring at the same rate, and the vapor pressure above a liquid reaches a constant value. The concept of dynamic equilibrium is useful in explaining many chemical phenomena, and we shall use it again.

8.3 MEASUREMENT OF VAPOR PRESSURE AND THE HEAT OF VAPORIZATION

There are many methods of measuring the vapor pressure of a liquid. A sketch of the apparatus for a straightforward method is shown in Fig. 8.2. A liquid is placed in a sealed glass vessel equipped with a mercury U-tube manometer.† The entire assembly

† A manometer is any device used for measuring gas or vapor pressure.

FIGURE 8.2

Measurement of the vapor pressure of a liquid.

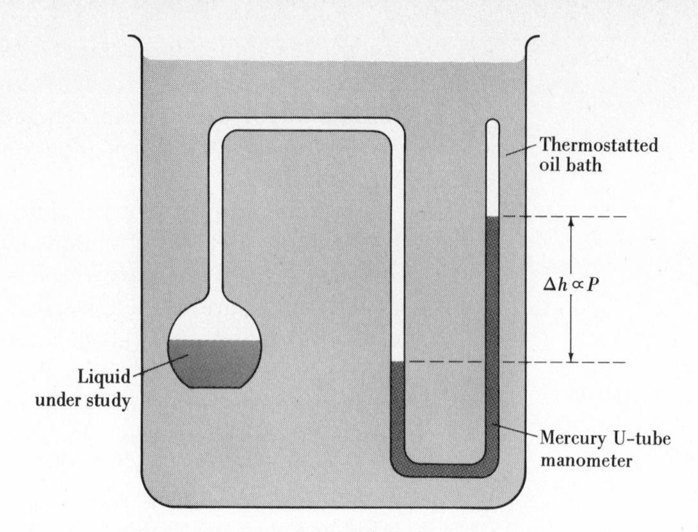

is immersed in a thermostated oil bath so that the temperature can be controlled. The vapor pressure of the liquid is determined by measuring the difference in the heights of the mercury columns in the U tube. Table 8.1 and Fig. 8.3 show the results of these kinds of measurements for a few liquids. It is obvious that the temperature has a very strong influence on the vapor pressure of a liquid. As a rough approximation the equilibrium vapor pressure doubles for each 10°C rise in temperature. Often the temperature dependence of the vapor pressure can be expressed quite accurately by logarithmic† function:

$$\ln p = \ln A - \frac{B}{T} \tag{8.1}$$

or

$$\log p = \log A - \frac{B}{2.303} \frac{1}{T} \tag{8.2}$$

The constants, A and B, depend on the nature of the liquid. As we shall see, B is related to the heat required to vaporize 1 mol of liquid. If we take $\log P$ and $1/T$ as dependent and independent variables, respectively, we see by comparison that Eq. (8.2) is in the form of the equation for a straight line:

$$\log p = \log A - \frac{B}{2.303} \frac{1}{T}$$

$y = b + mx$ equation for straight line

† This book follows the common practice of using log for logarithms to the base 10 and ln for natural logarithms, i.e., to the base e.

TABLE 8.1: THE EQUILIBRIUM VAPOR PRESSURE IN TORR FOR SOME COMMON LIQUIDS

	Temperature, °C						
	0	*20*	*40*	*60*	*80*	*100*	*Heat of vaporization,* ΔH, cal/mol
H_2O	4.579	17.535	55.324	149.38	355.1	760.00	9700
Ethanol, C_2H_5OH	12.2	43.9	135.3	352.7	812.6	169.0	9448
Benzene, C_6H_6	(Solid)	74.7	182.7	391.7	757.6	—	7360
Acetic acid, CH_3COOH	(Solid)	11.7	34.8	88.9	202.3	417.1	5810

A plot of log p, which corresponds to y, versus $1/T$, which corresponds to x, would yield a straight line with a slope m of $-B/2.303$, and an intercept b equal to log A. Figure 8.4 is a plot of log p versus $1/T$ for benzene using the data from Table 8.1. The plot is linear with a slope of -1632 and an intercept (not shown) of 7.551. Now we can substitute the values for the constants log A and $-B/2.303$ into Eq. (8.2) to obtain an equation which is specific for benzene:

$$\log p = 7.551 - \frac{1632}{T}$$

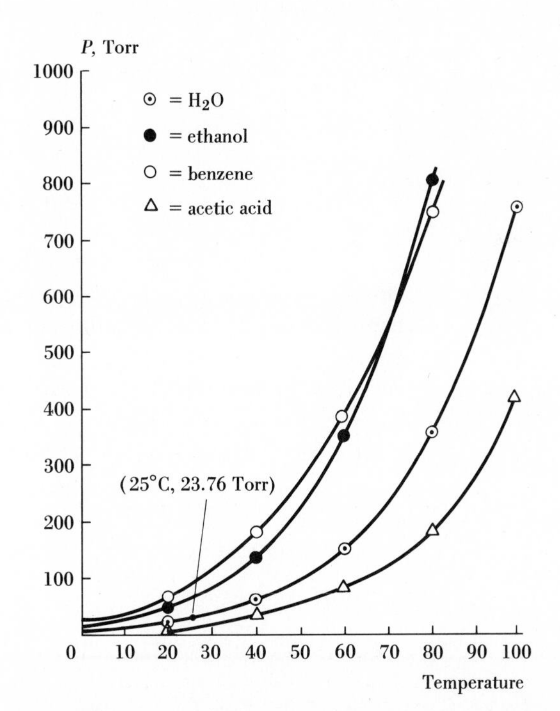

FIGURE 8.3

The influence of temperature on the equilibrium vapor pressure.

FIGURE 8.4

A plot of log *p versus 1/T for benzene.*

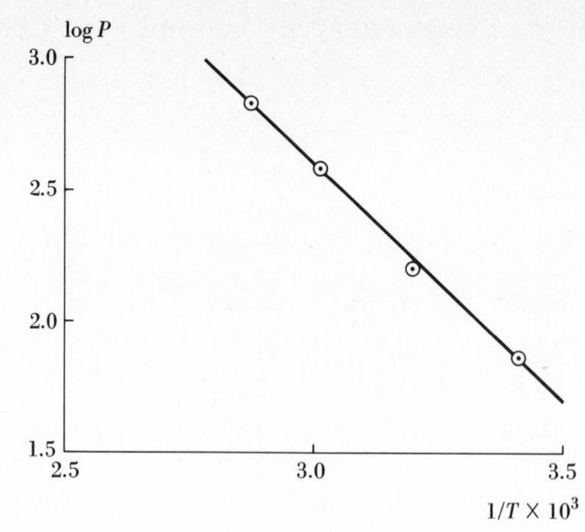

We can use this equation to estimate the vapor pressure of liquid benzene at any temperature between 0 and 80°C. It is really valid only in this temperature range because it was derived from the data in Table 8.1, which cover this range.

From thermodynamic theory it can be shown that B is simply related to the minimum amount of heat required to vaporize 1 mol of liquid at constant pressure. When water is vaporized at 100°C, that is, when water is boiled at atmospheric pressure, the process occurs at constant pressure. The boiling of any liquid at atmospheric pressure is essentially a vaporization process which occurs at constant pressure, and the heat associated with the process is called the *heat of vaporization*. The mathematical relationship between B and the heat of vaporization is

$$B = \frac{Q_p}{R} \tag{8.3}$$

Q is the symbol for heat; the subscript p indicates that we are considering heating at constant pressure. In Eq. (8.3) Q_p is the heat of vaporization, R is the gas-law constant, which must have units of energy per mole per kelvin. Since heat is usually measured in calories, the value of R we shall use in this discussion is 1.987 cal/mol-K.

The heat of vaporization Q_p is also called the *enthalpy change* ΔH. The symbol Δ, called *delta*, means "change in"; H is the symbol for enthalpy. ΔH_v will be termed the *heat* or *enthalpy of vaporization*. Equation (8.3) can be written

$$B = \frac{\Delta H_v}{R} \tag{8.4}$$

and Eq. (8.2) can be written

$$\log p = \log A - \frac{\Delta H}{2.303R}\frac{1}{T} \tag{8.5}$$

Equation (8.5) indicates that a plot of log p versus $1/T$ would give a straight line with a slope equal to $-\Delta H/2.303R$:

$$\text{Slope} = -\frac{\Delta H}{2.303R} \tag{8.6}$$

As we have noted, the plot of the benzene data in Fig. 8.4 gives a slope of −1632. From Eq. (8.5) we obtain an enthalpy of vaporization for benzene of 7480 cal/mol:

$$-1632 = -\frac{\Delta H}{2.303R}$$

$$\Delta H = 1632 \times 2.303R = 1632 \times 2.303 \times 1.987 = 7468 \text{ cal/mol}$$

The enthalpy of vaporization determined by measuring the heat required to vaporize 1 mol of benzene by direct heating at 80°C is 7360 cal/mol. In this direct determination of the enthalpy of vaporization the experiments are performed at a single temperature. When the enthalpy of vaporization is determined from vapor-pressure data, the value represents an average over the temperature range for which we have vapor-pressure measurements. This explains the difference in the results. Enthalpies of vaporization determined directly are given in the last column of Table 8.1.

EXAMPLE 8.1 Using the equilibrium vapor pressure at 0 and 80°C given in Table 8.1, calculate the enthalpy of vaporization for ethanol.

Solution: We need only use Eq. (8.5) to solve the example.

$$\log p = \log A - \frac{\Delta H_v}{2.303R}\frac{1}{T} \tag{8.5}$$

First, substitute the 0°C data where $p = 12.2$ Torr and $T = 273$ K:

$$\log 12.2 \text{ Torr} = \log A - \frac{\Delta H_v}{2.303R} \times \frac{1}{273 \text{ K}} \tag{a}$$

Then write another equation substituting the 80°C data where $p =$ 812.6 Torr and $T = 353$ K:

$$\log 812.6 \text{ Torr} = \log A - \frac{\Delta H_v}{2.303R} \times \frac{1}{353 \text{ K}} \tag{b}$$

Subtracting (a) from (b) eliminates the constant, log A, and the only unknown is ΔH, the enthalpy of vaporization:

$$\log 812.6 - \log 12.2 = \left(-\frac{\Delta H_v}{2.303R} \times \frac{1}{353 \text{ K}}\right) - \left(-\frac{\Delta H_v}{2.303R} \times \frac{1}{273 \text{ K}}\right)$$

$$\log \frac{812.6}{12.2} = -\frac{\Delta H_v}{2.303R}\left(\frac{1}{353 \text{ K}} - \frac{1}{273 \text{ K}}\right)$$

Substituting $R = 1.987$ cal/mol-K and $\log(812.6/12.2) = \log 66.6 = 1.823$, we can obtain ΔH_v:

$$\Delta H_v = -\frac{2.303 \times 1.987 \text{ cal/mol-K} \times 1.823}{1/353 \text{ K} - 1/273 \text{ K}} = 10{,}100 \text{ cal/mol}$$

Note that the *average* value of ΔH_v calculated here is slightly higher than the value in Table 8.1; the value in the table is the value of ΔH_v at the boiling point of ethanol.

EXAMPLE 8.2 Using the enthalpy of vaporization for water given in Table 8.1 and the normal boiling point, estimate the equilibrium vapor pressure of water at 90°C.

Solution: Again we need only use Eq. (8.5). First substitute $\Delta H_v = 9700$ cal/mol

$$\log p = \log A - \frac{\Delta H_v}{2.303R}\frac{1}{T} = \log A - \frac{9700 \text{ cal/mol}}{2.303 \times 1.987 \text{ cal/mol-K}}\frac{1}{T}$$

Then substitute the normal boiling point; i.e., substitute $T = 373$ K at $p = 760$ Torr:

$$\log 760 = \log A - \frac{9700 \text{ cal/mol}}{2.303 \times 1.987 \text{ cal/mol-K}} \times \frac{1}{373 \text{ K}}$$

The only unknown in the last equation is $\log A$. Solving the arithmetic, we find that $\log A$ equals 8.564.

Substituting $\log A$ and ΔH_v into Eq. (8.5) gives a function from which we can determine p at any value of T and vice versa:

$$\log p \text{ (Torr)} = 8.564 - \frac{2120}{T}$$

where $2120 = \Delta H_v/2.303R$.

Now to determine the vapor pressure at 90°C we substitute $T = 363$ K and solve for p.

$$\log p = 8.564 - \frac{2120}{363} = 2.724$$

$$p = 10^{2.724} = 10^{0.724} \times 10^2 = 5.30 \times 10^2 = 530 \text{ Torr}$$

The experimental value of the vapor pressure of water at 90°C is 526 Torr.

8.4 MOLECULAR IMPLICATIONS OF EQUILIBRIUM VAPOR PRESSURE AND THE HEAT OF VAPORIZATION

In the liquid phase the average separation between molecules is 2 to 3 Å, while in the vapor phase separation is 10 to 100 times greater. Intermolecular forces in the vapor are negligible. As a molecule escapes from the liquid phase into the vapor phase, it is actually escaping from the intermolecular attractive forces which exist in the liquid. The equilibrium vapor pressure is a direct measure of the number of molecules which have escaped from the liquid into a closed vapor space above it. When intermolecular forces are high, as they would be for polar molecules and for molecules which exhibit hydrogen bonding, the equilibrium vapor pressure is low. Compare polar water and nonpolar benzene, C_6H_6, in Table 8.1. The vapor pressure of H_2O is lower at all temperatures. This is explained by the fact that water molecules are electric tripoles and involved in hydrogen bonding (Fig. 8.5*a* and *b*). As a result, the intermolecular forces in liquid water are extraordinarily high, and the escaping tendency, as measured by the equilibrium vapor pressure, is quite low. The benzene molecule is electrically nonpolar and does not exhibit hydrogen bonding; the escaping tendency of benzene molecules as measured by the equilibrium vapor pressure is high when compared with water at the same temperature.

The enthalpy of vaporization is the energy or heat necessary to vaporize 1 mol of a liquid at constant pressure. On a molecular level we view it as the energy required to increase the average separation between molecules from 2 or 3 Å to 30 Å or more. When intermolecular forces are high, the energy required to increase the

FIGURE 8.5

The vapor pressure is influenced by intermolecular forces, i.e., the forces between molecules.

δ^- δ^+ H H δ^+

(a) The electric tripole molecule of water

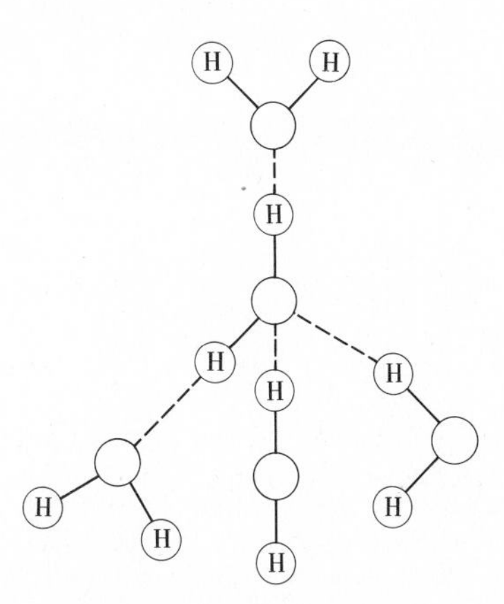

(b) Hydrogen bonding in the water structure

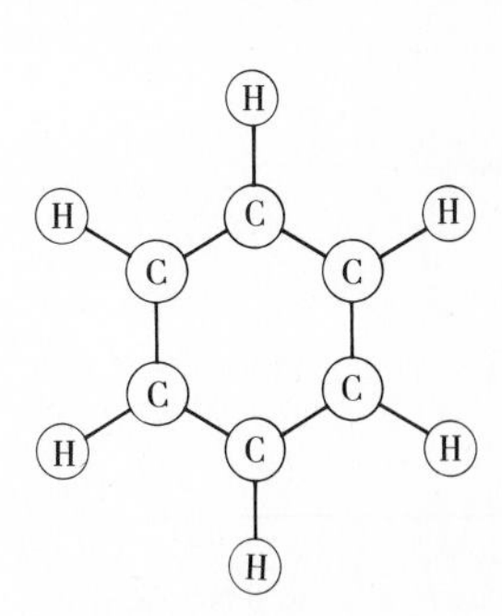

(c) Benzene is a symmetrical, nonpolar molecule

average separation is high. Therefore, polar liquids and liquids which exhibit hydrogen bonding have unusually high enthalpies of vaporization. For water the enthalpy of vaporization is 9700 cal/mol, whereas for benzene (even though it is composed of heavier molecules) it is only 7360 cal/mol.

8.5 PHASE DIAGRAMS

A plot of the equilibrium vapor pressure versus temperature is actually a plot of the conditions under which a liquid and its vapor can exist together at equilibrium. For instance, water and water vapor can coexist at a temperature of 25°C if the vapor pressure is 23.76 Torr, or at 40° if the pressure is 55.32 Torr, or at 100°C if the pressure is 760 Torr. These temperature-pressure pairs are points on the curve for water in Fig. 8.3. Any point on the curves in this figure represents a temperature-pressure condition under which the liquid and its vapor can coexist.

When a liquid and its vapor are at equilibrium at some fixed temperature, the vapor pressure is also fixed. The pressure of such a system cannot change unless either the liquid or the vapor completely disappears. Figure 8.6*a* shows that when water and its vapor are present at equilibrium in a closed vessel at 25°C, the vapor pressure is 23.76 Torr. If we decrease the volume of the system slightly while maintaining the temperature at 25°C, the vapor pressure remains 23.76 Torr. The decrease in volume causes some of the vapor to condense, but the pressure of the vapor does not change (Fig. 8.6*a* and *b*). If, while controlling the temperature at 25°C, we maintain an external pressure which is greater than 23.76 Torr, all

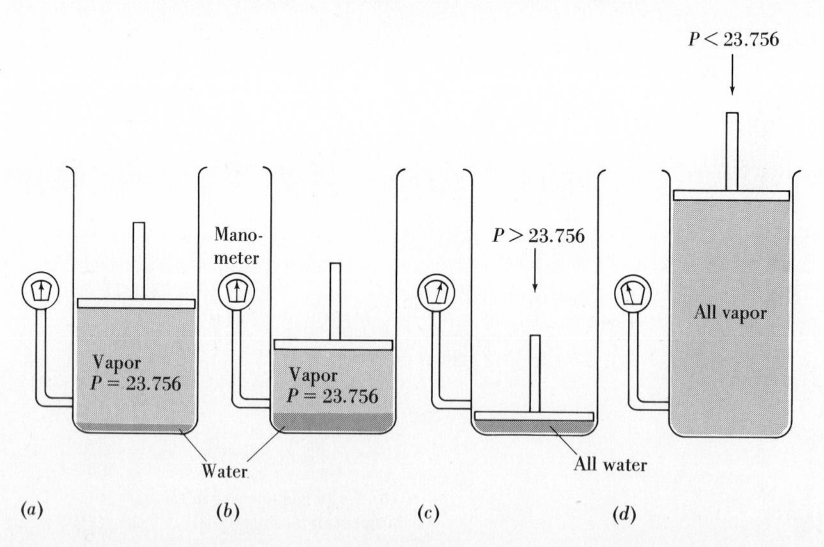

FIGURE 8.6

Effect of pressure on the state of H_2O at 25°C.

the vapor will condense and only liquid will be present; if we reduce the external pressure to a value less than 23.76 Torr, all the liquid will vaporize and only vapor will be present (Fig. 8.6*c* and *d*). Only at a pressure of 23.76 Torr can water and water vapor coexist at 25°C. For each temperature there is a unique value of the coexistence or equilibrium vapor pressure, and vice versa. For each value of the equilibrium vapor pressure there is a unique value of the temperature. If the temperature is held fixed, the equilibrium vapor pressure cannot change.

Curve *OA* in Fig. 8.7 represents the conditions under which water and water vapor can coexist in stable equilibrium. It is called the *boiling-point curve* because it represents the temperatures at which water boils at particular pressures. At 760 Torr water boils at 100°C. The boiling temperature at 760 Torr is called the *normal* boiling point. The boiling point at 25°C is 23.756 Torr.

Solids also have vapor pressure. Most solids have such low vapor pressures that their vaporization is not noticeable, but a few common solids vaporize quite rapidly. The vaporization of a solid is called *sublimation*. Snow and ice disappear from the ground due to sublimation on days when the temperature is below the freezing point. Paradichlorobenzene sublimes at an appreciable rate at room temperature, so that moth balls, which are made from this compound, grow small and disappear. Solid carbon dioxide, Dry Ice, is probably the most dramatic example of a material which undergoes rapid sublimation; a 50-lb block of Dry Ice can sublime in 8 h in an open room.

In a fashion analogous to the liquid, the solid form of a substance comes to equilibrium with its vapor in a sealed container if the temperature is constant; and we can obtain the set of temperature–

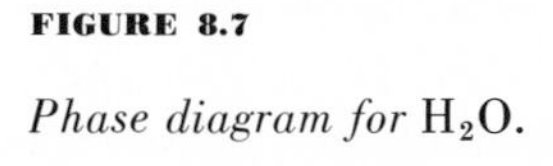
FIGURE 8.7

Phase diagram for H_2O.

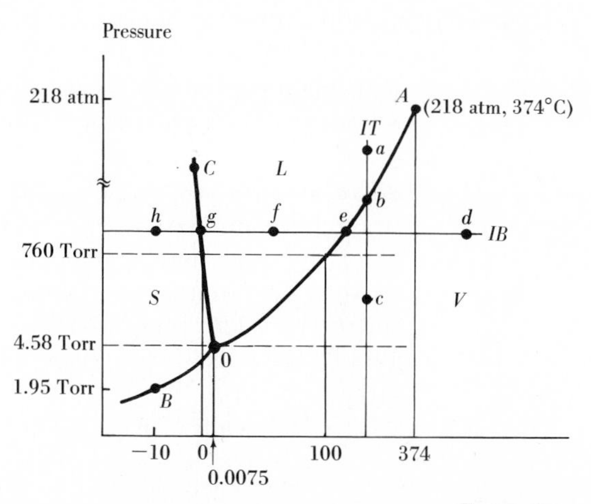

vapor-pressure pairs under which the solid and vapor can coexist in stable equilibrium. Curve *OB* in Fig. 8.7 represents the conditions under which ice and water vapor can coexist in stable equilibrium. The equilibrium vapor pressure of a solid has the same kind of temperature dependence as the vapor pressure of a liquid. Equation (8.5) can be used to analyze the data, and an enthalpy or heat of sublimation analogous to the enthalpy of vaporization can be calculated.

We have discussed the equilibria between vapor and liquid and between vapor and solid. The liquid and solid can also exist in equilibrium in the absence of vapor. The curve *OC* in Fig. 8.7 represents the set of temperatures and pressures under which the liquid-solid equilibrium can be maintained. The conversion of solid to liquid is called *fusion* or *melting*, so that each point on *OC* represents a *melting point*.

The melting-point temperature is not strongly affected by changes in pressure, as indicated by the fact that line *OC* is virtually vertical. The melting point of water occurs at 0.0075°C under its own vapor pressure of 4.58 Torr. When ice melts under the pressure of the atmosphere, 760 Torr, the melting point is 0.0°C, a change of only 0.0075°C. Compare this with effect of a pressure change on the boiling point: the same increase in pressure, from 4.58 to 760 Torr, causes the boiling point to go from 0 to 100°C; the change is 100° (see line *OA* in Fig. 8.7).

Point *O* in Fig. 8.7 represents the condition under which solid, liquid, and vapor can coexist; it is a *triple point*. For H_2O the triple point occurs at 0.0075°C and 4.58 Torr. The liquid and solid forms of H_2O have equal vapor pressures at 0.0075°C. At all other temperatures the vapor pressures of the solid and liquid are unequal.

Figure 8.7 is the phase diagram for water. A phase diagram is a convenient summary of data obtained from the study of the effects of temperature and pressure changes on the state of aggregation of a particular material. The lines *OA*, *OB*, and *OC* respectively represent conditions under which vaporization, sublimation, and fusion will occur. From another point of view these lines represent conditions of equilibrium between any two of the three possible phases. *OA* represents the temperatures and pressures under which vapor and liquid can coexist; *OB* and *OC* represent the conditions under which the vapor and solid and the liquid and solid, respectively, can coexist. The regions away from the lines represent the conditions under which only one phase can exist. Under the condition symbolized by the point *V* only vapor can exist; at *S* only solid exists; at *L* only liquid exists.

Line *IT* is called an *isotherm* because each point on the line has the same value of temperature. We can examine the effect of decreasing the pressure of water vapor while holding its temperature constant by going down the isotherm *IT*. Point *a* on *IT* represents a condition under which only the liquid can exist. As the pressure is decreased along *IT*, the point *b* is reached and vapor will begin to form because *b* lies on the boiling-point curve *OA*. There will be no further change in pressure as long as both phases coexist. When the liquid is completely vaporized, the pressure can drop below the value at *b*, say to *c* on *IT*.

Line *IB* is called an *isobar* since each point on the line represents the same pressure. At point *d* only vapor can exist. If the vapor is cooled along the isobar *IB*, liquid will eventually begin to form when a temperature corresponding to point *e* is reached. As long as the vapor and liquid coexist, no change in temperature will occur even though heat is removed. When all the vapor is converted into liquid, the temperature can drop again and we can reach a lower temperature on *IB*, say *f*, where only the liquid can exist. Continued cooling will cause ice to form at a temperature corresponding point *g*; again, the temperature will remain constant until all the liquid is converted to ice. The ice will then be able to cool below *g*, say to *h*.

It should be obvious that if two phases coexist, a knowledge of the temperature fixes the pressure and vice versa. For example, if water and water vapor coexist at 20°C, the pressure, read from the phase diagram, is 17.5 Torr; or if solid and vapor coexist at 3 Torr, the temperature, read from the phase diagram, is −5°C. If three phases coexist, the temperature *and* pressure are fixed: $t = 0.0075$°C, and $p = 4.58$ Torr. If only one phase is present, both the temperature and pressure are independently variable.

The conditions for the formation of snow are just the conditions for equilibrium between water vapor and ice as represented by curve *OB*. Therefore, the temperature must be below 0.0075°C and the vapor pressure must be below 4.58 Torr for snow to form. If water vapor in the atmosphere is cooled while its pressure is above 4.58 Torr, it forms raindrops, which upon further cooling produce hail (but no snow).

The liquid-solid equilibrium curve, or melting-point curve, *OC* for water has a negative slope. This means that as the pressure is increased, the temperature at which the two phases can exist at equilibrium decreases. In other words, the melting point is lower at higher pressures; this is very unusual behavior. For almost all other substances the melting point increases when the pressure increases. The phase diagram for carbon dioxide (Fig. 8.8) is normal in this respect; the melting-point curve has a positive slope; i.e., the melting point increases as the pressure increases. Note that liquid CO_2 is not stable below 5.2 atm and therefore, solid CO_2 cannot melt under atmospheric pressure; it sublimes.

FIGURE 8.8

Phase diagram for CO_2.

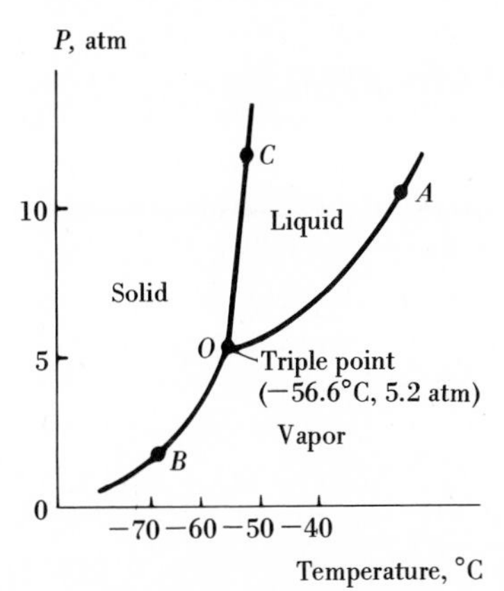

8.6 CRITICAL CONDITIONS

The liquid phase of a substance can often be produced by compression of the gas phase. As the phase diagram indicates, when water vapor is compressed at room temperature until the pressure exerted on it equals approximately 25 Torr, the vapor is converted to liquid; i.e., it *condenses*. Other gases and vapors can be condensed by compression at ordinary temperatures. These include NH_3, SO_2, and CO_2. Most atmospheric gases cannot be condensed by simply compressing them; their temperature must also be reduced before condensation can be achieved. For example, oxygen gas cannot be condensed above −118.8°C or nitrogen above −147.1°C. These temperatures are called the *critical temperatures* for the gases.

The critical temperature is often defined as the temperature above which it is impossible to condense or liquefy a gas, but this is an oversimplification. The critical temperature is also the temperature above which we cannot distinguish between a liquid and a vapor; no meniscus, or surface of demarcation, between the liquid and vapor will form no matter what pressure is exerted on a substance above its critical temperature. Only a single fluid phase is observable, which may resemble a liquid or a gas. At high pressures, when the fluid density is high, the fluid resembles a liquid; at low pressures the behavior of the fluid can be accurately predicted by the gas laws. Hence, by increasing the pressure on a gas which is above its critical point we can bring it to a condition which resembles a liquid yet no condensation will be observed; no meniscus separating the liquid from its vapor will form.

Carbon dioxide, CO_2, has a critical temperature of 31°C, which lies conveniently between room temperature, 20 to 25°C, and body temperature, 37.1°C. A sample of liquid CO_2 sealed in a glass capillary tube will show a sharp meniscus at room temperature. When the capillary is warmed by touch, the meniscus suddenly disappears. When the capillary is allowed to cool below 31°C, the meniscus reappears just as suddenly.

The minimum pressure required to cause the liquefaction of a gas and the formation of a meniscus when a gas is at its critical temperature is called the *critical pressure*. The volume of 1 mol of a substance at its critical temperature and pressure is called the *critical volume*. The critical temperature, pressure, and volume of a substance are collectively referred to as the *critical constants*. Table 8.2 shows that the critical temperatures for substances are very obviously related to intermolecular forces. When intermolecular forces are weak, as with helium, the critical temperature is low, reflecting the fact that the kinetic energy must be very drastically reduced before the weak intermolecular forces can bring about the formation of the liquid state. The critical temperature of water is very high for

TABLE 8.2: CRITICAL CONSTANTS

Substance	T_c, K	P_c, atm	V_c, ml
He	5.2	2.26	60
H_2	33.2	12.8	68
CO_2	304.2	73.0	95
NH_3	406.0	112.3	72
C_6H_6†	561.6	47.9	256
H_2O	647.3	217.7	57

† Benzene.

a compound with such a low molecular weight. This is consistent with our picture that the water molecule is involved in strong dipole-dipole interactions and hydrogen bonding.

8.7 SURFACE TENSION

All liquids resist expansion of their surfaces. A liquid droplet tends to become spherical because a sphere has a smaller surface-to-volume ratio than any other form; every other geometrical shape has more surface area per unit volume. (The surface area of a cube of volume 1 cm^3 is 6 cm^2 while the surface area of a sphere of the same volume is 4.56 cm^2.) In some cases the resistance to expansion is so great that dense objects do not penetrate a liquid but float on its surface instead. It is possible to float a clean steel needle or metal screen on pure water if it is carefully laid on the surface.

We believe that the resistance of a liquid surface to expansion is due to the fact that molecules at a surface, unlike molecules in the bulk, experience unsymmetrical intermolecular forces (Fig. 8.9). As a result, surface molecules are attracted downward toward the bulk

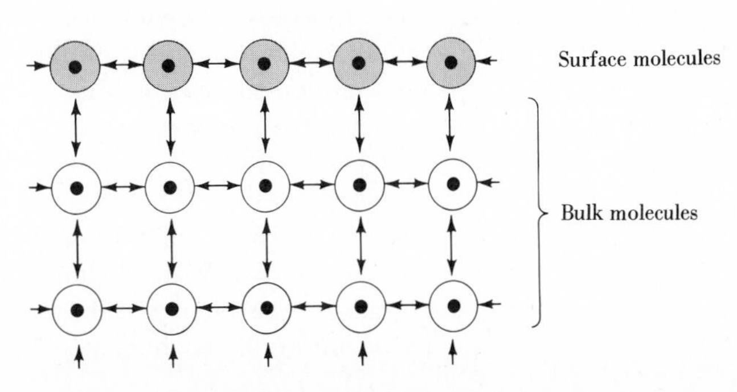

FIGURE 8.9

The molecules at the surface experience a net downward force; the molecules in the bulk experience uniform forces.

TABLE 8.3: SURFACE TENSION OF LIQUIDS AT 20°C

Substance	*Nature*	*Surface tension,* dyn/cm
Water	Very polar	72.75
Acetic acid	Polar	27.6
Acetone	Slightly polar	23.7
n-Hexane	Nonpolar	18.4
Ethyl ether	Slightly polar	17.0

of the liquid and there is a tightening of the surface molecules. We call this phenomenon *surface tension.*

The surface tension is a good measure of the intermolecular forces which are operative in liquids. Table 8.3 lists the surface tension for some common liquids. Note that polar liquids, in which intermolecular forces are high, have higher surface tensions than nonpolar liquids. The surface tension is defined as the force acting along a 1-cm length in the plane of the surface that resists the expansion of the surface; it is usually expressed in dynes per centimeter and is symbolized by γ.

The cleaning power of soap and other detergents is partially attributed to the fact that such reagents make big changes in the surface of liquid water. Greasy, nonpolar materials, which cannot penetrate the surface of pure water, can be brought through the surface of a soap solution and dissolved. The "tightness" of the liquid surface is reduced when soap or detergent is added, and dirt can be dissolved. The surface tension of a soap solution is about 35 dyn/cm, and that of a solution of a modern detergent is 25 dyn/cm, whereas it is $\sim$73 dyn/cm for pure water.

8.8 CAPILLARY ACTION

The surface tension of a liquid is responsible for capillary action or the spontaneous rise or fall of liquids in tubes of small diameter, e.g., blood vessels or plant stems. When a glass tube is immersed in a beaker of water, the level of the liquid inside the tube is higher than it is outside. The effect is really noticeable only in capillary tubes or tubes of small diameter. Hence, the phenomenon is called *capillary action.* Water exhibits a capillary *rise* in glass tubes and a capillary *fall* in silver tubes. Liquid mercury falls in a glass tube. These observations indicate that capillary action depends on the

FIGURE 8.10

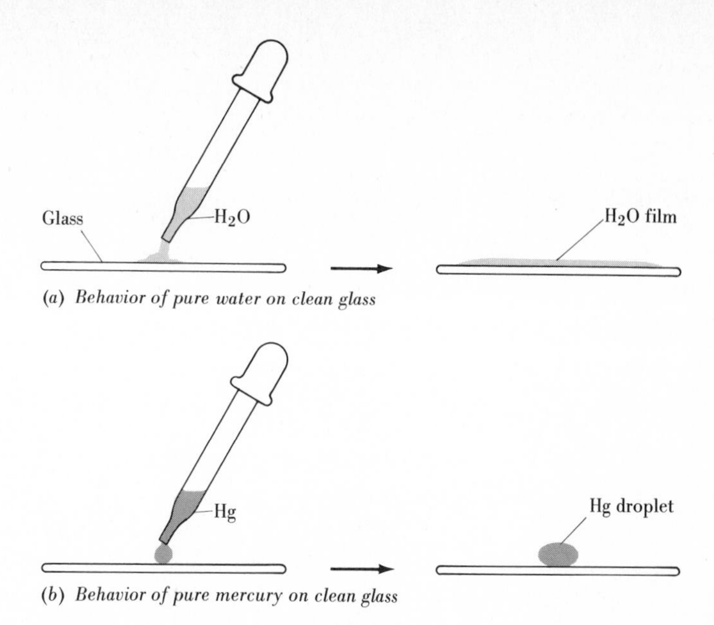

(*a*) *Behavior of pure water on clean glass*

(*b*) *Behavior of pure mercury on clean glass*

chemical nature of the liquid *and* of the capillary tube. We can explain the rise or fall of a meniscus inside a capillary in terms of the surface tension of the liquid and the interaction between the molecules of the liquid with the atoms and ions making up the surface of the tube.

Water rises in a glass capillary because the water molecules are strongly attracted by the glass surface. The attraction between the water molecules and the glass surface is stronger than the attraction of water molecules for themselves. We say that pure water *wets* a clean glass surface. When a drop of pure water is placed on clean glass, it spreads out and forms a film, but when it is placed on a wax surface it tends to "bead up" because water does not wet wax. Mercury does not wet glass, and it will produce rather rigid droplets on such a surface (see Fig. 8.10). The meniscus of mercury in glass is convex because it does not wet glass. The meniscus of water in glass capillaries is concave because water wets glass (see Fig. 8.11).

FIGURE 8.11

Meniscus shape in glass tubes.

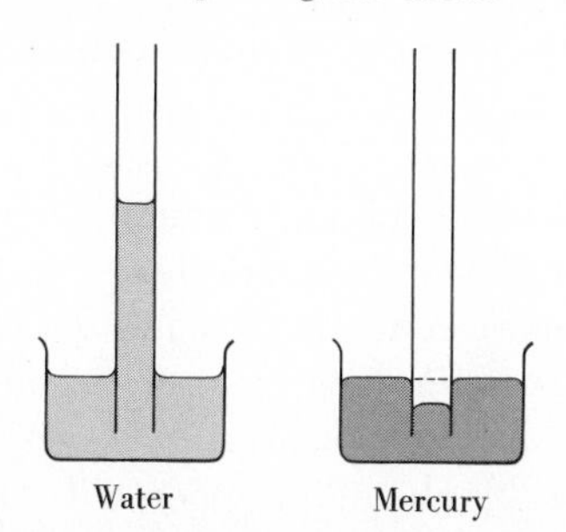

Let us examine how a combination of wetting and surface tension causes capillary rise. In Fig. 8.12*a* we imagine that a glass capillary has been placed in water and at the first instant the water level in the capillary is perfectly flat. However, wetting occurs and causes the liquid to creep up the walls (Fig. 8.12*b*). This results in a surface expansion. The surface tension, acting along the tube circumference, causes the liquid to rise in an attempt to reduce the surface area. The surface tension tends to flatten the surface (Fig. 8.12*c*). Again, the water wets the walls and creeps up, but the surface ten-

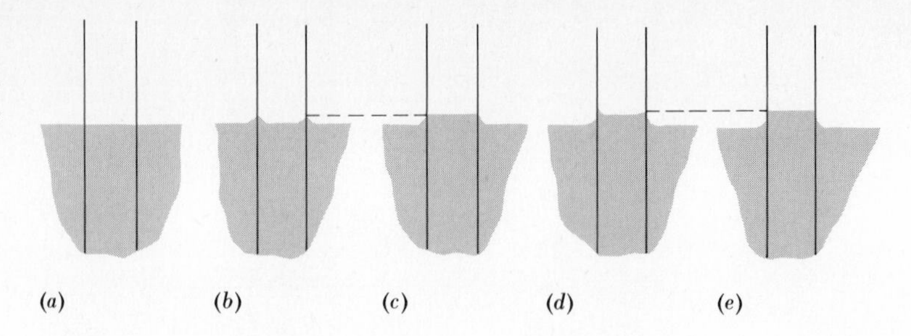

FIGURE 8.12

Capillary rise is due to combined action of surface wetting and surface tension.

sion, resisting surface expansion, causes the liquid to rise again (Fig. 8.12*d* and *e*). The capillary rise is due to wetting and surface tension. The whole process continues until the gravitational force acting down on the liquid in the column equals the force which causes the liquid to rise.

We have presented the mechanism for capillary rise as a stepwise process for convenience in explanation. Actually the wetting and surface-tension effects occur simultaneously, and a liquid rises smoothly up a capillary tube. Capillary fall can be explained in an analogous manner.

8.9 COMPARISON OF SOLIDS, LIQUIDS, AND GASES

It is very instructive to compare the three states of matter in terms of their (1) densities, (2) order, and (3) ability to undergo spontaneous mixing. In so doing a model for the liquid state will emerge.

Densities of gases, solids, and liquids

The densities of the solid and liquid states of a substance are high and almost equal, whereas the density of the vapor (under approximately the same conditions) is very low. This is apparent from the data in Table 8.4, where the densities of the solid, liquid, and vapor states of three common substances at their triple points are presented. The *specific volume* is the volume of 1 g of a substance, and is the reciprocal of the density. The specific volume shows in a dramatic way that the volume occupied by a vapor is much greater than an equal mass of either the liquid or the solid. At the triple point of water 1 g of vapor occupies ~206,000 times more space than 1 g of solid or liquid. This means that the molecules are much farther apart. The average separation between nearest molecules in water vapor is ~60 times greater than in either condensed phase.

Melting a solid to form a liquid is accompanied by an increase in volume of approximately 10 percent. This means that the average separation between particles increases by about 2 to 3 percent. This

TABLE 8.4: DENSITIES OF DIFFERENT STATES OF WATER, ACETIC ACID, AND BENZENE AT THEIR TRIPLE POINTS

Substance	*Triple point*, °C	*State*	*Density*, g/cm³	*Specific volume*, cm³/g	*Average separation between nearest molecules*, Å
H_2O	0.0075	Solid	0.917	1.09	3.20
		Liquid	0.998	1.002	3.11
		Vapor	0.00000485	206,000	183
Acetic acid, CH_3COOH	16.6	Solid	1.144	0.875	4.59
		Liquid	1.051	0.961	4.51
		Vapor	0.0000302	33,100	149
Benzene, C_6H_6	5.5	Solid	0.968	1.033	5.10
		Liquid	0.899	1.111	5.24
		Vapor	0.000148	8,710	96.7

is a very small, almost negligible, increase compared to the increase in average separation between particles which occurs when a liquid or a solid is transformed into a vapor under ordinary atmospheric conditions.

A comparison of the densities of the states leads us to the molecular picture that the average separation between particles increases as we go from solid to *liquid* to gas. This can be seen from Table 8.4 if we neglect water, which is anomalous. Our microscopic picture puts the liquid in an intermediate position. In terms of density, specific volume, and average separation between nearest molecules the liquid is seen to be more similar to a solid than to a gas.

Order in the solid, liquid, and gaseous states

We have already discussed x-ray analysis of crystalline solids and the high degree of order these experiments indicate. Some of the order of the crystal is preserved when it melts to form the liquid. The order in the liquid state decreases as the temperature increases. This is borne out by x-ray experiments, which show that liquids give diffuse diffraction patterns, indicating that some order remains. The patterns become more diffuse as the temperature of the liquid is increased, and finally, when the liquid vaporizes, all evidence of order is gone. Again, the intermediacy of the liquid state is evident. Order, or regularity, like separation of particles, relates the liquid more closely to the solid, crystalline state than to the gaseous state.

Spontaneous mixing

In general, solid materials retain their shape and appearance indefinitely. If two clean solid surfaces are placed in good contact, there is virtually no transport of matter across the boundary for years. When two metals are welded together, the sharp weld mark remains visible indefinitely; this must mean that net movement of the particles in a solid (ions, atoms, or molecules) is negligible. The particles of a solid may vibrate about fixed positions, but they undergo no net movement.

Mixing in the solid state does occur, but it is detectable only within a few atomic layers adjacent to the interface separating the different solids. Figure 8.13*a* is a two-dimensional picture of the contact between two crystalline solids, A and B. At the interface there is equal probability of finding particles of A or B. As we move away from the interface the probability of finding displaced particles decreases very rapidly. Once we have moved through 10 to 15 atomic or molecular layers, the probability of finding a displaced particle is virtually zero. Figure 8.13*b* is a graphical representation which shows how the number of displaced particles decreases as the distance from the interface increases.

Our picture emphasizes the virtual immobility of the particles of the solid state. In contrast it is impossible to maintain a sharp interface between two different gases for as much as a millisecond. Gases spontaneously mix or interdiffuse very rapidly. Spontaneous mixing in the liquid state is not nearly as rapid as in the gaseous state, but there is no difficulty in observing it in a relatively short time (minutes). In observations where radioactive, colored, or fla-

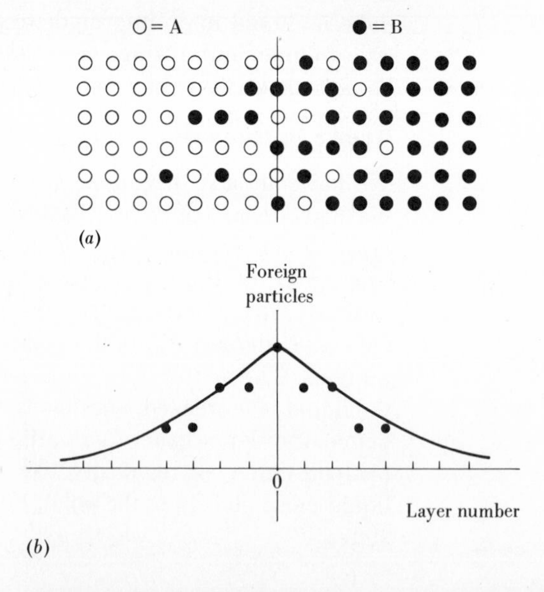

FIGURE 8.13

The particles of a solid are virtually immobile.

vored substances are initially restricted to a definite region in an unstirred liquid, spontaneous mixing eventually causes the liquid to become homogeneous (Fig. 8.14). These simple observations give us some insight into the molecular processes that occur in the three states of matter. We have concluded that the particles in a solid are virtually immobile or confined to very small regions of space. To a good approximation we may say the particles remain at a fixed point. The particles of gases and liquids have definite mobilities, otherwise we could not explain self-mixing in these phases. Gases undergo rapid self-mixing because of the great average separation between their molecules. Collisions are relatively infrequent, and gaseous molecules can move a long distance before being deflected or reflected. In the solid state there is virtually no room for movement. A particle can move only a fraction of an angstrom before it collides with a neighbor; or perhaps it is more accurate to say that a particle can move only a fraction of an angstrom before it experiences extremely high repulsive forces due to electron clouds of neighboring particles. These repulsive forces act as restoring forces, and the particle is moved back toward its original position. The particle moves in and out of its position continually and rapidly; essentially, then, a particle in the solid state undergoes oscillatory or vibrational motion about a fixed position and cannot escape the region to which it is confined by the repulsive forces of its very close neighbors (see Fig. 8.15*a*).

The density of a liquid is only slightly lower than the density of

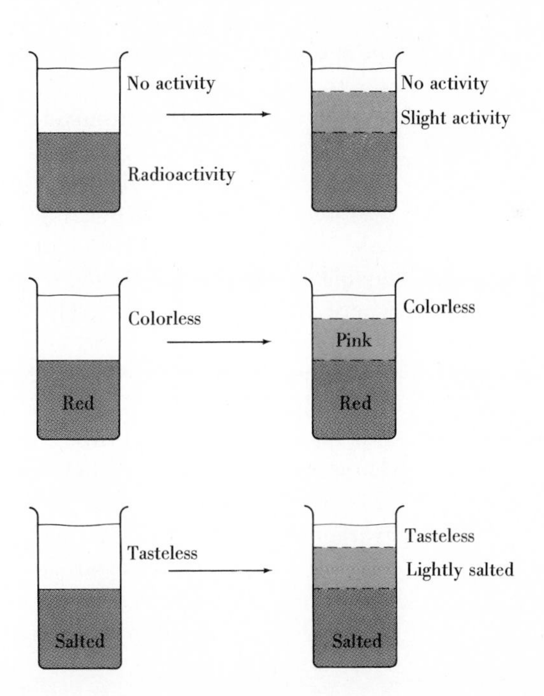

FIGURE 8.14

Mixing in the liquid state.

FIGURE 8.15

Diffusion in a liquid depends on the presence of vacancies, or holes.

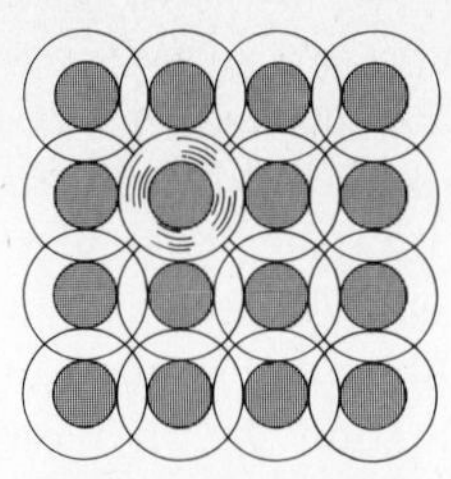

(*a*) *A particle of a solid undergoes oscillatory motion about a fixed position from which it cannot escape*

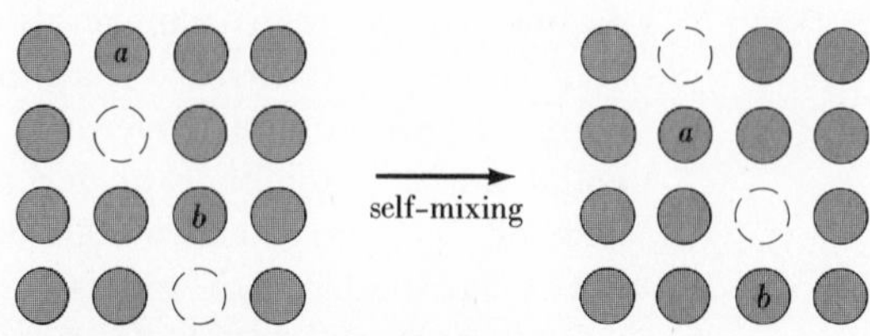

(*b*) *Liquid vacancies and molecules interchange positions resulting in mixing*

the corresponding solid phase. This indicates that the particles in a liquid are also quite close together. On a molecular level we may picture the liquid as a regular (crystalline) structure with a small fraction of the molecules missing from the lattice positions (see Fig. 8.15*b*). The unoccupied lattice sites are called *holes* or *vacancies*. The vacancies reduce the regularity of the structure and account for the diffuse x-ray diffraction patterns exhibited by liquids. It is obvious that the presence of vacancies reduces the density. Vacancies also account for the self-mixing of liquids. Molecules in the liquid can move into vacancies without experiencing strong repulsive forces; there is no tendency to restore the molecule to its original position. We envision a dynamic condition in the liquid as the molecules and vacancies continually interchange their positions. Self-mixing of liquids takes place even faster at high temperatures, where the density is lower. The decrease in density must mean that there are more vacancies at higher temperatures.

We know that the average velocity of gaseous molecules increases with temperature. If this is also true for liquids, self-mixing in liquids should be faster as the temperature increases for two reasons: (1) the molecules have higher velocities, and (2) more vacancies are present. Because the average translational kinetic energy and the average velocity of *gas* molecules increase with temperature, we guessed that an increase in the temperature would also cause the molecules of a liquid to move with greater velocities. Observations on small particles (diameter $\approx 10^{-4}$ cm) show that their movement in an unstirred liquid is continually undergoing a change in direction and speed. This observation was first made in 1827 by a botanist, Robert Brown, who used a microscope to view the particles. *Brownian motion* is attributed to bombardment of the particles by the molecules of the liquid, which also must be undergoing incessant changes in velocity. A study of brownian motion leads to the conclusion that the tiny particles *and* the molecules of the liquid which bombard them have values for the average kinetic energy and root-mean-square velocity equal to the values found for molecules of a gas. $\bar{\varepsilon} = \frac{3}{2}kT$ and $\bar{v}_{\text{rms}} = \sqrt{3RT/M}$. So the molecules of a liquid possess a very high average velocity. Self-mixing is slower in a liquid than in a gas not because the molecules have lower velocities but because the average distance a molecule travels between collisions is much smaller. In other words, a molecule in a liquid is always subject to the attractive and repulsive forces of its neighbors and is constantly deflected and reflected from a straight path.

8.10 GLASSES: THE NONCRYSTALLINE SOLIDS OR IMMOBILE LIQUIDS

A solid is a rigid substance which has a definite shape and volume and no tendency to flow under an applied stress. In Chap. 7, we discussed *crystalline* solids, which give well-defined diffraction patterns and sharp melting points. There are many examples of pure solids which give diffuse x-ray diffraction patterns and exhibit *softening* rather than melting. Tars, modern plastics, and ordinary glass are examples of such materials. When the temperature of these solids is increased, they are transformed from a solid into a liquid by softening over a wide temperature range of up to 100°C. These solids are called *glasses*. On the molecular level we picture

FIGURE 8.16

A comparison of a crystalline solid and a glass.

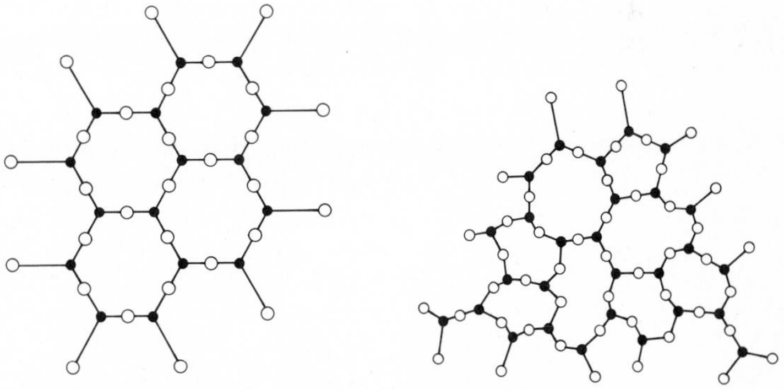

(*a*) *Quartz*
Crystalline SiO_2

(*b*) *Glass*
Noncrystalline SiO_2

them as resembling a liquid in terms of order and a solid in terms of mobility. The disorder of the liquid structure is maintained when a glass forms; the particles of a glass do not achieve the almost perfect order of the crystalline state. Glasses are often referred to as *supercooled liquids.*

Two common forms of silicon dioxide, SiO_2, demonstrate the difference between glass and a crystalline solid (see Fig. 8.16). The crystalline form is quartz; in this form there is a regular ring structure containing six oxygen and six silicon atoms, and all the Si—O bond lengths are equal. In the glass the rings are irregular and contain different numbers of atoms, and the Si—O bond lengths vary.

QUESTIONS AND PROBLEMS

8.1 (*a*) Why is the liquid state considered as being intermediate between the gaseous and solid crystalline states? (*b*) Cite data for some pure substances which show that the liquid state has intermediate properties. (Use information given in this chapter and in handbooks.)

8.2 When water and water vapor are enclosed in a vessel at 25°C, the vapor pressure of water reaches a constant value of 23.8 Torr. At this point we say that water and water vapor are in *dynamic* equilibrium although no change is apparently occurring within the vessel. Explain.

8.3 (*a*) Water evaporates completely when a full drinking glass is left on a kitchen table but not when the same full glass is put in a small, airtight box at the same temperature. Explain. (*b*) If $\frac{1}{2}$ l of water is left on the table in an airtight kitchen where the temperature is 27°C, will all the water evaporate? The dimensions of the kitchen are $4 \times 6 \times 2.5$ m. The equilibrium vapor pressure of water at 27°C is 26.5 Torr.

8.4 Using the equilibrium vapor pressures at 20 and 60°C given in Table 8.1, calculate the enthalpy of vaporization for ethanol. Compare your result with the value obtained in Example 8.1 and explain any differences.

8.5 The heat of vaporization of Freon 12 (an ordinary refrigerant, CF_2Cl_2) is 39.9 cal/g. Its normal boiling point is −29.4°C. Using Eq. (8.5), estimate the pressure in atmospheres which the cooling coils must withstand when the refrigerator is turned off on a warm summer day ($t \approx 30$°C).

8.6 (*a*) Using the H_2O data in Table 8.1, make a plot of $\log p$ versus $1/T$. (*b*) Determine the slope of the plot. (*c*) Calculate the average value of the enthalpy of vaporization from the slope.

8.7 Using the ethanol data in Table 8.1 make a plot of $\log p$ versus $1/T$ and determine the enthalpy of vaporization.

8.8 Make a sketch of the phase diagram for acetic acid using the data in Table 8.1. The triple point occurs at 16.6°C and 9.1 Torr. The normal boiling point is 118.5°C. Solid acetic acid is denser than the liquid.

8.9 What is the relationship between the enthalpy of vaporization and intermolecular forces? Refer to the data in Table 8.1.

8.10 A cylinder enclosed at the top with a movable piston at the height of 20 cm has a radius of 2 cm. It contains water to a height of 9.5 cm at 25°C. (*a*) What is the weight of H_2O present in the liquid space and in the vapor space above the liquid? (*b*) If the temperature is held constant, what will the pressure of the water vapor be if the piston is lowered to a height of 15.0 cm? (*c*) What weights of H_2O are present in the water and vapor spaces after the piston is lowered to 13.0 cm? [State assumptions that you might make to solve part (*c*).]

8.11 At 0 and −20°C the vapor pressures of ice are 4.579 and 0.776 Torr, respectively. Using Eq. (8.5), calculate the average enthalpy of sublimation in this temperature interval.

8.12 Dry Ice, solid CO_2, has a pressure of 1.00 atm at −78°C and a pressure of 56.6 atm at 20°C. Calculate the enthalpy of sublimation (see Prob. 8.11).

TABLE P8.13: VAPOR PRESSURE

Solid		*Liquid*	
t, °C	*P*, atm	*t*, °C	*P*, atm
−90	0.367	−50	6.74
−80	0.883	−40	9.92
−70	1.95	−30	14.1
−60	4.03	−20	19.4

8.13 (*a*) Using the data in Table P8.13 sketch the phase diagram for CO_2. The triple point occurs at −56.6°C and 5.13 atm. (*b*) From your phase diagram explain why Dry Ice sublimes under atmospheric conditions. (*c*) Under what conditions of temperature and pressure would it be possible for the liquid to be stable?

8.14 (*a*) Define critical temperature, critical pressure, and critical volume. (*b*) Why is it misleading to say that the critical temperature is the temperature above which it is impossible to liquefy a gas?

8.15 Is there any apparent relationship between attractive forces between molecules and the critical temperatures listed in Table 8.2? Explain.

8.16 (*a*) What is surface tension? (*b*) What is its exact definition? (*c*) What are the units of surface tension? (*d*) How does surface tension give rise to the capillary-tube effect? (*e*) Is there any apparent relationship between polarity of molecules and surface tension?

8.17 What properties of the liquid state are (*a*) more similar to gases than to solids and (*b*) more similar to solids than to gases?

8.18 (*a*) Explain why self-mixing occurs in gases and liquids. (*b*) Explain why self-mixing is negligible in the solid state.

8.19 (*a*) How is a glass similar to a liquid? (*b*) How is a glass similar to a crystalline solid? (*c*) Give three common examples of a glass (other than ordinary glass).

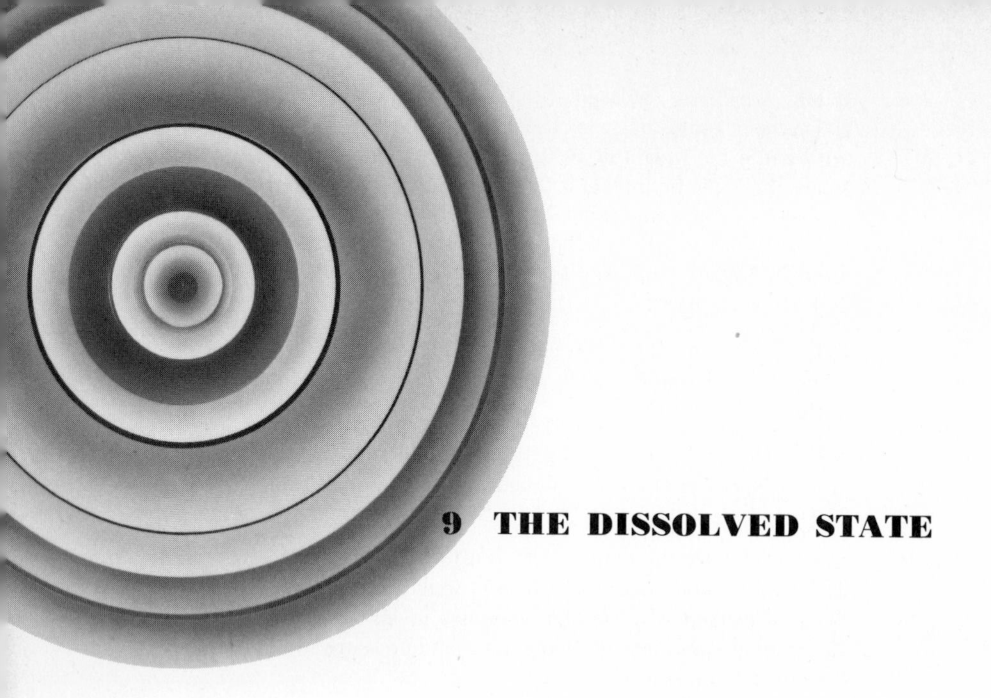

9 THE DISSOLVED STATE

9.1 INTRODUCTION

Solution is a term familiar to everyone. It is often just such a term that defies the precise definition required in science. Solutions are mixtures and can be solid (steel), liquid (brine), or gaseous (air). The physical state of a solution is not determined by the states of its pure components. Sodium and potassium are metallic solids at room temperature, but when they are mixed, they form a liquid solution. Often, however, the physical state of the solution is determined by the state of the major component, which is arbitrarily referred to as the *solvent*. The other component (or components) is called the *solute*. In this chapter we shall be primarily concerned with *binary* liquid solutions, i.e., liquid solutions containing two components.

9.2 THE DISTINCTIVE PROPERTIES OF A SOLUTION

When a small amount of benzene is added to a flask of water and the flask is shaken, the benzene, in the form of tiny globules, is dispersed throughout the water. The mixture is turbid, but on standing the lighter benzene separates and floats to the top; two liquid phases form. Benzene and water are immiscible; they do not form a solution. If we add some ethyl alcohol, C_2H_5OH, to water in a flask and shake the contents, the resulting mixture is clear and *chemically homogeneous.* There is no tendency for the mixture to separate into its components. Ethyl alcohol and water are *miscible.* Water forms a *solution* with alcohol but not with benzene.

There are many miscible and immiscible pairs of compounds. In addition there are pairs which are *partially* miscible. For instance, at room temperature a true solution of ethyl acetate, $C_2H_5—O—CO—CH_3$, and water can be formed provided that the mixture is less than 7.9 percent by weight ethyl acetate. Above this percentage, mixing will result in the formation of two liquid phases: a water-rich layer with only 7.9 percent ethyl acetate and an ethyl acetate–rich layer with 89.2 percent ethyl acetate.

Now let us get at the essential properties of a solution. When a crystal of blue copper(II) sulfate, $CuSO_4$, is carefully lowered into a flask of water, the liquid surrounding the crystal begins to take on the blue color. Shaking the flask causes the $CuSO_4$ to dissolve quickly; the resulting mixture is a clear and uniformly blue liquid. More $CuSO_4$ can be added and dissolved. The uniform color suggests that $CuSO_4$ is evenly distributed throughout the liquid, and analytical experiments bear this out. After the solution is well stirred, the ratio of the weight of $CuSO_4$ to the weight of water in any part of the mixture remains constant indefinitely. The mixture of $CuSO_4$ and water is homogeneous, and such a mixture is a solution.

The chemical and physical properties of the solution are a combination of the properties of the pure components. The water in solution can undergo chemical reactions that pure water would undergo; so can the $CuSO_4$. The dissolved $CuSO_4$ confers upon the solution a blue color, which of course is the color of pure $CuSO_4$. A dilute solution of $CuSO_4$ has vapor pressure and surface tension not much different from those of pure water. The solution conducts electricity as molten $CuSO_4$ would. We could prepare a long list of the properties of the solution which are similar to the properties of the individual pure components.

If the mixture is allowed to stand open to the atmosphere, the water evaporates and all the crystalline $CuSO_4$ is easily recovered. The mixture of $CuSO_4$ and water is a solution because water and $CuSO_4$ form homogeneous mixtures of various compositions which have no tendency to separate. The properties of the solution are a composite of the properties of the individual components, and, finally, the components are relatively easily recovered. In summary a mixture of two or more components is a solution if the mixture (1) is homogeneous, (2) has no tendency to separate into its components, (3) has a variable composition, and (4) has properties which are a composite of the properties of its individual components.

Solutions differ from mechanical mixtures in being homogeneous; they differ from pure compounds in having a variable composition and properties that are a composite. Mechanical mixtures, like powdered sulfur and iron filings, are not solutions because they are not homogeneous. Compounds are not solutions because compounds have a fixed composition and their properties can be very different from the properties of their components; e.g., water is not at all like its components, hydrogen and oxygen.

9.3 THE FORMATION OF A SOLUTION FROM A MOLECULAR POINT OF VIEW

The macroscopic properties of solutions have been discussed. These properties permit us to distinguish true solutions from other kinds of mixtures and from chemical compounds. Again, as always, the macroscopic properties have implications concerning the molecular level.

Since a well-stirred solution is chemically homogeneous, we must conclude that the solute is molecularly dispersed in the solvent. In a dilute solution of acetone, $CH_3—CO—CH_3$, in water the individual acetone molecules are randomly dispersed throughout the solution; each $CH_3—CO—CH_3$ molecule is surrounded predominantly by water molecules. The constant thermal motions cause the number of molecules in any small portion of the solution to fluctuate. But the fluctuations are so small that once a solution has been well stirred, the number of molecules in any volume large enough to be studied remains constant. When we say that the number of molecules remains constant, we mean that, within our ability to measure, the number of molecules remains constant. Remember that our most sensitive apparatus cannot detect changes of less than 10^{-9} g. This small mass corresponds to 3,300 trillion water molecules! In 1 cm^3 of water there are about 3.3×10^{22}, or 33 billion trillion molecules. A fluctuation of 3,300 trillion water molecules cannot be detected. Fortunately, compared to the total number of molecules in a cubic centimeter this is a small fluctuation; it corresponds to a change of only 0.0000001 percent.

That some materials are miscible and others immiscible can be understood in terms of intermolecular forces, van der Waals forces, and hydrogen bonding. In any pure liquid or solid the attractive forces between molecules must be quite strong, otherwise the molecules would fly apart from each other; i.e., immediate vaporization would occur. When different compounds are mixed, different molecules are forced to come into contact, and if a true solution is

FIGURE 9.1

A mechanism for the formation of a solution.

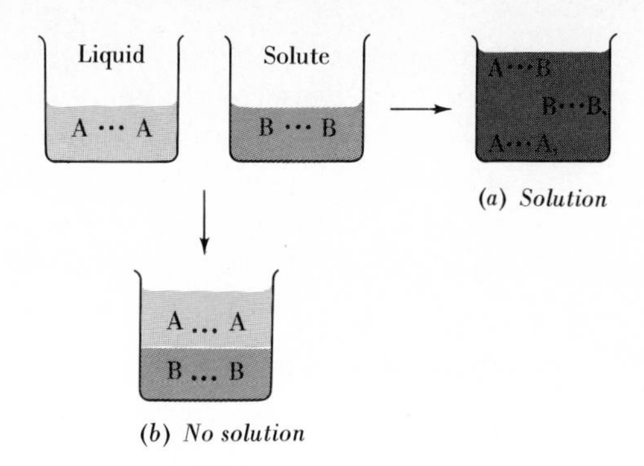

(*a*) *Solution*

(*b*) *No solution*

formed, intermolecular forces must be overcome both in the solute and in the solvent.

In Fig. 9.1 we see two beakers, one containing liquid A and the other a possible solute, B. If a solution is to form, the attractive force between the molecules A and B, represented as A····B, must be of the same magnitude as, or greater than, the attractive force between identical molecules A····A or B····B. Otherwise, the stronger forces between the molecules of a pure phase will preserve the purity of the phase; the A molecules will preferentially associate with other A molecules and the B molecules will preferentially associate with other B molecules. The foreign molecules will be squeezed out, and two separate phases will appear.

We see that a solution can form when there is an appreciable attraction between the molecules of the solute and solvent, but it must also be kept in mind that if the attractive force between two components is very high, a chemical reaction may occur. For example, when chlorine, Cl_2, is mixed with carbon monoxide, a new gaseous compound, phosgene, $COCl_2$, forms. The formation of a solution represents some intermediate stage in which molecules of the solvent and solute are bonded but in which no permanent or irreversible changes occur. If this were not true, it would be difficult to recover the original components from a solution.

9.4 MOLARITY, MOLALITY, AND MOLE FRACTION

The concentration of solutions is often expressed as a mass of solute per unit volume. It is more significant to express the concentration of a solution in terms of the number of molecules or moles present in a unit volume. The most important concentration terms used by chemists are *molarity*, *molality*, *normality*, and *mole fraction*. We can calculate the molarity and normality from the density and percentage composition of a solution. For dilute solutions the density of the solution is very close to the density of the pure solvent. To calculate the molality and mole fraction it is only necessary to know the percentage composition of the solution.

The *molarity* of a solution, abbreviated M, is the number of moles of solute per liter of solution. A solution whose molarity is 6 contains 6 mol of solute dissolved in 1 l of solution. It is called a 6 molar (6 M) solution.

EXAMPLE 9.1 An aqueous solution of $AgNO_3$ is 10 percent $AgNO_3$ by weight and has a density of 1.09 g/ml. Calculate the molarity of the solution.

Solution: We must transform the given data into the units of molarity, i.e., moles of $AgNO_3$ per liter of solution. The density has units of grams of solution divided by milliters of solution; 10 percent of the solution is $AgNO_3$ by weight. Multiplying the density by the weight fraction of $AgNO_3$ will give the weight of the $AgNO_3$ alone in 1 ml:

$$\frac{1.09 \text{ g soln}}{1 \text{ ml soln}} \times \frac{0.10 \text{ g AgNO}_3}{1 \text{ g soln}} = 0.109 \text{ g AgNO}_3\text{/ml soln}$$

The number of moles of $AgNO_3$ in each milliliter of solution can now be obtained; divide the weight of $AgNO_3$ in 1 ml by the gram-molecular weight of $AgNO_3$:

$$\frac{0.109 \text{ g AgNO}_3}{1 \text{ ml soln}} \times \frac{1 \text{ mol AgNO}_3}{170 \text{ g AgNO}_3} = 0.000641 \text{ mol AgNO}_3\text{/ml soln}$$

The number of moles of $AgNO_3$ in each liter of solution, i.e., the molarity, is now obtained by multiplying by 1000 since there are 1000 ml in each liter:

$$\frac{0.000641 \text{ mol AgNO}_3}{1 \text{ ml soln}} \times \frac{1000 \text{ ml soln}}{1 \text{ l soln}} = 0.641 \text{ mol AgNO}_3\text{/l soln}$$

$M = 0.641$

With a little practice the entire calculation can be done quickly in one equation:

$$\frac{1.09 \text{ g soln}}{1 \text{ ml soln}} \times \frac{0.10 \text{ g AgNO}_3}{1 \text{ g soln}} \times \frac{1 \text{ mol AgNO}_3}{170 \text{ g AgNO}_3} \times \frac{1000 \text{ ml soln}}{1 \text{ l soln}} = 0.641 \text{ mol AgNO}_3\text{/l soln}$$

The *molality* of a solution, abbreviated m, is the number of moles of solute per kilogram of solvent. The molality can be calculated from the percentage composition alone. The density is not necessary because the molality is not defined in terms of volume. From the weight percentage or weight fraction the relative weights of solute and solvent can be determined. Then with the molecular weight of the solute it is possible to determine the number of moles of solute dissolved in 1 kg of solvent.

EXAMPLE 9.2 An aqueous solution of NaOH is 10% NaOH by weight. Determine the molality of the solution.

Solution: The weight percentage permits us to calculate the weights of NaOH and H_2O in each gram of solution.

In 1 g of solution there are 0.10 g of NaOH and 0.90 g of H_2O; that is, there is 0.10 g NaOH per 0.90 g H_2O. We can use this unit, 0.10 g NaOH/0.90 g H_2O, as a starting point for the calculation of the molality. If we divide 0.10 g NaOH by 0.90 g H_2O, we obtain the grams of NaOH per gram of H_2O:

$$\frac{0.10 \text{ g NaOH}}{0.90 \text{ g H}_2\text{O}} = \frac{0.111 \text{ g NaOH}}{1 \text{ g H}_2\text{O}}$$

Next, division by the gram-molecular weight of NaOH gives us the number of moles of NaOH per gram of H_2O:

$$\frac{0.111 \text{ g NaOH}}{1 \text{ g H}_2\text{O}} \times \frac{1 \text{ mol NaOH}}{40 \text{ g NaOH}} = 0.00278 \text{ mol NaOH/g H}_2\text{O}$$

Now we can calculate the moles of NaOH per kilogram of solvent, i.e., the molality, if we simply multiply by 1000:

$$\frac{0.00278 \text{ mol NaOH}}{1 \text{ g H}_2\text{O}} \times \frac{1000 \text{ g H}_2\text{O}}{1 \text{ kg H}_2\text{O}} = 2.78 \text{ mol NaOH/kg H}_2\text{O}$$

$$m = 2.78$$

The entire calculation is more speedily performed in one operation:

$$\frac{0.10 \text{ g NaOH}}{0.90 \text{ g H}_2\text{O}} \times \frac{1 \text{ mol NaOH}}{40 \text{ g NaOH}} \times \frac{1000 \text{ g H}_2\text{O}}{1 \text{ kg H}_2\text{O}} = 2.78 \text{ mol NaOH/kg H}_2\text{O}$$

$$m = 2.78$$

Molarity and molality are basically different: molarity is a weight-volume concentration unit, and molality is a weight-weight concentration unit. The molarity is calculated from a knowledge of the weight of solute in a certain volume of solution. The molality depends only on the weight ratio of the solute and solvent. Since the volume of a solution changes as its temperature changes, the molarity depends on temperature. The molality of a solution is not affected by temperature changes.

The mole fraction of a component in solution, symbolized by X, is the number of moles of the component divided by the total number of moles in the solution. Since we are dividing moles by moles, the mole fraction is given without units. Like the molality, the mole fraction is a weight-weight concentration term and is not affected by temperature. It can be calculated from the percentage composition of the solution.

EXAMPLE 9.3 Calculate the mole fraction of the NaOH and of the H_2O in the solution described in Example 9.2.

Solution: By definition the mole fraction of NaOH is

$$X_{NaOH} = \frac{\text{moles of NaOH}}{\text{moles of NaOH} + \text{moles of H}_2\text{O}} = \frac{n_{NaOH}}{n_{NaOH} + n_{H_2O}}$$

where n is the number of moles.

Since the solution is 10 percent by weight NaOH, we have 10 g NaOH for each 90 g of H_2O. The number of moles of each is calculated in the usual way:

$$10 \text{ g NaOH} \times \frac{1 \text{ mol NaOH}}{40 \text{ g NaOH}} = 0.25 \text{ mol NaOH}$$

$$90 \text{ g H}_2\text{O} \times \frac{1 \text{ mol H}_2\text{O}}{18 \text{ g H}_2\text{O}} = 5 \text{ mol H}_2\text{O}$$

The mole fraction of NaOH is:

$$X_{NaOH} = \frac{0.25 \text{ mol}}{0.25 \text{ mol} + 5 \text{ mol}} = \frac{0.25}{5.25} = 0.048$$

Likewise the mole fraction of H_2O is

$$X_{H_2O} = \frac{n_{H_2O}}{n_{NaOH} + n_{H_2O}} = \frac{5}{5.25} = 0.952$$

It should be noted that since the sum of the mole fractions for all the components in a system equals unity, $\sum_i X_i = 1$, we could have calculated the mole fraction of H_2O, X_{H_2O}, by subtracting the mole fraction of NaOH, X_{H_2O}, from unity:

$$X_{H_2O} = 1 - X_{NaOH} = 1 - 0.048 = 0.952$$

9.5 FACTORS AFFECTING SOLUBILITY

Solubility is defined as the concentration of a substance in a solution which is in contact with an excess of the substance. The solubility represents the maximum concentration which can be achieved. The solubility of table salt in water at 20°C is 36 g per 100 g H_2O. If 40 g of NaCl is added to 100 g of H_2O, 4 g of salt will *not* dissolve. This will be excess solute in contact with the solution. The solution is *sat-*

urated with NaCl, and the concentration of such a solution is called the *saturation concentration*, which is synonymous with solubility.

The solubility of a solid is very sensitive to changes in temperature; it is not appreciably affected by ordinary pressure changes. Figure 9.2 shows how the temperature influences the solubility of some common solids in water. The behavior of other solid-liquid pairs is similar. It is evident from Fig. 9.2 that the solubility of a solid in a liquid generally increases with increasing temperature. Sucrose, NH_4NO_3, and KOH demonstrate this very nicely; the solubility of NaCl, however, is hardly affected by increasing the temperature.

It is possible in some instances to achieve concentrations which are in excess of the saturation concentration. A solution in which this is true is called a *supersaturated solution*. A supersaturated solution may be produced by chilling a hot, concentrated solution. A supersaturated solution is metastable; i.e., stirring, introducing a tiny crystal of solute, or standing often makes the excess solute crystallize out, reducing the concentration to the normal solubility.

Ionic compounds are soluble in water, methyl alcohol, and liquid ammonia; they are insoluble in octane, benzene, and carbon tetrachloride. The molecules of water, methyl alcohol, and ammonia are polar; there are centers of positive and negative electric charge

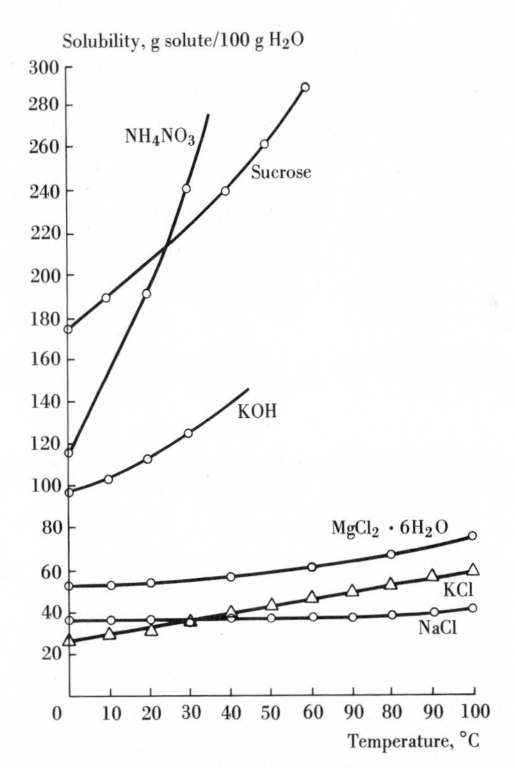

FIGURE 9.2

Solubility of some common solids in water at various temperatures.

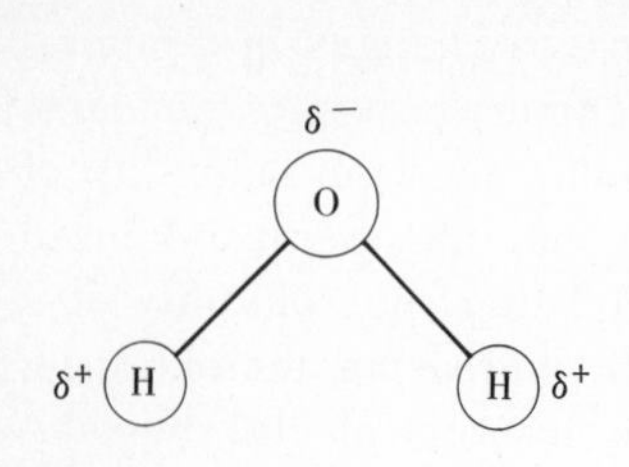

Water

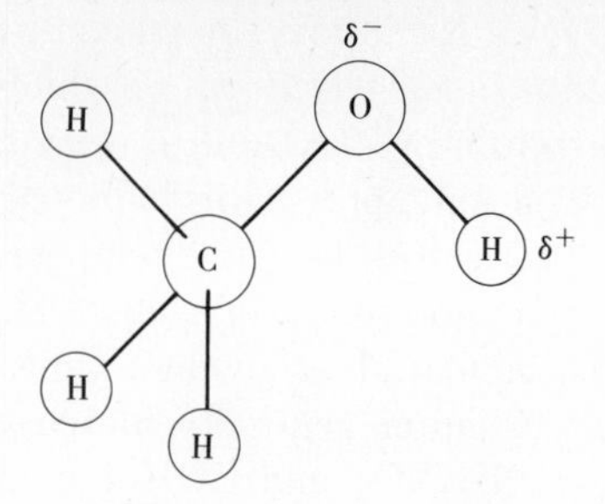

Methyl alcohol (a water derivative)

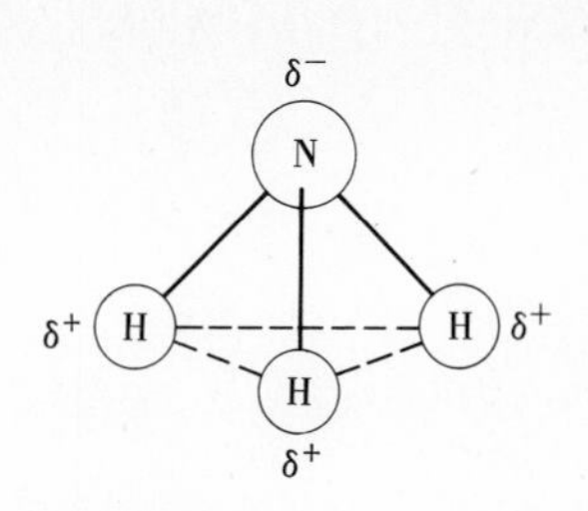

Ammonia

FIGURE 9.3

Some polar molecules.

in each molecule (see Fig. 9.3). Other polar liquids are also effective in dissolving ionic compounds. The molecules of octane and benzene are nonpolar (see Fig. 9.4). Since there is only a slight difference in the electronegativities of carbon and hydrogen, any bond between these atoms is nonpolar. The carbon chlorine bond *is* polar, as we would guess from electronegativities in Table 5.3, p. 109. Nevertheless, the CCl_4 molecule is nonpolar because the four chlorine atoms in the CCl_4 molecule are arranged symmetrically about the central carbon atom; the effect of the individual bonds is thereby canceled. Nonpolar liquids are generally ineffective in dissolving ionic compounds.

We also find that a polar liquid will dissolve other polar compounds. Ammonia and methyl alcohol dissolve in water. Sucrose, a polar (nonionic) solid, is soluble in water and methyl alcohol; it is insoluble in benzene, octane, and carbon tetrachloride. The nonpolar solvents dissolve nonpolar compounds. Greases and oils, which are nonpolar hydrocarbons, will dissolve in any of the three nonpolar solvents we have specifically discussed but are virtually insoluble in the three polar solvents. The rule usually followed in choosing solvents is *like dissolves like*. Polar solvents dissolve ionic

FIGURE 9.4

Some nonpolar molecules.

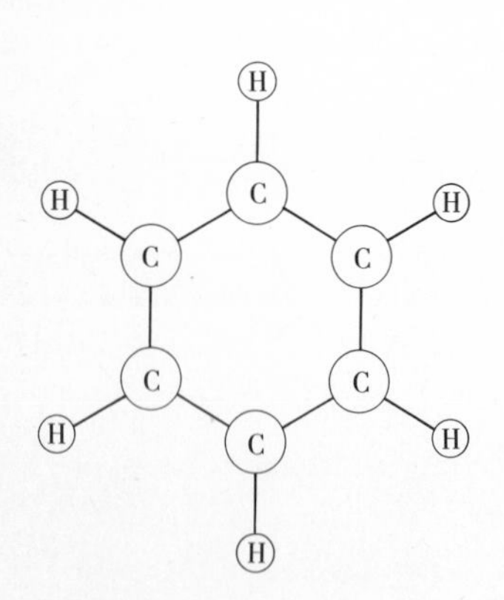

Benzene, C_6H_6
(Planar hexagon)

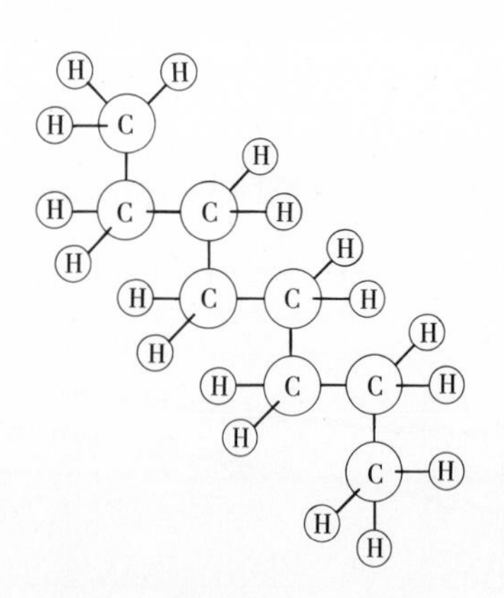

Octane, C_8H_{18}
(A staggered chain of 8 carbon atoms)

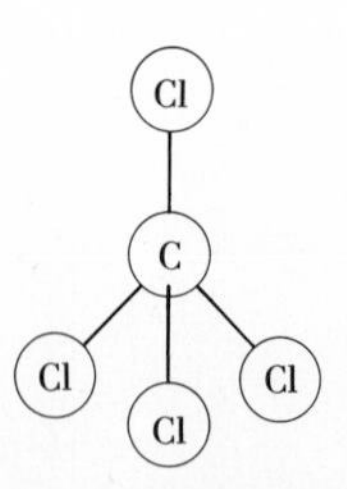

Carbon tetrachloride, CCl_4
(A regular tetrahedron of chlorine atoms)

TABLE 9.1: DIELECTRIC CONSTANTS FOR COMMON LIQUIDS AT 25°C

Compound	ϵ
Nonpolar:	
Cyclohexane, C_6H_{12}	2.015
Carbon tetrachloride, CCl_4	2.228
Benzene, C_6H_6	2.274
Chlorobenzene, C_6H_5Cl	5.621
Polar:	
Methanol, CH_3OH	32.63
Nitrobenzene, $C_6H_5NO_2$	34.82
Water, H_2O	78.54
Liquid ammonia, NH_3	22.4†

† At −33°C.

and polar compounds; nonpolar solvents dissolve nonpolar compounds.

The molecular process of dissolution can be interpreted in terms of the *electric forces* which operate between molecules (and ions). We shall try to explain the dissolution of Na^+Cl^- in these terms.

It is well known that charged bodies exert a mutual force. Coulomb found the quantitative relationship for this force: *two electric charges repel or attract each other with a force which is proportional to the product of their charges and inversely proportional to the square of the distance between them*

$$F = \frac{q_1 q_2}{\epsilon r^2}$$

q_1 and q_2 are the charges on the bodies and ϵ, called the *dielectric constant*, is a unitless constant dependent on the medium in which the charges exist. For vacuum $\epsilon = 1$. Table 9.1 gives the dielectric constants for some liquids often used as solvents. It is important to see the correlation between the dielectric constant and polarity. Liquids of low dielectric constant are nonpolar, and liquids of high dielectric constant are polar. We would choose a liquid of high dielectric constant as a solvent for ionic or polar compounds.

Now consider the interface between solid ionic Na^+Cl^- and water during dissolution. We know from our discussions of ionic bonds (page 99) and ionic solids (page 213) that bonds between ions are the strongest known. In order to remove an ion from the crystal structure the strong attractive forces must be reduced. The polar water molecule can do this at the crystal surface. Figure 9.5*b* pictures the polar H_2O molecules attacking individual ions at the surface. The negative oxygen end of the water dipole is attracted by the positive Na^+ ion. As it approaches, it neutralizes some of the charge on the ion and the coulomb attractive force which operates between

FIGURE 9.5

Mechanism for dissolution of an ionic crystal by a polar solvent.

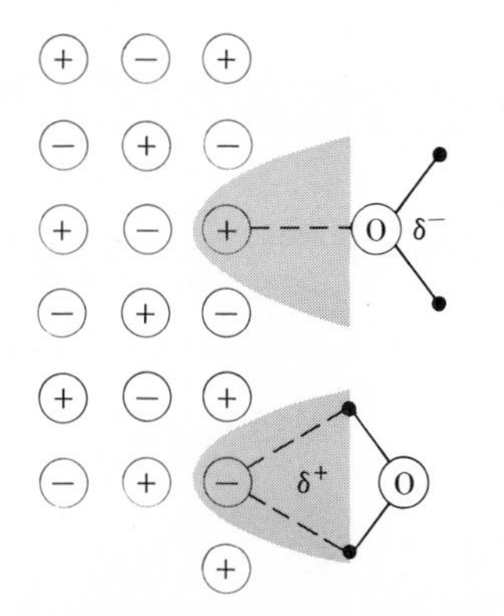

(*a*) H_2O *molecules neutralize ionic charge and "loosen" the ions at the surface.*

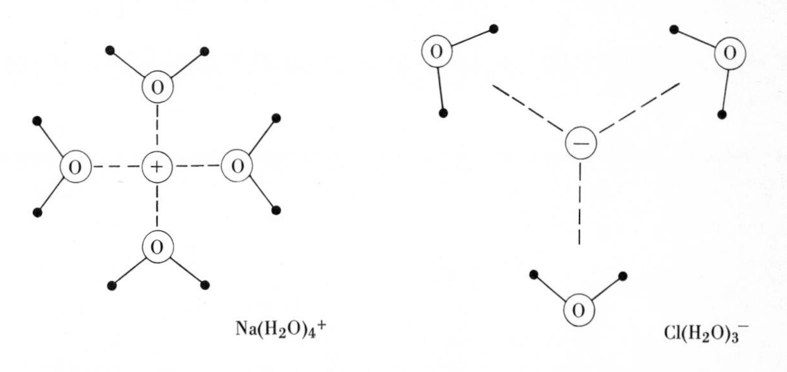

(*b*) H_2O *molecules hydrate ions in solution and partially neutralize their charge. The tendency for recombination of* Na^+ and Cl^- *is thereby reduced.*

the cation under attack and the surrounding anions is considerably reduced. The ion can be dislodged from the crystal surface and moved into the water phase. Once in solution, the ion is hydrated (bonded to more water molecules), as shown in Fig. 9.5*c*. This further neutralizes the ionic charges and reduces the tendency for recombination with oppositely charged ions.

We can explain the successful dissolution of ionic compounds in polar solvents in terms of the dielectric constant of the solvent. Coulomb's law also states that the electric force between charged bodies is *inversely* proportional to the dielectric constant of the medium. Then the electric force between Na^+ ions and Cl^- ions is low in water because the dielectric constant is high. The coulomb force, which tends to preserve the crystal, is appreciably reduced, and a solution can form.

The dissolution of polar compounds by polar solvents can be explained similarly. The molecules of a polar compound are held in lattice positions by electric attractions between dipoles or by hydrogen bonding. The molecules of a polar solvent can reduce such forces and allow the polar compound to dissolve.

Solubility of gases

Gases are only slightly soluble in water and other common liquids. The solubility of a gas can be increased by raising the pressure on the gas above the solution. The solubility of oxygen at various pressures is shown in Table 9.2. It is obvious from the last column that the solubility is directly proportional to the gas pressure. This behavior is normal for solutions of gases in liquids.

The solubility of gases in liquids, unlike that of solids (or liquids), decreases as the temperature increases. A chemist will boil a sample of water to reduce the concentration of dissolved atmospheric gases. Because the solubility of oxygen in water is greatly reduced by heating, the discharge of hot water into natural streams may cause serious damage to aquatic life.

The effect of temperature on the solubility of gases is typified by

TABLE 9.2: SOLUBILITY OF OXYGEN IN WATER AT 25°C

P, Torr	*Solubility, moles* O_2 *per liter of* $H_2O = M$	$\frac{M}{P} \times 10^6 \approx$ const
175	0.000307	1.75
300	0.000500	1.67
414	0.000688	1.66
610	0.00100	1.64
760	0.00128	1.68

TABLE 9.3: EFFECT OF TEMPERATURE ON THE SOLUBILITY† OF GASES IN WATER‡

	Temperature, °C				
Gas	*0*	*10*	*25*	*50*	*100*
H_2	0.000960	0.000873	0.000783	0.000717	0.000715
N_2	0.00105	0.000830	0.000647	0.000485	0.000423
O_2	0.00212	0.00170	0.00128	0.000933	0.000758
CO_2	0.0765	0.0533	0.0343	0.0194	

† Solubility is given in moles of gas dissolved per liter of water.

‡ Gas pressure above the solution is 1 atm.

the data in Table 9.3. The solubility of CO_2 is much higher than that of the other three because CO_2 reacts with water to form carbonic acid:

$$CO_2 + H_2O \rightleftharpoons \underset{\text{Carbonic acid}}{H_2CO_3}$$

Hydrogen chloride and ammonia, which are gases under ordinary conditions, also react with water, and are therefore very soluble.

9.6 COLLIGATIVE PROPERTIES

We noted (Sec. 9.2) that solution properties are a composite of the properties of the solvent and the solute. Surprisingly, in relatively dilute solutions of any solvent certain physical properties are almost independent of the nature of the solute. Such properties, called *colligative properties*, depend on the molar concentration of solute but not on its nature. For instance, all solutes cause the freezing point of a solution to be lower than the freezing point of the pure solvent. The freezing points of a 0.01 m sucrose solution and of a 0.01 m urea, $CO(NH_2)_2$, solution are both −0.0186°C. Not only is the freezing point of water depressed by sucrose and urea, but the depression is directly proportional to the number of moles of solute present. In 1 kg of water the freezing point is lowered 1.86°C per mole of solute dissolved, regardless of the nature of the solute. There are three other colligative properties which we shall study in this chapter: the elevation of the boiling point, the relative lowering of the vapor pressure, and osmotic pressure. One of the important applications of colligative properties is in molecular-weight determination. The osmotic pressure was the first colligative property to be studied thoroughly, but the lowering of the vapor pressure has a special significance. It is easy to demonstrate that if the vapor pressure of a solution is lower than the vapor pressure of the pure solvent, the solution freezing point must be lower than the solvent freezing point and the solution boiling point must be higher.

We shall limit our first consideration to solutions in which the solute is *nonvolatile* and *nonionic*. Such a solution is simple because the vapor pressure above it is due only to the solvent and the ultimate solute particle is the uncharged molecule, not an ion. The complicating effects of volatile solutes will be discussed at the end of the chapter; the colligative properties of ionic solutions are discussed in Chap. 14.

Osmosis and osmotic pressure

The cell walls of plants and animals are *semi*permeable; they permit the free flow of water but the passage of other materials is restricted. This phenomenon, the transport of solvent (but not of solute) through a membrane, was observed in the middle of the eighteenth century, and is called *osmosis*. It was carefully investigated in the 1830s by a French physiologist, Dutrochet, who named the phenomenon. If a pig's bladder is partially filled with a sugar solution and placed in a beaker of water, the solution will be diluted by a net transfer of water into the bladder. The pressure inside will increase, and if the sucrose concentration is high enough, the bladder may burst (see Fig. 9.6). Dutrochet found that the pressure produced is proportional to the concentration of the solution in the bladder. It is known today that the pressure is proportional to the *difference* in concentration: if sucrose solutions are placed in the bladder and beaker, no net transfer of water is observed. When a dilute sucrose solution is placed in the beaker and a more concentrated one in the bladder, water is transferred into the bladder until the concentrations are equal. If a sucrose solution is placed in the beaker and pure water in the bladder, water is transferred into the beaker. This demonstrates that the solvent can move in either direction across the semipermeable membrane.

When a solvent and a solution are separated by a semipermeable membrane, there is a net transfer of solvent into a solution. The tendency is toward increased dilution of the solution. A little consideration will convince you that this is the expected result under the circumstances. Picture a vessel which has two compartments; place

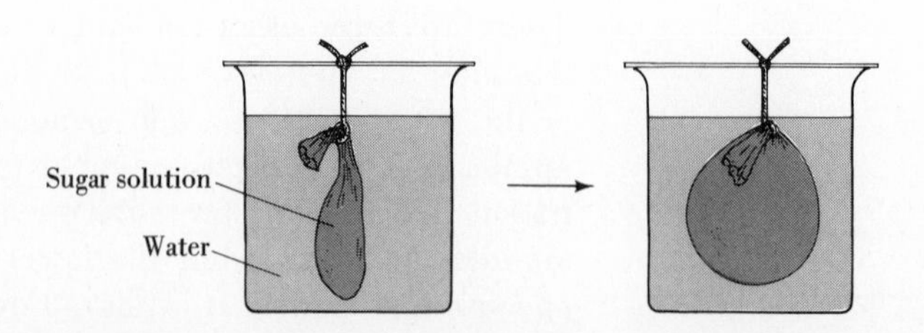

FIGURE 9.6

Osmosis: water passes through the pig's bladder, a semipermeable membrane.

FIGURE 9.7

The natural tendency to mix.

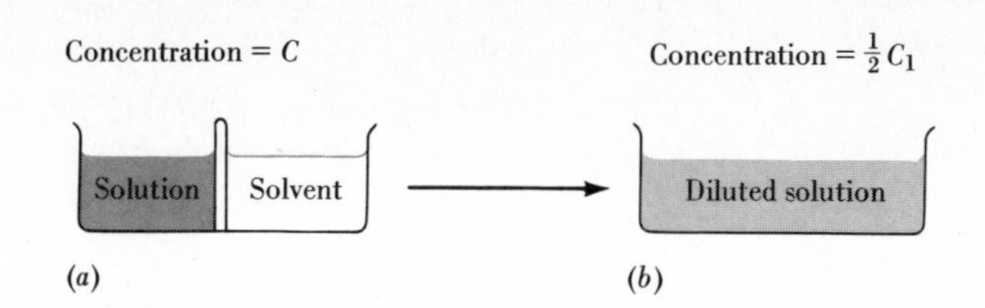

pure solvent in one compartment and a volume of a colored solution of the same solvent in the other. Carefully withdraw the wall which separates the solvent and the solution (see Fig. 9.7) and wait; eventually the pure solvent and the solution are thoroughly mixed, resulting in a solution more dilute than the original one. *The tendency is for the system to become thoroughly mixed.* How could the solvent and the original solution mix if they were separated by a semipermeable membrane, where the only mechanism for mixing is solvent transfer? The answer is clear: pure solvent must pass through the membrane into the solution. The opposite process, transfer of solvent from the solution into the pure solvent, could never lead to thorough mixing.

In order to define *osmotic pressure* it is convenient to consider a different kind of experiment. Figure 9.8 depicts an osmotic-pressure cell separated into two compartments by a semipermeable membrane. The solution side is equipped with a tube carrying a piston, which can be used to increase the pressure on the solution. Fused to the solvent compartment is a vertical capillary tube. A small change in solvent volume will cause a big change in the height of the liquid level in the capillary. The solvent will tend to flow into the solution. By placing weights on the piston the pressure on the solution can be increased until no transfer of solvent is observed. This point is determined by observing the solvent level in the capillary. The level becomes constant when there is no net transfer of solvent. *Osmotic pressure π is defined as the minimum pressure which must be applied to a solution to stop the net transfer of solvent into a solution through a semipermeable membrane.*

FIGURE 9.8

An osmotic-pressure cell.

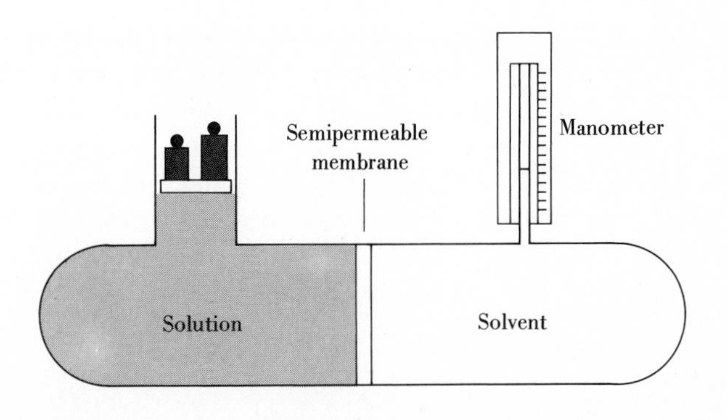

TABLE 9.4: EFFECT OF CONCENTRATION ON π†

π, atm	c, mol/l	$\frac{\pi}{c}$, l-atm/mol
1.34	0.0588	22.8
2.00	0.0809	24.7
2.75	0.119	23.1
4.04	0.179	22.6

† Aqueous sucrose solutions at 15°C (π/c should equal RT, which equals 23.6 l-atm/mol at 15°C).

The work of the German botanist Wilhelm Pfeffer in the 1870s and 1880s represents the first quantitative study of osmotic pressure. An examination of Pfeffer's findings (Tables 9.4 and 9.5) shows that π is directly proportional to concentration and temperature in dilute solutions:

$$\pi \propto cT$$

Inserting a proportionality constant k gives

$$\pi = kcT \tag{9.1}$$

It was the Dutch theoretical chemist and first Nobel laureate, J. H. van't Hoff, who first pointed out (1885) the similarity of the behavior of the osmotic pressure of dilute solutions and the pressure of ideal gases. For an ideal gas, the pressure is directly proportional to concentration and temperature; in a dilute solution, the osmotic pressure π is directly proportional to concentration and temperature. Compare:

For solutions: $\pi = kcT$

and

for ideal gases: $P = RcT \qquad c = \frac{n}{V}$

When the proportionality constant k in Eq. (9.1) is evaluated, it is found to be equal to R, the ideal-gas-law constant, so that the analogy is almost perfect.

EXAMPLE 9.4 The osmotic pressure developed by 0.0588 M sucrose solution at 14°C is 1.34 atm (Table 9.4). Evaluate the proportionality constant k in Eq. (9.1).

Solution: $\pi = kcT$

$$k = \frac{\pi}{cT} = \frac{1.34 \text{ atm}}{0.0588 \text{ mol/l} \times 287 \text{ K}} = 0.0810 \text{ l-atm/mol K} \approx R$$

TABLE 9.5: EFFECT OF TEMPERATURE ON π†

π, atm	T, K	$\frac{\pi}{T}$, atm/K
0.649	273	0.00237
0.664	280	0.00237
0.684	289	0.00237
0.721	295	0.00244
0.746	309	0.00241

† Aqueous sucrose solution at 0.0294 M (π/T should equal Rc, which equals 0.00241 atm/K at 0.0294 M).

Equation (9.1) is a summary of Pfeffer's experimental studies of the osmotic pressure of dilute solutions. It is an empirical equation but one which van't Hoff successfully derived from thermodynamic theory. His derivation demonstrates an important point: Eq. (9.1) is only an approximation, applicable to dilute solutions; more complicated formulas must be used to obtain accurate values of π at higher concentrations. It may be noted that when an empirical equation is successfully derived from physical theory, approximations must invariably be made to alleviate the mathematical difficulties, and these approximations limit the conditions under which the resulting equations are applicable. A study of the theoretical derivation alerts us to any limitations in the empirical equations and shows us how to make meaningful measurements and better analyses of data. For instance, since we know that Eq. (9.1) applies more exactly as a solution becomes more dilute, molecular weights calculated by applying Eq. (9.1) to osmotic-pressure data should be most reliable in dilute solutions.

EXAMPLE 9.5 Pfeffer found that the osmotic pressure developed by an aqueous solution containing 0.994 g of sucrose in 100 ml of solution is 0.704 atm at 15°C, whereas π for a solution containing 28.2 g in 100 ml is 25.2 atm at 15°C. Calculate the molecular weight of sucrose from both sets of data.

Solution: Using Eq. (9.1) with $k = R$ and $c = n/V$, we obtain

$$\pi V = nRT$$

But n = number of moles = g/M; finally we obtain

$$\pi V = \frac{g}{M} RT$$

or

$$M = \frac{gRT}{\pi V} \tag{9.2}$$

Substitution of the first set of data into Eq. (9.2) gives for a molecular weight of sucrose:

$$M = \frac{0.994 \text{ g} \times 0.082 \text{ l-atm/mol-K} \times 288 \text{ K}}{0.704 \text{ atm} \times 0.100 \text{ l}} = 333 \text{ g/mol}$$

Since the accepted molecular weight for $C_{12}H_{22}O_{11}$ is 342 g, the error in this determination is 9 parts in 342, or about 2.6 percent. Use of the data for the more concentrated solution gives a molecular weight of 264 g, which is equivalent to a 22.8 percent error.

Raoult's law

In the 1870s and 1880s François Marie Raoult studied the effect a solute has on the freezing point and vapor pressure of a pure solution, and in so doing discovered a basic law governing the behavior of dilute solutions. He found that the vapor pressure P of a solvent in a solution is always lower than the vapor pressure P^0 of the pure solvent at the same temperature. He expressed his findings concisely in what has come to be called *Raoult's law: The relative lowering of the vapor pressure,* $(P^0 - P)/P^0$, *is equal to the mole fraction of the solute* X_2 *in a dilute solution.* An equivalent mathematical expression of the law is

$$\frac{P^0 - P}{P^0} = \text{relative lowering of vapor pressure} = X_2 \tag{9.3}$$

Often it is more convenient to express Raoult's law as

$$P = X_1 P^0 \tag{9.4}$$

where X_1 is the mole fraction of the solvent. Equation (9.4) can be obtained from Eq. (9.3) by substituting $1 - X_1$ for X_2.

EXAMPLE 9.6 In one experiment Raoult observed that when a 27.6-g sample of benzaldehyde, C_7H_6O, was dissolved in 128.8 g of pure ethyl ether, $C_4H_{10}O$, at 18°C, the vapor pressure of ether above the solution was 352 Torr. The vapor pressure of pure ether is 404 Torr at this temperature. Is this observation consistent with Eq. (9.3)?

Solution: To answer this question we shall independently calculate the mole fraction of benzaldehyde and the relative lowering of the vapor pressure. If Raoult's observation is consistent with Eq. (9.3), these two quantities will be equal.

$$\text{Mole fraction of benzaldehyde} = X_2$$

$$= \frac{\text{mol benzaldehyde}}{\text{mol benzaldehyde} + \text{mol ether}} = \frac{n_2}{n_1 + n_2}$$

$$n_2 = \text{mol benzaldehyde} = \frac{g_2}{M_2} = \frac{27.6 \text{ g}}{106 \text{ g/mol}} = 0.260 \text{ mol}$$

$$n_1 = \text{mol ether} = \frac{g_1}{M_1} = \frac{128.8 \text{ g}}{74 \text{ g/mol}} = 1.74 \text{ mol}$$

$$X_2 = \frac{0.260}{1.74 + 0.260} = 0.130$$

$$\text{Relative lowering of vapor pressure} = \frac{P^0 - P}{P^0}$$

$$= \frac{404 - 352}{404} = 0.129$$

Therefore, $X_2 = (P^0 - P)/P^0$, and the result is consistent with Eq. (9.3).

It will become clear from Example 9.7 that it is possible to estimate the molecular weight of the solute by measuring the relative lowering of the vapor pressure. It is not possible to obtain accurate values of the molecular weight by this method because in dilute solutions where Raoult's law holds best, the vapor pressure of the solvent is not appreciably different from the vapor pressure of the pure solvent, and it is difficult to obtain a precise value of $P^0 - P$.

EXAMPLE 9.7 When a 3.50-g sample of a nonvolatile solute is dissolved in 100 g of H_2O, the solution has a vapor pressure of 31.71 Torr at 30°. Calculate the gram-molecular weight of the solute. The equilibrium vapor pressure of pure water at this temperature is 31.82 Torr.

Solution: From Raoult's law, Eq. (9.3), $(P^0 - P)/P^0 = X_2$, we write

$$\frac{P^0 - P}{P^0} = \frac{n_2}{n_1 + n_2} = \frac{g_2/M_2}{g_1/M_1 + g_2/M_2}$$

where M_2 = g mol wt of solute
M_1 = g mol wt of H_2O = 18 g

Substituting all data into the last equation and solving for M_2, we obtain

$$\frac{31.82 - 31.71}{31.82} = \frac{3.50/M_2}{\frac{100}{18} + 3.50/M_2}$$

$M_2 = 182$ g

Note that although the vapor pressures are known to four significant figures, the difference $P^0 - P$, which is crucial in the molecular-weight determination, is known to only two significant figures and there really are not three significant figures in the resultant value of M_2.

Freezing-point depression

Raoult's initial experiments on solutions involved the investigation of a phenomenon which was well known, that the freezing of a dilute solution is lower than the freezing point of a pure solvent. By expressing concentrations in terms of moles of solute (rather than in terms of mass), he showed that the freezing-point depression was a colligative property. The freezing-point depression ΔT_f of a dilute solution depends on the molar concentration of the solute and the nature of the *solvent;* it does not depend on the nature of the solute. *The freezing-point depression of a dilute solution is directly proportional to the solute molality:*

$$\Delta T_f = k_f m \tag{9.5}$$

k_f is a proportionality constant called the molal freezing-point-depression constant. It is the freezing-point depression per mole of solute dissolved in 1 kg of solvent. Note its units and that it varies from solvent to solvent (see Table 9.6).

TABLE 9.6: MOLAL FREEZING-POINT-DEPRESSION CONSTANTS

Solvent	*Freezing point, °C*	k_f deg-kg solvent/mol solute
Benzene	5.5	5.12
Camphor	178.4	37.7
Cyclohexane	6.5	20.0
Naphthalene	80.2	6.9
Water	0.00	1.86

FIGURE 9.9

Comparison of freezing-point data with predictions; $A = CCl_4$ in benzene; $B =$ sucrose in water.

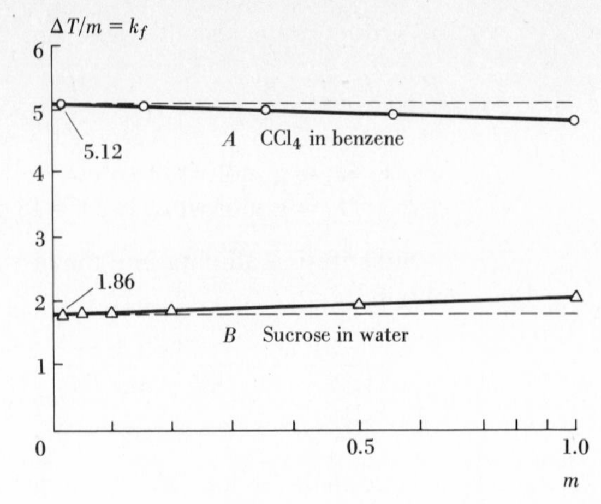

The molal freezing-point constant corresponds to the freezing-point depression that a 1.00 m solution would exhibit *if* the freezing-point depression followed Eq. (9.5) up to 1.00 m. Actually, deviations are generally observed below this concentration. Figure 9.9 shows a plot of $\Delta T/m$ versus m for two solutions, carbon tetrachloride in benzene and sucrose in water. By Eq. (9.5) $\Delta T/m$ should be a constant, equal to k_f; a plot of $\Delta T/m$ versus m should yield a horizontal line. The plot shows that both solutions exhibit deviations from the behavior predicted by Eq. (9.5); this deviation is negligible at concentrations below 0.1 m, and the molecular weight of solutes determined from freezing-point experiments carried out below this concentration should be reliable.

EXAMPLE 9.8 The freezing point of a solution made up by dissolving 0.644 g of acetone in 100 g of H_2O is -0.21°C. Calculate the molecular weight of acetone.

Solution: The calculation is made by substitution into Eq. (9.5):

$$\Delta T = k_f m$$

The absolute value of the freezing-point depression, or $|\Delta T|$, is used. The freezing point of pure water is 0.00°C, and therefore ΔT is 0.21 deg. From Table 9.5 we find that k_f is 1.86.

The molality is defined as moles of solute per kilogram of solvent. By this definition, for a solution which contains g_2 g of solute of molecular weight M_2 dissolved in g_1 g of solvent, the number of moles of solute is g_2/M_2, the number of kilograms of solvent is $g_1/1000$, and the molality is

$$m = \frac{\text{mol solute}}{\text{kg solvent}} = \frac{g_2/M_2}{g_1/1000}$$

or

$$m = \frac{1000g_2}{g_1 M_2} \tag{9.6}$$

Substituting $g_2 = 0.644$ g acetone and $g_1 = 100$ g H_2O into Eq. (9.6) gives

$$m = \frac{6.44}{M_2} \frac{\text{g solute}}{\text{kg } H_2O}$$

Use of Eq. (9.5) enables us to calculate M_2

$$\Delta T = k_f m = k_f \frac{6.44}{M_2}$$

$$0.21 \text{ deg} = 1.86 \frac{\text{deg-kg } H_2O}{\text{mol solute}} \frac{6.44}{M_2} \frac{\text{g solute}}{\text{kg } H_2O}$$

or

$$M_2 = 57 \text{ g/mol} \qquad \text{(acetone)}$$

To solve this type of problem it is more convenient to substitute Eq. (9.6) into Eq. (9.5) and obtain an equation for M_2:

$$\Delta T = k_f \frac{1000g_2}{g_1 M_2}$$

or

$$M_2 = \frac{1000 g_2 k_f}{g_1 \, \Delta T} \tag{9.7}$$

Substituting the data into Eq. (9.7), we have

$$M_2 = \frac{1000 \times 0.644 \times 1.86}{100 \times 0.21} \text{ g/mol} = 57 \text{ g/mol}$$

The true molecular weight of acetone is 58 g. In order to obtain a better value of the molecular weight a more precise value of the freezing-point depression is required. In this concentration range the freezing-point-depression technique is capable of giving molecular weights with an uncertainty of less than 1 percent.

Boiling-point elevation

The presence of a solute not only depresses the freezing point; it elevates the boiling point. The boiling-point elevation in dilute solutions, like the freezing-point depression, is directly proportional to the solute concentration, which is conventionally expressed as molality. The equation for the boiling-point elevation has a form identical to Eq. (9.5):

$$\Delta T_b = k_b m \tag{9.8}$$

k_b is the molal boiling-point-elevation constant. It is analogous to k_f but is generally smaller.

By substituting the mathematical definition of molality, Eq. (9.6), into Eq. (9.8) and rearranging we obtain an equation which is very useful for molecular-weight determinations from measurement of ΔT_b:

$$M_2 = \frac{1000 g_2 k_b}{g_1\, \Delta T_b} \tag{9.9}$$

Values of k_b for various solvents are listed in Table 9.7. Generally speaking, at identical concentrations, in a given solvent, freezing-point depressions are absolutely greater than boiling-point elevations. This follows from the fact that k_f is generally greater than k_b.

EXAMPLE 9.9 A solution consisting of 5.50 g of an organic substance in 27.50 g of CCl_4 boils at 81.5°C. What is (*a*) the molality of the solution and (*b*) the molecular weight of the solute?

Solution: (*a*) For carbon tetrachloride, CCl_4, $k_b = 5.0$, and the boiling point is 76.8°C. The boiling-point elevation of the solution is therefore 81.5 − 76.8°C, or 4.7°C. By substitution into Eq. (9.8) we can obtain m:

$$\Delta T_b = k_b m$$

$$4.7 \text{ deg} = \frac{5 \text{ deg-kg solvent}}{\text{mol solute}} \times m$$

$$m = 0.94 \text{ mol solute/kg solvent}$$

(*b*) It is most convenient to use Eq. (9.9) to obtain M_2:

$$M_2 = \frac{1000 g_2 k_b}{g_1\, \Delta T_b} = \frac{1000 \times 5.50 \times 5.0}{27.50 \times 4.7} \text{ g/mol} = 213 \text{ g/mol}$$

(There are really only two significant figures in this calculation. Why?)

TABLE 9.7: MOLAL BOILING-POINT-ELEVATION CONSTANTS

Solvent	*Normal boiling point,* °C	k_b, deg-kg solvent/mol solute
Carbon tetrachloride	76.8	5.0
Benzene	80.1	2.57
Ethyl alcohol	78.4	1.20
Water	100.0	0.52

Interrelationships of the colligative properties

In the last part of the nineteenth century van't Hoff showed that the existence of one colligative property implies the existence of all the others. He was able to do so by the application of rigorous thermodynamic theory; we shall show the interrelationships between the colligative properties in a more qualitative way.

Let us pretend that we know only Raoult's law:

$$P = P^0X_1$$

This statement of Raoult's law says that the vapor pressure of the solvent in solution is always lower than the vapor pressure of a pure solvent because X_1 is always less than 1. Suppose that we plot the vapor pressure of pure water at −25, 0, 25, 50, 75, and 100°C. We would obtain curve A in Fig. 9.10 (this is part of the phase diagram for water shown in Fig. 8.7). Now suppose we plot the vapor pressures of a dilute aqueous solution at the same temperatures; by Raoult's law the solution vapor pressures at each of these temperatures will be lower than the solvent vapor pressure. We obtain curve B in Fig. 9.10.

When pure water is at 0°C, the liquid and solid phases can coexist in equilibrium; 0°C is the freezing point.

Figure 9.10 shows that the solution must be below 0°C to be in equilibrium with solid; i.e., the solution freezing point is below 0°C. Because the vapor pressure of the solution is depressed, the solution freezes at a lower temperature than the solvent.

At the other end of curve A we see that the vapor pressure of water reaches 760 Torr at 100°C. But the solution vapor pressure is lower than 760 Torr at 100°C. The solution must be heated above 100°C before it begins to boil. Because the vapor pressure of the solution is depressed, the solution boils at a higher temperature than the solvent.

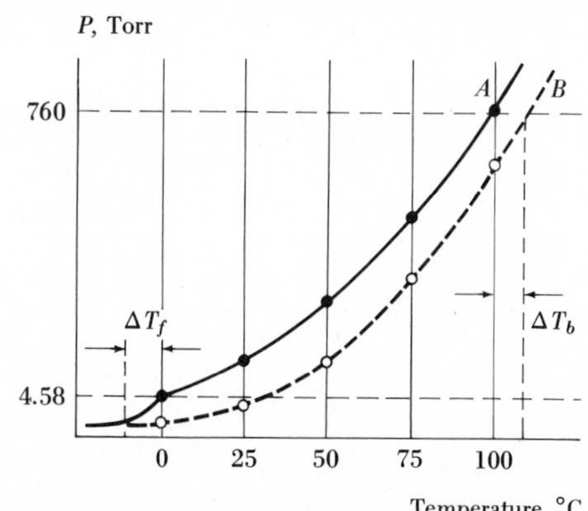

FIGURE 9.10

Plot of the vapor pressure of pure water and of a dilute solution versus temperature.

To complete the picture consider the osmotic-pressure cell of Fig. 9.8. A semipermeable membrane separates pure solvent from solution. Only solvent can pass through the membrane. By Raoult's law the pure solvent has a higher vapor pressure than the solvent in solution, and therefore the solvent will be forced through the semipermeable membrane into the solution. Osmosis will occur.

We have qualitatively deduced three other properties starting from one, Raoult's law. With the use of thermodynamic theory, it is possible to derive from Raoult's law the mathematical equations which were introduced in this chapter as empirical laws: $\pi = cRT$, $\Delta T_f = k_f m$, and $\Delta T_b = k_b m$. This is what van't Hoff did. His treatment included the deduction of more accurate equations for the colligative properties of solutions. He showed that the simple colligative-property equations we have discussed are approximations to the more accurate equations.

9.7 RAOULT'S LAW AND SOLUTIONS OF VOLATILE SOLUTES

In solutions in which both the solvent and solute have appreciable vapor pressure Raoult's law can be applied to both components. Using Eq. (9.4) for solvent and solute, we write

$$P_1 = \text{vapor pressure of solvent above solution} = X_1 P_1^0$$
$$P_2 = \text{vapor pressure of solute above solution} = X_2 P_2^0$$

Dalton's law of partial pressures allows us to write an equation for the total pressure above the solution:

$$P_{\text{total}} = P_1 + P_2 = X_1 P_1^0 + X_2 P_2^0$$

EXAMPLE 9.10 Calculate the partial pressures of water and ethyl alcohol and the total vapor pressure above a solution which contains 1.00 g of ethyl alcohol per 10.0 g of solution at 19°C. The vapor pressures of pure H_2O and ethyl alcohol are 16.3 and 40.0 Torr, respectively, at 19°C.

Solution: For water

$$P_1 = X_1 P_1^0 = \frac{n_1}{n_1 + n_2} P_1^0 = \frac{\frac{9}{18}}{\frac{9}{18} + \frac{1}{46}} \times 16.3 \text{ Torr} = 15.6 \text{ Torr}$$

For ethyl alcohol, C_2H_6O (molecular weight 46 g/mol)

$$P_2 = X_2 P_2^0 = \frac{n_2}{n_1 + n_2} P_2^0 = \frac{\frac{1}{46}}{\frac{9}{18} + \frac{1}{46}} \times 40.0 \text{ Torr} = 1.7 \text{ Torr}$$

The total pressure is

$$P_{\text{total}} = P_1 + P_2 = 15.6 + 1.7 \text{ Torr} = 17.3 \text{ Torr}$$

9.8 IDEAL SOLUTIONS

When a solution is formed from two volatile and chemically similar compounds, Raoult's law is obeyed by both components over the entire concentration range. More often there are deviations, but Raoult's law has gained a special significance in solution properties. A solution is considered an *ideal solution* if the solvent obeys Raoult's law. Deviations from ideal behavior are interpreted in terms of the interactions between the molecules (and ions) making up the solution. If the solution exhibits a vapor pressure lower than that predicted by Raoult's law (negative deviation), the intermolecular forces existing in the solution must be greater than the intermolecular forces in the separate pure components; the molecules have a reduced tendency to escape the solution (and the vapor pressure of the solution is lower than expected). If the solution exhibits a vapor pressure higher than that predicted by Raoult's law (positive deviation), the intermolecular forces existing in the solution must be weaker than those in the separate pure components; the molecules have a high tendency to escape the solution (and the vapor pressure of the solution is higher than expected). Figure 9.11 shows plots of solutions which obey Raoult's law (curve *a*), exhibit a negative deviation from Raoult's law (curve *b*), and exhibit a positive deviation from Raoult's law (curve *c*). Solutions which exhibit deviations from

FIGURE 9.11

Vapor pressure above solutions of completely miscible volatile liquid pairs.

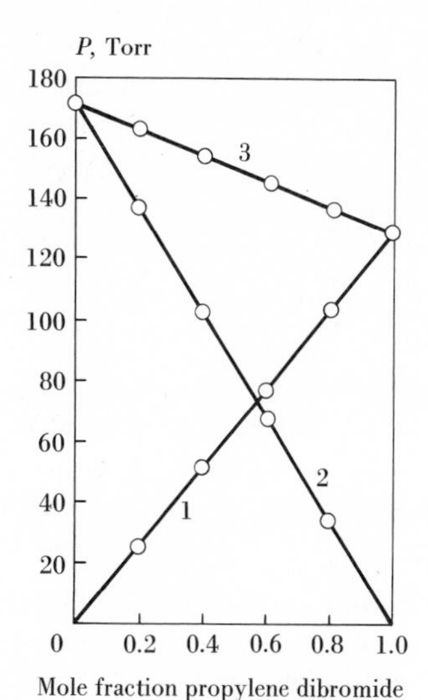

(*a*) *Vapor pressure of an ethylene dibromide–propylene dibromide solution at 85° C. 1 Partial pressure of propylene dibromide; 2 ethylene dibromide; 3 total pressure*

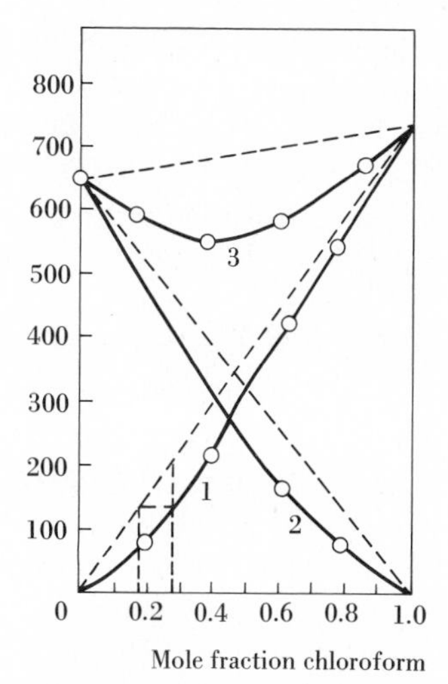

(*b*) *Vapor pressure of an acetone-chloroform solution at 55°C. 1 Partial pressure of chloroform; 2 acetone; 3 total pressure*

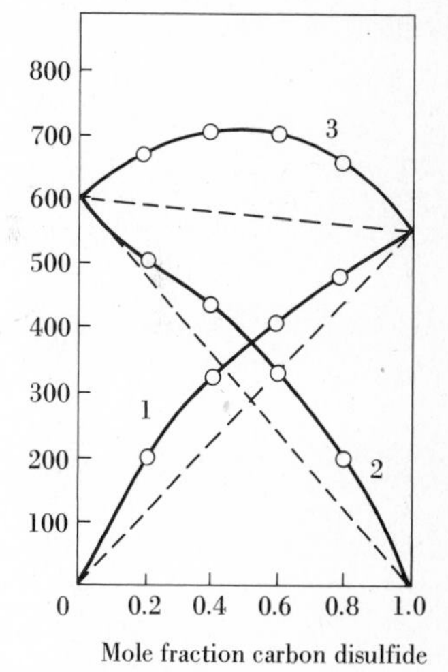

(*c*) *Vapor pressure of methanol carbon disulfide solutions at 35°C*

Raoult's law show corresponding deviations from other colligative property laws; this is to be expected in view of the fact that these properties are interdependent.

9.9 DEVIATIONS FROM IDEAL-SOLUTION BEHAVIOR

Figure 9.11*b* shows the vapor pressures above solutions of acetone, $CH_3—CO—CH_3$, and chloroform, $CHCl_3$, at various concentrations. At all concentrations the vapor pressures of the components are lower than the Raoult's law predictions. For instance, at a chloroform mole fraction of 0.26 the vapor pressure of chloroform is 130 Torr, considerably lower than the predicted 210 Torr. The solution behaves as if the chloroform concentration were much lower than 0.26 mole fraction. Under ideal solution conditions the chloroform partial pressure would reach 130 Torr at a mole fraction of only 0.15. The *effective concentration* of chloroform is 0.15; its measured concentration is 0.26.

The difference between effective concentration and the measured concentration is observed in other colligative properties. To illustrate this, freezing-point depressions for solutions of nitrobenzene in benzene are given in Table 9.8. At a molality of 0.137 the actual depression differs from the depression calculated using Eq. (9.5) by only 1 percent, but at all higher concentrations the depression is appreciably lower. Again, we can view these data in terms of the effective concentration of the solutions. The effective concentration is the value of the concentration calculated from the simple colligative-property laws using an experimental value of the colligative property. For example, the effective concentration of a nitrobenzene-benzene solution which has a freezing-point depression of 2.727°C is 0.535 mol of nitrobenzene per kilogram of benzene:

$$m = \frac{\Delta T_f}{k_f} = \frac{2.727}{5.10} = 0.535$$

The actual concentration is 0.589 *m* (see second row, Table 9.8).

The effective concentration of a solution is usually called the *activity a*, but it is often more convenient to discuss the behavior of a component in a solution in terms of its activity coefficient *f*. The activity coefficient is the ratio of the experimental value of a colligative property to its value as calculated from the simple colligative-property equations. The activity and activity coefficient are related by the simple equation

$$a = f \times \text{concentration} \tag{9.10}$$

The following equations define the activity coefficient in terms of any colligative property:

TABLE 9.8: FREEZING-POINT DEPRESSIONS FOR SOLUTIONS OF NITROBENZENE, $C_6H_5NO_2$, IN BENZENE C_6H_6

	ΔT_f, °C		
Concentration, m	*Observed* (1)	*Calculated from Eq. (9.5)* (2)	*Effective concentration* $\frac{\Delta T_{f,\text{obs}}}{k_f}$ (3)
0.137	0.689	0.699	0.135
0.589	2.727	3.01	0.535
0.898	3.945	4.58	0.775
2.04	8.015	10.4	1.57

$$f = \frac{\Delta T_{f,\text{obs}}}{\Delta T_{f,\text{calc}}} = \frac{\Delta T_{b,\text{obs}}}{\Delta T_{b,\text{calc}}} = \frac{\text{vapor pressure}_{\text{obs}}}{\text{vapor pressure}_{\text{calc}}} = \frac{\pi_{\text{obs}}}{\pi_{\text{calc}}} \tag{9.11}$$

EXAMPLE 9.11 Calculate the activity coefficient and the activity of a 0.898 m solution of nitrobenzene in benzene. Use the data in Table 9.8. k_f for benzene is 5.10.

Solution: The activity coefficient is

$$f = \frac{\Delta T_{f,\text{obs}}}{\Delta T_{f,\text{calc}}} = \frac{\Delta T_{f,\text{obs}}}{k_f m} = \frac{3.945}{4.58} = 0.863$$

The activity is

$$a = fm = 0.863 \times 0.898 = 0.775$$

and by our mathematical definitions the activity and effective concentration are identical (see last column of Table 9.8).

The molecular interpretation of the deviation of solution behavior from the simple colligative-property laws is discussed further in Chap. 14, which deals with ionic solutions.

9.10 COLLOIDS

When you turn on the headlights of a car to guide you on a foggy night, much of the light is scattered by the water droplets. The droplets of a very fine fog are small masses kept from falling to the ground by collisions with air molecules.

Fog is an example of the dispersion of liquid droplets in a gas. The dispersed phase is not dispersed as individual molecules, as in a true solution, but nevertheless the fog is quite stable and homoge-

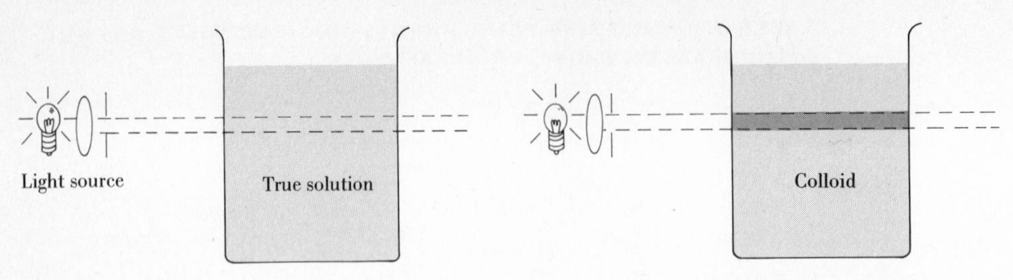

(a) Beam is invisible when viewed at 90° as it passes through a true solution

(b) The beam is visible when viewed at 90° as it passes through a colloid, since the beam is scattered in all directions by the colloidal particles

FIGURE 9.12

The Tyndall effect.

neous. The dispersion phase of liquid or solid particles in a gas is called an *aerosol.* The dispersion of particles of one substance in another is generally referred to as a *colloid.* An aerosol is a special kind of colloid. Colloids may result from the dispersion of a solid, a liquid, or a gas in a solid or liquid; there have been no examples of colloids resulting from dispersion of gases in gases, but every other combination is known. The most common colloids are *sols,* in which solid particles are dispersed in liquid; *gels,* in which liquid particles are suspended in a solid; *emulsions,* in which liquid particles are suspended in another immiscible liquid; and *aerosols,* in which solid or liquid particles are suspended in a gas. Examples of each type are shown in Table 9.9.

Colloids may look like true solutions but the two are easily distinguished by shining a beam of light through them. Viewed from the side, the beam is invisible as it passes through a true solution, which contains individually dispersed molecules. However, a colloid, which contains aggregates of thousands of molecules, scatters the light beam in all directions. The scattering of light by colloids is called the *Tyndall effect;* it is shown for a sol of soap in water in Fig. 9.12.

The particles of a colloid have a very high surface-to-mass ratio; this stabilizes the colloids and keeps the dispersed particles from

TABLE 9.9: TYPES OF COLLOIDS

Type	*Dispersed phase*	*Medium*	*Example*
Sol	Solid	Liquid	Soap in water
Gel	Liquid	Solid	Jello, jellies, opal
Emulsion	Liquid	Liquid	Salad dressing (oil and vinegar)
Aerosol	Liquid	Gas	Fog, mist, clouds
	Solid	Gas	Smoke, coal dust in air of coal mines

coagulating. The surfaces of the colloidal particles are charged and repel each other, preventing them from forming larger aggregates which might grow into separate phases. It is well to keep in mind that all surfaces are charged, but for large bodies, where the surface-to-mass ratio is small, the effect is not noticeable. The high ratio of surface to mass in tiny colloidal particles allows them to build up a relatively high charge.

Soap and detergent "solutions" are actually *sols*. Sodium stearate is a typical soap. The stearate ion is a long hydrocarbon-chain molecule ionized at one end, as shown in Fig. 9.13*a*. The ionized end, —COO^-, called the *carboxylate end*, attracts water molecules, which are very polar; it is said to be *hydrophilic*. The hydrocarbon chain is nonpolar and repels water molecules; it is said to be *hydrophobic*. When sodium stearate is mixed with water, the stearate ions form tiny collodial spheres with the carboxylate end at the surface and the hydrocarbon chains pointing toward the center (Fig. 9.13*b*). As many as 100 molecules may be present in the aggregate, which is called a *micelle*.

FIGURE 9.13

Soap forms micelles.

H_3C—CH_2—CH_2—CH_2—CH_2—CH_2—CH_2—CH_2—CH_2—CH_2—CH_2—CH_2—CH_2—CH_2—CH_2—CH_2—CH_2—C(=O)O^-

or

COO^-

(*a*) *Two representations of a stearate ion,* $C_{17}H_{35}COO^-$

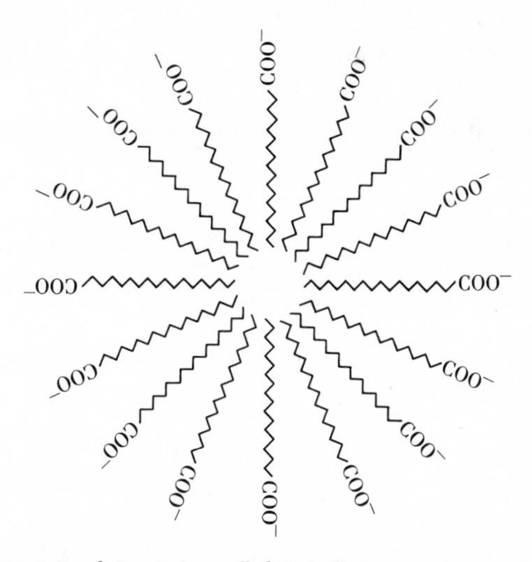

(*b*) *A spherical aggregate of stearate ions called a micelle*

Modern synthetic detergents, like sodium lauryl sulfonate, also form micelles. The detergents, like soaps, have a nonpolar hydrocarbon chain attached to a polar end. Whenever a compound is made up of molecules which contain a long or large nonpolar part bonded to an ionic or polar part, the molecules tend to aggregate and the result is a colloidal dispersion.

Macromolecules in "solution"

We have said that the difference between a true solution and a colloid is that the dispersed particles of a true solution are individual molecules whereas the dispersed particles of a colloid are large aggregates. When we consider macromolecules, synthetic polymers (polyethylene, polyvinyl chloride, etc.), and rubber, along with biopolymers (proteins, DNA, RNA, and polysaccharides), we have systems in which the dispersed particles may be individual molecules but have the physical properties of colloids. A single giant molecule can be as large as a micelle; some are much larger. Human hemoglobin is a macromolecule 75 Å long and 40 Å thick. It has a gram-molecular weight of 68,000. Many other biological molecules are quite large (Table 9.10) and form dispersions with the properties of colloidal systems rather than true solutions. A knowledge of the properties of colloids is extremely important in understanding the manufacture of industrial polymers and in the study of biochemical processes.

The osmotic pressure of colloidal dispersions can be used to determine the molecular weights of the dispersed aggregates or macromolecules. No other colligative property is sensitive enough for the purpose. This point is best shown by example.

EXAMPLE 9.12 Determine the osmotic pressure and the lowering of the vapor pressure at 25°C for a solution of 5.00 g of egg albumin in 150.0 g of water.

TABLE 9.10: MOLECULAR WEIGHTS OF SOME PROTEINS

Protein	*Dimensions,*† Å	*mol wt,* g/mol
Ribonuclease (beef pancreas)	. . .	14,000
Myoglobin (horse muscle)	. . .	17,000
β-Lactoglobulin (milk protein)	130 × 35	38,500
Albumin (egg protein)	120 × 30	45,700
Hemoglobin (horse red blood cells)	75 × 40	68,000
Fibrinogen (beef plasma)	700 × 30	340,000
Bushy stunt virus, rod-shaped (tomato)	3000 × 150	8,000,000

† Most of these giant molecules are ellipsoidal. The dimensions given are for the length and thickness of the ellipsoid of revolution.

Solution: From Table 9.10 we find that the gram-molecular weight of egg albumin is 45,700 g. We shall assume that the volume of the solution of 5 g of egg albumin and 150 g of water is 155 ml since the albumin has a density near unity. Using Eq. (9.2), we can obtain the osmotic pressure:

$$\pi = \frac{gRT}{VM} = \frac{5 \text{ g} \times 0.082 \text{ l-atm/mol-K} \times 298 \text{ K}}{0.155 \text{ l} \times 45{,}700 \text{ g}}$$

$$\approx 0.017 \text{ atm or } 13.1 \text{ Torr}$$

The lowering of the vapor pressure is too small to measure, as can be seen from the calculation:

$$\frac{\Delta P}{P^0} = X_2 = \frac{n_2}{n_1 + n_2}$$

$$\Delta P = P^0 \frac{n_2}{n_1 + n_2}$$

or

$$\Delta P = 23.8 \text{ Torr} \times \frac{5 \text{ g}/45{,}700 \text{ g}}{5 \text{ g}/45{,}700 \text{ g} + 150 \text{ g}/18 \text{ g}} = 1 \times 10^{-5} \text{ Torr}$$

$$P^0 - P \approx 0$$

The albumin lowers the pressure by about 0.00001 Torr, but in effect there is no vapor-pressure lowering because it cannot be measured with any accuracy or precision.

QUESTIONS AND PROBLEMS

9.1 (*a*) List the tests you would perform to prove or disprove that a homogeneous liquid mixture of 1 g of sugar and 100 g of methyl alcohol forms a true solution and not a compound. (*b*) What tests would you perform to help you decide whether steel is a solution or a compound?

9.2 If 75 ml of alcohol is dissolved in enough water to make 100 ml of solution, which liquid is the solvent and which is the solute? Explain how you made your choice.

9.3 Explain how polar solvents easily dissolve polar and ionic solids. Use sketches.

9.4 In terms of Coulomb's law explain why solvents of high dielectric constant are best for dissolving salts.

9.5 Two compounds, A and B, form a solution. Does this mean that the attraction between a molecule of A and a molecule of B is *much* greater than the attraction between two A molecules? Explain.

9.6 Calculate the molality and mole fractions of solute and solvent in the following aqueous solutions: (*a*) 11.5 percent by weight $KMnO_4$, (*b*) 5.00 g NaCl in 17.5 g H_2O, (*c*) 50.0 g of sucrose, $C_{12}H_{22}O_{11}$, in 1.5 kg of H_2O.

9.7 Is it possible to calculate the exact molarity of the solutions described in Prob. 9.6? If not, what further information is required?

9.8 Calculate the molality, mole fractions of solute and solvent, and molarity of the following solutions: (*a*) an aqueous H_2SO_4 solution which is 50.0 percent by weight H_2SO_4 and has a density of 1.40 g/ml, (*b*) an aqueous sucrose, $C_{12}H_{22}O_{11}$, solution which is 19.0 percent sucrose by weight and has a density of 1.08 g/ml, and (*c*) a solution made up of 24.4 g of NaOH and 97.6 g of H_2O which has a volume of 100 ml.

9.9 Define (*a*) solubility, (*b*) saturated solution, and (*c*) supersaturated solution.

9.10 The solubility of a solid in a liquid generally increases with temperature. Explain this in terms of kinetic-molecular theory.

9.11 At room temperature aniline, $C_6H_5NH_2$, and ethylene glycol, $C_2H_6O_2$, have dielectric constants of 6.89 and 37.7, respectively. Choose a good solvent from Table 9.1 for each of these compounds.

9.12 William Henry (1774–1836), a friend of Dalton's, discovered that the solubility of a gas is directly proportional to the pressure of the gas above a solution:

$M = k_H P$ Henry's law

where k_H is called the Henry's law constant. (*a*) From Table 9.2 determine an average value of the Henry's law constant for solutions of O_2 in H_2O at 25°C. (*b*) k_H for N_2 in water is 6.4×10^{-5} mol/l-atm at 25°C. What is the solubility of N_2 in moles per liter under atmospheric conditions? (*c*) What volume of N_2 (measured at STP) will dissolve in 1 ml of water at 25°C? (*d*) Determine the Henry's law constants at 0°C in moles per liter-Torr for H_2, N_2, O_2, and CO_2 from the data in Table 9.3.

9.13 (*a*) What is a colligative property? (*b*) Name and define the four colligative properties discussed in this chapter. (*c*) Name four solution properties which are *not* colligative.

9.14 What will happen if a semipermeable bag containing pure water is immersed in a stream of concentrated salt solution?

9.15 (*a*) Using the data in Table 9.4, calculate π/T for each concentration, keeping in mind that $T = 288$ K in each case. Plot π/T versus c. (*b*) What is the average value of the slope $\Delta(\pi/T)/\Delta c$? (*c*) What are the units of the slope? (*d*) In terms of Eq. (9.1) what is the significance of the slope?

9.16 (*a*) Using the data in Table 9.5, calculate the value of π/c for each temperature, keeping in mind that $c = 0.0294$ M in each case. Plot π/c versus T. (*b*) What is the average value of the slope $\Delta(\pi/c)/\Delta T$? (*c*) What are the units of the slope? In terms of Eq. (9.1) what is the significance of the slope?

9.17 Calculate the osmotic pressure of a 0.0294 M aqueous sucrose solution at 0°C (see Table 9.5).

9.18 Calculate the osmotic pressure of a 0.179 M aqueous sucrose solution at 15°C (see Table 9.4).

9.19 How do we know that Eqs. (9.1) and (9.2) are valid only at low concentration? (Consider the work of van't Hoff and the results of Example 9.5.)

9.20 In terms of a kinetic-molecular explanation why do you suppose that the vapor pressure of a pure liquid is greater than the vapor pressure of the same liquid in a solution? (Consider the molecules at the surface of the pure liquid and at the surface of a solution.)

9.21 What is the strict definition of the *relative lowering of the vapor pressure?* Define all terms.

9.22 Show that $(P^0 - P)/P^0 = X_2$ and $P = X_1P^0$ are equivalent forms of Raoult's law.

9.23 Determine the vapor pressure at 25°C of an aqueous solution consisting of 10.0 g of sucrose, $C_{12}H_{22}O_{11}$, and 75.0 g of H_2O.

9.24 (*a*) The water vapor pressure of a solution containing 1.7930 g of the sugar mannitol, ($C_6H_{12}O_6$), in 100 g of water at 20°C is 0.0307 Torr lower than the vapor pressure of pure water. Calculate the molecular weight of mannitol from its formula *and* from the vapor-pressure data. At 20°C the equilibrium vapor pressure of water is 17.535 Torr. How many significant figures do you have in the answer obtained using vapor-pressure data?

9.25 (*a*) The vapor pressure of pure benzene, C_6H_6, at 20°C is 74.66 Torr. When 2.00 g of a nonvolatile hydrocarbon is dissolved in 100 g of benzene, the vapor pressure is reduced to 74.01 Torr. Calculate the molecular weight of the hydrocarbon. (*b*) The hydrocarbon is 94.4 percent carbon and 5.6 percent hydrogen by weight; determine the empirical formula. (*c*) What is the molecular formula?

9.26 Calculate the water vapor pressure of a solution which contains 10.0 g of glycerol, $C_3H_8O_3$, in 100 g of water at 30°C. The vapor pressure of pure water at 30°C is 31.824 Torr.

9.27 A solution containing 8.00 g of a nonelectrolyte dissolved in 350 g of water freezes at −0.820°C. Calculate the molecular weight of the solute.

9.28 Calculate the molecular weights of the solutes in the solutions of nonelectrolytes shown (Table P9.28). The freezing point of pure benzene is 5.48°C, and the molal freezing-point constant of benzene is 5.12°C.

TABLE P9.28

Nonelectrolyte

Solute, g	*Benzene*, g	*Freezing point*, °C
(*a*) 20	250	3.5
(*b*) 25	200	0.20
(*c*) 80	400	2.2

9.29 Calculate the freezing point of a 20% by weight solution of ethyl alcohol, C_2H_6O, in water.

9.30 An aqueous solution of sucrose, $C_{12}H_{22}O_{11}$, freezes at −0.93°C. Calculate the percentage of sugar in this solution.

9.31 An aqueous solution of ethyl alcohol, C_2H_6O, is 4.01 M and has a specific gravity of 0.970. Calculate the freezing point of the solution.

9.32 A solution containing 52.5 g of a solute in 1 kg of water boils at 100.39°C. What is the molecular weight of the solute?

9.33 A solution containing 25.0 g of a solute in 200 g of water has a density of 1.08 g/ml at 25°C. Its boiling point is 100.26°C. Calculate (*a*) the molecular weight of the solute, (*b*) the water vapor pressure of the solution, (*c*) the freezing point of the solution, and (*d*) the osmotic pressure of the solution. The vapor pressure of pure water at 25°C is 23.756 Torr.

9.34 When 0.05844 g of NaCl is dissolved in 100 g of water, the freezing point of the solution is −0.0372°C. (*a*) Calculate the "molecular weight" of NaCl from the freezing point of the solution *and* from its formula. (*b*) Explain the results. (*c*) What would the freezing point of a 0.001 *m* Na_2SO_4 solution be?

9.35 A compound on analysis is found to have the following percentage composition: 40.7% carbon, 5.09% hydrogen, and 54.2% oxygen. When 1.18 g of the compound is dissolved in 100 g H_2O, the solution boils at 100.052°C. (*a*) Determine the empirical formula, (*b*) calculate the molecular weight, and (*c*) determine the molecular formula.

9.36 A compound containing only carbon, hydrogen, and oxygen is 26.65 percent carbon and 2.22 percent hydrogen by weight. When 4.50 g of this compound is dissolved in 250 g of benzene, the freezing point of the solution is 4.47°C. (*a*) Determine the molecular formula of the solute and (*b*) the boiling point of the solution.

9.37 Van't Hoff showed that one colligative property implies the existence of all the others. How does the relative lowering of the vapor pressure imply a depression of the freezing point, an elevation of boiling point, and the existence of osmotic pressure?

9.38 Chlorobenzene and bromobenzene are both volatile and completely miscible at 140°C. Assuming that they form an ideal solution, calculate the vapor pressure of each component of a solution which has a chlorobenzene mole fraction of 0.655 at 140°C. At this temperature the vapor pressures of the pure liquids are 939.4 and 495.8 Torr for chlorobenzene and bromobenzene, respectively.

9.39 A solution of two volatile liquids, A and B, obeys Raoult's law. At 25°C the vapor pressures are 100 and 250 Torr for A and B, respectively. (*a*) Calculate the vapor pressures of A and B in solutions which have mole fractions of B equal to 0.0, 0.2, 0.4, 0.6, 0.8, and 1.0. (*b*) Make a plot of P_A, versus X_B; P_B versus X_B; and P_t versus X_B. Use Fig. 9.11 as a model.

9.40 How do you distinguish between a true solution and a colloid (*a*) in the laboratory and (*b*) by definition?

9.41 Give an example of a common sol, gel, aerosol, and emulsion. Do not use the examples given in Table 9.9.

9.42 What is the most important factor in stabilizing colloidal dispersions?

9.43 Using an ultramicroscope, it is possible to count the number of colloidal particles in a definite volume. Suppose that 5.00 g of cottonseed oil, which has a density of 0.926 g/ml, is emulsified with 100 g of H_2O. Counting with an ultramicroscope shows that there are about 1000 microglobules of oil present in 1 μl, or 10^{-6} l. (*a*) Calculate the volume and diameter of the globules assuming they are uniform spheres. (*b*) Using a molecular weight of 680 g/mol for cottonseed oil, determine the number of moles and *molecules* present in each microglobule.

9.44 Are dispersions of macromolecules in liquids true solutions or colloids? Consider the definitions of solution and colloid and support your contention.

9.45 (*a*) Determine the vapor pressure, freezing point, boiling point, and osmotic pressure of a dispersion of 1.00 g of ribonuclease (Table 9.10) in 100 g of H_2O at 25°C. (*b*) Which colligative property is most suited for the estimation of the molecular weights of macromolecules?

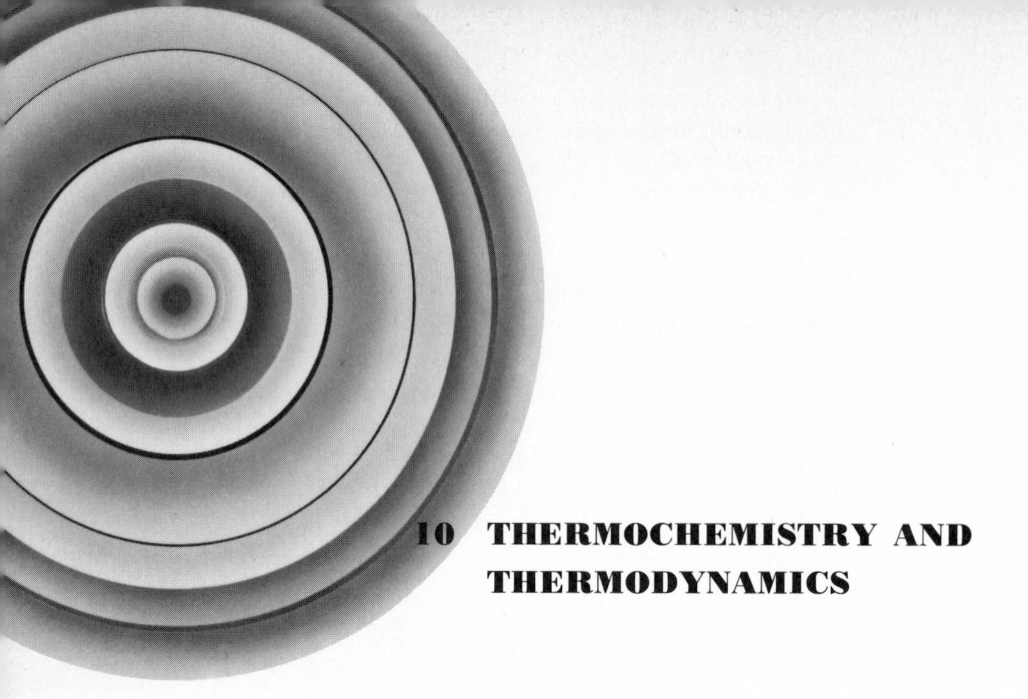

10 THERMOCHEMISTRY AND THERMODYNAMICS

10.1 INTRODUCTION

Chemists have long searched for a principle for determining whether a particular chemical reaction will occur. If without recourse to direct experimentation we could accurately decide whether or not two or more materials would spontaneously react, a great deal of time, effort, and money could be saved. We shall be concerned in this chapter with the factors which govern the *spontaneity* of chemical reactions (and other physical changes). Two broad approaches to the problem have been made, one resting on the principles of molecular structure and the other on two experimental laws of *thermodynamics*. We have a firm belief that we shall be able to predict the reactions or other changes molecules will undergo if we can obtain a good understanding of their structure and bonding. This theoretical approach has high promise, and we have already discussed its basis in Chap. 5, but so far it has not produced good results except for the simplest reactions, like $H + H \rightarrow H_2$. Note that in this theoretical approach we state that if we understand the invisible molecular world, we shall be able to predict a macroscopic observable result, in this case a chemical reaction.

The thermodynamic approach to chemical reactions is philosophically quite different from the molecular approach. It is highly developed and very successful. Using only two principles, the first and second laws of thermodynamics, a whole science of chemical reactions has been developed. Using thermodynamics, the chemist can decide not only whether a chemical reaction will occur but to what extent it will occur. For instance, ammonia is produced industrially by the reaction of N_2 and H_2:

$$N_2(g) + 3H_2(g) \rightarrow 2NH_3(g)$$

By use of the thermochemical tables we deduce that when a mixture of $H_2(g)$ and $N_2(g)$ is prepared in a molar ratio of 3:1 at 450°C and a total pressure of 10 atm, the partial pressure of ammonia will reach 0.204 atm exactly and 13.37 kcal of heat will be evolved.

The conclusions of thermodynamics are reached without recourse to the hypothesis of the existence of atoms and molecules. We can make accurate predictions about chemical reactions even though we do not know the structure of the molecules involved; but, unfortunately, from the science of thermodynamics alone we cannot decide how much time will be required for a chemical transformation. For example, we can show that the oxidation of diamond (a crystalline form of pure carbon) by oxygen under ordinary conditions is *thermodynamically* favorable, or should *occur spontaneously*. This means that thermodynamics predicts that the following reaction will occur at room temperature:

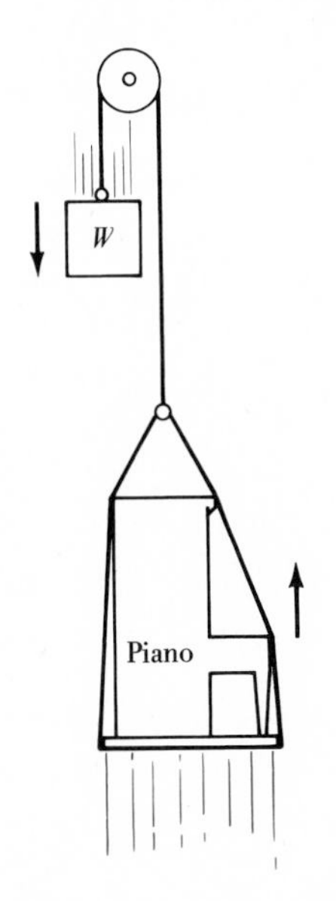

FIGURE 10.1

The body W possesses energy because it is capable of doing work; it lifts the piano.

$$C(\text{diamond}) + O_2 \rightarrow CO_2$$

but everyone knows that the reaction does not take place at an *observable rate* under ordinary circumstances. Thermodynamics also predicts that gaseous hydrogen and oxygen will react to produce water, but mixtures of these two gases can be kept in glass vessels indefinitely; yet we know that the mixture is dangerously reactive. A spark or the presence of certain materials like palladium will cause the gases to undergo an explosive transformation into water. The energy liberated by the reaction and the amount of water formed can be exactly calculated by the methods of thermodynamics, but the time required for the transformation cannot be determined.

10.2 THE FIRST LAW OF THERMODYNAMICS

The capacity of a body to do work is called its *energy*. For instance, the body in Fig. 10.1 possesses energy because it can lift a piano; i.e., it has the capacity to do work. As we shall see, the total energy of a body depends on its position, state of motion, internal condition, chemical composition, and mass.

Position and potential energy

Any body on the surface of the earth has energy by virtue of the fact that it exists in the earth's gravitational field. The body shown in Fig. 10.1 can lift the piano because it is in the earth's gravitational field. When properly harnessed, the fall of a body in a gravitational field can be used to do work: water falling over a dam can be used to generate electricity, and water falling over a waterwheel can power machinery.

There are force fields other than gravitational fields, and each kind can be used to do work. A charged body will tend to move in an electric field, and a magnetic body will tend to move in a magnetic field. We can obtain work by harnessing such bodies. The energy a body possesses by virtue of its position in a field is called its *potential energy* (PE), and we can speak of gravitational potential energy, electric potential energy, magnetic potential energy, etc. The potential energy is often symbolized by V.

State of motion and kinetic energy

We have already discussed kinetic energy (KE) in Sec. 6.7. It is the energy a body possesses by virtue of its motion and can be calculated from the defining formula

$$\text{KE} = T = \tfrac{1}{2}mv^2$$

where m = mass of body
v = velocity of body

A body which has potential energy because of its position in an electric or magnetic field may also have kinetic energy if it is moving.

Einstein's mass-energy relationship

Einstein predicted that mass and energy are interconvertible and that the relationship between the two is given by a simple equation. If a body of mass m is converted into energy, the energy produced E_m will be

$$E_m = mc^2$$

where c is the velocity of light in a vacuum, 3×10^{10} cm/s. If m is in grams, E_m will be in g-cm^2/s^2, or ergs. (Remember: 8.314×10^7 ergs = 1.987 cal = 0.08203 l atm.)

With Einstein's equation it can be shown that 1 g of mass converted into energy is 9×10^{20} ergs, which is the equivalent of 11,000 tons of TNT. The correctness of the Einstein equation was dramatically demonstrated in August 1945, when atomic bombs were exploded over Hiroshima and Nagasaki.

The internal condition of a body and internal energy

The internal condition of a body refers to its physical and chemical condition as described by its temperature, density, vapor pressure, index of refraction, etc., and its composition. The energy of a body depends upon its internal condition as well as on its position in a field and its velocity. The gravitational and kinetic energies of two bodies may be equal, but if one body is very hot, it can be used as an energy source to drive a heat engine; a cool body would be less effective. The energy of a body depends on the temperature. The energy of a body also depends on its composition. Burning of a kilogram of coal delivers much more heat than burning of a kilogram of carbon monoxide; therefore more work can be obtained from coal than from an equal weight of carbon monoxide. The energy of a body depends on all the other internal properties we have mentioned previously. The energy which a body has by virtue of its internal condition is called *internal energy* U. The internal energy is of utmost importance in the thermodynamics of chemical reactions.

Total energy E

The total energy of a body may be taken as a sum of the different kinds of energy discussed in the previous sections:

$$E = \text{PE} + \text{KE} + \text{internal energy} + \text{energy equivalent of mass}$$

$$E = V + T + U + E_m \tag{10.1}$$

The first law of thermodynamics can be stated in many ways, but the broadest is that *the total energy of an isolated body or system is constant*. Then we can write

$$E = V + T + U + E_m = \text{const} \tag{10.2}$$

An isolated system is defined as one which is completely insulated from its surroundings; neither mass nor energy can move across the boundary between an isolated system and its surroundings. The system has no effect on the surroundings, and vice versa (see Fig. 10.2). This statement does not mean that the potential energy or the internal energy must remain constant. The potential energy can increase in an isolated system provided there is a compensating decrease in the internal energy, kinetic energy, or in the mass-energy.† For instance, if some mass disappears in an isolated system, there is an accompanying change in the internal conditions (temperature, pressure, etc.). The loss of mass-energy would be compensated by an increase in the internal energy; the total energy would remain constant.

FIGURE 10.2

Neither mass nor energy can enter or leave an isolated system.

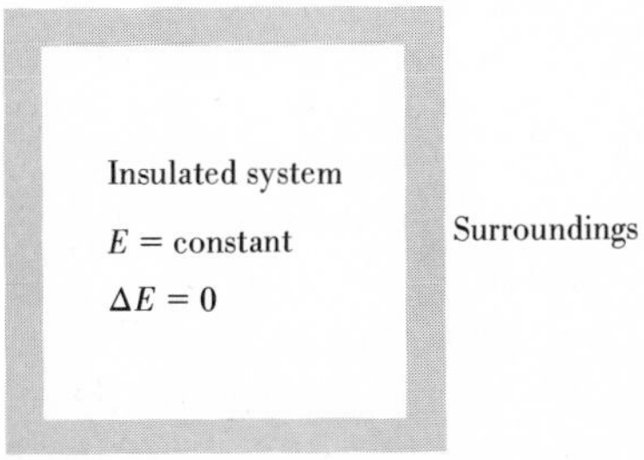

Insulated system

E = constant

$\Delta E = 0$

Surroundings

† The hyphen emphasizes the equivalence of the two halves of this term, which is simpler than writing "mass and its equivalent energy."

10.3 THE IMPORTANCE OF INTERNAL ENERGY IN THE STUDY OF CHEMICAL REACTIONS

The first law of thermodynamics allows us to come to some definite conclusions about isolated systems, but unfortunately we can never make observations on truly isolated systems. In order to study any system we must be able to examine its properties. This always requires an exchange of energy between the system and the surroundings. Even a temperature measurement requires that some energy pass from the system to the surroundings via the thermometer. Therefore, we can never investigate a truly isolated system although we try to approximate such systems in the calorimetric study of chemical reactions.

In most experiments chemists are not concerned with isolated systems. We often study chemical reactions in a constant-temperature oil bath (Fig. 10.3). As the reaction proceeds, heat may be given up or absorbed by an oil bath that is maintained at a constant temperature by a thermostat. Because the oil bath maintains the internal temperature of the system, the other internal properties (pressure, composition, index of refraction, etc.) of the reaction system are the ones that change. This means that the internal energy changes. Since the reaction system is not in motion before, during, or after the reaction and its position does not change, its kinetic and potential energies remain constant. There is no measurable change in mass, and therefore the energy equivalent of the mass of the system remains constant. It is obvious from Eq. (10.1) that the change in *total* energy just equals the change in the *internal* energy. This is true in most changes of chemical interest. The total energy of the system is

$$E = V + T + U + E_m$$

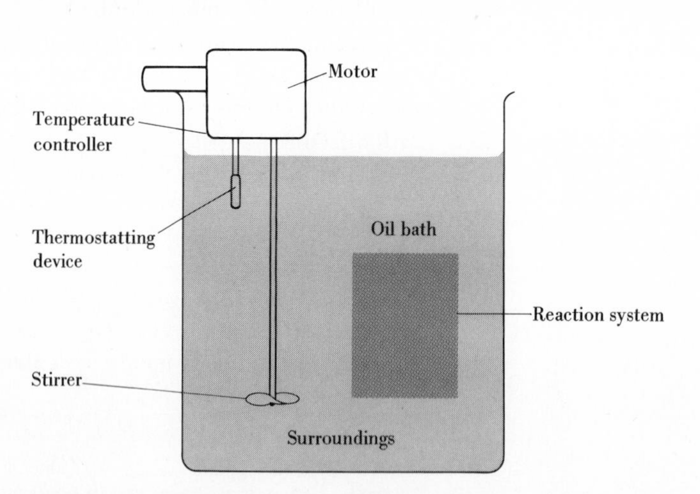

FIGURE 10.3

Reaction studied in a thermostatted oil bath.

and the change in energy is equal to the sum of each energy change,

$$\Delta E = \Delta V + \Delta T + \Delta U + \Delta E_m$$

but in the system we have described, ΔV, ΔT, and ΔE_m equal zero. Therefore,

$$\Delta E = \Delta U = \text{change in internal energy}$$

To summarize, in chemical reactions and other processes of chemical interest only the internal energy of the system ordinarily undergoes any change.

10.4 EFFECT OF WORK AND HEAT ON THE INTERNAL CONDITION OF A SYSTEM

Since we do not conduct our experiments or make our observations on isolated systems, the first law of thermodynamics as given in Eq. (10.2) is not very useful for chemical problems. During ordinary experiments the temperature, pressure, and other properties of a system do change, and hence the internal energy generally undergoes a net change.

We can change the internal conditions of a system in nonchemical ways. We can add or extract heat; we can perform work on the system or cause the system to perform work. In each instance the internal conditions change, and therefore the internal energy of the system changes. Consider a fixed amount of water. By heating it, we can raise its temperature and vapor pressure and decrease its index of refraction and density. By performing work on the water we can cause the same changes; e.g., agitating the water with a motorized stirrer will cause the temperature and vapor pressure to rise and the density and index of refraction to decrease. The internal energy U of a body can be changed by a heat transfer or a work transfer. The change in internal energy of a body will be equal to the amount of heat *added* q and the amount of work *done on* it w. This is summarized

$$\Delta U = q + w$$

This is another mathematical statement of the first law of thermodynamics but is more useful for solving chemical problems than Eq. (10.2). Because of the historical development of the first law it is more conventionally written

$$\Delta U = q - w \qquad (10.3)$$

By normal convention if work is *done on* a system by its surroundings, the work is considered to be negative; if work is *done by* a system on its surroundings, the work is considered to be positive. We shall use this convention with Eq. (10.3) as the mathematical statement of the first law of thermodynamics.

10.5 WORK, HEAT, AND ENTHALPY

Suppose we enclose two reactants in a vessel having *rigid*, heat-conducting walls. As a reaction occurs, heat may pass from the system into the surroundings (or vice versa). The chemical composition inside the reaction vessel changes, and all the internal properties (pressure, temperature, density, etc.) may change; therefore, the internal energy changes. But no work is done, either *on* or *by* the reaction system, because, by definition, the performance of work requires that some body move or undergo a change of position.† Work is defined as the product of the force exerted on a body and the distance through which the body moves:

$$\begin{aligned} W &= \text{force} \times \text{distance through which body moves} \\ &= F \times s \end{aligned} \tag{10.4}$$

When a reaction takes place within a rigid vessel, no motion is permitted; neither the vessel nor any body in the surroundings will move. The distance through which a body moves is zero, and by Eq. (10.4) no work is done.

The process we have described is a *constant-volume*, or *isochoric*, process. In such a process no work is done. Hence, by the first law of thermodynamics, Eq. (10.3), the internal energy change for a constant-volume process must equal the heat evolved or absorbed:

$$\Delta U = q - w \qquad w = 0$$

and

$$\Delta U_v = q_v \tag{10.5}$$

which means that the heat associated with a constant-volume process equals the internal-energy change.

Processes of interest to chemists often occur at constant pressure. Chemical reactions and physical processes (like fusion, vaporization, and sublimation) are often studied in open vessels, under the constant pressure of the atmosphere. During constant-pressure processes, work is done and heat is transferred. The change in internal energy is $q - w$ [from Eq. (10.3)]. If a chemical reaction occurs in such a way that the pressure of the reaction system remains constant, the volume of the system must, in general,

† In everyday usage, work may mean mental or muscular exertion. In chemistry and physics the term has a much more restricted definition.

FIGURE 10.4

When $COCl_2$ *decomposes, the piston moves up and work is done on the surroundings.*

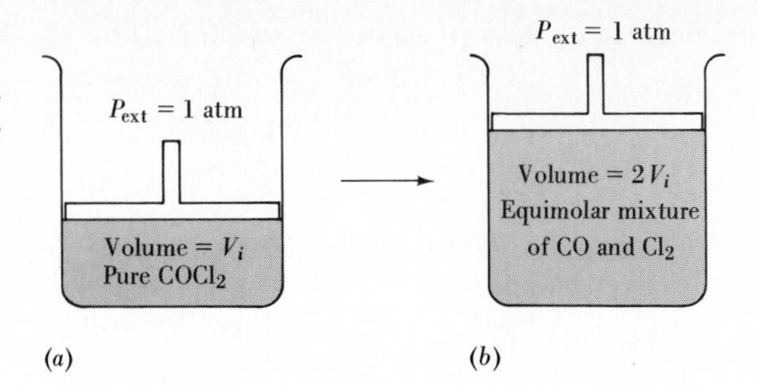

change. For example, the war gas phosgene, $COCl_2$, decomposes into two gases, CO and Cl_2, according to the equation

$$COCl_2(g) \rightarrow CO(g) + Cl_2(g)$$

If the total pressure is to remain constant and equal to the initial pressure, the volume of the reaction system must be exactly doubled. But this would require work. If the reaction were allowed to take place in a cylinder equipped with a free-moving piston,† the piston would rise against the external pressure of the surroundings; a body moves against a force, and, by definition, work is done by the system on the surroundings (see Fig. 10.4). If the reaction involved a decrease in the number of moles of gas, the piston would have to move down in order to maintain a constant pressure. The surroundings would do work on the system.

Very often the constant pressure at which reactions occur is the pressure of the atmosphere. A chemical-reaction system does work on the atmosphere when the volume of the system increases; the atmosphere does work on the system when the volume of the system decreases. From Eq. (10.3) the heat absorbed or evolved in such a process is given by

$$q_p = \Delta U + w_{\text{atm}} \tag{10.6}$$

where w_{atm} refers to the work due to a volume change of the reaction system. When a system changes volume at the constant, external pressure of the surroundings, it can be shown from the definition of work, Eq. (10.4), that

$$\text{Work} = P_{\text{ext}}\, \Delta V \tag{10.7}$$

This equation shows that the work involved in changing the volume of a system against a constant pressure is equal to the product of the pressure and the volume change ΔV. If the work is done at the constant pressure of the atmosphere P_{atm}, the equation is

$$w = P_{\text{atm}}\, \Delta V$$

† Here we use the kind of piston made famous by physics departments the world over, i.e., the weightless, frictionless variety.

EXAMPLE 10.1 Calculate the work done when 1 g of ice melts and forms 1 g of water at 0°C under 1 atm of pressure.

$$1 \text{ g ice} \xrightarrow{0°\text{C}} 1 \text{ g water}$$

Solution: The volume of 1 g of ice is about 1.1 ml; the volume of 1 g of water is 1.0 ml. The volume change on melting is equal to the final volume V_f minus the initial volume V_i:

$$\Delta V = V_f - V_i = 1.0 - 1.1 = -0.1 \text{ ml} = -0.0001 \text{ l}$$
$$w = P_{atm}\, \Delta V = (1 \text{ atm})(-0.0001 \text{ l}) = -0.0001 \text{ l-atm}$$

The work in liter-atmospheres can be converted into calories by use of the ideal-gas-law constant in liter-atmospheres and in calories:

$$R = 0.082 \text{ l-atm/mol-K} = 1.99 \text{ cal/mol-K}$$

Hence, the relationship between liter-atmospheres and calories is

$$0.082 \text{ l-atm} = 1.99 \text{ cal}$$

and we can use this equality to calculate w in calories:

$$w = -0.0001 \text{ l-atm} \times \frac{1.99 \text{ cal}}{0.082 \text{ l-atm}} = -0.00243 \text{ cal}$$

The negative sign indicates that the system is being compressed; i.e., work is being done *on* the system *by* the surroundings.

When a process occurs at constant external pressure, as in Example 10.1, the work is given by $P_{ext}\, \Delta V$, and Eq. (10.6) becomes

$$q_p = U + P_{ext}\, \Delta V \tag{10.8}$$

If a gaseous system expands rapidly against a constant external pressure, the gas pressure in the system is not equal to the external pressure. If a gaseous system in a cylinder equipped with a free-moving piston is heated very slowly, the external pressure and the internal pressure are almost equal at every moment. Then we can substitute the internal pressure of the system for external pressure in Eq. (10.8). For very slow processes $P_{ext} = P_{int}$, and

$$q_p = \Delta U + P_{int}\, \Delta V \qquad \text{for very slow processes} \tag{10.9}$$

The advantage of Eq. (10.9) is that P_{int} can be approximated by an equation of state *which describes a system at equilibrium.* During a very slow compression of oxygen gas, the oxygen is in virtual equilibrium and its pressure can be calculated from the ideal-gas law, $PV = nRT$.

The heat evolved in a very slow process occurring at constant pressure q_p is called the *enthalpy change* ΔH. The enthalpy is actually defined in terms of the internal energy, pressure, and volume *of a system;* specifically

$$H = U + PV \qquad \text{definition of enthalpy} \tag{10.10}$$

From this definition we can easily show that q_p for a very slow process equals ΔH. The enthalpy change occurring during any process is

$$\Delta H = \Delta U + \Delta(PV) \tag{10.11}$$

If the internal pressure of the system is constant during a process, Eq. (10.11) becomes

$$\Delta H = \Delta U + P_{\text{int}}\,\Delta V \tag{10.12}$$

Comparing the right-hand members of Eqs. (10.9) and (10.12), we see that the constant-pressure heat q_p *for a slow process* equals the enthalpy change†

$$\Delta H = q_p \tag{10.13}$$

It should be understood that the enthalpy was intentionally defined so that this would be true.

EXAMPLE 10.2 Suppose that 1 mol of oxygen at 25°C and atmospheric pressure is enclosed in a cylinder equipped with a free-riding piston. It is heated very slowly to 50°C, and it expands against the constant pressure of the atmosphere. (*a*) Calculate the initial and final volumes of the system. (*b*) Calculate the work in calories. (*c*) The heat required for this process is 175 cal. Calculate ΔU and ΔH.

Solution: (*a*) Oxygen obeys the ideal-gas law, $PV = nRT$. The initial volume V_i for 1 mol at 298 K and 1.00 atm is

$$V_i = \frac{nRT_i}{P} = \frac{1\text{ mol} \times 0.082\text{ l-atm/mol-K} \times 298\text{ K}}{1\text{ atm}} = 24.4$$

The final volume V_f after heating to 323 K is

$$V_f = \frac{nRT_f}{P} = \frac{1\text{ mol} \times 0.082\text{ l-atm/mol-K} \times 323\text{ K}}{1\text{ atm}} = 26.5$$

(*b*) The work is $P_{\text{ext}}\,\Delta V$ or $P_{\text{int}}\,\Delta V$ since in a slow process $P_{\text{ext}} = P_{\text{int}}$.

$$w = P_{\text{ext}}\,\Delta V = P_{\text{int}}\,\Delta V = P(V_f - V_i)$$

Note that ΔV always means final volume minus initial volume; it is not just the absolute difference. ΔV can be positive or negative.

† No matter how a process occurs, the enthalpy change at constant pressure is always equal to $\Delta U + P\,\Delta V$, but this is equal to the heat q only for very slow isobaric processes. Most authors refer to very slow processes as reversible processes. It is important to realize that we are referring to processes during which the system is practically in equilibrium.

$$w = P(V_f - V_i) = (1 \text{ atm})(26.5 - 24.4 \text{ l}) = 0.21 \text{ l-atm}$$

$$= 2.1 \text{ l-atm} \times \frac{1.99 \text{ cal}}{0.082 \text{ l-atm}} = 51 \text{ cal}$$

(*c*) To calculate the internal-energy change we use the first law of thermodynamics [Eq. (10.3)]

$$\Delta U = q - w$$

Since $q = 175$ cal and $w = 51$,

$$\Delta U = 175 - 51 = 124 \text{ cal}$$

Realizing that the heat in a slow constant-pressure process equals ΔH, we have

$$q_p = \Delta H = 175 \text{ cal}$$

If we use Eq. (10.12), we obtain the same result:

$$\Delta H = \Delta U + P\,\Delta V = 124 + 51 = 175 \text{ cal}$$

10.6 THE RELATIONSHIP BETWEEN ΔU AND ΔH

The heat liberated by a chemical reaction is an important quantity to a chemical engineer who may be designing equipment for a chemical plant. The amount of heat liberated will affect his design and choice of materials. If the heat of a reaction is very high, the engineer will have to incorporate a special cooling system and choose materials which will withstand the high temperatures. It is obviously important to know the heat liberated by burning fuels or the detonation of explosives. As we have seen, the heat of a reaction at constant volume equals ΔU, and the heat of a reaction at constant pressure equals ΔH. These quantities can be obtained from calorimetric experiments described later in this chapter.

In an isothermal chemical reaction which involves no appreciable volume change, the difference between ΔU and ΔH is virtually unmeasurable, but when we consider reactions of gases involving a large volume change, there is a possibility of an appreciable difference between ΔU and ΔH. We can demonstrate this point by analysis of the definition of H [Eq. (10.10)]:

$$\text{Enthalpy} = H = U + PV$$

$$\text{Change in enthalpy} = \Delta H = \Delta U + \Delta(PV) \qquad (10.14)$$

Equation (10.14) shows that ΔH equals ΔU when $\Delta(PV)$ is zero. $\Delta(PV)$ is very small when no gases are consumed or produced during a chemical reaction. Many solution reactions fall into this category.

A useful expression which is the equivalent of Eq. (10.14) can be developed for gaseous reactions. It shows that the difference between ΔH and ΔU is $RT\,\Delta n_g$, where Δn_g is the change in the number of moles of gas during an *isothermal* chemical reaction. Consider a gaseous system with an initial pressure-volume product P_iV_i; after a reaction has occurred, the pressure-volume product is P_fV_f. Each pressure-volume product can be expressed in terms of the ideal-gas law: †

$$P_fV_f = n_fRT \qquad P_iV_i = n_iRT$$

During the reaction the number of moles changes from n_i to n_f, but since the process is isothermal, the temperature remains constant.

The change in the pressure-volume product is

$$\Delta(PV) = P_fV_f - P_iV_i = n_fRT - n_iRT$$

When this is substituted into Eq. (10.14), we have

$$\Delta H = \Delta U + (n_fRT - n_iRT) = \Delta U + RT(n_f - n_i)$$

$$\Delta H = \Delta U + RT\,\Delta n_g \tag{10.15}$$

Equation (10.15) shows that ΔH and ΔU differ only if the number of moles of gas changes during a chemical reaction. This equation is also applicable to physical transformations (like vaporization, condensation, and fusion) and heterogeneous chemical reactions which involve solids, liquids, and solutions as well as gases.

EXAMPLE 10.3 Will ΔH equal ΔU for the following reaction?

$$COCl_2(g) \rightarrow CO(g) + Cl_2(g)$$

Solution: In this case the change in moles of gas which occurs if the reaction goes to completion is

$$\Delta n_g = n_f - n_i = 2 - 1 = 1$$

and

$$\Delta H \neq \Delta U \qquad \Delta H = \Delta U + RT\,\Delta n_g = \Delta U + RT$$

EXAMPLE 10.4 Will ΔH equal ΔU for the reaction

$$H_2(g) + Cl_2(g) \rightarrow 2HCl(g)$$

Solution: In this case $\Delta n_g = 2 - 2 = 0$; the number of moles of gas does not change, and ΔH equals ΔU.

EXAMPLE 10.5 Will ΔH equal ΔU for the reaction

$$CaO(s) + CO_2(g) \rightarrow CaCO_3(s)$$

† We could use more complicated gas laws if we felt it was necessary, but at low pressure most gases obey the ideal-gas law.

Solution: The change in the number of *gaseous* moles Δn_g is $0 - 1$, or -1.

$$\Delta H \neq \Delta U \qquad \Delta H = \Delta U + RT\,\Delta n_g = \Delta U - RT$$

Note that we count only the moles of gas (CaO and $CaCO_3$ are solids) and that Δn_g may be negative.

EXAMPLE 10.6 Will ΔH equal ΔU for

$$1 \text{ mol ice} \xrightarrow[1 \text{ atm}]{0°\text{C}} 1 \text{ mol water} \qquad \text{fusion}$$

Solution: ΔH is 80 cal/g for this process. There are no gases involved; therefore, $\Delta n_g = 0 - 0 = 0$, and

$$\Delta H = \Delta U = 80 \text{ cal/g}$$

We can look at the fusion of water in another way. In Example 10.1 we calculated the work associated with this process:

$$w = P_{\text{ext}}\,\Delta V = -0.00243 \text{ cal}$$

Since the process occurs at constant pressure, the change in the PV product will equal the work, $P\,\Delta V$:

$$\Delta(PV) = P\,\Delta V = -0.00243 \text{ cal}$$

By Eq. (10.12) for very slow constant-pressure processes

$$\Delta H = \Delta U + P_{\text{int}}\,\Delta V$$

or

$$\Delta U = \Delta H - P_{\text{int}}\,\Delta V = 80 - (-0.00243) \text{ cal} \approx 80 \text{ cal}$$

We have shown that in a process which does not involve a gas, $\Delta(PV)$ is negligible compared to ΔH; that is, $\Delta H = \Delta U$, which is a conclusion more easily reached by the application of Eq. (10.15).

EXAMPLE 10.7 Will ΔH equal ΔU for

$$1 \text{ mol water} \xrightarrow[1 \text{ atm}]{100°\text{C}} 1 \text{ mol steam} \qquad \text{vaporization}$$

or

$$H_2O(l) \xrightarrow[1 \text{ atm}]{100°\text{C}} H_2O(g)$$

Solution: In this case $\Delta n_g = 1 - 0 = 1$ and $\Delta H \neq \Delta U$; $\Delta H = \Delta U + RT$.

$$\Delta H - \Delta U = RT = 1.99 \text{ cal/mol-K} \times 373 \text{ K} = 743 \text{ cal/mol}$$

The difference between ΔH and ΔU lies in the change of the PV

product. During vaporization the change in the PV product equals

$$\Delta(PV) = (PV \text{ product of vapor}) - (PV \text{ product of liquid})$$
$$= P_gV_g - P_lV_l$$

But $P_gV_g \gg P_lV_l$. (Compare the volumes of 1 mol of water, 0.018 l, and 1 mol of steam, 33.6 l at 100°C.) Therefore, the change in the PV product equals P_gV_g, but by the ideal-gas law this simply equals RT for 1 mol. The difference between ΔH and ΔU for the vaporization of 1 mol of a liquid is RT, which is the conclusion we reached by the application of Eq. (10.15). This reasoning is summarized mathematically:

$$\Delta H = \Delta U + \Delta(PV)$$
$$\Delta H - \Delta U = \Delta(PV) = P_gV_g - P_lV_l = P_gV_g \qquad \text{since } P_gV_g \gg P_lV_l$$

If the gas formed by vaporization obeys the ideal-gas law,

$$P_gV_g = RT \qquad \text{and} \qquad \Delta H - \Delta U = RT$$

By the same reasoning we can show that during the sublimation of 1 mol of a solid, ΔH again differs from ΔU by RT. During the reverse of vaporization (condensation) or sublimation the difference, $\Delta H - \Delta U$, is $-RT$.

10.7 SPECIFIC HEAT AND MOLAR HEAT CAPACITY

When 1 g of water at 9.5°C absorbs 1.00129 cal of heat, its temperature rises to 10.5°C; it goes from 49.5 to 50.5°C by absorption of 0.99854 cal and from 97.5 to 98.5°C by absorption of 1.00640 cal. The exact amount of heat required to make a change of 1 Celsius degree is different at each water temperature. As a good approximation, however, we say that the temperature of 1 g of water increases by 1 Celsius degree for each calorie of heat absorbed; it is also true that the temperature of 1 g decreases by 1 Celsius degree for each calorie evolved or lost.

The amount of heat required to raise the temperature of 1 g of a substance by 1 Celsius degree is called the *specific heat*† c. It has the units of calorie per gram. From the above paragraph it should be clear that the specific heat of water is 1.00129 at 10°C, 0.99854 at 50°C, and 1.00640 cal/g at 98°C, but it is very nearly 1 cal/g over its entire liquid range from 0 to 100°C.

EXAMPLE 10.8 How many calories are required to raise the temperature of 15.5 g of mercury from 25 to 100°C? The specific heat of mercury is 0.0333 cal/g-deg.

† Specific heat could be defined as the heat gained *or lost* by 1 g of a substance during a 1 Celsius degree temperature change.

Solution: Heat = mass × specific heat × temperature change

$$q = m \times \text{specific heat} \times \Delta t$$
$$= 15.5 \text{ g} \times 0.0333 \text{ cal/g-deg} \times (100 - 25 \text{ deg})$$
$$= 38.7 \text{ cal}$$

EXAMPLE 10.9 When a 10.0-g sample of Ni at 99.90°C is placed in an insulated vessel containing 80.0 g of water at 20.00°C, the system reaches temperature equilibrium at 21.11°C. Calculate the specific heat of nickel.

Solution: The heat loss by the nickel is equal *in magnitude* to the heat gained by the water:

Heat lost by nickel = −heat gained by water

The negative sign indicates that one body cools down as the other body warms up.

$$(m \times \text{specific heat} \times \Delta t)_{\text{Ni}} = -(m \times \text{specific heat} \times \Delta t)_{\text{H}_2\text{O}}$$
$$10.0 \text{ g} \times \text{specific heat} \times (21.11 - 99.90)$$
$$= -[80.0 \text{ g} \times 1 \text{ cal/g-deg} \times (21.11 - 20.00) \text{ deg}]$$

Specific heat of Ni = 0.113 cal/g-deg

The specific heats of substances cannot always be meaningfully compared; they measure the heat *per gram*, but 1-g samples of different substances do not contain the same number of molecules. It is more significant to compare the properties of samples which contain equal numbers of molecules. To do this we multiply the specific heat by the gram-molecular weight. This gives the heat required to raise the temperature of 1 mol of a substance (N_A molecules) by 1 Celsius degree. The resulting quantity is the *molar heat capacity* C.

$$C = \text{specific heat} \times \text{gram-molecular weight}$$

C has the units of calories per mole-degree. To obtain the heat involved in heating or cooling 1 mol of a substance we simply multiply the heat capacity by the temperature change ΔT:

$$q(1 \text{ mol}) = C\ \Delta T$$

If we are dealing with n mol, the heat involved is

$$q(n \text{ mol}) = nC\ \Delta T \qquad (10.16)$$

Two types of heat capacities are important: the constant-volume heat capacity C_v and the constant-pressure heat capacity C_p. C_v is the heat required to raise the temperature of a 1-mol sample by 1 Celsius degree when the sample is held at constant volume. C_p is the heat required to raise the temperature of a 1-mol sample by 1 Celsius degree when the sample is held at constant pressure. C_p will always be greater than C_v because during the isobaric or constant-

FIGURE 10.5

The difference between C_p and C_v lies in the work of expansion during an isobaric process.

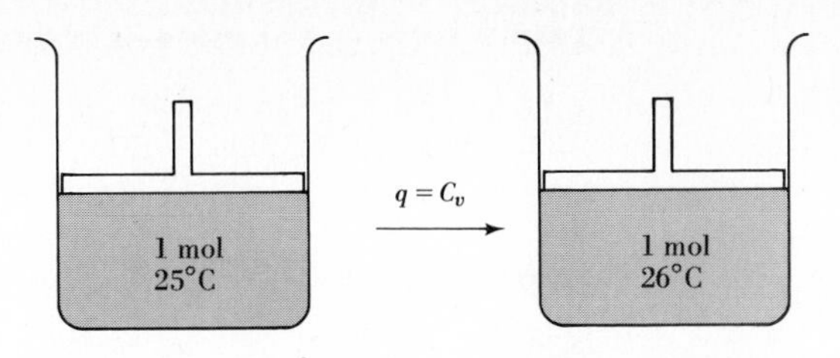

(*a*) *When a gas is heated at constant volume no work is done*

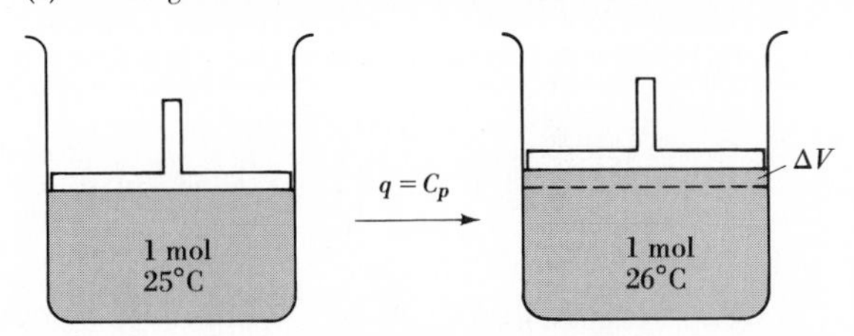

(*b*) *When a gas is heated at constant pressure work is done against the atmosphere*

pressure heating the system must also expand; the heat absorbed is used to raise the temperature *and* to do work against the constant pressure of the surroundings (see Fig. 10.5). For liquids and solids C_p and C_v are approximately equal since the thermal expansion of these condensed phases is usually very small. For example, the C_p for ice at $-5°C$ is 8.879 cal/mol whereas C_v is 8.788 at the same temperature. Also C_p for water at 25°C is 17.969 cal/(mol-deg); C_v is 17.850. This is true because solids and liquids undergo only slight expansion when they are warmed, and even during isobaric heating almost all the heat goes into raising the temperature. For gases the situation is quite different. Gases undergo large volume expansions during isobaric heating; C_p and C_v are very different. For instance, at room temperature C_p for $O_2(g)$ is 7.05 cal/mol whereas C_v is 5.06 cal/mol-deg; C_p is greater by 40 percent. In many instances C_p for a gas is approximately 2 units greater than C_v. We can summarize what we have been discussing by writing two simple empirical† equations:

For solids and liquids: $C_p - C_v \approx 0$
For gases: $C_p - C_v = 2$ cal/mol-deg

From the thermodynamic theory it can be proved that the difference between the heat capacities for an *ideal gas* is exactly equal to the gas-law constant R which in calories per mol-K equals 1.987.

Table 10.1 lists the constant-pressure molar heat capacities for various substances. Note that C_p is approximately 7 cal/mol-K for the common diatomic gases. For the monatomic rare gases it is ex-

† An empirical equation is one which is based on observation or experimental results.

TABLE 10.1: C_p **VALUES FOR VARIOUS SUBSTANCES**

Substance	*Temperature range*, K	C_p, cal/mol-K
Gases, $C_p - C_v = 2$		
$H_2(g)$	200–700	6.82
$O_2(g)$	200–700	6.94
$N_2(g)$	200–400	6.94
$CO(g)$	200–400	6.96
$He(g)$	200–1000	4.968
$Ne(g)$	200–1000	4.968
$Ar(g)$	200–1000	4.968
$CO_2(g)$	200–400	8.811
$NH_3(g)$	200–400	8.859
$CH_4(g)$	200–400	8.54
$H_2O(g)$	200–400	8.04
$HCl(g)$	200–700	6.81
$Cl_2(g)$	200–700	7.71
Liquids, $C_p \approx C_v$		
$H_2O(l)$	273–373	18.04
$C_2H_5OH(l)$, ethyl alcohol	250–350	27.1
$C_6H_6(l)$, benzene	280–353	34.2
$CH_3COOH(l)$, acetic acid	290–385	31.8
Solids, $C_p \approx C_v$		
C, graphite	250–1000	2.07
Diamond	250–1000	1.45
H_2O, ice	220–273	8.87
NaCl	200–400	11.88

actly 4.968 cal/mol-deg. When the actual value of the heat capacity is unknown, it is quite common to use 7 and 5 cal/mol-K for diatomic and monatomic gases, respectively. The values given in Table 10.1 are average values for the temperature ranges shown. The heat capacities of the listed substances can be quite different outside the given temperature ranges.

EXAMPLE 10.10 Calculate the heat absorbed when 2.00 mol of gaseous oxygen contained in a cylinder equipped with a weightless, frictionless piston is slowly heated from 5°C (*a*) isobarically to 35°C, and (*b*) isochorically to 35°C. (*c*) Show that the difference between these heats is due to work done in the isobaric expansion. (*d*) What is the relationship of these two heats to ΔH and ΔU? From Table 10.1 the average value of C_p in this temperature interval is 6.94 cal/mol-K.

Solution: (*a*) To calculate the heat absorbed in the isobaric process we use Eq. (10.16), and the constant-pressure heat capacity C_p.

$$q_p = nC_p\,\Delta T = 2.00 \text{ mol} \times 6.94 \text{ cal/mol-K} \times (308 - 278) \text{ K} = 417 \text{ cal}$$

(*b*) In the isochoric process we use C_v. For gases $C_v = C_p - 2$, and therefore C_v for O_2 is 4.94.

$$q_v = nC_v\,\Delta T = 2.00 \text{ mol} \times 4.94 \text{ cal/mol-K} \times (308 - 278) \text{ K} = 298 \text{ cal}$$

(*c*) During the isobaric process the gas expands against the constant pressure of the atmosphere. The work is $P_{\text{ext}}\,\Delta V$. We must calculate the volume change ΔV and then the work in *calories*.

$$\Delta V = V_f - V_i$$

At this low pressure oxygen obeys the ideal-gas law:

$$V_f = \frac{nRT_f}{P} = \frac{2 \text{ mol} \times 0.082 \text{ l-atm/mol-K} \times 308 \text{ K}}{1 \text{ atm}} = 50.5 \text{ l}$$

$$V_i = \frac{nRT_i}{P} = \frac{2 \text{ mol} \times 0.082 \text{ l-atm/mol-K} \times 278 \text{ K}}{1 \text{ atm}} = 45.6 \text{ l}$$

$$w = P_{\text{ext}}\,\Delta V = (1 \text{ atm})(50.5 - 45.6 \text{ l}) = 4.9 \text{ l atm}$$

Next we use the equality 0.082 l-atm = 1.99 cal to calculate w in calories:

$$w = 4.9 \text{ l-atm} \times \frac{1.99 \text{ cal}}{0.082 \text{ l-atm}} = 119 \text{ cal}$$

In parts (*a*) and (*b*) of this problem we showed that $q_p = 417$ cal and $q_v = 298$ cal. The difference, 119 cal, is due to the work done in the isobaric expansion.

(*d*) By definition, the isobaric heat q_p is equal to ΔH, and the isochoric heat q_v is equal to ΔU [see Eqs. (10.5) and (10.13)]. Hence, we have equations for calculating ΔH and ΔU for the heating or the cooling of a system:

$$q_p = \Delta H = nC_p\,\Delta T \tag{10.17}$$

$$q_v = \Delta U = nC_v\,\Delta T \tag{10.18}$$

Molecular interpretation of heat capacities

The variation of molar heat capacities from substance to substance is quite large. The heat capacity of ice near its melting point is 9 cal/mol-K whereas the heat capacity of heme, the pigment molecule of hemoglobin, is almost 170 cal/mol-K. The capacity to absorb heat has been interpreted in terms of the kinds of motions a mole-

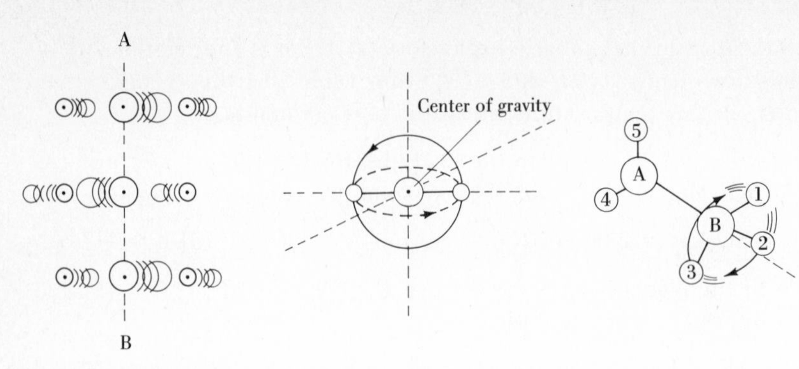

(a) Vibration. The molecules have equilibrium positions in the solid but they vibrate back and forth through these positions.

(b) Rotation. Atoms of a molecule may rotate about the center of gravity.

(c) Internal rotation. Atoms 1, 2, and 3 rotate about the axis joining atoms A and B. Atoms A, B, 4 and 5 are stationary relative to 1, 2, and 3.

FIGURE 10.6

Molecular motions in a crystal lattice.

cule can undergo. In the solid crystalline state a molecule can vibrate about its equilibrium position; as the temperature is raised, it may begin to rotate "in place" before it melts. These motions are induced and enhanced by the absorption of heat (see Fig. 10.6). As we approach the melting point, the amplitude of the vibration of the molecule becomes larger. At the melting point the amplitude may become infinite, which is a way of saying that the molecule escapes its lattice position.

When a crystal melts, the molecules are no longer restricted to definite positions in the lattice. The heat absorbed by the molecules of a liquid can be taken up not only as vibrational and rotational motion but as a translational motion. The molecules can vacate positions, leaving "holes," which may or may not be occupied by incoming molecules. As the temperature of a liquid rises, the number of holes increases. In order to escape its neighbors and leave a hole a

TABLE 10.2: COMPARISON OF MOLAR HEAT CAPACITIES C_p**, cal/mol-K**

	Liquids and vapors			*Solids and liquids*		
Substance	*t, °C*	*Liquid*	*Vapor*	*t, °C*	*Solid*	*Liquid*
H_2O	100	18	8.6	0	9.0	18.0
Chlorine, Cl_2	−33	16	7.9	. . .	. . .	. . .
Xenon, Xe	−108	10.7	5.0	. . .	. . .	. . .
Sulfur dioxide, SO_2	0	20.7	9.3	−75	16.5	21.0
Carbon tetrachloride, CCl_4	25	31.49	19.96	. . .	. . .	. . .
Benzene, C_6H_6	. . .	. . .	. . .	5	21.8	30.3
Carbon monoxide, CO	. . .	. . .	. . .	−205	12.0	14.0

molecule must overcome the high intermolecular attractive forces. A large fraction of the heat absorbed by a liquid is used in the creation of holes. The heat capacity of a liquid is therefore greater than the heat capacity of a solid because the molecules of a liquid have definite, though limited, mobility.

When a liquid is vaporized, the resulting gaseous molecules are separated by relatively large distances. The heat absorbed by a gas is taken up as rotational, vibrational, and translational motion of the molecules; virtually no heat is required to overcome the weak intermolecular forces. Therefore, the capacity of a substance to absorb heat, on a molar basis, decreases when liquid is converted to vapor (see Table 10.2).

10.8 CALORIMETRY: THE MEASUREMENT OF HEAT AND HEAT CAPACITY

The heat absorbed (or evolved) during a chemical reaction is, in principle, a very simple quantity to obtain. The chemical reaction is allowed to proceed in a thermally insulated vessel called a *calorimeter* the heat capacity of which is accurately known.

Figure 10.7 shows a simplified version of an adiabatic† reaction calorimeter. The reactants are placed in a stainless-steel bomb A, which is immersed in a weighed amount of water. When the reaction occurs, heat is evolved and is transferred to the water; the

FIGURE 10.7

An adiabatic calorimeter.

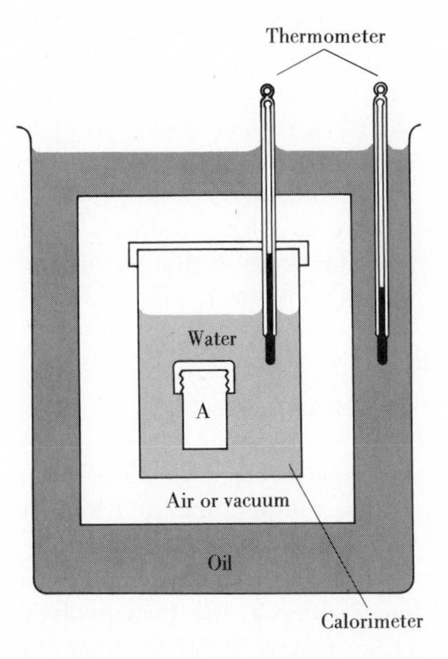

† *Adiabatic* means thermally insulated so that no heat can leave or enter.

water temperature changes and finally becomes constant after the reaction is complete. The surrounding oil bath of the calorimeter is automatically controlled so that its temperature always equals the temperature of the water in which the bomb is immersed. This minimizes the heat loss by the calorimeter water since a transfer of heat requires a temperature difference.

The temperature change of the calorimeter is used to calculate the heat of reaction. We can apply Eq. (10.14*a*) for this purpose:

$$q = c\ \Delta T$$

$$\text{Heat of reaction} = \text{heat capacity of calorimeter} \times \text{temperature change of calorimeter}$$

Since the reaction takes place in a rigid bomb, heat is generated at *constant volume* and is equal to the change in internal energy ΔU accompanying the chemical reactions:

$$\text{Heat of reaction} = q_v = \Delta U_{\text{reac}}$$

Using Eq. (10.13), we can calculate ΔH_{reac} from ΔU_{reac}. Hence, through calorimetric measurements we can obtain q_v and ΔU, ΔH, and q_p.

The specific heat of a material is measured directly by noting the temperature rise in a sample caused by transferring a definite amount of energy to it. This is a very tedious experiment, especially if we need to determine the heat capacity of a material over wide ranges of temperature. Fortunately, the recent development of scanning calorimeters with electronic control permits rapid determination of accurate heat capacities. In these instruments the temperature rise is set at some constant rate, for example, 10 deg/min, and the energy input is automatically controlled so that the temperature rise is kept at this rate. At the same time the instrument produces an instantaneous recording of the energy input and the temperature of the sample. Under ordinary conditions the samples are heated at constant pressure so that we obtain values of C_p directly from the scanning calorimeters.

10.9 STANDARD STATE

The heat evolved (or absorbed) during a chemical reaction depends on the temperature at which the reaction occurs. When 1 mol of ammonia is produced from its elements at room temperature, 11 kcal of heat is generated. At 450°C, the industrial temperature for the synthesis, the heat evolved is 14 kcal per mole of ammonia. Other factors affect the heat of a reaction. The pressure has a small effect, which we shall neglect. The heat will depend on the physical state of the reactants and products; if 1 mol of H_2O *liquid* is produced from its elements, $H_2(g)$ and $O_2(g)$, the heat evolved is approximately −68 kcal. If, however, the product is H_2O vapor, the heat is −58 kcal. To complicate the picture further, different amounts of heat are evolved when different crystalline forms react; the combustion of 1 mol of carbon as diamond produces −94.50 kcal, whereas the combustion of 1 mol as graphite produces −94.05 kcal. If the reaction occurs in solution, the heat will depend on the concentration of reactants and the nature of the solvent. Since the heat of a chemical reaction is very sensitive to the exact conditions under which the reaction proceeds, we must be careful to indicate these conditions when we specify the heat. We use *thermochemical equations* for this purpose. Thermochemical equations give the temperature, pressure, physical state of the reactants and products, and the enthalpy change, in addition to the stoichiometry. The reactions we have discussed above can be described by the following thermochemical equations:

$$\underset{1\text{ atm}}{\tfrac{1}{2}N_2(g)} + \underset{1\text{ atm}}{\tfrac{3}{2}H_2(g)} \xrightarrow{298\text{ K}} \underset{1\text{ atm}}{NH_3(g)}$$

$$q_p = \Delta H^\circ_{298} = -11.04 \text{ kcal/mol} \tag{10.19a}$$

$$\underset{1\text{ atm}}{\tfrac{1}{2}N_2(g)} + \underset{1\text{ atm}}{\tfrac{3}{2}H_2(g)} \xrightarrow{723\text{ K}} \underset{1\text{ atm}}{NH_3(g)}$$

$$q_p = \Delta H^\circ_{723} = -13.78 \text{ kcal/mol} \tag{10.19b}$$

$$\underset{1\text{ atm}}{H_2(g)} + \underset{1\text{ atm}}{\tfrac{1}{2}O_2(g)} \xrightarrow{298\text{ K}} H_2O(l)$$

$$q_p = \Delta H^\circ_{298} = -68.32 \text{ kcal/mol} \tag{10.20a}$$

$$\underset{1\text{ atm}}{H_2(g)} + \underset{1\text{ atm}}{\tfrac{1}{2}O_2(g)} \xrightarrow{298\text{ K}} \underset{1\text{ atm}}{H_2O(g)}$$

$$q_p = \Delta H^\circ_{298} = -57.80 \text{ kcal/mol} \tag{10.20b}$$

$$\text{C(diamond)} + \underset{1\text{ atm}}{O_2(g)} \xrightarrow{298\text{ K}} \underset{1\text{ atm}}{CO_2(g)}$$

$$q_p = \Delta H^\circ_{298} = -94.50 \text{ kcal/mol} \tag{10.21a}$$

$$\text{C(graphite)} + \underset{1\text{ atm}}{O_2(g)} \xrightarrow{298\text{ K}} \underset{1\text{ atm}}{CO_2(g)}$$

$$q_p = \Delta H^\circ_{298} = -94.05 \text{ kcal/mol} \tag{10.21b}$$

TABLE 10.3: STANDARD ENTHALPIES OF FORMATION AT 25°C

Substance	$\Delta H_{f,298}$ *(kcal)*
All elements	0.0†
$H_2O(g)$	−57.80
$H_2O(l)$	−68.32
$HCl(g)$	−22.06
$SO_2(g)$	−70.96
$SO_3(g)$	−94.45
$H_2S(g)$	−4.82
$H_2SO_4(l)$	−193.91
$NO(g)$	21.60
$NO_2(g)$	8.09
$N_2O_4(g)$	2.31
$NH_3(g)$	−11.04
$PCl_3(g)$	−73.22
$PCl_5(g)$	−95.35
$CO(g)$	−26.42
$CO_2(g)$	−94.05
Methane, $CH_4(g)$	−17.89
Ethane, $C_2H_6(g)$	−20.24
Propane, $C_3H_8(g)$	−24.82
n-Butane, $C_4H_{10}(g)$	−24.82
Isobutane, $C_4H_{10}(g)$	−31.45
n-Pentane, $C_5H_{12}(g)$	−35.00
n-Hexane, $C_6H_{14}(g)$	−39.96
n-Heptane, $C_7H_{16}(g)$	−44.89
n-Octane, $C_8H_{18}(g)$	−49.82
Ethylene, $C_2H_4(g)$	12.50
Propylene, $C_3H_6(g)$	4.88
Acetylene, $C_2H_2(g)$	54.19
Carbon tetrachloride, $CCl_4(l)$	−33.30
Methanol, $CH_3OH(l)$	−57.02

† The standard enthalpy of formation of an element *must be zero* by the very definition of the enthalpy of formation. Why?

Note that the physical state of each substance is indicated: $g =$ gas, $l =$ liquid, and $s =$ solid; c is used for crystal, but if more than one crystalline solid form is possible, it must be given explicitly. The temperature is written over the arrow and as a subscript on the symbol for the enthalpy change. All heats are given as constant-pressure heats and are therefore equal to the enthalpy changes.

Because the heat or enthalpy change is so sensitive to reaction conditions, a set of *standard conditions* has been internationally adopted and all enthalpy of reaction data obtained by thermochemists are reported for these standard conditions. If a substance is present under standard conditions, it is said to be in its *standard state.*

Standard state for various substances

Gases: A gas is in standard state when it is at 298.15 K (25°C) and at a pressure of 1 atm.

Liquids: A liquid is in standard state when it is under 1 atm at 298.15 K.

Solids: A solid is in standard state when it is in its most stable crystalline form under 1 atm and at 298.15 K. (The standard state of carbon is graphite.)

The standard temperature is really arbitrary, but in most cases 298.15 K is chosen and we shall use this convention.

By our definition of standard conditions the reactants and products shown in Eqs. (10.19*a*) and (10.20*a*) to (10.21*b*) are in standard state. The enthalpy change associated with processes in which reactants and products are in standard state is called the *standard enthalpy change* $\Delta H°$. The superscript, read "super zero," always refers to a process which occurs under standard-state conditions $\Delta U°$ means *standard* internal-energy change.

When a substance in its standard state is formed *from its elements* in their standard states, the associated enthalpy change is called the *standard enthalpy of formation,* $\Delta H^\circ_{f,298}$. Methanol, CH_3OH, can be produced by the reaction of carbon monoxide and hydrogen. The equation for the process under standard conditions is

$$\underset{1\text{ atm}}{CO(g)} + \underset{1\text{ atm}}{2H_2(g)} \xrightarrow{25°C} CH_3OH(l) \qquad \Delta H° = -30.60 \text{ kcal/mol}$$

The enthalpy change shown is *not* the standard enthalpy of formation because in this process methanol is not formed from its elements. The enthalpy change for the following process is the standard enthalpy of formation for CH_3OH:

$$\text{C(graphite)} + \tfrac{1}{2}O_2(g) + 2H_2(g) \xrightarrow{25°C} CH_3OH(l)$$

$$\qquad 1\ \text{atm} \qquad 1\ \text{atm} \qquad\qquad \Delta H^\circ_{f,298} = -57.02\ \text{kcal/mol}$$

The standard enthalpies of formation have been measured and tabulated for many substances, a brief list being given in Table 10.3. By use of the $\Delta H^\circ_{f,298}$ data it is possible, as we shall learn, to calculate the standard-enthalpy changes for many reactions. Note that all the values given in Table 10.3 are for compounds. The standard enthalpy of formation of any element *must* be zero.

10.10 THERMOCHEMICAL MANIPULATIONS

The enthalpy change for a process *must be* equal and opposite to the enthalpy change for the reverse process. For example, the slow vaporization of water at its normal boiling point requires 9700 cal/mol. This is a constant-pressure process and $q_p = \Delta H$.

$$H_2O(l) \xrightarrow[1\ \text{atm}]{100°C} H_2O(g) \qquad \Delta H = +9700\ \text{cal/mol}$$

The reverse process, the slow condensation of water vapor under the same conditions, *must* generate 9700 cal/mol

$$\underset{1\ \text{atm}}{H_2O(g)} \xrightarrow{100°C} H_2O(l) \qquad \Delta H = -9700\ \text{cal/mol}$$

If the enthalpy changes for the two processes were not equal and opposite, we could create energy. To illustrate this point, suppose the condensation liberated 10,000 cal/mol and the vaporization required 9000 cal/mol. We could take 1 mol of liquid and vaporize it at 100°C by transferring only 9000 cal into it; and then, simply by condensing the vapor, we could obtain 10,000 cal. This would be a net increase in energy of 1000 cal; no other compensating changes would have occurred in the universe. This would be a clear violation of the first law of thermodynamics (the total energy of the universe must be constant) and is therefore impossible. In chemical processes the same principle must hold. For instance, the combustion of carbon produces CO_2 and 94.05 kcal of heat. The conversion of CO_2 into carbon and oxygen absorbs 94.05 kcal:

$$\text{C(graphite)} + \underset{1\ \text{atm}}{O_2(g)} \xrightarrow{25°C} \underset{1\ \text{atm}}{CO_2(g)} \qquad \Delta H^\circ = -94.05\ \text{kcal/mol}$$

$$\underset{1\ \text{atm}}{CO_2(g)} \xrightarrow{25°C} \text{C(graphite)} + \underset{1\ \text{atm}}{O_2(g)} \qquad \Delta H^\circ = +94.05\ \text{kcal/mol}$$

By this line of reasoning we must also conclude that *the enthalpy change for the chemical reaction is the same whether it occurs in one or several steps*. This principle, called *Hess's law of constant heat summation*, is also a requirement of the first law of thermodynamics. We can use Hess's law to calculate enthalpy changes for chemical reactions from the enthalpy-of-formation data in Table 10.3.

EXAMPLE 10.11 Calculate ΔH_f° for carbon monoxide, CO, at 25°C.

Solution: It is experimentally impossible to determine ΔH_f° directly for carbon monoxide because whenever carbon is reacted with limited oxygen, a mixture of carbon oxides form:

$$\mathrm{C} + \text{limited } \mathrm{O_2} \rightarrow \mathrm{CO_2} + \mathrm{CO} + \text{carbon suboxides}$$

However, it is possible to separate pure CO from a mixture, and this can be combusted alone. The following information is available:

$$\underset{1\text{ atm}}{\mathrm{CO}(g)} + \underset{1\text{ atm}}{\tfrac{1}{2}\mathrm{O_2}(g)} \xrightarrow{25^\circ\mathrm{C}} \underset{1\text{ atm}}{\mathrm{CO_2}(g)} \qquad \Delta H^\circ = -67.62 \text{ kcal} \qquad (a)$$

$$\mathrm{C(graphite)} + \underset{1\text{ atm}}{\mathrm{O_2}(g)} \xrightarrow{25^\circ\mathrm{C}} \underset{1\text{ atm}}{\mathrm{CO_2}(g)} \qquad \Delta H^\circ = -94.05 \text{ kcal} \qquad (b)$$

From these two thermochemical equations we can devise a two-step process for the formation of CO from its elements. Step 1 is just reaction (*b*), this gives us $CO_2(g)$.

$$\text{Step 1:} \quad \mathrm{C(graphite)} + \underset{1\text{ atm}}{\mathrm{O_2}(g)} \xrightarrow{25^\circ\mathrm{C}} \mathrm{CO_2}(g) \qquad \Delta H^\circ = -94.05 \text{ kcal}$$

By reversing reaction (*a*) we can convert the $CO_2(g)$ into $CO(g)$.

$$\text{Step 2:} \quad \underset{1\text{ atm}}{\mathrm{CO_2}(g)} \xrightarrow{25^\circ\mathrm{C}} \underset{1\text{ atm}}{\tfrac{1}{2}\mathrm{O_2}} + \underset{1\text{ atm}}{\mathrm{CO}(g)} \qquad \Delta H^\circ = +67.62 \text{ kcal}\dagger$$

$$\text{Net process:} \quad \mathrm{C(graphite)} + \underset{1\text{ atm}}{\tfrac{1}{2}\mathrm{O_2}(g)} \xrightarrow{25^\circ\mathrm{C}} \underset{1\text{ atm}}{\mathrm{CO}(g)}$$

The net process is the chemical reaction for the formation of CO from its elements. By Hess's law the enthalpy change for the two-step process must equal the enthalpy of formation of CO at 25°C. Hence, ΔH_f° for CO is

$$-94.05 + 67.62 = -26.42 \text{ kcal}$$

The complete thermochemical equation for the process is

† Note change from −67.62 kcal to +67.62 kcal; i.e., this is the reverse of Eq. (*a*).

$$\text{C(graphite)} + \tfrac{1}{2}\text{O}_2(g) \xrightarrow{25^\circ\text{C}} \text{CO}(g) \qquad \Delta H^\circ = -26.42 \text{ kcal/mol}$$

$$\text{1 atm} \qquad\qquad \text{1 atm}$$

Hess's law can be applied to any reaction provided that the enthalpies of formation ΔH_f° are known for all compounds involved. It is easily shown that Hess's law reduces to

$$\Delta H^\circ_{\text{reac}} = \Sigma\ \Delta H^\circ_{f,\text{prod}} - \Sigma\ \Delta H^\circ_{f,\text{reac}} \tag{10.22}$$

This represents a very convenient form for the calculation of reaction enthalpies.

EXAMPLE 10.12 Using data from Table 10.3, calculate the standard enthalpy change ΔH° and the standard internal-energy change ΔU° at 25°C for the hydrogenation of propylene gas, C_3H_6; propane, C_3H_8, is the product:

$$\text{CH}_2\text{=CH—CH}_3(g) + \text{H}_2(g) \rightarrow \text{CH}_3\text{—CH}_2\text{—CH}_3(g)$$

Solution: Use Eq. (10.22):

$$\Delta H^\circ_{\text{reac}} = \Sigma\ \Delta H^\circ_{f,\text{prod}} - \Sigma\ \Delta H^\circ_{f,\text{reac}}$$

The standard enthalpy of formation of the product, propane, is −24.82 kcal; the standard enthalpy of formation for the reactant, propene, is +4.88 kcal, and for H_2 it is 0:

$$\Delta H^\circ = -24.82 - (4.88 + 0) = -29.70 \text{ kcal}$$

In this reaction $\Delta n_g = 1 - 2 = -1$. Therefore, since

$$\Delta H = \Delta U + RT\ \Delta n_g$$

we have

$$\begin{aligned}\Delta U^\circ &= \Delta H^\circ - RT\ \Delta n_g \\ &= -29{,}700 \text{ cal} - [(1.99 \text{ cal/mol-K})(298 \text{ K})(-1 \text{ mol})] \\ &= -29{,}110 \text{ cal} = -29.11 \text{ kcal}\end{aligned}$$

EXAMPLE 10.13 Calculate ΔH° and ΔU° at 25°C using the data in Table 10.3 for the reaction

$$\text{CH}_4(g) + 4\text{Cl}_2(g) \rightarrow \text{CCl}_4(l) + 4\text{HCl}(g)$$

Solution: Using Eq. (10.22), we have

$$\Delta H^\circ = -33.30 + 4(-22.06) - [17.89 + (4 \times 0)] = -103.65 \text{ kcal}$$

In this reaction $\Delta n_g = 4 - 5 = -1$ and $\Delta U^\circ = -103{,}650 + RT$. Note that we must multiply the ΔH_f° values for Cl_2 and HCl by their stoichiometric coefficients in order to obtain the correct enthalpy change. Remember that the data in Table 10.4 are given in kilocalories *per mole*. ($\Delta U = -103.058$ kcal)

EXAMPLE 10.14 At 100°C the standard enthalpy of formation of water vapor is −60.27 kcal/mol, while ΔH_f° for liquid water at this temperature is −69.97 kcal/mol. Calculate the heat (enthalpy change) of vaporization at 100°C.

Solution: Hess's law applies to physical as well as chemical transformation. Therefore, we can use Eq. (10.22) to solve this problem. The vaporization of water can be represented as

$$H_2O(l) \xrightarrow[760\ \text{Torr}]{100°C} H_2O(g)$$

By Eq. (10.22) we have

$$\Delta H^\circ = H_{f,\text{vap}}^\circ - H_{f,\text{liq}}^\circ = -60.27 - (-69.97) = +9.70\ \text{kcal/mol}$$

Note that the ΔH_f° values for $H_2O(g)$ and $H_2O(l)$ at 100°C are slightly different from the values at 25°C (Table 10.3).

10.11 THE VARIATION OF ΔH WITH TEMPERATURE

With the aid of data for the enthalpy of formation like those presented in Table 10.3, we can obtain the standard enthalpies for many reactions *but* only at 25°C. By combining ΔH_f° and C_p data we can obtain the standard enthalpy changes for reactions over wide ranges of temperature. We shall demonstrate this with a specific example. The principle upon which the calculation is based is Hess's law: the enthalpy change is the same whether it occurs in one or several steps.

EXAMPLE 10.15 Calculate the standard-enthalpy change for the formation of HCl gas from its elements at 600 K.

Solution: We must determine ΔH° for the reaction

$$\underset{1\ \text{atm}}{\tfrac{1}{2}H_2(g)} + \underset{1\ \text{atm}}{\tfrac{1}{2}Cl_2(g)} \xrightarrow{600\ \text{K}} \underset{1\ \text{atm}}{HCl(g)} \tag{10.23}$$

According to Hess's law, the enthalpy change for this process will be the same whether it occurs in one or several steps. The following three-step process converts the gases H_2 and Cl_2 at 600 K into HCl gas at 600 K:

1. Cooling H_2 and Cl_2 from 600 to 298 K

$$\underset{\substack{1\ \text{atm}\\600\ \text{K}}}{\tfrac{1}{2}H_2(g)} + \underset{\substack{1\ \text{atm}\\600\ \text{K}}}{\tfrac{1}{2}Cl_2(g)} \xrightarrow[\substack{\text{constant}\\\text{pressure}}]{\text{cool}} \underset{\substack{1\ \text{atm}\\298\ \text{K}}}{\tfrac{1}{2}H_2(g)} + \underset{\substack{1\ \text{atm}\\298\ \text{K}}}{\tfrac{1}{2}Cl_2(g)}$$

2. The reaction of H_2 and Cl_2 at 298 K

$\frac{1}{2}H_2(g) + \frac{1}{2}Cl_2(g) \rightarrow HCl(g)$

1 atm	1 atm	1 atm
298 K	298 K	298 K

3. Warming HCl from 298 to 600 K

$$HCl(g) \xrightarrow[\text{constant pressure}]{\text{warm}} HCl(g)$$

1 atm	1 atm
298 K	600 K

The net result of steps 1 to 3 is the formation of HCl gas under standard conditions at 600 K. This is equivalent to the transformation described in Eq. (10.23), and the associated enthalpy change is the standard enthalpy of formation of HCl gas at 600 K.

By application of Eq. (10.17), we can determine the enthalpy change for cooling the reactants (step 1) and heating the product (step 3). The enthalpy change for step 2 is just the standard enthalpy of formation for HCl at 25°C as given in Table 10.3.

Step 1
From Table 10.1, C_p for H_2 is 6.82 cal/mol-K and C_p for Cl_2 is 7.71 cal/mol-K in this temperature range. Cooling H_2:

$$\Delta H = nC_p\ \Delta T = (\tfrac{1}{2}\text{ mol})(6.82\text{ cal/mol-K})(298 - 600\text{ K}) = -1030\text{ cal}$$

Cooling Cl_2:

$$\Delta H = (\tfrac{1}{2}\text{ mol})(7.71\text{ cal/mol-K})(298 - 600\text{ K}) = -1160\text{ cal}$$

Step 2
From Table 10.3, $\Delta H = \Delta H^\circ_{f,298} = -22.06$ kcal $= -22{,}060$ cal.

Step 3
From Table 10.1, C_p for HCl is 6.81 cal/mol-K.

$$\Delta H = nC_p\ \Delta T = (1\text{ mol})(6.81\text{ cal/mol-K})(600 - 298\text{ K}) = +2060$$

The total enthalpy change is equal to the algebraic sum for steps 1 to 3:

$$\begin{aligned}\Delta H_{\text{total}} &= -1030 - 1160 - 22{,}060 + 2060 \\ &= -22{,}190\text{ cal} = -22.19\text{ kcal}\end{aligned}$$

Therefore, the standard-enthalpy change for the formation of HCl at 600 K is -22.19 kcal/mol. $\Delta H^\circ_{f,600} = -22.19$ kcal/mol.

The procedure we used to calculate the standard enthalpy of formation of HCl at 600 K can serve as a model for the calculation of the enthalpy change for any reaction at a temperature other than 25°C.

Because so many of the reactions studied by chemists were exothermic,† the principle that all spontaneous reactions are accompanied by a negative ΔH was generally accepted. (In fact, it is still used as an approximation.) Unfortunately, the problem of spontaneity is not quite so simple. Spontaneous chemical reactions have been discovered which are endothermic (ΔH is positive). The following endothermic reaction is spontaneous:

$$Ag(s) + \tfrac{1}{2}Hg_2Cl_2(s) \xrightarrow{20°C} AgCl(s) + Hg(l) \qquad \Delta H = +1280 \text{ cal/mol}$$

Obviously, the enthalpy change is not the criterion by which we judge whether a reaction is spontaneous or not. We must search for another principle.

At all levels of human experience we note that natural processes are accompanied by an increase in disorder. Unattended flower beds become overgrown with weeds and grass, while flower seeds spill into the grass and take root. A house constantly requires straightening out, or ordering. If we place a small amount of potassium permanganate, $KMnO_4$, in a beaker of water, the purple crystals slowly dissolve; eventually the water and the $KMnO_4$ are thoroughly mixed. *The reverse process of unmixing never occurs.*

Our experience shows without exception that there is a natural tendency toward mixing or chaos in all systems. This may not seem true in biological systems, which are highly structured and increase in order as they grow. But it must be remembered that for the trivial increase in order which occurs in our biosphere there is a fantastic increase in chaos on the sun; and the sun must be included when we consider the total biological process because it is the ultimate source of energy.

In view of these considerations we can write one of the several equivalent statements of the second law of thermodynamics: *All isolated systems tend toward maximum disorder.* Every spontaneous process which occurs in an isolated system is accompanied by an increase in disorder. Since the universe is an isolated system, any change occurring in the universe must be accompanied by a net increase in disorder, or chaos, or mixed-upness. When an isolated system is at a point of maximum disorder, it will cease to change on a macroscopic level; the pressure, temperature, index of refraction, and all other observable properties become constant. We say such a system is at equilibrium. It is important to associate the equilibrium condition with the concept of maximum disorder.

The degree of disorder has been placed on a quantitative scale by

† *Exothermic* means heat is evolved; q and ΔH are negative. *Endothermic* means heat is absorbed.

devising a quantity called the *entropy* S. *The entropy of an isolated system tends toward a maximum;* this is another statement of the second law of thermodynamics. When an isolated system is at equilibrium, it is at maximum disorder, or entropy. *Spontaneous reactions in isolated systems must be accompanied by an increase in entropy.* This is the thermodynamic criterion for spontaneity, and therefore it is important to learn how to calculate entropy changes for physical and chemical processes. By such calculations we can decide without recourse to experimentation whether processes can occur.

10.13 THE THIRD LAW OF THERMODYNAMICS

If the entropy of a substance is a measure of its degree of disorder, the entropy of a pure, perfectly ordered substance must be zero. Only perfect crystals at the absolute zero of temperature can be perfectly ordered, and such materials are assigned an absolute entropy of zero: *The third law of thermodynamics states that the entropy of a perfectly crystalline solid at* 0 K *is zero; or* $S_0^\circ = 0$.†

As the temperature of a pure crystalline material is raised, its entropy increases. This can be understood if we consider the state of the atoms or molecules in the lattice of a crystal. At the absolute of

FIGURE 10.8

Order and disorder in a crystal.

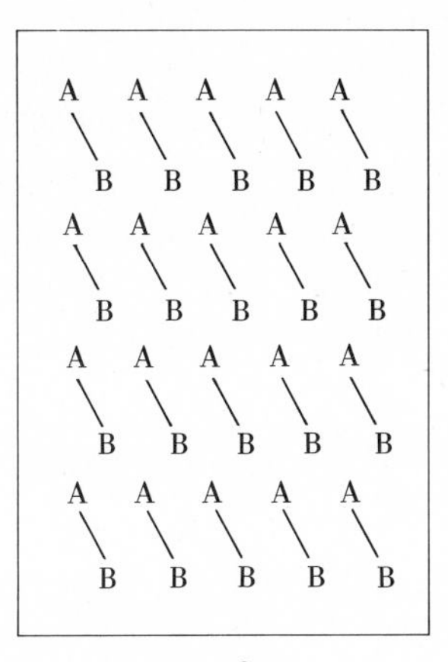

Perfect order at 0°K

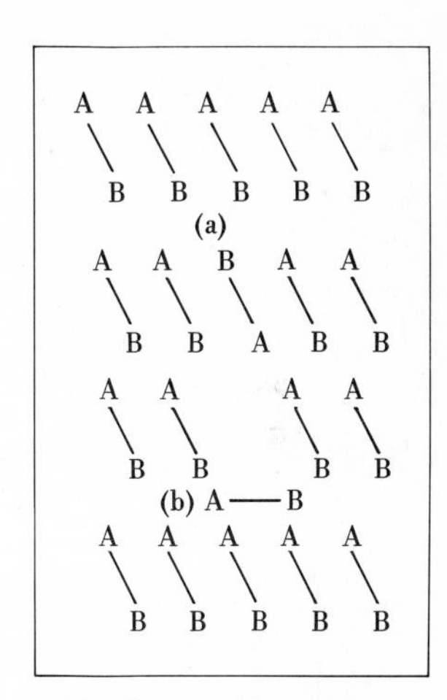

Disorder caused by increase in temperature. Molecule a *is rotated out of perfect alignment. Molecule* b *has migrated from its lattice position.*

† Note that the third law is not a law in the same sense as the first and second laws. The third law is not experimentally verifiable because we cannot reach the absolute of zero.

zero there can be perfect order. As we raise the temperature, the kinetic energy of the atoms and molecules increases. Some molecules may be displaced from their normal lattice positions by undergoing vibrations of high amplitudes; others may rotate, causing a nonuniform alignment of the molecules. On a molecular level the disorder increases as the temperature increases, and the entropy must increase (see Fig. 10.8) The entropy of a substance which is above 0 K must always be positive.

10.14 THE ABSOLUTE ENTROPY $S°$

In our molecular interpretation on page 293 we related the heat capacity to the kinds of molecular motions a molecule can undergo. Like the heat capacity, the entropy increases as the freedom of molecular motion increases; an increase in molecular motion is equivalent to an increase in disorder, or entropy. There is a direct relationship between the heat capacity and the entropy, and it should be suspected that the entropy of a substance can be calculated from its heat capacity. The absolute entropies $S°$ for some common substances in their standard states at 25°C are given in Table 10.4. It was necessary to measure the heat capacities in order

TABLE 10.4: ABSOLUTE ENTROPIES $S°$ AT 25°†

Substance	$S°$, cal/mol-K		
$H_2O(g)$	45.11	$O_2(g)$	49.00
$H_2O(l)$	16.72	Methane, $CH_4(g)$	44.50
$HCl(g)$	44.62	Ethane, $C_2H_6(g)$	54.85
$SO_2(g)$	59.40	Propane, $C_3H_8(g)$	64.51
$SO_3(g)$	61.24	*n*-Butane, $C_4H_{10}(g)$	74.10
$H_2S(g)$	49.15	Isobutane, $C_4H_{10}(g)$	70.42
$NO(g)$	50.34	*n*-Pentane, $C_5H_{12}(g)$	83.27
$NO_2(g)$	57.47	*n*-Hexane, $C_6H_{14}(g)$	92.45
$N_2O_4(g)$	72.73		
$NH_3(g)$	46.01	*n*-Heptane, $C_7H_{16}(g)$	101.64
$PCl_3(g)$	74.49	*n*-Octane, $C_8H_{18}(g)$	110.82
$PCl_5(g)$	84.3	Ethylene, $C_2H_4(g)$	52.45
$CO(g)$	47.30	Propene, $C_3H_6(g)$	63.80
$CO_2(g)$	51.06	Acetylene, $C_2H_2(g)$	48.00
$N_2(g)$	45.77	Carbon tetrachloride,	
$H_2(g)$	31.21	$CCl_4(l)$	51.25
$Cl_2(g)$	53.29	Methanol, $CH_3OH(l)$	30.3

† All substances in standard state.

to obtain the absolute entropies of these materials. Note that $S°$ for the elements is *not* zero.

We used the molecular theory in the discussion above because it is easier to relate entropy changes to chemical and physical processes in this light; but the actual values of the absolute entropies $S°$ presented in Table 10.4 were calculated from an entropy definition which does not depend on the existence of molecules:

$$dS = \frac{dq(\text{slow process})}{T}$$

where dS is the infinitesimal change in the entropy which accompanies a transfer of an infinitesimal amount of heat dq at a temperature T. Under these conditions the heat transfer must occur in a slow process. You have not been given enough information in this text to calculate the absolute entropy of a substance from its heat capacity. You will learn how to do this in advanced courses. At this point you should realize that there is a simple relationship between C_p and $S°$. It is more important to learn how to use the absolute entropies to calculate entropy changes accompanying chemical and physical transformations.

10.15 THE CALCULATION OF STANDARD-ENTROPY CHANGES

To calculate the standard-entropy change $\Delta S°$ for a reaction we use the standard absolute entropies $S°$. $\Delta S°$ for a reaction is given by

$$\Delta S° = \Sigma S°_{\text{prod}} - \Sigma S°_{\text{reac}} \qquad (10.24)$$

EXAMPLE 10.16 Calculate the standard-entropy change for the hydrogenation of propene gas at 25°C

$$\underset{\substack{1\text{ atm}\\ \text{Propene}}}{C_3H_6(g)} + \underset{1\text{ atm}}{H_2(g)} \xrightarrow{25°C} \underset{\substack{1\text{ atm}\\ \text{Propane}}}{C_3H_8(g)}$$

$$\Delta S° = S°_{(C_3H_8)} - (S°_{(C_3H_6)} + S°_{(H_2)})$$

Solution: From Table 10.4, we obtain the $S°$ values for each compound:

$$\Delta S° = 64.51 - (63.80 + 31.21) = -30.50 \text{ cal/K}$$

Note that $S°$ for the *element* hydrogen is *not* zero.

Entropy changes at constant temperature

When a substance melts or boils in the open atmosphere, it does so at constant temperature. Everyone knows that H_2O (ice) melts at 0°C and H_2O (water) boils at 100°C; other pure substances also

TABLE 10.5: COMPARISON OF ENTROPY CHANGES FOR FUSION AND VAPORIZATION

Process equation	$\Delta H = q_p$, cal/mol	$\Delta S = \frac{\Delta H}{T}$, cal/mol-K
Fusion $H_2O(s) \xrightarrow{0°C} H_2O(l)$	+1140	+5.28
Vaporization $H_2O(l) \xrightarrow{100°C} H_2O(g)$	+9720	+26.1

have sharp melting points and boiling points. In view of the fact that entropy is a measure of disorder, it should not be surprising that fusion and vaporization cause large increases in the entropy of a substance. For such processes the entropy change is easily calculated since the isothermal entropy change is simply

$$\Delta S = \frac{q(\text{slow process})}{T} \tag{10.25}$$

where q = heat absorbed
T = absolute temperature at which process occurs

If the process also occurs at constant pressure, as ordinary melting and boiling do, the heat is the constant-pressure heat which equals the enthalpy change; and entropy change is calculated from

$$\Delta S = \frac{q_p}{T} = \frac{\Delta H}{T} \tag{10.26}$$

Table 10.5 shows that the entropy changes accompanying the fusion of ice and the vaporization of water are consistent with our interpretation that the entropy is directly related to disorder. The large increase in entropy in going from liquid to vapor is expected. Vaporization causes a greater increase in disorder or randomness than fusion, consistent with our earlier discussions of gases, liquids, and solids (Chaps. 6 to 8).

10.16 THE CHEMICAL POTENTIAL ENERGY *G*

The second law of thermodynamics tells us that the entropy change in an isolated system is a criterion for spontaneity; that is, ΔS must be positive for a spontaneous process which occurs in an isolated system. Then ΔS calculations using data from Table 10.4 enable us to decide whether a hypothetical process is possible in an *isolated system*. Obviously, chemical studies are not carried out in isolated systems. In fact, most chemical studies are carried out in systems held at constant pressure and temperature by external control. Under these conditions it can be proved from thermodynamic theory

TABLE 10.6: THERMODYNAMIC CRITERIA FOR SPONTANEITY

Quantity	*Condition of system*	*Change in quantity*
Entropy S	Isolated	$\Delta S > 0$
Chemical potential energy G	Constant temperature and pressure	$\Delta G < 0$

that the criterion for spontaneity depends on the enthalpy change ΔH as well as on ΔS. If the quantity

$$\Delta H - T\,\Delta S$$

is negative for a process occurring at constant temperature and pressure, the process is spontaneous. A new quantity, called the *chemical potential energy* G, has been defined so that a change in G will equal the quantity $\Delta H - T\,\Delta S$ at constant temperature. By definition, $G = H - TS =$ chemical potential energy. A change in G is given by

$$\Delta G = \Delta(H - TS) = \Delta H - \Delta(TS) = \Delta H - (T_2S_2 - T_1S_1)$$

At constant temperature, $T_2 = T_1$. This gives

$$\Delta G = \Delta H - T\,\Delta S \tag{10.27}$$

Finally, we say that the criterion for a spontaneous process at constant temperature and pressure is that ΔG be negative. When a chemical reaction occurs in a system at constant temperature and pressure, G must decrease; ΔG must be negative. When the reaction comes to equilibrium, G is constant and $\Delta G = 0$ (see Table 10.6).

In Chap. 11 we expand the concept of the chemical potential energy, show its relationship to the state of chemical equilibrium, and demonstrate how to calculate equilibrium constants from thermochemical data.

QUESTIONS AND PROBLEMS

10.1 (*a*) Explain what is meant by the statement: Time is not a legitimate variable in thermodynamics. (*b*) Must all thermodynamically spontaneous reactions occur? Explain.

10.2 (*a*) What is the definition of energy? (*b*) Distinguish between potential energy and kinetic energy. (*c*) Give examples of bodies which possess different kinds of potential energy.

10.3 (*a*) What does *internal* energy refer to? (*b*) What is the broadest statement of the first law of thermodynamics? (*c*) Why is it possible to consider only the internal energy U in the study of ordinary chemical reactions?

10.4 (*a*) A body expands from 5.00 to 10.0 l at atmospheric pressure.

Calculate the work in calories done *by* the body. (*b*) If the body absorbs 1500 cal during the expansion, what is the change in internal energy ΔU of the body? (*c*) If the body, which weighs 5.00 g, is simultaneously raised 100 ft above the surface of the earth [while it expands, part (*a*), and absorbs heat, part (*b*)] what is the change in the total energy ΔE_{total} of the body ($g = 980$ cm/s^2)?

10.5 A sample of 2.00 mol of oxygen in a cylinder equipped with a free-moving weightless piston is initially at 1.00 atm and 25°C. It is slowly cooled at the constant pressure of the atmosphere until its volume is reduced by a factor of $\frac{1}{2}$. Using the ideal-gas law calculate (*a*) the initial volume, (*b*) the final volume, (*c*) the work done on the gas by the atmosphere. (*d*) What is the final temperature of the gas? (*e*) If 2100 cal of heat is lost by the oxygen, what is the internal energy change ΔU?

10.6 Do ΔH and ΔU differ in the following processes?

(*a*) $NH_4HS(s) \xrightarrow{25°C} NH_3(g) + H_2S(g)$

(*b*) $H_2(g) + Cl_2(g) \xrightarrow{25°C} 2HCl(g)$

(*c*) $CO_2(s) \xrightarrow{-78°C} CO_2(g)$
Dry Ice

(*d*) $AgNO_3(aq) + NaCl(aq) \xrightarrow{25°C} AgCl(s) + NaNO_3(aq)$

10.7 (*a*) How many calories are required to raise 40 g of iron from 20.0 to 85°C? (The specific heat of iron is 0.106 cal/g-deg.) (*b*) How many calories are given off when 200 g of water cools from the normal boiling point to 20°C?

10.8 How many calories are necessary to convert 30.0 g of ice at −10.0°C to steam at 110°C?

10.9 Determine the resulting temperature when 1000 g of ice at 0°C is mixed with 200 g of steam at 100°C.

10.10 The specific heat of tin is 0.0510 cal/g-deg. Determine the final temperature when 30.0 g of Sn at 95.0°C is placed in an insulated vessel with 75.0 g of water at 16.3°C.

10.11 When 5.63 g of Ag at 96.71°C is placed in an insulated vessel with 61.22 g of H_2O at 20.00°C, the final temperature is 20.39°C. (*a*) Calculate the specific heat of silver and (*b*) estimate the atomic weight of silver using the law of Dulong and Petit.

10.12 Using the equations given in Prob. 10.6, calculate the *difference* between ΔH and ΔU for (*a*) the decomposition of 1.00 mol of $NH_4HS(s)$, (*b*) the formation of 2.00 mol of HCl, (*c*) the sublimation of 5.00 mol of Dry Ice, and (*d*) the precipitation of 2.00 mol of AgCl(*s*).

10.13 (*a*) Use the data in Table 10.1 to calculate ΔU for the following processes: (1) heating 25.0 g of Ar(*g*) from the freezing point to the boiling point of water at constant volume, (2) cooling 100 g of diamond from 75 to 10°C at constant volume. (*b*) Calculate q and ΔH for the following processes: (1) heating 2.5 mol of $N_2(g)$ from 25 to 100°C at constant pressure,

(2) cooling 100 g of liquid ethyl alcohol from 80 to 20°C at constant pressure. (*c*) Calculate ΔH for the two processes given in part (*a*). (*d*) Calculate ΔU for the two processes given in part (*b*).

10.14 The heat capacity of a calorimeter is 1508 cal/K near room temperature. When 5.000 g of certain hydrocarbon with a molecular weight of 117 g/mol is completely combusted in the calorimeter, the temperature rises from 25.015 to 27.513°C. (*a*) Calculate the heat produced per gram of hydrocarbon. (*b*) Calculate the heat of combustion per mole. (*c*) The combustion chamber of the calorimeter is a rigid vessel. Is the heat of combustion equal to ΔU or ΔH? (*d*) When 1 mol of the hydrocarbon burns, the number of gaseous moles increases by 3.5. What are ΔH and ΔU for the combustion of 1 mol at the temperature of the experiment, 25°C.

10.15 (*a*) What is a thermochemical equation? (*b*) Write thermochemical equations for (1) the formation of liquid benzene, C_6H_6, from its elements under standard conditions, (2) the formation of nitric acid, HNO_3, from its elements under standard conditions, and (3) the reaction of iron and chlorine to produce iron(III) chloride ($P_{Cl_2} = 10$ atm, $t = 25$°C).

10.16 Why was it necessary for thermochemists to agree on a set of standard conditions?

10.17 In order for 1.00 g of solid benzene to melt in a slow process, the benzene must absorb 30.3 cal. Why is it impossible for more than 30.3 cal to be given up by 1.00 g of liquid benzene when it freezes under the same conditions?

10.18 Prove that the following information cannot be completely correct:

$$H_2(g) + \tfrac{1}{2}O_2(g) \xrightarrow[\text{spark}]{\text{electric}} H_2O(l) \qquad \Delta H = -100 \text{ kcal}$$

$$H_2(g) + \tfrac{1}{2}O_2(g) \xrightarrow[\text{metal}]{\text{palladium}} H_2O(l) \qquad \Delta H = -60 \text{ kcal}$$

(Assume that the heat generated by electric spark is 10.0 cal.)

10.19 Show that Hess's law reduces to Eq. (10.22).

10.20 In view of the definition of the standard enthalpy of formation ΔH_f°, show that ΔH_f° must be zero for an element.

10.21 (*a*) Calculate the enthalpy of formation of ethanol, $C_2H_5OH(l)$, at 25°C from the data in Table 10.3 and the fact that the standard enthalpy of combustion for ethanol is −327 kcal/mol:

$$C_2H_5OH(l) + 3O_2(g) \rightarrow 2CO_2(g) + 3H_2O(l) \qquad \Delta H = -327 \text{ kcal/mol}$$

(*b*) Calculate the enthalpy of formation of naphthalene, $C_{10}H_8(s)$, at 25°C; its standard enthalpy of combustion is −1228 kcal/mol.

10.22 Using the data in Table 10.3, calculate the standard enthalpy changes for the following reactions at 25°C:

(*a*) $PCl_5(g) \rightarrow PCl_3(g) + Cl_2(g)$
(*b*) $SO_2(g) + \tfrac{1}{2}O_2(g) \rightarrow SO_3(g)$
(*c*) $H_2S(g) + \tfrac{3}{2}O_2(g) \rightarrow H_2O(l) + SO_2(g)$

(d) $NH_3(g) + \frac{3}{4}O_2(g) \rightarrow N_2(g) + \frac{3}{2}H_2O(g)$

(e) $H_2(g) + C_2H_2(g) \rightarrow C_2H_4(g)$

Acetylene Ethylene

(f) $CH_4(g) + 2O_2(g) \rightarrow CO_2(g) + 2H_2O(g)$

(g) $CH_4(g) + 4Cl_2(g) \rightarrow CCl_4(l) + 4HCl(g)$

(h) $C_4H_{10}(g) + \frac{13}{2}O_2(g) \rightarrow 4CO_2(g) + 5H_2O(l)$

Isobutane

10.23 Use the data in Tables 10.1 to 10.3 to calculate the enthalpy change at 500 K for reactions (f) and (g) in Prob. 10.22. For reaction (g) you must know that ΔH°_{vap} for CCl_4 is 7551 cal/mol at its normal boiling point, 76.75°C. Use Example 10.15 as a model for this calculation.

TABLE P10.24

Compound	$\Delta H^\circ_{f,298}$
$H_2O_2(g)$	−31.83
$H_2O_2(l)$	−44.84
$CHCl_3(g)$	−24.0
$CHCl_3(l)$	−31.5

10.24 Given the data in Table P10.24, what is the enthalpy change accompanying (a) the vaporization of 1.00 mol of hydrogen peroxide under standard conditions at 25°C and (b) the condensation of 3.00 mol of chloroform, $CHCl_3$, at 25°C?

10.25 (a) Make a rough plot of the entropy S versus time for an isolated system which approaches and finally reaches equilibrium. (b) On the same graph show a plot of the total energy E versus time in the same time interval. (Reread the first and second laws of thermodynamics before you make the plots.)

10.26 Using the data in Table 10.4, calculate the standard-entropy change at 25°C for the reactions given in Prob. 10.22.

10.27 Using the data in Table 10.4, calculate the entropy of vaporization of $H_2O(l)$ at 25°C:

$$H_2O(l) \xrightarrow{25^\circ C} H_2O(g)$$

10.28 Under the same conditions, for the same substances, the entropy of vaporization (liquid → gas) is greater than the entropy of fusion (solid → liquid). Explain.

10.29 (a) The enthalpy change accompanying the vaporization of 1 mol of propionic acid, CH_3CH_2COOH, at its normal boiling point, 141°C, is 10,100 cal/mol. Calculate the entropy of vaporization for this process. (b) When ammonia, $NH_3(l)$, freezes isobarically at −78°C, it liberates 108 cal/g. What is the entropy change accompanying the freezing of 1 mol of NH_3? Is your sign correct? Does your calculation show that freezing causes an increase or decrease in order?

10.30 (a) What is the relationship between the chemical potential energy G and spontaneous processes? (b) Calculate ΔG° for the reactions given in (a) to (g) in Prob. 10.22. Use the answers to Probs. 10.22 and 10.26 in this calculation.

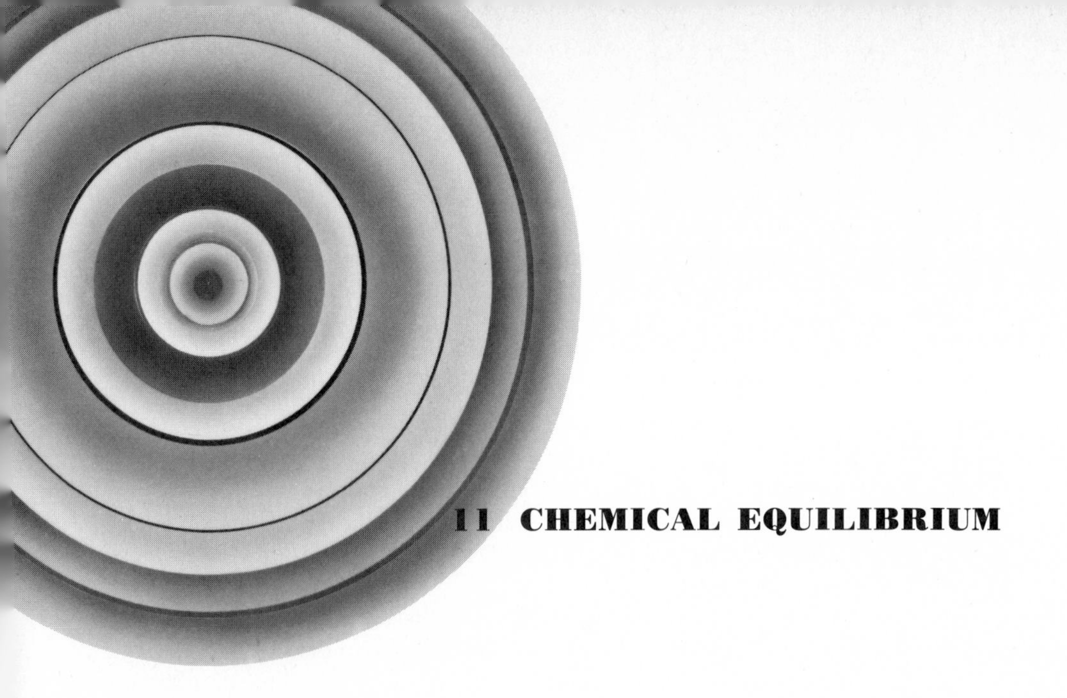

11 CHEMICAL EQUILIBRIUM

11.1 THE EQUILIBRIUM STATE

In general, the things all about us, i.e., physical and biological systems, are in a constant process of change. Some bodies change slowly or to only a slight extent, so that the change is detectable only after long periods of observation or with very sensitive apparatus; in other cases, the change is very rapid and continuous and can be observed in a very short time. In special cases, as when we make an effort to hold a body or system at constant temperature and pressure, we can observe that the physical properties become constant after a period of change (see Fig. 11.1). We then assume that as long as such a system is maintained at constant temperature and pressure, no further change will occur no matter how long we monitor the system. To an observer, the system now appears to be in a static condition: all chemical forces are in balance, and all physical properties such as density, color, index of refraction, and all the concentrations of the chemical species remain constant in magnitude. The system has no further tendency to undergo a chemical change, and we say that such a system is in *chemical equilibrium*.

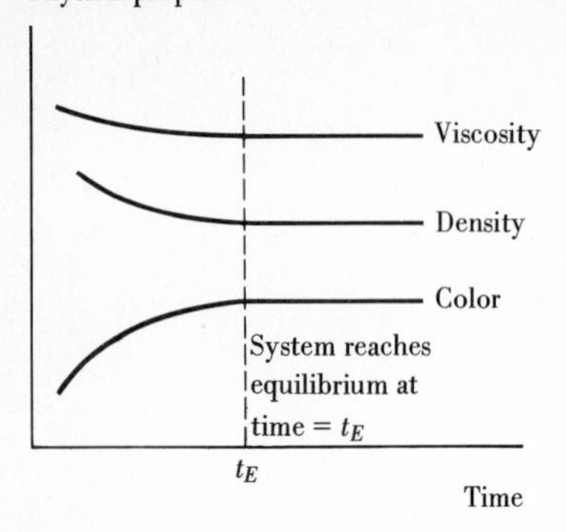

FIGURE 11.1

All the physical properties of a system become constant when a system reaches chemical equilibrium.

Our senses report to us that the properties of a chemical system at equilibrium (or simply equilibrium system) do not change, but if we really accept the atomic and molecular theory of matter, it will be impossible to admit that a truly static condition exists. To understand why, consider the gaseous mixture consisting of the chemically reactive species H_2 and I_2. It has been learned from experimentation that when these two gases come together at 700°C, a reaction occurs and HI is produced:

$$H_2(g) + I_2(g) \xrightarrow{700°C} 2HI(g)$$

The reactants do not disappear completely, and eventually the concentrations of the three gases reach constant values. Thereafter, as long as the temperature is constant, there is no further measurable change in the system.

Can we be led to believe that processes involving the different molecules have ceased? In view of the well-founded atomic theory and kinetic-molecular gas theory, the answer to this question must be no. The molecules of the mixture are moving rapidly and randomly and are continually colliding. Collisions between H_2 and I_2 molecules are likely to lead to HI formation; collision of HI with other molecules or with the container walls could lead to decomposition into elements. In other words, although a chemical equilibrium system is unchanging as far as observable properties are concerned, it is reasonable to believe that the invisible molecules are undergoing chemical transformations. In order to accommodate these apparently contradictory concepts of macroscopic constancy and molecular change we must admit that the microscopic processes occur so that no net macroscopic change results: each process is balanced by its reverse. In the above example this could mean that under equilibrium conditions, the formation of HI proceeds at the same rate as its decomposition into elements. Chemical reactions which come to equilibrium with all reactants and products present are conventionally symbolized by writing equations using the double arrow; the above reaction is symbolized as

$$H_2(g) + I_2(g) \xrightleftharpoons{700°C} 2HI(g)$$

In summary, we can say that a chemical-equilibrium system is a static system at the macroscopic level and a dynamic system at the microscopic level. (The student should consider possible laboratory tests of the hypothesis that a chemical-equilibrium system involves motion and change at the molecular level.)

11.2 THE EQUILIBRIUM CONSTANT

The concentrations of reactants and products in a chemical system reach a constant value at equilibrium; obviously, then, any quantity which is solely dependent upon these equilibrium concentrations will also be a constant. To demonstrate, again consider the reaction involving H_2, I_2, and HI:

$$H_2(g) + I_2(g) \xrightleftharpoons{700°C} 2HI(g)$$

At equilibrium $[H_2]$, $[I_2]$, and $[HI]$ are constant.† Hence, the quantities

$$[H_2] \times [I_2] \times [HI] \quad \text{and} \quad \frac{[H_2]}{[I_2] \times [HI]}$$

and

$$\frac{[H_2]^2 \times [I_2]^3}{[HI]^{1/2}} \quad \text{and} \quad \frac{[HI]^2}{[H_2] \times [I_2]}$$

are constants. But the analysis of chemical-equilibrium data has led to a very important discovery involving the last quotient,‡

$$\frac{[HI]_{eq}^2}{[H_2]_{eq}[I_2]_{eq}}$$

In this case, the concentration of the product raised to a power equal to its stoichiometric coefficient appears in the numerator; the concentration of each reactant raised to a power equal to its stoichiometric coefficient, i.e., unity, appears in the denominator. The value of this quotient for an equilibrium system is constant regardless of the initial amounts of H_2, I_2, and HI which are mixed. For instance, we could start a reaction with equal numbers of moles of all three species, with a ratio of 2:1 mol of H_2 to I_2 and no HI; or with pure HI and no H_2 or I_2. Still, after equilibrium is reached, the same value of the quotient, $[HI]_{eq}^2/[H_2]_{eq}[I_2]_{eq}$, is obtained. This is obvious from the last column in Table 11.1: regardless of the initial concentrations, the value of $[HI]_{eq}^2/[H_2]_{eq}[I_2]_{eq}$ is a constant. Other quotients or products are not constant; see the next to the last column in Table 11.1.

More important, it has been found that for *any* reaction which comes to equilibrium with all reactants and products present the analogous quotient has a characteristic equilibrium value independent of the initial amounts of reactants and products present. Since

† Brackets indicate molar concentrations; that is, $[H_2]$ means the molarity, or number of H_2 moles per liter.

‡ The subscript eq after the brackets emphasizes that we are dealing with *equilibrium* concentrations.

TABLE 11.1: $H_2 + I_2 \xrightleftharpoons{700°C} 2HI$

Initial concentration			Equilibrium concentration				
$[H_2]$	$[I_2]$	$[HI]$	$[H_2]$	$[I_2]$	$[HI]$	$[H_2]_{eq}[I_2]_{eq}[HI]_{eq}$	$\frac{[HI]^2_{eq}}{[H_2]_{eq}[I_2]_{eq}}$
11.148	9.948	0	2.907	1.707	16.482	81.8	54.7
0	0	4.489	0.479	0.479	3.531	0.808	54.4
11.337	7.510	0	4.565	0.738	13.544	45.7	54.5
11.337	7.510	2.533	4.769	0.942	15.669	70.39	54.6

the quotient has the same constant value at equilibrium, it is called the *equilibrium constant* K_{eq}.

The constancy of K_{eq} can also be shown by considering the gaseous system consisting of N_2, H_2, and NH_3; the following reactions occur:

$$\tfrac{1}{2}N_2 + \tfrac{3}{2}H_2 \rightarrow NH_3$$

and its opposite

$$NH_3 \rightarrow \tfrac{1}{2}N_2 + \tfrac{3}{2}H_2$$

usually combined as

$$\tfrac{1}{2}N_2 + \tfrac{3}{2}H_2 \rightleftharpoons NH_3$$

to indicate an equilibrium reaction. Chemical analysis of this system at equilibrium would yield the equilibrium concentrations of the three gases. Regardless of initial concentrations, whenever this system is allowed to reach equilibrium, the following quotient is constant if the temperature is constant

$$\frac{[NH_3]_{eq}}{[H_2]^{3/2}_{eq}[N_2]^{1/2}_{eq}} = K_{eq} = 0.712 \qquad \text{at } 400°C$$

TABLE 11.2: $\tfrac{1}{2}N_2 + \tfrac{3}{2}H_2 \xrightleftharpoons{400°C} NH_3$

P_T, atm	NH_3 *at equilibrium, %*	K_{eq}
10	3.85	0.712
30	10.09	0.712
50	15.11	0.712

If a mixture of N_2 and H_2 in a mole ratio of 1:3 is brought to equilibrium at 400° under various total pressures, the data in Table 11.2 would be obtained.

Notice that the K_{eq} remains constant even though the total pressure P_T increases fivefold and the percentage of ammonia increases approximately fourfold. This demonstrates clearly that although there may be variations in equilibrium concentrations or equilibrium partial pressures, the equilibrium constant does not change. As long as the temperature is constant, the value of K_{eq} is constant: K_{eq} is a function of temperature.

11.3 THE EQUILIBRIUM CONSTANT IN TERMS OF PARTIAL PRESSURES

The equilibrium constant K_{eq} is a special quotient of the equilibrium molar concentrations of the reactants and products. The concentration terms in this quotient are raised to powers (exponents) equal to the stoichiometric coefficients in the balanced equation for the reaction. In gaseous reactions, the equilibrium partial pressures (which are proportional to concentrations) may be used in place of the molar concentrations in the equilibrium-constant expression.

For the reaction $N_2O_4(g) \rightleftharpoons 2NO_2(g)$, the equilibrium-constant expressions are

$$K_{eq} = K_C = \frac{[NO_2]^2_{eq}}{[N_2O_4]_{eq}} \qquad \text{and} \qquad K'_{eq} = K_P = \frac{P^2_{NO_2}}{P_{N_2O_4}}$$

In these expressions the brackets refer to the equilibrium values of the molar concentrations. The P's refer to the equilibrium values of the partial pressures. K_P and K_C will be numerically unequal but will give equivalent information concerning the equilibrium system.

To obtain the exact relationship between K_P and K_C assume that the individual gases in this equilibrium system $N_2O_4 \rightleftharpoons 2NO_2$ obey the ideal-gas law, $PV = nRT$. Rewriting the ideal-gas law as

$$\frac{n}{V} = \frac{P}{RT}$$

we recognize the quantity n/V as the molar concentration (moles per liter) of the gas. Hence, the molar concentration of the gas is equal to its partial pressure divided by the product of the gas constant and the absolute temperature, or $C = P/RT$. Since

$$K_C = \frac{[NO_2]^2_{eq}}{[N_2O_4]_{eq}} \tag{11.1}$$

and

$$[NO_2] = C_{NO_2} = \frac{P_{NO_2}}{RT} \tag{11.2}$$

and

$$[N_2O_4] = C_{N_2O_4} = \frac{P_{N_2O_4}}{RT} \tag{11.3}$$

we can convert the concentration terms in Eq. (11.1) to pressure:

$$K_C = \frac{P^2_{NO_2}/(RT)^2}{P_{N_2O_4}/RT} = \frac{P^2_{NO_2}}{P_{N_2O_4}}\frac{1}{RT} = \frac{P^2_{NO_2}}{P_{N_2O_4}}(RT)^{-1}$$

or

$$K_C = K_P(RT)^{-1}$$

for the N_2O_4-NO_2 equilibrium system. By applying the above reasoning to any equilibrium-constant expression in terms of concentrations, we can show in general that

$$K_C = K_P(RT)^{-\Delta n_g} \quad (11.4)$$

where Δn_g is the stoichiometric change in gaseous moles. For example, in the reaction, $N_2O_4(g) \rightleftharpoons 2NO_2(g)$, the initial stoichiometric number of moles is 1, the final stoichiometric number of moles is 2, and Δn_g equals 1:

$$\Delta n_g = 2 - 1 = 1$$
$$K_C = K_P(RT)^{-\Delta n_g} = K_P(RT)^{-1}$$

For the reaction

$$H_2(g) + I_2(g) \rightleftharpoons 2HI(g)$$

when $\Delta n_g = 0$, the numerical values of K_P and K_e will be identical:

$$K_C = K_P(RT)^{-\Delta n_g} = K_P(RT)^0 = K_P$$

For the reaction

$$NH_4HS(s) \rightleftharpoons NH_3(g) + H_2S(g)$$
$$\Delta n_g = 2 - 0 = 2$$

and

$$K_C = K_P(RT)^{-2}$$

since the initial number of gaseous moles is zero. (NH_4HS is a solid.)

For the reaction: $PCl_3(g) + Cl_2(g) \rightarrow PCl_5(g)$ the equilibrium-constant expressions are

$$K_C = \frac{[PCl_5]_{eq}}{[PCl_3]_{eq}[Cl_2]_{eq}}$$

$$K_P = \frac{P_{PCl_5}}{P_{PCl_3}P_{Cl_2}}$$

In this case Δn_g is negative; $\Delta n_g = 1 - 2 = -1$. Therefore, $K_C = K_P(RT)^{-\Delta n} = K_P RT$.

11.4 RULES FOR WRITING THE EQUILIBRIUM-CONSTANT EXPRESSION

Stoichiometry

The primary information we must obtain for a system which undergoes a chemical change includes the formulas for all reactants and products. This enables us to write the balanced chemical equation. Without this information, no meaningful calculations can be performed. In obtaining this information for new chemical reactions we are aided considerably by the fact that the law of conservation of mass is valid within experimental error. Also, the systematic studies of many chemical reactions provide the insight for fairly accurate predictions of the products of new, untried reactions.

Four rules must be observed when calculations involving the equilibrium constant are performed:

1. The chemical process must be clearly described in terms of a balanced chemical equation. The equilibrium-constant expression and its numerical value depend on the way the chemical equation is written. For the reaction

$$H_2(g) + \tfrac{1}{2}O_2(g) \rightleftharpoons H_2O(g)$$

the equilibrium-constant expression is

$$K_C = \frac{[H_2O]_{eq}}{[H_2]_{eq}[O_2]_{eq}^{1/2}}$$

If we choose to write the equation as

$$2H_2(g) + O_2(g) \rightleftharpoons 2H_2O$$

the equilibrium-constant expression is

$$K_C' = \frac{[H_2O]_{eq}^2}{[H_2]_{eq}[O_2]_{eq}^{1/2}}$$

By inspection we see that

$$K_C' = (K_C)^2$$

If we write the first reaction in reverse, i.e.,

$$H_2O(g) \rightleftharpoons H_2(g) + \tfrac{1}{2}O_2(g)$$

the equilibrium constant is

$$K_C'' = \frac{[H_2]_{eq}[O_2]_{eq}^{1/2}}{[H_2O]_{eq}}$$

Obviously,

$$K_C = \frac{1}{K_C''}$$

2. The equilibrium concentrations of the products appear in the numerator, and the equilibrium concentrations of the reactants appear in the denominator.

3. The stoichiometric coefficients in the chemical equation become the exponents of the concentration terms in the equilibrium-constant expression; e.g., for the reaction

$$A + 2B \rightleftharpoons C + 3D$$

the equilibrium-constant expression is

$$K_C = \frac{[C]_{eq}[D]^3_{eq}}{[A]_{eq}[B]^2_{eq}}$$

4. The concentration of a pure solid or of a pure liquid (not a solution) does not appear in the equilibrium-constant expression because the molar concentration of a pure phase does not change at constant temperature.† For example,

$$NH_4HS(s) \rightleftharpoons NH_3(g) + H_2S(g)$$

$$K_C = [NH_3]_{eq}[H_2S]_{eq} \qquad \text{or} \qquad K_P = P_{NH_3}P_{H_2S}$$

The following points should be kept in mind regarding the equilibrium constant for a specific chemical reaction:

1. It is a special quotient of concentrations *under equilibrium* conditions.

2. It is an approximate constant except for a change in temperature.

3. Its magnitude is an indicator of the extent to which a reaction can proceed.

If K_{eq} is very large, the formation of products is highly favored; the concentration of products at equilibrium will be high. If K_{eq} is small, the concentration of products will be low.

11.5 DERIVATION OF K_C FROM THE LAW OF MASS ACTION

The *law of mass action* states that the rate of a chemical reaction is directly proportional to the concentrations of the reacting species. For example, in the formation of HI from H_2 and I_2 described by the equation $H_2 + I_2 \rightarrow 2HI$, the rate of the reaction is directly proportional to the concentrations of the reactants, H_2 and I_2, or

$$\text{Rate} \propto [H_2][I_2]$$

This proportionality is transformed into an equation by introducing a proportionality constant, k_r:

$$\text{Rate} = k_r[H_2][I_2]$$

† NaCl, a solid, has a density of 2.16 g/ml. One liter of pure NaCl weighs 2160 g. The number of moles in 1 l is 2160/58.5, or 37.0. This is the molar concentration of pure NaCl. If we consider 0.250 l of NaCl, we have only 540 g of NaCl. The number of moles is 540/58.5, or 9.25; but this is the number of moles per 0.25 l. The molar concentration is 9.25/0.250, or 37.0 *M*. We see that the molar concentration of a pure solid (or liquid) is independent of the actual amount present.

The proportionality constant k_r is called the *rate constant* of the reaction; when k_r is large, the reaction rate is high.

Using these ideas, K_{eq} can be derived for certain simple reaction systems in the following way. Consider the equilibrium reaction

$$A + B \underset{r_2}{\overset{r_1}{\rightleftharpoons}} C + D$$

where r_1 = rate of reaction between A and B
r_2 = rate of reaction between C and D

From the law of mass action we can write equations for the rates of the opposing reactions:

$$r_1 = k_1[A][B] \qquad \text{and} \qquad r_2 = k_2[C][D]$$

At equilibrium all concentrations become invariant; that is, $[A]_{eq}$ = const; $[B]_{eq}$ = const, etc. This means that the opposing reactions occur at the same rate:

$$r_1 = r_2 \qquad \text{at equilibrium}$$

and

$$k_1[A]_{eq}[B]_{eq} = k_2[C]_{eq}[D]_{eq}$$

Therefore

$$\frac{k_1}{k_2} = \frac{[C]_{eq}[D]_{eq}}{[A]_{eq}[B]_{eq}}$$

The ratio of the constants, k_1/k_2, must be a constant, and this is identified as the equilibrium constant K_{eq}, as defined above. Hence,

$$\frac{k_1}{k_2} = K_{eq} = K_C = \frac{[C]_{eq}[D]_{eq}}{[A]_{eq}[B]_{eq}}$$

The derivation of the equilibrium constant from the law of mass action can be generalized so that for the reaction

$$aA + bB + \cdots \rightleftharpoons mM + nN + \cdots$$

where the small letters represent stoichiometric coefficients and the capital letters represent chemical compounds

$$K_{eq} = K_C = \frac{[M]^m_{eq}[N]^n_{eq}}{[A]^a_{eq}[B]^b_{eq}}$$

11.6 DISPLACEMENT OF EQUILIBRIUM: LE CHÂTELIER'S PRINCIPLE

Le Châtelier's principle states that whenever an external stress is imposed on a system in equilibrium, the system reacts to the stress in such a way that the effect is minimized. Using this principle, it is possible to predict the effect caused by changes in the conditions of a system in equilibrium.

Before specific examples are discussed, we shall devote a moment to the specialized language chemists use when they discuss the application of Le Châtelier's principle. When an equilibrium system is disturbed, i.e., a stress is imposed on the system, the system is, at least momentarily, no longer in equilibrium. The system will undergo a chemical reaction until it is again at equilibrium. Chemists say that "the stress has caused the equilibrium to shift to the left (or right)." Shift in this context means that when the system is in equilibrium, a disturbance throws the system out of equilibrium and then the system undergoes a chemical change, eventually reaching a new equilibrium condition.

In terms of Le Châtelier's principle, we shall consider the effects of various stresses on a chemical system at equilibrium.

Concentration changes

In the reaction $PCl_5(g) \rightleftharpoons PCl_3(g) + Cl_2(g)$, if the concentration of Cl_2 is increased, a stress is imposed on the system. The system will react to alleviate the applied stress. The added Cl_2 must be removed. This can be achieved through reaction with PCl_3; at the same time, PCl_5 must form. Therefore, addition of Cl_2 will decrease the number of moles of PCl_3 and increase the number of moles of PCl_5. Some of the added Cl_2 will react with some of the PCl_3 already present, to form more PCl_5. This will continue until a new equilibrium condition is attained. We say the equilibrium shifts to the left. (This does not mean the K_{eq} changes. It does not!)

Conversely, if the concentration of PCl_5 is increased, some of the added PCl_5 will dissociate and thus form more PCl_3 and Cl_2. The equilibrium shifts to the right.

If a decrease in concentration is effected, e.g., if some PCl_3 is removed from the system, the reaction *to the right* becomes favored as the system tends to replace some of the removed PCl_3.

Changes in pressure

A change in applied pressure affects only reaction systems in which some or all of the compounds are in the gaseous state and only those gaseous reactions in which the number of moles of gas changes during a chemical reaction. Thus, the gaseous reaction at equilibrium

$$H_2(g) + I_2(g) \rightleftharpoons 2HI(g)$$

is unaffected by changes in applied pressure, since the total number of moles of gas does not change during a reaction; 1 mol of H_2 plus 1 mol of I_2 will produce 2 mol of HI.

However, the gaseous system

$$2NO(g) + O_2(g) \rightleftharpoons 2NO_2(g)$$

2 mol   1 mol   2 mol

will be affected by a change in applied pressure. Since 2 mol of NO_2 exerts a lower pressure than 2 mol of NO plus 1 mol of O_2 taken at the same temperature and volume, the equilibrium system reacts to the stress of increased pressure by favoring the formation of NO_2; the stress of increased pressure is thereby reduced.

Conversely, a decrease in applied pressure on this system at equilibrium will favor the reaction *to the left* so as to increase the number of molecules present in the system and thus relieve the effect of the reduced pressure.

Changes in temperature

Changing the temperature of an equilibrium system has an effect significantly different from that of concentration or pressure changes. These latter changes affect only the equilibrium concentrations of the species present; a temperature change alters the value of the equilibrium constant as well.

Whenever the temperature of a chemical-equilibrium system is raised, the reaction which proceeds with the absorption of heat, i.e., the *endothermic* reaction, is favored. This is the only way a chemical system can minimize a stress of increased temperature.

The reaction

$$\tfrac{1}{2}N_2 + \tfrac{3}{2}H_2 \xrightarrow{400°C} NH_3 \qquad \Delta H = -12{,}350 \text{ cal} \qquad (a)$$

is exothermic; when $\frac{3}{2}$ mol of hydrogen combines with $\frac{1}{2}$ mole of nitrogen to form 1 mol of ammonia, 12,350 cal of heat is evolved.

The reverse of this reaction,

$$NH_3 \xrightarrow{400°C} \tfrac{1}{2}N_2 + \tfrac{3}{2}H_2 \qquad \Delta H = +12{,}350 \text{ cal} \qquad (b)$$

involves an absorption of 12,350 cal of heat for each mole of ammonia which dissociates. Thus, when the temperature of this reaction system at equilibrium is raised, the decomposition reaction (*b*), which requires heat, is favored. This results in the production of more H_2 and N_2 and the disappearance of NH_3. On the other hand, if the temperature of this system is lowered, the formation reaction (*a*) is favored.

TABLE 11.3: $\frac{1}{2}N_2 + \frac{3}{2}H_2 \rightleftharpoons NH_3$

t, °C	K_{eq}
350	1.359
400	0.712
450	0.390

The effect of a temperature change on the value of the equilibrium constant is shown in Table 11.3. Obviously, K_{eq} is greatly affected by a change in temperature. In this case K_{eq} decreases as the temperature increases. The temperature effect on the K_{eq} value depends on the sign of $\Delta H°$. If the reaction is exothermic, $\Delta H°$ is negative and the K_{eq} decreases as the temperature increases. If $\Delta H°$ is positive, K_{eq} increases as the temperature increases.

There is a simple empirical relationship between the equilibrium constant in terms of partial pressures K_P and temperature. It has been observed for almost all reactions that K_P is an exponential function of the absolute temperature:

$$K_P = Ae^{-B/T}$$

or, in equivalent logarithmic form:

$$\log K_P = \log A - \frac{B}{T} \qquad (11.5)$$

A and B are empirical constants; B may be positive or negative and has a physical significance which we shall discuss. A plot of log K_P versus $1/T$ yields a straight line with a slope equal to negative B. Figure 11.2 shows such a plot for the equilibrium reaction

$$N_2(g) + O_2(g) \rightleftharpoons 2NO(g)$$

From the theory of chemical thermodynamics it can be shown that the slope of the line is related to the standard enthalpy change $\Delta H°$ for the reaction by

$$\text{Slope} = -B = -\frac{\Delta H°}{2.303R} \qquad (11.6)$$

where R is the gas-law constant. Substitution of Eq. (11.6) into Eq. (11.5) gives

$$\log K_P = \log A - \frac{\Delta H°}{2.303RT} \qquad (11.7)$$

In Fig. 11.2 the plot of log K_P versus $1/T$ gives a slope of −9500 K. By Eq. (11.6)

$$\Delta H° = -2.303R(\text{slope}) = (-2.303 \times 1.987 \text{ cal/mol-K})(-9500 \text{ K})$$
$$= +44{,}000 \text{ cal}$$

EXAMPLE 11.1 Calculate the standard-enthalpy change for the reaction

$$CO(g) + H_2O(g) \rightleftharpoons CO_2(g) + H_2(g)$$

K_P equals 10.0 at 690 K and 3.60 at 800 K.

Solution: Substitute both sets of data into Eq. (11.7), eliminate log A, and solve the resulting equation for $\Delta H°$:

$$\log 10.0 = \log A - \frac{\Delta H°}{2.303(1.987 \text{ cal/mol-K})(690 \text{ K})} \qquad (a)$$

$$\log 3.60 = \log A - \frac{\Delta H°}{2.303(1.987 \text{ cal/mol-K})(800 \text{ K})} \qquad (b)$$

FIGURE 11.2

A plot of log K_p versus $1/T$ in the temperature range 1900 to 2500 K for the reaction $N_2(g) + O_2(g) \rightarrow 2NO(g)$.

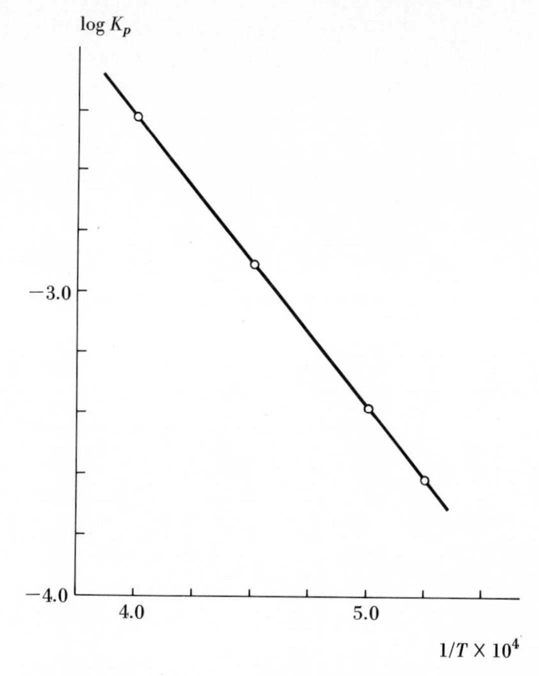

Subtraction of Eq. (*b*) from (*a*) eliminates log *A*, giving

$$\log 10.0 - \log 3.60 = -\frac{\Delta H^\circ}{2.303 \times 1.987}\left(\frac{1}{690} - \frac{1}{800}\right)$$

Finally,

$$\Delta H^\circ = -10{,}200 \text{ cal}$$

The average value of ΔH° in the temperature interval 690 to 800 K is −10,200 cal.

Thus, we have an alternate, noncalorimetric method for obtaining ΔH° for chemical reactions. The values of ΔH° calculated in this way agree with those obtained from calorimetry studies, but they are not as precise.

The effect of a catalyst

A catalyst is an agent which when placed in a reaction system in relatively small amounts alters the rate at which equilibrium is attained. The catalyst does not undergo any net chemical transformation. A catalyst has no effect when added to a system at equilibrium. The equilibrium concentrations of the species of the reaction system remain the same: *only the speed with which equilibrium is attained is affected.*

Consider the reaction system

$$\underset{\text{Acetic acid}}{CH_3COOH} + \underset{\text{Ethyl alcohol}}{C_2H_5OH} \overset{25°C}{\rightleftharpoons} \underset{\text{Ethyl acetate}}{CH_3COOC_2H_5} + H_2O$$

$$K_{eq} = \frac{[CH_3COOC_2H_5]_{eq}[H_2O]_{eq}}{[CH_3COOH]_{eq}[C_2H_5OH]_{eq}}$$

This reaction may require many weeks before equilibrium is attained. Addition of HCl causes the system to come to equilibrium in a much shorter period of time but does not affect the equilibrium concentrations of the reactants or products; nor will it affect the value of the equilibrium constant.

The HCl used as a catalyst in the above reaction is called a *homogeneous* catalyst because it is present in the same phase as the reactants. Very often a gaseous or solution reaction can be catalyzed by the presence of certain solids. The reaction actually occurs on the surface of the solid catalyst. The dissociation of NH_3 into its elements is very slow under ordinary conditions. On a tungsten metal surface the reaction proceeds at a much faster rate. Tungsten in this case is an example of a *heterogeneous* catalyst.

Many important industrial processes are carried out in the presence of heterogeneous catalysts. Probably all biological reactions are catalyzed by enzymes. The field of catalysis is very important; we shall say more about it in the chapter on chemical kinetics (Chap. 17).

11.7 CALCULATIONS WITH THE EQUILIBRIUM CONSTANT

The equilibrium constants for chemical reactions are expressed in terms of the molar concentrations or the partial pressures of the reactants. Since the calculations involved are slightly different in the two cases, they are considered separately and then related.

EXAMPLE 11.2 Calculate K_C for the reaction

$$2HI(g) \rightleftharpoons H_2(g) + I_2(g)$$

if an equilibrium mixture contains 1.25×10^{-3} mol/l of I_2, 3.56×10^{-3} mol/l of H_2, and 1.56×10^{-2} mol/l of HI.

Solution: Since the equilibrium concentrations are given, the problem is straightforward. Simply substitute into the correct equilibrium expression and solve.

$$K_C = \frac{[H_2]_{eq}[I_2]_{eq}}{[HI]_{eq}^2} = \frac{(3.56 \times 10^{-3})(1.25 \times 10^{-3})}{(1.56 \times 10^{-2})^2} = 0.0183$$

EXAMPLE 11.3 The equilibrium constant K_C for the reaction

$$\underset{\text{Ethyl alcohol}}{C_2H_5OH} + \underset{\text{Acetic acid}}{CH_3COOH} \rightleftharpoons \underset{\text{Ethyl acetate}}{CH_3COOC_2H_5} + H_2O$$

is 4.00. Calculate the concentration of ethyl acetate at equilibrium if the concentration of each of the other species in the reaction is 1.50 mol/l at equilibrium.

Solution:

$$K_C = \frac{[CH_3COOC_2H_5]_{eq}[H_2O]_{eq}}{[C_2H_5OH]_{eq}[CH_3COOH]_{eq}}$$

$$4.00 = \frac{[CH_3COOC_2H_5]_{eq}(1.5)}{1.5 \times 1.5}$$

$$[CH_3COOH_2H_5]_{eq} = 1.50 \times 4.00 = 6 \text{ mol/l}$$

EXAMPLE 11.4 Consider the hypothetical equilibrium reaction $2A \rightleftharpoons B + C$. When a 1.00-l vessel is charged with 1.00 mol of A, the number of moles of A is reduced to 0.250 at equilibrium. Calculate K_C.

Solution: The equilibrium-constant expression is

$$K_C = \frac{[B]_{eq}[C]_{eq}}{[A]^2_{eq}}$$

The difficult part of this problem is determining the equilibrium concentrations, $[A]_{eq}$, $[B]_{eq}$, and $[C]_{eq}$. Then the calculation of K_C is simple.

The number of moles of A which has reacted equals the original number of moles minus the number of moles present at equilibrium, or $1.000 - 0.250 = 0.750$.

From the stoichiometry (2 mol of A produces 1 mol of B and 1 mol of C), the number of moles of B or C produced is exactly one-half of the amount of A which reacted, or one-half of 0.75, which is 0.375 mol.

Since the volume of the vessel is 1.00 l, the number of moles of each species equals its concentration: $[A]_{eq} = 0.25$, $[B]_{eq} = 0.375$, and $[C]_{eq} = 0.375$. Substitution of the equilibrium concentrations into the proper equilibrium-constant expression gives the correct answer:

$$K_C = \frac{[B]_{eq}[C]_{eq}}{[A]^2_{eq}} = \frac{0.375^2}{0.250^2} = 2.25$$

It is a good idea to tabulate the given concentrations of the species involved in the chemical equilibrium under the equation for the reaction. In the case we have just discussed we would have the following setup:

	$2A \rightleftharpoons$	B +	C
Original concentration, mol/l:	1	0	0
Equilibrium concentration, mol/l:	0.250	——	——

The concentrations of species B and C at equilibrium are not given. The original concentrations are the concentrations before the reaction takes place or at some time before equilibrium is attained. If the original concentrations and the stoichiometry are known, it is possible to deduce all the equilibrium concentrations from the equilibrium concentration of only one of the species. In this example we were given the chemical equation, the original concentrations, $[A]_{\text{initial}} = 1.00\ M$, $[B]_{\text{initial}} = 0$, and $[C]_{\text{initial}} = 0$, and the equilibrium concentration of A. From $[A]_{\text{initial}}$ and $[A]_{\text{eq}}$ it was possible to determine the amount of A which reacted; this is simply the difference

$$[A]_{\text{initial}} - [A]_{\text{eq}} = 1.000 - 0.250 = 0.750 \text{ mol}$$

The chemical equation says that if 2 mol of A reacts, 1 mol each of A and B will be produced. Then, if 0.750 mol A reacts, 0.375 mol each of A and B will be produced. Our completed concentration data will be

	$2A \rightleftharpoons$	B +	C
Original concentration, mol/l:	1	0	0
Equilibrium concentration, mol/l:	0.250	0.375	0.375

And then the equilibrium constant is calculated from the concentrations in the second row.

EXAMPLE 11.5 When $H_2(g)$ and $I_2(g)$ are placed in a reaction vessel at 700 K with initial concentrations of 0.1110 and 0.0995 mol/l, respectively, the concentration of H_2 at equilibrium is 0.0288 mol/l. Calculate K_C for the reaction

$$H_2(g) + I_2(g) \rightleftharpoons 2HI(g) \qquad K_C = \frac{[HI]^2_{\text{eq}}}{[H_2]_{\text{eq}}[I_2]_{\text{eq}}}$$

Solution: The difficult part of this problem is to determine the equilibrium concentrations of all species. From the stoichiometry we can say that equal numbers of moles of H_2 and I_2 will react and for each mole of H_2 (or I_2) which reacts, 2 mol of HI must form. Therefore, let us calculate the number of moles of H_2 which have reacted. Then we can calculate the equilibrium concentrations of I_2 and HI, which, along with the known equilibrium concentration of H_2, is all that is required for the calculations of K_C.

1. Number of moles per liter of H_2 which have reacted equals original H_2 concentration minus equilibrium H_2 concentration

$$0.1110 \text{ mol/l} - 0.0288 \text{ mol/l} = 0.0822 \text{ mol/l}$$

2. Number of moles per liter of H_2 which have reacted equals number of moles of I_2 which have reacted, or 0.0822 mol/l.

3. Equilibrium concentration of I_2 equals original $[I_2]$ minus number of moles per liter which have reacted

$$0.0995 \text{ mol/l} - 0.0822 \text{ mol/l} = 0.0173 \text{ mol/l} = [I_2]_{eq}$$

4. Number of moles of HI which have formed equals 2 times the number of moles of H_2 which have reacted

$$[HI]_{eq} = 2(0.0822 \text{ mol/l}) = 0.1644 \text{ mol/l}$$

First let us tabulate the given concentration data:

	$H_2(g)$ +	$I_2(g) \rightleftharpoons$	$2HI(g)$
Original concentration, M:	0.1110	0.0995	0
Equilibrium concentration, M:	0.0288	——	——

With the original and equilibrium concentrations of all species a complete tabulation can be made beneath the equation for the reaction:

	$H_2(g)$ +	$I_2(g) \rightleftharpoons$	$2HI(g)$
Original concentration, M:	(0.1110)	(0.0995)	(0)
Equilibrium concentration, M:	(0.0288)	0.0173	0.1644

The concentrations in parentheses are given in the example; the other concentrations are deduced from the stoichiometry of the reaction.

The equilibrium constant can now be calculated by simple substitution into the correct equilibrium-constant expression:

$$K_C = \frac{[HI]_{eq}^2}{[H_2]_{eq}[I]_{eq}} = \frac{0.1644^2}{0.0288 \times 0.0173} = 54.4$$

EXAMPLE 11.6 In the gaseous equilibrium $A + B \rightleftharpoons 2C + D$, all species have partial pressures of 0.75 atm at equilibrium. Calculate K_P.

Solution:

$$K_P = \frac{P_C{}^2 \times P_D}{P_A \times P_B} = \frac{0.75^2 \times 0.75}{0.75 \times 0.75} = 0.75$$

EXAMPLE 11.7 When N_2O_4 is sealed in a reaction vessel and brought to equilibrium at 25°C, it dissociates according to the equation $N_2O_4(g) \rightleftharpoons 2NO_2(g)$. K_P for the reaction at this temperature is 0.140. If the total pressure in the vessel is 2.00 atm, calculate the partial pressure of each gas.

Solution: To solve this problem, we note that the partial pressures of the two gases must obey two mathematical laws:

The equilibrium law: $K_P = \dfrac{P^2_{NO_2}}{P_{N_2O_4}}$

Dalton's law of partial pressures:

$$P_{total} = \sum_{i=1}^{i=n} P_i \qquad P_{total} = P_{NO_2} + P_{N_2O_4}$$

Simultaneous solution of these equations yields P_{NO_2} and $P_{N_2O_4}$.

$K_P = 0.140$; $P_{total} = 2.00$ atm.

$$0.140 = \frac{P^2_{NO_2}}{P_{N_2O_4}} \qquad (a)$$

$$2 \text{ atm} = P_{NO_2} + P_{N_2O_4} \qquad \text{or} \qquad P_{N_2O_4} = 2 \text{ atm} - P_{NO_2} \qquad (b)$$

Substitution of (*b*) into (*a*) yields

$$0.140 = \frac{P^2_{NO_2}}{2 - P_{NO_2}} \qquad \text{or} \qquad P^2_{NO_2} + 0.140P_{NO_2} - 0.280 = 0$$

Using the general solution for the quadratic equation,

$$P_{NO_2} = \frac{-0.140 \pm \sqrt{0.140^2 + (4 \times 0.280)}}{2} = 0.46 \text{ atm}†$$

$$P_{N_2O_4} = 2.00 \text{ atm} - P_{NO_2} = 2.00 - 0.46 \text{ atm} = 1.54 \text{ atm}$$

EXAMPLE 11.8 Solid NH_4HS is placed in an evacuated, sealed vessel at 25°C. Dissociation takes place according to the equation

$NH_4HS(s) \rightleftharpoons NH_3(g) + H_2S(g)$

The equilibrium constant for the reaction K_P (in atmospheres) equals 0.111 at 25°C. Calculate the partial pressures of the two gases in Torr.

Solution: NH_4HS is a pure solid of constant composition and concentration; it is not included in the expression for K_P, which is

$K_P = P_{NH_3}P_{H_2S}$

† Note that two values of P_{NO_2} are actually obtained by solution of the quadratic equation. One value is negative and has no physical significance.

Tabulate the given information under the chemical equation:

	$NH_4HS(s) \rightleftharpoons$	$NH_3(g)$ +	$H_2S(g)$
Original pressure:	Present	0	0
Equilibrium pressure:	Present	___	___

Note that we simply indicate that $NH_4HS(s)$ is present since it is a pure phase. It is either present or absent. There is no $NH_3(g)$ or $H_2S(g)$ present originally before dissociation of $NH_4HS(s)$ occurs. Since NH_3 and H_2S are formed in equimolar amounts, the partial pressures of the two gases at equilibrium will be equal:

$$P_{NH_3} = P_{H_2S}$$

We can substitute this equality into the K_P expression:

$$K_P = P_{NH_3}P_{H_2S} = P^2_{NH_3} = 0.111$$

and

$$P_{NH_3} = \sqrt{0.11} = 0.33 \text{ atm} = P_{H_2S}$$

The pressure in Torr for either gas is

$$0.33 \text{ atm} \times 760 \text{ Torr/atm} = 250 \text{ Torr}$$

EXAMPLE 11.9 Solid NH_4HS is placed in a flask at 25°C containing 0.50 atm of ammonia. What are the partial pressures of ammonia and hydrogen sulfide when equilibrium between the gaseous and solid phases is reached? $K_P = 0.11$ at 25°C. The equation for the reaction is

$$NH_4HS(s) \rightleftharpoons NH_3(g) + H_2S(g)$$

If the flask did not already contain some NH_3, the P_{NH_3} would be equal to the P_{H_2S}. In the present case, however, the ammonia pressure will always be 0.50 atm greater than H_2S pressure, since the stoichiometry indicates that both gases are formed in equal amounts from the dissociation. The relationship between P_{NH_3} and P_{H_2S} is

$$P_{NH_3} = 0.50 + P_{H_2S} \qquad (a)$$

Let us tabulate the original and equilibrium partial pressures in terms of the partial pressure of H_2S:

	$NH_4HS(s) \rightleftharpoons$	$NH_3(g)$ +	$H_2S(g)$
Original pressure, atm:	Present	0.50	0
Equilibrium pressure, atm:	Present	$0.5 + P_{H_2S}$	P_{H_2S}

Now we write the equilibrium constant in terms of partial pressures

$$K_P = P_{NH_3} P_{H_2S} = 0.111 \qquad (b)$$

Substitution of the equilibrium concentrations into (*b*) gives

$$K_P = (0.50 + P_{H_2S})(P_{H_2S}) = 0.111$$
$$P_{H_2S} = 0.170 \text{ atm}$$
$$P_{NH_3} = 0.50 + 0.17 \text{ atm} = 0.67 \text{ atm}$$

EXAMPLE 11.10 For the Haber process ($\frac{3}{2}H_2 + \frac{1}{2}N_2 \rightleftharpoons NH_3$) K_P is 0.0266 at 350°C and 0.00659 at 450°C. Calculate (*a*) K_C at each temperature and (*b*) $\Delta H°$ for the reaction.

Solution: (*a*) The relationship between K_P and K_C for gaseous reactions is

$$K_C = K_P (RT)^{-\Delta n_g}$$

where R is in liter-atmospheres. For the above reaction, $\Delta n_g = 1 - (\frac{3}{2} + \frac{1}{2}) = -1$. Therefore

$$K_C = K_P (RT)^{-(-1)} = K_P RT$$

At 350°C or 623 K: $K_C = 0.0266 \times 0.082 \times 623 = 1.36$
At 450°C or 723 K: $K_C = 0.00659 \times 0.082 \times 723 = 0.391$

(*b*) Equation (11.7) relates K_P and $\Delta H°$:

$$\log K_P = \log A - \frac{\Delta H°}{2.303RT} \qquad \text{where } R \text{ is in calories.}$$

At 623 K,

$$\log 0.0266 = \log A - \frac{\Delta H°}{2.303 \times 1.987 \times 623}$$

At 723 K,

$$\log 0.00659 = \log A - \frac{\Delta H°}{2.303 \times 1.987 \times 723}$$

By subtraction, we obtain

$$\log 0.0266 - \log 0.00659 = -\frac{\Delta H°}{2.303 \times 1.987 \times 623} + \frac{\Delta H°}{2.303 \times 1.987 \times 723}$$

which is equivalent to

$$\log \frac{0.0266}{0.00659} = \frac{\Delta H°}{2.303 \times 1.987} \left(\frac{1}{723} - \frac{1}{623} \right)$$

$$\Delta H° = -12{,}495 \text{ cal}$$

We see that the above analysis will yield the general equation

$$\log \frac{K_{P,2}}{K_{P,1}} = \frac{\Delta H^\circ}{2.303R}\left(\frac{1}{T_1} - \frac{1}{T_2}\right) \tag{11.7a}$$

where $K_{P,2}$ = equilibrium constant at temperature T_2
$K_{P,1}$ = equilibrium constant at temperature T_1

11.8 DRIVING FORCES AND THE CHEMICAL POTENTIAL ENERGY

The forces which cause systems to change, or react, are called *driving forces*. For example, bodies tend to fall when acted upon by a gravitational driving force; they lose gravitational potential energy. Charged bodies move together or apart depending on whether they have opposite or like charges. In doing so, they lose electric potential energy. The north pole of a magnet attracts the south pole of another magnet, and the magnets may move toward each other. In so doing, they lose magnetic potential energy. All systems tend toward a minimum in potential energy.† It should not be surprising that a concept of chemical potential energy has been developed in analogy to other kinds of potential energy, so that a reaction system loses chemical potential energy during a chemical process.

When chemical reactions occur at constant temperature and pressure,‡ the chemical potential energy of the reacting system is falling. The chemical potential energy is symbolized by G and the change in the chemical potential energy by ΔG. In terms of quantities we have studied previously,

$$G = H - TS \tag{11.8}$$

where H = enthalpy
T = absolute temperature
S = entropy

The change in chemical potential energy is then equal to the change in $H - TS$:

$$\Delta G = \Delta(H - TS) \tag{11.9}$$

or

$$\Delta G = \Delta H - \Delta(TS) \tag{11.9a}$$

† Implicitly, during all these processes we consider that the temperature remains constant.
‡ G, the chemical potential, decreases during a chemical reaction when the initial and final temperatures and pressures are equal. They need not remain constant during the process.

Since we shall want to calculate ΔG only when T is constant, Eqs. (11.9) and (11.9a) become

$$\Delta G = \Delta H - T\,\Delta S \tag{11.10}$$

For reactions which occur so that the initial and final temperatures and pressures are equal ($T_i = T_f$, and $P_i = P_f$), G can only decrease. After a reaction attains equilibrium under these conditions, G remains constant (as do all macroscopic properties of an equilibrium system). Therefore, the change in the chemical potential energy can be only negative or zero; ΔG (or $\Delta H - T\,\Delta S$) is negative for chemical reactions which occur at constant temperature and pressure. ΔG is zero for a chemical system at equilibrium.

It should be noted that the change in chemical potential energy is related to the change in entropy. We have learned that the entropy change must be positive for any process occurring in an isolated system. Now we learn that the chemical potential-energy change must be negative for chemical processes which occur at constant temperature and pressure. The fact that ΔG and ΔS are related should not be surprising: they both serve as criteria of spontaneity.

1. If a process occurs in an isolated system, S increases (see Chap. 10).

2. If a process occurs in a system held at constant temperature and pressure, G decreases.

The chemical potential energy G is a particularly convenient quantity because many observations are made at constant temperature and pressure.

The change in chemical potential energy ΔG which accompanies a chemical reaction at constant pressure can be calculated from an equation derived from thermodynamic principles. This equation is called the *reaction isotherm*.

$$\Delta G = \Delta G^\circ + RT \ln Q \tag{11.11}$$

In Eq. (11.11) ΔG° is the *standard* change in the chemical potential energy for the reaction under consideration. It is analogous to the *standard*-enthalpy change. It is the change in chemical potential energy which accompanies the transformation of reactants in their standard states into products in their standard states at some specific temperature. R is the gas constant, T is the absolute temperature, and Q is the pressure or concentration quotient.† In the general reaction

$$a\mathrm{A} + b\mathrm{B} + \cdots \rightleftharpoons m\mathrm{M} + n\mathrm{N} + \cdots$$

† To be exact, activities instead of pressures or concentrations should be employed. Under ordinary conditions ($P = 1$ atm, $T = 298$ K) gas pressures are nearly equal to activities, but the concentrations of compounds in solutions may be appreciably different from their activities (see Chaps. 9 and 14).

where a, b, m, and n are the stoichiometric coefficients of the chemical species, the Q's are defined as

$$Q_P = \frac{P_M{}^m P_N{}^n \cdots}{P_A{}^a P_B{}^b \cdots}$$

and

$$Q_C = \frac{[M]^m [N]^n \cdots}{[A]^a [B]^b \cdots}$$

The partial pressures shown in Q_P are the pressures of the reactants at the beginning of the reaction and the pressures of the products at the end of the reaction. Hence, Q_P has the form of the equilibrium constant K_P, but it is not the same quantity; only equilibrium pressures are used to evaluate K_P, and K_P has a unique value at a particular temperature. The value of Q_P can be varied by changing initial concentrations of the reactants. As an example of the use of the reaction isotherm, consider the following problem.

EXAMPLE 11.11 Calculate the decrease in the chemical potential energy per mole of N_2O_4 when NO_2 at 10 atm is converted at 25°C into N_2O_4 at 10 atm

$$\underset{10\text{ atm}}{2NO_2(g)} \xrightarrow{25^\circ C} \underset{10\text{ atm}}{N_2O_4(g)}$$

Solution: Remember that

$$\Delta G = \Delta G^\circ + RT \ln Q = \Delta G^\circ + RT \ln \frac{P_{N_2O_4}}{P_{NO_2}^2}$$

ΔG° equals $\Delta H^\circ - T\,\Delta S^\circ$ and can be calculated from ΔH° and S° values in Tables 10.3 and 10.4. For this reaction, $\Delta G^\circ = -1291$ cal; therefore

$$\Delta G = -1291 + \left(1.987 \times 298 \ln \frac{10}{10^2}\right)$$

Note that Q is *not* the equilibrium constant! It has the form of the equilibrium constant, but the partial pressures used in its evaluation are not the *equilibrium* partial pressures.

$$\begin{aligned}\Delta G &= -1291 + (1.987 \times 298 \ln 0.1)\\ &= -1291 + (1.987 \times 298 \times 2.303 \log 0.1)\\ &= -1291 - (1.987 \times 298 \times 2.303) = -2655 \text{ cal}\end{aligned}$$

Let us further analyze the reaction isotherm in view of the above example. Because a reaction did occur at constant temperature and pressure, ΔG had to be negative. At any given temperature (in this case 25°C), ΔG°, R and T of Eq. (11.11) are constants. The decrease in the chemical potential energy is affected only by the value of Q.

The smaller the value of Q, the more favorable the reaction under consideration. If the final pressure of N_2O_4 had been 1 atm instead of 10 atm, Q would have been 0.01, which would make ln Q equal to -4.6. ΔG would have a value of -3850 cal, which is a greater drop in chemical potential energy than when the pressure of N_2O_4 was 10 atm.

It is possible to predict the effect a change in pressure will have simply by looking at the chemical equation for the reaction. When pressures (or concentrations) of reactants are high relative to the products, the reaction from left to right is more favorable. In terms of the chemical potential energy, this means that there would be a greater drop, that is, ΔG would be more highly negative, which is exactly what Eq. (11.11) bears out.

11.9 THE RELATIONSHIP BETWEEN $\Delta G°$ AND K_{eq}

When a chemical reaction is at equilibrium, G is constant and ΔG is zero. At equilibrium, then, Eq. (11.11) becomes

$$0 = \Delta G° + RT \ln Q_{eq}$$

or

$$\Delta G° = -RT \ln Q_{eq}$$

or

$$\Delta G° = -2.303RT \log Q_{eq} \qquad (11.12)$$

where Q is calculated from the equilibrium value of the concentrations or partial pressures and is therefore, by definition, the equilibrium constant K

$$\Delta G° = -RT \ln K$$

or

$$\Delta G° = -2.303RT \log K \qquad (11.13)$$

This equation expresses one of the most important principles in chemistry. From $\Delta G°$ the value of the equilibrium constant can be calculated. The *standard* chemical-potential-energy change $\Delta G°$ can be calculated from *standard* enthalpies and entropies of reaction, which we have learned to calculate from thermochemical data. Thermochemical data such as heats of combustion and heat capacities are tabulated for thousands of compounds at many temperatures. One of the reasons for their tabulation is obvious now: from them one can determine equilibrium constants without performing any additional experiments.

EXAMPLE 11.12 For the reaction $2NO_2(g) \rightleftharpoons N_2O_4(g)$ at 25°C, $\Delta G°$ is −1291 cal. Calculate K_P.

Solution:

$$\Delta G° = -RT \ln K_P$$

or

$$\log K_P = -\frac{\Delta G°}{2.303RT} = -\frac{(-1291)}{2.303 \times 1.987 \times 298}$$

$$\log K_P = 0.947$$
$$K_P = 8.85$$

or

$$K_P = \frac{P_{N_2O_4}}{P^2_{NO_2}} = 8.85$$

K_p is the pressure quotient at equilibrium.

EXAMPLE 11.13 From thermochemical data, calculate $\Delta G°$ and K_P for the reaction

$$NO(g) + \tfrac{1}{2}O_2(g) \xrightleftharpoons{25°C} NO_2(g)$$

Relationships to be used are $\Delta G° = \Delta H° - T\,\Delta S°$ and $\Delta G° = -2.303RT \log K_P$.

Solution:
To calculate $\Delta H°$, we have, from Table 10.3, that the standard enthalpies of formation of NO and NO_2 are +21.60 and +8.09 kcal/mol. Of course, $\Delta H_f°$ for O_2, an element, is zero. The standard-enthalpy change for the above reaction is

$$\Delta H° = \Sigma\ \Delta H°_{prod} - \Sigma\ \Delta H°_{reac}$$
$$= 8.09 - 21.60 = -13.51 \text{ kcal} = -13{,}510 \text{ cal}$$

To calculate $\Delta S°$, we have, from Table 10.4, that the absolute entropies of NO, NO_2, and O_2 are 50.34, 57.47, and 49.00 cal/mol-K, respectively. The standard-entropy change for the above reaction is

$$\Delta S° = \Sigma S°_{prod} - \Sigma S°_{reac}$$
$$= 57.57 - [50.34 + (\tfrac{1}{2} \times 49.00)] = -17.27 \text{ cal/K}$$

To calculate $\Delta G°$

$$\Delta G° = \Delta H° - T\,\Delta S° = -13{,}510 - [298 \times -(17.27)] = -8364 \text{ cal}$$

To calculate K_P

$$\Delta G° = -RT \ln K_P$$

or

$$\Delta G° = -2.303RT \log K_P$$

$$\log K_P = -\frac{\Delta G^\circ}{2.303RT} = -\frac{(-8364)\ \text{cal/mol}}{2.303 \times 1.987\ \text{cal/mol-deg} \times 298.2\ \text{K}}$$

$$\log K_P = 6.13 = 6 + 0.13$$
$$K_P = 10^{0.13} \times 10^6 = 1.35 \times 10^6$$

Since K_P is very large, we can state that the products are strongly favored in this reaction. In fact, K_P is so large in this case that it would probably be impossible to detect reactants at equilibrium if they are initially mixed in their stoichiometric ratios.

QUESTIONS AND PROBLEMS

11.1 State (*a*) the law of mass action and (*b*) Le Châtelier's principle.

11.2 (*a*) How can one determine *experimentally* that a chemical reaction is at equilibrium? (*b*) What is true of the chemical potential energy G of a system at equilibrium?

11.3 How would you prove that when the reaction $H_2 + I_2 \rightleftharpoons 2HI$ comes to equilibrium, I_2 and H_2 continue to react? (That is, how would you prove that the equilibrium is dynamic?)

11.4 (*a*) Define catalyst. (*b*) In view of this definition why would it be impossible for a catalyst to affect the concentrations or pressures of the species present in an equilibrium system?

11.5 Write the equilibrium-constant expressions K_C and K_P for the following reactions:

(*a*) $H_2S(g) + NH_3(g) \rightleftharpoons NH_4HS(s)$
(*b*) $CaCO_3(s) \rightleftharpoons CaO(s) + CO_2(g)$
(*c*) $CaCO_3(s) + H^+(aq) \rightleftharpoons HCO_3^-(aq) + Ca^{2+}(aq)$
(*d*) $PCl_5(g) \rightleftharpoons PCl_3(g) + Cl_2(g)$
(*e*) $N_2O_4(g) \rightleftharpoons 2NO_2(g)$

11.6 The density of pure solid ammonium hydrosulfide, NH_4HS, is 1.17 g/cm^3. (*a*) Calculate the volume of 100 g of NH_4HS. (*b*) Determine the number of moles in 100 and 200 g. (*c*) Divide the number of moles in 100 g by the volume of 100 g; do the same for 200 g. (*d*) Does this exercise prove that the concentration of a pure phase is independent of the actual mass of the phase? (*e*) Explain why the concentration of NH_4HS does not appear in the equilibrium-constant expression for $NH_4HS(s) \rightleftharpoons NH_3(g) + H_2S(g)$.

11.7 For the gaseous reaction $CO(g) + 2H_2(g) \rightleftharpoons CH_3OH(g)$, $\Delta H^\circ = -20.2$ kcal. In terms of Le Chatelier's principle explain what happens to the equilibrium concentrations of the species and the equilibrium constant if (*a*) the pressure is increased, (*b*) the temperature is raised, (*c*) the CO partial pressure is increased, (*d*) the CH_3OH concentration is increased, (*e*) a catalyst is added.

11.8 Consider the equilibrium system $2HI(g) \rightleftharpoons H_2(g) + I_2(g)$; ΔH is negative. What will happen to the concentrations and the equilibrium constant if (*a*) the pressure is slowly decreased, (*b*) the concentration of I_2 is decreased, (*c*) the temperature is raised, (*d*) a catalyst is added?

11.9 Consider the equilibrium system $CaO(s) + CO_2(g) \rightleftharpoons CaCO_3(s)$; $\Delta H° = +43.1$ kcal/mol. (*a*) Write the equilibrium-constant expression. (*b*) What happens to the partial pressure of CO_2 if the volume of the system is decreased slowly? (*c*) What happens to the *concentrations* of the *solids* CaO and $CaCO_3$ if the volume of the system is slowly increased? (*d*) What is the effect on P_{CO_2} and K_P of increasing the temperature?

11.10 The equilibrium concentrations (in moles per liter) are listed for the species involved in particular stoichiometries. Write the equilibrium-constant expression and calculate K_C for each:

(*a*) A + 2B ⇌ C

1.00 2.00 0.500

(*b*) 2M + 2N ⇌ 2P + Q

0.500 2.75 3.50 1.00

(*c*) A + 3B ⇌ 2C + 2D

5.00×10^{-5} 3.00×10^{-3} 2.50 1.00

(*d*) A + B + 2C ⇌ 3D

0.0115 1.05 2.05 0.0525

(*e*) 2E + 3F ⇌ G + 2H

10.5 5.4 0.115 0.0267

11.11 Consider the equilibrium A + B ⇌ C. In one experiment A and B each have initial concentrations of 0.500 *M*; the equilibrium concentration of B is 0.400 *M*. (*a*) Calculate the equilibrium concentration of each species. (*b*) Calculate the equilibrium constant K_C.

11.12 Consider the gaseous equilibrium 2M + 2N ⇌ 2P + D. Initially, 0.100 mol each of M and N are placed in a 1.00-l flask. At equilibrium 0.0250 mol of D is present in the flask. Calculate (*a*) the equilibrium concentrations of all species and (*b*) the equilibrium constant K_C.

11.13 Consider the gaseous equilibrium A + B ⇌ 2C at 25°C. In one experiment the initial partial pressures of A and B are 0.375 atm; at equilibrium the partial pressure of C is 0.500 atm. Calculate (*a*) the equilibrium partial pressures of all species, (*b*) the equilibrium constant K_P, and (*c*) the equilibrium constant K_C [Eq. (11.4)]. (*d*) What happens to the pressure as the reaction proceeds?

11.14 Consider the gaseous equilibrium 2A + B ⇌ C + 2D at 50°C. In one experiment the initial pressures of A and B were 0.100 and 1.00 atm, respectively. The equilibrium partial pressures of D was 0.0375 atm. Calculate (*a*) the equilibrium partial pressures of all species, (*b*) K_P, and (*c*) K_C.

11.15 Consider the gaseous equilibrium A + B ⇌ C + D. Initially the concentrations of all reactants and products are 0.100 *M*. At equilibrium the concentration of D is 0.123 *M*. (*a*) Calculate the concentrations of all species and (*b*) K_C.

11.16 Acetic acid, CH_3COOH, and ethyl alcohol, CH_3CH_2OH, are allowed to react, whereupon ethyl acetate, $CH_3CH_2OCOCH_3$, and water are produced according to the equation

$$CH_3COOH + CH_3CH_2OH \rightleftharpoons CH_3CH_2OCOCH_3 + H_2O$$

In an experiment at room temperature the molar concentrations of species at equilibrium were $[CH_3COOH] = 1.00$, $[CH_3CH_2OH] = 0.250$, $[CH_3CH_2OCOCH_3] = 3.00$, and $[H_2O] = 0.333$. Calculate K_C for this reaction.

11.17 For the gaseous equilibrium $N_2O_4 \rightleftharpoons 2NO_2$, K_P is 0.141 at 25°C and 0.628 at 45°C. (*a*) Using Eq. (11.7), calculate the standard-enthalpy change for this temperature interval. (*b*) Using Eq. (11.13), calculate the standard chemical potential-energy change at each temperature.

11.18 K_C for the dissociation of N_2O_4 is 0.00577 at 25°C. [$N_2O_4(g) \rightleftharpoons 2NO_2(g)$; $K_C = 0.00577$.] A 0.0500-g sample of gaseous N_2O_4 is introduced into a 200-ml flask and thermostated at 25°C. (*a*) Calculate the number of moles of N_2O_4 introduced into the flask. (*b*) What would the concentration of N_2O_4 be if it did not dissociate? (*c*) Calculate the equilibrium concentrations and partial pressures of each gas. (*d*) Calculate K_P from the equilibrium partial pressures *and* from Eq. (11.4).

11.19 The partial pressures of SO_2 and O_2 in a reaction vessel are both 0.100 atm before the reaction between the two gases begins at 1000°C [$2SO_2(g) + O_2(g) \rightleftharpoons 2SO_3(g)$]. At equilibrium the total pressure is 0.189 atm. (*a*) Calculate the initial total pressure. (*b*) What are the partial pressures of the three gases at equilibrium? (*c*) What is the value of K_P?

11.20 For the reaction $H_2(g) + I_2(g) \rightleftharpoons 2HI(g)$, $K_P = 66.9$ at 448°C, and $K_P = 50.0$ at 350°C. (*a*) Using Eq. (11.7), calculate $\Delta H°$ for this temperature range. (*b*) Is the sign of $\Delta H°$ consistent with predictions based on Le Châtelier's principle for this reaction? Explain. (*c*) Calculate $\Delta G°$ for this reaction at 448 and 350°C. (*d*) Calculate ΔG (not $\Delta G°$) for the following transformation at 350°C:

$$\underset{10\ \text{atm}}{H_2(g)} + \underset{10\ \text{atm}}{I_2(g)} \rightarrow \underset{5\ \text{atm}}{2HI(g)}$$

[Use Eq. (11.11).]

11.21 For the $N_2O_4 \rightleftharpoons 2NO_2$ equilibrium the values of K_P are given in Table P11.21. (*a*) Plot log K_P versus $1/T$. (*b*) From the value of the slope calculate $\Delta H°$ for this temperature interval. (*c*) Calculate $\Delta G°$ at each temperature. (*d*) Using Eq. (11.10), calculate $\Delta S°$ at each temperature.

TABLE P11.21

K_P	t, °C
0.141	25
0.303	35
0.628	45

11.22 Using the $\Delta H°$ and $S°$ values tabulated in Tables 10.3 and 10.4 calculate $\Delta G°$ and K_P for the following reactions at 25°C:

(*a*) $CH_4(g) + 4Cl_2(g) \rightleftharpoons CCl_4(l) + 4HCl(g)$
(*b*) $NO(g) + \frac{1}{2}O_2(g) \rightleftharpoons NO_2(g)$
(*c*) $SO_3(g) \rightleftharpoons SO_2(g) + \frac{1}{2}O_2(g)$
(*d*) $C_2H_2(g) + 2H_2 \rightleftharpoons C_2H_6(g)$

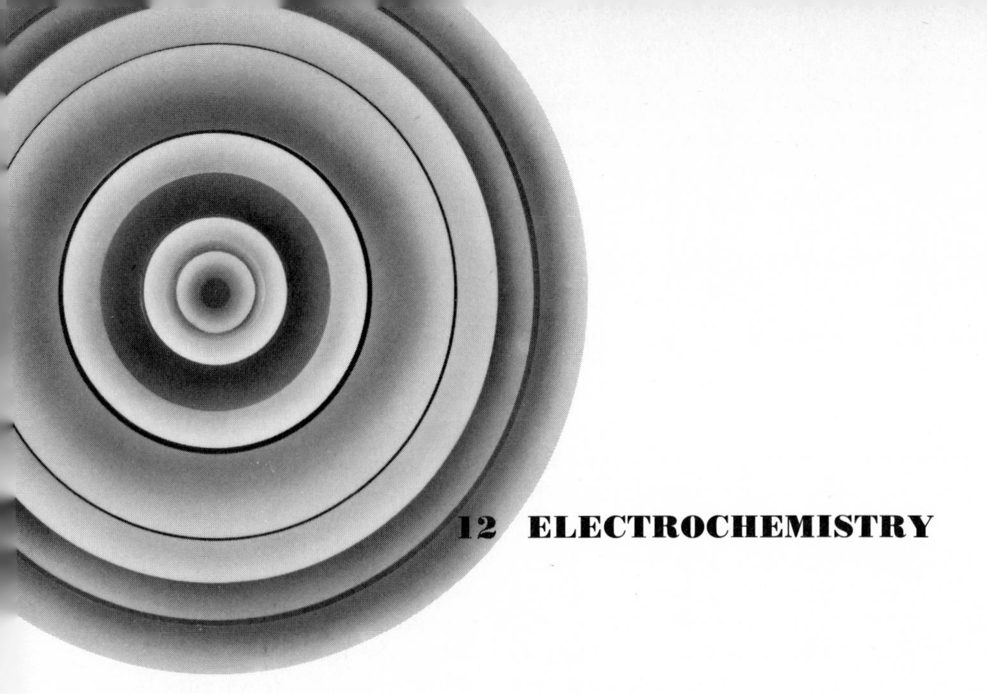

12 ELECTROCHEMISTRY

12.1 INTRODUCTION

Electrochemistry might better be called *electron chemistry* because it is a study of the chemical reactions involving electrons. As the chapter progresses, this will become increasingly evident.

When Mendeleev learned of J. J. Thomson's conclusions about the nature of cathode rays, he attempted to fit the electron into his periodic classification of the elements, insisting that the electron was the lightest of all elements. This may seem absurd today, but electrons do have chemical characteristics which make them similar to atoms. Although electrons are extremely reactive, so that ordinarily they cannot be found at very high concentrations in any chemical system, liquid ammonia, the amines, certain alcohols, and ethers solvate electrons (just as ions are solvated) and stabilize them in solution. The study of solvated electrons is a very active field of research. The consequences of these investigations are of great importance to the development of nuclear reactors; radiation caused by atomic fission produces electrons, which are solvated in coolant liquids. The study of the properties of solvated electrons has also shed light on the nature of electronic conduction in nonmetallic conductors.

In this chapter we deal with the reactions of electrons at interfaces, where materials of different chemical composition are in contact. Here spontaneous chemical reactions can cause a flow of electrons through an electric circuit which includes the interfaces. Conversely, by using a battery to force an electron current to pass through an interface, it is possible to cause a reaction involving electrons to occur. This type of reaction is called *electrolysis*.

12.2 THE DISCOVERY OF A SOURCE OF CONTINUOUS ELECTRIC CURRENT

Until the 1790s the study of electrical effects was severely limited. The only source of electricity was the electrophorus (Fig. 12.1), a frictional machine incapable of sustaining electric currents for more than a fraction of a second. In 1780, when the Italian physiologist Luigi Galvani was conducting some routine experiments on muscle preparations, he accidentally left a dissected frog near an electrophorus. When one of his assistants touched the nerve endings with a knife, the legs performed violent contractions. Another assistant noticed that the frog's legs also contracted when the electrophorus was discharged. Continued investigation of this effect showed that the contraction could be caused by lightning from an electric storm. More important, Galvani found that the leg muscle contracted when the nerve was placed in contact with two different metals, such as copper and zinc, provided that the metals were also in contact at another point. He was probably reminded of the action of the electric eel and explained his frog's-legs experiments in terms of "animal electricity." He believed that nerve fibers produced electricity which caused the muscles to contract.

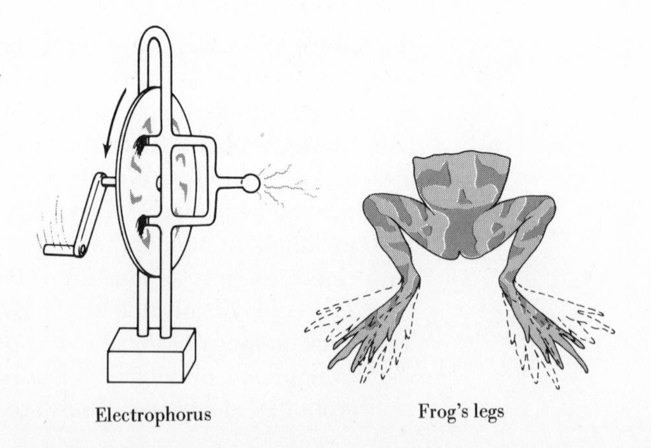

FIGURE 12.1

Electrophorus. (Courtesy of Burndy Library.)

FIGURE 12.2

Equivalent circuits.

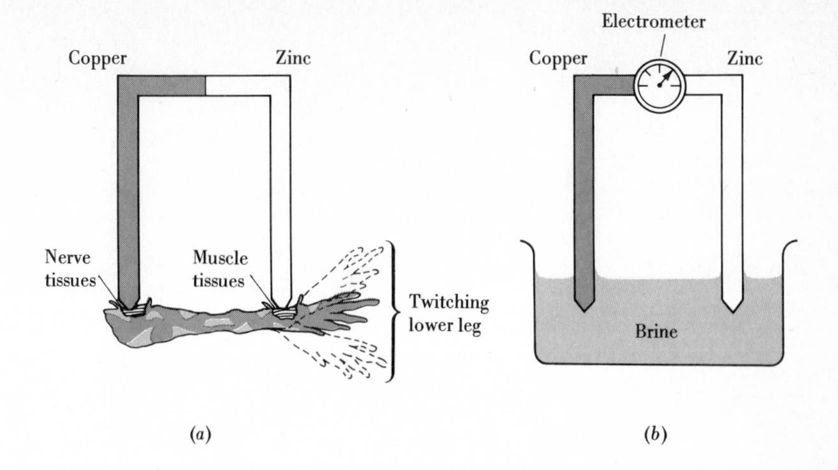

Galvani's frog's-legs experiments intrigued Alessandro Volta, a physicist who had already worked with the electrophorus and electric condensers and invented an electrometer capable of measuring very small quantities of electricity. Volta eventually determined that the source of electricity in the frog's-leg phenomenon is the contact of the two dissimilar metals. The frog's leg serves two purposes: (1) it is a sensitive instrument for the detection of an electric current, and (2) because of the presence of salts in the animal fluids, the leg is a good conductor of electricity. The dissimilar metals and the frog's leg make a complete electric circuit (see Fig. 12.2*a*). As shown in Fig. 12.2*b*, when Volta substituted the combination of a brine solution and his electrometer for the frog's leg, he also detected a current. Volta devised a continuous source of electric current by making a *pile* of alternate disks of silver, blotting paper soaked with brine, and zinc (see Fig. 12.3). Volta's pile produced a continuous, feeble shock when he placed one finger on the top of the pile and another in the brine solution. Today, Volta's pile would be called a *battery;* each unit of a battery is called a *cell.* In Volta's pile a cell consists of one zinc disk, or electrode, and one silver disk, or electrode, separated by a brine-soaked paper disk. Volta's work resulted in one of the most significant experimental discoveries in history. Modern civilization could not have developed or be sustained without the energy supplied by a continuous electric current. All sciences, but especially physics and chemistry, developed rapidly as soon as a source of continuous current was available. Volta's discovery changed the course of history.

It is important to keep in mind the state of natural science at the time of Volta's work. Thomson's identification of the electron was a century off; electricity was considered to be a weightless fluid; Dalton had not published his atomic hypothesis. The theoretical basis for understanding an electric cell or pile was undeveloped.

FIGURE 12.3

Volta's pile, or battery.

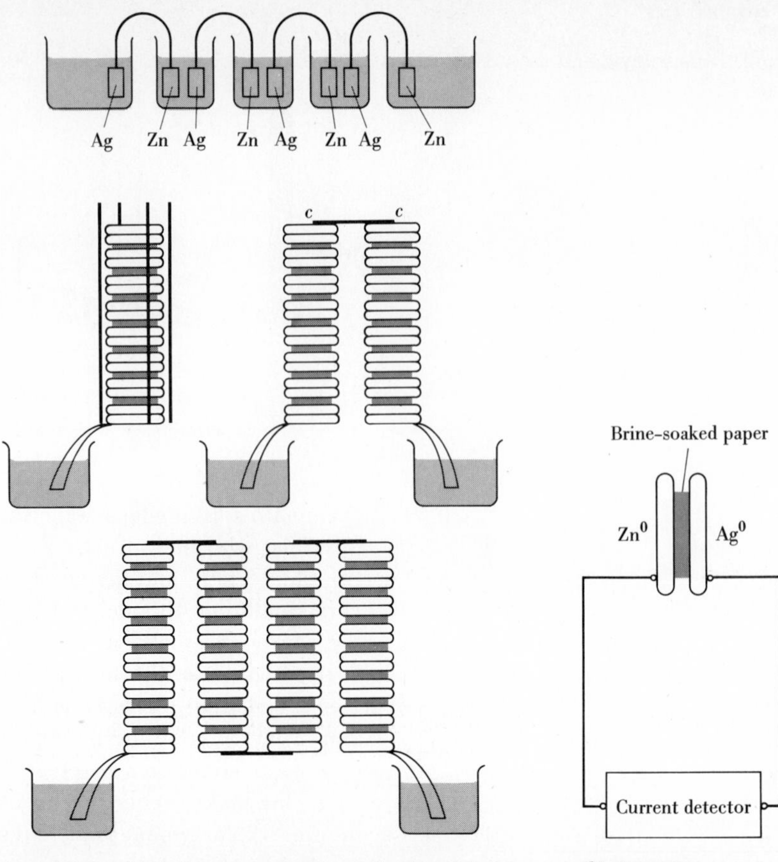

(*a*) *Diagrams of Volta's apparatus for the production of electric current*

(*b*) *One cell*

Early workers had no conception that ions carry an electric current through salt solution and that electrons are responsible in the metallic conductors. In fact, the concept of stable ions did not exist. There was no need for such a concept until scientists began to investigate the chemical effects produced when an electric current passed through salt solutions.

12.3 VOLTAIC CURRENT IS PRODUCED BY OXIDATION-REDUCTION REACTIONS

Volta had not noticed (or at least never recorded) that chemical reactions at the zinc and silver electrodes accompany the generation of current in his battery. It is well established today that the source of the current is a chemical reaction at each electrode surface. Figure 12.4 depicts what is equivalent to a cell in Volta's battery. The reaction at the silver electrode produces hydrogen gas and hydroxyl ions:

$$2HOH(l) + 2e^- \rightarrow H_2(g) + 2OH^-(aq) \tag{12.1}$$

FIGURE 12.4

The equivalent of one cell from Volta's pile.

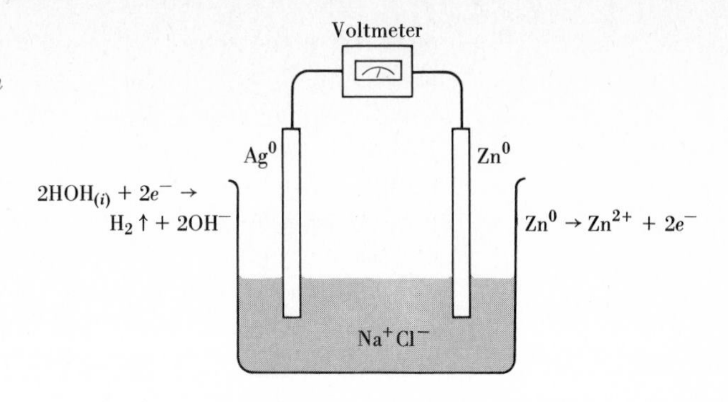

The zinc electrode is gradually consumed, producing zinc ions in the other reaction:

$$Zn(s) \rightarrow Zn^{2+}(aq) + 2e^- \tag{12.2}$$

The reaction described by Eq. (12.1) is a reduction reaction because electrons are gained by a reactant. The silver electrode is a cathode in Volta's battery; a *cathode* is an electrode at which reduction occurs. Equation (12.2) represents an oxidation reaction because electrons are lost by a reactant. The zinc electrode is an anode in Volta's battery; an *anode* is an electrode at which oxidation occurs. Oxidation and reduction reactions are always paired and occur simultaneously, often being referred to as *half reactions* to emphasize this fact. When they can occur at different points, their concerted action produces an electron current. One half reaction (reduction) pulls electrons, and the other half reaction pushes electrons. This is the principle of voltaic or galvanic cells. Because an electric current is produced by a chemical reaction in such cells, they are often called *electrochemical cells.*

12.4 ELECTRICAL UNITS

Before our discussion can proceed, we must discuss how quantities of electricity are defined, i.e., electrical units. The unit of electricity or electric charge the chemist uses most often is the *coulomb* (C). In terms of the charge on the electron it is a very large unit, the total charge on 6.24×10^{18} electrons being equal to 1 C. The *electric current* measures the quantity of charge passing through a conductor in unit time. If we choose the coulomb for the unit of charge and the second for the unit of time, the current is in coulombs per second, or *amperes* (A). When 1 C passes through a conductor in 1 s, the current is said to be 1 A. The current is also called the *amperage.*

The only other electrical unit we shall need is the *volt* (V), named for Volta. It is more difficult to define precisely at this level than the coulomb or the ampere. From ordinary usage we know that the voltage has to do with the force under which a current will flow. That is why the voltage of a cell is often called the *electromotive force* (emf). However, the voltage is not a force. It is a measure of the work required to move a quantity of charge from one point to another; nevertheless, it is convenient to think of the emf as the cause of an electric current, and ordinarily this will cause no errors. If 1 joule (J) (or 0.239 cal) must be expended in order to move a charge of 1 C from a point A to a point B, then an electric battery or other source of electric current must be used which has an emf of 1 V.

12.5 WHAT PRODUCES VOLTAGE

The voltage of an electrochemical cell depends only on the materials from which it is constructed. Volta's pile generates approximately 0.1 V for each cell in the pile. In the Daniell cell (Fig. 12.5*a*), a copper electrode is immersed in a 1 M $CuSO_4$ solution in one compartment of the cell; a zinc electrode is immersed in a 1 M $ZnSO_4$ solution in another compartment of the cell. The electrode compartments are separated by a porous membrane. When the circuit is closed and a current flows, Cu^{2+} ions are reduced at the copper electrode:

$$Cu^{2+}(1\ M) + 2e^- \rightarrow Cu^0(s)$$

and Zn is converted to Zn^{2+} ions at the zinc electrode:

$$Zn^0(s) \rightarrow Zn^{2+}(1\ M) + 2e^-$$

The direction of the electron flow is toward the copper electrode. Since reduction occurs at the copper electrode, it is the cathode and zinc is the anode. The emf of the Daniell cell is approximately 1.1 V.

When the zinc-electrode compartment is replaced by a silver-electrode compartment (Fig. 12.5*b*), copper becomes the anode and electrons flow away from the copper electrode; the emf is approximately 0.46 V.

To understand these observations, consider what might occur in a copper-electrode compartment with the cell circuit open so that no current can flow. We know that copper ions can be consumed (Fig. 12.5*a*, Daniell cell) or generated (Fig. 12.5*b*, copper-silver cell), depending on the nature of the opposing electrode. It is reasonable to suspect that in an open circuit when the copper electrode is not joined to another electrode, the copper surface can be involved in a reduction and oxidation reaction:

$$\text{Reduction: } Cu^{2+} + 2e^- \rightarrow Cu^0 \qquad (12.3)$$

$$\text{Oxidation: } Cu^0 \rightarrow Cu^{2+} + 2e^- \qquad (12.4)$$

Neither reaction can proceed to any great extent because individu-

FIGURE 12.5

Voltaic cells with copper electrodes.

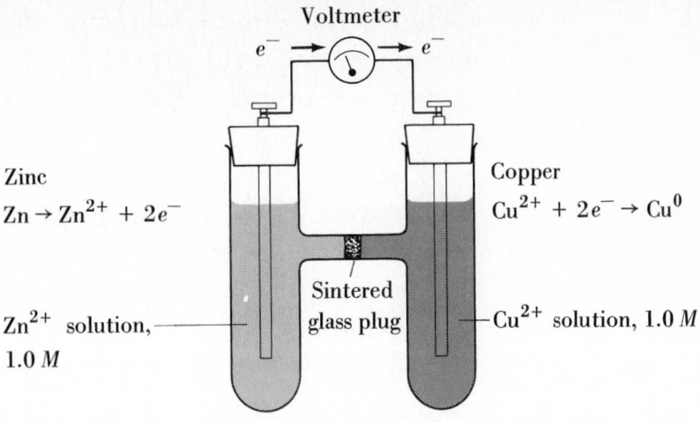

Cell reaction: $Cu^{2+} + Zn^0 \rightarrow Zn^{2+} + Cu^0$; $\mathcal{E}^0 = 1{,}099$ V

(*a*) *Copper is the cathode*

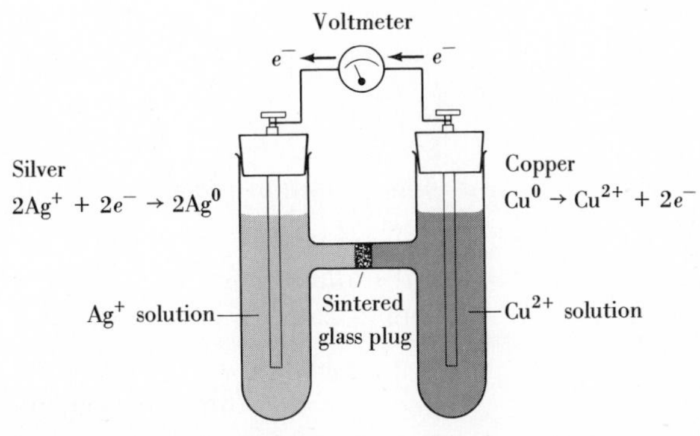

Cell reaction: $2Ag^+ + Cu^0 \rightarrow 2Ag^0 + Cu^{2+}$; $\mathcal{E}^0 = 0.46$ V

(*b*) *Copper is the anode*

ally each would generate a highly charged electrode, a result which is not observed. We conclude that in an open circuit the interface between copper metal and $CuSO_4$ solution quickly comes to equilibrium with respect to the reaction

$$Cu^{2+} + 2e^- \rightleftharpoons Cu^0$$

and no net reduction or oxidation is observed.

When the copper electrode is coupled with another electrode, it may become the anode or the cathode. With a zinc electrode, as in the Daniell cell, copper is the cathode because under these conditions Cu^{2+} ions have a greater affinity for electrons than Zn^{2+} ions do. As the reduction of Cu^{2+} occurs [Eq. (12.3)], the electrons are consumed at the copper electrode; they are replaced by the electrons flowing from the zinc electrode as they are given up by zinc atoms which are converted into Zn^{2+} ions. As a result, there is no buildup of charge at either electrode, and the electrode reactions can proceed until the copper-ion concentration becomes very low or until the zinc electrode is consumed.

When the copper electrode is coupled with a silver electrode (Fig. 12.5*b*), copper is the anode; Ag^+ ions have a greater affinity for electrons than Cu^{2+} ions. Electrons are generated at the copper surface and consumed at the silver surface. These concerted chemical reactions result in a continuous flow of electrons from the copper electrode to the silver electrode.

When the copper electrode is coupled with another identical copper electrode, no current flows because the identical interfaces in the circuit have equal affinities for electrons (see Fig. 12.6*a*). If we change the concentration of the $CuSO_4$ solution in one compartment, a current can be detected and will flow in the direction tending to equalize the $CuSO_4$ concentration in the two compartments. Cells which generate a current because of a difference in the *concentrations* of electrolyte in the two compartments are called *concentration cells* (see Fig. 12.6*b*).

Summary:

1. *Essential condition for voltaic cell:* two dissimilar interfaces in an electric circuit constitute a voltaic cell.

2. *Condition for high voltage:* the voltage depends on the difference in the affinity for electrons at the two interfaces; the greater this difference the higher the voltage.

3. *Condition for high amperage:* if the interface materials are good electric conductors, the amperage can be high.

FIGURE 12.6

Concentration differences cause a current flow.

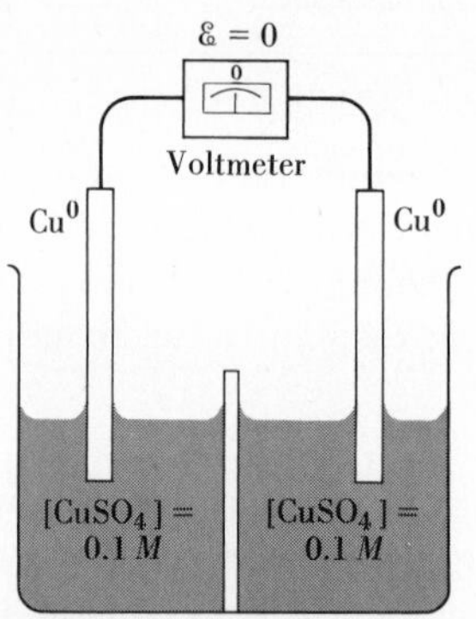

(*a*) *Identical interfaces*
When both electrodes are made from the same metal and both solutions are identical in composition and concentration no voltage is generated

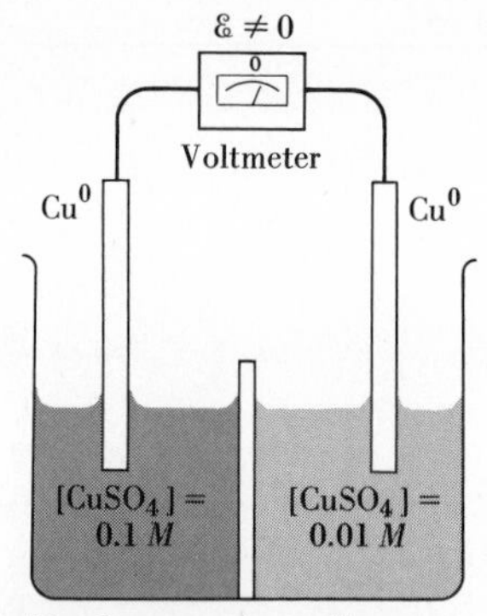

(*b*) *Concentration cell*
Although both electrodes are made from the same metal and both compartments contain a $CuSO_4$ *solution, a flow of current is detected because the solution concentrations and, hence, the electrode interfaces, are different*

12.6 OXIDATION-REDUCTION REACTIONS

Chemical reactions involving the transfer of electrons from one species to another produce electric currents in voltaic cells. Reactions in which electrons are transferred are called *oxidation-reduction* or *redox reactions.* In this section the rules for balancing the equations describing oxidation-reduction reactions are presented.

Oxidation is the loss of electrons in a chemical reaction. *Reduction* is the gain of electrons in a chemical reaction. In the reaction

$$Na^0(s) + \frac{1}{2}Cl_2{}^0(g) \rightarrow Na^+Cl^-$$

sodium is oxidized; i.e., sodium loses an electron, $Na^0 \rightarrow Na^+ + e^-$. Chlorine is reduced; i.e., chlorine gains an electron, $\frac{1}{2}Cl_2 + e^- \rightarrow Cl^-$. Oxidation is always accompanied by reduction; the loss of electrons (oxidation) by one species must always equal the gain of electrons (reduction) by another. In the sodium-chlorine reaction, sodium metal is called the reducing agent or *reductant* because it causes the reduction of chlorine; chlorine is called the oxidizing agent or *oxidant* because it causes the oxidation of sodium.

Oxygen gas is the most common oxidizing agent. All combustion reactions are oxidation-reduction reactions in which oxygen is the oxidizing agent:

$$H_2(g) + \frac{1}{2}O_2(g) \rightarrow H_2O$$

$$\underset{\text{Butane}}{C_4H_{10}(g)} + \frac{13}{2}O_2(g) \rightarrow 4CO_2(g) + 5H_2O$$

In all these reactions O_2 changes its oxidation number from 0 to -2. This does not necessarily mean that each oxygen atom gains two electrons during reaction. The oxidation number is a convenient formalism, but only when ionic compounds are formed can we relate it to the actual number of electrons transferred from one species to another. In the reaction

$$2Na^0(s) + Br_2(l) \rightarrow 2NaBr(s)$$

the product NaBr, sodium bromide, is an ionic solid; x-ray data indicate a transfer of the electrons from sodium atoms to bromine atoms. In the reaction

$$2H_2(g) + O_2(g) \rightarrow 2H_2O$$

the product is covalent. Since oxygen is more electronegative than hydrogen (Table 5.3), the electrons of the covalent bond have a higher probability of being at the oxygen atom than at the hydrogen atoms. There is no permanent transfer of electrons from hydrogen to oxygen. Nevertheless, we say that the oxidation states of oxygen and hydrogen are -2 and $+1$, respectively. As we shall see, this formalism is convenient for balancing redox equations.

To determine whether a reaction is an oxidation-reduction reaction we must show that there is a change in oxidation number or state (see Tables 2.4 and 2.5 for the common oxidation numbers). If there is no change in oxidation number for any of the species, the reaction is not a redox reaction. Consider the reaction between copper metal and concentrated nitric acid, HNO_3. The unbalanced equation is

$$Cu^0 + HNO_3 \rightarrow Cu(NO_3)_2 + NO_2(g) + H_2O$$

From Tables 2.4 and 2.5, the oxidation numbers of hydrogen and oxygen are +1 and −2, respectively. Their oxidation numbers do not change in this reaction. All species in their elemental states have oxidation numbers of 0; Cu^0 has an oxidation number of 0. The product $Cu(NO_3)_2$ contains copper in the +2 oxidation state. To deduce this we note that the species $Cu(NO_3)_2$ has no net charge; each NO_3^- ion has an oxidation number of −1. For $Cu(NO_3)_2$ to be electrically neutral copper must be in the +2 state.

HNO_3 is an electrically neutral molecule; hydrogen is in the +1 state and oxygen is in the −2 state. To have overall neutrality nitrogen must be in the +5 state; the net formal charge on one hydrogen, one nitrogen, and three oxygen atoms must be zero:

$$+1 + x + (-6) = 0$$
$$x = +5$$

In a neutral molecule of nitrogen dioxide, NO_2, nitrogen is in the +4 state; the net formal charge on one nitrogen and two oxygen atoms must be zero:

$$x + (-4) = 0$$
$$x = +4$$

Therefore, when HNO_3 reacts and produces NO_2, the oxidation number of nitrogen changes from +5 to +4, a formal gain of one electron; HNO_3 is the oxidizing agent. Copper undergoes a change from 0 to +2, which is a loss of two electrons; copper is the reducing agent. Balancing an equation for a redox reaction means choosing stoichiometric coefficients so that the loss of electrons equals the gain of electrons.

EXAMPLE 12.1 Balance the equation for the oxidation of copper by HNO_3:

$$Cu + HNO_3 \rightarrow Cu(NO_3)_2 + NO_2 + H_2O$$

Solution: Each copper atom loses two electrons:

$$Cu^0 \rightarrow \overset{2+}{Cu}(NO_3)_2$$

and each nitrogen atom formally gains one electron:

$$H\overset{+5}{N}O_3 \rightarrow \overset{+4}{N}O_2$$

To balance the loss and the gain we must use a Cu/N ratio of 1:2:

Preliminary balance:

$$\underline{1}Cu^0 + \underline{2}HNO_3 \rightarrow \underline{1}Cu(NO_3)_2 + \underline{2}NO_2 + ?H_2O$$

There are two remaining problems. HNO_3 acts as an oxidizing agent, and in this role it produces two molecules of NO_2 for each copper atom oxidized; but HNO_3 also supplies two NO_3^- ions for each Cu^{2+} ion formed. Therefore, for each copper atom oxidized we require four molecules of HNO_3, two for oxidation and two for electric neutrality. The equation is now

$$\underline{1}Cu^0 + (\underline{2+2})HNO_3 \rightarrow \underline{1}Cu(NO_3)_2 + \underline{2}NO_2 + ?H_2O$$

To determine the stoichiometric coefficient of H_2O we count the oxygen atoms on both sides of the equation; they must be equal in a completely balanced equation (Dalton's atomic theory). On the left side oxygen is present only in HNO_3. Four molecules of HNO_3 contain 12 oxygen atoms. On the right side, not counting H_2O, which does not yet have a coefficient, there are 10 oxygen atoms:

Left side: $4 \times 3 = 12$
Right side (not counting O in H_2O): $3 \times 2 + 2 \times 2 = 10$

The right side is deficient by 2 oxygen atoms, and therefore the stoichiometric coefficient of H_2O is 2. The completely balanced chemical equation is

$$Cu + 4HNO_3 \rightarrow Cu(NO_3)_2 + 2NO_2 + 2H_2O$$

Check: To ensure that no errors are present count the number of hydrogen atoms on both sides: their number should be equal.

Rules for balancing redox equations

1. Write the unbalanced equation including all reactants and products.

2. Determine the changes in oxidation state.

3. Determine the ratio of the stoichiometric coefficients for the oxidizing and reducing agents only and make a preliminary balance.

4. Inspect the equation and decide whether the oxidant or reductant has any other chemical role. If it does, its stoichiometric coefficient must be adjusted to account for this role. In first learning to balance redox equations it is wiser to write the formula of a species twice if it has two roles.

5. Balance the oxygen atoms.

6. Check the balance by counting the hydrogen atoms.

EXAMPLE 12.2 Potassium permanganate reacts with HCl to produce manganous chloride, chlorine gas, potassium chloride, and water. (*a*) What is the oxidant? (*b*) What is the reductant? (*c*) Write the balanced chemical equation for the reaction.

Solution: In answering these questions we shall follow the rules for balancing redox equations given above.

1. Write the unbalanced equation; be sure to include all reactants and products:

$$KMnO_4 + HCl \rightarrow MnCl_2 + Cl_2 + KCl + H_2O$$

2. Using the assignments in Tables 2.4 and 2.5, we find that the only changes in oxidation state involve

$KMnO_4$(Mn in +7 state) $\rightarrow$
$MnCl_2$(Mn in +2 state) 5 electrons gained per Mn

and

HCl(Cl in −1 state) $\rightarrow$
Cl_2(Cl in 0 state) 1 electron lost per Cl

Answer: (*a*) $KMnO_4$ is the oxidant (gains electrons). (*b*) HCl is the reductant (loses electrons).

3. A preliminary balance is obtained by using the $KMnO_4$/HCl ratio of 1:5. This will make the electron gain by $KMnO_4$ equal to the electron loss by HCl. This also means that the coefficients of the products, $MnCl_2$ and Cl_2, must be 1 and $\frac{5}{2}$ respectively.

$$\underline{1}KMnO_4 + \underline{5}HCl \rightarrow \underline{1}MnCl_2 + \underline{\tfrac{5}{2}}Cl_2 + ?KCl + ?H_2O$$

4. The coefficients of the species involved in the change in oxidation state are underlined.

Since the coefficient of the reactant $KMnO_4$ is 1, the coefficient of product KCl must also be 1; these are the only compounds which contain potassium. We can now write

$$\underline{1}KMnO_4 + \underline{5}HCl \rightarrow \underline{1}MnCl_2 + \underline{\tfrac{5}{2}}Cl_2 + 1KCl + ?H_2O$$

Note that HCl is the reductant since it produces Cl_2, but HCl must also supply Cl^- ions for manganous chloride and potassium chloride. This requires 3 additional moles of HCl:

$$1KMnO_4 + (\underline{5}HCl + 3HCl) \rightarrow \underline{1}MnCl_2 + \underline{\tfrac{5}{2}}Cl_2 + 1KCl + ?H_2O$$

5. There are four oxygen atoms on the left side and none on the right (ignoring water, for which the coefficient has not been chosen). Therefore the stoichiometric coefficient of H_2O must be 4.

Answer (*c*): The final equation is

$$KMnO_4 + 8HCl \rightarrow MnCl_2 + \tfrac{5}{2}Cl_2 + KCl + 4H_2O$$

6. The number of hydrogen atoms (eight) on both sides is equal.

The above equation is easier to balance if we consider only the species involved in the chemical reaction. The reaction between $KMnO_4$ and HCl is usually carried out in water. When $KMnO_4$ dissolves in this solvent, it is present as K^+ and MnO_4^- ions; HCl is present as H^+ (or H_3O^+) and Cl^- ions. The products $MnCl_2$ and KCl are also ionic. Taking these facts into consideration in writing the equation, we have

$$K^+ + MnO_4^- + H^+ + Cl^- \rightarrow Mn^{2+} + 2Cl^- + Cl_2 + K^+ + Cl^- + H_2O$$

Comparison of reactants and products now enables us to simplify the equation. Since K^+ ion appears on both sides of the equation, it does not undergo any net chemical change and is not written in the *simple ionic equation* for the reaction. Since Cl^- ion is a reactant, the presence of Cl^- in the product does not represent a chemical change and in the simple ionic equation Cl^- is not written on the product side. The simple ionic equation is

$$MnO_4^- + H^+ + Cl^- \rightarrow Mn^{2+} + Cl_2 + H_2O$$

To balance the equation we use the rules given above, but we do not have to contend with the double role of HCl; we only consider its role as the reductant. The ratio of the coefficients for the oxidant, MnO_4^-, and the reductant, Cl^-, remains the same at 1:5. A preliminary balance of the equation gives

$$\underline{1}MnO_4^- + ?H^+ + \underline{5}Cl^- \rightarrow \underline{1}Mn^{2+} + \underline{\tfrac{5}{2}}Cl_2 + ?H_2O$$

Examination of the preliminary balance reveals that the number of oxygen atoms on the reactant side is fixed at 4 because in the reactants oxygen occurs only in MnO_4^-, the coefficient of which is fixed. Therefore, the coefficient of H_2O must be 4. This in turn means that the number of H^+ ions on the reactant side must be 8. The final equation is

$$MnO_4^- + \underline{8}H^+ + \underline{5}Cl^- \rightarrow Mn^{2+} + \underline{\tfrac{5}{2}}Cl_2 + \underline{4}H_2O$$

To check the final balance of a simple ionic equation sum up the ionic charge on both sides of the equation. If it is the same, the equation is balanced. In this case the charge on both sides is +2. On the reactant side we have $-1+8-5$, or $+2$, and on the product side we have simply +2.

12.7 CHEMICAL EQUIVALENT WEIGHTS

When the stoichiometry of a chemical reaction is known, the weights of reactants which will combine can be calculated. These weights are "chemically" equal. For example, the stoichiometry of the reaction of H_2SO_4 with KOH is

$$H_2SO_4 + 2KOH \rightarrow K_2SO_4 + 2H_2O$$

Since 1 mol, or 98 g, of H_2SO_4 reacts with 2 mol, or 112 g, of KOH, 98 g of H_2SO_4 and 112 g of KOH are equivalent chemically. Furthermore, since 1 mol, or 98 g, of H_2SO_4 will produce 2 mol, or 36 g, of H_2O, 98 g of H_2SO_4 and 36 g of H_2O are equivalent chemically. Also, one could say that 49 g of H_2SO_4 and 18 g of H_2O are equivalent chemically. It is obvious that unless some standard is chosen, the term *chemical equivalent* is arbitrary. If, however, a weight of a particular element is established as standard equivalent weight, then the weight of any material which can combine with, produce, or be produced from an equivalent weight of the standard material will also be an equivalent weight. Equivalent weight is abbreviated equiv.

Hydrogen has been selected as the standard material. Its gram-atomic weight, 1.008 g, is chosen as the standard equivalent weight. The weight of any other element or compound which reacts with, is produced from, or produces 1.008 g of hydrogen is also an equivalent weight. Consider the reaction:

$$H_2 + Cl_2 \rightarrow 2HCl$$

Since 35.45 g of chlorine reacts with 1.008 g of hydrogen, a chemical equivalent weight of chlorine is 35.45 g. By considering other reactions, it will be seen that, ordinarily, 8 g of oxygen, 19 g of fluorine, and 23 g of sodium are chemical equivalent weights based on the definition of 1.008 g of hydrogen.

However, it is important to realize that the chemical reaction must be known if one is to calculate equivalent weights. For example, the gram equivalent weight of oxygen is ordinarily taken as 8 g. This is demonstrated by the reaction

$$H_2 + \tfrac{1}{2}O_2 \rightarrow H_2O$$

In the following reaction, the chemical equivalent weight of oxygen is 16.00 g:

$$H_2 + O_2 \rightarrow H_2O_2$$

The weight of oxygen which combines with 1.008 g of hydrogen is 16 g. The point of this discussion is that since many elements and compounds exhibit more than one equivalent weight, the stoichiometry of the reaction must be known before the calculation of an equivalent weight is begun.

When hydrogen is involved in an oxidation-reduction reaction, it gains or loses one electron per atom, or Avogadro's number N_A of electrons per g atom. In the reaction

$$3H_2 + Fe_2O_3 \rightarrow 3H_2O + 2Fe$$

one atom of hydrogen formally loses one electron. However, in the reaction

$$2HCl + Zn \rightarrow ZnCl_2 + H_2$$

one atom of hydrogen gains one electron. We see that one hydrogen atom is associated with an electron change of +1 or −1. One gram-atomic weight of hydrogen (the chosen standard equivalent weight) is associated with a change of $+N_A$ or $-N_A$ electrons.

For oxidation-reduction reactions, it is practical to calculate the gram equivalent weight by dividing the gram formula weight by the number of moles of electrons gained or lost by one gram formula weight:

$$\text{Equiv} = \frac{\text{formula weight}}{\text{moles of electrons gained or lost}}$$

In the reaction

$$H_2S + HNO_3 \rightarrow H_2SO_4 + NO_2 + H_2O$$

the equivalent weight of H_2S is 34/8 = 4.25 g, and the equivalent weight of HNO_3 is 63/1 = 63 g. In the reaction

$$As_2S_3 + HNO_3 \rightarrow H_3AsO_4 + S + NO$$

the equivalent weight of As_2S_3 is 246/10 = 24.6 g, since each arsenic atom and each sulfur atom lose two electrons. The loss per formula weight is 10 electrons. The equivalent weight of HNO_3 is 63/3 = 21 g. The equivalent weight of HNO_3 depends upon the reaction. Many other compounds and elements fall in this same category.

As further examples, consider the reactions

$$K_2Cr_2O_7 + 3H_2S + 4H_2SO_4 \rightarrow 3S + Cr_2(SO_4)_3 + K_2SO_4 + 7H_2O \quad (12.5)$$

$$NaOH + Br_2 \rightarrow NaBrO_3 + NaBr + H_2O \quad (12.6)$$

In Eq. (12.5), the equivalent weights of $K_2Cr_2O_7$ and H_2S are calculated in the usual way. The equivalent weight of $K_2Cr_2O_7$ is 294/6 = 49 g. The equivalent weight of H_2S is 34/2 = 17 g. However, the calculation of the equivalent weight of H_2SO_4 for this reaction requires some consideration. Since we have already established that the equivalent weight of H_2S is 17 g, we must say that the equivalent weight of H_2SO_4 is that weight of H_2SO_4 which will react with 17 g of H_2S (or 49 g of $K_2Cr_2O_7$). Whereas it was not necessary to balance the equation to determine the equivalent weights of the oxidizing and reducing agents, it is necessary to balance the equation to determine the equivalent weight of H_2SO_4 since it is not involved in the oxidation-reduction part of the reaction. The balanced equation shows that the 4 mol of H_2SO_4 react for every 3 mol of H_2S. By the following analysis, we calculate the equivalent weight of H_2SO_4

to be 65.33 g:

$$\frac{17\text{ g } \cancel{H_2S}}{1\text{ equiv}} \times \frac{1\text{ mol } \cancel{H_2S}}{34\text{ g } \cancel{H_2S}} \times \frac{4\text{ mol } \cancel{H_2SO_4}}{3\text{ mol } \cancel{H_2S}} \times \frac{98\text{ g } H_2SO_4}{1\text{ mol } \cancel{H_2SO_4}} = \frac{65.33\text{ g } H_2SO_4}{1\text{ equiv}}$$

In Eq. (12.6) the concept of equivalent weight loses its usefulness, and it is better to use mole-mole relationships. However, the equivalent weights of the oxidizing and reducing agents can be calculated. In this case, bromine is both the oxidizing agent and the reducing agent:

Reducing agent: $Br_2 \rightarrow 2BrO_3^-$ $\quad$ $\text{equiv} = \frac{160\text{ g}}{10} = 16\text{ g}$

Oxidizing agent: $Br_2 \rightarrow 2Br^-$ $\quad$ $\text{equiv} = \frac{160\text{ g}}{2} = 80\text{ g}$

The equivalent weight of NaOH can be calculated from the balanced equation. It is the weight associated with 16 g of the reducing agent, bromine, or 80 g of the oxidizing agent, bromine.

Normality

The normality N is a designation of the concentration of a solution. It is defined as the number of equivalents of solute per liter of solution.

EXAMPLE 12.3 In terms of Eq. (12.5), what is the normality of a solution which contains 100 g of $K_2Cr_2O_7$ per liter?

Solution: By Eq. (12.5) the equivalent weight of $K_2Cr_2O_7$ is 49 g since in the reaction with H_2S this compound gains 6 mol of electrons per formula weight:

$$\text{Equiv} = \tfrac{294}{6} = 49\text{ g}$$

The normality can now be calculated:

$$N = \text{equiv/l of solution}$$

$$\frac{100\text{ g } \cancel{K_2Cr_2O_7}}{1\text{ l}} \times \frac{1\text{ equiv}}{49\text{ g } \cancel{K_2Cr_2O_7}} = 2.04\text{ equiv/l}$$

$$N = 2.04$$

EXAMPLE 12.4 A solution containing 15 g of H_2SO_4 per 75 ml of solution is used to neutralize a KOH solution:

$$H_2SO_4 + 2KOH \rightarrow K_2SO_4 + 2HOH$$

Calculate the normality of the acid solution.

Solution: The equivalent weight of sulfuric acid is one-half the gram-molecular weight because each mole supplies 2 mol of hydrogen ion.

$$\text{equiv} = \tfrac{98}{2} = 49 \text{ g}$$

The equivalent weight is known and it is possible to calculate the normality:

$$\frac{15 \text{ g } \cancel{H_2SO_4}}{75 \cancel{\text{ml}}} \times \frac{\text{equiv}}{49 \text{ g } \cancel{H_2SO_4}} \times \frac{1000 \cancel{\text{ml}}}{1} = 4.08 \text{ equiv/l}$$

$$N = 4.07$$

12.8 ELECTROLYSIS

As the knowledge of Volta's battery spread through Europe, many scientists began investigating *electrolysis*, chemical reactions caused by passing an electric current through substances. Cells in which chemical reactions are caused by passage of an electric current are called *electrolytic cells*. Among the first reports was a study

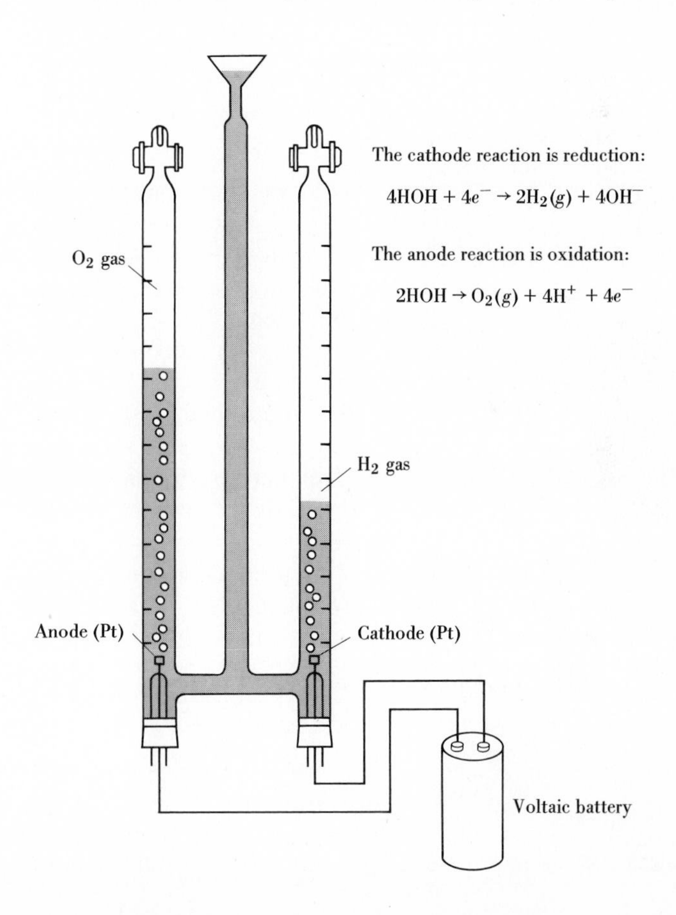

FIGURE 12.7

Electrolysis of water.

Michael Faraday.

Michael Faraday at work in the Royal Institution Laboratory. (Courtesy of the Edgar Fahs Smith Collection, Van Pelt Library, University of Pennsylvania, Philadelphia.)

by two Englishmen, Nicholson and Carlisle, describing the electrolysis of water, which produced hydrogen and oxygen gases (1800) (see Fig. 12.7). At about the same time Sir Humphry Davy isolated the very active metals sodium and potassium by electrolysis of their molten hydroxides. In addition, he similarly obtained all the alkaline-earth metals except beryllium and radium. The science of chemistry was given a great impetus as the number of known elements rapidly increased.

Davy's work in itself represents a very important contribution to science, but it has often been said that his greatest discovery was Michael Faraday (1791–1867). Faraday apprenticed himself to Davy in 1813 and began general investigations in chemistry and physics. He initiated experiments in electricity and continually returned to this inquiry after excursions into other fields. He discovered quantitative relationships between the electric charge and the weights of substances consumed and produced by electrolysis. These findings are summarized in Faraday's first and second laws of electrolysis.

First law: The mass of a substance involved in an electrode reaction is directly proportional to the quantity of charge, or electricity, which passes through the solution.

Second law: The ratio masses of different substances produced during electrolysis in the same cell are equal to the ratio of their equivalent weights.

Today we know that these are not independent laws. Faraday performed his electrolysis experiments before the electron was identified. With the concept of the electron as the entity common to all electrolysis reactions we can reduce Faraday's electrolysis laws to a single statement.

Consider the electrolysis of a sulfuric acid solution. Figure 12.7 depicts an electrolytic cell designed so that gases produced at electrode surfaces can be separately collected for volume measurement. Essentially, there are two inert platinum-foil electrodes connected to a voltaic battery. When current is passing, hydrogen gas is evolved at the cathode and oxygen gas at the anode. For each coulomb passed through the cell 1.0447×10^{-5} g of hydrogen forms. Therefore, to produce exactly 1 gram-atomic weight, or 1.00797 g, of hydrogen 96,487 C is required, as can be seen from a simple calculation:

$$\frac{1\text{ C}}{1.0447 \times 10^{-5}\text{ g H}_2} \times 1.00797\text{ g H}_2 = 96{,}487\text{ C}$$

Hydrogen gas is often produced at the cathode during the electrolysis of aqueous solutions of many mineral acids and salts. It is always found that 96,487 C produces exactly 1.00797 g of this gas. If we remember that a gram atom of hydrogen contains N_A (6.02×10^{23})

hydrogen atoms, we can calculate the charge on the electron by assuming that one electron reacts with one hydrogen ion in the electrolysis:

$$H^+ + e^- \rightarrow \tfrac{1}{2}H_2(g) \tag{12.7}$$

By this stoichiometry the ratio of H^+ ions to electrons is 1:1. Since 96,487 C is required to neutralize 6.0225×10^{23} hydrogen ions, 96,487 C must be the charge on 6.0225×10^{23} electrons. The electronic charge ϵ is

$$\epsilon = \frac{96{,}487 \text{ C}}{6.0225 \times 10^{23} \text{ electrons}}$$

$$= 1.6021 \times 10^{-19} \text{ C/electron}$$

Whenever 1.008 g of H_2 gas is produced in the electrolysis of H_2SO_4 solution, 8.00 g of oxygen gas is also produced; 8.00 g of oxygen gas must also be associated with Avogadro's number of electrons. The anode reaction written for one electron is

$$\tfrac{1}{2}HOH \rightarrow \tfrac{1}{4}O_2 + H^+ + e^-$$

Avogadro's number, or 1 mol, of electrons is generated along with each $\frac{1}{4}$ mol, or 8.00 g, of oxygen.

The weight of a substance produced by passage of 96,487 C at an electrode during electrolysis is an electrochemical equivalent weight. It is the weight of a chemical substance which is associated with Avogadro's number of electrons. In the electrolysis of sulfuric acid solution 1.008 g of hydrogen and 8.00 g of oxygen are *electrochemical equivalent weights.* The unit of charge, 96,487 C, is very significant in electrochemistry; it is called the faraday, $\mathcal{F}$; 1 $\mathcal{F}$ = 96,487 C.

Table 12.1 lists the electrolysis products of five electrolytes. Comparison of columns 3 and 4 shows that the weight of each product is directly proportional to the charge passing through the cell. This is in accordance with Faraday's first law of electrolysis. In four cases (HCl, H_2SO_4, dilute NaCl, and concentrated NaCl) H_2 gas is the cathode product; in each case 1.008 g of H_2 is produced by 1 $\mathcal{F}$. In three cases (HCl, concentrated NaCl, and molten NaCl) Cl_2 gas is the anode product; in each instance 35.5 g of Cl_2 is produced by 1 $\mathcal{F}$. By comparison of columns 3 and 5 we see that 96,487 C, or 1 $\mathcal{F}$, produces one electrochemical equivalent weight of each product; column 5 shows the cell reactions as one-electron or 1-$\mathcal{F}$ reactions, and therefore the amount of each substance involved must be one electrochemical equivalent weight. This is in accordance with Faraday's second law. Faraday's laws can be combined in a single, more convenient statement: *The passage of 96,487 C through an electrolyte solution will decompose, liberate, or deposit one electrochemical equivalent of material at each electrode.* For the pur-

TABLE 12.1: ELECTROLYSIS PRODUCTS IN AQUEOUS SOLUTION†

Electrolyte	*Cathode product* (*1*)	*Anode product* (*2*)	*Weight produced*, g: *By 96,487 C* (*3*)	*Weight produced*, g: *By 192,974 C* (*4*)	*Cell reaction based on 1 equiv* (*5*)
$AgNO_3$	Ag	O_2	108 Ag, 8 O_2	216 Ag, 16 O_2	$AgNO_3(aq) + \frac{1}{2}HOH \rightarrow Ag^0(s) + \frac{1}{4}O_2(g) + HNO_3(aq)$
HCl	H_2	Cl_2	1.008 H_2, 35.5 Cl_2	2.016 H_2, 71.0 Cl_2	$HCl(aq) \rightarrow \frac{1}{2}H_2(g) + \frac{1}{2}Cl_2(g)$
H_2SO_4	H_2	O_2	1.008 H_2, 8 O_2	2.016 H_2, 16 O_2	$\frac{1}{2}HOH \rightarrow \frac{1}{2}H_2(g) + \frac{1}{4}O_2(g)$
$CuSO_4$	Cu	O_2	63.5 Cu, 8 O_2	127.0 Cu, 16 O_2	$\frac{1}{2}CuSO_4(aq) + \frac{1}{2}HOH \rightarrow \frac{1}{2}H_2SO_4(aq) + \frac{1}{2}Cu^0(s) + \frac{1}{4}O_2(g)$
Dilute NaCl	H_2	O_2	1.008 H_2, 8 O_2	2.016 H_2, 16 O_2	$\frac{1}{2}HOH \rightarrow \frac{1}{2}H_2(g) + \frac{1}{4}O_2(g)$
Concentrated NaCl	H_2	Cl_2	1.008 H_2, 35.5 Cl_2	2.016 H_2, 71.0 Cl_2	$\frac{1}{2}HOH + NaCl(aq) \rightarrow \frac{1}{2}H_2(g) + \frac{1}{2}Cl_2(g) + NaOH(aq)$
Molten pure NaCl	Na	Cl_2	23 Na, 35.5 Cl_2	46 Na, 71.0 Cl_2	$NaCl(aq) \rightarrow Na^0(s) + \frac{1}{2}Cl_2(g)$

† All electrodes are platinum.

pose of calculations it is important to keep in mind the relationships between coulombs, ampere-seconds, the faraday, and the electrochemical equivalent weight:

$$96{,}487\ \text{C} = 96{,}487\ \text{A-s} = 1\ \mathscr{F} = 1\ \text{equiv}$$

These equations indicate that one equivalent weight of any substance is chemically equal to Avogadro's number of electrons, which has a charge of approximately 96,500 C or A-s.

Table 12.1 reveals that the electrolysis products depend on the electrolyte concentration. Before electrochemical calculations can be performed, the stoichiometry of the reactions must be determined by experiment or from published results.

The electrolysis of NaCl is extremely important from an industrial viewpoint. It may yield the gases H_2, Cl_2, O_2, metallic sodium, and NaOH. The products, all of which have industrial value, are controlled by regulating the NaCl concentration. Electrolysis is used in the production of many elements and in plating and refining metals. Copper, aluminum, zinc, cadmium, and many other metals are produced by the electrolysis of salts.

EXAMPLE 12.5 What amperage is required to pass 1.5 $\mathscr{F}$ through a solution in 2 h?

Solution: Since 1 $\mathscr{F}$ equals 96,500 A-s, we need 1.5 × 96,500 A-s to pass in 2 × 3600 s. If we divide the number of ampere-seconds required by the number of seconds the current flows, we obtain the required amperage:

$$\frac{1.5 \times 96{,}500 \text{ A-}\cancel{\text{s}}}{2 \times 3600 \ \cancel{\text{s}}} = 20.1 \text{ A}$$

EXAMPLE 12.6 Calculate (*a*) the number of grams of Cl_2 liberated when 96,500 C passes through an HCl solution and (*b*) the volume of H_2 liberated at STP; 96,500 C will liberate 1 equiv of H_2 at the cathode and 1 equiv of Cl_2 at the anode:

Cathode: $H^+ + e^- \rightarrow \frac{1}{2}H_2$ Anode: $Cl^- \rightarrow \frac{1}{2}Cl_2 + e^-$

From these equations it is obvious that 1 g atom of hydrogen and 1 g atom of chlorine are each associated with Avogadro's number of electrons (96,500 C). Therefore, the equivalent weight of hydrogen is 1.01 g, and the equivalent weight of chlorine is 35.5 g.

Solution: (*a*) 96,500 C will liberate 1 equiv, or 35.5 g, of Cl_2. (*b*) 1.01 g of H_2 will be produced; this is the equivalent weight and one-half the molecular weight. The liberated gas occupies 22.4 l/mol at STP. Therefore, the volume of 1.01 g of H_2, which is $\frac{1}{2}$ mol, is 11.2 l at STP.

EXAMPLE 12.7 A current of 3.00 A is passed through an aqueous H_2SO_4 solution for 2.00 h. Calculate (*a*) the weight of oxygen liberated and (*b*) the volume of hydrogen liberated at STP.

Solution: The number of equivalents of oxygen produced at the anode will equal the number of equivalents of hydrogen at the cathode:

Anode: $\frac{1}{2}HOH \rightarrow \frac{1}{4}O_2 + H^+ + e^-$
Cathode: $H^+ + e^- \rightarrow \frac{1}{2}H_2$

The number of equivalents depends only on the number of coulombs or ampere-seconds; 96,500 C (or A-s) will produce 1 equiv each of oxygen and of hydrogen.

$$\text{Number of A-s} = 3 \text{ A } (2 \text{ h} \times 3600 \text{ s/h}) = 21{,}600 \text{ A-s}$$

$$\text{Number of equiv} = 21{,}600 \text{ A-s} \times \frac{1 \text{ equiv}}{96{,}500 \text{ A-s}} = 0.112 \text{ equiv}$$

Therefore, 0.224 equiv of O_2 and of H_2 is produced.

(*a*) From the anode reaction, it is obvious that only $\frac{1}{4}$ mol, or 8.00 g of O_2, is produced by Avogadro's number of electrons (96,500 C), and therefore 8.00 g is an equivalent weight of O_2.

$$\text{Weight of } O_2 = 0.224 \text{ equiv} \times \frac{8.00 \text{ g } O_2}{1 \text{ equiv}} = 1.79 \text{ g } O_2$$

(*b*) The volume of 1 equiv of H_2 at STP is 11.2 l (see Example 12.2).

$$\text{Volume of } H_2 = 0.224 \text{ equiv} \times \frac{11.2 \text{ l } H_2}{1 \text{ equiv}} = 2.51 \text{ l } H_2$$

EXAMPLE 12.8 Calculate the number of hours required for a current of 4.00 A to deposit 127 g of copper from a solution of $CuSO_4$. The cathode reaction is $Cu^{2+} + 2e^- \rightarrow Cu^0$.

Solution: The equation for the cathode shows that the equivalent is one-half the atomic weight because 2 times Avogadro's number of electrons produces 1 g atom of Cu^0.

$$\text{equiv} = \frac{63.5}{2} = 31.75 \text{ g}$$

The solution is obtained by first determining the number of ampere-seconds required to deposit 127 g of copper. Then we divide this number of ampere-seconds by the amperage to be used (4 A) to determine the time required to deposit 127 g of copper. First, convert the weight, 127 g of copper, to equivalents and then convert the equivalents to ampere-seconds:

$$127 \text{ g Cu} \times \frac{\text{equiv}}{31.75 \text{ g Cu}} \times \frac{96{,}500 \text{ A-s}}{\text{equiv}} = 386{,}000 \text{ A-s}$$

386,000 A-s is required to deposit 127 g copper, and the time required for 386,000 A-s to pass when the current is 4 A is

$$\frac{386{,}000 \text{ A-s}}{4 \text{ A}} \times \frac{1 \text{ h}}{3600 \text{ s}} = 26.8 \text{ h}$$

EXAMPLE 12.9 What number of electrons are involved in the electrolysis described in Example 12.8?

Solution: Since 127 g of copper represent 4 equiv (because an equivalent weight of copper is 31.75 g), 4 $\mathscr{F}$ passes through the cell. Each faraday represents 6.02×10^{23} electrons:

$$4 \mathscr{F} \times \frac{6.02 \times 10^{23} \text{ electrons}}{\mathscr{F}} = 2.41 \times 10^{24} \text{ electrons}$$

Therefore 2.41×10^{24} electrons are involved at each electrode. The solution could have been obtained in a single-line calculation:

$$127 \text{ g Cu} \times \frac{1 \mathscr{F}}{31.75 \text{ g Cu}} \times \frac{6.02 \times 10^{23} \text{ electrons}}{1 \mathscr{F}} = 2.41 \times 10^{24} \text{ electrons}$$

EXAMPLE 12.10 If 96.5 C is required to produce 0.0120 g of metallic magnesium by electrolysis of a molten magnesium salt, what is the equivalent weight of magnesium?

Solution: We know that 0.0120 g magnesium is produced per 96.5 C. We must calculate the grams of magnesium produced per 96,500 C because this quantity of charge produces 1 equiv of a substance.

$$\frac{0.0120 \text{ g Mg}}{96.5 \cancel{\text{C}}} \times \frac{96{,}500 \cancel{\text{C}}}{1 \text{ equiv}} = \frac{12.0 \text{ g Mg}}{1 \text{ equiv}}$$

Equiv wt = 12.0 g

12.9 REVERSIBLE AND IRREVERSIBLE CELLS

When a voltaic cell is no longer capable of producing a voltage, it is said to be dead. This means either that the electrochemical reactions responsible for generating the voltage have gone to completion or have come to equilibrium. If a cell is dead because the electrochemical reactions have come to equilibrium, it is possible to regenerate, or recharge, it. The electrochemical reactions in such a cell are reversed when a slightly higher external voltage is connected in opposition, so that the current direction is the reverse of the direction during spontaneous discharge. In the Daniell cell (Fig. 12.5*a*) electrons flow toward the copper electrode during the spontaneous discharge of the cell. The emf of the cell is 1.099 V; the reactions which occur are shown. When we apply an external emf slightly greater than 1.099 V, the electrons flow toward the zinc electrode and the electrochemical reactions are reversed:

Cathode: $Zn^{2+} + 2e^- \rightarrow Zn^0$
Anode: $Cu^0 \rightarrow Cu^{2+} + 2e^-$
Cell: $Zn^{2+} + Cu^0 \rightarrow Zn^0 + Cu^{2+}$

All substances consumed during the spontaneous operation of the cell are regenerated. Since we can recharge a dead Daniell cell, it is said to be *reversible*.

TABLE 12.2: ELECTROCHEMICAL REACTIONS IN VOLTA'S PILE

Spontaneous electrode reactions	*Reactions under influence of opposing external emf*
Zn electrode (anode): $\frac{1}{2}Zn^0 \rightarrow \frac{1}{2}Zn^{2+} + e^-$	Zn electrode (cathode): $H_2O + e^- \rightarrow \frac{1}{2}H_2 + OH$
Ag electrode (cathode): $H_2O + e^- \rightarrow \frac{1}{2}H_2 + OH^-$	Ag electrode (anode): $Cl^- \rightarrow \frac{1}{2}Cl_2 + e^-$

Volta's pile goes dead because the electrochemical reactions go to completion. When an external emf is applied to Volta's pile and the current direction is reversed, the reactions at the electrode surfaces are *not* reversed. Instead, completely different reactions occur. It is obvious from Table 12.2 that Volta's pile cannot be regenerated or recharged by the application of an external emf; it is therefore said to be *irreversible*.

12.10 EXAMPLES OF IMPORTANT VOLTAIC CELLS

The familiar dry cell is depicted in Fig. 12.8. Dry cells are discarded when they are dead. The spontaneous reactions given in Fig. 12.5 are irreversible. A good dry cell generates approximately 1.6 V.

In the common lead storage battery used in automobiles each cell consists of a lead metal plate (anode) and a lead dioxide plate (cathode) immersed in a 15% aqueous sulfuric acid solution. A good lead storage battery yields 2 V per cell. The common automobile battery has six cells and can generate 12 V. The lead storage battery can be charged when it is dead because the electrode reactions responsible for the emf are reversible. The anode reaction is

$$Pb^0 + H_2SO_4 \rightarrow PbSO_4(s) + 2e^- + 2H^+$$

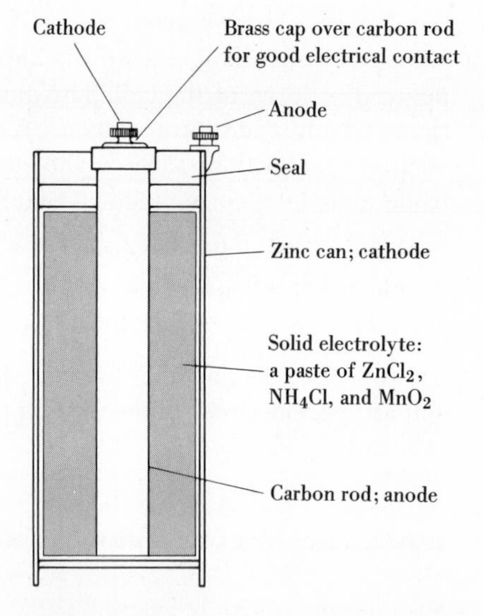

FIGURE 12.8

Leclanche dry cell.

Cathode reaction:

$$2NH_4^+ + MnO_2 + 2e^- \rightarrow 2NH_3 + MnO + H_2O$$

Anode reaction: $Zn^0 \rightarrow Zn^{2+} + 2e^-$

Cell reaction:

$$2NH_4^+ + MnO_2 + Zn^0 \rightarrow 2NH_3 + MnO + H_2O + Zn^{2+}$$

As $PbSO_4$ is formed, it adheres to the lead plate. The cathode reaction is

$$PbO_2 + H_2SO_4 + 2e^- \rightarrow PbSO_4 + 2OH^-$$

The net cell reaction is

$$Pb^0 + PbO_2 + 2H_2SO_4 \rightarrow 2PbSO_4 + 2HOH$$

It is obvious that H_2SO_4 is consumed during the spontaneous operation of a lead storage battery. This causes an appreciable decrease in the density of the sulfuric acid solution, a fact used to determine the condition of the battery.

Another type of cell which has received a great deal of attention is the fuel cell. With these cells it is possible to avoid the waste of energy which occurs when the heat produced by the combustion of coal or other fuels is used to do work. Fuel cells produce energy from the same oxidation-reduction reaction which occurs in combustion. The fuel-cell oxidant is usually air, and the reductants are ordinary fuels like coal, oil, hydrogen, propane, and butane. What is the advantage of a fuel cell?

It is the *heat* of a combustion reaction which is used as energy to drive the most common high-power engines, e.g., gasoline, diesel, and steam engines. From thermodynamic theory we know that it is impossible for a heat engine to convert heat energy completely into work. Some heat is always lost to the surroundings. Under ordinary operating conditions the theoretical limit of heat conversion is about 35 percent, and in practice it is much less. If the combustion reaction could be harnessed as the electrochemical reaction of a cell, we could convert chemical energy directly into work without the intermediacy of heat. There is no theoretical limit to the conversion of chemical energy into work, and hence were it not for frictional losses, we could convert chemical energy completely into useful work.

Consider the combustion of coal (graphite):

$$C(\text{graphite}) + O_2(g) \rightarrow CO_2(g) \qquad \Delta G^\circ = -94 \text{ kcal}$$

The chemical potential-energy change which accompanies this reaction is 94 kcal per mole of carbon. It is an oxidation-reduction reaction, and it has been found possible to separate the oxidation half reaction from the reduction half reaction (see Fig. 12.9).

The carbon or coal anode is oxidized to CO_2, and the OH^- ions of the electrolyte, NaOH, are used up. The OH^- ions in turn are replenished by the reduction of the O_2 as air is bubbled through the NaOH solution over the inert silver electrode. In a later section of this chapter we shall show that the theoretical limit of the voltage for this cell is approximately 1.01 V. In practice the voltage of the carbon-air fuel cell approaches the theoretical limit only during ini-

FIGURE 12.9

A carbon-air fuel cell.

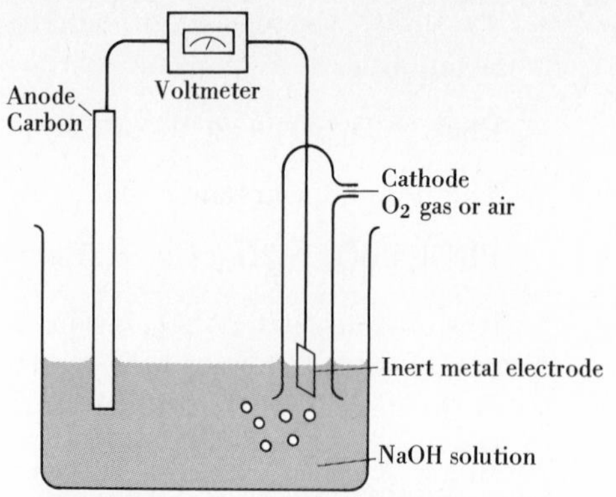

Anode reaction: $C(s) + 4OH^- \rightarrow CO_2(g) + 2H_2O + 4e^-$
Cathode reaction: $O_2(g) + 2H_2O + 4e^- \rightarrow 4OH^-$
Cell reaction: $C(s) + O_2(g) \rightarrow CO_2(g)$

tial discharge and then falls off to very low values. If this cell could be perfected, it would cause a revolution in energy utilization but thus far scientists have not been able to overcome the technical difficulties. Many other fuel cells have been studied. Almost all involve oxygen (air) as the *oxidant* and various fuels as reductants. The hydrogen-oxygen fuel cell has been most thoroughly studied (Fig. 12.10), and hydrogen-oxygen cells have actually been used in space vehicles.

We have discussed some of the important *practical* voltaic cells. Other cells have been developed primarily as standards for voltage measurements. The most common cell in this category is the Weston cell (Fig. 12.11). Properly constructed, it generates exactly 1.01810 V at 25°C.

FIGURE 12.10

Hydrogen-oxygen fuel cell.

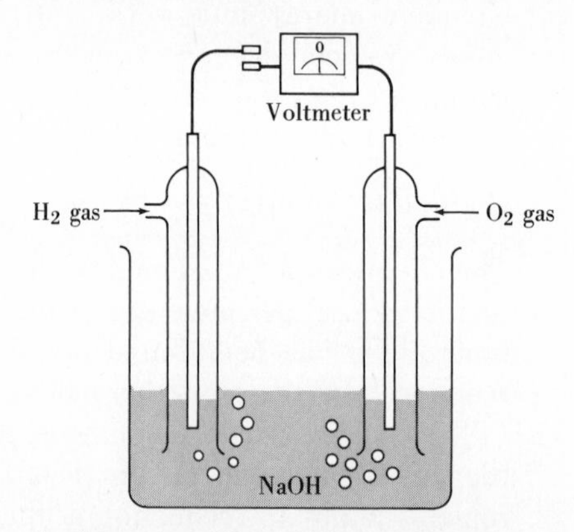

Anode reaction: $2H_2 + 4OH^- \rightarrow 4H_2O(l) + 4e^-$

Cathode reaction: $O_2 + 2H_2O(l) + 4e^- \rightarrow 4OH^-$

Cell reaction:

$2H_2(g) + O_2(g) \rightarrow 2H_2O(l)$; $\Delta G° = -56.7$ kcal/mol

FIGURE 12.11

Weston standard cell.

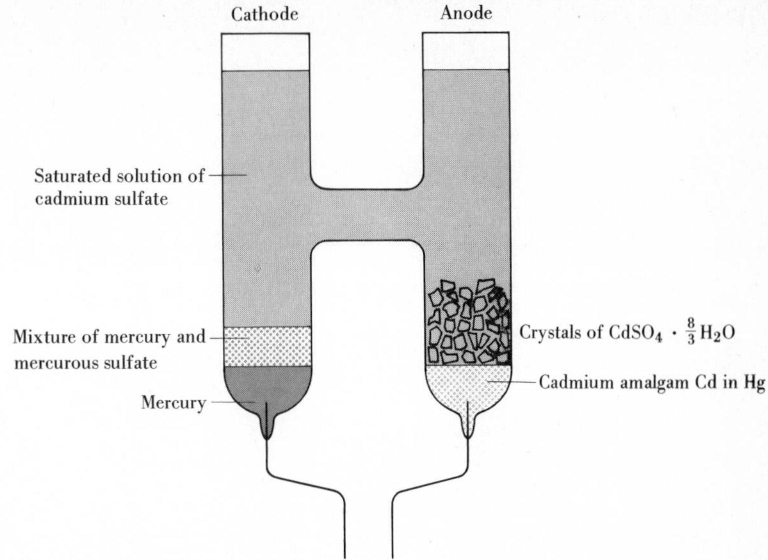

Anode: $\frac{1}{2}Cd(Hg) \rightarrow \frac{1}{2}Cd^{2+} + e^-$

Cathode: $\frac{1}{2}Hg_2SO_4 + e^- \rightarrow Hg^0 + \frac{1}{2}SO_4{}^{2-}$

$$\text{Cell:} \begin{cases} \frac{1}{2}Cd(Hg) + \frac{1}{2}Hg_2SO_4 \rightarrow Hg^0 + \frac{1}{2}CdSO_4 \\ \text{and} \\ CdSO_4 + \frac{8}{3}H_2O \rightarrow CdSO_4 \cdot \frac{8}{3}H_2O \end{cases}$$

12.11 STANDARD ELECTRODE POTENTIALS

An electrode is said to be a *standard* electrode if all substances involved in the electrode reaction are in their standard states. The standard states for pure gases, liquids, and solids are defined on page 298. The standard state for a solute in a solution is the solute at a concentration of unit molarity.†

In the standard hydrogen electrode (SHE) depicted alone in Fig. 12.12 and coupled with other standard electrodes in Fig. 12.13, pure hydrogen gas in its standard state (1 atm, 25°C) passes over a platinum-foil electrode so that the foil is alternately bathed with the gas and a hydrochloric acid solution (1 *M*, 25°C). When the electrode is isolated, the reaction

$$\tfrac{1}{2}H_2(g) \rightleftharpoons H^+(aq) + e^-$$

quickly comes to equilibrium at the platinum surface. When the SHE is coupled to another electrode, H_2 may be oxidized or H^+ may be reduced. In Fig. 12.13 the SHE is shown coupled to standard cadmium, zinc, and copper electrodes. The SHE is the cathode in the cells with cadmium and zinc electrodes; it is the anode in the cell with the copper electrode. The reactions and experimental voltages

FIGURE 12.12

Standard hydrogen electrode.

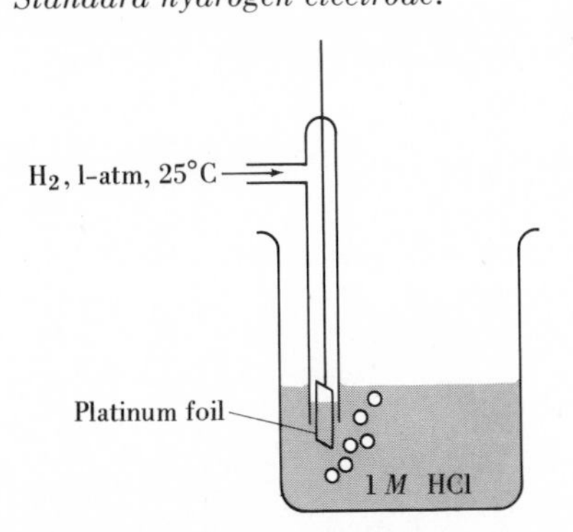

As an anode: $\frac{1}{2}H_2 \rightarrow H^+ + e^-$

As a cathode: $H^+ + e^- \rightarrow \frac{1}{2}H_2$

† Actually, the standard state for the solute is *unit activity*, which may be considerably different from the molar concentration, especially for ionic solutes.

FIGURE 12.13

Standard cells.

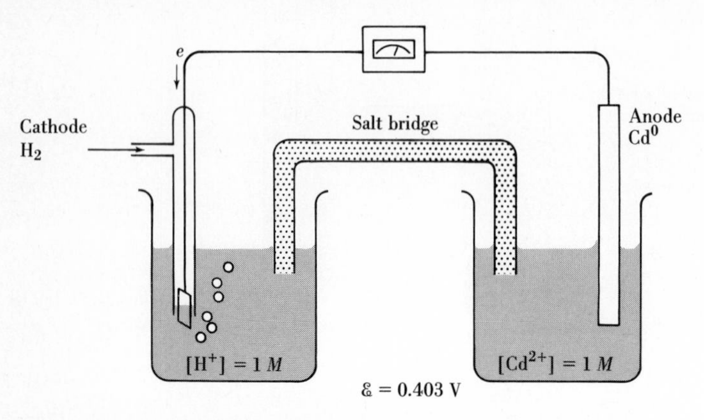

Anode: $\frac{1}{2}\ Cd^0 \rightarrow \frac{1}{2}\ Cd^{2+} + e^-$

Cathode: $H^+ + e^- \rightarrow \frac{1}{2}\ H_2(g)$

Cell: $\frac{1}{2}\ Cd^0 + H^+ \rightarrow \frac{1}{2}\ Cd^{2+} + \frac{1}{2}\ H_2(g)$

(*a*) *Hydrogen–cadmium cell*

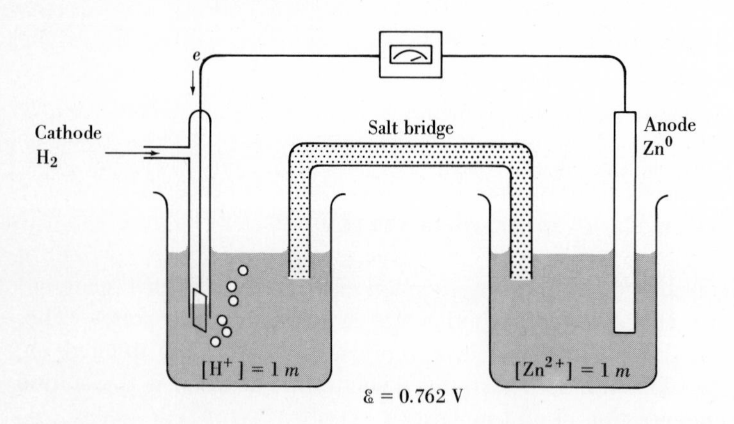

Anode: $\frac{1}{2}\ Zn^0 \rightarrow \frac{1}{2}\ Zn^{2+} + e^-$

Cathode: $H^+ + e^- \rightarrow \frac{1}{2}\ H_2$

Cell: $\frac{1}{2}\ Zn^0 + H^+ \rightarrow \frac{1}{2}\ Zn^{2+} + \frac{1}{2}\ H_2$

(*b*) *Hydrogen–zinc cell*

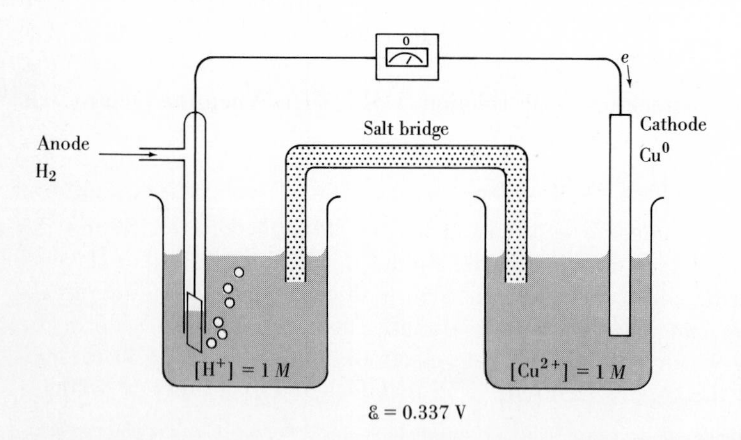

Anode: $\frac{1}{2}\ H_2(g) \rightarrow H^+ + e^-$

Cathode: $\frac{1}{2}\ Cu^{2+} + e^- \rightarrow \frac{1}{2}\ Cu^0$

Cell: $\frac{1}{2}\ H_2 + \frac{1}{2}\ Cu^{2+} \rightarrow H^+ + \frac{1}{2}\ Cu^0$

(*c*) *Hydrogen–copper cell*

generated by these cells are also shown in Fig. 12.13. The electrode compartments are joined by a *salt bridge*, a tube containing a salt, such as KCl or NH_4NO_3, dissolved in a gel. The bridge is a good conductor, yet it keeps the two solutions separated.

The SHE has arbitrarily been assigned an emf of exactly 0 V. The emf generated by a combination of the SHE and any other standard electrode is just equal to standard potential of the other electrode, because the value for the SHE is zero by definition. The symbol most often used to denote the voltage, or the emf, or the potential of an electrode or a cell is $\mathscr{E}$; for a standard electrode we use $\mathscr{E}^0$. Referring to Fig. 12.13 we can write

$$\mathscr{E}^0_{\mathrm{Cd}} = 0.403\ \mathrm{V} \qquad \mathscr{E}^0_{\mathrm{Zn}} = 0.762\ \mathrm{V} \qquad \mathscr{E}^0_{\mathrm{Cu}} = -0.337\ \mathrm{V}$$

Note that the standard oxidation potential $\mathscr{E}^0$ for the copper electrode is negative, which means that the copper electrode will be the cathode when it is coupled with the SHE. The $\mathscr{E}^0$ values are values of the oxidation potential with respect to the SHE. If, in fact, an electrode is the site of reduction, i.e., it is the cathode when coupled with the SHE, its $\mathscr{E}^0$ value is listed as negative. A whole series of standard potentials have been measured by comparing all standard electrodes to the SHE. The data in Table 12.3 are called *standard oxidation potentials* because all electrode reactions are written as oxidation reactions. If the electrodes are actually sites for oxidation

TABLE 12.3: STANDARD OXIDATION POTENTIALS AT 25°C

Electrode	*Electrode reaction*	$\mathscr{E}^0$, *V*
$Zn^0 \mid Zn^{2+}$	$Zn^0 \rightleftharpoons Zn^{2+} + 2e^-$	0.762
$Fe^0 \mid Fe^{2+}$	$Fe^0 \rightleftharpoons Fe^{2+} + 2e^-$	0.441
$Cd^0 \mid Cd^{2+}$	$Cd^0 \rightleftharpoons Cd^{2+} + 2e^-$	0.403
$Pb^0 \mid PbSO_4(s), SO_4^-$	$Pb(s) + SO_4^{2-} \rightleftharpoons PbSO_4(s) + 2e^-$	0.355
$Ni^0 \mid Ni^{2+}$	$Ni^0 \rightleftharpoons Ni^{2+} + 2e^-$	0.236
$Ag^0 \mid AgI(s), I^-$	$Ag^0 + I^- \rightleftharpoons AgI + e^-$	0.152
$Pb^0 \mid Pb^{2+}$	$Pb^0 \rightleftharpoons Pb^{2+} + 2e^-$	0.1265
$H_2^0 \mid H^+$	$H_2(g) \rightleftharpoons 2H^{+}$† $+ 2e^-$	0.000
$Ag^0 \mid AgBr(s), Br^-$	$Ag^0 + Br^- \rightleftharpoons AgBr(s) + e^-$	−0.0711
$Ag^0 \mid AgCl(s)Cl^-$	$Ag^0 + Cl^- \rightleftharpoons AgCl(s) + e^-$	−0.223
$Cu^0 \mid Cu^{2+}$	$Cu^0 \rightleftharpoons Cu^{2+} + e^-$	−0.337
$I_2^0 \mid I^-$	$2I^- \rightleftharpoons I_2 + 2e^-$	−0.536
$Ag^0 \mid Ag^+$	$Ag^0 \rightleftharpoons Ag^+ + e^-$	−0.799
$Br \mid Br^-$	$2Br^- \rightleftharpoons Br_2(l) + 2e^-$	−1.06
$Cl_2 \mid Cl^-$	$2Cl^- \rightleftharpoons Cl_2(g) + 2e^-$	−1.36

† Activity = 1.

reactions when coupled with the SHE, the potentials are positive; if the electrodes are sites for reduction, the oxidation potentials are negative.

The potentials listed in Table 12.3 are a measure of the tendency toward oxidation relative to H_2 in aqueous solutions under standard conditions. Cadmium has a much greater tendency toward oxidation than hydrogen under these conditions, and copper has lower tendency.

The standard oxidation potentials can be used to calculate the voltage of a cell and equilibrium constants for cell reactions and to predict the direction of chemical reactions; we shall learn to do all three in this chapter.

EXAMPLE 12.11 The standard silver and zinc electrodes are coupled to make a cell. (*a*) What are the electrode reactions? (*b*) What is the cell reaction? (*c*) What is the cell voltage? (*d*) Will the cell be affected by increasing the Ag^+ concentration; if so, how?

Solution: (*a*) From Table 12.3 we note that zinc has a much higher standard oxidation potential than silver; i.e., it has a much greater potential to undergo oxidation under standard-state conditions. Therefore, the zinc electrode is the anode, and the silver electrode is the cathode. The electrode reactions must be

Anode: $Zn^0 \rightarrow Zn^{2+} + 2e^-$
Cathode: $Ag^+ + e^- \rightarrow Ag^0$

(*b*) The cell reaction is the *sum* of the electrode reactions, but electrode reactions cannot be added unless they involve the same number of electrons. We can multiply the equation for the anode reaction by $\frac{1}{2}$ or the equation for the cathode reaction by 2. Choosing the latter alternative, we obtain

Anode: $Zn^0 \rightarrow Zn^{2+} + 2e^-$
Cathode: $2Ag^+ + 2e^- \rightarrow 2Ag^0$
Cell reaction: $Zn^0 + 2Ag^+ \rightarrow Zn^{2+} + 2Ag^0$

(*c*) The cell voltage can be calculated from the data in Table 12.3, where the potentials are listed for the *oxidation* reactions which would occur if all the electrodes were *anodes*. If in a particular case an electrode acts as a cathode, the sign of the voltage given in the table must be reversed. In this example, the anode is $Zn \mid Zn^{2+}$, and the cathode is $Ag^0 \mid Ag^+$. The voltage listed for the $Ag^0 \mid Ag^+$, -0.799 V, must be reversed in sign and added to the voltage given for the $Zn \mid Zn^{2+}$ electrode:

Anode: $Zn^0 \rightarrow Zn^{2+} + 2e^-$ $\mathscr{E}^0_{anode} = 0.762$ V
Cathode: $2Ag^+ + 2e^- \rightarrow 2Ag^0$ $\mathscr{E}^0_{cathode} = 0.799$ V
Cell: $Zn^0 + 2Ag^+ \rightarrow Zn^{2+} + 2Ag^0$ $\mathscr{E}^0_{cell} = 1.561$ V

Note that no matter how we write the stoichiometric equation, we cannot change the voltage. If we choose the electrode reactions

Anode: $\frac{1}{2}Zn^0 \rightarrow \frac{1}{2}Zn^{2+} + e^-$
Cathode: $Ag^+ + e^- \rightarrow Ag^0$
Cell: $\frac{1}{2}Zn^0 + Ag^+ \rightarrow \frac{1}{2}Zn^{2+} + Ag^0$

the cell emf will nevertheless be 1.561 V. The cell voltage cannot be affected by the way we choose to represent the cell reaction on paper.

A calculation giving a negative cell voltage means that we have chosen the anode and the cathode incorrectly and that the direction of the cell reaction is just the opposite of the one we have chosen. If we had incorrectly made the silver electrode the anode and the zinc electrode the cathode, our cell reaction would have been $2Ag^0 + Zn^{2+} \rightarrow 2Ag^+ + Zn^0$ and the calculated emf would have been -1.561 V.

(*d*) To deduce the effect, if any, of increasing $[Ag^+]$ we can use the principle of Le Châtelier:

Cell reaction: $Zn^0 + Ag^+(aq) \rightarrow Zn^{2+}(aq) + 2Ag^0$

An increase in $[Ag^+]$ would favor the cell reaction, and therefore the cell voltage would be augmented. A decrease in Zn^{2+} would also increase the cell voltage. In this reaction we can only change the concentrations of the Ag^+ and Zn^{2+} ions; we cannot affect the concentrations of pure solid phases like Zn^0 or Ag^0.

12.12 THERMODYNAMICS OF VOLTAIC CELLS

Voltaic cells are easily operated at constant temperature and pressure. Under such conditions the electric work w_{elec} obtained from the discharge of the cell equals the fall or decrease in chemical potential energy (see pages 309 and 333).

$$w_{elec} = -\Delta G$$

By definition, electric work is just equal to the voltage times the charge Q which moves under the influence of the voltage:

$$w_{elec} = \mathscr{E}Q$$

The work will be in *joules* if $\mathscr{E}$ is in volts and Q is in coulombs. In other words a joule can be considered to be a *volt-coulomb:*

$$w_{elec}(\text{joules}) = \mathscr{E}Q(\text{volt-coulombs})$$

In electrochemistry it is more convenient to express the charge passing during the operation of a cell in terms of equivalents. When n equiv pass, the charge is $n\mathscr{F}$, where $\mathscr{F}$ is the faraday, or 96,500 C:

$$Q = n\mathscr{F}$$

and

$$w_{\text{elec}} = n\mathscr{F}\mathscr{E} \tag{12.8}$$

During the operation of a voltaic cell the relationship between the decrease in chemical potential energy and the voltage is

$$\Delta G = -n\mathscr{F}\mathscr{E} \tag{12.9}$$

We can use Eq. (12.9) to calculate ΔG for a cell reaction when the cell voltage is known, and, conversely, we can use Eq. (12.9) to estimate the voltage of a cell reaction if the thermochemical data are available for it.

EXAMPLE 12.12 Calculate the change in chemical potential energy for spontaneous operation of a cell consisting of an SHE and a standard cadmium electrode.

Solution: Since the oxidation potential of the $Cd \mid Cd^{2+}$ electrode is higher, it is the anode:

Anode: $Cd \rightarrow Cd^{2+}(aq) + 2e^-$ $\qquad \mathscr{E}_{Cd}{}^0 = 0.403$ V
Cathode: $2H^+(aq) + 2e^- \rightarrow H_2(g)$ $\qquad \mathscr{E}_{H_2}{}^0 = 0.000$ V

The cell reaction and the cell voltage are

$$2H^+(aq) + Cd^0 \rightarrow Cd^{2+}(aq) + H_2(g)$$

$$\mathscr{E}^0 = 0.403 + 0.000 = +0.403 \text{ V}$$

From Eq. (12.9), the change in chemical potential energy is

$$\Delta G = -n\mathscr{F}\mathscr{E}$$

Since the cell is a standard cell and 0.403 is the standard voltage, the calculated value of ΔG will be the standard change in chemical potential energy $\Delta G°$.

Because of the way we have written the electrode and cell reactions, the number of equivalents n is 2. Substituting into Eq. (12.9), we obtain

$$\Delta G° = -2 \text{ equiv} \times \frac{96{,}500 \text{ C}}{\text{equiv}} \times 0.403 \text{ V} = -77{,}700 \text{ V-C}$$

Since 1 V-C is equal to 1 J,

$$\Delta G° = -77{,}700 \text{ V-C} = -77{,}700 \text{ J}$$

ΔG values are ordinarily given in calories; the relationship between calories and joules is

$$1.987 \text{ cal} = 8.314 \text{ J}$$

$$\Delta G° = -77{,}700 \text{ J} \times \frac{1.987 \text{ cal}}{8.314 \text{ J}} = -18{,}600 \text{ cal}$$

EXAMPLE 12.13 At 25°C, the chemical potential energy change for the reaction between H_2 and O_2 under standard conditions is −56.7 kcal per mole of water formed. Estimate the voltage of the hydrogen fuel cell shown in Fig. 12.10.

Solution: The reaction which generates the current in the hydrogen fuel cell is identical with the ordinary combustion of H_2:

$$H_2(g) + \tfrac{1}{2}O_2(g) \rightarrow H_2O(l)$$

and therefore ΔG for the hydrogen-fuel-cell reaction and hydrogen combustion must be equal. Using Eq. (10.27), we can calculate $\Delta G°$ with thermochemical data

$$\Delta G° = \Delta H° - T\,\Delta S°$$

This demonstrates the power of the thermodynamic theory. It allows us to calculate the voltage from data that have no obvious relationship to it. $\Delta H_f°$ values come from combustion experiments; $\Delta S°$ values come from specific-heat measurements. But by calculating the thermochemical quantities $\Delta H°$ and $\Delta S°$ for a reaction which can occur in a cell we can also estimate the voltage. Using the $\Delta H_f°$ data given in Table 10.3, and the $S°$ data from Table 10.4, we obtain

$$\Delta G° = -68{,}320 \text{ cal} - 298 \text{ K}\left(-39.0\,\frac{\text{cal}}{\text{mol-K}}\right)$$
$$= -56{,}700 \text{ cal}$$

To calculate the voltage of the hydrogen fuel cell we use Eq. (12.8), but $\Delta G°$ must first be converted into joules:

$$1.987 \text{ cal} = 8.314 \text{ J}$$

$$\Delta G° = -56{,}700 \text{ cal} \times \frac{8.31 \text{ J}}{1.99 \text{ cal}} = -237{,}000 \text{ J}$$

Rearrangement of Eq. (12.8) gives

$$\mathscr{E} = -\frac{\Delta G}{n\mathscr{F}}$$

or for standard cells in which the cell reactions proceed under standard conditions

$$\mathscr{E}^0 = -\frac{\Delta G°}{n\mathscr{F}}$$

Substitution of $\Delta G° = -237{,}000$ J and $n = 2$ equiv gives

$$\mathscr{E}^0 = -\frac{(-237{,}000 \text{ J})}{2 \text{ equiv} \times 96{,}500 \text{ C/equiv}} = 1.23 \text{ J/C} = \frac{1.23 \text{ V-C}}{\text{C}} = 1.23 \text{ V}$$

12.13 EFFECT OF CONCENTRATION ON CELL VOLTAGE: THE NERNST EQUATION

Since the current of a voltaic cell is generated by a chemical reaction, it should not be surprising that the voltage depends upon the concentration of the species in the reaction. In Example 12.11, we applied Le Châtelier's principle to deduce the effect that a change in concentration would have on the voltage of a silver-zinc cell. In that case the reaction which caused the current to flow was

$$Zn^0(s) + 2Ag^+(aq) \rightarrow Zn^{2+}(aq) + 2Ag^0(s)$$

Any change which gives more impetus to this reaction will increase the cell voltage. From Le Châtelier's principle we concluded that increasing the Ag^+ concentration and decreasing the Zn^{2+} concentration would give more impetus to the cell reaction and would increase the cell voltage.

The conclusion reached by the application of Le Châtelier's principle is correct, but it is only qualitative. We cannot calculate the actual voltage of a cell by application of Le Châtelier's principle. To do this we must return to the results of thermodynamics. From our study of the thermodynamics of chemical reactions we learned that the relationship between the change in chemical potential and concentration is given by the reaction isotherm:

$$\Delta G = \Delta G^\circ + RT \ln Q \qquad (11.11)$$

where Q is the concentration or pressure coefficient. If the change in chemical potential energy ΔG° is known for a chemical reaction, the change in chemical potential energy under different concentration conditions can be calculated from Eq. (11.11).

By substituting Eq. (12.8) into Eq. (11.11) we can obtain an equation which relates the voltage to the concentrations of the reactants and products in a cell reaction:

$$-n\mathcal{F}\mathcal{E} = -n\mathcal{F}\mathcal{E}^0 + RT \ln Q$$

or

$$\mathcal{E} = \mathcal{E}^0 - \frac{RT}{n\mathcal{F}} \ln Q \qquad (12.10)$$

This is known as the *Nernst equation*, and it is the most important mathematical equation of electrochemistry.

For calculations of cell voltages at 25°C, Eq. (12.10) reduces to

$$\mathcal{E} = \mathcal{E}^0 - \frac{0.05916}{n} \log Q \qquad (12.11)$$

To obtain Eq. (12.11) from Eq. (12.10) substitute $R = 8.314$ J/mol-K, $T = 298.16$ K, $\mathcal{F} = 96{,}487$ C, and $\ln Q = 2.303 \log Q$. In Eq. (12.11), the term, $0.05916/n \log Q$, appears as a correction term to the standard voltage $\mathscr{E}^0$ and accounts for the fact that a cell may not be a standard cell in which the voltage is $\mathscr{E}^0$. This correction term is zero when the cell operates under standard-state conditions.

EXAMPLE 12.14 In Example 12.11, the standard voltage $\mathscr{E}^0$ for the silver-zinc cell was determined to be 1.561 V. It was concluded from Le Chatelier's principle that an increase of silver-ion concentration would cause an increase in voltage. Test this conclusion by calculating the $\mathscr{E}$ value of a silver-zinc cell at 25°C which differs from a standard cell only by the fact that $[Ag^+] = 1.50\ M$.

Cell reaction: $Zn^0(s) + 2Ag^+(aq) \rightarrow Zn^{2+}(aq) + 2Ag^0(s)$

Solution: The concentration quotient for the reaction as described by the above equation is

$$Q = \frac{[Zn^{2+}][Ag^0]^2}{[Zn^0][Ag^+]^2} = \frac{[Zn^{2+}]}{[Ag^+]^2} = \frac{1}{1.50^2} = \frac{1}{2.25}$$

As written, the equation represents a reaction involving two electrons; therefore, $n = 2$. Substituting into the Nernst equation,

$$\mathscr{E} = \mathscr{E}^0 - \frac{0.0592}{\mathrm{n}} \log Q$$

we obtain

$$\mathscr{E} = 1.561\ \mathrm{V} - \frac{0.0592}{2} \log \frac{1}{2.25} = 1.561\ \mathrm{V} + 0.0296 \log 2.25$$

$$= 1.561\ \mathrm{V} + 0.0105\ \mathrm{V} = 1.572\ \mathrm{V}$$

Increasing $[Ag^+]$ from 1.00 to 1.50 M causes a voltage increase or $\sim$0.01 V. A simple inspection of the Nernst equation will reveal that a decrease in Zn^{2+} concentration will also cause an increase in the voltage.

Note that the value of Q for a standard silver-zinc cell would be $1/1^2$, or 1; $\log 1 = 0$, and the Nernst equation says that the cell voltage $\mathscr{E}$ equals $\mathscr{E}^0$ when the cell is a standard cell.

EXAMPLE 12.15 In a hydrogen-gas–cadmium cell the following conditions prevail: $T = 298.16$ K, $P_{H_2} = 1.50$ atm, $[H^+] = 1.20$, $[Cd^{2+}] = 0.500$. (*a*) Write the electrode reactions. (*b*) Write the cell reaction and calculate the voltage of a standard hydrogen-gas–cadmium cell. (*c*) Calculate the cell voltage under the conditions described above.

Solution: Since the Cd-Cd^{2+} electrode has a higher standard oxidation potential, it would be the anode under standard-state conditions. We shall guess that it is also the anode under the above conditions; if our guess is incorrect, the calculated voltage will be negative but the magnitude will be correct.

(*a*) The reactions based on 1 equiv are

Anode: $\frac{1}{2}Cd \rightarrow \frac{1}{2}Cd^{2+} (M = 0.500) + e^-$ $\qquad \mathscr{E}^0 = 0.403$ V

Cathode: $H^+ (M = 1.20) + e^- \rightarrow$

$\frac{1}{2}H_2(g)(P = 1.50 \text{ atm})$ $\qquad \mathscr{E}^0 = 0.000$ V

(*b*) The cell reaction is the sum of the electrode reactions.

Cell: $\frac{1}{2}Cd + H^+ \ (M = 1.20) \rightarrow$

$\frac{1}{2}Cd^{2+} (M = 0.500) + \frac{1}{2}H_2 (g) (P = 1.50 \text{ atm})$ $\qquad \mathscr{E}^0 = 0.403$

(*c*) From the cell reaction we can write the expression for Q:

$$Q = \frac{[Cd^{2+}]^{1/2} P_{H_2}{}^{1/2}}{[H^+]} = \frac{0.500^{1/2} \times 1.50^{1/2}}{1.20}$$

Substitution into the Nernst equation (12.11) gives

$$\mathscr{E} = 0.403 \text{ V} - \frac{0.0592}{1} \log \frac{0.500^{1/2} \times 1.50^{1/2}}{1.20}$$

$$= 0.403 + 0.00839 = 0.411 \text{ V}$$

The terms in the quotient Q may be confusing because we have molar concentrations and pressures together. The quotient should actually be calculated from the activities of the species involved. The molar concentration of solutes and the pressures of gases are only approximations of the activities, which have no units.

An analysis of the cell in terms of Le Châtelier's principle would give a qualitative result in agreement with the calculation from Nernst's equation.

12.14 THE EQUILIBRIUM CONSTANT FOR CELL REACTIONS

When a cell is dead its voltage is zero; $\mathscr{E} = 0$. If the dead cell is reversible, the reaction responsible for the voltage is at equilibrium and the concentration quotient Q is equal to the equilibrium constant [see Eq. (11.12)]. From an analysis of the Nernst equation we can see that it is possible to determine equilibrium constants from $\mathscr{E}^0$ data.

$$\mathscr{E} = \mathscr{E}^0 - \frac{0.0592}{n} \log Q \qquad \text{Nernst equation}$$

For a dead reversible cell, $\mathscr{E} = 0$, and Q equals the equilibrium constant K. When these substitutions are made, the Nernst equation becomes

$$0 = \mathscr{E}^0 - \frac{0.0592}{n} \log K$$

or

$$\mathscr{E}^0 = \frac{0.0592}{n} \log K \tag{12.12}$$

Equation (12.12) allows us to calculate equilibrium constants for oxidation-reduction reactions from standard oxidation potentials.

EXAMPLE 12.16 Calculate the equilibrium constant for the reaction

$$Cu^0 + 2Ag^+(aq) \rightarrow Cu^{2+}(aq) + 2Ag^0$$

Solution: We can recognize this reaction as the cell reaction in the copper-silver cell; the electrode reactions are

Anode: $Cu^0 \rightarrow Cu^{2+}(aq) + 2e^-$	$\mathscr{E}^0 = -0.337$ V
Cathode: $2Ag^+(aq) + 2e^- \rightarrow 2Ag^0$	$\mathscr{E}^0 = +0.799$ V
Cell: $Cu + 2Ag^+(aq) \rightarrow Cu^{2+}(aq) + 2Ag^0$	$\mathscr{E}^0 = 0.462$ V

Substitution into Eq. (12.12) gives

$$\log K = \frac{n\,\mathscr{E}^0}{0.0592} = \frac{2.0 \times 0.462}{0.0592} = 15.61$$

$$K = 10^{15.61} = 10^{0.61} \times 10^{15} = 4.07 \times 10^{15}$$

This is an extremely large equilibrium constant, and it means the reaction very strongly favors the formation of products Ag^0 and Cu^{2+}. When a silver salt solution is placed in contact with metallic copper, all the silver will be reduced. No silver salt will be detectable in the solution.

12.15 CONCLUSION

Electrochemical studies have been extremely important in the development of modern theory, technology and the quality of life. Try to imagine what life would be like without electric power. Today's scientists and engineers are very much concerned with the development of new sources of energy and different methods of energy storage. To avoid the pollution of our atmosphere and still provide the needs of the people, electrochemists are investigating new methods of converting chemical energy into work. The quality and even the continuance of human life depend on the success of these attempts.

QUESTIONS AND PROBLEMS

12.1 What is the essential difference between Volta's and Galvani's interpretations of the frog's-legs experiments?

12.2 Give your reasons for supporting or for not supporting the contention that Volta's pile is the most important *experimental* discovery in the last 300 years.

12.3 Figure 12.6*b* depicts a concentration cell. If the force which generates the current is due to a tendency to equalize the Cu^{2+} concentration in each compartment, which side is the anode?

12.4 What are the essential features of a voltaic cell capable of producing a high voltage and a high current?

12.5 Define (*a*) oxidation, (*b*) oxidant, and (*c*) reductant.

12.6 Balance the following equations. Identify the oxidant and the reductant in each case:

(*a*) $Cu + HNO_3 \rightarrow Cu(NO_3)_2 + NO_2 + H_2O$
(*b*) $HI + H_2SO_4 \rightarrow I_2 + H_2S + H_2O$
(*c*) $KMnO_4 + HBr \rightarrow Br_2 + MnBr_2 + KBr + H_2O$
(*d*) $K_2Cr_2O_7 + H_2S + H_2SO_4 \rightarrow S + Cr_2(SO_4)_3 + K_2SO_4 + H_2O$
(*e*) $As_2O_3 + HNO_3 + H_2O \rightarrow H_3AsO_4 + NO$
(*f*) $KOH + CrCl_3 + Cl_2 \rightarrow K_2CrO_4 + KCl + H_2O$
(*g*) $Cu_2S + HNO_3 \rightarrow Cu(NO_3)_2 + H_2SO_4 + NO_2 + H_2O$

Note that both elements in Cu_2S undergo a formal change of oxidation state.

(*h*) $FeS + HNO_3 \rightarrow Fe(NO_3)_3 + NO + S + H_2O$
(*i*) $KMnO_4 + H_2O_2 + H_2SO_4 \rightarrow MnSO_4 + O_2 + K_2SO_4 + H_2O$
(*j*) $NaBrO_3 + NaBr + H_2SO_4 \rightarrow Br_2 + Na_2SO_4 + H_2O$

In example (*j*) the product of oxidation is identical with the product of reduction.

(*k*) $Cr_2O_7^{2-} + H^+ + S^{2-} \rightarrow S^0 + Cr^{3+} + H_2O$
(*l*) $Cl_2 + OH^- \rightarrow ClO_3^- + Cl^- + H_2O$
(*m*) $MnO_4^- + NO_2 + OH^- \rightarrow NO_3^- + MnO_2 + H_2O$
(*n*) $Zn^0 + NO_3^- + OH^- \rightarrow ZnO_2^{2-} + NH_3 + H_2O$

12.7 Calculate the equivalent weights of the oxidant and reductant for each reaction in Prob. 12.6.

12.8 Calculate the normality of (*a*) 10.0 g of HNO_3 in 500 ml of solution reacting according to reaction (*a*) in Prob. 12.6, (*b*) 22.5 g of HBr in 1100 ml of solution reacting according to reaction (*c*) in Prob. 12.6, (*c*) 2.50 g of H_2O_2 in 15 ml of solution reacting according to reaction (*i*) in Prob. 12.6, (*d*) 10.0 g of H_2SO_4 in 50 ml of solution reacting according to reaction (*i*) in Prob. 12.6.

12.9 Calculate the normality of the following solutions; reference is made to the appropriate part of Prob. 12.6 for the reaction in which the solute is involved:

(*a*) 25.0 g of HNO_3 in 500 ml of solution [part (*a*)]
(*b*) 25.0 g of HNO_3 in 500 ml of solution [part (*e*)]
(*c*) 30.0 g of $K_2Cr_2O_7$ in 250 ml of solution [part (*d*)]
(*d*) 50.0 g of $CrCl_3$ in 750 ml of solution [part (*f*)]

(*e*) 10.0 g of H_2O_2 in 2.50 l of solution [part (*i*)]
(*f*) 24.0 g of $NaBrO_3$ in 1.75 l of solution [part (*j*)]

12.10 What number of faradays $\mathscr{F}$ is represented by (*a*) 17,500 C, (*b*) a 5-A current passing for $\frac{1}{2}$ day, (*c*) 100,000 A-s?

12.11 An aqueous $AgNO_3$ solution has a concentration 0.1 *N*. One liter of the solution is electrolyzed by a constant current of 0.100 A for 10.0 h. (*a*) Write the anode, cathode, and cell reactions. (*b*) What weight of silver metal is deposited at the cathode? (*c*) What volume of dry oxygen gas, measured at STP, is liberated at the anode? (*d*) What is the normality of the silver ion in the solution at the end of the electrolysis?

12.12 An aqueous solution of $CuSO_4$ electrolyzed 8.00 h produces 100 g of Cu metal. (*a*) Write the anode, cathode, and cell reactions (see Table 12.1). (*b*) What is the average current? (*c*) What weight and volume (STP) of oxygen are produced? What weight of $CuSO_4$ is decomposed?

12.13 A dilute aqueous H_2SO_4 solution is electrolyzed using a steady current of 0.150 A for 1.25 h. The hydrogen collected over the solution has a volume of 87.0 ml. The total pressure of the gas above the solution is 755 Torr, and its temperature is 22.0°C. The vapor pressure of the solution is 18.4 Torr. Assuming the cathode stoichiometry to be $H^+(aq) + e^- \rightarrow \frac{1}{2}H_2(g)$, calculate the charge on the electron in coulombs.

12.14 A steady current is passed for 3.00 h through solutions of $AgNO_3$ and $CuSO_4$ arranged in series. It is found that 0.555 g of metallic silver is deposited at the cathode from the $AgNO_3$ solution. (*a*) How many equivalents and coulombs of electricity pass through the two solutions? (*b*) How many grams of metallic copper are produced? (*c*) Calculate the volume of dry oxygen produced at 25°C and 770 Torr. (*d*) What was the amperage during the electrolysis?

12.15 (*a*) In terms of chemical reactions, what is the difference between reversible and irreversible cells? (*b*) Why is Volta's pile irreversible? (*c*) Why is the dry cell (Fig. 12.8) irreversible?

12.16 The combustion of graphite under standard-state conditions is accompanied by a chemical-potential-energy decrease of −94 kcal per mole of graphite. What is the maximum voltage of a carbon-oxygen fuel cell operating under standard-state conditions? [See Eq. (12.9).]

12.17 Define (*a*) standard electrode and (*b*) standard cell; (*c*) sketch a carefully labeled standard cell using any two electrodes from Table 12.3. (*d*) Write equations for the anode, cathode, and cell reactions.

12.18 (*a*) Write the anode, cathode, and cell reactions for voltaic cells formed from the pairs of electrodes given below. (*b*) Calculate the cell voltage for each. (*c*) Calculate the change in chemical potential energy which accompanies the passage of 1.00 $\mathscr{F}$ of charge under standard-state conditions in each case.

(1) $Zn^0 \mid Zn^{2+}(aq)(1.00\ M)$ versus $Cd^0 \mid Cd^{2+}(aq)(1.00\ M)$
(2) $H_2(P = 1.00\ \text{atm}) \mid H^+(aq)(1.00\ M)$ versus $Br_2(l) \mid Br^-(aq)(1.00\ M)$

(3) $Ni^0 \mid Ni^{2+}(aq)(1.00\ M)$ versus $Cu^0 \mid Cu^{2+}(aq)(1.00\ M)$
(4) $Pb^0 \mid Pb^{2+}(aq)(1.00\ M)$ versus $Ag^0 \mid AgCl(s),\ Cl^-(aq)(1.00\ M)$

12.19 The change in standard chemical potential energy $\Delta G°$ has been measured by calorimetric methods for the oxidation-reduction reactions listed below. (*a*) If voltaic cells could be devised in which these reactions were responsible for the emf, what would the voltage be?

(1) $H_2(g) + \frac{1}{2}O_2(g) \xrightarrow{25°C} H_2O(l) \qquad \Delta G° = -54.6$ kcal

(2) $S(\text{rhombic}) + O_2(g) \xrightarrow{25°C} SO_2(g) \qquad \Delta G° = -71.8$ kcal

(3) $\frac{1}{2}N_2(g) + \frac{3}{2}H_2(g) \xrightarrow{25°C} NH_3(g) \qquad \Delta G° = -3.98$ kcal

(*b*) Make a sketch of a possible voltaic cell in which reaction (2) is the cell reaction; label the anode and cathode.

12.20 A voltaic cell consists of a silver electrode, $Ag^0 \mid Ag^+$, in which the Ag^+ concentration is 0.25 *M*, and $Ni^0 \mid Ni^{2+}$ electrode in which the Ni^{2+} concentration is 1.50 *M*. (*a*) Make a sketch of the cell. (*b*) Write the anode, cathode, and cell reactions based on 1 equiv. (*c*) Calculate the standard emf of the cell. (*d*) Calculate the equilibrium constant for the cell reaction. (*e*) Calculate the emf of the cell as described. (*f*) Calculate the change in chemical potential energy based on the equations as written in part (*b*). (*g*) Are the values of $\mathscr{E}^0$ and $\mathscr{E}$ obtained in parts (*c*) and (*e*), respectively, consistent with Le Châtelier's principle?

12.21 Repeat Prob. 12.20 for the following cells:

(*a*) $Cu^0(s) \mid Cu^{2+}(aq)(1.00\ M)$ versus $H_2(g)(P = 1.00\ \text{atm}) \mid H^+(aq)(1.50\ M)$
(*b*) $Ni^0(s) \mid Ni^{2+}(aq)(2.50\ M)$ versus $Cd^0(s) \mid Cd^{2+}(aq)(0.750\ M)$
(*c*) $Fe^0(s) \mid Fe^{2+}(aq)(0.750\ M)$ versus $Zn^0(s) \mid Zn^{2+}(aq)(0.750\ M)$
(*d*) $Ag^0(s) \mid AgBr(s),\ Br^-(aq)(0.445\ M)$ versus $Pb^0(s) \mid Pb^{2+}(aq)(1.50\ M)$

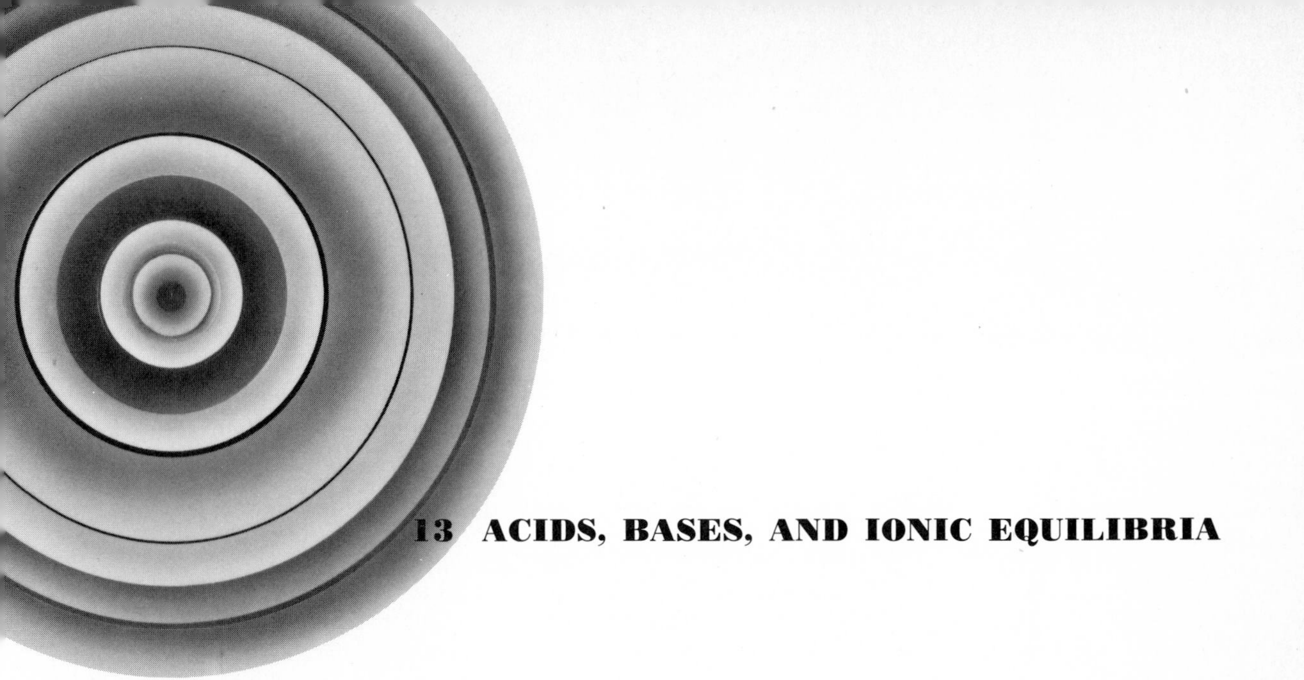

13 ACIDS, BASES, AND IONIC EQUILIBRIA

13.1 THE PROPERTIES OF ACIDS AND BASES

Acids have long been recognized as unusual compounds. Their best-known property is sour taste, but many more "scientific" characteristics have been discovered. For example, acids "dissolve" iron, tin, zinc, and other metals causing the evolution of hydrogen gas:

$$Zn(s) + \text{acid} \rightarrow \text{salt of zinc} + H_2(g)$$

Acids react with carbonates causing the evolution of carbon dioxide gas. The reaction of vinegar (acetic acid) and baking soda, $NaHCO_3$, is a good example of this:

$$\underset{\text{Acetic acid}}{C_2H_3O_2H(aq)} + NaHCO_3(aq) \rightarrow \underset{\text{Sodium acetate}}{C_2H_3O_2Na(aq)} + H_2O(l) + CO_2(g)$$

Acids cause certain organic compounds to change color and for this reason such compounds are used as indicators of the presence of acid. Litmus changes from blue to red and phenolphthalein from pink to colorless when acids are present. Acids often act as *catalysts* for reactions in solution; i.e., acids often increase the rate of a reaction even though they are not consumed by the reaction. The catalytic power of an acid is a measure of its strength as an acid.

Compounds which counteract, or neutralize, the effects of acids were first discovered in the ashes of plants. They were called *alkali* or alkaline (from the Arabic *al kali*, ashes of salt-marsh plants). Today we refer to these substances as *bases*. Aqueous solutions of bases have a slimy feeling and a bitter taste, like soap. The most common bases are ammonia, soap, and sodium bicarbonate. Bases react with aqueous solutions of many heavy-metal salts to produce insoluble hydroxides:

$$Fe(NO_3)_2(aq) + \text{base} \rightarrow Fe(OH)_3(s) + \text{a nitrate salt}$$

Since bases reverse the effects of acids, colored indicators change color as the base "uses up" the acid present. Bases cause red litmus to become blue and colorless phenolphthalein to turn pink.

When equivalent amounts of an acid and base are mixed together, the resulting solution is neither acidic or basic. It does not cause the evolution of H_2 or CO_2 when zinc or baking soda is added; it has no effect on the color of litmus or phenolphthalein; it does not feel slimy; and it does not taste sour or bitter. The solution may taste salty because when acids and bases react they produce salts.

13.2 THEORIES OF ACIDS AND BASES

Early speculations

Beginning with Lavoisier's speculations at the end of the eighteenth century, chemists have tried to interpret the properties of acids and bases in terms of elemental composition and molecular structure. During his study of dephlogisticated air, which he later renamed oxygen, Lavoisier observed that many common acidic substances result from the union of oxygen with nonmetals:

$$C(\text{graphite}) + O_2(g) \rightarrow CO_2(g)$$
$$P_4(s) + 5O_2(g) \rightarrow P_4O_{10}(s)$$
$$S_8(s) + 8O_2(g) \rightarrow 8SO_2(g)$$

These oxides, CO_2, P_4O_{10}, and SO_2 give acidic solutions when they are dissolved in water. Lavoisier inferred *incorrectly* that all acids contain oxygen, a name he coined from Greek roots: it means *acid former*. In 1808 Sir Humphry Davy's careful experiments showed that hydrogen chloride gas, HCl, which forms hydrochloric acid when it is dissolved in water, contains no oxygen, and Lavoisier's oxygen theory of acids began to lose support. By 1816 Davy finally concluded that hydrogen, not oxygen, is the element peculiar to all acid compounds. Davy's conclusion was not accepted immediately, but gradually, as it withstood the test of experiments, chemists accepted the hydrogen theory of acids.

The Arrhenius theory of acids and bases

Svante Arrhenius. (Courtesy of the Edgar Fahs Smith Collection, Van Pelt Library, University of Pennsylvania, Philadelphia.)

In the 1880s Svante Arrhenius developed a theory for the behavior of aqueous electrolyte solutions which included acids and bases (see Chap. 14 for a fuller discussion). His main thesis was that when a salt, acid, or base is dissolved in water, the molecules of the solute partially ionize, or split into ions, which can then carry an electric current. Thus, these solutions possess an extraordinarily high electric conductivity.

Arrhenius defined an acid as a hydrogen-containing compound which forms hydrogen ions, H^+, in aqueous solution and a base as a hydroxy-containing compound which forms hydroxyl ions, OH^-, in aqueous solution. Thus, he wrote

$HCl \rightleftharpoons H^+ + Cl^-$ hydrochloric acid†
$NaOH \rightleftharpoons Na^+ + OH^-$ sodium hydroxide†

According to Arrhenius, neutralization resulted from the combination of H^+ and OH^- to produce water:

$H^+ + OH^- \rightarrow HOH$ neutralization

For example,

$$\underset{\text{Base}}{NaOH} + \underset{\text{Acid}}{HCl} \rightarrow \underset{\text{Water}}{HOH} + \underset{\text{Salt}}{NaCl}$$

$$\underset{\text{Base}}{KOH} + \underset{\text{Acid}}{HNO_3} \rightarrow \underset{\text{Water}}{HOH} + \underset{\text{Salt}}{KNO_3}$$

$$\underset{\text{Base}}{Ba(OH)_2} + \underset{\text{Acid}}{H_2SO_4} \rightarrow \underset{\text{Water}}{2HOH} + \underset{\text{Salt}}{BaSO_4}$$

Arrhenius made acid-base theory quantative. The equilibrium constants for the ionization reactions are a direct measure of the strength of an acid. The higher the equilibrium constant the higher the hydrogen concentration:

$$\underset{\text{Acetic acid}}{CH_3COOH} \rightleftharpoons \underset{\text{Acetate ion}}{CH_3COO^-} + H^+ \qquad K_{eq} = K_i = \frac{[CH_3COO^-][H^+]}{[CH_3COOH]} = 1.8 \times 10^{-5}$$

† We know today that HCl and NaOH molecules do not exist in aqueous solutions, but in the 1880s the concept of stable ions was new; it was not realized that some substances could be completely ionized in solution.

$$HNO_2 \rightleftharpoons H^+ + NO_2^- \qquad K_{eq} = K_i = \frac{[H^+][NO_2^-]}{HNO_2} = 4.6 \times 10^{-4}$$

Nitrous acid — Nitrite ion

$$HCl \rightleftharpoons H^+ + Cl^- \qquad K_{eq} = K_i = \frac{[H^+][Cl^-]}{[HCl]} = \text{very large}$$

Hydrochloric acid

The equilibrium constant for the ionization of nitrous acid is about 25 times greater than that for the ionization of acetic acid. Nitrous acid is stronger than acetic acid. The equilibrium constant for the ionization of an acid or a base is called the *ionization constant* K_i.

The ion product for water

Arrhenius knew that the purest obtainable water has some ability to conduct electricity. To account for this observation he proposed that water molecules ionize slightly into H^+ and OH^-:

$$HOH \rightleftharpoons H^+ + OH^- \tag{13.1}$$

The equilibrium constant for this ionization is

$$K_{eq} = \frac{[H^+][OH^-]}{[HOH]} \tag{13.2}$$

The concentrations of H^+ and OH^- must be equal in pure water because they form in equimolar amounts. At 25°C the concentrations of H^+ and OH^- are both 10^{-7} mol/l:

$$[H^+] = [OH^-] = 1 \times 10^{-7} \text{ mol/l}$$

We can substitute this value in Eq. (13.2)

$$K_{eq} = \frac{(1 \times 10^{-7})(1 \times 10^{-7})}{[HOH]} \tag{13.3}$$

The concentration of water in moles per liter can be determined from the stoichiometry of the ionization, Eq. (13.1). If 1×10^{-7} mol/l of H^+ (or OH^-) is formed, then 1×10^{-7} mol/l of HOH is consumed. One liter, or 1000 g, of water represents 55.55 mol; if only 1×10^{-7} mol of the 55.55 mol is consumed, the concentration of water is constant for all practical purposes and we can write:

$$K_{eq}[HOH] = [H^+][OH^-] = \text{constant}$$

The ion product of water, $[H^+][OH^-]$, is constant. Using the concentrations of H^+ and OH^- for pure water at 25°C, we find that the ion product of water K_w is 10^{-14} at 25°C:

$$K_w = K_{eq}[HOH] = [H^+][OH^-]$$
$$K_w = [H^+][OH^-] = (1 \times 10^{-7})(1 \times 10^{-7}) = 1 \times 10^{-14} \qquad \text{at 25°C}$$

Note that since the product $[H^+][OH^-]$ can never be zero, neither H^+ nor OH^- can ever be zero. In a basic solution $[H^+]$ is low but never zero; in an acidic solution $[OH^-]$ is low but never zero.

EXAMPLE 13.1 Calculate the concentration of OH^- in a 3.00 M hydrochloric acid solution at 25°C.

Solution: Since HCl is completely ionized, the H^+ concentration is 3.00 M; 3 mol of HCl will produce 3 mol of H^+: $HCl \rightarrow H^+ + Cl^-$. (The contribution of H^+ from the ionization of water is negligible compared to the 3 mol produced by the HCl.) The ion product for water is constant:

$$K_w = [H^+][OH^-] = 1 \times 10^{-14}$$

Substituting $[H^+] = 3.00$ mol/l, we have

$$3.00[OH^-] = 1 \times 10^{-14} \qquad \text{or} \qquad [OH^-] = 3.33 \times 10^{-15} \text{ mol/l}$$

Questions unanswered by the Arrhenius theory

There are obvious shortcomings in the Arrhenius theory, of which three are outstanding: (1) the theory is limited to aqueous solutions; (2) there are substances, like ammonia and other nitrogen-containing compounds, which contain no hydroxy groups but which form basic solutions in water;† (3) the theory cannot explain why some salts do not form neutral solutions. The salt of a weak acid like sodium acetate, $NaC_2H_3O_2$, which does not contain a hydroxyl group, forms a basic solution and the salt of a weak base, like ammonium chloride, NH_4Cl, which contains no ionizable hydrogen, forms an acid solution.

The solvent-system theory of acids and bases

The solvent-system theory generalizes the Arrhenius theory and makes it applicable to solvents other than water. In this theory an acid is defined as any substance that gives the cation characteristic of the solvent and a base as any substance that gives the anion characteristic of the solvent. The solvents other than water in which studies have been carried out include liquid ammonia, NH_3, and liquid sulfur dioxide, SO_2. Each of these solvents can undergo ionization to produce cations and anions:

$$2NH_3 \rightleftharpoons \underset{\text{Ammonium ion}}{NH_4^+} + \underset{\text{Amide ion}}{NH_2^-}$$

† Aqueous solutions of ammonia are called ammonium hydroxide solutions, and at one time chemists believed that ammonia, NH_3, was present as ammonium hydroxide, NH_4OH. The formation of NH_4OH in water could occur simply: $NH_3 + HOH \rightarrow NH_4OH$, but no one has ever found any evidence for the existence of NH_4OH molecules.

$$2SO_2 \rightleftharpoons \underset{\text{Thionyl ion}}{SO^{2+}} + \underset{\text{Sulfite ion}}{SO_3^{2-}}$$

We can represent the ionization of water in analogous manner:

$$2HOH \rightleftharpoons \underset{\text{Hydronium ion}}{H_3O^+} + \underset{\text{Hydroxide ion}}{OH^-}$$

The hydronium ion is a *hydrated* hydrogen ion or a hydrated proton, a proton bonded to a water molecule; it is the cation of the solvent water. According to the solvent-system theory of acids and bases, neutralization involves combining the cation and anion to produce the solvent and a salt:

$$\underset{\text{Ammonium chloride}}{NH_4Cl} + \underset{\text{Potassium amide}}{KNH_2} \xrightarrow[NH_3]{\text{in}} \underset{\text{Solvent}}{2NH_3} + \underset{\text{Salt}}{KCl}$$

$$\underset{\text{Thionyl chloride}}{SOCl_2} + \underset{\text{Sodium sulfite}}{NaSO_3} \xrightarrow[SO_2]{\text{in}} \underset{\text{Solvent}}{2SO_2} + \underset{\text{Salt}}{2NaCl}$$

$$\underset{\text{Hydrochloric acid}}{H_3O^+Cl^-(aq)} + Na^+OH^-(aq) \xrightarrow[H_2O]{\text{in}} \underset{\text{Solvent}}{2H_2O} + \underset{\text{Salt}}{NaCl(aq)}$$

Note that NH_4^+ and H_3O^+ are *solvated* protons; in liquid ammonia the proton is solvated by NH_3 ($H^+ + NH_3 \rightarrow NH_4^+$), and in water the proton is solvated by H_2O ($H^+ + H_2O \rightarrow H_3O^+$).

Although the solvent-system theory generalizes the Arrhenius acid-base theory so that it can be applied to nonaqueous solutions, it still does not overcome the objection that substances which contain no hydroxy groups act like bases in water. In addition, there are substances which contains no NH_2^- ions but act like bases in ammonia. Nevertheless the solvent-system theory is very useful in comparing and classifying reactions in nonaqueous media.

The Brönsted-Lowry proton theory of acids and bases

Almost simultaneously in the 1920s J. N. Brönsted (1879–1947), of Denmark, and T. M. Lowry (1874–1936), of England, independently suggested that an acid can be defined as a proton donor and a base as a proton acceptor. By these newer definitions an Arrhenius acid is still an acid and an Arrhenius base is still a base. For example, in the reaction of acetic acid and sodium hydroxide the acetic acid donates a proton to the NaOH:

$$\text{(H)}C_2H_3O_2 + OH^- \rightarrow C_2H_3O_2^- + HOH$$

(NaOH) (NaC_2H_3O)

According to the Brönsted-Lowry theory, NaOH is a base because it accepts a proton, not because it produces OH^- ions.

The Brönsted-Lowry theory of acids and bases emphasized the role of the solvent as an acid or base. When the covalent compound HCl is dissolved in most solvents, it is a weak acid and its solutions have a low electric conductivity. In water, however, an HCl solution is strongly acidic and has a high electric conductivity. Evidently, an aqueous HCl solution, commonly called hydrochloric acid, is highly ionized. These observations suggest that the solvent is more than a matrix for the ionization of the acid; the solvent must be intimately involved. By the Arrhenius theory we would write

$$HCl \rightarrow H^+ + Cl^-$$

which ignores any possible role of the solvent, but by the Brönsted-Lowry theory we would write

$$HCl + H_2O \rightarrow H_3O^+ + Cl^-$$

In other words an aqueous HCl solution is a strong acid because water readily accepts a proton from HCl to form H_3O^+. In other solvents proton transfer does not occur to such a great extent, and HCl solutions are weak acids.

By the Brönsted-Lowry definition, HCl, the proton donor, is an acid and H_2O, the proton accepter, is a base. As a result of the reaction between HCl and H_2O a new acid and a new base are formed, called the *conjugate acid* and the *conjugate base*. Every acid and every base has its conjugate in the Brönsted-Lowry theory:

$$\underset{\text{Acid}}{HCl} + \underset{\text{Base}}{H_2O} \rightarrow \underset{\text{Conjugate acid}}{H_3O^+} + \underset{\text{Conjugate base}}{Cl^-} \tag{13.4}$$

The acid HCl donates a proton and is transformed into the conjugate base, Cl^-; the base H_2O accepts a proton and is transformed into the conjugate acid, H_3O^+. HCl and Cl^- are called a *conjugate acid-base* pair; H_3O^+ and H_2O are also a conjugate acid-base pair.

The reaction represented by Eq. (13.4) goes completely to the right; no HCl molecules can be detected in an aqueous HCl solution. One mole of HCl will produce 1 mol of H_3O^+ and 1 mol of Cl^- in aqueous solution. Equation (13.4) is written with a single arrow to indicate that the reaction is not reversible. Many of the common mineral acids, for example, HNO_3, H_2SO_4, and $HClO_4$ (perchloric acid), are completely ionized in water. These are called *strong acids.*

When acetic acid is dissolved in water, the electric conductance of the resulting solution is low, indicating that not many ions are present. Most of the dissolved acetic acid remains molecular. The acid-base reaction by which acetic acid molecules donate protons to water molecules comes to equilibrium with most of the solute present as molecules. The equation for the reaction is written with a double arrow to indicate reversibility:

$$\underset{\text{Acid}}{HC_2H_3O_2} + \underset{\text{Base}}{H_2O} \rightleftharpoons \underset{\text{Conjugate acid}}{H_3O^+} + \underset{\text{Conjugate base}}{C_2H_3O_2^-} \qquad (13.5)$$

The conjugate base in this case is the acetate ion, $C_2H_3O_2^-$.

In Eq. (13.4) the conjugate base is the chloride ion, Cl^-. The chloride ion is a weaker base, i.e., a weaker proton acceptor, than the acetate ion. There is no measurable tendency for the chloride to accept a proton from H_3O^+ in water solution to produce molecular HCl.

In the Brönsted-Lowry theory water can act as an acid or a base. When ammonia gas, NH_3, is dissolved in water, the solution is basic. By the Brönsted-Lowry theory we can indicate this as

$$\underset{\text{Base}}{NH_3} + \underset{\text{Acid}}{\text{(H)}OH} \rightleftharpoons \underset{\text{Conjugate acid}}{NH_4^+} + \underset{\text{Conjugate base}}{OH^-}$$

The acid, HOH, donates a proton to the base, NH_3. The ammonium ion, NH_4^+, is the conjugate acid. This reaction does not go to completion; the conductance of an ammonia solution is quite low, indicating that few ions are present and ammonia is a weak base. Substances like water which can donate or accept a proton are called *amphiprotic*.

Arrhenius explained the slight electric conductance of a pure water by simple ionization:

$$HOH \rightleftharpoons H^+ + OH^-$$

By the Brönsted-Lowry theory we explain this property by proton transfer, which emphasizes the amphiprotic character of water:

$$HOH + HOH \rightleftharpoons H_3O^+ + OH^-$$

The water product in terms of the Brönsted-Lowry theory would involve $[H_3O^+]$ instead of $[H^+]$:

$$K_w = [H_3O^+][OH^-] = 1 \times 10^{-14} \qquad \text{at } 25°C$$

The chief difference between the Brönsted-Lowry and Arrhenius theories is in the definition of a base. The Brönsted-Lowry theory allows us to account for the basicity of many compounds (like NH_3, other nitrogen compounds, and certain ions) which do not contain hydroxyl groups. When such compounds are dissolved, they accept a proton from water and produce hydroxide ions. The following equations show how these substances act as Brönsted-Lowry bases by accepting a proton from H_2O:

$$\underset{\text{Base}}{NH_3} + \underset{\text{Acid}}{H_2O} \rightleftharpoons \underset{\text{Conjugate acid}}{NH_4^+} + \underset{\text{C. base}}{OH^-} \tag{13.6}$$

$$\underset{\text{Acetate (base)}}{(Na^+)C_2H_3O^-} + \underset{\text{(acid)}}{H_2O} \rightleftharpoons \underset{\text{Acetic acid (conjugate acid)}}{HC_2H_3O} + \underset{\text{(conjugate base)}}{(Na^+)OH^-} \tag{13.7}$$

The Brönsted-Lowry theory is applicable in any "protonic" solvent system, i.e., in any solvent containing molecules that can donate or accept a proton. These include many common solvents such as ammonia, ethyl alcohol, and acetic acid, in addition to water. Proton transfer in these solvents is represented by the following equations:

$$\underset{\text{Ammonia}}{NH_3} + \underset{\text{Ammonia}}{(H)NH_2} \rightleftharpoons \underset{\text{Ammonium ion}}{NH_4^+} + \underset{\text{Amide ion}}{NH_2^-} \tag{13.8}$$

$$\underset{\text{Ethyl alcohol}}{CH_3CH_2OH} + \underset{\text{Ethyl alcohol}}{CH_3CH_2O(H)} \rightleftharpoons \underset{\text{Ethonium ion}}{CH_3CH_2OH_2^+} + \underset{\text{Ethoxide ion}}{CH_3CH_2O^-} \tag{13.9}$$

Since ethyl alcohol is very weak acid and the ethoxide ion is a very strong base, the equilibrium represented by Eq. (13.9) lies far to the left.

The Lewis theory of acids and bases

G. N. Lewis, one of the pioneers in the electron theory of bonding (see Chap. 5), considered acids and bases in terms of the electronic structures of the molecules. According to Lewis, a base is a species that can donate a pair of electrons in the formation of a covalent bond, and an acid is a species that can accept a pair of electrons in the formation of a covalent bond. All the species previously classified as acids or bases by the Arrhenius and Brönsted-Lowry theories are still acids or bases by the Lewis theory. However, the list of acids and bases is extended by this more general theory. In the Arrhenius theory an acid-base reaction is simply a combination of H^+ and OH^- ions to form H_2O. By the Lewis theory this is an acid-base reaction because the acid H^+ accepts an electron pair from OH^- and a covalent bond is formed:

$$H^+ + :\ddot{\underset{..}{O}}:H^- \rightarrow H:\ddot{\underset{..}{O}}:H$$

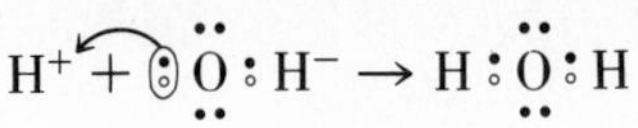

By the Brönsted-Lowry theory the reaction between NH_3 and acetic acid, $HC_2H_3O_2$, is an acid-base reaction because the acid donates a proton to ammonia; but by the Lewis theory this is an acid-base reaction because it involves the formation of a covalent bond as the ammonia molecule donates an electron pair:

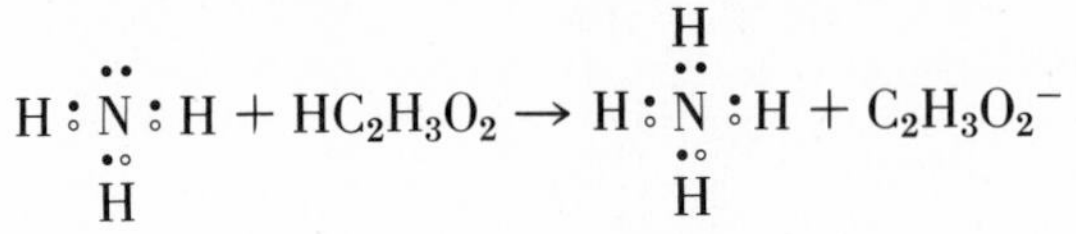

$$H:\underset{\underset{H}{:}}{\ddot{N}}:H + HC_2H_3O_2 \rightarrow H:\overset{\overset{H}{:}}{\underset{\underset{H}{:}}{N}}:H + C_2H_3O_2^-$$

Ammonium acetate

The Lewis theory extends the acid-base concept beyond those compounds which can be involved in protonic transfers. Boron trifluoride, BF_3, is an electron-deficient compound; in this compound boron has only six electrons in its valence shell. Ammonia and many other compounds in which nitrogen is trivalent possess an unshared pair of electrons. By the Lewis theory BF_3 and NH_3 undergo an acid-base reaction:

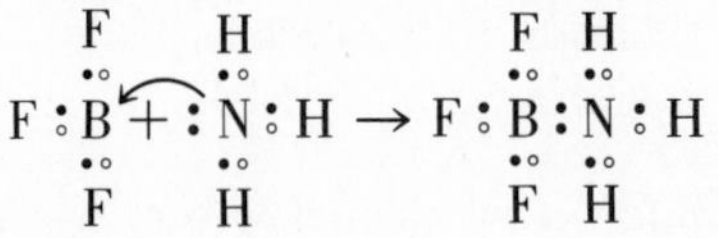

$$F:\overset{\overset{F}{:}}{\underset{\underset{F}{:}}{B}} + :\overset{\overset{H}{:}}{\underset{\underset{H}{:}}{N}}:H \rightarrow F:\overset{\overset{F}{:}}{\underset{\underset{F}{:}}{B}}:\overset{\overset{H}{:}}{\underset{\underset{H}{:}}{N}}:H$$

Acid Base

The acid, BF_3, accepts an electron pair from the base NH_3, and a new covalent bond is formed.

All Lewis bases have one fundamental property in common. They have available a *lone pair*, or unshared pair, of electrons. All the acids have one fundamental property in common: they have an empty molecular orbital available for the acceptance of an electron pair. The following reactions are given as examples of Lewis acid-base reactions:

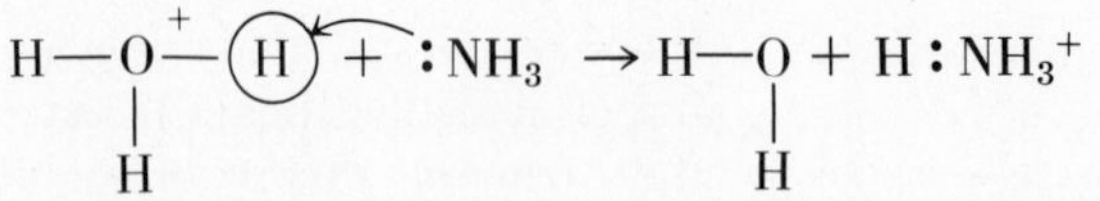

$$H-\overset{+}{\underset{\underset{H}{|}}{O}}-H + :NH_3 \rightarrow H-\underset{\underset{H}{|}}{O} + H:NH_3^+$$

Hydronium ion (Hydrochloric acid) | Ammonia | Ammonium ion (Ammonium chloride)

$$\mathrm{F\!:\!\underset{\displaystyle F}{\overset{\displaystyle F}{\ddot{\underset{\cdot\cdot}{B}}}} + :\!\ddot{O}\!-\!C_2H_5 \; (\text{with } C_2H_5 \text{ on O}) \rightarrow F\!:\!\underset{\displaystyle F}{\overset{\displaystyle F}{\ddot{\underset{\cdot\cdot}{B}}}}\!:\!\ddot{O}\!-\!C_2H_5 \; (\text{with } C_2H_5 \text{ on O})}$$

Boron trifluoride | Diethyl ether | Boron trifluoride ether (addition compound)

$$\mathrm{F\!:\!\underset{\displaystyle F}{\overset{\displaystyle F}{\ddot{\underset{\cdot\cdot}{B}}}} \;+\; :\!\ddot{\underset{\cdot\cdot}{F}}\!:^{-} \;\rightarrow\; \left[F\!:\!\underset{\displaystyle F}{\overset{\displaystyle F}{\ddot{\underset{\cdot\cdot}{B}}}}\!:\!F\right]^{-}}$$

Boron trifluoride | Fluoride ion (from KF) | Tetrafluoroboride ion (KBF_4)

13.3 CHEMICAL EQUILIBRIUM AND THE IONIZATION OF WEAK ACIDS AND BASES

In this section we use numerical examples to show that the principles of chemical equilibrium presented in Chap. 11 are also applicable to the ionization reactions of weak acids and bases. We shall use the Brönsted-Lowry acid-base definitions.

When a weak acid like acetic acid, $HC_2H_3O_2$ (or HAc), is dissolved in water, hydronium ions, H_3O^+, are formed in a reaction that quickly comes to equilibrium.

$$\mathrm{HAc + HOH \rightleftharpoons H_3O^+ + Ac^-}$$

The equilibrium-constant expression for this reaction is

$$K_{eq} = \frac{[\mathrm{H_3O^+}][\mathrm{Ac^-}]}{[\mathrm{HAc}][\mathrm{HOH}]}$$

Since the number of moles of water hardly changes during the reaction, its concentration is virtually constant and we can write

$$K_i = K_{eq}[\mathrm{HOH}] = \frac{[\mathrm{H_3O^+}][\mathrm{Ac^-}]}{[\mathrm{HAc}]}$$

The product $K_{eq}[\mathrm{HOH}]$ is a constant which we call the *ionization constant* K_i. It is the constant which measures the extent to which the dissolved acid will form hydronium ions, and therefore the constant is a measure of the strength of the acid. Many chemists simply write H^+ for hydronium ion; it is understood that the proton, H^+, is hydrated in an aqueous solution. Hence, we can write the ionization-constant expression as

$$K_i = \frac{[\mathrm{H^+}][\mathrm{Ac^-}]}{[\mathrm{HAc}]}$$

which is the way we would write the ionization-constant expression if we were using the Arrhenius theory, where the ionization of acetic acid would be represented as

$$HAc \rightleftharpoons H^+ + Ac^-$$

EXAMPLE 13.2 At 25°C a 0.100 *M* HAc solution is 1.34 percent ionized. Calculate $[H^+]$, $[Ac^-]$, and K_i.

Solution: Write the equation for the reaction and under it tabulate the facts known about the species involved.

	HAc	+ HOH ⇌	H_3O^+	+ Ac^- (a)
Initial concentration, *M*:	0.100	——	~0.0	0.0
Equilibrium concentration, *M*:	0.100 − (0.100 × 0.0134) = 0.0987	——	0.100 × 0.0134 = 0.00134	0.100 × 0.0134 = 0.00134

In the line called initial concentration write down what the concentrations of the different species in the reaction would be *if* none of the dissolved HAc underwent ionization. In this case, there would be exactly 0.100 mol of HAc in each liter of solution, and its concentration would be 0.100 *M*. The concentration of water does not appear in the ionization-constant expression, and so we do not consider it. The concentration of H_3O^+ is not zero because there is always some H_3O^+ present in water, but it is very small compared to the amount that would form from the ionization of HAc. Therefore, without the ionization of HAc the concentration of H_3O^+ is virtually zero. The concentration of Ac^- would also be zero if no HAc ionized. Therefore, as shown in the first line under reaction (*a*) the concentrations of HAc, H_3O^+, and Ac^- are 0.100, 0, and 0 *M*, respectively.

In actuality, the acetic acid does undergo a rapid ionization reaction in aqueous solution, and the reaction quickly comes to equilibrium. What are the concentrations of the reactive species when the reaction is at equilibrium? We deduce these equilibrium concentrations from the initial concentrations and the chemical equation for the ionization. Then we write the equilibrium concentrations in the second line under reaction (*a*).

According to the information given in Example 13.2, 1.34 percent of the HAc ionized. The acetic acid concentration is decreased by 1.34 percent, it goes from 0.100 *M* to 0.100 − (0.0134 × 0.100) *M* or to 0.0987 *M*. The equation for the reaction indicates that H_3O^+ and Ac^- will form in equimolar amounts and that 1 mol of HAc will produce 1 mol each of H_3O and Ac^-. In this case 1.34 percent of the acetic acid, which is 0.00134 mol of acetic acid, reacts, and 0.00134 mol each of H_3O^+ and Ac^- form (in each liter). Therefore, as shown in the second line under reaction (*a*), the equilibrium concentrations of HAc, H_3O^+, and Ac^- are 0.0987, 0.00134, and 0.00134 *M*, respectively.

The equilibrium-constant expression is

$$K_i = \frac{[H_3O^+][Ac^-]}{[HAc]}$$

Substituting the equilibrium concentration of the species involved, we have

$$K_i = \frac{0.00134^2}{0.0987} = 1.82 \times 10^{-5}$$

Note that the equilibrium concentration of HAc is very near the initial concentration. No serious error would result if we used the initial concentration of HAc instead of its equilibrium concentration to determine the value of the equilibrium constant:

$$[HAc]_{eq} \cong [HAc]_{initial} = 0.100$$

$$K_i = \frac{[H_3O^+][Ac^-]}{[HAc]} = \frac{0.00134^2}{0.100} = 1.80 \times 10^{-5}$$

EXAMPLE 13.3 Ammonia, NH_3, is the most common weak base. By coincidence, at 25°C, its ionization constant is equal to the ionization constant of acetic acid, the most common weak acid.

$$NH_3 + HOH \rightleftharpoons NH_4^+ + OH^-$$

$$K_i = \frac{[NH_4][OH^-]}{[NH_3]} = 1.8 \times 10^{-5} \qquad \text{at } 25°C$$

Calculate the concentration of hydroxide ion, ammonium ion, and ammonia and the percentage ionization of NH_3 in a 0.300 *M* solution at 25°C.

Solution: Again we make a table under the equation for the reaction and consider what the concentration of the species would be if no ionization occurred.

	NH_3	+ HOH ⇌	NH_4^+	+ OH^-
Initial concentration, *M*:	0.300	——	0.0	~0.0
Equilibrium concentration, *M*:	$0.300 - 0.300\alpha$	——	0.300α	0.300α

The concentration of NH_3 would be 0.300 *M*; the concentration of water is not required. The concentration of NH_4^+ would be zero, and the concentration of OH^- would be *virtually* zero. Therefore, as shown in the first line, the concentrations of NH_3, NH_4^+, and OH^- without ionization are 0.300, 0.0, and 0.0 M, respectively. When equilibrium is reached, a certain fraction of the NH_3 will have ionized; let this fraction equal α. The concentration of NH_3 will diminish by 0.300α, so that the equilibrium concentration of NH_3 will be $0.300 - 0.300\alpha$. This is simply the initial amount minus the amount used in the reaction. The stoichiometry indicates that 1 mol of NH_3 will produce 1 mol each of NH_4^+ and OH^-. Since 0.300α mol/l of NH_3 has reacted, 0.300α mol of NH_4^+ and of OH^- will have been produced; therefore, the concentrations of NH_4^+ and OH^- at equilibrium are 0.300α M. The equilibrium concentrations of NH_3, NH_4^+, and OH^- are shown in the second line.

Substitution of the equilibrium concentrations of NH_3, NH_4^+, and OH^- into the ionization-constant expression will yield α, the fraction of ionization, since K_i is known:

$$K_i = \frac{[NH_4^+][OH^-]}{[NH_3]}$$

$$1.8 \times 10^{-5} = \frac{(0.300\alpha)^2}{0.300 - 0.300\alpha} = \frac{0.300^2\alpha^2}{0.300(1-\alpha)} = \frac{0.300\alpha^2}{1-\alpha}$$

The result is a quadratic equation in α

$$\alpha^2 + \frac{1.8 \times 10^{-5}}{0.300}\alpha - \frac{1.8 \times 10^{-5}}{0.300} = 0$$

or

$$\alpha^2 + 6 \times 10^{-5}\alpha - 6 \times 10^{-5} = 0$$

The general solution to a quadratic equation is

$$\alpha = \frac{-b \pm \sqrt{b^2 - 4ac}}{2a}$$

In our case $a = 1$, $b = 6 \times 10^{-5}$, and $c = -6 \times 10^{-5}$. Therefore,

$$\alpha = \frac{-6 \times 10^{-5} \pm \sqrt{(6 \times 10^{-5})^2 - 4(1)(-6 \times 10^{-5})}}{2(1)}$$

$$= \frac{-6 \times 10^{-5} \pm \sqrt{(36 \times 10^{-10}) + (24 \times 10^{-5})}}{2}$$

Under the square-root sign we can neglect 36×10^{-10} since $|24 \times 10^{-5}| \gg |36 \times 10^{-10}|$, and we have

$$\alpha = \frac{-6 \times 10^{-5} \pm \sqrt{2.4 \times 10^{-4}}}{2} = \frac{-6 \times 10^{-5} \pm 1.55 \times 10^{-2}}{2}$$

Now we can neglect -6×10^{-5} since $|6 \times 10^{-5} \ll |1.55 \times 10^{-2}|$. Finally, we have

$$\alpha = \frac{\pm 1.55 \times 10^{-2}}{2} = \pm 0.775 \times 10^{-2}$$

Of the two mathematical solutions we consider only the positive value, as there is no physical significance to a *negative* fraction of ionization.

With the value of α it is possible to calculate the equilibrium concentrations of NH_3, NH_4^+, and OH^-. From the second line under reaction (a)

$$[NH_3] = 0.300 - 0.300\alpha = 0.300 - (0.300 \times 0.00775)$$
$$= 0.300 - 0.00232 = 0.298\ M$$
$$[NH_4^+] = [OH^-] = 0.300\alpha = 0.300 \times 0.00775 = 2.32 \times 10^{-3}\ M$$

Note that the equilibrium concentration of NH_3 is only slightly less than the initial concentration given in the first line under reaction (a):

$$[NH_3]_{eq} = 0.297 \qquad [NH_3]_{initial} = 0.300$$

If we keep in mind the fact that the concentration of a weak acid or weak base is hardly affected by ionization, we can make an assumption which simplifies the arithmetic in these calculations, namely that the equilibrium concentration of ammonia, $[NH_3]_{eq}$, equals the initial concentration of ammonia

$$[NH_3]_{eq} = [NH_3]_{initial} = 0.300$$

We have neglected the ionization of NH_3. Substitution of the 0.300 for NH_3 in the ionization-constant expression now yields

$$K_i = \frac{[NH_4^+][OH^-]}{[NH_3]}$$

$$1.8 \times 10^{-5} = \frac{(0.300\alpha)^2}{0.300} = 0.300\alpha^2$$

$$\alpha = \pm 0.00775$$

This is the same result we obtained when we used the quadratic formula, so that we are justified in making the simplifying assumption.

As a rule of thumb we can neglect the decrease in the concentration of the molecular form of a weak acid or a weak base caused by ionization provided that the molar concentration M divided by the ionization constant is equal to or greater than 1000. If $M/K_i > 1000$, neglect the effect of ionization. Note that we must consider the concentration as well as the ionization constant. At extremely low concentrations the fraction of ionization can be quite high, even for extremely weak acids and bases.

Let us apply our rule to the solved examples above. In Example 13.2, $K_i = 1.82 \times 10^{-5}$, $M = 0.100$, and

$$\frac{M}{K_i} = \frac{0.100}{1.82 \times 10^{-5}} = 5.49 \times 10^3$$

Since M/K_i is greater than 1000, we could have neglected the decrease in the acetic acid concentration due to its ionization. This is obvious from the examination of reaction (*a*), which shows $[HAc]_{initial} \approx [HAc]_{eq}$. In Example 13.3 involving NH_3, $K_i = 1.8 \times 10^{-5}$, and the concentration of ammonia is initially 0.300; $M/K_i = 0.300/(1.8 \times 10^{-5}) = 16{,}700$. Our rule shows, as we proved in the solution of Example 13.3, that we could have neglected the decrease of the ammonia concentration due to ionization.

The ionization constants for many weak acids and bases have been determined. These are presented in Table 13.1. The significance of the numbers in the columns under K_2 and K_3 is discussed in Sec. 13.8.

EXAMPLE 13.4 Calculate the hydrogen-ion concentration in a 0.100 *M* solution of hydrocyanic acid, HCN, at 25°C; K_i for HCN is 7.2×10^{-10}.

Solution: In the solution of this problem we shall calculate $[H^+]_{eq}$ directly. It is not necessary to calculate the fraction of ionization to obtain $[H^+]_{eq}$. A summary of the initial and equilibrium concentrations is shown:

	HCN	+ HOH	⇌ H_3O^+	+ CN^-
Initial concentration:	0.100	——	~0	0
Equilibrium concentration:	$0.100 - x \approx 0.100$	——	x	x

In this case we let x equal the concentration of H_3O^+ at equilibrium. Since CN^- is formed in equimolar amounts, its equilibrium concentration is also x. The stoichiometry indicates that if 1 mol of H_3O^+ is produced, 1 mol of HCN is consumed. If x mol of H_3O^+ is produced, then, x mol of HCN must have been consumed. The initial concentration was 0.100 *M*, and equilibrium concentration is $0.100 - x$, but since $M/K_i = 0.100/7.2 \times 10^{-10} \gg 10^3$, we can neglect x with respect to 0.100 *M*; the equilibrium concentration of HCN is very nearly 0.100 *M*.

Substitution of the equilibrium concentrations into the ionization-constant expression yields

$$K_i = \frac{[H_3O^+][CN^-]}{[HCN]} \qquad \text{or} \qquad 7.2 \times 10^{-10} = \frac{x^2}{0.100}$$

or

$$x = 8.5 \times 10^{-6}\ M = [H_3O^+]_{eq}$$

It is obvious that our neglect of x with respect to 0.100 was justified:

$$0.100 - x = 0.100 - 0.0000085 \approx 0.100$$

TABLE 13.1: IONIZATION CONSTANTS FOR WEAK ACIDS AND BASES AT 25°C

	Formula	K_1	K_2	K_3
Acid:				
Formic	$HCOOH$	1.77×10^{-4}		
Acetic	CH_3COOH	1.75×10^{-5}		
Propionic	C_2H_5COOH	1.34×10^{-5}		
Benzoic	C_6H_5COOH	6.29×10^{-5}		
Hydrofluoric	HF	7.2×10^{-4}		
Phosphoric	H_3PO_4	7.52×10^{-3}	6.23×10^{-8}	4.8×10^{-13}
Phosphorous	H_3PO_3	1.6×10^{-2}	7×10^{-7}	
Sulfuric	H_2SO_4	Strong	1.01×10^{-2}	
Sulfurous	H_2SO_3	1.72×10^{-2}	6.24×10^{-8}	
Arsenic	H_3AsO_4	5.0×10^{-3}	8.3×10^{-8}	6×10^{-10}
Boric	H_3BO_3	5.80×10^{-10}		
Carbonic	H_2CO_3	4.52×10^{-7}	4.69×10^{-11}	
Hydrocyanic	HCN	7.2×10^{-10}		
Iodic	HIO_3	1.67×10^{-1}		
Hydrosulfuric	H_2S	7.9×10^{-8}	2.0×10^{-15}	
Base:				
Ammonia	NH_3	1.76×10^{-5}		
Methylamine	CH_3NH_2	4.40×10^{-4}		
Aniline	$C_6H_5NH_2$	3.83×10^{-10}		
Pyridine	C_5H_5N	1.56×10^{-9}		

13.4 THE HYDROGEN-ION INDEX, pH

The ion product K_w for water is 1×10^{-14} at 25°C. In pure water the concentration of hydrogen and hydroxide ions must be equal because of the stoichiometry of the dissociation process:

Arrhenius: $H_2O \rightleftharpoons H^+ + OH^-$

Brönsted-Lowry: $H_2O + H_2O \rightleftharpoons H_3O^+ + OH^-$

$K_w = [H_3O^+][OH^-] = 1 \times 10^{-14} \qquad \text{at } 25°\text{C}$

and

$$[H_3O^+] = [OH^-] = \sqrt{1 \times 10^{-14}} = 1 \times 10^{-7}\ M$$

TABLE 13.2: K_w FOR WATER AT VARIOUS TEMPERATURES

t, °C	$K_w = [H_3O^+]\ [OH^-]$
0	1.14×10^{-15}
10	2.92×10^{-14}
25	1.01×10^{-14}
50	5.46×10^{-14}
60	0.961×10^{-13}

The ion product changes with temperature. At 0°C it is very close to 10^{-15}, and at 60°C it is almost 10^{-13}. Values of K_w are given for several temperatures in Table 13.2.

The hydrogen-ion concentration of aqueous solutions is conventionally given in terms of pH. The pH of a solution is defined as the negative logarithm (to the base 10) of the hydrogen-ion activity (see page 267 for discussion of activity). In our work we shall assume that the activity and the concentration are equal so that we can consider the pH to be the negative logarithm of the hydrogen-ion *con-*

centration:

$$pH = -\log [H_3O^+] \tag{13.10}$$

The symbol p represents "negative log of," and we can speak of pOH, pK_w and pK_i, where

$$pOH = -\log [OH^-] \tag{13.11}$$

$$pK_w = -\log K_w \qquad pK_i = -\log K_i$$

At 25°C the concentration of hydronium ion in pure water is 1×10^{-7} *M*. The pH is very simply calculated:

$$pH = -\log [H_3O^+] = -\log (1 \times 10^{-7}) = -(0 - 7) = 7$$

The pH of a neutral solution at ordinary temperatures is 7. The pOH is also 7 and the pK_w is 14 as can be seen from the following manipulation:

$$K_w = [H_3O^+][OH^-] = 1 \times 10^{-14} \tag{13.12}$$

$$\log K_w = \log [H_3O^+] + \log [OH^-] = \log (1 \times 10^{-14})$$
$$-\log K_w = -\log [H_3O^+] - \log [OH^-] = 14$$

By the definition of p, we have

$$pK_w = pH + pOH = 14 \tag{13.13}$$

We shall now show the solution to some example problems involving the pH concept.

EXAMPLE 13.5 Calculate the pH of an acidic solution at 25°C if the hydrogen-ion concentration is 5.00×10^{-4}.

Solution

$$pH = -\log [H_3O^+] = -\log (5.00 \times 10^{-4})$$
$$= 4 - \log 5.00 = 4 - 0.699 = 3.30$$

EXAMPLE 13.6 Calculate the pH and the pOH of a 0.001 *M* HCl solution at 25°C.

Solution: HCl is completely ionized in water, so that $[H_3O^+] = 0.001$.

$$pH = -\log 0.001 = -\log 10^{-3} = -(-3) = 3$$

We can use Eq. (13.12) or (13.13) to calculate the pOH. Using Eq. (13.13), we have

$$pH + pOH = 14 \qquad \text{or} \qquad pOH = 14 - pH = 14 - 3 = 11$$

Using Eq. (13.12), we have

$$[H_3O^+][OH^-] = 1 \times 10^{-14}$$

from which we can calculate the hydroxide-ion concentration:

$$[OH] = \frac{1 \times 10^{-14}}{[H_3O^+]} = \frac{1 \times 10^{-14}}{10^{-3}} = 1 \times 10^{-11}$$

By definition

$$pOH = -\log [OH^-] = -\log (1 \times 10^{-11}) = -(0 - 11) = 11$$

EXAMPLE 13.7 The pH of normal blood plasma measured at 25°C is 7.4. Calculate $[H_3O^+]$ and $[OH^-]$. Is blood plasma acidic, basic, or neutral?

Solution:

$$pH = 7.4 \qquad \text{and} \qquad pH = -\log [H_3O^+]$$

Therefore,

$$\log [H_3O^+] = -7.4$$
$$[H_3O^+] = 10^{-7.4} = 10^{0.6} \times 10^{-8} = 3.9 \times 10^{-8}$$

We can calculate $[OH^-]$ by using the water product:

$$K_w = [H_3O^+][OH^-]$$
$$1 \times 10^{-14} = 3.9 \times 10^{-8}[OH^-]$$
$$[OH^-] = 2.6 \times 10^{-7}$$

Since $[OH^-]$ is slightly greater than $[H_3O^+]$, blood plasma is slightly basic.

EXAMPLE 13.8 Calculate the pH of a 0.0200 M base solution if its ionization constant is 2.00×10^{-7}.

Solution: As before, we write the equation representing the ionization process and then write the initial and equilibrium concentrations of all the species. Since the key to the solution of the problem is to obtain the hydroxide-ion concentration at equilibrium, we let $x = [OH^-]_{eq}$ in the analysis of the problem.

	B	+ HOH	⇌ BH^+	+ OH^-
Initial concentration, M:	0.0200	——	0.0	~0.0
Equilibrium concentration, M:	$0.0200 - x \approx 0.0200$	——	x	x

Since $M/K_i = 0.0200/(2 \times 10^{-7}) = 100{,}000 > 1000$, we can neglect x with respect to 0.0200 and the equilibrium concentration of the base is approximately equal to the initial concentration. Next we substitute the equilibrium concentrations into the ionization-constant expression and solve for x:

$$K_i = \frac{[BH^+][OH^-]}{[B]} \qquad \text{or} \qquad 2.00 \times 10^{-7} = \frac{x^2}{0.0200}$$

$x^2 = 40.0 \times 10^{-10}$
$x = 6.33 \times 10^{-5}$
$[OH^-] = x = 6.33 \times 10^{-5}\ M$

With $[OH^-]$ it is possible to calculate the pOH and the pH:

$$pOH = -\log [OH^-] = -\log (6.33 \times 10^{-5})$$
$$= 5 - \log 6.33 = 5 - 0.801 = 4.20$$

From Eq. (13.13)

$$pH + pOH = 14 \qquad \text{and} \qquad pH = 14 - pOH$$

Substituting the value for pOH calculated above, we have

$$pH = 14 - pOH = 14 - 4.20 = 9.80$$

13.5 HYDROLYSIS OF SALTS

When hydrochloric acid reacts with sodium hydroxide, a salt, NaCl, is produced. When this salt is dissolved in pure water, the solution is neutral; at 25°C its pH is 7. This is true for any salt produced from a neutralization reaction of a *strong* acid and *strong* base.

The salts formed in the reaction of weak acids and strong bases produce aqueous solutions which are basic; the salts of weak bases and strong acids produce aqueous solutions that are acidic. It is not quite as simple to predict whether the salt of a weak acid and a weak base will form an acidic or basic solution. In this case, as we shall see below, both ionization constants must be considered. These observations, summarized in Table 13.3, are easily understood in terms of the Brönsted-Lowry theory of acids and bases.

When HCl is dissolved in water, the resulting hydrochloric acid solution contains H_3O^+ and Cl^- ions:

$$\underset{\text{Acid}}{HCl(g)} + \underset{\text{Base}}{HOH} \rightarrow \underset{\text{Conjugate acid}}{H_3O^+} + \underset{\text{C. base}}{Cl^-(aq)} \tag{13.14}$$

TABLE 13.3: THE pH OF SALT SOLUTIONS

Salts derived from	pH *of aqueous solution*	*Example*
Strong acid and strong base	~7 (neutral)	NaCl (NaOH + HCl) KNO_3 (KOH + HNO_3)
Weak acid and strong base	>7 (basic)	NaAc (NaOH + HAc) KCN (KOH + HCN)
Strong acid and weak base	<7 (acidic)	NH_4Cl (NH_3 + HCl)
Weak acid and weak base	Varies	NH_4CN (NH_3 + HCN) NH_4Ac (NH_3 + HAc)

This is not a reversible reaction because Cl^- is an extremely weak base. It cannot abstract a proton from H_2O or H_3O^+.

$$Cl^-(aq) + H_2O \text{ or } H_3O^+ \rightarrow \text{no reaction} \qquad (13.15)$$

HCl is strong acid because Cl^- does not react with water and the reaction given by Eq. (13.14) cannot "go to the left." Reaction with H_2O is called *hydrolysis*, and we could say that HCl is a strong acid in water because the chloride ion does not *hydrolyze* [Eq. (13.15)]; 1 mol of HCl will produce 1 mol of H_3O^+ in aqueous solutions.

When NaOH dissolves in pure water, the sodium and hydroxide ions present in the pure solid are able to move through the solution with some independence. NaOH is a strong base because Na^+ is a very weak acid and does not hydrolyze; 1 mol of NaOH produces 1 mol of OH^- in aqueous solutions.

NaCl is the salt of NaOH and HCl. Neither Na^+ nor Cl^- hydrolyzes, and a solution of NaCl is neutral.

Sodium acetate forms a basic solution because the anion of the acid, the acetate ion, is the Brönsted-Lowry base. It can accept a proton. The anions of all weak acids hydrolyze, or accept protons from the solvent, in solution. Sodium acetate is made up of Na^+ ions, which do not hydrolyze, and Ac^- ions, which do. Since the acetate ion is a Brönsted-Lowry base, sodium acetate forms a basic solution:

$$(Na^+) + Ac^- + HOH \rightleftharpoons (Na^+) + HAc + OH^-$$

Ammonium chloride forms an acidic solution because the ammonium ion, NH_4^+, is a Brönsted-Lowry acid. The cations of all weak bases hydrolyze, or donate protons to the solvent. Ammonium chloride consists of NH_4^+ ions, which do hydrolyze, and Cl^- ions, which do not. Since the ammonium ion is a Brönsted-Lowry acid, NH_4Cl produces an acidic solution:

$$NH_4^+ + (Cl^-) + HOH \rightleftharpoons NH_3 + (Cl^-) + H_3O^+$$

For the salt of a weak acid and a weak base both ions hydrolyze, and a knowledge of the ionization constants is required to predict whether the solution is acidic or basic. Ammonium cyanide, NH_4CN, is the salt of NH_3 and hydrocyanic acid, HCN. Both ions hydrolyze; NH_4^+ donates protons to the solvent, and CN^- accepts protons from the solvent:

$$H_3NH^+ + HOH \rightarrow NH_3 + H_3O^+$$
$$CN^- + HOH \rightarrow HCN + OH^-$$

Since the CN^- is more extensively hydrolyzed, the resulting solution of NH_4CN is basic. Other salts of weak acids and weak bases form acidic solutions. If the K_i of the weak base is greater than the K_i of the weak acid from which the salt is formed, the salt is basic. (K_i for

NH_3 is 1.8×10^{-5}, and K_i for HCN is 7.2×10^{-10}; NH_4CN is basic.) The converse of this statement is also true.

The hydrolysis reactions of the salts are equilibrium reactions, and we can analyze solutions of the salts of weak acids and bases in terms of the principles of chemical equilibrium. Again, the application of these principles will be demonstrated by specific numerical examples.

EXAMPLE 13.9 Calculate the hydroxide-ion concentration of a 0.100 M NaAc solution.

Solution: In this case we need only consider the hydrolysis reaction of the acetate ion because the sodium ion does not hydrolyze.

$$Ac^- + HOH \rightleftharpoons HAc + OH^-$$

The equilibrium constant for this reaction is

$$K_{eq} = \frac{[HAc][OH^-]}{[Ac^-][HOH]}$$

Since the extent of the hydrolysis is slight, the water concentration is approximately constant and it is convenient to rewrite the expression as

$$K_h = K_{eq}[HOH] = \frac{[HAc][OH^-]}{[Ac^-]} \tag{13.16}$$

K_h is called the *hydrolysis constant*. Since in an aqueous solution $[OH^-]$ equals $K_w/[H_3O^+]$, the hydrolysis constant is simply equal to K_w/K_i as can be seen from

$$K_h = \frac{[HAc][OH^-]}{[Ac^-]} = \frac{[HAc](K_w/[H_3O^+])}{[Ac^-]} = \frac{[HAc]K_w}{[H_3O^+][Ac^-]}$$

The quotient $[HAc]/([H_3O^+][Ac^-])$ is just equal to the reciprocal of K_i for HAc and

$$K_h = \frac{K_w}{K_i} \tag{13.17}$$

Equation (13.17) holds for any weak acid or weak base. It is left as an exercise to show that Eq. (13.17) holds for NH_3 and other weak bases (see Prob. 13.31).

The hydrolysis constant for NaAc at 25°C is

$$K_h = \frac{K_w}{K_i} = \frac{1 \times 10^{-14}}{1.82 \times 10^{-5}} = 5.49 \times 10^{-10}$$

Now let us systematically attack Example 13.9. Tabulate the initial and equilibrium concentration of all species under an equation for the reaction.

	Ac^- +	$HOH \rightleftharpoons$	HAc +	OH
Initial concentration, M:	0.100	—	0.0	~0.0
Equilibrium concentration, M:	$0.100 - x \approx 0.100$	—	x	x

Let x equal the hydroxide-ion concentration at equilibrium; by the stoichiometry of the reaction this must also equal the HAc concentration. In view of the very low value of K_h the hydrolysis of Ac^- is only slight, and the concentration of Ac^- is hardly affected; we neglect x with respect to 0.100, and the equilibrium concentration of Ac^- is approximately equal to the initial concentration.

For hydrolysis reactions we can use the same criterion given for ionization of weak acids or bases. If the molar concentration of the salt divided by K_h is equal to or greater than 1000, the diminution of the salt concentration due to hydrolysis is negligible compared to the initial concentration of the salt. In this case M/K_h is much greater than 1000:

$$\frac{M}{K_h} = \frac{0.100}{5.49 \times 10^{-10}} = 1.82 \times 10^8 \gg 1000$$

and we are justified in neglecting x with respect to 0.100. Substitution of the equilibrium concentrations into the hydrolysis-constant expression gives

$$K_h = \frac{[HAc][OH^-]}{[Ac^-]} \qquad \text{or} \qquad 5.49 \times 10^{-10} = \frac{x^2}{0.100}$$

$$x = 7.41 \times 10^{-6} = [OH^-]$$

The hydroxide-ion concentration is 7.41×10^{-6} M.

EXAMPLE 13.10 An aqueous NH_4Cl solution must be prepared so that the pH is 5.00 at 25°C. What must the molar concentration of NH_4Cl be?

Solution: NH_4Cl is the salt of a weak base and strong acid. Only NH_4^+ will hydrolyze:

$$NH_4^+ + HOH \rightleftharpoons NH_3 + H_3O^+$$

Since the pH is specified, the concentration of H_3O^+ can be calculated:

$$pH = -\log [H_3O^+] \qquad 5 = -\log [H_3O^+]$$

or

$$[H_3O^+] = 1 \times 10^{-5}\ M$$

The NH_3 concentration must also be 1×10^{-5} M since H_3O^+ and NH_3 are formed in equimolar amounts by the hydrolysis of NH_4^+.

We shall assume that the NH_4^+ concentration is not appreciably diminished by hydrolysis since the hydrolysis constant for NH_4Cl is so small.

$$K_h = \frac{K_w}{K_i} = \frac{1.0 \times 10^{-14}}{1.80 \times 10^{-5}} = 5.55 \times 10^{-10}$$

All these considerations are tabulated under the equation for the hydrolysis; M represents the NH_4Cl concentration.

	NH_4^+	+ HOH $\rightleftharpoons$	NH_3	+ H_3O^+
Initial concentration:	M	—	0	~ 0
Equilibrium concentration:	$M - (1 \times 10^{-5}) \approx M$	—	1×10^{-5}	1×10^{-5}

Next we substitute into the hydrolysis constant expression and solve for M:

$$K_h = \frac{[NH_3][H_3O^+]}{[NH_4^+]} \qquad 5.55 \times 10^{-10} = \frac{(1 \times 10^{-5})^2}{M}$$

$M = 0.180$ = molar concentration of NH_4Cl.

Examples 13.9 and 13.10 dealt with NaAc and NH_4Cl, salts of a weak acid and strong base and of a strong acid and weak base, respectively. For the salt of a weak acid and a weak base, for example, NH_4CN or NH_4Ac, the hydrolysis depends on the ionization constants for the acid and the base since two separate hydrolysis reactions occur; e.g., consider NH_4CN:

$$NH_3\text{: } NH_4^+ + HOH \rightleftharpoons NH_3 + H_3O^+ \qquad K_h = \frac{K_w}{K_i} \tag{13.18}$$

$$HCN\text{: } CN^- + HOH \rightleftharpoons HCN + OH^- \qquad K_h = \frac{K_w}{K_i} \tag{13.19}$$

These reactions do not proceed to the same extent, so that the resulting solution is not neutral. The hydrolysis constant for CN^- is greater than the hydrolysis constant for NH_4^+:

$$CN^-\text{: } K_h = \frac{K_w}{K_i} = \frac{1.00 \times 10^{-14}}{7.2 \times 10^{-10}} = 1.4 \times 10^{-5}$$

$$NH_4^+\text{: } K_h = \frac{1.00 \times 10^{-14}}{1.8 \times 10^{-5}} = 5.5 \times 10^{-10}$$

Therefore, the reaction given by Eq. (13.19) predominates, and the solution is basic.

The concentration of the various species in a solution of a salt of a weak acid and a weak base, like NH_4CN, can be determined, but the calculation is complicated and we shall be satisfied to know only whether the solution of such a salt is acidic or basic.

An ionic equilibrium in solution, like any other chemical equilibrium, can be displaced by a change in concentration of any of the species involved. Let us examine the effect of adding NH_4Cl to a solution of NH_3.

In the ammonia solution the equilibrium is

$$NH_3 + HOH \rightleftharpoons NH_4^+ + OH^- \tag{13.20}$$

The addition of NH_4Cl will increase the ammonium-ion concentration, and the system will be momentarily out of equilibrium. By the principle of Le Châtelier the system will shift so as to relieve the stress; the system will react to consume the added NH_4^+ ion, which means that the addition of NH_4Cl to an NH_3 solution would cause the system to shift to the left decreasing $[OH^-]$ and increasing $[NH_3]$. We can demonstrate this *common-ion effect* quantitatively by the use of a numerical example.

EXAMPLE 13.11 Calculate (*a*) the OH^- concentration and (*b*) the pH of a 0.100 *M* NH_3 solution which is also 0.200 *M* in NH_4Cl.

Solution: (*a*) Again we tabulate the initial and equilibrium concentration of the species involved in the reaction under the equation for the reaction.

	NH_3	+	HOH	$\rightleftharpoons$	NH_4^+	+	OH^-
Initial concentration, *M*:	0.100		——		0.200		~0.0
Equilibrium concentration, *M*:	$0.100 - x \approx 0.100$		——		$0.200 + x \approx 0.200$		x

The initial concentration of NH_4^+ is just equal to the concentration of NH_4Cl since this salt is completely ionic and will contribute 1 mol of NH_4^+ ions for each mole of salt. Let x equal the moles per liter of OH^- formed at equilibrium. By the stoichiometry the concentration of NH_3 will be diminished by x and the concentration of the NH_4^+ ions will be augmented by x. Next, we substitute the equilibrium concentrations into the equilibrium-constant expression:

$$K_i = \frac{[NH_4^+][OH^-]}{[NH_3]} \qquad 1.8 \times 10^{-5} = \frac{(0.200 + x)x}{0.100 - x} \tag{13.21}$$

We would have a quadratic equation to solve, but again simplifying approximations can be made. The addition of NH_4Cl suppresses the ionization of NH_3. Therefore, if the extent of ionization is negligible in the absence of NH_4Cl, it is more negligible in the presence of NH_4Cl. Since x is the number of moles of NH_3 which ionize, x can be neglected with respect to 0.100, or $0.100 - x \approx 0.100$. If we neglect x with respect to 0.100, we must also neglect it with respect to 0.200. Therefore, the concentration of NH_4^+ is approximately 0.200 M. What we are saying is that the increase in the ammonium-ion concentration due to the hydrolysis of NH_3 is negligible. With these simplifications Eq. (13.21) reduces to

$$1.8 \times 10^{-5} = \frac{0.200x}{0.100} \qquad \text{or} \qquad x = 9 \times 10^{-6}$$

$$[OH^-] = 9 \times 10^{-6}\ M$$

The hydroxide-ion concentration in a 0.100 M NH_3 solution with no NH_4Cl added is 1.34×10^{-3} M, so that the concentration of OH^- ion is greatly depressed by the addition of NH_4Cl. This is the conclusion we reached qualitatively by the application of the principle of Le Châtelier.

(*b*) We can obtain the pH easily since $[OH^-]$ is now known:

$$K_w = [H_3O^+][OH^-] \qquad \text{or} \qquad [H_3O^+] = \frac{K_w}{[OH^-]}$$

$$[H_3O^+] = \frac{1.00 \times 10^{-14}}{9 \times 10^{-6}} = 1.11 \times 10^{-9}\ M$$

$$\text{pH} = -\log\ [H_3O^+] = -\log\ (1.11 \times 10^{-9}) = 9 - \log 1.11 = 8.95$$

A solution which contains a weak base and its salt (or a weak acid and its salt) is referred to as a *buffer* solution because it resists changes in pH. Examples of such solution are HAc and NaAc, NH_3 and NH_4Cl, and HCN and NaCN. Addition of a small amount of acid or base to a buffer solution causes only a slight change in pH relative to the unbuffered solutions. Again we can understand this behavior in terms of the Brönsted-Lowry theory. In a buffer solution containing HAc and NaAc, HAc acts as an acid and Ac^- acts as a base. Suppose we add HCl to such a solution. The Ac^- will react to reduce the H_3O^+ concentration:

$$\underset{\text{Acid}}{H_3O^+Cl^-} + \underset{\text{Base}}{Ac^-} \rightarrow HAc + HOH + Cl^-$$

If we add NaOH to the solution, the HAc will react to reduce the OH^- concentration:

$$\underset{\text{Base}}{Na^+OH^-} + \underset{\text{Acid}}{HAc} \rightarrow Na^+ + HOH + Ac^-$$

In a buffer system of NH_3 and NH_4^+, the NH_3 would counteract the effect of added acids, and the NH_4^+ ion would counteract the effect of added bases:

$$\underset{\text{Base}}{NH_3} + \underset{\text{Acid}}{H_3O^+Cl} \rightarrow NH_4^+ + HOH + Cl^-$$

$NH_4^+ + Na^+OH^- \rightarrow NH_3 + Na^+ + HOH$

Acid Base

Examples 13.12 through 13.15 demonstrate that a buffer solution resists changes in pH.

EXAMPLE 13.12 Calculate the pH of a 0.200 *M* HAc solution.

Solution: From the ionization-constant expression for HAc we can obtain H_3O^+ and then the pH:

$$K_i = \frac{[H_3O^+][Ac^-]}{[HAc]} \quad \text{or} \quad 1.8 \times 10^{-5} = \frac{x^2}{0.200}$$

$$x = [H_3O^+] = 1.9 \times 10^{-3}\ M$$
$$pH = -\log\ [H_3O^+] = -\log\ (1.9 \times 10^{-3}) = 3 - \log 1.9 = 2.72$$

EXAMPLE 13.13 Calculate the pH of the solution which results when 10.0 ml of 0.200 *M* NaOH solution is added to 90.0 ml of a 0.200 *M* HAc solution.

Solution: When we mix the HAc and NaOH solutions, a partial neutralization occurs. To determine the extent of the neutralization we must determine the number of moles of the acid and base:

$$\text{Moles acid} = \frac{0.200\ \text{mol}}{1000\ \text{ml}} \times 90\ \text{ml} = 0.018\ \text{mol}$$

$$\text{Moles base} = \frac{0.200\ \text{mol}}{1000\ \text{ml}} \times 10\ \text{ml} = 0.002\ \text{mol}$$

There are 0.002 mol of NaOH and 0.018 mol of HAc. When they are mixed, the NaOH neutralizes 0.002 mol of acid:

$$Na^-OH^+ + HAc \rightarrow Na^+Ac^- + HOH$$

	Na^-OH^+	HAc	Na^+Ac^-	HOH
Initial, mol:	0.002	0.018	0.0	____
Final, mol:	0.000	0.016	0.002	____

After the neutralization there will be 0.016 mol of HAc left, and 0.002 mol of Ac^- ion will have been produced. These substances are present in a total of 100 ml of solution, so that the concentrations are

$$[HAc] = \frac{0.016\ \text{mol}}{0.100\ \text{l}} = 0.16\ M \quad \text{and} \quad [Ac^-] = \frac{0.002\ \text{mol}}{0.1\ \text{l}} = 0.02\ M$$

By reasoning analogous to that given in Example 13.12 we write

$$K_i = \frac{[H_3O^+][Ac^-]}{[HAc]} \quad \text{or} \quad 1.8 \times 10^{-5} = \frac{0.02x}{0.16}$$

$$x = 1.44 \times 10^{-4} = [H_3O^+]$$
$$pH = -\log\ [H_3O^+] \doteq 4 - \log 1.44 = 3.84$$

Hence, comparing Examples 13.12 and 13.13, the pH changes from 2.72 to 3.84 when the NaOH is added to an unbuffered HAc solution. Now let us see what happens in a buffer solution of the same HAc concentration when the same amount of NaOH is added.

EXAMPLE 13.14 Calculate the pH of a buffer solution which is 0.200 *M* in HAc and 0.200 *M* in NaAc.

Solution: The pH of the buffer solution is calculated by recalling that NaAc supplies a species involved in the equilibrium ionization of HAc. This is the common-ion effect.

Make a tabulation of the initial and equilibrium concentration.

	HAc	+ HOH ⇌	H_3O^+	Ac^-
Initial concentration, *M*:	0.200	—	~0.0	0.200
Equilibrium concentration, *M*:	$0.200 - x$	—	x	$0.200 + x$

The equilibrium values of the concentration are substituted into the ionization-constant expression:

$$K_i = \frac{[H_3O^+][Ac^-]}{[HAc]} = \frac{x(0.200 + x)}{0.200 - x}$$

The extent of ionization is negligible in a 0.200 *M* HAc solution; $M/K_i \gg 1000$. When the common ion is added, the extent of ionization is even lower and so we may neglect x with respect to 0.200.

$$K_i = \frac{0.200x}{0.200} = x$$

$$x = K_i$$

but since $x = [H_3O^+]$

$$[H_3O^+] = K_i$$

We have obtained a very interesting result; when a weak acid and its salt are present in equimolar amounts, the hydrogen-ion concentration is equal to the ionization constant for the acid (or the pH equals the pK_i):

$$[H_3O^+] = K_i = 1.8 \times 10^{-5}$$
$$-\log\,[H_3O^+] = -\log K_i$$

By the definition of p,

$$pH = pK_i = -\log\,(1.8 \times 10^{-5}) = 5 - \log 1.8 = 4.74$$

The pH of the buffer solution is higher than the pH of the 0.200 *M*

HAc solution. This is to be expected since the buffering agent, Na^+Ac^-, is a Brönsted-Lowry base.

In Example 13.15 we shall see that the pH of the buffer solution undergoes only a relatively small change when base is added.

EXAMPLE 13.15 Calculate the pH of the solution which results when 10.0 ml of 0.200 *M* NaOH solution is added to 90.0 ml of the buffer solution described in Example 13.14.

Solution: When base is added to the buffer solution, an equivalent amount of the acid in the solution is neutralized. From Example 13.13 we know that 10.0 ml of 0.200 *M* NaOH contains 0.002 mol of NaOH and 90 ml of 0.200 *M* HAc contains 0.018 mol of HAc. After neutralization all the NaOH is gone, and 0.016 mol of HAc is left; at the same time 0.002 mol of Ac^- ions is formed. In 90 ml of the buffer solution 0.018 mol of Ac^- is present from the beginning. After neutralization the number of moles is increased to 0.020 mol. Hence, the "initial" concentrations of HAc and Ac^- are 0.016 mol per 0.100 l, or 0.16 *M*, and 0.020 mol per 0.100 l, or 0.200 *M*, respectively. Again we have a buffer solution, and it can be treated as such. Let x equal the number of moles of H_3O^+ formed by the ionization of HAc, and tabulate the information under the equation for the ionization of HAc:

	HAc	+ HOH ⇌	H_3O^+	+ Ac^-
Initial concentration, *M*:	0.16	——	~0.0	0.200
Equilibrium concentration, *M*:	$0.16 - x \approx 0.16$	——	x	$0.20 + x \approx 0.20$

Again we neglect the addition and subtraction of x to and from the much larger numbers, and the ionization-constant expression is

$$K_i = \frac{[H_3O^+][Ac^-]}{[HAc]} \qquad 1.8 \times 10^{-5} = \frac{0.20x}{0.16}$$

$$x = [H_3O^+] = \frac{0.16}{0.20} \times 1.8 \times 10^{-5}$$

$$[H_3O^+] = 1.43 \times 10^{-5}$$

The H_3O^+ concentration before addition of acid was 1.8×10^{-5}, so that only a small change occurred. The pH of the solution is

$$pH = -\log H_3O^+ = -\log (1.43 \times 10^{-5}) = 5 - \log 1.43 = 4.85$$

The pH of the buffer solution before the addition of base was 4.74. Therefore, the addition of base caused a change of only 0.11 pH unit, whereas in the unbuffered 0.200 *M* solution (see Examples

13.12 and 13.13) the pH change was 1.12 units. Nature has taken advantage of this property of buffer solution. All living organisms maintain a constant internal pH by buffer action. The pH of human blood is 7.35, faintly basic. Since a change of only ±0.2 pH unit can cause death, buffering action is vital.

13.7 THE SOLUBILITY-PRODUCT CONSTANT

So far we have considered only very soluble electrolytes. When a slightly soluble salt is added to water, an equilibrium is established between the undissolved solute and the solute present in the solution. The solution is then said to be *saturated.* We have a heterogeneous equilibrium involving ions in solution and in the undissolved salt. When AgCl, which is only slightly soluble, is added to water, the solid AgCl comes to equilibrium with Ag^+ and Cl^- in solution:

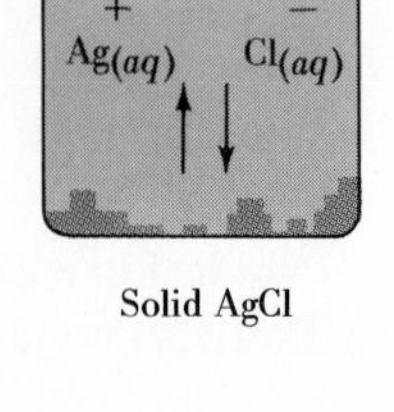

Solid AgCl

$$Ag^+Cl^-(s) \rightleftharpoons Ag^+(aq) + Cl^-(aq)$$

The equilibrium-constant expression is

$$K_{eq} = \frac{[Ag^+][Cl^-]}{[AgCl(s)]}$$

Since the AgCl is present as a solid, its concentration is constant and the equilibrium expression can be rewritten

$$K_{eq}[AgCl(s)] = [Ag^+][Cl^-]$$

The product, $K_{eq}[AgCl(s)]$, is a constant, known as the *solubility-product constant,* K_{sp}.

$$K_{sp}(\text{for AgCl}) = [Ag^+][Cl^-]$$

At 25°C, K_{sp} for AgCl is 1.6×10^{-10}. This represents the upper limit of the product of the concentrations of the two ions; that is, $[Ag^+][Cl^-] \leqslant 1.6 \times 10^{-10}$. When a solution is saturated with AgCl, the product equals 1.6×10^{-10}.

Other examples are

$$AgBr(s) \rightleftharpoons Ag^+(aq) + Br^-(aq) \qquad K_{sp} = [Ag^+][Br^-] = 7.7 \times 10^{-13}$$

$$BaSO_4(s) \rightleftharpoons Ba^{2+}(aq) + SO_4^{2-}(aq)$$
$$K_{sp} = [Ba^{2+}][SO_4^{2+}] = 1.1 \times 10^{-10}$$

$$Ag_2CrO_4(s) \rightleftharpoons 2Ag^+(aq) + CrO_4^{2-}(aq)$$
$$K_{sp} = [Ag^+]^2[CrO_4^{2-}] = 1.1 \times 10^{-12}$$

$$CaF_2(s) \rightleftharpoons Ca^{2+}(aq) + 2F^-(aq)$$
$$K_{sp} = [Ca^{2+}][F^-]^2 = 4.0 \times 10^{-11}$$

EXAMPLE 13.16 The solubility of AgBr is 1.65×10^{-4} g per liter of solution. Calculate the solubility-product constant.

Solution: The solubility of AgBr in moles per liter is

$$\frac{1.65 \times 10^{-4}\ \text{g/l}}{188\ \text{g/mol}} = 8.8 \times 10^{-7}\ \text{mol/l}$$

$$\text{AgBr}(s) \rightleftharpoons \text{Ag}^+(aq) + \text{Br}^-(aq)$$

For every mole of dissolved AgBr, there is 1 mol of Ag^+ and 1 mol of Br^- in solution. Thus, since 8.8×10^{-7} mol of AgBr dissolve, the concentrations of Ag^+ and Br^- are both equal to 8.8×10^{-7}.

$$K_{sp} = [\text{Ag}^+][\text{Br}^-] = (8.8 \times 10^{-7})(8.8 \times 10^{-7}) = 7.7 \times 10^{-13}$$

EXAMPLE 13.17 K_{sp} for PbI_2 is 1.39×10^{-8} at 25°C. Calculate the solubility of PbI_2 in grams per liter at 25°C. The formula weight is 461 g.

Solution: Since 1 mol of PbI_2 yields 1 mol of Pb^{2+} and 2 mol of I^-, upon dissolution, we have:

$$\text{PbI}_2(s) \rightleftharpoons \text{Pb}^{2+}(aq) + 2\text{I}^-(aq)$$

Hence, the solubility of PbI_2 equals the molarity of Pb^{2+}, and the molarity of I^- will be twice the molarity of Pb^{2+}. Let $x = [\text{Pb}^{2+}] =$ molar solubility of PbI_2. Then,

$$2x = [\text{I}^-]$$
$$K_{sp}(\text{for PbI}_2) = [\text{Pb}^{2+}][\text{I}^-]^2$$
$$1.39 \times 10^{-8} = x(2x)^2$$
$$4x^3 = 1.39 \times 10^{-8}$$
$$x^3 = 3.47 \times 10^{-9}$$
$$x = 1.52 \times 10^{-3}\ \text{mol/l} = \text{molar solubility of PbI}_2$$

The solubility in grams per liter is

$$(1.52 \times 10^{-3}\ \text{mol/l})(461\ \text{g/mol}) = 7.0 \times 10^{-1}\ \text{g/l} = 0.70\ \text{g/l}$$

The equilibria between slightly soluble salts and their dissolved ions are sensitive to addition of very soluble salts and acids containing a common ion. The common-ion effect is often utilized to reduce the solubility of slightly soluble compounds still further.

EXAMPLE 13.18 What will be the concentration of Ag^+ left in solution if enough HBr is added to a solution of $AgNO_3$ to make the final Br^- concentration 0.10 *M*?

Solution:

$$\text{AgBr}(s) \rightleftharpoons \text{Ag}^+(aq) + \text{Br}^-(aq)$$
$$K_{sp}(\text{for AgBr}) = 7.7 \times 10^{-13}$$
$$[\text{Ag}^+][\text{Br}^-] = 7.7 \times 10^{-13}$$
$$[\text{Ag}^+](0.10) = 7.7 \times 10^{-13}$$

$$[Ag^+] = \frac{7.7 \times 10^{-13}}{0.10} = 7.7 \times 10^{-12} \text{ mol/l}$$

(which is also the solubility of AgBr under these conditions).

Note that the solubility of AgBr is very substantially reduced compared to that of a solution of AgBr which does not contain a common ion (Example 13.16).

The solubility-product constants for several salts are given in Table 13.4.

13.8 POLYPROTIC ACIDS

In Secs. 13.3 to 13.6 we applied the principles of chemical equilibrium to solutions containing weak *monoprotic* acids and bases. In these cases only one proton is exchanged during the acid-base reaction:

$$HOH + HAc \rightleftharpoons H_3O^+ + Ac^-$$
$$HOH + NH_3 \rightleftharpoons OH^- + NH_4^+$$

TABLE 13.4: SOLUBILITY PRODUCTS OF SOME SLIGHTLY SOLUBLE SALTS AT 25°C

Salt	K_{sp}
AgCl	1.6×10^{-10}
AgBr	7.7×10^{-13}
AgI	1.6×10^{-16}
Ag_2S	1.6×10^{-49}
BaF_2	1.7×10^{-6}
$BaSO_4$	1.1×10^{-10}
$CaCO_3$	5×10^{-9}
CaF_2	3.8×10^{-11}
$PbCrO_4$	1.8×10^{-14}
PbI_2	1.4×10^{-8}
PbS	2.3×10^{-27}
$PbSO_4$	1.1×10^{-8}
MnS	8×10^{-17}
FeS	1.3×10^{-17}
CoS	3×10^{-26}
NiS	1.5×10^{-24}
CuS	8.5×10^{-45}
ZnS	2.4×10^{-24}

Many of the common acids are polyprotic; they can donate more than one proton. Examples include carbonic acid, H_2CO_3; hydrosulfuric acid, H_2S; phosphoric acid, H_3PO_4; and sulfuric acid, H_2SO_4. All these acids undergo stepwise ionization reactions. Consider carbonic acid:

$$HOH + H_2CO_3 \rightleftharpoons H_3O^+ + HCO_3^- \qquad K_1 = \frac{[H_3O^+][HCO_3^-]}{[H_2CO_3]}$$

The ionization constant for the first ionization is symbolized as K_1. The bicarbonate ion, HCO_3^-, can act as an acid by donation of a second proton:

$$HOH + HCO_3^- \rightleftharpoons H_3O^+ + CO_3^{2-} \qquad K_2 = \frac{[H_3O^+][CO_3^{2-}]}{[HCO_3^-]}$$

K_2 is the ionization constant for the second ionization step. A polyprotic acid is characterized by an ionization constant for each ionization step. The ionization constant always becomes progressively smaller; for H_2CO_3,

$$K_1 = 4.52 \times 10^{-7} \qquad \text{and} \qquad K_2 = 4.69 \times 10^{-11}$$

This is reasonable in terms of simple electrostatic considerations. After the first ionization occurs, the negatively charged HCO_3^- ion is formed. Removal of a second positively charged proton from a negatively charged ion would be impeded by electrostatic attraction between the two particles. Hence, $K_1 \gg K_2$; furthermore, the bicarbonate ion can accept a proton and, like the ions formed by all acids, it can act as a base.

$$HCO_3^- + HOH \rightleftharpoons H_2CO_3 + OH^-$$

Sodium bicarbonate, $NaHCO_3$, is widely used as a stomach neutralizer because the HCO_3^- ion can counteract excess acidity or basicity.

Sulfuric acid is a diprotic acid since it can donate two protons per molecule:

$$HOH + H_2SO_4 \rightarrow H_3O^+ + HSO_4^-$$

$$HOH + HSO_4^- \rightleftharpoons H_3O^+ + SO_4^{2-} \qquad K_i = \frac{[H_3O^+][SO_4^{2-}]}{[HSO_4^-]}$$

The first ionization reaction goes to completion; since no H_2SO_4 molecules exist in aqueous solutions, no equilibrium constant is given for the first step. The bisulfate, or hydrogen sulfate, ion, HSO_4^-, does not ionize completely, but it is a stronger acid than HAc. The acid ionization constant for HSO_4^- is 1.01×10^{-2}. In a 0.200 M solution of $NaHSO_4$ at room temperature the HSO_4^- ion is 6 percent ionized; HAc is 1 percent ionized at the same concentration. The ionization constants for some of the common polyprotic acids are listed in Table 13.1.

Hydrogen sulfide, H_2S, is a gas under ordinary conditions. It has the odor of rotten eggs and is a deadly poison. It dissolves in water to form a hydrosulfuric acid solution which is an important reagent for the chemical identification of metals. Hydrosulfuric acid is a diprotic acid, ionizing two steps:

$$HOH + H_2S \rightleftharpoons H_3O^+ + HS^- \qquad K_1 = \frac{[H_3O^+][HS^-]}{[H_2S]} = 7.9 \times 10^{-8}$$

$$HOH + HS^- \rightleftharpoons H_3O^+ + S^{2-} \qquad K_2 = \frac{[H_3O^+][S^{2-}]}{[HS^-]} = 2.0 \times 10^{-15}$$

Note that $K_1 > K_2$, but in fact K_1 is small, so that H_2S is a very weak acid. This means that an aqueous H_2S solution is essentially a molecular solution. A saturated solution of H_2S is less than 0.1 percent ionized. The concentrations of the H_3O^+ and HS^- ions are low, and the concentration of S^{2-} is extremely low. We shall now apply the principles of chemical equilibrium to an H_2S solution to show how a polyprotic/acid solution is analyzed.

EXAMPLE 13.19 Calculate pH, $[HS^-]$, and $[S^{2-}]$ for a saturated aqueous H_2S solution. At 25°C a saturated solution has a concentration of 0.10 M.

Solution: Whenever K_1 is equal to or greater than 1000 times K_2, as in this case, the effect of the second ionization step in producing H_3O^+ may be neglected. The first ionization alone may be used, and for the calculation of $[H_3O^+]$ the acid is treated as a monoprotic acid like HAc. Tabulate the initial and equilibrium concentrations under the chemical equation:

	H_2S +	$HOH \rightleftharpoons$	H_3O^+ +	HS^-
Initial concentration, M:	0.10	—	~0	0
Equilibrium concentration, M:	$0.10 - x \approx 0.10$	—	x	x

We let x equal the number of moles of H_3O^+ formed per liter. In the second ionization, $HS^- + HOH \rightleftharpoons H_3O^+ + S^{2-}$, some of the HS^- is consumed and more H_3O^+ is formed, but the extent of this second step is neglected; the concentrations $[H_3O^+]$ and $[HS^-]$ are almost exactly equal: $x = [H_3O^+] = [HS^-]$. Substitution in the ionization-constant expression gives

$$K_1 = \frac{[H_3O^+][HS^-]}{[H_2S]} \qquad \text{or} \qquad 7.9 \times 10^{-8} = \frac{x^2}{0.1}$$

$$x = [H_3O^+] = [HS^-] = 8.9 \times 10^{-5}$$

Then the pH can be calculated:

$$pH = -\log [H_3O^+] = -\log (8.9 \times 10^{-5}) = 5 - 0.95 = 4.05$$

To calculate the sulfide-ion concentration the second step must be considered. The H_3O^+ and HS^- ions are involved, but, as we said above, their concentrations are not appreciably affected because the ionization of HS^- is very slight. Let y represent the number of moles of S^{2-} ion formed per liter and then tabulate all the concentration data as usual:

	HS^-	$+ HOH \rightleftharpoons$	H_3O^+ +	S^{2-}
Initial concentration, M:	8.9×10^{-5}	—	8.9×10^{-5}	0
Equilibrium concentration, M:	$(8.9 \times 10^{-5}) - y \approx 8.9 \times 10^{-5}$	—	$(8.9 \times 10^{-5}) + y \approx 8.9 \times 10^{-5}$	y

Substitution into the ionization-constant expression gives

$$K_2 = \frac{[H_3O^+][S^{2-}]}{HS^-}$$

$$2.0 \times 10^{-15} = \frac{8.9 \times 10^{-5} y}{8.9 \times 10^{-5}} \qquad y = [S^{2-}] = 2.0 \times 10^{-15}$$

The sulfide-ion concentration is just equal to the ionization constant for the second step.

Analogous reasoning processes are used to solve problems involving other polyprotic acids.

Salts of polyprotic acids undergo hydrolysis just as salts of monoprotic acids do. With a polyprotic acid, however, more than one salt is possible. For instance, H_2S gives NaHS and Na_2S, which are called sodium bisulfide and sodium sulfide. Phosphoric acid gives rise to $NaHPO_4$, Na_2HPO_4, and Na_3PO_4. Species like NaHS, NaH_2PO_4, and Na_2HPO_4, which contain ionizable hydrogen, are called *acid salts*. All these salts can hydrolyze acting either as acids or bases.

The actual calculation of $[H_3O^+]$ for acid-salt solutions is complicated, but in the special case of the acid salts of a weak diprotic acid we can estimate the hydrogen-ion concentration easily provided that the solution is neither very dilute nor very concentrated:

$$[H_3O^+] = \sqrt{K_1K_2} \quad \text{for acid-salt concentration between 0.01 and 1.0 } M \qquad (13.22)$$

EXAMPLE 13.20 Calculate the pH of 0.250 M NaHS solution.

Solution: Taking the logarithm of both sides of Eq. (13.22), we have

$$\log\,[H_3O^+] = \log \sqrt{K_1K_2}$$
$$pH = -\log \sqrt{K_1K_2} = -\log\,(K_1K_2)^{1/2}$$
$$pH = -\tfrac{1}{2}\log K_1K_2$$

For hydrosulfuric acid $K_1 = 7.9 \times 10^{-8}$ and $K_2 = 2.0 \times 10^{-15}$. Substitution into the above equation gives

$$pH = -\tfrac{1}{2}\log\,[(7.9 \times 10^{-8})(2.0 \times 10^{-15})] = 10.9$$

13.9 COLORED INDICATORS AND pH

The aqueous solutions of many organic compounds are *color-sensitive* to the pH. These compounds are called *visual* indicators of the pH. Indicators are actually Brönsted-Lowry acids or bases. We can designate the Brönsted-Lowry form of the indicator in acidic solution as HIn and in basic solution as In^-:

$$HIn + OH^- \rightleftharpoons HOH + In^-$$

The color of the acid form of the indicator, HIn, is different from the color of the basic form. For a very good indicator the color of the HIn species is in deep contrast to the In^- species, and the conversion from HIn to In^- occurs very sharply as the pH changes. Table 13.5 lists some of the common indicators used to estimate the pH of a solution.

With a single indicator one cannot determine the actual pH, but by using a combination it is possible to limit the pH to a very narrow range, which is often all that is required. For example, suppose a solution tested with three indicators gives the following results: the

TABLE 13.5: VISUAL INDICATORS

Indicator	pH *range*	*Color change*
Picric acid	0–2.2	Colorless–yellow
Thymol blue	1.2–2.8	Red–yellow
Methyl red	4.2–6.3	Red–yellow
Bromothymol blue	6.0–7.6	Yellow–blue
Litmus	5.5–8.2	Red–blue
Phenolphthalein	8.3–10.0	Colorless–pink
Alizarin yellow	10.1–12.0	Yellow–violet

solution causes litmus to turn red (pH < 5.5), thymol blue to turn yellow (pH > 2.8), and methyl red to turn red (pH < 4.2). These observations mean that the pH of the solution lies between 2.8 and 4.2.

Titration

Using a solution of accurately known concentration, i.e., a standard solution, for analyzing another solution of unknown concentration is called *titration*. The volumes of the solutions are accurately measured in burettes, or tubes of uniform diameter graduated in 0.1-ml divisions, so that volumes can be measured to approximately 0.01 ml. In acid-base titrations, a volume of acid is measured into a flask, and a standard solution of base is added dropwise from a burette (see Fig. 13.1). The point at which the base exactly neutralizes the acid, called the *end point*, can be conveniently determined by an indicator.

In titrations of acids with strong bases the end point, or the point at which equivalent amounts of acid and base are present, lies between a pH of 7.0 and 9.0. Phenolphthalein is an excellent indicator for these titrations because it has a sharp change, from colorless to brillant pink, in this pH range. A base solution of known concentration is added slowly and carefully to an acid solution whose concentration is to be determined until the faint pink color of phenolphthalein persists. At this point virtually equivalent amounts of acid and base are present.

Since 1 equiv of any acid will neutralize 1 equiv of any base, it follows that any number of equivalents of acid will exactly neutralize the same number of equivalents of a base:

equiv acid = equiv base
N = number of equiv/l

it follows that

N (equiv/l) $\times$ V (l) = number of equiv

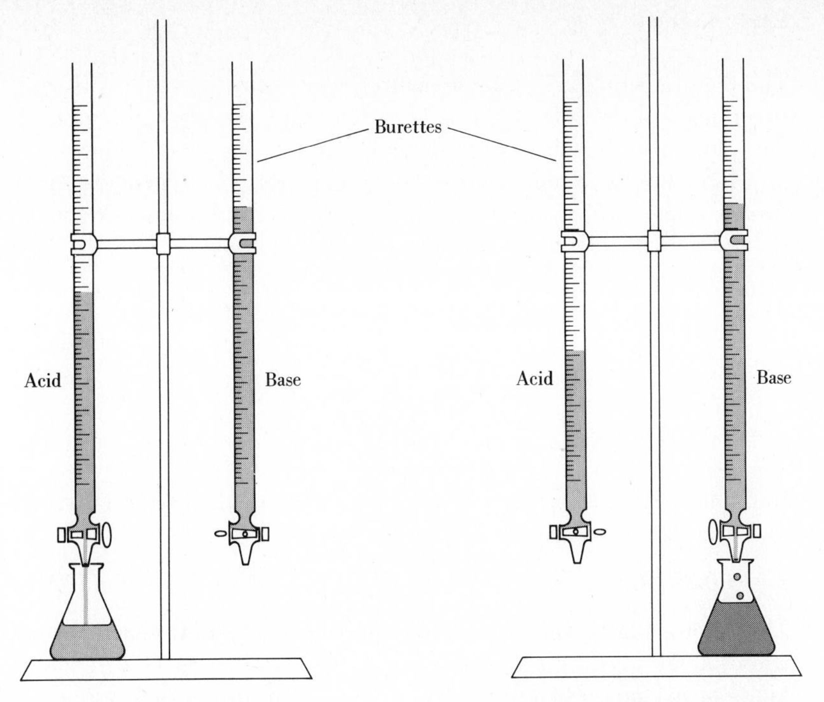

(*a*) *Measure out a volume of acid*

(*b*) *Titrate with base to the faint pink end-point color of phenolphthalein*

FIGURE 13.1

Acid-base titration.

Then, since the number of equivalents of acid and base must be equal at the end point of a titration, the following expression holds:

$$N \text{ acid} \times V(\text{l}) \text{ acid} = N \text{ base} \times V(\text{l}) \text{ base}$$

or

$$N \text{ acid} \times V(\text{ml}) \text{ acid} = N \text{ base} \times V(\text{ml}) \text{ base}$$

or

$$N_A V_A = N_B V_B$$

EXAMPLE 13.21 What volume of 0.30 N HCl will be necessary to neutralize 30 ml of 0.20 N NaOH?

Solution:

$$HCl + NaOH \rightleftharpoons NaCl + H_2O$$
$$V_A N_A = V_B N_B$$
$$V_{HCl} \times 0.30 = 30 \text{ ml} \times 0.20$$
$$V_{HCl} = 20 \text{ ml}$$

The pH of a solution can be measured more conveniently by using a reversible voltaic cell which has one electrode sensitive to hydrogen ions. The emf of such an electrode changes as the pH of the solution in which the electrode is immersed changes. Hydrogen-gas electrodes (page 367) can be used because the reversible chemical reaction at this electrode involves the hydrogen ion:

$$\tfrac{1}{2}H_2(g) \rightleftharpoons H^+(aq) + e^- \qquad \text{or} \qquad \tfrac{1}{2}H_2(g) + H_2O \rightleftharpoons H_3O^+(aq) + e^-$$

By the Nernst equation (12.11), the emf of the hydrogen electrode at 25°C is

$$\mathscr{E} = -0.05916 \log \frac{[H_3O^+]}{P_{H_2}{}^{1/2}}$$

But since $pH = -\log [H_3O^+]$ and, if we keep the hydrogen pressure at 1 atm during the experiment, the equation becomes

$$\mathscr{E} = +0.05916 \text{ pH} \tag{13.23}$$

This shows that a very simple relationship exists between the voltage of a cell and the pH. Analytical chemists have developed many kinds of pH meters calibrated to give a rapid, direct determination of the pH of a solution. Two sturdy electrodes are immersed in the solution, and a voltage readout corresponding to a particular pH is produced immediately.

There are many experimental disadvantages in using a hydrogen-gas electrode to determine pH: (1) there is a need for a supply of very pure hydrogen gas, (2) the gas must be kept at constant pressure in the electrode compartment, and (3) its pressure must be measured accurately. For pH measurements the hydrogen-gas electrode has been replaced by the *glass electrode*, a compact, sealed unit which contains an acidic solution within a special glass bulb.

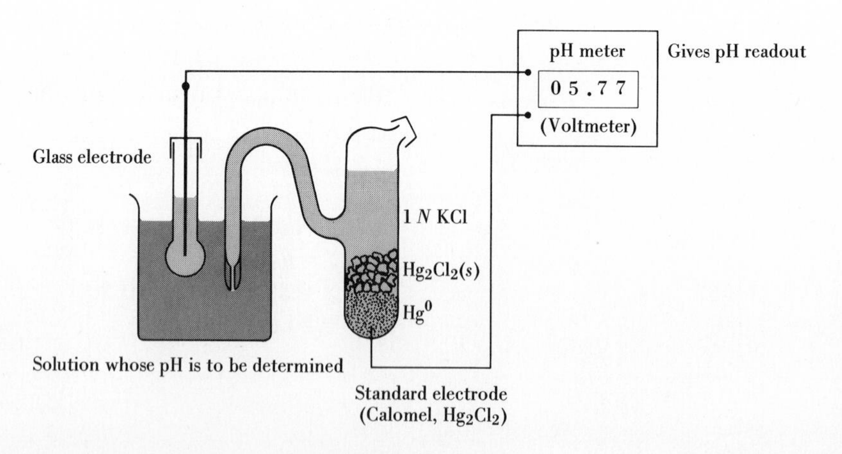

FIGURE 13.2

The glass electrode is used to measure pH.

When the glass electrode is coupled with another reversible electrode and immersed in a solution, it forms a voltaic cell very sensitive to pH changes. Figure 13.2 shows a voltaic cell made from the glass electrode and a standard electrode with constant emf. Under ideal conditions changes in the pH of the solution will affect only the glass electrode, and the measured voltage is directly proportional to the pH.

13.11 ACID-BASE TITRATIONS; DETERMINATIONS OF K_i

With a series of visual indicators it is possible to trace the change in pH as a base is gradually added to an acid solution or vice versa. With the use of a pH meter it is possible to determine the pH continuously as acid is neutralized by the addition of base. Table 13.6 shows the variations of the pH of a 0.100 *M* HCl solution as NaOH is added. This is an example of the titration of a strong acid with a strong base. The pH increases from 1.00 in the pure acid solution to 7.00 when equivalent amounts of acid and base are present. The H_3O^+ concentration of 0.100 *M* HCl is also 0.100 *M* because HCl is a strong, completely ionized acid. The $pH = -\log 0.1 = 1$. When 0.05 mol of NaOH has been added, 0.05 mol of HCl per liter remains unneutralized. At this point the HCl and the H_3O^+ concentration are 0.05 *M* and the $pH = -\log 0.05 = 1.31$, as shown in Table 13.6. As the equivalence point is approached, the change in pH per mole of NaOH added rises very rapidly. Note that at the beginning, the addition of 0.01 mol of NaOH causes the pH to increase from 1.00 to 1.05, which is a change of a 0.05 pH unit per 0.01 mol of NaOH, or 5 pH units per mole. When 0.09 mol of NaOH is added, 90 percent of the HCl has been neutralized. The addition of 0.005 mol at this point causes the pH to rise from 2.00 to 2.31. This is a change of 0.31 pH unit per 0.005 mol of added NaOH, or 62 pH units per mole. At the equivalence point the pH changes at a rate of 30,000 pH units per mole of added NaOH. After neutralization addition of NaOH does not cause such a large change in pH. Because of the large change in pH near the equivalence point it is very easy to determine the equivalence point from a plot of pH versus moles of added base. The data presented in Table 13.6 are plotted in Fig. 13.3. The solid line represents the variation of the pH as NaOH is added. At the equivalence point the curve is almost vertical so that the slope is virtually infinite. The dotted line in Fig. 13.3, the *derivative curve*, represents a plot of the change in pH per mole of added base, Δ. There is a sharp maximum in this curve at the equivalence point. It is easy to determine the equivalence point from the derivative curve.

TABLE 13.6: CHANGES IN THE pH OF HCl SOLUTION AS NaOH IS ADDED TO A SOLUTION CONTAINING 0.100 mol OF HCl PER LITER OF SOLUTION

Moles NaOH *added*	pH	*Comment*
0.00	1.00	Pure acid solution
0.01	1.05	
0.03	1.16	
0.05	1.31	Half neutralized
0.07	1.52	
0.09	2.00	
0.095	2.31	
0.099	3.00	
0.0999	4.00	
0.100	7.00	Neutral
0.1001	10.00	Excess base
0.101	11.00	
0.110	12.00	
0.130	12.48	
0.150	12.70	

It is important to note that the equivalence point for the titration

FIGURE 13.3

Titration curve: change in pH *as* NaOH *is added to a solution which contains 0.1 mol of* HCl *per liter of solution.*

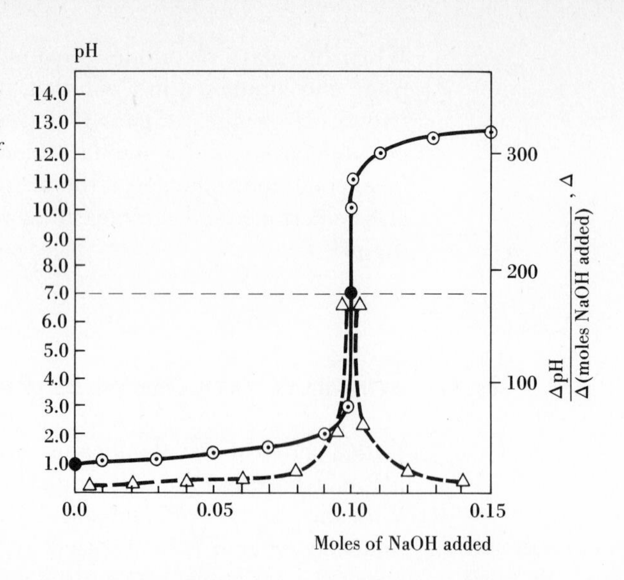

TABLE 13.7: CHANGES IN pH OF HAc SOLUTION AS NaOH IS ADDED TO A SOLUTION CONTAINING 0.100 mol PER LITER OF SOLUTION

Moles NaOH *added*	pH	*Comments*
0.00	2.87	Pure acid solution
0.01	3.79	
0.03	4.38	
0.05	4.74	$pH = pK_i$
0.07	5.11	
0.09	5.70	
0.095	6.02	
0.099	6.74	
0.09945	7.00	Neutral
0.0999	7.74	
0.100	8.87	HAc = NaOH; pH > 7
0.1001	10.00	Excess base
0.101	11.00	
0.110	12.00	
0.130	12.48	
0.150	12.70	

of a strong acid with a strong base occurs at a pH of 7 (see Fig. 13.3). When a weak acid is titrated with a strong base, the equivalence point occurs at a pH greater than 7. We can understand this behavior in terms of the salts which form in the two cases. The salt of a strong acid and strong base, for example, NaCl, does not hydrolyze, and its solution is neutral; the pH is 7. The salt of weak acid and a strong base, for example NaAc, is basic. The acetate ion hydrolyzes; $Ac^- + HOH \rightleftharpoons HAc + OH^-$ produces a basic solution with a pH > 7.

The data for the titration of a 0.100 *M* HAc solution with NaOH are shown in Table 13.7. The pH of the solution can be calculated at various points during the titration by the methods demonstrated in the examples above. The pH of the solution before the addition of NaOH is calculated using the ionization-constant expression:

$$HAc + HOH \rightleftharpoons H_3O^+ + Ac^-$$

$$K_i = \frac{[H_3O^+][Ac^-]}{[HAc]}$$

The hydronium-ion concentration of a 0.100 *M* HAc solution is 1.34×10^{-3} *M*, and the pH is 2.87. When small amounts of NaOH are added, some of the HAc is neutralized and the resulting solution contains HAc and NaAc. The pH can be determined by considering the solution to be a buffer solution; it contains a weak acid and its salt. For example, when 0.03 mol of NaOH is added to 1 l of solution which was initially 0.100 *M* HAc, approximately 0.03 mol of NaAc will form and 0.07 mol of HAc will remain. Using the ionization-constant expression, we can determine the pH of this solution, but for convenience we shall show the simple relationship which exists between the pH and the $[Ac^-]/[HAc]$ ratio before we do any calculations. For HAc solutions:

$$K_i = \frac{[H_3O^+][Ac^-]}{[HAc]} = [H_3O^+] \frac{[Ac^-]}{[HAc]}$$

Taking the logarithm of both sides of the equation and multiplying by -1, we have

$$-\log K_i = -\log [H_3O^+] - \log \frac{[Ac^-]}{[HAc]}$$

Remembering that p means negative logarithm, we have

$$pK_i = pH - \log \frac{[Ac^-]}{[HAc]}$$

or

$$pH = pK_i + \log \frac{[Ac^-]}{[HAc]} \tag{13.24}$$

This equation is convenient for the calculations of the pH of the solution when less than equivalent amounts of NaOH are also present.

When equivalent amounts of HAc and NaOH are present, we have a virtual NaAc solution. Therefore, we must simply calculate the pH of a 0.100 M NaAc solution which is slightly basic due to hydrolysis:

$$Ac^- + HOH \rightleftharpoons HAc + OH^-$$

$$K_h = \frac{K_w}{K_i} = \frac{[HAc][OH^-]}{[Ac]}$$

$$\frac{10^{-14}}{1.8 \times 10^{-5}} = \frac{x^2}{0.1}$$

$x = [OH^-] = 7.45 \times 10^{-6}$, $[H^+] = 1.34 \times 10^{-9}$, pH = 8.87

Above the equivalence point the pH of the solution is determined by the excess NaOH present. For instance, when 0.101 mol of NaOH has been added, there is an excess of 0.001 mol/l. The pOH is 3, and the pH is 14 − 3 or 11 (see Table 13.7).

The data presented in Table 13.7 are plotted in Fig. 13.4. In this plot the point at which half the acid has been neutralized has a special significance. At this point the Ac^- and HAc concentrations are equal, and by Eq. (13.24) we can see that the pH equals the pK_i:

$$pH = pK_i - \log \frac{[Ac^-]}{[HAc]}$$

That is, when $[Ac^-] = [HAc]$, $[Ac^-]/[HAc] = 1$ and log

FIGURE 13.4

Titration curve: change in pH *as* NaOH *is added to a solution which contains 0.1 mol of* HAc *per liter of solution.*

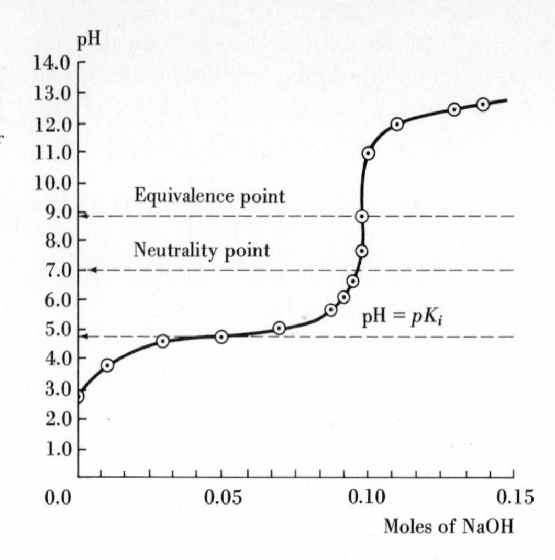

$([Ac^-]/[HAc]) = 0$. Therefore, when the acid solution is half neutralized, pH = pK_i. The value of K_i for a weak acid or base can be determined from the analysis of titration, which is one of the most common methods employed.

Figure 13.5 shows the HCl and HAc titration curves. Initially the increase in pH as NaOH is added is greater for the HAc solution. The pH of the HAc solution is always higher before the equivalence point is reached. After the equivalence point of HAc is reached, the plots are collinear. Note that the HAc solution reaches neutrality (pH of 7) before an equivalent amount of NaOH has been added; at the equivalence point the HAc solution is basic (pH = 8.87).

FIGURE 13.5

Comparison of the titration curves for a weak acid and strong acid.

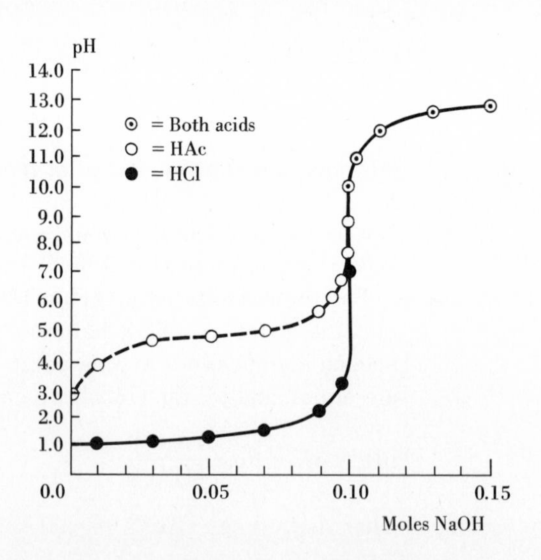

Titration curves for monoprotic acids will resemble the HCl and HAc curves. The titration of a weak base with a strong acid produces curves which are mirror images of the weak-acid–strong-base curves we have examined.

13.12 A MECHANISTIC PICTURE OF ACIDITY

From the Brönsted-Lowry point of view the strength of an acid depends on the favorability of the reaction of the type given by

$$HA + HOH \rightarrow A^- + H_3O^+ \tag{13.25}$$

A^- represents an anion which may be a single atom like Cl^- or a radical like ClO_3^-. The favorability of such a reaction depends on the strength of the H—A bond, which must be broken, and on the stability of the anion A^- which is formed. The process of forming H_3O^+ ions from an acid requires sufficient energy to break the H—A bond. As we shall see, some of this energy comes from the formation of the hydrated anion and the hydrated proton.

The H—A bond in an acid molecule is polar. Polar bonds are relatively strong, and at first thought might seem to be the least likely to be severed. What we should not forget is that the strength of an acid depends on a reaction with solvent molecules. If we dissolve HCl

FIGURE 13.6

The role of water in the ionization of an acid.

(a) Polar HCl *bond is weakened by interactions with water molecules*

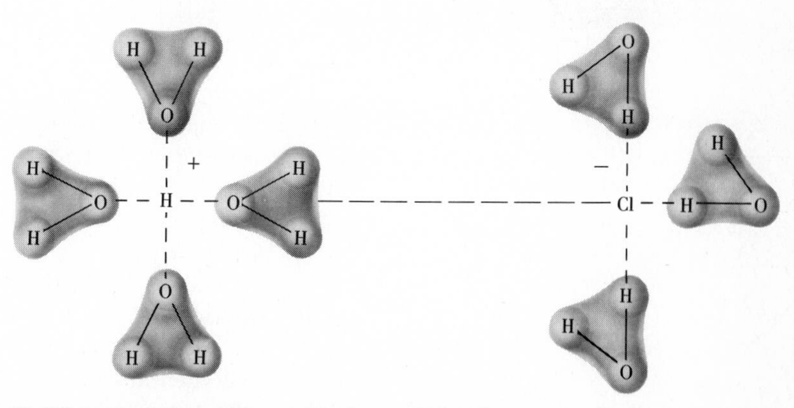

(b) HCl *bond is broken and the resulting ions are hydrated*

gas in benzene, which is a nonpolar solvent, the solution is nonacidic. When HCl is dissolved in water, the H_2O molecules, which are very polar, are attracted by the polar H—Cl bond (see Fig. 13.6*a*). The negative, oxygen end of the water molecule neutralizes. the partial positive charge of the hydrogen end of the HCl molecule. The positive hydrogen end of the water molecules neutralizes the partial negative charge of the chlorine end. As a result, the H—Cl bond is less polar and is weakened; it can be broken by a collision with an energetic water molecule. As soon as ions are formed, they are bonded strongly to a number of water molecules. This is called *hydration*, and it is an exothermic process (see Fig. 13.6*b*). Hydration stabilizes the ions because it disperses the electric charge. The attraction between a pair of hydrated ions is much less than that between nonhydrated ions.

We see that the strength of an acid depends on the interaction of the acid and solvent molecules. Nonpolar benzene molecules do not attack the polar H—Cl bond; polar water molecules do. A strong interaction between solvent molecules and acid molecules weakens the strong H—A bond, and then collisions with energetic solvent molecules are sufficient to sever it. Still, energy is required and this comes from the final process, hydration, which also stabilizes the ions formed.

Our Brönsted-Lowry picture of the solvent-acid interaction explains the trends observed in the strengths of the binary acids within a family in the periodic table. For example, the ionization constants for the binary hydrohalic acids increase continuously from HF down through HI. Hydrofluoric acid is a weak acid in water; K_i for HF is 7×10^{-4}. The H—X bond strength decreases as we descend in the halogen family. At the same time the anionic radius increases, making the anion a poorer base. These factors explain why HF is the weakest and HI the strongest binary halogen acid.

13.13 OXYACIDS

Many acidic compounds contain hydrogen and oxygen along with another nonmetallic element. Examination of the molecules of these acids shows that hydrogen atoms which can ionize are always bonded to oxygen atoms. Consider the two acids phosphoric, H_3PO_4, and phosphorous, H_3PO_3; both contain three hydrogen atoms per molecule, but phosphoric is a *tri*protic acid and phosphorous is *di*protic. A comparison of structures shows that all three hydrogen atoms of H_3PO_4 are bonded to oxygen atoms. In H_3PO_3 only two hydrogen atoms are bonded to oxygen atoms; the third is bonded to the central phosphorus atom.

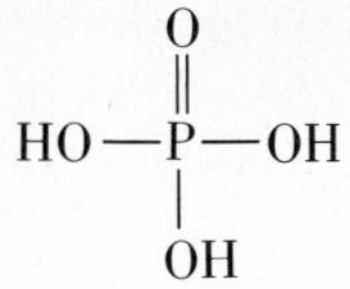

$$\begin{array}{c} O \\ \| \\ HO-P-OH \\ | \\ OH \end{array}$$

Phosphoric acid, H_3PO_4

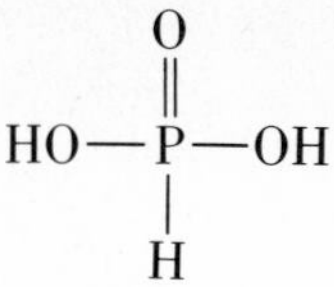

$$\begin{array}{c} O \\ \| \\ HO-P-OH \\ | \\ H \end{array}$$

Phosphorous acid, H_3PO_3

In the oxyacids, the ionizable protons are always bonded to oxygen atoms, which in turn are all bonded to a single central atom of a nonmetallic element, like phosphorus, sulfur, nitrogen, or chlorine.

Most organic acids contain acidic and nonacidic hydrogen atoms. Acetic acid, $C_2H_4O_2$, contains one acidic hydrogen atom and three nonacidic hydrogen atoms:

$$H-\overset{\displaystyle H}{\underset{\displaystyle H}{\overset{|}{\underset{|}{C}}}}-C\begin{array}{l} \nearrow\!\!\!\!\!= O \\ \searrow O-H \end{array}$$

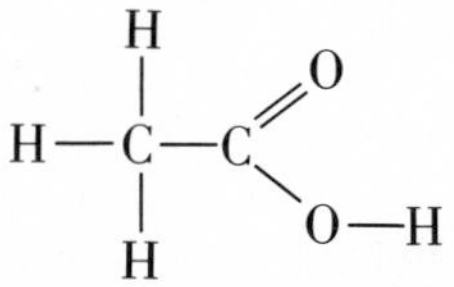

Acetic acid, $C_2H_4O_2$

The acidic hydrogen is bonded to oxygen. To emphasize the fact there is only one acidic hydrogen atom in acetic acid we write the formula as CH_3CO_2H or simply HAc, where Ac^- stands for the acetate ion, CH_3COO^-.

From the ionization constants for a large number of acids Linus Pauling concluded that the strength of oxyacids depends on the number of *nonhydrogenated* oxygen atoms. Thus, boric acid, H_3BO_3, and hypochlorous acid, HClO, are weak since they contain

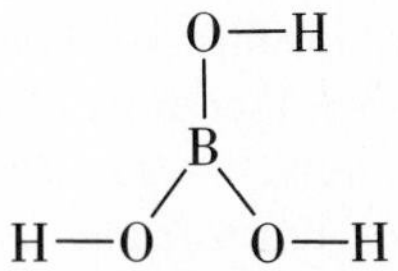

$$\begin{array}{c} O-H \\ | \\ B \\ \diagup \quad \diagdown \\ H-O \qquad O-H \end{array}$$

Boric acid, $B(OH)_3$ or H_3BO_3; $K_1 = 0.64 \times 10^{-9}$

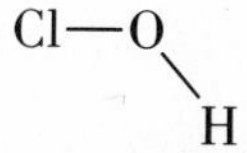

$$\begin{array}{l} Cl-O \\ \qquad\quad \diagdown \\ \qquad\qquad H \end{array}$$

Hypochlorous acid, HClO or Cl(OH); $K_1 = 37 \times 10^{-9}$

no nonhydrogenated oxygen atoms. Acids with no nonhydrogenated oxygen atoms have ionization constants of the order of 10^{-9}. With a single nonhydrogenated oxygen atom the acid strength is considerably higher. Sulfurous acid, H_2SO_3, and phosphoric acid, H_3PO_4, are in this category, and their ionization constants are near 10^{-2}.

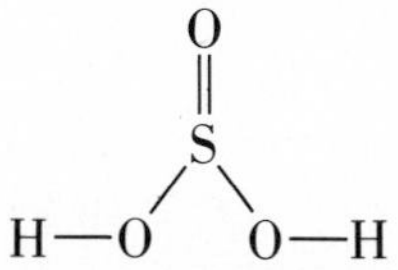

$$\begin{array}{c} O \\ \| \\ S \\ \diagup \quad \diagdown \\ H-O \qquad O-H \end{array}$$

Sulfurous acid, H_2SO_3 or $SO(OH)_2$; $K_1 = 1.7 \times 10^{-2}$

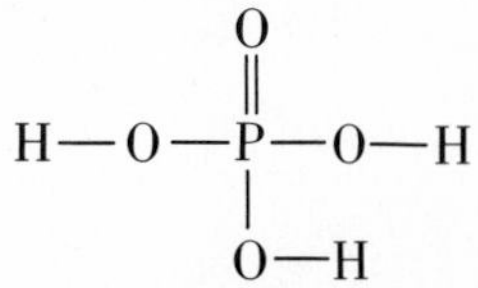

$$\begin{array}{c} O \\ \| \\ H-O-P-O-H \\ | \\ O-H \end{array}$$

Phosphoric acid, H_3PO_4 or $PO(OH)_3$; $K_1 = 1.1 \times 10^{-2}$

Acids with two or more nonhydrogenated oxygen atoms are very strong. Sulfuric, nitric, chloric, and perchloric acids are examples. Such acids are completely ionized in aqueous solutions.

```
        O                       O
        ↑                       ‖
H—O—S—O—H                       N
        ↓                      ↙ \
        O                     O   O—H
```

Sulfuric acid,
H_2SO_4 or $SO_2(OH)_2$

Nitric acid,
HNO_3 or $NO_2(OH)$

```
     O                    O
     ↑                    ↑
     Cl              O←—Cl—O—H
    ↙ ·· \                ↓
   O     O—H              O
```

Chloric acid,
$HClO_3$ or $ClO_2(OH)$

Perchloric acid,
$ClO_3(OH)$

The strength of oxyacids can also be related to the oxidation state of the central atom. The series of *oxychlorine* acids demonstrates the point nicely (Table 13.8). The strength of an acid increases as the oxidation state of the central atom increases. This is another way of saying that the acidity increases as the number of nonhydrogenated oxygen atoms increases.

If we consider a group of oxyacids in a single family of elements in the periodic table in which the central atoms are in the same oxidation state, we shall find that the strength of the acids decreases as we go down the family. This is understandable if we recall that the elements become more metallic as we go down a family and the metallic elements tend to form basic hydroxides. Using the same line of reasoning, we should expect the strength of the oxyacids to increase from left to right across a period since nonmetallic character increases. The oxyacids of the second period do increase in strength from H_4SiO_4 to H_3PO_4 to H_2SO_4 and to $HClO_4$. This is what is observed.

The strength of the oxygen acids can be related very nicely to the

TABLE 13.8: ACID STRENGTH AND OXIDATION STATE OF THE CENTRAL ATOM

Acid	*Formula*	*Oxidation state of central atom*	K_i
Hypochlorous	$HClO$	+1	3×10^{-7}
Chlorous	$HClO_2$	+3	1×10^{-2}
Chloric	$HClO_3$	+5	10
Perchloric	$HClO_4$	+7	10^{11}

electronic structures of their conjugate bases. For example, the conjugate base of nitric acid is the nitrate ion:

$$\underset{\text{Acid}}{HNO_3} + HOH \rightarrow \underset{\text{Conjugate base}}{NO_3^-} + H_3O^+$$

The nitrate ion is a very weak base; in water it has no measurable tendency to accept a proton. This behavior is partially accounted for by the fact that the negative charge on this ion is dispersed or delocalized by *resonance* (see page 113). Three equivalent Lewis structures can be written for the nitrate ion:

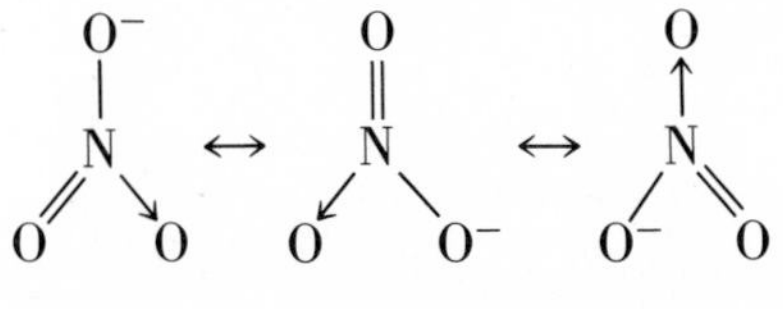

```
  O⁻            O            O
  |             ‖            ↑
  N      ↔      N      ↔     N
 // \↘         ↙ \          / \\
O    O       O    O⁻      O⁻    O
```

A better picture of the bonding would be a composite of all three structures so that the negative charge is dispersed over four atoms. As a result the ion does not have a great affinity for protons and is a weak base.

The conjugate base of nitrous acid, HNO_2, the nitrite ion, is definitely basic. Two equivalent Lewis structures can be drawn for the nitrite ion:

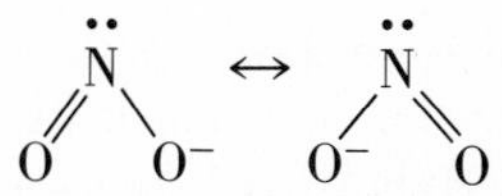

```
  ••            ••
  N      ↔      N
 // \          / \\
O    O⁻      O⁻    O
```

Note the presence of a nonbonded pair of electrons. The negative charge is dispersed but not as much as in the NO_3^- ion. The nitrite ion is stabilized to a smaller degree by resonance, and it can act as a weak base, accepting protons from water. Bonding theory explains the relationship between acid strength and the number of nonhydrogenated oxygen atoms. As the number of nonhydrogenated oxygen atoms increases, the extent of delocalization of charge also increases. This reduces the base strength of the oxyanions. The order of increasing basicity in the chlorine oxyanions is ClO_4^-, ClO_3^-, ClO_2^-, and ClO^-. The order of increasing acidity of the corresponding acids is $HClO$, $HClO_2$, $HClO_3$, and $HClO_4$.

13.14 THE RELATIVE STRENGTHS OF STRONG ACIDS

Acids like HCl, HI, H_2SO_4, HNO_3, and $HClO_4$ are called *strong acids* because they are 100 percent ionized in aqueous solution. Since the degree of ionization is the same in all these cases, we must conclude that these acids are of equal strength. This is surprising because a different bond and different bond strength are involved in each case. The fact that these acids appear to be of equal strength is due to the fact that the solvent, water, is a fairly strong base and is able to effect the complete transfer of protons. This is called the *leveling effect* of water. The acid-base reactions between the *strong* acids and water are therefore irreversible. For example, when nitric acid is dissolved, the reaction with HOH goes completely to the right:

$$\underset{\text{Acid}}{HNO_3} + \underset{\text{Base}}{HOH} \rightarrow NO_3^- + H_3O^+$$

What we are saying is that *acid strength* is relative; it depends on the strength of the base with which the acid reacts.

If we wish to determine the order of acid strength, we must study the strong acids in a solvent which is a weaker base than water. Methanol, CH_3OH, is a very weak base. The reaction of dissolved acids with methanol is analogous to the reaction with water except that it does not go to completion. For example, when HNO_3 is dissolved in CH_3OH, the *methonium ion* forms:

$$\underset{\text{Acid}}{HNO_3} + \underset{\text{Base}}{CH_3OH} \rightleftharpoons NO_3^- + \underset{\substack{\text{Methonium ion} \\ \text{(strong acid)}}}{CH_3OH_2^+}$$

but the above reaction is an equilibrium reaction and an ionization constant can be determined. As a result of studies in nonaqueous solvents the relative strengths of the strong acids have been determined. Of the common acids $HClO_4$ is the strongest; this is not surprising since there are three nonhydrogenated oxygen atoms in the molecule. The order of acid strength of acids 100 percent ionized in water is $HClO_4 > HI > HBr > HCl > HNO_3 > H_2SO_4$. In solvents which are weaker bases than water the leveling effect is not present. The differences in acid strengths due to the internal structure of the acid molecules are manifested.

QUESTIONS AND PROBLEMS

13.1 List the physical and chemical properties which are peculiar to acids and bases.

13.2 (*a*) Before 1780 oxygen was called "dephlogisticated air"; why did Lavoisier rename it "oxygen"? (*b*) How did Davy show that the name oxygen is a misnomer?

13.3 Define acid in terms of: (*a*) Arrhenius theory, (*b*) the solvent-system theory, (*c*) the Brönsted-Lowry theory, and (*d*) the Lewis theory. (*e*) Which of these definitions includes the greatest number of compounds?

13.4 Show how the reaction of NaOH and HCl in a water solution is an acid-base reaction by (*a*) the Arrhenius theory, (*b*) the solvent-system theory, (*c*) the Brönsted-Lowry theory, and (*d*) the Lewis theory.

13.5 Can the reaction of HNO_3 with pyridine, C_5H_5N:, in H_2O be considered an acid-base reaction by (*a*) the Arrhenius theory, (*b*) the solvent-system theory, (*c*) the Brönsted-Lowry theory, and (*d*) the Lewis theory.

13.6 Write equations for Brönsted-Lowry acid-base reactions; indicate the acid and its conjugate base and the base and its conjugate acid.

13.7 What is the common-ion effect? Use a chemical equation for a specific case to explain it.

13.8 What is a buffer solution? Using a specific example, explain how a buffer solution works.

13.9 With the help of Table 13.5 estimate the pH range of the following solutions: (*a*) picric acid is yellow and litmus is red in solution A; (*b*) thymol blue and methyl red are yellow, while phenolphthalein is colorless in solution B; (*c*) bromothymol blue is blue and alizarin yellow is yellow in solution C.

13.10 Calculate the normality of a KOH solution if 55.35 ml is brought to a phenolphthalein end point with 61.66 ml of 0.1003 *N* HCl.

13.11 (*a*) Why can a hydrogen-gas electrode be used to determine pH? (*b*) What is a glass electrode and what is it used for?

13.12 Put the following groups of acids in order of increasing strength: (*a*) H_2O, H_2S, H_2Se, and H_2Te; (*b*) H_2SO_4, H_2SeO_4, and H_2TeO_4; (*c*) H_4GeO_4, H_3AsO_4, H_2SeO_4, and $HBrO_4$. (*d*) State the rule which guides you in each case.

13.13 A solution contains 0.0155 mol/l of a completely ionized monoprotic acid. Determine $[H_3O^+]$, $[OH^-]$, and the pH.

13.14 Calculate (*a*) $[OH^-]$, (*b*) $[H_3O^+]$, and (*c*) the pH of a 0.0250 *M* aqueous KOH solution.

13.15 The pH of 0.500 *M* of a weak monoprotic acid is 4.735. Calculate the ionization constant.

13.16 Benzoic acid is a weak monoprotic acid. In a 0.100 *M* solution it is 2.51 percent ionized at 25°C. Calculate the $[H_3O^+]$ and pH of this solution; calculate K_i for benzoic acid at 25°C.

13.17 Calculate the percentage ionization and $[OH^-]$ in ammonia solutions which are (*a*) 1.00 *M*, (*b*) 0.100 *M*, and (*c*) 0.0100 *M*; $K_i = 1.8 \times 10^{-5}$. (*d*) How is the percentage of ionization affected as the solution becomes more dilute? (*e*) How is $[OH^-]$ affected as the solution becomes more dilute? (*f*) Do the answers to parts (*d*) and (*e*) contradict each other? Explain.

13.18 Calculate the pH and pOH of a 0.350 *M* HCN solution at 25°C. See Table 13.1 for the ionization constant.

13.19 Calculate the pH and pOH of a 1.50 M solution of the base pyridine at 25°C. $K_i = 1.56 \times 10^{-9}$.

13.20 Calculate the pH and pOH of a 0.125 M acetic acid solution at 25°C.

13.21 Calculate the degree of ionization of a 0.001 M NH_3 solution. (Can ionization of NH_3 be neglected in determining its concentration in this solution?)

13.22 Calculate the pH and pOH of buffer solutions which contain (*a*) 0.500 M HAc and 0.500 M NaAc and (*b*) 0.350 M NH_3 and 0.350 M NH_4Cl. (*c*) What is the relationship between pH and pK_i for the buffer solution in part (*a*)? (*d*) What is the relationship between pOH and pK_i for the buffer solution in part (*b*)?

13.23 Calculate the pH before and after the addition of 0.0500 mol of HCl to 1 l of a buffer solution which is 0.445 M in NH_3 and 0.200 M in NH_4Cl. (Neglect the volume change due to addition of the HCl.)

13.24 Calculate the pH, pOH, $[HS^-]$, and $[S^{2-}]$ of a solution in which 0.10 mol per liter of H_2S has been dissolved.

13.25 Calculate $[S^{2-}]$ in a solution saturated with H_2S at pH = 1.00. The saturation concentration of H_2S is 0.10 M.

13.26 Estimate the pH of a 0.115 M $NaHCO_3$ solution.

13.27 Calculate the K_{sp} for each of the following salts; solubilities at room temperature are given in moles per liter: (*a*) $PbCl_2$, 1.6×10^{-2}; (*b*) Ag_2S, 3.64×10^{-17}; and (*c*) $BaSO_4$, 1.1×10^{-5}.

13.28 Using the K_{sp} values in Table 13.4, determine the solubility in grams per liter for (*a*) AgCl, (*b*) BaF_2, and (*c*) PbI_2.

13.29 A solution is 0.0010 M in NaBr and 0.010 M in NaCl. If a concentrated $AgNO_3$ solution is added dropwise, will AgBr or AgCl precipitate first? Assume no volume change occurs upon addition of $AgNO_3$ solution.

13.30 Calculate the minimum concentration of Fe^{2+} in moles per liter that will cause the precipitation of FeS from a solution saturated with H_2S at a pH equal to 1.0.

13.31 Show that the hydrolysis constant for weak bases like ammonia is given by Eq. (13.17).

13.32 Calculate the pH of the following solutions at 25°C: (*a*) 0.100 M NaCl, (*b*) 0.215 M NaCN, and (*c*) 0.500 M NH_4Cl.

FIGURE P13.33

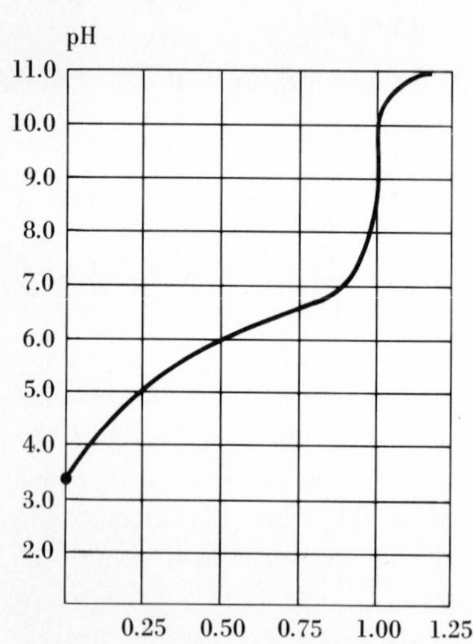

13.33 Shown in Fig. P13.33 is a curve for the titration of a weak, monoprotic acid with NaOH. The ordinate shows the pH, and the abscissa shows the ratio of the equivalents of NaOH to equivalents of acid in solution. (*a*) What is the pH at the end point? (*b*) Estimate the NaOH/acid ratio when the solution is neutral (pH = 7). (*c*) What is the K_i for the acid? (*d*) What is the initial pH? (*e*) What is the initial concentration of weak acid?

13.34 Using the values of K_w given in Table 13.2, estimate the enthalpy change for the ionization of water. [Refer to Eq. (11.7).]

13.35 The K_i of HCN is given in Table 13.1. (*a*) Calculate the pH of 0.500 *M* solution after the following percentages of acid have been neutralized by addition of KOH: 0, 10, 20, 50, 70, 90, 95, 99, 100, 101, and 110 percent. Assume no volume change occurs when KOH is added. (*b*) Plot pH versus equivalents of KOH added. (*c*) Draw a smooth curve through your data and estimate the number of equivalents of KOH which must be added to make the solution neutral (pH = 7). (*d*) At the midpoint of the titration, i.e., when half the HCN is neutralized, pH = p*K*. Is your plot in accordance with this fact?

13.36 (*a*) Make a plot like the one in Prob. 13.33 which shows the variation of the pH of a 0.300 *M* methylamine solution as HCl is added (see Table 13.1 for K_i). (*b*) How many equivalents of HCl are required to neutralize the solution? (*c*) What is the pH at the end point?

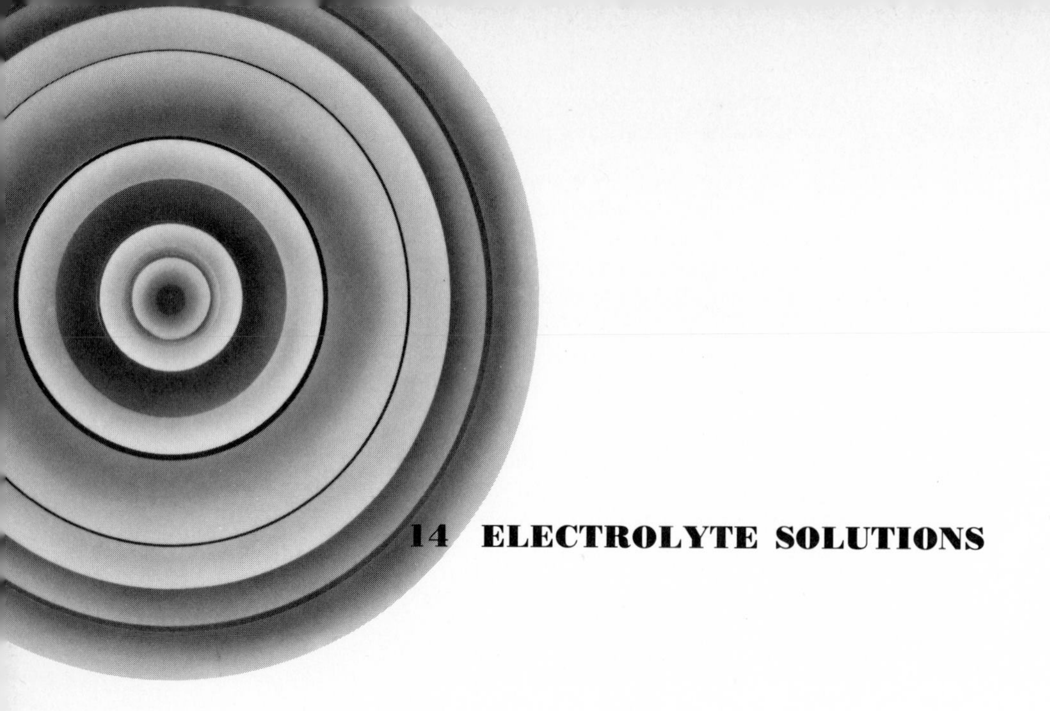

14 ELECTROLYTE SOLUTIONS

14.1 INTRODUCTION

We have studied electrochemistry and the theory of acids and bases. At many points these studies overlapped, and so they should, since in both cases we were dealing with electrolyte solutions, i.e., solutions which conduct electricity. In Chap. 12 we learned that chemical reactions occur at electrode surfaces when a current passes through electrolyte solutions. In this chapter we shall show how conductance and colligative-property studies have led to our modern picture of electrolyte solutions. We shall focus our attention on the nature of the electrolyte solution rather than the electrodes.

When a small amount of an electrolyte (salt, acid, or base) is added to water or some other solvent, most of the physical properties of the solvent are hardly affected; however, the conductance seems to increase out of proportion. This is demonstrated by the data in Table 14.1, which compares the conductance of typical electrolyte and nonelectrolyte solutions, insulators, and metals. The conductance of the nonelectrolyte solutions is not very different

TABLE 14.1: ELECTRIC CONDUCTANCE OF SOLUTIONS, METALS, AND INSULATORS

Solvent	Solute	Weight % of 0.01 M solute	Specific conductance† at 25°C, $(\Omega\text{-cm})^{-1}$ Pure solvent	Solution
Electrolyte solutions:				
H_2O	HCl	0.037	7.5×10^{-8}	4.12×10^{-3}
	HAc	0.06	7.5×10^{-8}	1.63×10^{-4}
	KCl	0.07	7.5×10^{-8}	1.14×10^{-3}
Nonelectrolyte solution:				
H_2O	Sucrose	17	7.5×10^{-8}	7.0×10^{-8}

Metal	Specific conductance† at 20°C, $(\Omega\text{-cm})^{-1}$	Insulator	Specific conductance† at 20°C, $(\Omega\text{-cm})^{-1}$
Silver	6.1×10^{5}	Glass	2×10^{-14}
Copper	5.8×10^{5}	Hard rubber	3×10^{-15}
Gold	4.2×10^{5}	Wood	$\sim 10^{-10}$
Aluminum	3.5×10^{5}	Pure water	7.5×10^{-8}

† Specific conductance is defined in Sec. 14.2.

from that of the pure solvent. The conductance of a metal is more than a million times greater than an electrolyte solution, but in comparison with electric insulators, which includes most solvents, electrolytes have a high conductance. A comparison of the conductance of pure water with the conductance of a solution which contains only 0.037 percent by weight HCl shows that the solution conductance is about 50,000 times greater. The addition of the nonelectrolyte sucrose to the extent of 17 percent by weight hardly affects the conductance.

14.2 SPECIFIC CONDUCTANCE AND EQUIVALENT CONDUCTANCE

The electric conductance of materials is generally reported in terms of the *specific conductance* κ, but when dealing with electrolyte solutions it is more meaningful to discuss the data in terms of the *equivalent conductance* Λ.

The *specific conductance* is the conductance of a cube of material 1 cm on the side placed between electrodes which are in complete contact with any pair of opposite faces of the cube. The method of measuring the specific conductance is indicated in Fig. 14.1*a*. The specific conductance κ has units of $(\Omega\text{-cm})^{-1}$, read "reciprocal ohm-centimeters."

FIGURE 14.1

Specific conductance and equivalent conductance.

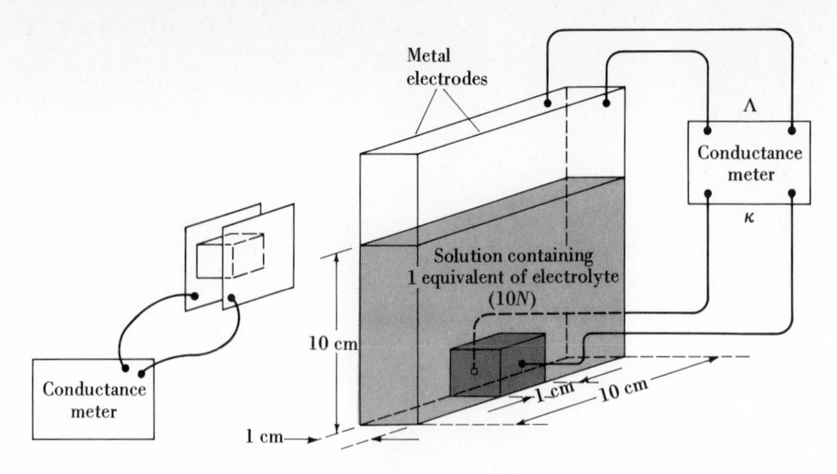

(a) The conductance of a cube, 1 cm on the side, is the specific conductance, κ

(b) 100 cm³ or 100 ml of a 10-N solution contains 1 equivalent weight; the conductance of the entire solution between parallel electrodes, 1 cm apart, is the equivalent conductance Λ

The equivalent conductance Λ is the conductance of a volume of solution which has 1 equiv of electrolyte dissolved in it. There is a simple relationship between Λ and κ: κ is the conductance per cubic centimeter (or milliliter); multiplication of κ by the volume of a solution which contains 1 equiv would give Λ by definition:

$$\text{Equivalent conductance} = \frac{\text{conductance}}{\text{equiv}}$$

$$\text{Equivalent conductance} = \frac{\text{conductance}}{\text{ml}} \times \frac{\text{ml}}{\text{equiv}} = \kappa \times V \qquad (14.1)$$

The volume per equivalent V can be calculated from the normality N of the solution, i.e., the number of equivalents per liter of solution. The reciprocal, $1/N$, is the number of *liters* per equivalent; $1000(1/N)$ is the number of milliliters per equivalent, which is V in Eq. (14.1).

$$V = \frac{1000}{N} \qquad (14.2)$$

Therefore, by substitution of Eq. (14.2) into Eq. (14.1), we obtain

$$\Lambda = \frac{1000}{N}\,\kappa = \frac{1000\,\kappa}{N} \qquad (14.3)$$

The unit of equivalent conductance is $\text{cm}^2/\text{equiv-}\Omega$, called the *kohlrausch unit* (ku). Figure 14.1*b* shows the distinction between κ and Λ.

FIGURE 14.2

The equivalent conductance of a typical strong electrolyte, KCl, *and a typical weak electrolyte,* HAc.

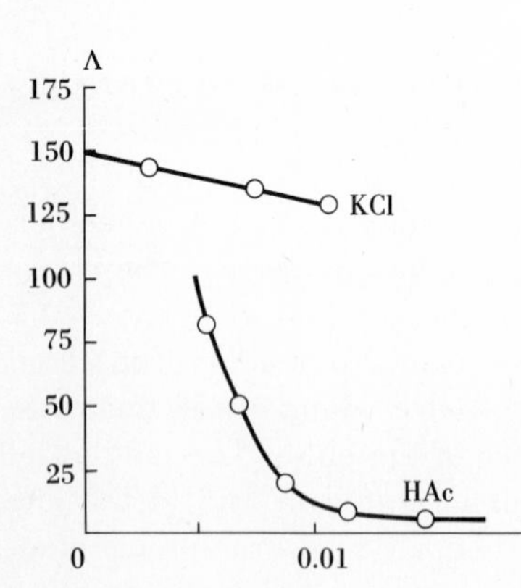

In Fig. 14.2 the equivalent conductances of two typical electrolytes are plotted versus normality. The equivalent conductance of the aqueous KCl solutions is very high at all concentrations; it increases slightly as the concentration of the solution is decreased. The equivalent conductance of acetic acid solutions is low at all concentrations above 0.01 N. The equivalent conductance of acetic acid rises sharply as the concentration decreases below 0.01 N. Electrolytes like KCl, which exhibit high conductance, are called *strong* electrolytes. Those like HAc are called *weak* electrolytes.

When the conductance curves are extrapolated to zero normality, the intercept is called the equivalent conductance at infinite dilution Λ_0. Λ_0 is interpreted as the hypothetical value which the equivalent conductance would have if the electrolyte particles were separated by infinite distance. For practical purposes when the solute particles in an aqueous electrolyte solution are separated by an average distance of 200 Å (0.0001 N), the solution is infinitely dilute and the equivalent conductance Λ equals the equivalent conductance at infinite dilution Λ_0. For example, the value of Λ for a 0.0001 N solution of KCl is 148.9, whereas the value extrapolated to infinite dilution Λ_0 is 149.8, a difference of only 0.67 percent.

14.3 THE GROTTHUS EXPLANATION FOR IONIC CONDUCTANCE

In 1805, shortly after Volta published his discovery of a cell which generated continuous current and long before the discovery of the electron, the mechanism of transport of electricity (electric current) was very poorly understood. In this year C. J. D. von Grotthus proposed an explanation for the high electric conductance of electrolyte solutions. He hypothesized that the solute molecules, which are the cause of the increased electric conductance, are electrically neutral but polar: the center of positive charge does not coincide with the center of negative charge. When solutions of these molecules are placed in an electric field, the solute molecules align themselves as shown in Fig. 14.3*a*. Then the molecule, labeled A (or B), is ruptured due to the force of the electric field. The negative electrode attracts the positive part of molecule A, and repels the negative part, thus splitting the molecule. The positive residue moves to the electrode, where it is neutralized. An opposite effect is produced on molecule B (see Fig. 14.3*b*). The negative residue of A then combines with the positive part of D forming a new molecule. The recombination process is propagated down the length of the chain until the positive residue of B is combined with the negative part of molecule C. The final result is shown in Fig. 14.3*c*. Next the molecules rotate to realign themselves in the field, and then the whole process is repeated.

FIGURE 14.3

The Grotthus mechanism.

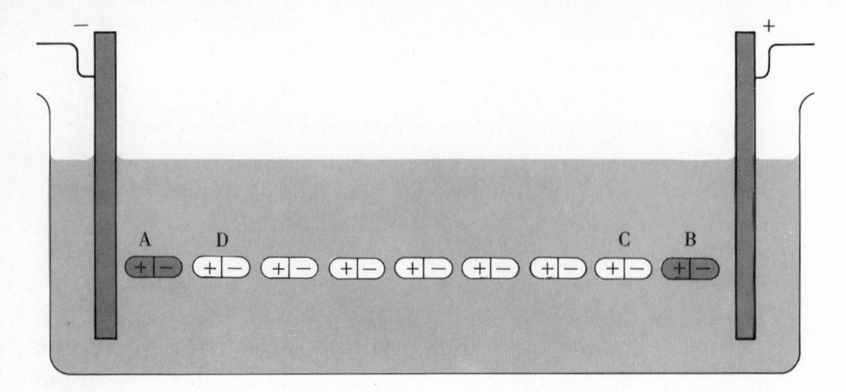

(a) Grotthus: First, the electric field causes polar molecules to align

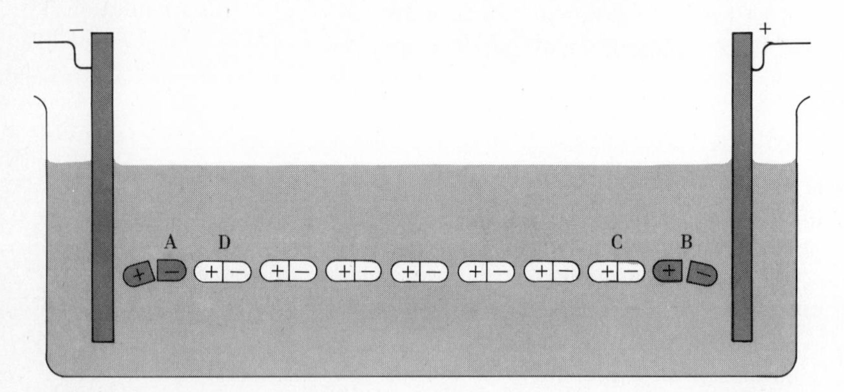

(b) The electric field splits the polar molecules

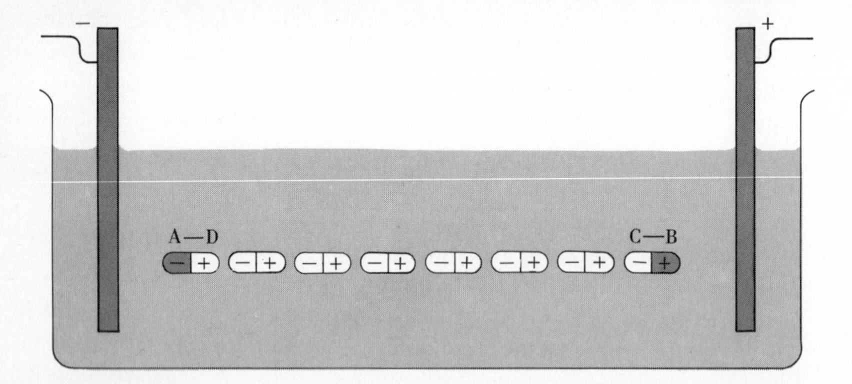

(c) The charged residues recombine before realigning

This conduction mechanism (model) presupposes that no charged particles, i.e., ions, normally exist in a solution. The polar solute dissociates only when acted upon by an external electric field. Herein lay Grotthus's error.

14.4 CLAUSIUS SEES A FLAW IN THE GROTTHUS MECHANISM

According to Grotthus's mechanism, the electric field generates the charged particles by rupturing polar molecules. If this is true, then at very low voltage, when the force is not great enough to break the chemical bonds in the polar solute molecules, no charged species would be generated and current would not flow. As the voltage between the electrodes is increased, a point would be reached at which current would just begin to flow. Hence, a plot of current versus voltage would look like Fig. 14.4*a*, which shows that the current is zero until the voltage reaches a certain minimum value; then the solute molecules are ruptured, ions are produced, and a current flows.

R. Clausius analyzed the data available for electrolyte solutions and in 1857 pointed out that when the current is plotted versus voltage, the result is a straight line which extrapolates to zero current at zero applied voltage (see Fig. 14.4*b*). This fact is inconsistent with the Grotthus hypothesis because it means that ions are present at the lowest voltages and even at zero voltage. The external electric field does not generate ions. Clausius hypothesized that the charged particles were produced by energetic collisions between electrolyte molecules and solvent molecules. Forceful collisions caused dissociation of the electrolyte molecules into ions. At any concentration of electrolyte a small equilibrium fraction of ions must exist. Nevertheless, even in the face of the strong evidence supporting the existence of ions, the concept found very few followers.

FIGURE 14.4

Comparison of the Grotthus mechanism with experiment.

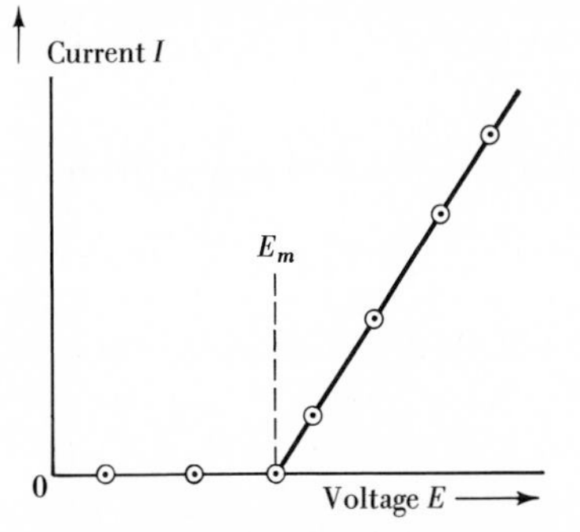

(*a*) *The hypothetical plot of current vs. voltage based on the Grotthus mechanism*

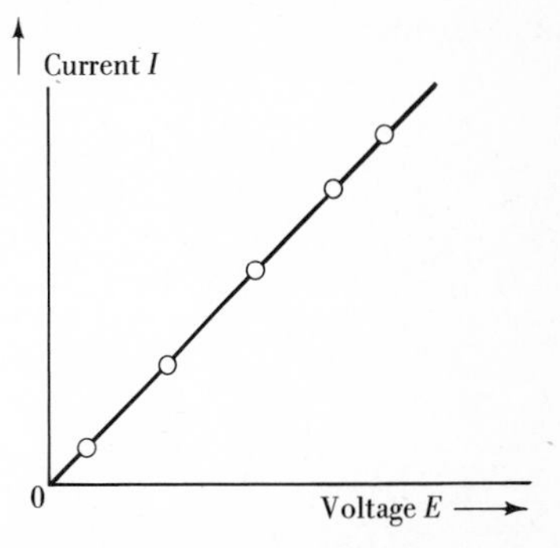

(*b*) *Experiments showed that there was a current even at very low voltage*

In his doctoral thesis (1884) the young Swedish chemist Svante Arrhenius attempted to explain the high conductance of electrolyte solutions by the postulate that ions formed from a partial dissociation of the solute are *always* present in electrolyte solutions. His examining professors were not impressed by his speculations, and the candidate was passed only with reservations. Nevertheless, his ideas were finally to gain wide acceptance.

Arrhenius's conclusions were based primarily on an interpretation of the effect dilution has on the value of the equivalent conductance Λ (see Fig. 14.2). Noting that Λ increases as the concentration decreases, he concluded that a greater fraction of the electrolyte is in ionic form in the more dilute solutions. With more ions present the conductance would be higher. He proposed that electrolyte molecules are in equilibrium with ions, as

$$MA \rightleftharpoons M^+ + A^- \tag{14.4}$$

and that such equilibria are shifted to the right by dilution.

Let us examine Arrhenius's proposal in view of Le Châtelier's principle. According to this principle, an equilibrium system reacts to a stress so as to minimize it. If we dilute a chemical-equilibrium system, we tend to decrease the number of solute particles per unit volume. To alleviate this stress an electrolyte will ionize and thereby increase the number of particles. This minimizes the stress of dilution; hence, Arrhenius's proposal is consistent with Le Châtelier's principle.

Of course, as soon as a chemical-equilibrium process is hypothesized, an equilibrium-constant expression can be written. For the ionization process given in Eq. (14.4)

$$K = \frac{[M^+][A^-]}{[MA]}$$

FIGURE 14.5

Λ_0 is obtained by extrapolation to $C = 0$.

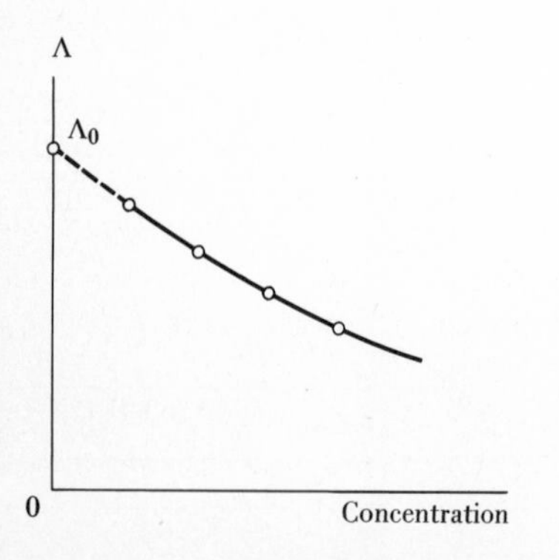

The equilibrium constant for an ionization process, called an *ionization* constant K_i, was discussed in Chap. 13. Arrhenius had to find a way of evaluating K_i to test his theory.

The fraction of ionization

Another important conclusion reached by Arrhenius concerns the equivalent conductance at infinite dilution. At very low concentrations, i.e., as a solution becomes infinitely dilute, the dissociation of the electrolyte into ions becomes complete. Then the equivalent conductance should have its maximum value at infinite dilution. Figure 14.5 shows that Λ increases steadily as a solution is diluted. Λ_0 can be obtained by extrapolating a plot of Λ versus concentration back to a concentration of zero, as shown in Fig. 14.5. The value of

Λ at any concentration above zero is lower than Λ_0 because the solute is not completely ionized. Now consider the two extremes: when the solute is completely molecular, there is no conductance and $\Lambda = 0$; when it is completely ionized, $\Lambda = \Lambda_0 =$ maximum value. Accordingly, Arrhenius suggested that the fraction α of solute which is ionized is given by Λ/Λ_0. When the solute is completely molecular, $\alpha = \Lambda/\Lambda_0 = 0/\Lambda_0 = 0$; when it is completely ionized, $\alpha = \Lambda/\Lambda_0 = \Lambda_0/\Lambda_0 = 1$.

To summarize, Arrhenius's ideas are stated succinctly:

1. An equilibrium exists between ions and their parent molecules in electrolyte solution.

2. At infinite dilution the solute is completely dissociated into ions.

3. The fraction of ions present is proportional to the equivalent conductance.

4. The fraction or degree of dissociation α of an electrolyte is Λ/Λ_0.

There are two major flaws in the Arrhenius theory: (1) he neglected the role of the solvent in ionization, and (2) he did not realize that some electrolytes are completely ionized at all concentrations. In the last chapter we explicitly indicated the role of solvent when we wrote equations for the ionization of acids and bases:

$$HAc + HOH \rightleftharpoons H_3O^+ + Ac^-$$
$$NH_3 + HOH \rightleftharpoons NH_4^+ + OH^-$$

For ionic materials like KCl and $NaNO_3$ a modern chemist would never write an equation indicating an equilibrium between molecules and ions:

$$\underset{\text{Molecule}}{KCl} \rightleftharpoons \underset{\text{Ions}}{K^+ + Cl^-}$$

KCl is ionic as a pure crystal and when it is dissolved. No molecules of KCl exist in the pure crystal or in solution, and it would be impossible to have an equilibrium between KCl molecules and ions.

14.6 EVALUATION OF α AND K_i FROM CONDUCTANCE DATA

If Arrhenius's ideas are valid, we can compare different solutes at the same concentrations and determine which is the most highly dissociated using conductance measurements alone. For instance, acetic acid and monochloroacetic acid have similar structures, but monochloroacetic acid is much stronger. This is understandable in terms of the conductance data in Table 14.2. The two acids have approximately the same Λ_0 value, 390 ku. At 0.001 N, Λ is 48 ku for acetic acid and 263 ku for monochloroacetic acid. Hence, the frac-

TABLE 14.2: COMPARISON OF TWO ACIDS AT 25°C AND AT A CONCENTRATION OF 0.001 N

Electrolyte	*Dissociation equilibrium*	Λ, ku	$\alpha = \dfrac{\Lambda}{\Lambda_0}$
Acetic acid	$HCH_2{-}C(=O){-}O{-}H \rightleftharpoons HCH_2{-}C(=O){-}O^- + H^+$	48	$\frac{48}{390} = 0.123$
Monochloroacetic acid	$ClCH_2{-}C(=O){-}O{-}H \rightleftharpoons ClCH_2{-}C(=O){-}O^- + H^+$	263	$\frac{263}{390} = 0.674$

tion of monochloroacetic ionized is much greater, and the concentration of H^+ is higher.

The fraction of ionization can be used to calculate the ionization constant, and this is best demonstrated by an example.

EXAMPLE 14.1 The equivalent conductance of a 0.05 N HAc solution is 7.4 ku, while Λ_0 is 390.6 ku at 25°C. Calculate (*a*) the fraction of dissociation and (*b*) the ionization constant. (Note that the normality and molarity are equal in this case.)

Solution: The fraction of dissociation α is just Λ/Λ_0.

$$\alpha = \frac{\Lambda}{\Lambda_0} = \frac{7.4}{390.6 \text{ ku}} = 0.0189$$

To determine K_i write the equation for the ionization and tabulate concentration data under it:

	HAc $\rightleftharpoons$	H^+ +	Ac^-
Original concentration, N:	0.05	~ 0	0
Equilibrium concentration, N:	$0.05 - 0.05\alpha$	0.05α	0.05α

The solution is 0.05 N in HAc "before" ionization occurs. At equilibrium a certain fraction has ionized. If 0.05 mol of acid is present in each liter before ionization and α is the fraction which ionizes, the amount of HAc present after ionization is $0.05 - 0.05\alpha$; that is, 0.05α mol ionizes. For each mole of acid which ionizes, 1 mol each of H^+ and Ac^- will form; at equilibrium 0.05α mol/l of H^+ and Ac^- will be present.

The ionization-constant expression is

$$K_i = \frac{[H^+][Ac^-]}{[HAc]}$$

Substituting the equilibrium concentrations from the table, we have

TABLE 14.3: TEST OF ARRHENIUS'S HYPOTHESIS†

M	Λ	$\alpha \times 10^2$	$K_i \times 10^5$
0.2529	3.221	0.838	1.759
0.03162	9.260	2.389	1.846
0.003952	25.60	6.605	1.843
0.001000	48.00	12.3	1.844
0.000494	68.22	17.60	1.853

† D. A. MacInnes and T. Shedlovsky, *J. Am. Chem. Soc.*, **54:** 1429 (1932).

$$K_i = \frac{0.05\alpha \times 0.05\alpha}{0.05 - 0.05\alpha} = \frac{0.05\alpha^2}{1 - \alpha}$$

Substitution of $\alpha = \Lambda/\Lambda_0 = 0.0189$ gives

$$K_i = \frac{0.05 \times 0.0189^2}{1 - 0.0189} = 1.82 \times 10^{-5}$$

In general the ionization-constant expression for a monoprotic acid like HAc is

$$K_i = \frac{C\alpha^2}{1 - \alpha} = \frac{C\,(\Lambda/\Lambda_0)^2}{1 - \Lambda/\Lambda_0} \tag{14.5}$$

where C is the concentration. Using Eq. (14.5), we can calculate the ionization constant of acetic acid at other concentrations. This has been done, and the results are shown in Table 14.3. The fact that K_i is fairly constant supports Arrhenius's hypothesis. Note also that α increases in value as the concentration of HAc decreases.

14.7 KOHLRAUSCH DISCOVERS STRONG ELECTROLYTES

Between 1869 and 1880, the German physicist Friedrich Kohlrausch (1840–1910) performed the experimental work which led to a clear distinction between weak and strong electrolytes. This work was performed when it was not known that salts such as NaCl are ionic in the solid state. Most scientists believed all materials to be predominantly molecular. Ions were believed to exist only under very special circumstances and then only at very low concentrations.

The difference between strong and weak electrolytes is illustrated by plots of equivalent conductance versus concentration. In Fig. 14.2, KCl represents a typical strong electrolyte. Its conductance is high at all concentrations, and as the concentration of KCl increases, the equivalent conductance decreases only slightly below

TABLE 14.4: THERE IS NO IONIZATION CONSTANT FOR STRONG ELECTROLYTES

Electrolyte	*Concentration*	"K_i" $= \dfrac{C\alpha^2}{1-\alpha}$
HCl	0.003	0.15
	0.001	0.07
	0.0001	0.03
KCl	0.01	0.17
	0.001	0.051
	0.0001	0.016

the infinite dilution value Λ_0. Acetic acid is representative of weak electrolytes; its conductance is relatively low at all concentrations. Below 0.01 N the equivalent conductance of HAc rises sharply as the concentration decreases. A strong electrolyte is completely dissociated at all concentrations. There is no equilibrium between molecules and ions in this case. The application of Arrhenius's hypothesis to the conductance data for the strong electrolytes studied by Kohlrausch yields ionization "constants" which vary considerably (see Table 14.4).

Kohlrausch also noted that the equivalent conductance of the strong electrolytes is a linear function of the *square root of the concentration* in dilute solutions:

$$\Lambda = \Lambda_0 - A\sqrt{C} \tag{14.6}$$

A plot of Λ versus $\sqrt{C}$ for strong electrolytes gives a linear plot, as shown in Fig. 14.6.

Equation (14.5), which is based on the Arrhenius hypothesis, predicts that Λ is a linear function of the concentration. This is shown by rearrangement of Eq. (14.5), which yields

$$\Lambda = \Lambda_0 - \frac{\Lambda_0}{K_i}\frac{\Lambda}{\Lambda_0}C$$

In a very dilute solution of strong electrolytes $\Lambda \sim \Lambda_0$ or $\Lambda/\Lambda_0 \sim 1$,

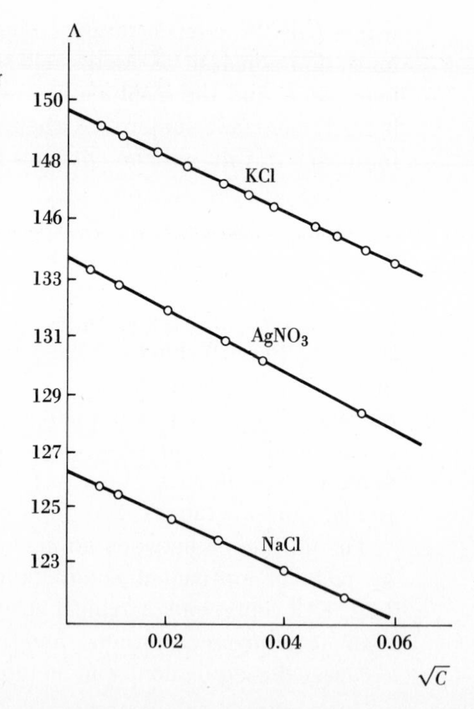

FIGURE 14.6

A plot of Λ versus $\sqrt{C}$ for strong electrolytes in water.

and we have

$$\Lambda = \Lambda_0 - \frac{\Lambda_0}{K_i} C \qquad (14.7)$$

Since Λ_0 and K_i are constants, Arrhenius's theory predicts that Λ is a linear function of the concentration. But Kohlrausch found that Λ is a linear function of the *square root* of the concentration. Hence, Arrhenius theory is not valid for strong electrolytes. The conclusion is that there exist two kinds of electrolytes in aqueous solutions.

1. Strong electrolytes
  a. The magnitude of Λ is approximately 100 ku at ordinary concentrations.
  b. Λ is not strongly affected by changes in concentrations. These electrolytes obey Kohlrausch's law ($\Lambda = \Lambda_0 - A\sqrt{C}$) and are *not* amenable to Arrhenius's theory.
  c. The solute is completely ionic, and there is no equilibrium between molecules and ions; no neutral solute molecules exist in strong electrolyte solutions.

2. Weak electrolytes
  a. The magnitude of Λ is below 15 ku at ordinary concentrations.
  b. Λ rises very sharply in very dilute solutions.
  c. These electrolytes are amenable to Arrhenius's theory but do *not* obey Kohlrausch's law [Eq. (14.6)].
  d. The solute is predominantly molecular, but ions are present in equilibrium with molecules, $MA \rightleftharpoons M^+ + A^-$. This equilibrium shifts to the right with increasing dilution and causes a sharp increase in Λ in dilute solutions.

14.8 COLLIGATIVE PROPERTIES OF STRONG ELECTROLYTE SOLUTIONS

If strong electrolytes are completely ionic and do not participate in an equilibrium with molecules, why does Λ decrease as the concentration is increased? This question is answered best by interpreting the results of colligative-property experiments; 1 mol of a solute such as NaCl, which is believed to be completely ionic in solution, should cause a freezing-point depression 2 times greater than 1 mol of a solute such as sucrose, which is molecular. Recall, from page 259 that the freezing-point depression ΔT_f is related to concentration by the equation

$$\Delta T_f = k_f m \qquad (14.8)$$

where k_f = molal freezing-point-depression constant
m = molality

For strong electrolytes we should be able to calculate the freezing-point depression by incorporating into Eq. (14.8) a factor i which accounts for the number of ions formed per mole of electrolyte.

$$T = ik_f m \qquad i = \begin{cases} 2 & \text{for NaCl} \\ 3 & \text{for } CaF_2 \end{cases}$$

Ideally, a NaCl solution should have twice the depression of a sucrose solution at the same concentration because each mole of NaCl gives 2 mol of ions, Na^+ and Cl^-. Table 14.5 shows experimental and theoretical ΔT_f values for sucrose and NaCl solutions at several concentrations. (For water at 0°C k_f is 1.86 deg/mol.) Also included are the freezing-point depression per mole of solute $\Delta T_f/m$. This value should be constant, independent of concentration. For sucrose it should be equal to the molal freezing-point constant:

$$\Delta T_f = k_f m \qquad \text{and} \qquad \frac{\Delta T_f}{m} = k_f = 1.86$$

For NaCl it should be equal to twice k_f:

$$\Delta T_f = ik_f m = 2k_f m \qquad \text{and} \qquad \frac{\Delta T_f}{m} = 2k_f = 3.72$$

It is evident from Table 14.5 that although the values for sucrose solutions are in excellent agreement with theory, the freezing-point depressions of the NaCl solutions deviate considerably from those predicted by theory. All the values are lower than the theoretical values. Also, the freezing-point depression per mole $\Delta T_f/m$ decreases steadily with increasing concentration.

The results of freezing-point-depression experiments on NaCl solutions are typical of the results obtained for all colligative-property experiments on strong-electrolyte solutions. All colligative-property values for strong-electrolyte solutions are lower than the

TABLE 14.5: ΔT_f FOR AQUEOUS SOLUTIONS OF SUCROSE AND NaCl

molality, m	*0.01*		*0.1*		*0.2*		*0.5*	
	Calc.	*Obs.*	*Calc.*	*Obs.*	*Calc.*	*Obs.*	*Calc.*	*Obs.*
Sucrose:								
ΔT	0.0186	0.0186	0.186	0.186	0.372	0.372	0.93	0.92
$\frac{\Delta T_f}{m}$	1.86	1.86	1.86	1.86	1.86	1.86	1.86	1.84
NaCl:								
ΔT	0.0372	0.0340	0.372	0.290	0.744	0.543	1.86	1.26
$\frac{\Delta T_f}{m}$	3.72	3.40	3.72	2.90	3.72	2.72	3.72	2.52

values predicted from the simple laws studied in Chap. 9:

1. Relative lowering of vapor pressure:

$$\frac{P^0 - P}{P^0} = X(\text{solute})$$

2. Osmotic pressure:

$$\pi = cRT$$

3. Freezing-point depression:

$$\Delta T_f = k_f m$$

4. Boiling-point elevation:

$$\Delta T_b = k_b m$$

14.9 EXPLANATION OF THE OBSERVED DEVIATIONS OF COLLIGATIVE PROPERTIES

The colligative properties depend only on the concentration of the solute, not on the nature of the solute. If we accept the colligative-property equations as accurate, we conclude that the effective concentration of the solute in electrolyte solutions is less than the actual concentration *m*. We can explain the experimental deviations from theory by mechanisms which effectively reduce the number of ions in solution without causing molecules to form.

The electrostatic forces of attraction among ions are strong even at long distances. For instance, when two uncharged sucrose molecules are separated by a distance of 100 Å, there is an attractive force between them amounting to only 10^{-37} dyn. However, when a sodium ion is separated from a chloride ion by the same distance, the electrostative attractive force is 2×10^{-7} dyn! Obviously forces operating between ions are enormous compared to forces operating between uncharged molecules. As a result of these forces, ions do not move independently; they are influenced by neighboring ions. The electrostatic forces of attraction may cause ions separated by small distances (<8 Å) to move as a unit, even though there is no evidence of covalent bonds between them. These units, called *aggregates*, move together until they are separated by forceful collisions with other particles. Therefore, the effective concentration of ions may be less than the actual concentration, and colligative properties of electrolyte solutions are lower than predicted from simple theory.

As the concentration of electrolyte is decreased and the solution becomes very dilute, the experimental values agree more closely with the theoretical values (see Table 14.3) because the ions have larger average separations and the attractive forces between them are reduced. This explanation in terms of microscopic particles, reached by analysis of colligative-property data, also explains the experimental observation that the equivalent conductance Λ for strong electrolytes decreases as the concentration of solute increases, even though no molecules form. As shown in Fig. 14.7, $\Delta T_f/m$ and Λ decrease as the electrolyte concentration increases. This is attributed to ionic aggregation, which reduces not only the number of particles but also the number of charged particles. An aggregate such as $Na^+ \cdots Cl^-$ is neutral and does not conduct electricity.

14.10 THE DEBYE-HÜCKEL THEORY

In 1923 Peter Debye and Erich Hückel gave a theoretical explanation of deviations of the colligative-property values in dilute solutions of strong electrolytes. Their model, the basis of all their calculations, was very simple. Focusing their attention on a cation in solution, they envisioned it as being surrounded by other anions and cations. However, because ions of like charges repel one another while ions of unlike charges attract one another, there would be a preponderance of anions in the vicinity of a central cation (see Fig. 14.8*a*). Likewise, they pictured a central anion as being surrounded by many cations and other anions, but on the average there would be a preponderance of cations about the anion. To simplify their model they looked upon the ions which surrounded the central ion as being part of a continuous *ionic cloud*, as shown in Fig. 14.8*b*. And then by rather sophisticated thermodynamic and statistical methods Debye and Hückel estimated the electrostatic force which acts between every possible pair of ions in solution. They reasoned that the electrostatic force is related to the effective concentration of the electrolyte in solution. The higher the force the lower the effective concentration of the electrolyte since the force would cause aggregates to form. Their analysis of the electrostatic forces made it possible to calculate the activity, or effective concentration, of the electrolyte in solution. (Remember that $a = fC$; see page 267.)

FIGURE 14.7

Aggregation of ions reduces the equivalent conductance and the freezing-point depression per mol, $\Delta T/m$.

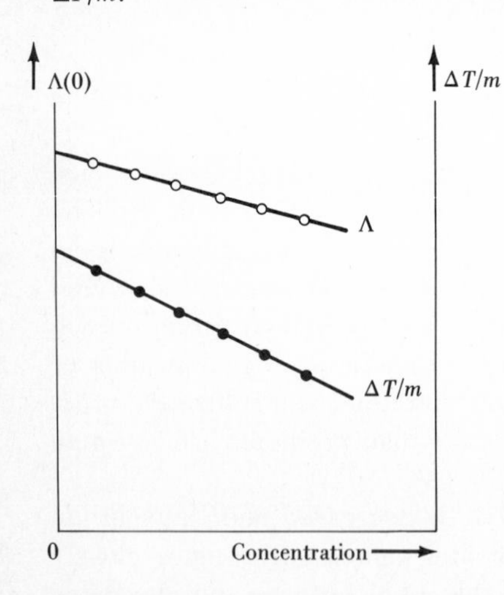

The equation which bears their name is somewhat complicated, but for aqueous electrolyte solutions at 25°C it reduces to

$$\log f = -0.509|Z_+Z_-|\sqrt{I} \qquad (14.9)$$

where f = activity coefficient
Z_+ = charge on cation
Z_- = charge on anion
I = ionic strength

FIGURE 14.8

The Debye-Hückel model for an electrolyte solution.

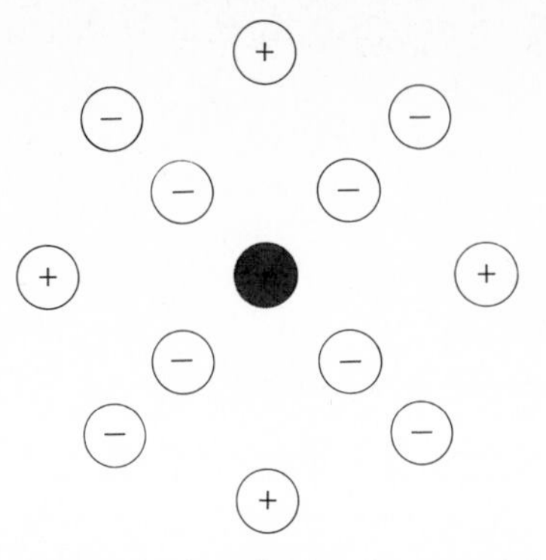

(*a*) *A cation is predominantly surrounded by anions*

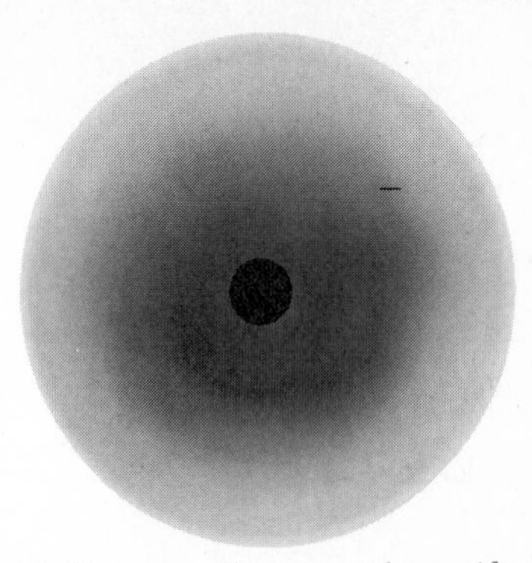

(*b*) *The surrounding ions can be considered as an "ionic cloud" of opposite sign*

The ionic strength is equal to one-half the sum of the product of the molarity and the charge squared for each ion in the solution:

$$I = \frac{1}{2}\sum_{i=1}^{i=n} M_i Z_i^2 \tag{14.10}$$

The ionic strength of a one-one electrolyte, like Na^+Cl^- or H^+Cl^-, is equal to the molar concentration.

EXAMPLE 14.2 Calculate the (*a*) ionic strength, (*b*) activity coefficient, and (*c*) activity of a 0.01 *M* NaCl solution.

Solution: (*a*)

$I = \frac{1}{2}\Sigma M_i Z_i^2$

The concentrations of the Na^+ and Cl^- ions are both 0.01 *M*:

$$I = \tfrac{1}{2}[0.01(+1)^2 + 0.01(-1)^2] = 0.01$$

(*b*)
$$\log f = -0.509|Z_+Z_-|\sqrt{I}$$
$$= -0.509|(+1)(-1)|\sqrt{0.01} = -0.0509$$
$$f = 10^{-0.0509} = 10^{0.9491} \times 10^{-1} = 0.889$$

(*c*) Activity = activity coefficient × concentration

$a = fC = 0.889 \times 0.01 = 0.00889$

EXAMPLE 14.3 Calculate the (*a*) ionic strength and (*b*) activity coefficient for K_2SO_4 in a 0.15 *M* solution. (In this case *I* will not equal the concentration of the electrolyte.)

Solution: (*a*) The molar concentration of potassium ion is twice the concentration of the salt, whereas the concentration of SO_4^{2-} is just equal to the concentration of the salt.

$$I = \frac{1}{2}\Sigma M_i Z_i^2 = \frac{1}{2}[\underbrace{0.30(+1)^2}_{K^+} + \underbrace{0.15(-2)^2}_{SO_4^{2-}}] = 0.45$$

$$(b)\ \log f = -0.509|Z_+Z_-|\sqrt{I} = -0.509|1(-2)|\sqrt{0.45} = -0.687$$
$$f = 10^{-0.687} = 10^{0.313} \times 10^{-1} = 2.05 \times 10^{-1} = 0.205$$

With the activity coefficient it is possible to make good estimates for the values of the colligative properties of electrolyte solutions.

EXAMPLE 14.4 Calculate the osmotic pressure of a 0.02 M aqueous NaCl solution at 25°C (*a*) using the ionic concentration and (*b*) using the activity instead of concentration.

Solution: (*a*) The concentration of particles in a 0.20 M NaCl solution is 2 times the molar concentration:

$$C = 2 \times 0.02 = 0.040\ M$$

and

$$\pi = CRT = 0.040 \times 0.082 \times 298 = 0.975 \text{ atm}$$

(*b*) Actually we should use the activity to obtain a better estimate of π.

$$\pi = aRT = fCRT$$

Using the Debye-Hückel equation, and remembering that the ionic strength equals the molarity for a one-one electrolyte, we have

$$\log f = -0.509|Z_+Z_-|\sqrt{C} = -0.509|1(-1)|\sqrt{0.02}$$

which reduces to

$$\log f = -0.071$$

$$f = 10^{-0.071} = 10^{+0.929} \times 10^{-1} = 8.5 \times 10^{-1} = 0.85$$

The osmotic pressure is then

$$\pi = fCRT$$

$$= 0.85(0.040 \text{ mol/l})(0.082 \text{ l-atm/mol-K})(298 \text{ K}) = 0.83 \text{ atm}$$

The experimental value of π is 0.85 atm. Obviously this value is much closer to the experimental value than the value obtained from the simple theory. In general, when activities are used instead of concentrations in the colligative-property equations, the results are in much better agreement with the experiment.

The colligative-property experiments lead us to the conclusion that aggregation of ions occur in solutions, and at the same time we begin to understand why the *equivalent* conductance decreases as the concentration of electrolyte increases.

Starting with the Debye-Hückel model for electrolyte solutions, Lars Onsager was able to derive Kohlrausch's experimental law [Eq. (14.6)] from theory. Onsager predicted two conductance-reducing effects, the *asymmetry* effect and the *electrophoretic* effect. Neither effect involves the formation of molecules.

Asymmetry effect

Onsager considered the effect of the ionic cloud on a moving ion. In an equilibrium system the ionic cloud would be symmetrical about the central ion (see Fig. 14.9). In the presence of an electric field, a central cation would move toward the cathode (negative electrode), while the negative ionic cloud would move toward the anode (positive electrode). Thus, when the field is acting on the system, the ion and its ionic cloud move in opposite directions, disrupting the symmetry of the ionic cloud about the central ion, as shown in Fig. 14.9*b*. The center of charge of the negative ionic cloud no longer coincides with the central cation. Because of *asymmetry* the ionic cloud exerts an attractive force on the cation and impedes its flow toward the negative electrode. The conductance is therefore reduced by the *asymmetry effect*.

The electrophoretic, or drag, effect

Onsager predicted another conductance-reducing effect. Each ion, in addition to having an ionic cloud, is solvated. Thus when an anion moves, its sheath of solvent molecules moves with it. At the same time a "stream" of solvated cations is moving in the opposite direction. The bulky solvated anions are "swimming" against a stream of cations and vice versa. The motion of the ions is mutually impeded, and the conductance is reduced by this *electrophoretic*, or drag, ef-

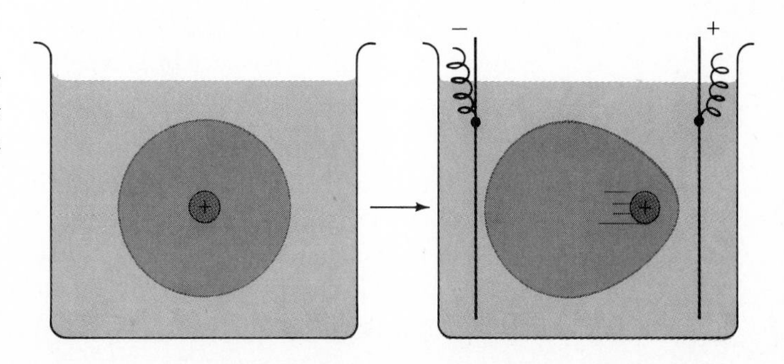

FIGURE 14.9

The asymmetry effect. The ionic cloud becomes unsymmetrical about the central ion when the central ion moves in an electronic field.

FIGURE 14.10

The electrophoretic, or drag effect.

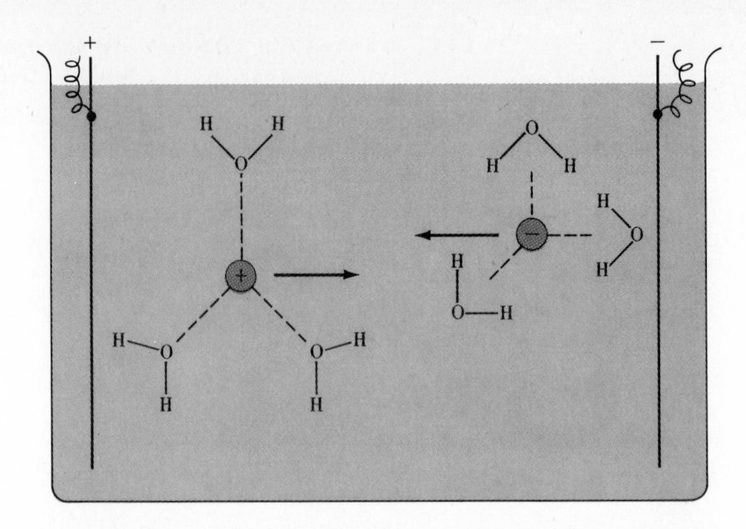

fect (see Fig. 14.10). As the concentration of the electrolyte is reduced, the asymmetry and the electrophoretic effects should diminish. The charge of the ionic cloud depends on the electrolyte concentration. At infinite dilution the ionic cloud does not exist. The electrophoretic effect is unimportant in very dilute solution since the oppositely charged ions are not near enough to cause any resistive drag on each other. At the hypothetical condition of infinite dilution the electrophoretic effect disappears. Since both effects are diminished by decreasing concentration, the Onsager picture implies that the equivalent conductance increases with decreasing concentration, in agreement with experiment.

Considering these two effects and *not* hypothesizing an equilibrium between ions and molecules as Arrhenius did, Onsager was able to derive the equation which predicts that Λ should decrease as the concentration increases. Onsager's equation has the same form as Kohlrausch's experimental equation:

$$\Lambda = \Lambda_0 - A\sqrt{C} \qquad \text{Kohlrausch's equation} \tag{14.6}$$

$$\Lambda = \Lambda_0 - \left[\frac{82.48}{(\epsilon T)^{1/2}\eta} + \frac{8.20 \times 10^5}{(\epsilon T)^{3/2}}\Lambda_0\right]\sqrt{C} \qquad \text{Onsager's equation} \tag{14.11}$$

where ϵ = dielectric constant of solvent
η = viscosity of solvent
T = temperature

Both equations show that Λ is a decreasing, linear function of the square root of the concentration. Therefore, Onsager's theoretical equation has the correct form. But Onsager's equation also predicts the slope of a plot of Λ versus $\sqrt{C}$. Comparing Eqs. (14.6) and (14.11) shows that the experimental slope A should be equal to the bracketed term in Eq. (14.11). At a fixed temperature in a given solvent the term within the brackets, the *Onsager slope*, is constant. In hundreds of experimental tests of the theory the experimental slope has been found to be equal to the Onsager slope.

14.12 CATEGORIES OF ELECTROLYTES

Mainly on the basis of conductance measurements, but in conjunction with other physical and chemical measurements, it is possible to categorize electrolytes as true, potential, strong, and weak. *True electrolytes* are salts like NaCl, KNO_3, CsI, etc., which are ionic in the crystalline and molten state. In the molten state, where the mobility of the ionic particles is high, these salts conduct electricity.

X-ray measurements can be used to establish the existence of ions in the crystalline state. The data from X-ray experiments show that, relative to the neutral atom, a cation has a deficiency of electrons and an anion has an excess of electrons. For example, in solid crystalline NaCl the number of electrons at Na^+ is 10.02 and at Cl^- is 17.96. This indicates an almost complete transfer of one electron from ${}_{11}Na$ to ${}_{17}Cl$. NaCl is a true electrolyte because it is intrinsically ionic.

Potential electrolytes are covalent compounds like HCl, HAc, etc., which have the ability to react with solvents to produce charged species. For example, in the reaction

$$HCl + H_2O \rightarrow H_3O^+ + Cl^-$$

HCl, which is covalently bonded, reacts with water to form a hydrated proton plus the chloride ion. Pure liquid or gaseous HCl is neither ionic nor acidic. Only when HCl reacts with water is an acid produced. Hence, pure HCl should not be referred to as hydrochloric acid. Hydrochloric acid is an aqueous solution of HCl.

Strong electrolytes are solutes which produce solutions having high equivalent conductances ($\Lambda > 10$) at ordinary concentrations. A strong electrolyte may be a true or potential electrolyte. Here the *solvent* must be considered. For example, NaCl is a strong electrolyte in water but a weak electrolyte in nitrobenzene. Some other examples of strong electrolytes are NaCl in H_2O; HCl in H_2O; tetramethylammonium hydroxide, $(CH_3)_4NOH$, in H_2O; and potassium amide, KNH_2, in liquid NH_3.

Weak electrolytes are solutes which produce solutions having low equivalent conductances ($\Lambda < 10$) at ordinary concentrations. Usually the solute is molecular or highly aggregated. Again, the solvent must be considered. Even a true electrolyte can be a weak electrolyte in a particular solvent. Some examples of weak electrolytes are HCl in ethanol, HAc in H_2O, NaCl in nitrobenzene, NH_3 in H_2O, and alcohols in H_2O.

QUESTIONS AND PROBLEMS

14.1 (*a*) Do electrolyte solutions have a high conductance compared to metals? Support your answer with data. (*b*) Why do we consider that electrolyte solutions have a high conductance?

14.2 (*a*) Distinguish between specific conductance and equivalent conductance. (*b*) What information is required for the calculation of the equivalent conductance of an electrolyte solution?

14.3 Calculate the equivalent conductances of the four electrolyte solutions described in Table 14.1. In each case 1 mol of electrolyte equals 1 equiv.

14.4 How is the equivalent conductance at infinite dilution Λ_0 obtained for a strong electrolyte, and what is its physical significance? Why is it not possible to obtain a value of Λ_0 for a weak electrolyte like HAc in the same way as for a strong electrolyte like KCl? (Look at Fig. 14.2.)

14.5 Why did Clausius decide that the Grotthus mechanism for conductance was incorrect?

14.6 (*a*) What are the four salient features of Arrhenius's electrolyte theory? (*a*) Why was it considered revolutionary?

14.7 Using Fig. 14.2, make an estimate of K_i for HAc. Λ_0 for HAc is 391 ku. Use the Λ value at 0.01 N for your calculation.

14.8 Using the data in Table 14.2, calculate K_i for monochloroacetic acid at 25°C.

14.9 Is the trend in α shown for HAc in Table 14.3 contrary to that expected from Arrhenius's theory? Explain.

14.10 (*a*) Make plots of Λ versus $\sqrt{N}$ for LiCl and $AgNO_3$ using the data in Table P14.10, which were obtained at 25°C. Put both plots on the same graph. (*b*) By extrapolation estimate Λ_0 for each salt.

TABLE P14.10

	Λ, ku	
Concentration N	LiCl	$AgNO_3$
0.0005	113.2	131.4
0.001	112.4	130.5
0.005	109.4	127.2
0.01	107.3	124.8
0.02	104.7	121.4
0.05	100.1	115.2
0.1	95.9	109.1

14.11 Why is the Arrhenius electrolyte theory inapplicable to the strong electrolytes studied by Kohlrausch?

14.12 Assuming that K_2SO_4 is completely dissociated in an aqueous solution calculate ΔT_f and ΔT_b for a solution which is 20 percent by weight K_2SO_4. (Let concentration equal activity.)

14.13 Calculate the osmotic pressure at 25°C of a 15 percent by weight NaCl solution. The density of the solution is 1.11 g/ml, and the activity coefficient of NaCl in the solution is 0.714.

14.14 Calculate the ionic strengths of the following solutions: (*a*) 0.100 M CsCl, (*b*) 0.100 M Cs_2SO_4, (*c*) 0.250 M Na_3PO_4, (*d*) 0.210 M $LaCl_3$, (*e*) 0.100 M NaCl and 0.100 M $CaSO_4$, (*f*) 0.250 M KNO_3 and 0.010 M $Cu(NO_3)_2$.

14.15 Using Eq. (14.9), calculate the activity coefficient of (*a*) KCl in a 0.01 M solution, (*b*) Na_2SO_4 in a 0.005 M solution, and (*c*) $AgNO_3$ in a 0.002 M solution.

14.16 Calculate the osmotic pressure at 25°C of a 0.0050 M KNO_3 solution (*a*) assuming that the activity equals the concentration and (*b*) using the activity (as determined from the Debye-Hückel theory) in place of concentration.

14.17 As the concentration of weak electrolyte increases, the fraction of solute in the form of molecules increases and the equivalent conductance drops. Onsager showed how the equivalent conductance of a strong electrolyte can decrease without supposing that strong electrolytes form neutral molecules. Explain in terms of the asymmetry effect and the electrophoretic effect.

14.18 (*a*) Define true, potential, strong, and weak electrolytes. (*b*) Can a true electrolyte be a weak electrolyte? If so, give an example. (*c*) Can a potential electrolyte be a strong electrolyte? If so, give an example.

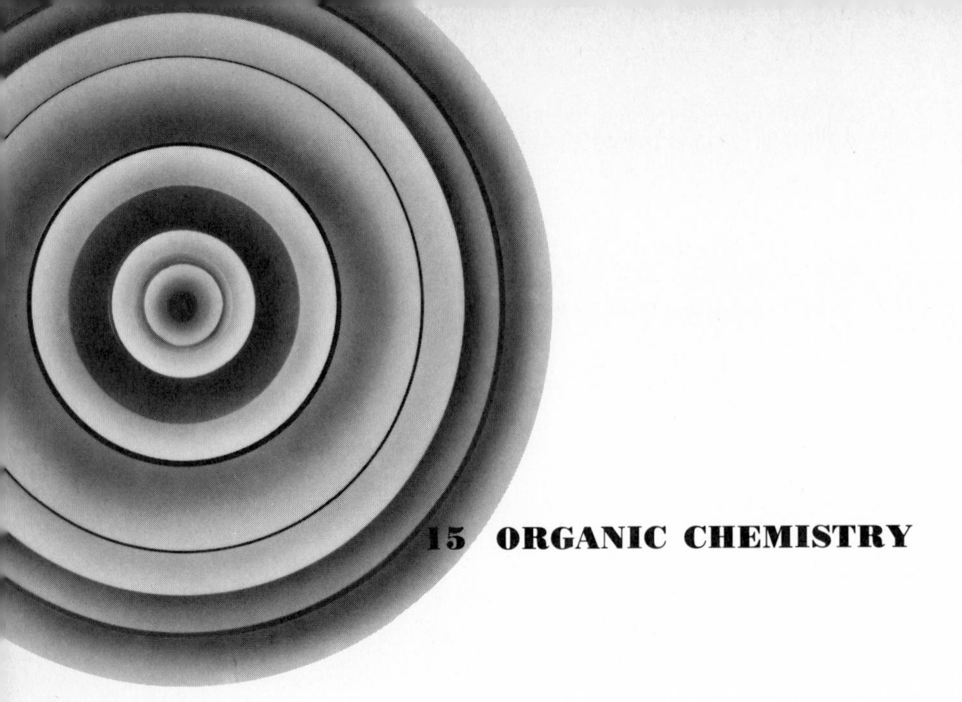

15 ORGANIC CHEMISTRY

15.1 INTRODUCTION

Before 1828 organic chemistry was a study of the compounds produced by living organisms, plants and animals. Chemists believed that an intangible *vital force* was required for the synthesis of such compounds, but in 1828 the German chemist Friedrich Wohler was studying the reactions of the cyanates and attempting to synthesize ammonium cyanate when instead he obtained urea, a compound found in the urine of mammals.

Predicted reaction:
$$\underset{\text{Ammonium chloride}}{NH_4Cl} + \underset{\text{Silver cyanate}}{AgCNO} \rightarrow \underset{\text{Ammonium cyanate}}{NH_4CNO} + \underset{\text{Silver chloride}}{AgCl}$$

Actual reaction:
$$NH_4Cl + AgCNO \rightarrow \underset{\text{Urea}}{CO(NH_2)_2} + AgCl$$

(Note that ammonium cyanate and urea have the same molecular formula, CON_2H_4.) The synthesis of urea did not signal an instant death of *vitalism*, but it did begin the erosion of this concept. Several other organic compounds were produced synthetically in the next 20 years, notably the synthesis by Hermann Kolbe (1818–1884) of acetic acid and the synthesis by Marcelin Berthelot (1827–1907) of methane, acetylene, and several other organic compounds. Chemists began to realize that the compounds produced by plants and animals can also be produced in the laboratory under vastly different conditions. By 1850 the principle of vitalism was abandoned, and synthetic organic chemistry became a well-defined field of research.

Organic chemistry is no longer a study restricted to *natural products*, which are compounds produced by living organisms. It is a study of the compounds of carbon. Carbon compounds can undergo acid-base, oxidation-reduction, polymerization, and a multitude of other reactions. The range of their physical properties is vast. Some carbon compounds have the tensile strength of steel. Some are hallucinogens, poisons, vitamins, pigments; some are liquids very similar to water; some conduct electricity, while others are excellent insulators. The modern organic chemist has learned the relationship between the physical properties of the carbon compounds and their molecular structure. Since he also understands the reactions of these compounds, he can synthesize materials to fit particular needs. The industries of the world rely heavily on the ingenuity of the organic chemist.

15.2 PETROLEUM

The starting materials with which the organic chemist practices his magic usually come from petroleum, a mixture of gaseous, liquid, and solid carbon compounds found in sedimentary rock deposits throughout the world. Some scientists believe that petroleum reserves were formed from the remains of aquatic plants and animals which lived billions of years ago. After they died their remains accumulated and were covered over by layers of sedimentary mud and sand. As the earth's crust buckled and heaved, high pressures and temperatures were created and oil was formed from the organic remains. Petroleum contains thousands of different chemical compounds, but the major components are the *hydrocarbons*, compounds of hydrogen and carbon only. The hydrocarbons of lowest molecular weight are gases and are used as fuel for heating, kitchen ranges, and welding. The heavier compounds, with molecules containing from 4 to 10 carbon atoms, are used as gasoline fuel for internal-combustion engines. Still heavier compounds are used as cruder fuels, lubricants, and asphalt. These materials are separated by distillation, and organic chemists can convert the various components of petroleum into compounds which are used to make synthetic rubber, plastics, pigments, textiles and a host of other materials.

15.3 STRUCTURE OF THE ALKANES

The alkanes are compounds of hydrogen and carbon only. Methane, ethane, and propane are the alkanes with the lowest molecular weights. They are gases, and their formulas (CH_4, C_2H_6, C_3H_8, respectively) fit the general formula for the alkanes, C_nH_{2n+2}. In methane, $n = 1$, and $2n + 2 = 4$; therefore the formula for methane is CH_4. In propane, $n = 3$, and $2n + 2 = 8$; therefore the formula of propane is C_3H_8.

In the alkane series each carbon atom in a molecule is bonded to four other atoms, as shown in Table 15.1 for methane, ethane, and propane. In methane the single carbon atom is bonded to four hydrogen atoms. In ethane each carbon atom is bonded to three hydrogen atoms and the other carbon atom. In propane each terminal carbon atom is bonded to three hydrogen atoms and to one other carbon atom while the interior carbon is bonded to two hydrogen atoms and two carbon atoms. Since carbon can form bonds with a *maximum* of four other atoms, we say that in the alkanes the bonding power of carbon is saturated and the alkanes are called the *saturated hydrocarbons*.

There is no known limit to the number of carbon atoms in an alkane molecule. Alkanes with more than 100 carbon atoms are known, and with our understanding of organic chemistry we could synthesize compounds containing even greater numbers if we had any reason to do so. Carbon is a natural chain former, and this property makes it unique among the elements.

On page 141 we discussed the structure and the nature of the bonding in the methane molecule. The hydrogen atoms are arranged tetrahedrally about a carbon atom, as shown in Table 15.1; all C—H bond lengths are 1.09 Å, and all H—C—H bond angles are 109°28′. We explained the valence of carbon in methane and the structure of the molecule in terms of hybridization theory. By sp^3 hybridization the carbon atom can have four equivalent orbitals available for bonding, and these orbitals have a tetrahedral orientation in space. The structures of all other alkanes are patterned after methane

TABLE 15.1: ALKANE STRUCTURE

Compound	*Formula*	*Structure*
Methane	CH_4	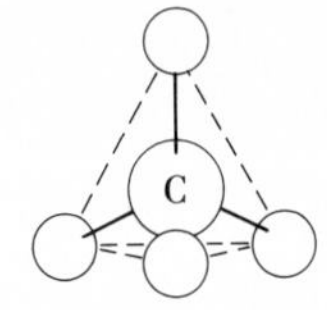
Ethane	C_2H_6 or $CH_3—CH_3$	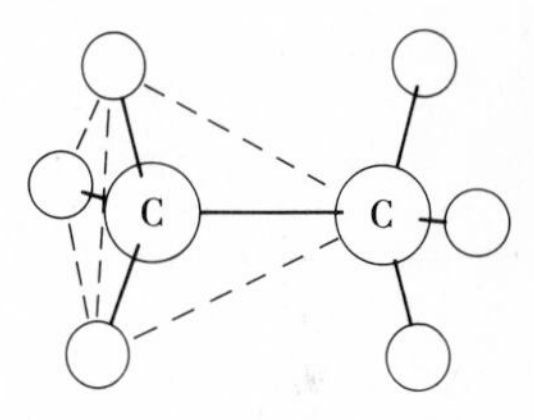
Propane	C_3H_8 or $CH_3—CH_2—CH_3$	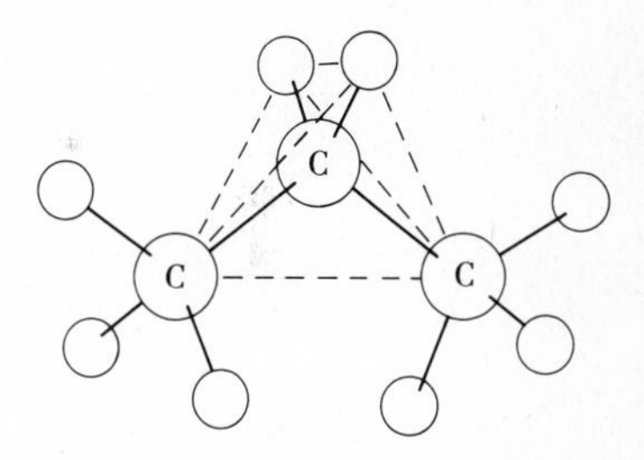

and can be explained in terms of sp^3 hybridization of the carbon atom.

Ethane, C_2H_6, can be considered a derivative of methane. Conceptually remove one of the hydrogen atoms of methane

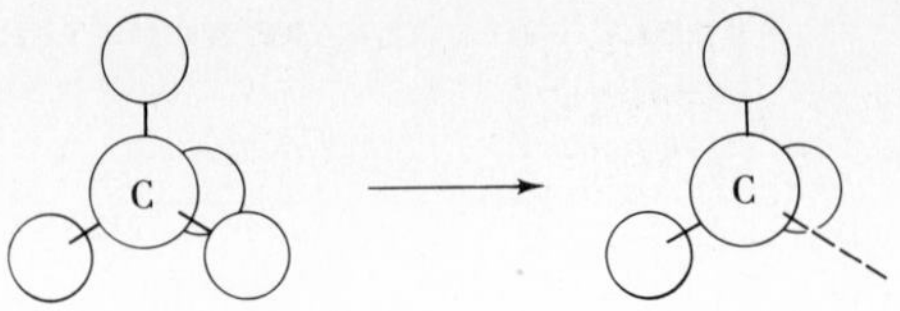

CH_4 → CH_3

put a carbon atom in its place

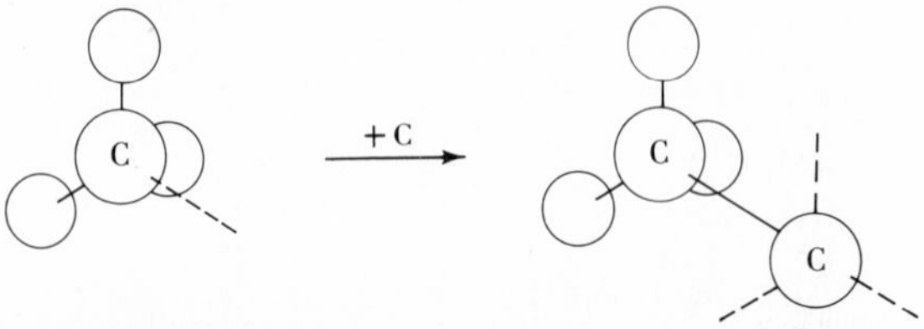

CH_3 +C → C_2H_3

and then bond three hydrogen atoms tetrahedrally around the new carbon atom

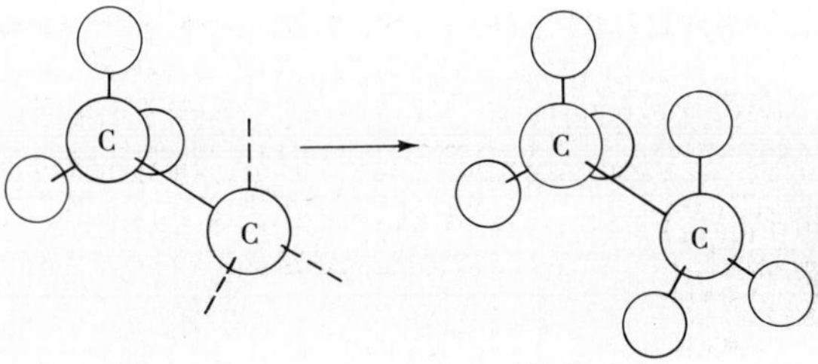

C_2H_3 → C_2H_6, ethane

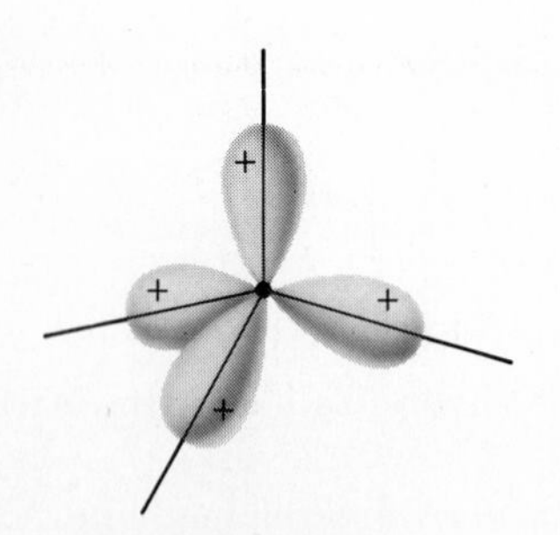

The four sp^3 *orbitals on* C

Hence, ethane is a derivative of methane in which one of the hydrogen atoms has been replaced by a CH_3 group. The CH_3 group is called a *methyl group* or *methyl radical*. Each carbon atom in ethane can be considered to reside at the center of a tetrahedron, as shown in Table 15.1. The four vertices about each central carbon atom are occupied by three hydrogen atoms and one carbon atom. All C—H bond lengths are 1.09 Å, and the C—C bond length is 1.54 Å. All bond angles are 109°28′.

The molecular structure of ethane can be explained by sp^3 hybridization. The ground state of the carbon atom is $_6C\ 1s^2, 2s^2, 2p_x^1, 2p_y^1, 2p_z^0$, and the sp^3-hybridized state is $_6C\ 1s^2, (sp^3)_1^1,$

FIGURE 15.1

The σ bond between carbon atoms in ethane.

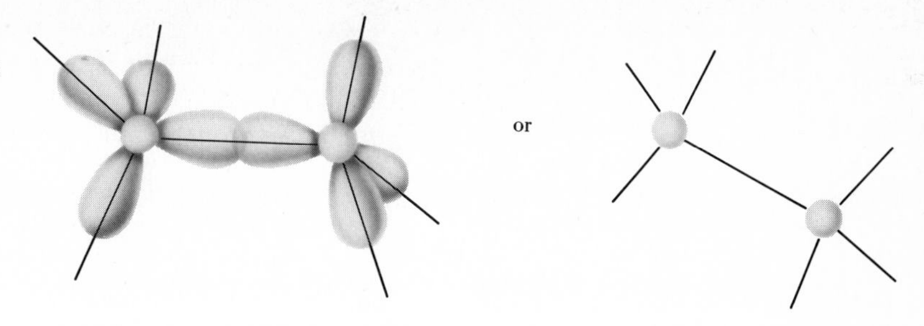

$(sp^3)_2{}^1$, $(sp^3)_3{}^1$, $(sp^3)_4{}^1$. A σ bond forms between the two carbon atoms by overlap of sp^3 (see Fig. 15.1). Each of the remaining six sp^3 orbitals overlaps with the 1*s* orbital of a hydrogen atom, and six C—H bonds are formed (see Fig. 15.2).

The next higher-molecular-weight alkane is propane,

CH_3—CH_2—CH_3

which differs from ethane by a —CH_2 group in the middle of the molecule. But each carbon atom is at the center of a tetrahedron; the central carbon atom is bonded to two carbon and two hydrogen atoms. The two terminal carbon atoms are bonded to three hydrogen atoms and one carbon atom.

As a result of the tetrahedral pattern, the three carbon atoms of propane are not collinear (see Table 15.1 and Fig. 15.3). The

FIGURE 15.2

The complete bonding in ethane in terms of hybridization theory.

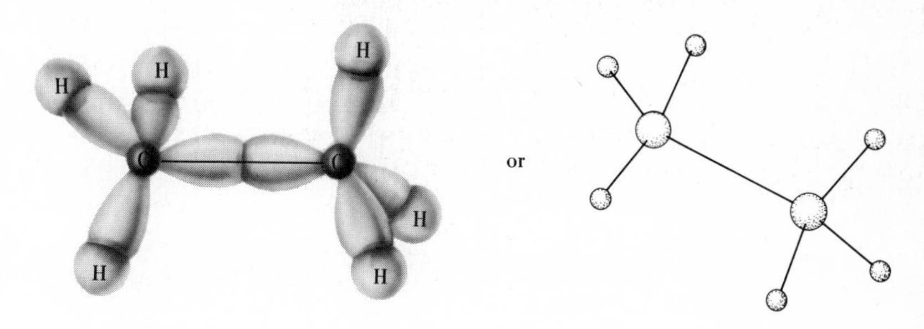

FIGURE 15.3

The tetrahedral pattern of propane.

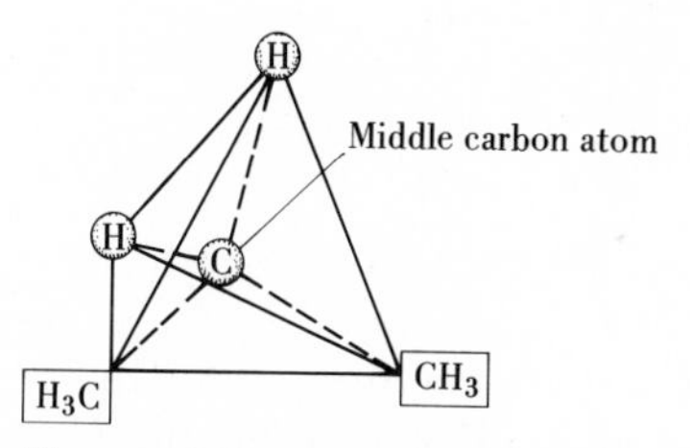

The central C atom in propane is bonded to two other C atoms and two H atoms

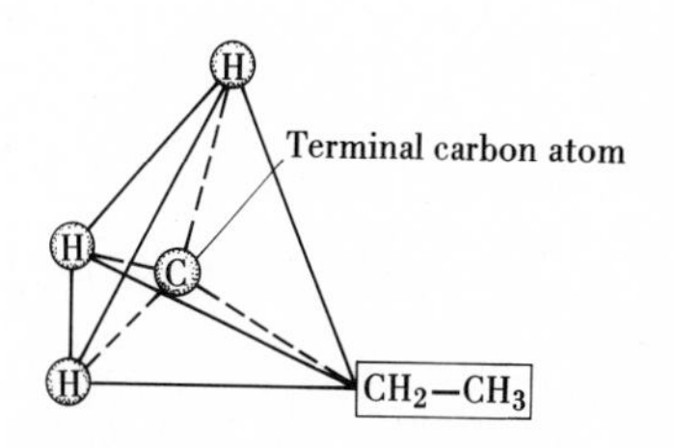

The terminal C atoms in propane are bonded to one other C atom and three H atoms

C—C—C bond angle is close to 109°, which is the expected value for a regular tetrahedron. The H—C—H and C—C—H bond angles are 109°28′; each C—H bond length is 1.09 Å, and each C—C bond length is 1.54 Å, just as in ethane.

15.4 BUTANE AND ISOMERISM

Propane

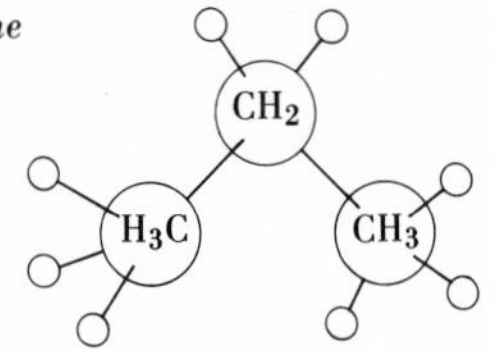

n-Butane

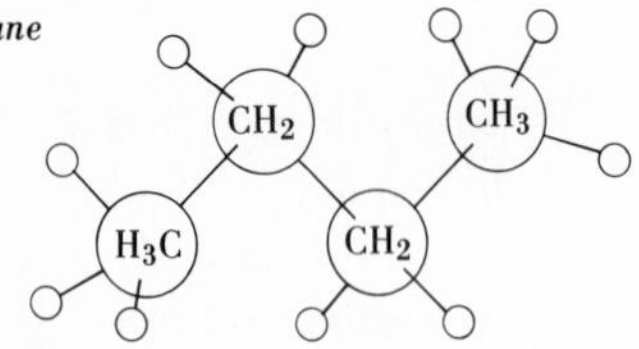

Isobutane

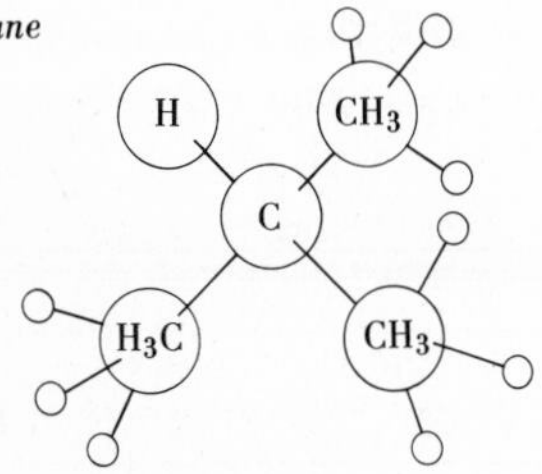

The next member of the alkane series is *butane;* its formula is C_4H_{10}. However, there are *two* known compounds which have this molecular formula; there are two butane *isomers.* One boils at −10.2°C and the other at −0.6°C, so that each is a gas at ordinary temperature. One of these compounds is called *normal butane* (*n*-butane) and the other is called *isobutane.* In *n*-butane the four carbon atoms are strung out in a chain; it can be considered as propane with a methyl group bonded at a terminal position. Isobutane can be considered as propane with a methyl group bonded at the middle position. *n*-Butane is referred to as a *straight-chain* hydrocarbon, while isobutane is a *branched-chain* hydrocarbon. It is understood that the carbon atoms of a straight-chain hydrocarbon do not lie in a straight line.

As we increase the number of carbon atoms in the alkane molecule, the tetrahedral pattern of the low-molecular-weight alkanes persists. All bond angles are close to 109°, all C—C bond lengths are 1.54 Å, and all C—H bond lengths are 1.09 Å. However, the possibility for branching and the number of possible isomers increase rapidly. Butane, C_4H_{10}, exhibits two isomers, and pentane, C_5H_{12}, exhibits three (see Fig. 15.4). There are 35 isomers of nonane, C_9H_{20}, and about 70 million *possible* isomers of tetracontane, $C_{40}H_{82}$.

Once the structure of the alkanes is understood, it is convenient to use more compact, faster-to-write formulas, for instance, *n*-butane, $CH_3—(CH_2)_2—CH_3$; isobutane, $CH_3—CH—(CH_3)_2$; and isopentane is $CH_3—CH_2—CH—(CH_3)_2$.

15.5 NOMENCLATURE

Since the structure of organic molecules can become very complex, it was realized at an early stage that a systematic nomenclature would be required. The names of many organic compounds are based on the names for the *normal,* or *straight-chain,* alkanes. Table 15.2 shows the names for the first 10 alkanes. After butane, the

FIGURE 15.4

The isomers of pentane.

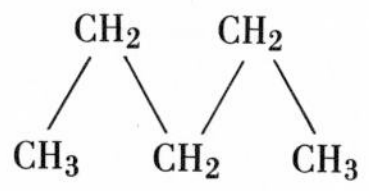

$$\begin{array}{ccccc} & CH_2 & & CH_2 & \\ / & & \backslash \; / & & \backslash \\ CH_3 & & CH_2 & & CH_3 \end{array}$$

Normal pentane

$$\begin{array}{ccc} & CH_2 & CH_3 \\ / & \backslash & / \\ CH_3 & CH & \\ & | & \\ & CH_3 & \end{array}$$

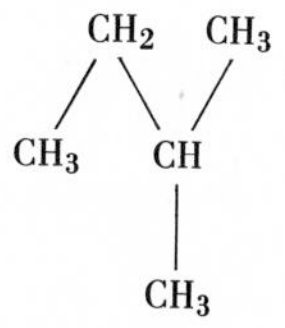

Isopentane (2-Methylbutane)

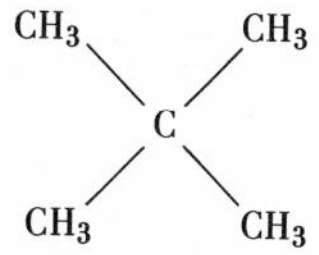

$$\begin{array}{ccc} CH_3 & & CH_3 \\ & \diagdown \; \diagup & \\ & C & \\ & \diagup \; \diagdown & \\ CH_3 & & CH_3 \end{array}$$

Neopentane (2,2-Dimethylpropane)

TABLE 15.2: NAMES OF THE FIRST 10 ALKANES

Compound	*Formula*
Methane	CH_4
Ethane	C_2H_6
Propane	C_3H_8
Butane	C_4H_{10}
Pentane	C_5H_{12}
Hexane	C_6H_{14}
Heptane	C_7H_{16}
Octane	C_8H_{18}
Nonane	C_9H_{20}
Decane	$C_{10}H_{22}$

prefix of each name is derived from the Greek or Latin number indicating the number of carbon atoms in the molecule. The suffix *-ane* indicates that the compound is an alkane. Thus octane is a compound whose molecules contain eight carbon atoms in a chain. The names of the first 10 alkanes must be memorized because the nomenclature of organic chemistry is largely derived from them.

In the systematic nomenclature the name of an organic compound depends on the number of carbon atoms in the longest carbon chain. Figure 15.5 shows the carbon skeleton of the pentane isomer which we call *isopentane*, but isopentane is not a systematic name. In its longest chain there are four carbon atoms; hence, it is a butane derivative with a *methyl group*, $—CH_3$, substituted on the chain. The name must indicate where the methyl group is attached. If we number the carbon atoms of the chain from left to right, the methyl group is bonded to the carbon atom numbered 2; from right to left it is bonded to the carbon atom numbered 3. By convention we number the chain in such a way that the methyl group is bonded to the carbon atom bearing the smaller number. Hence, we number this example from left to right, and the correct name of the compound in Fig. 15.5 is 2-methylbutane. This name means that the compound is an alkane (*-ane*), that the longest chain of carbon atoms

FIGURE 15.5

Two representations of 2-methylbutane, C_5H_{12}. The hydrogen atoms are omitted intentionally.

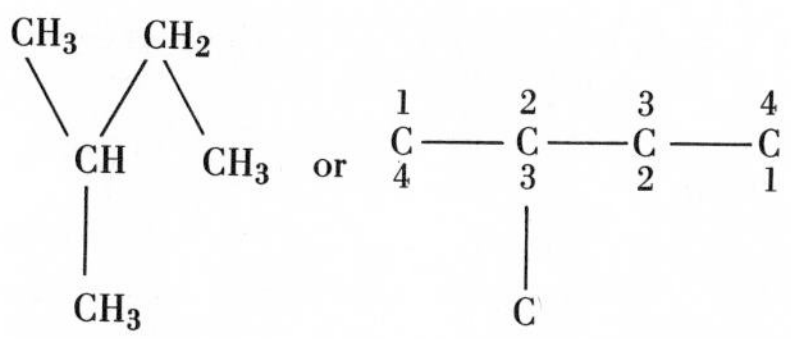

$$\begin{array}{ccc} CH_3 & & CH_2 \\ \backslash & / & \backslash \\ & CH & & CH_3 \\ & | & \\ & CH_3 & \end{array} \quad \text{or} \quad \begin{array}{ccccccc} \overset{1}{\underset{4}{C}} & — & \overset{2}{\underset{3}{C}} & — & \overset{3}{\underset{2}{C}} & — & \overset{4}{\underset{1}{C}} \\ & & | & & & & \\ & & C & & & & \end{array}$$

is four (*but-*), and that there is a methyl substituent on the chain occurring on the number 2 carbon atom (*2-methyl-*).

Actually, in this case the systematic name, 2-methylbutane, is redundant because there can be no methylbutane other than 2-methylbutane. The two interior carbon atoms are equivalent, and each must be considered to be a number 2 carbon atom. The terminal carbon atoms, numbers 1 and 4, are also equivalent. If we substitute a methyl group on either of them, we do not have 1-methylbutane or 4-methylbutane but *n*-pentane, because the longest chain now includes five carbon atoms; so the only methylbutane possible is 2-methylbutane.

Figure 15.6 shows an 11-carbon atom alkane. To name this compound we trace its structural formula, looking for the longest carbon chain. This is just seven, so that the compound is a heptane. If we

FIGURE 15.6

Two representations of 2,5-dimethyl-4-ethylheptane. The hydrogen atoms are not shown below.

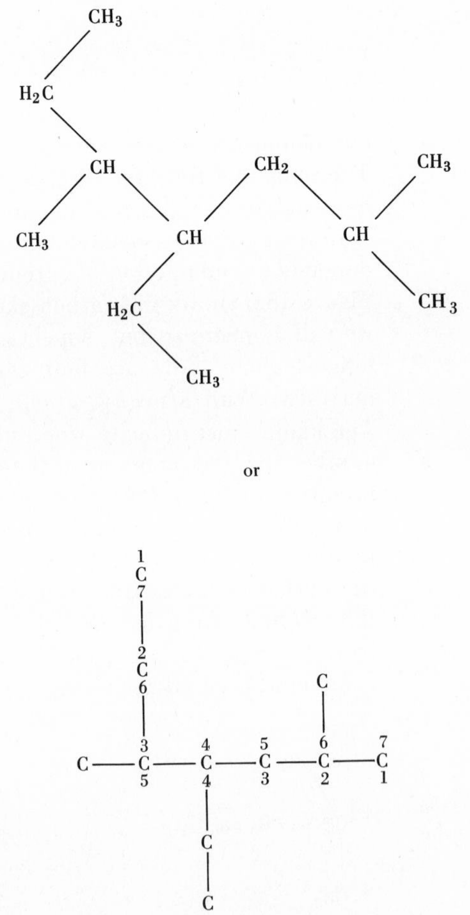

CH_3
H_2C
CH CH_2 CH_3
CH_3 CH CH
H_2C CH_3
CH_3

or

1 C 7
2 C 6
C
C — 3 C 5 — 4 C 4 — 5 C 3 — 6 C 2 — 7 C 1
C
C

number the chain from left to right, we have substituents on the number 3, number 4, and number 6 carbon atoms. If we number from right to left, we have substituents on the 2, 4, and 5 carbon atoms. Since the smaller set of numbers is 2, 4, and 5, we number the chain from right to left. The substituents on the number 2 and number 5 carbon atoms are methyl groups. The substituent on the number 4 carbon atom has two carbon atoms. Its name is *ethyl* (from ethane, C_2H_6). The name of the compound is 2,5-dimethyl-4-ethylheptane. Note that the number of times the methyl substituent appears is indicated by the prefix *di-*. It is not sufficient to write 2,5-methyl; the positions and the number of times a substituent occurs must be explicitly indicated. The name 4-ethyl-2,5-dimethylheptane would be satisfactory except that it is customary to name the smaller substituents first.

We have seen methyl, $—CH_3$, and ethyl, $—C_2H_5$, groups as substituents. Other organic radicals can act as substituents, and their names are derived from the alkane with the same number of

FIGURE 15.7

Some alkane structures with the systematic names.

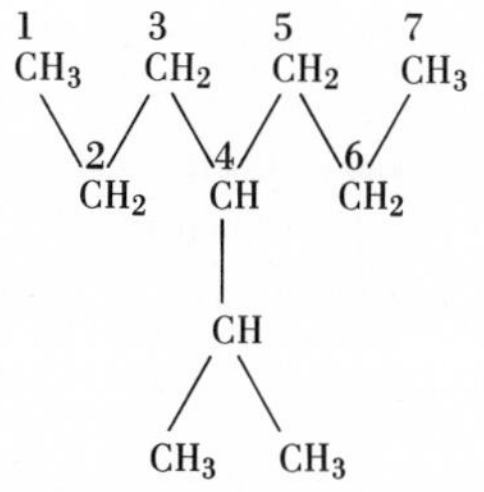

```
 1     3     5     7
CH3   CH2   CH2   CH3
  \2/   \4/   \6/
  CH2   CH    CH2
         |
         CH
        /  \
     CH3    CH3
```

(*a*) *4-Isopropylheptane*

```
          CH3
           |
 1         |3      5
CH3       CH      CH2
   \2/      \4/      \6
   CH        CH      CH3
  /         /
CH3       CH2
             \
              CH3
```

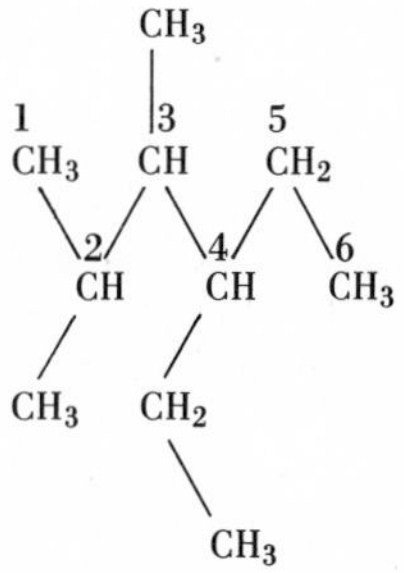

(*b*) *2,3-Dimethyl-4-ethylhexane*

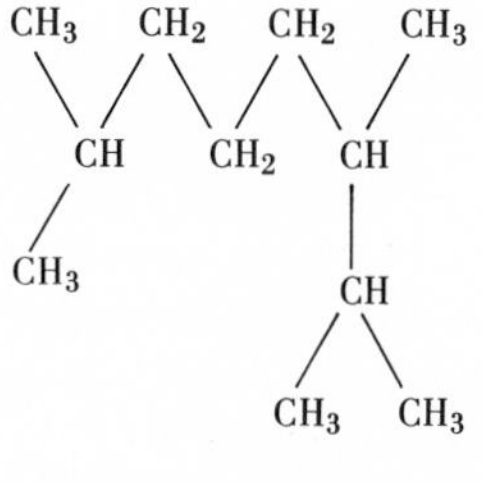

```
CH3   CH2   CH2   CH3
   \ /   \ /   \ /
   CH    CH2   CH
  /             |
CH3             CH
               /  \
            CH3    CH3
```

(*c*) *2,3,7-Trimethyloctane*

```
          CH3
           |
 1         |3
CH3       CH
   \2/      \4
   CH        CH3
    |
   CH3
```

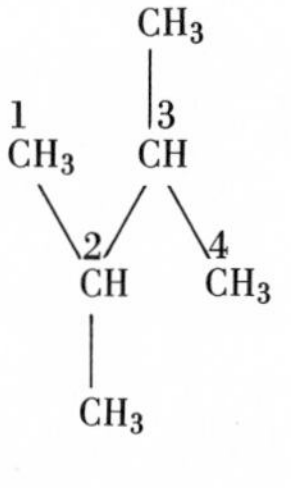

(*d*) *2,3-Dimethylbutane*

carbon atoms: —C_3H_7 is propyl, —C_4H_9 is butyl, etc. There are two radicals derived from propane; one is the *n*-propyl radical, —CH_2—CH_2—CH_3, and the other is the isopropyl radical,

$$\overset{|}{CH_3—CH—CH_3}$$

Figure 15.7 shows other alkanes and their systematic names.

15.6 ALKENES, OR OLEFINS

The gaseous hydrocarbon commonly called *ethylene* has the molecular formula C_2H_4. It is the first member of a series of hydrocarbons with the general formula C_nH_{2n} called the *alkenes* or *olefins*.

From molecular-structure studies it is known that all the atoms of the ethylene molecule lie in the same plane. In addition, all the bond angles (H—C—H, H—C—C, and C—C—H) are 120° (see Fig. 15.8).

FIGURE 15.8

Ethylene is a planar symmetrical molecule.

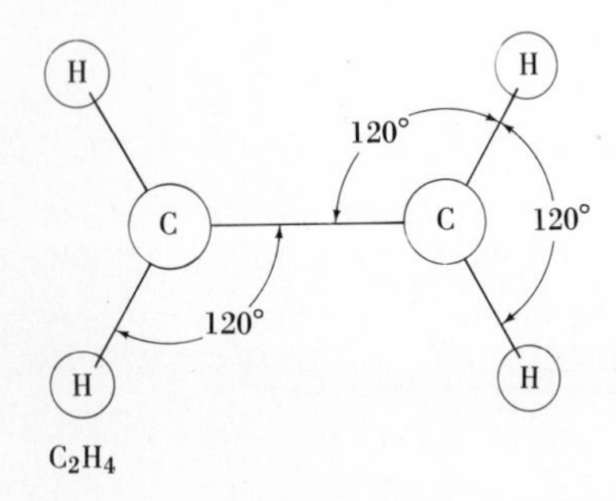

The geometry of the ethylene molecule suggests that sp^2 or trigonal hybridization (page 144) is involved in the bonding. The electronic ground state of the carbon atom is $_6C\ 1s^2, 2s^2, 2p_x{}^1, 2p_y{}^1, 2p_z{}^0$. The sp^2 hybrid state is $_6C\ 1s^2(sp^2)^1, (sp^2)^1, (sp^2)^1, 2p_x{}^1$. The sp^2 orbitals are at approximately the same energy as the unhybridized $2p_z$. Therefore, by Hund's rule, the electrons fill the three sp^2 orbitals and the $2p_z$ orbital one at a time, and there are four half-filled orbitals in the hybrid state. The axes of the sp^2 orbitals are coplanar, and the angles between adjacent axes are 120° (see Fig. 15.9*a*). The axis of the unhybridized $2p_z$ orbital is perpendicular to the plane of the sp^2 orbitals (Fig. 15.9*b*). As we shall see, the sp^2 orbitals will be available to form σ bonds, and the $2p_z$ orbital will be involved in a π bond. Figure 15.10*a* shows how the sp^2 orbitals are involved in the bonding in the ethylene molecule. Each carbon atom forms σ bonds with two hydrogen atoms by overlap of carbon sp^2 orbitals with the hydrogen 1*s* orbitals. The two carbon atoms are also held together by a σ bond formed by overlap of sp^2 orbitals on each carbon. Because of the half-filled $2p_z$ orbitals, the carbon atoms are also involved in a π bond (Fig. 15-10*b*). There is a double bond between the carbon atoms, and we usually write the formula for ethylene as $H_2C{=}CH_2$, but all the details shown in Fig. 15.10 are implied.

Ethylene is the simplest member of the alkene series. It is considered to be *un*saturated because each carbon atom is bonded to only three other atoms; the maximum number of atoms to which carbon can be bonded is four, as in the alkane series.

The next member of the alkene series is commonly called propylene, C_3H_6. As its name suggests, there are three carbon atoms in the molecule and a double bond. It can be considered as ethylene in which one of the hydrogen atoms has been replaced by a methyl

FIGURE 15.9

Trigonal, or sp^2, hybridization.

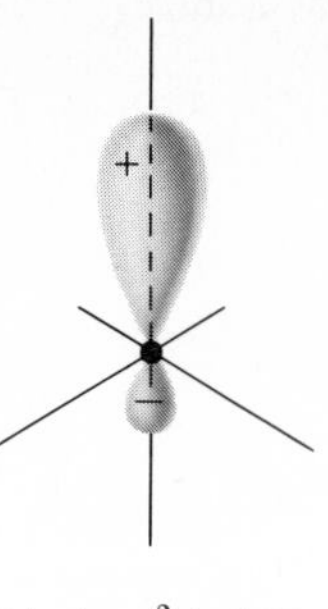

(*a*) *An sp^2-hybrid orbital*

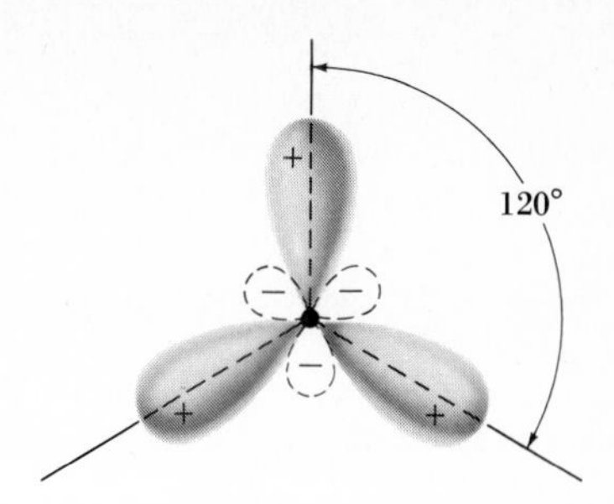

(*b*) *The three sp^2 hybrids*

FIGURE 15.10

The bonds in the ethylene molecule.

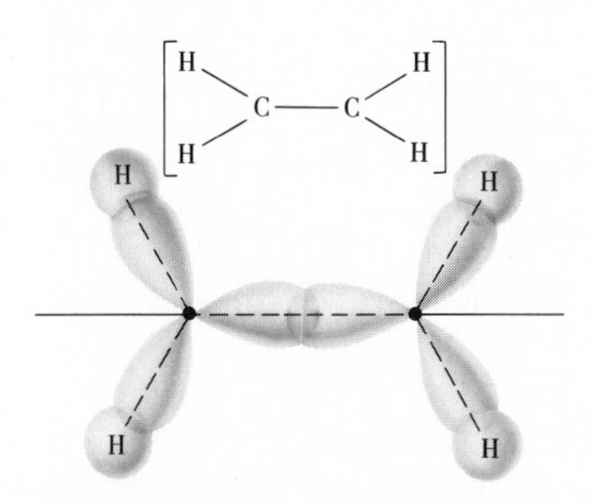

(*a*) *σ-bond structure of ethylene*

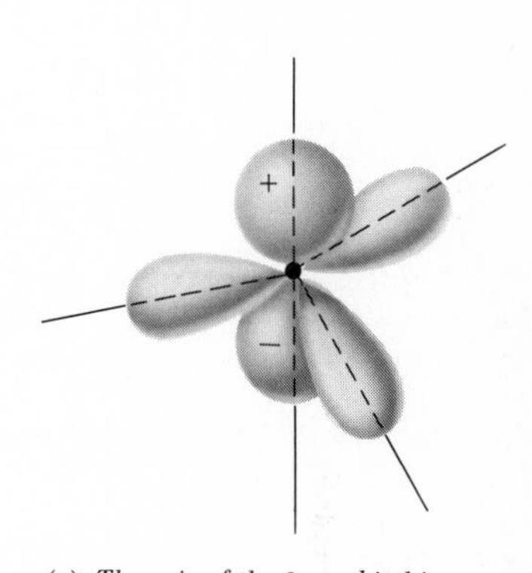

(*c*) *The axis of the $2p_z$ orbital is perpendicular to the plane of the sp^2 orbitals*

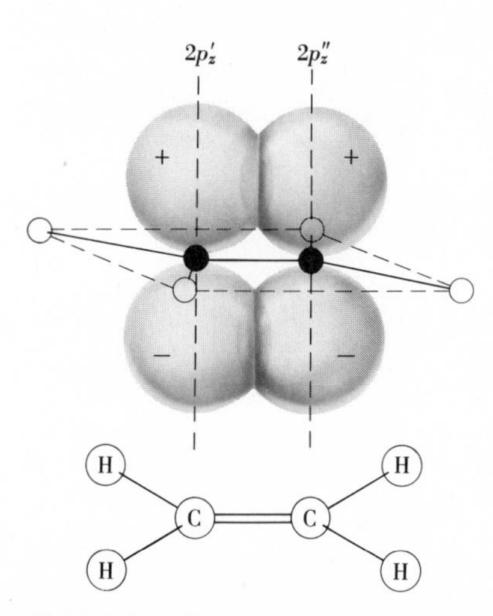

(*b*) *The π bond forms by overlap of the parallel $2p_z$ orbitals on adjacent carbon atoms*

group. The structure of the molecule around the carbon double bond can be explained in terms of sp^2 hybridization (see Fig. 15.11). This means that all the atoms of the propylene molecule are in the same plane except the protons of the methyl group. The methyl group maintains the tetrahedral structure of the alkanes.

As you might guess, the next member of the alkene series is butylene, C_4H_8. There are three different alkene structures corresponding to the formula C_4H_8. If we consider butylene to be a derivative of propylene, we can see how the three isomers are possible. We can replace a hydrogen atom by a methyl group on any one of the three carbon atoms of propylene. This is shown in Fig. 15.12. Two of the resulting butylene isomers (*a* and *c*) have straight chains of four carbon atoms, whereas one of the isomers (*b*) is branched and contains only three carbon atoms in its longest chain.

The names of higher-molecular-weight alkenes follow the nomenclature pattern developed for the alkanes. Thus pentene is the five-carbon alkene, C_5H_{10}; hexene is the six-carbon alkene, C_6H_{12}, etc. If you take out a pencil and paper and start drawing some structures, you will find that the number of possible isomers of the

FIGURE 15.11

Propylene (propene).

$$H_2C = CH - CH_3$$

or

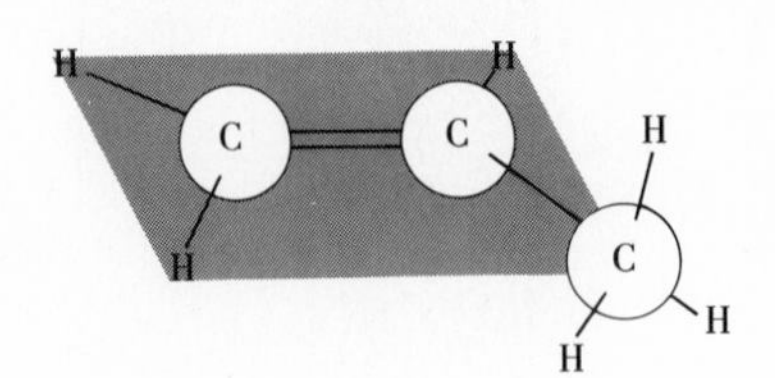

alkenes increases rapidly with molecular weight. Again the resulting structural complexities demand a systematic nomenclature.

To decide the proper name for an alkene you must (1) find the longest chain of atoms which contains the double bond, (2) number the carbon atoms so that the double-bond position is indicated by the smallest possible number, and (3) indicate the names and positions of all substituents.

Ethylene is not the systematic name for the first member of the alkene series, but it is the name which chemists ordinarily use. The correct name is derived from the alkane which has the same number of carbon atoms; this is ethane. To the stem *eth-* add the family designation *-ene* as a suffix, and we have *ethene* as the systematic name for C_2H_4. Likewise, the systematic names for propylene and butylene are propene and butene, respectively. It is acceptable to use the *trivial names*,† like ethylene and propylene for the lower-

† A trivial name may be common or scientific, but it differs from the systematic name because it is not explicitly related to the structure.

FIGURE 15.12

The butylene (butene) isomers.

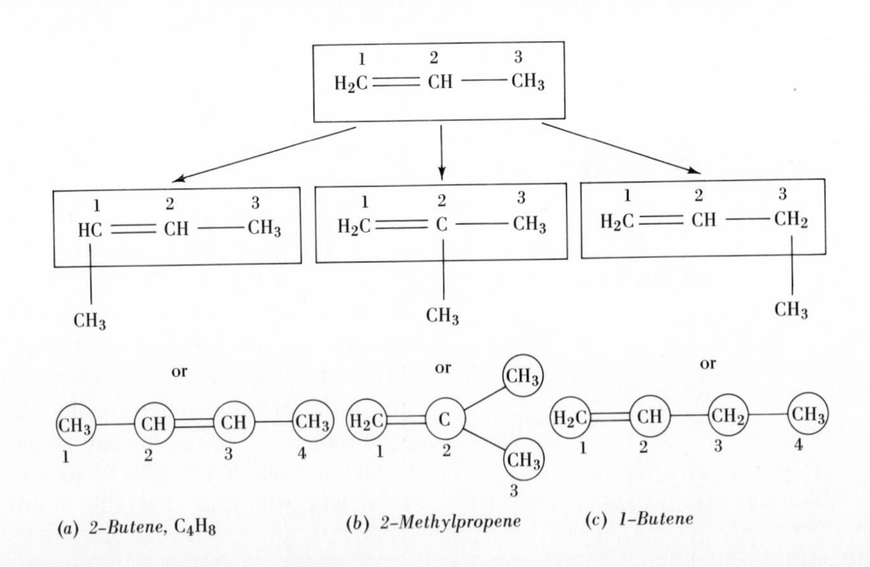

(*a*) *2-Butene*, C_4H_8 (*b*) *2-Methylpropene* (*c*) *1-Butene*

molecular-weight alkenes, but for the higher-molecular-weight compounds, the systematic nomenclature must be used.

The systematic names of the C_4H_8 isomers are given in Fig. 15.12. We see that 1-butene (Fig. 15.12*c*) has four carbon atoms in the chain which contains the double bond; the prefix 1 means that the double bond joins carbon atoms 1 and 2. If the chain had been numbered from right to left the name would have been 3-butene. 1-Butene and 3-butene are names for identical structures, but, by the convention of selecting the smallest possible number, 3-butene is incorrect.

Figure 15.12*a* shows 2-butene, which contains a chain of four carbon atoms with a double bond between carbon atoms 2 and 3. The longest chain in 2-methylpropene contains three carbon atoms with a double bond joining a terminal and the middle (number 2) carbon atoms. There is a methyl group on the number 2 carbon atom. (The prefix 2- does not refer to the position of the double bond. In propene there is no need to indicate the double-bond position.) The structures and names of more complicated alkenes are given in Fig. 15.13.

FIGURE 15.13

The structure and nomenclature of some specific alkenes.

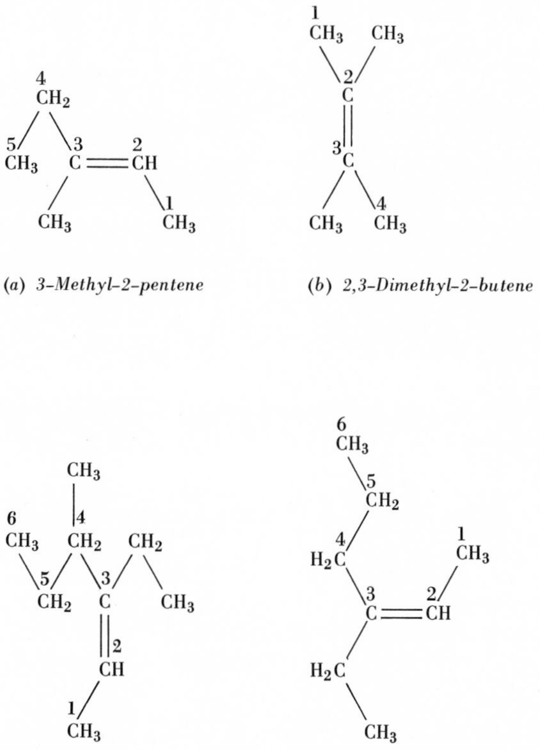

(*a*) *3-Methyl-2-pentene*

(*b*) *2,3-Dimethyl-2-butene*

(*c*) *4-Methyl-3-ethyl-2-hexene*

(*d*) *3-Ethyl-2-hexene*

15.7 ALKYNES, OR ACETYLENES

Acetylene is a combustible hydrocarbon commonly used for welding. Its molecular formula is C_2H_2; it is the first member of a series of compounds called the *acetylenes* or *alkynes* which have the general formula C_nH_{2n-2}.

Molecular structure studies show that C_2H_2 is a linear molecule, which suggests that the carbon atom might be involved in *sp* hybridization as in the case of BeF_2 and $MgCl_2$ (page 146).

Let us compare the electronic configurations of the ground state and *sp*-hybridized states of carbon:

Ground state: ${}_6C\ 1s^2, 2s^2, 2p_x{}^1, 2p_y{}^1, 2p_z{}^0$
sp-hybrid state: ${}_6C\ 1s^2, (sp)^1, (sp)^1, 2p_y{}^1, 2p_z{}^1$

The two hybrid *sp* orbitals are collinear, as shown in Fig. 15.14, and their orientation is ideal for explaining the structure of the acetylene molecule. In addition there are two unhybridized 2*p* orbitals on each atom. These can form π bonds by parallel overlap. As we shall see, the carbon atoms of acetylene are joined by one σ bond and two π bonds; chemists say that acetylenes (or alkynes) have triple bonds, H—C≡C—H.

FIGURE 15.14

The bonding orbitals in sp-hybridized carbon.

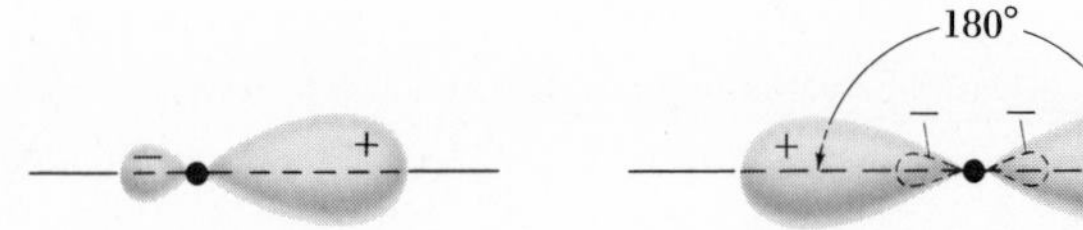

(*a*) *An sp–hybrid orbital*

(*b*) *The sp–hybrid orbitals of* C;

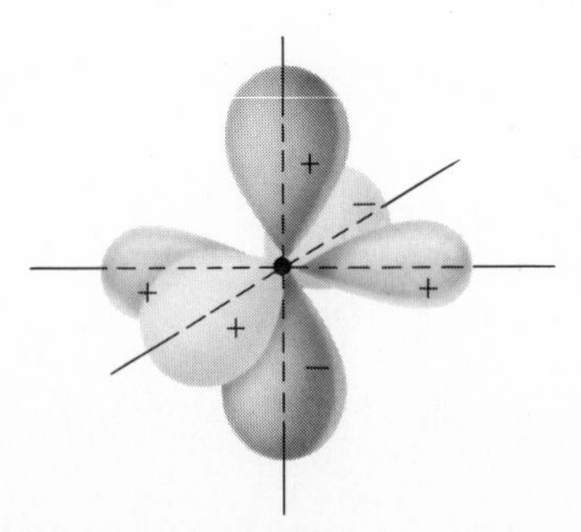

(*c*) *A composite of the bonding orbitals; the negative lobes of the sp–hybrids are omitted.*

In Fig. 15.15*a* we see how the two *sp* orbitals of each carbon atom form the linear σ-bond structure of acetylene. The two carbon atoms form a carbon-carbon σ bond by overlap of *sp* orbitals; each carbon is σ-bonded to hydrogen by overlap of a carbon *sp* and hydrogen 1*s*. Figure 15.15*b* and *c* demonstrates how the $2p_y$ and $2p_z$ orbitals overlap to form π bonds.

The characteristic of the acetylenic, or alkyne, compounds is the carbon-carbon triple bond. The nomenclature of this series of compounds is analogous to the nomenclature of the alkenes. The systematic name for acetylene is based on the corresponding alkane, ethane. The suffix used to indicate that a compound is an alkyne is *-yne*. Hence, the systematic name for acetylene is ethyne. Names for some higher-molecular-weight alkynes are given with the corresponding structures in Fig. 15.16.

FIGURE 15.15

The bonding in acetylene, C_2H_2.

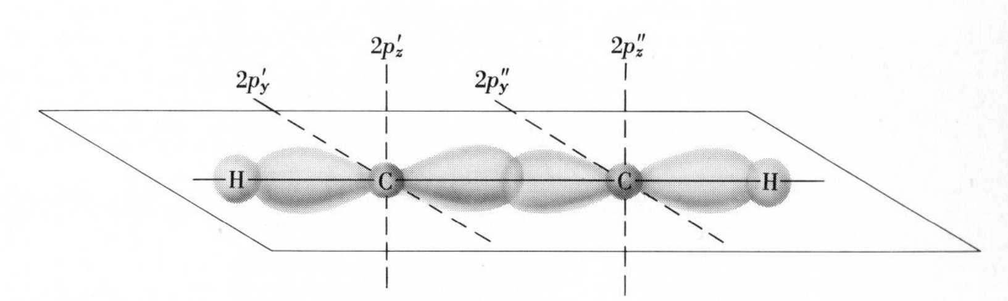

(*a*) *σ bonding from overlap of* C_{sp} *and* H *(1s)*

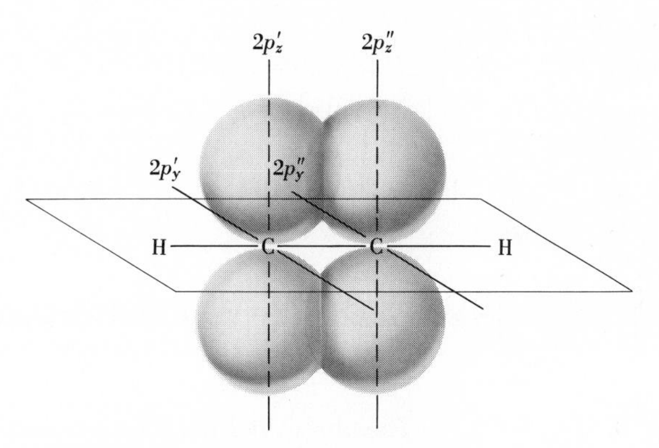

(*b*) *π bonding from overlap of* $2p_z$ *with* $2p_z$

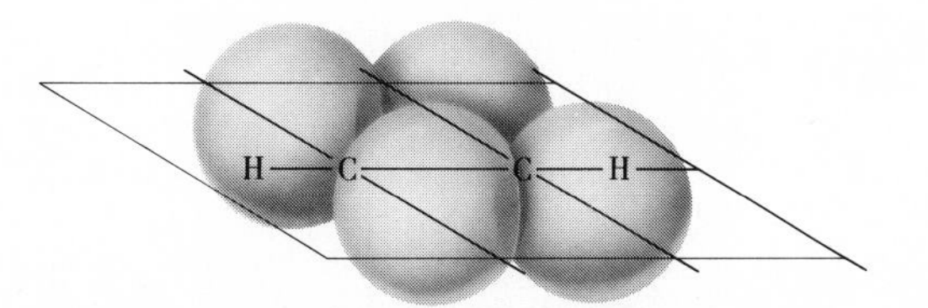

(*c*) *π bonding from overlap of* $2p_y$ *with* $2p_y$

FIGURE 15.16

The structure and nomenclature of some alkynes. The linear part of each molecule is boxed.

$$\mathrm{H{-}C{\equiv}C{-}CH_3}$$

(Structure: H—C≡C—C—H, with H above and H below the last C)

(*a*) *Propyne*

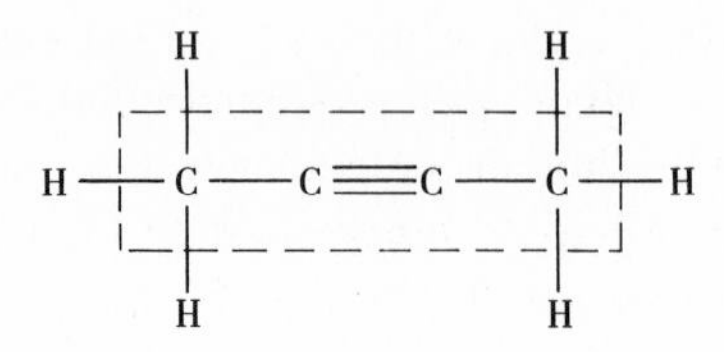

(*b*) *2-Butyne*

7 CH$_3$—6 CH$_2$—5 C(CH$_3$)(CH$_2$CH$_3$)—4 C≡3 C—2 C(H)(CH$_3$)—1 CH$_3$

(*c*) *2,5-Dimethyl-5-ethyl-3-heptyne*

15.8 COMPARATIVE CHEMISTRY OF THE ALKANES, ALKENES, AND ALKYNES

The three hydrocarbon series of the alkanes, alkenes, and alkynes differ considerably in their chemical properties. The alkanes are termed saturated and are often referred to as the *paraffins*, which means "small affinity." We can guess from these names that the alkanes are relatively unreactive.

Reactions of the alkanes

Since the alkanes are commonly used as fuels, one of the important reactions of this class of compounds is combustion. The products are $CO_2(g)$ and $H_2O(g)$, provided that there is sufficient oxygen:

$$CH_4(g) + 2O_2(g) \rightarrow CO_2(g) + 2H_2O(g)$$

Methane

$$C_8H_{18}(l) + 12\tfrac{1}{2}O_2(g) \rightarrow 8CO_2(g) + 9H_2O(g)$$

Octane

These are the kinds of reactions which occur in a gasoline engine. Unfortunately, the conditions in such an engine do not permit the complete combustion of the fuel, so that poisonous carbon monoxide, CO, always forms and escapes into our atmosphere. In addition because of the high temperatures achieved and because air is used for the combustion, nitrogen oxides, NO, N_2O, and NO_2, are produced during the internal combustion of the hydrocarbons. These oxides are highly reactive acid gases which are very damaging to living organisms and are one cause of the corrosive effects of our city environments. These gases contaminate the air and the waterways.

The alkanes are used to produce halogenated solvents. For instance, methane reacts with chlorine to produce four very useful derivatives. The reaction between methane and chlorine is an example of a substitution reaction which is typical of the alkanes:

$$\mathrm{H-\overset{\displaystyle H}{\underset{\displaystyle H}{\overset{|}{\underset{|}{C}}}}-H + Cl-Cl}(g) \xrightarrow{\Delta} \mathrm{H-\overset{\displaystyle H}{\underset{\displaystyle H}{\overset{|}{\underset{|}{C}}}}-Cl + HCl}(g)$$

Methane, CH_4 — Chloromethane, CH_3Cl

Note that the reaction involves replacing a hydrogen atom by a chlorine atom. In order for this reaction to occur a C—H bond and a Cl—Cl bond must be severed. A substitution reaction is generally slow and includes a by-product gas, like HCl(*g*).

The reaction between methane and Cl_2 can proceed until all the hydrogen atoms have been replaced. We have learned to control the reaction so that we can produce any desired product:

$$CH_4 + Cl_2 \rightarrow CH_3Cl(g) + HCl(g)$$

Chloromethane

$$CH_3Cl + Cl_2 \rightarrow CH_2Cl_2(l) + HCl(g)$$

Dichloromethane

$$CH_2Cl + Cl_2 \rightarrow CHCl_3(l) + HCl(g)$$

Trichloromethane
(chloroform)

$$CHCl_3 + Cl_2 \rightarrow CCl_4(l) + HCl(g)$$

Tetrachloromethane
(carbon tetrachloride)

All four chlorinated methanes are useful solvents for dissolving oils and tars. In addition chloroform has anesthetic properties and until recently was used extensively for medical purposes. Ethane and propane also produce useful solvents when they are halogenated.

High-octane gasoline consists of a mixture of branched-chain alkanes and unsaturated hydrocarbons containing from 4 to 12 carbon atoms. A large fraction of natural petroleum consists of alkanes with much higher molecular weights, which are useless for automobile fuel. Petroleum chemists convert the higher-molecular-weight fractions of petroleum into branched-chain and unsaturated lower-molecular-weight materials by *pyrolysis*, the chemical breakdown, or degradation, produced by heating a material to high temperatures. Pyrolysis of high-molecular-weight alkanes in the absence of air or oxygen converts them into more powerful fuels for gasoline engines and into valuable raw materials for the rubber, plastics, and textile industries.

In general, the hot vapors of high-molecular-weight alkanes are passed over catalysts like silica, SiO_2, or alumina, Al_2O_3. A complicated series of reactions occurs on the catalytic surfaces, severing chemical bonds, rearranging atoms, and abstracting protons. The end product is a mixture of alkenes, branched-chain alkanes, methane, and hydrogen:

$$\text{High-molecular-weight alkanes} \xrightarrow[450\text{–}550^\circ\text{C}]{SiO_2,\ Al_2O_3} \text{alkenes} + \text{branched alkanes} + CH_4 + H_2$$

By selecting the proper conditions for pyrolysis it is possible to enhance the production of one of the components. Since large molecules are broken up into smaller ones, pyrolysis of petroleum is commonly referred to as *cracking*. A specific example of this reaction involving $C_{14}H_{30}$ is

$$\underset{\text{Tetradecane}}{C_{14}H_{30}(l)} \xrightarrow[SiO_2]{\Delta} \underset{\text{Hexene}}{2C_6H_{12}(l)} + \underset{\text{Ethylene}}{C_2H_4(g)} + H_2(g)$$

Ethylene, the most important raw material of the plastics industry, is produced by *steam cracking*, in which the hydrocarbon vapor is diluted with steam and heated to approximately 900°C. Other chemicals of value to the chemical industry produced by the same process include propylene, butadiene, and isoprene. Ethylene and propylene are used for the manufacture of the plastics polyethylene and polypropylene. Butadiene and isoprene are used in the manufacture of synthetic rubber. These materials are discussed in Sec. 15.14.

In summary there are three important reactions which the alkanes undergo:

Combustion: $C_3H_8(g) + 5O_2(g) \rightarrow 3CO_2(g) + 4H_2O$

Substitution: $C_3H_8(g) + Cl_2(g) \rightarrow C_3H_7Cl(l) + HCl(g)$

Pyrolysis: $C_3H_8(g) \xrightarrow[\text{cat.}]{} \underset{\text{Ethylene}}{C_2H_4(g)} + CH_4(g)$

Reactions of the alkenes and alkynes

The characteristic reaction of the alkanes is substitution. The characteristic reaction of the unsaturated hydrocarbons, including the alkenes and alkynes, is addition. Compare the reactions of ethane, ethylene, and acetylene with Cl_2:

Substitution: $CH_3—CH_3 + Cl_2 \rightarrow CH_3—CH_2Cl + HCl$
Addition: $CH_2═CH_2 + Cl_2 \rightarrow CH_2Cl—CH_2Cl$
Addition: $CH≡CH + Cl_2 \rightarrow CHCl═CHCl$

We say that chlorine adds across the double bond:

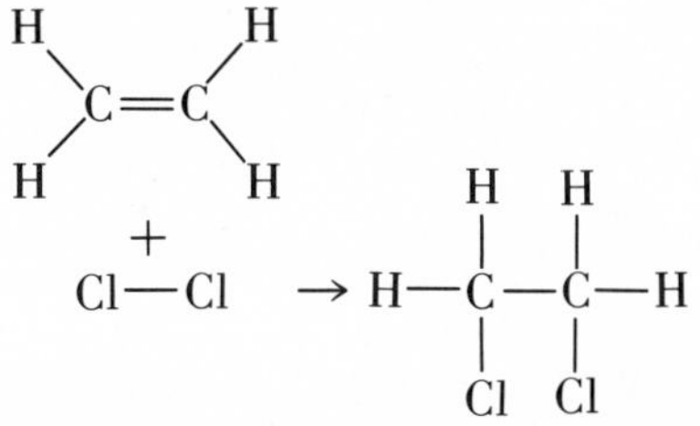

$$\begin{matrix} H & & & & H \\ & \diagdown & & \diagup & \\ & & C{=}C & & \\ & \diagup & & \diagdown & \\ H & & & & H \end{matrix} + Cl—Cl \rightarrow H—\underset{\underset{Cl}{|}}{\overset{\overset{H}{|}}{C}}—\underset{\underset{Cl}{|}}{\overset{\overset{H}{|}}{C}}—H$$

Addition reactions of the unsaturated hydrocarbons yield no gaseous by-product. The substitution reactions of the alkanes do form by-products. Halogens, hydrohalides, and hydrogen can add across double and triple bonds:

$$CH_2═CH_2 + HBr \rightarrow \underset{\text{Bromoethane}}{CH_3—CH_2Br}$$

$$CH_2═CH—CH_3 + I_2 \rightarrow \underset{\text{1,2-Diiodopropane}}{CH_2I—CHI—CH_3}$$

$$CH_3—CH═CH—CH_3 + H_2 \rightarrow CH_3—CH_2—CH_2—CH_3$$

Hydrogenation of olefins is an important process in the food industry. Although the hydrogenation of ethylene is not important in itself, it demonstrates the reaction:

$$\underset{\text{Ethylene}}{CH_2═CH_2} + H_2(g) \rightarrow \underset{\text{Ethane}}{CH_3—CH_3}$$

Vegetable oils, which are mixtures of unsaturated liquid compounds, are converted into saturated, solid compounds by catalytic addition of H_2 at high pressure. A common component of vegetable oils is glyceryl trioleate. More highly unsaturated oils present are commonly called *polyunsaturates*. The hydrogenation of glyceryl trioleate is shown:

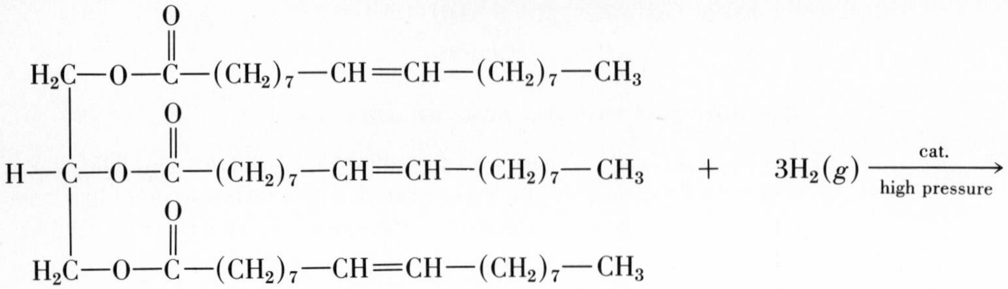

$$
\begin{array}{l}
H_2C-O-\overset{\overset{\Large O}{\|}}{C}-(CH_2)_7-CH=CH-(CH_2)_7-CH_3 \\
\quad | \\
H-C-O-\overset{\overset{\Large O}{\|}}{C}-(CH_2)_7-CH=CH-(CH_2)_7-CH_3 \quad + \quad 3H_2(g) \xrightarrow[\text{high pressure}]{\text{cat.}} \\
\quad | \\
H_2C-O-\overset{\overset{\Large O}{\|}}{C}-(CH_2)_7-CH=CH-(CH_2)_7-CH_3
\end{array}
$$

Glyceryl trioleate, a component of vegetable oil

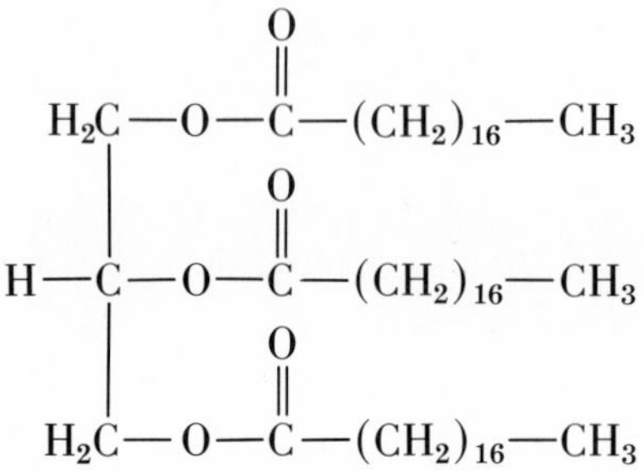

$$
\begin{array}{l}
H_2C-O-\overset{\overset{\Large O}{\|}}{C}-(CH_2)_{16}-CH_3 \\
\quad | \\
H-C-O-\overset{\overset{\Large O}{\|}}{C}-(CH_2)_{16}-CH_3 \\
\quad | \\
H_2C-O-\overset{\overset{\Large O}{\|}}{C}-(CH_2)_{16}-CH_3
\end{array}
$$

Glyceryl tristearate, a solid fat

Without hydrogenation the peanut oil in peanut butter would separate and come to the top of the jar. Hydrogenation of the oil helps keep the peanut butter homogenous (homogenized). Hydrogenated vegetable oils, e.g., Crisco and Spry, were very popular until biochemists found a relationship between saturated fats and circulatory diseases. At present the cholesterol-conscious housewife uses more of the liquid, polyunsaturated fats (oil) for cooking.

Addition to triple bonds can be controlled so that it can occur stepwise:

$$HC \equiv CH + HCl \rightarrow H_2C=CHCl$$

Acetylene — Vinyl chloride (chloroethene)

$$H_2C=CHCl + HCl \rightarrow H_3C=CHCl_2$$

1,1-Dichloroethane

The most obvious chemical difference between the saturated and unsaturated hydrocarbons lies in the ability of unsaturated hydrocarbons to undergo addition reactions, which can be used to distinguish alkanes from alkenes or alkynes. When a red solution of bromine, Br_2, in carbon tetrachloride is treated with an unsaturated hydrocarbon (alkene or alkyne), the bromine color disappears quickly because of the rapid addition of Br_2 across the double or triple bond:

$$CH_3—CH=CH_2 + Br_2 \xrightarrow{\text{fast}} CH_3—CHBr—CH_2Br$$

Propene (Red) (Colorless)

$$CH_3—CH_2—C\equiv CH + Br_2 \xrightarrow{\text{fast}}$$

1-Butyne (Red)

$$CH_3—CH_2—CBr=CHBr \xrightarrow[\text{addition of } Br_2]{\text{continued}} CH_3—CH_2—CBr_2—CHBr_2$$

(Colorless)

In these addition reactions there is only one product, a colorless oil. When the same bromine solution is treated with an alkane, a very slow substitution reaction usually occurs, so that the red color of bromine lingers. In bright light certain alkanes like isobutane can react rapidly with Br_2. The accompanying evolution of HBr gas, which is not very soluble in carbon tetrachloride, is proof that bromine is reacting with an alkane. For example,

$$C_6H_{14}(l) + Br_2 \xrightarrow[\text{Br}_2\text{ color lingers}]{\text{slow}} C_6H_{13}Br(l) + \underline{HBr}(g)$$

Hexane

$$CH_3—\overset{\displaystyle CH_3}{\underset{\displaystyle CH_3}{\overset{|}{\underset{|}{C}}}}—H + Br_2 \xrightarrow[\text{color disappears fast}]{\text{in light}} CH_3—\overset{\displaystyle CH_3}{\underset{\displaystyle CH_3}{\overset{|}{\underset{|}{C}}}}—Br(l) + \underline{HBr}(g)$$

Isobutane (2-methylpropane) 2-Bromo-2-methylpropane

The first reaction is the normal slow reaction between an alkane and Br_2; in the second, which occurs rapidly in the presence of bright light, the evolution of gas bubbles indicates the presence of the alkane.

15.9 MORE ON MOLECULAR STRUCTURE

The methane, ethane, and propane structures are basically tetrahedral, the relative positions of the atoms being fixed by this pattern. When we go to butane and higher alkanes, there is more than one possible arrangement of the atoms within the tetrahedral pattern. More succinctly we say that such molecules have more than one *conformation*. The different arrangements are called *conformers*. Figure 15.17 shows three *conformers* of *n*-butane and *n*-pentane. The butane (and pentane) conformers are not different compounds because conversion from one to the other occurs easily; the energy these

FIGURE 15.17

Conformations of n-butane and n-pentane.

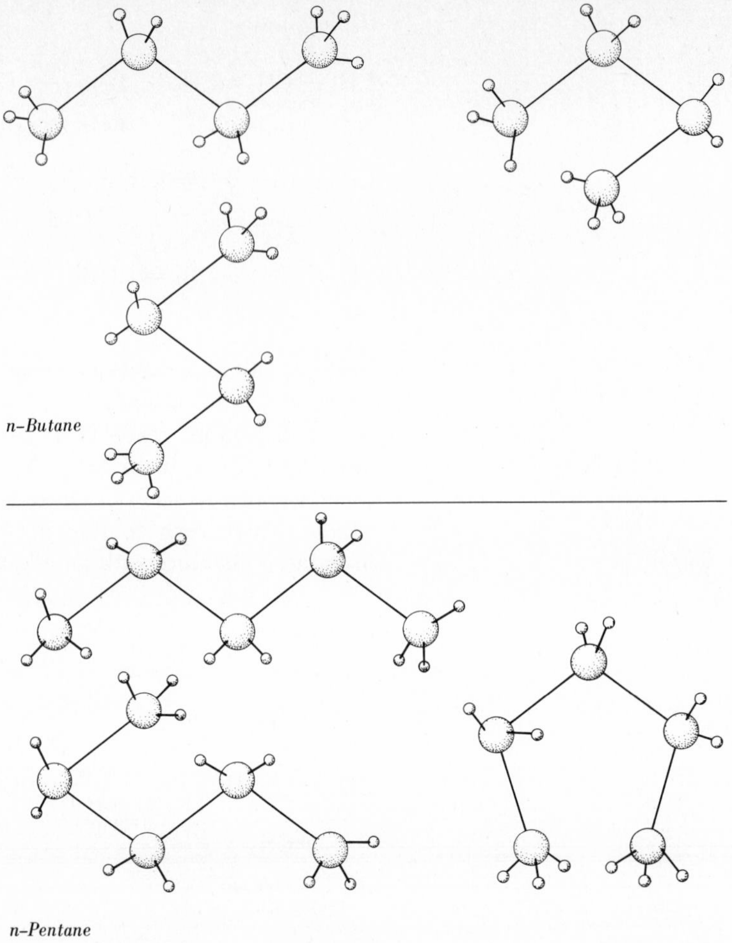

n–Butane

n–Pentane

molecules possess causes them to alternate constantly among the possible conformations. Since the alkane molecules are continually flip-flopping, a long alkane chain could become very twisted within the tetrahedral pattern of such molecules.

Conversion from one conformer to another amounts to rotation about a carbon-carbon bond. This is shown in Fig. 15.18, where rotation about a C—C bond in butane changes the conformation of the molecule. These kinds of change occur very rapidly, and we say that rotation about a C—C single bond is unrestricted. The unrestricted rotation about the C—C bond of an alkane means that the hydrogen atoms are not fixed in position, as demonstrated for ethane and propane in Fig. 15.19. Actually the rotation about any bond meets with some resistance. At extremely low temperatures, near 0 K, when the molecules have very little energy, the molecule could be

FIGURE 15.18

Interconversion of conformers of n-butane.

2 CH_2 4 CH_3 1 CH_3 3 C H_a H_b

rotation about the 2—3 bond

2 CH_2 H_b 1 CH_3 3 C 4 CH_3 H_a

FIGURE 15.19

"Free" rotation about the C—C *bond in the alkanes.*

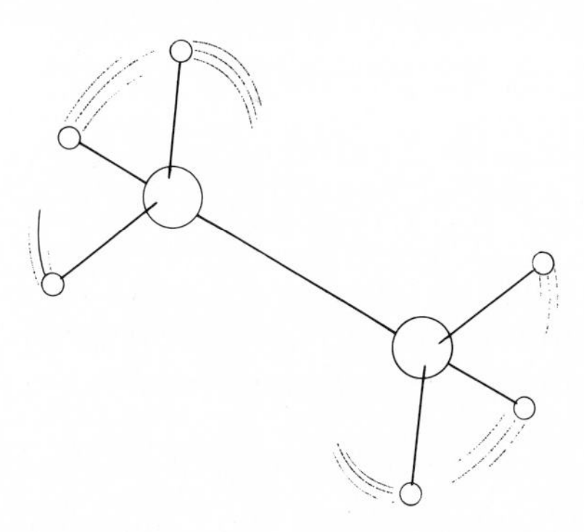

(*a*) *Ethane*

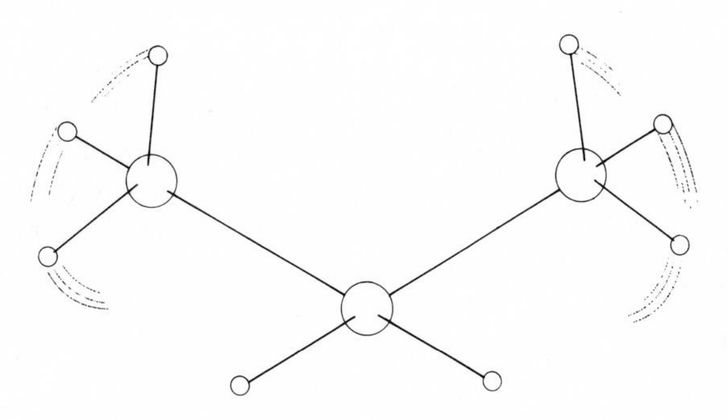

(*b*) *Propane*

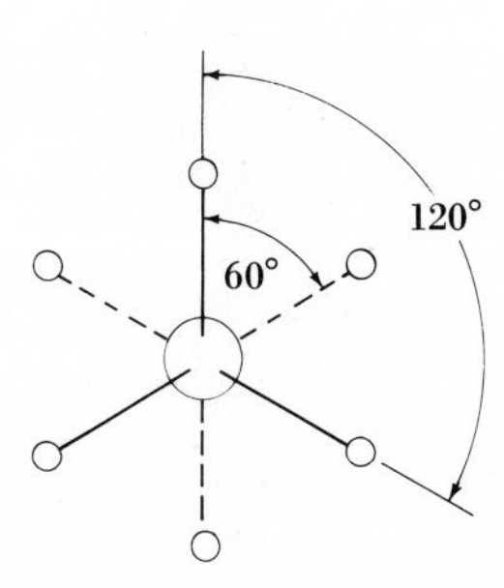

(*c*) *Ethane*

FIGURE 15.20

Cis-trans, or geometric, isomerism.

```
H       H            Cl       H
 \     /               \     /
  C = C                 C = C
 /     \               /     \
Cl      Cl            H       Cl
```

(*a*) *cis-1,2-Dichloroethylene, bp 60.1°C, density 1.29 g/cm³*

(*b*) *trans-1,2-Dichloroethylene, bp 48.4°C, density 1.26 g/cm³*

"frozen" in the preferred conformation. But since the energy required for rotation is so small, there is virtually free rotation about the C—C single bond at ordinary temperatures. Figure 15.19*c* shows a model of ethane as it might appear near 0 K if we sighted down the C—C bond. We would see only one carbon atom; the other would be hidden. The six hydrogen atoms, which would be visible, are shown in the preferred conformation, as far apart as possible within the tetrahedral structure of the molecule.

Geometric isomerism

The introduction of a double bond into a molecule reduces the possible position interchanges that the atoms of a molecule can undergo because rotation about a C—C double bond is restricted. Thus molecules with identical molecular formulas and internal bonding can have different structures. For example, there are *two* compounds named 1,2-dichloroethylene (Fig. 15.20). When the two chlorine atoms are on the same side of the double bond, we have a *cis isomer;* when they are on opposite sides of the double bond we have a *trans isomer.* The cis and trans isomers have different physical properties. As we shall see, cis-trans isomerism, also called *geometric isomerism*, has important consequences for the properties of rubber and other essential materials.

The restriction to rotation about a double bond can be explained

FIGURE 15.21

When a C═C double bond is twisted so that the molecule is no longer coplanar, the overlap between the p orbitals is diminished and the π bond is broken.

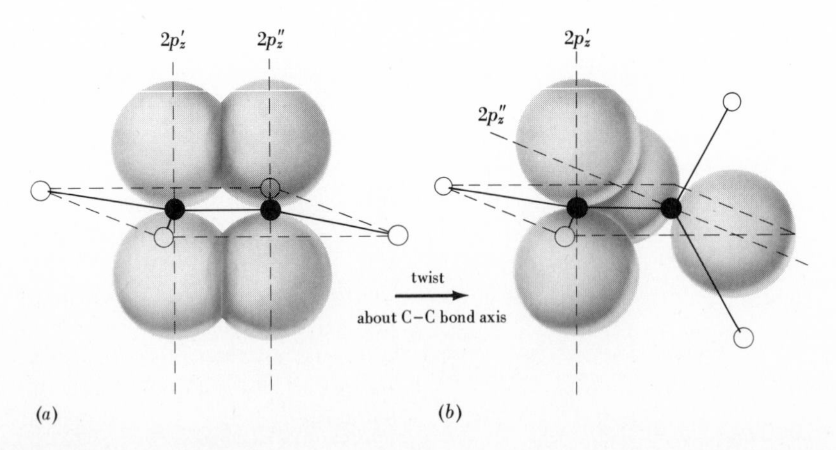

FIGURE 15.22

Geometric isomers of 2-butene. Rectangles include atoms of the molecule which lie in a plane due to restricted rotation about the C=C bond.

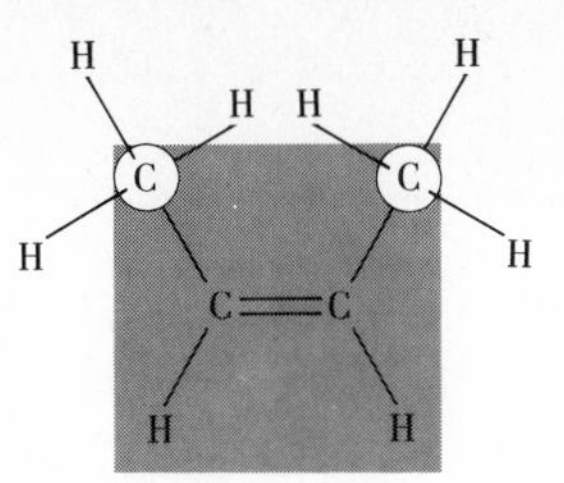

cis-2-Butene

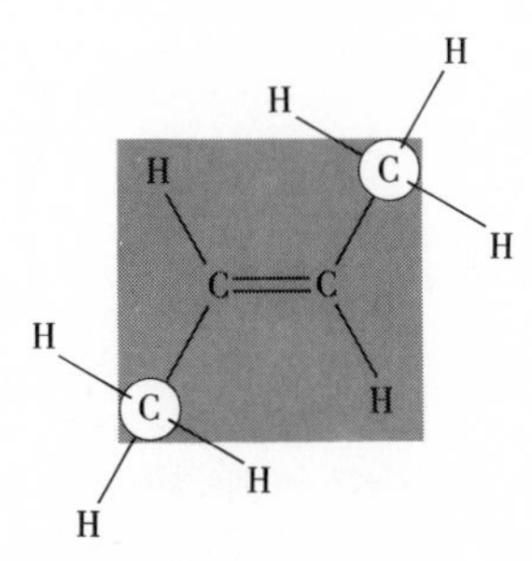

trans-2-Butene

in terms of the bonding theory given in Sec. 15.6 for ethylene. Figure 15.21*a* shows the unhybridized $2p_z$ orbitals on the carbon atoms of ethylene. Since their axes are parallel, they overlap appreciably and form a π bond. If the molecule were slowly twisted about the C—C bond, the overlap between the $2p_z$ orbitals would be diminished and at 90°C the overlap would be at a minimum; the π bond would no longer be present. This is shown in Fig. 15.21*b*. The most stable configuration of atoms in a molecule permits the maximum overlap of bonding orbitals. Hence, the most stable configuration of ethylene is planar and any attempt to twist it out of the plane is strongly resisted. As a result, it is possible to have substituted ethylenes which exhibit cis-trans isomerism.

Whenever there is a carbon-carbon double bond in a molecule, that segment of the molecule has a fixed, planar structure. This is demonstrated in Fig. 15.22 with 2-butene, which exhibits cis-trans isomerism.

Optical isomerism

As a result of the tetrahedral structure of the alkane derivatives another kind of isomerism is possible. Whenever a carbon atom is bonded to four different atoms or radicals, two different structures are possible. Figure 15.23 depicts the two isomers of CHFClBr, fluorochlorobromomethane. They are *nonsuperimposable* mirror images. A carbon atom attached to four different atoms or groups in this way is called an *asymmetric* carbon atom.

FIGURE 15.23

Optical isomers of CHFClBr, *fluorochlorobromomethane.*

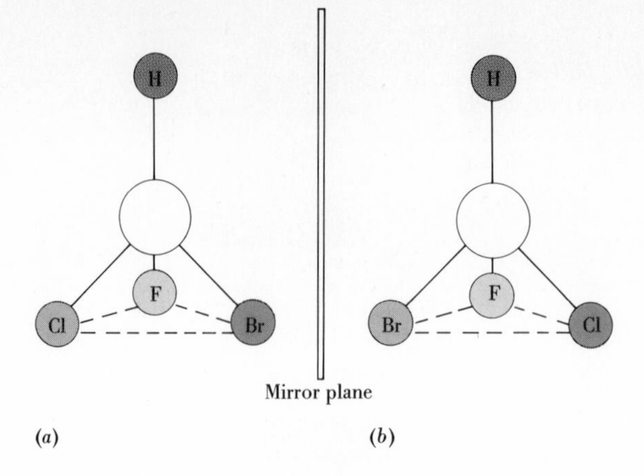

It may be difficult to see that the structures in Fig. 15.23 are different. It is much easier to see the difference and to understand this kind of isomerism with atomic models because the effect is due to the three-dimensionality of the molecule. Remember that the key to this kind of isomerism is that the molecules cannot be superimposed. We can make atomic models of CH_2FBr or CHF_2Cl which are mirror images, but such mirror images are superimposable and do not represent different molecules.

The isomerism due to an asymmetric carbon atom is called *optical isomerism* because the only difference in the physical properties of the isomers is the effect they have on light. As shown in Fig. 15.24, light produces mutually perpendicular electric and magnetic field vectors as it moves through space (page 188). When a light ray passes through a sample of an optical isomer, the directions of the electric and magnetic fields are rotated (see Fig. 15.24). One isomer will rotate the fields clockwise; the other will rotate them counterclockwise, but to the same extent. This property of optical isomers is called *optical rotatory power*. Table 15.3 shows some of the properties of optically active isomers of 1-hydroxy-2-methylbutane. The structure of one isomer is shown. The asymmetric carbon atom is designated by an asterisk; the symbol (+) preceding the name signifies that the isomer possesses a clockwise rotatory power.

Table 15.3 shows that the properties are identical except for optical rotatory power, which is equal and opposite for the two isomers.

The 20-or-so α-amino acids from which living cells produce proteins possess an asymmetric carbon atom. The formulas of four of them are shown in Fig. 15.25. For some unknown reason only one of the two possible optical isomers for each amino acid is synthesized by living cells and used to make proteins. We shall return to amino acids and proteins in Sec. 16.5.

FIGURE 15.24

Rotation of the electric and magnetic fields by an optical isomer.

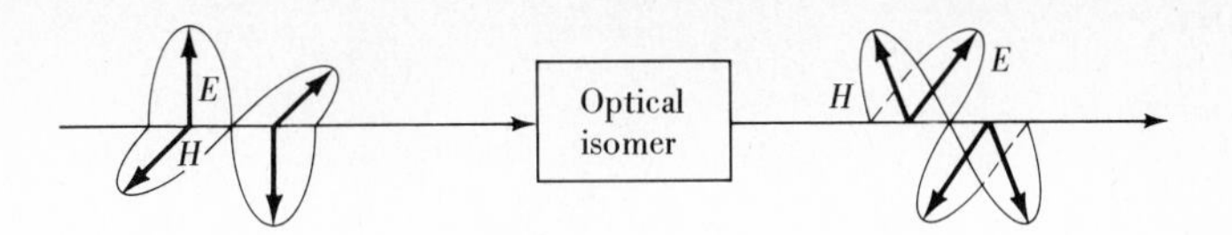

(*a*) *When a light wave passes through a sample of an optical isomer the directions of the electric and magnetic fields are rotated*

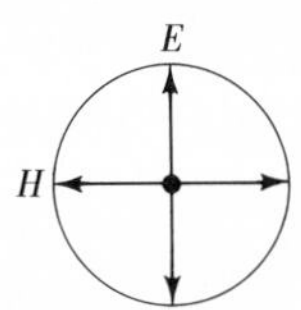

(*b*) *The oscillation of the electric field E occurs in the vertical plane before the wave passes through the sample*

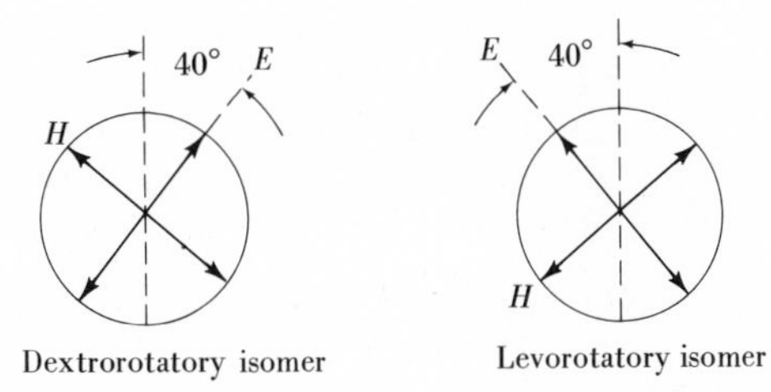

(*c*) *After the light wave passes through the sample the plane in which the electric field oscillates has been rotated*

TABLE 15.3: PROPERTIES OF THE OPTICAL ISOMERS OF 1-HYDROXY-2-METHYLBUTANE

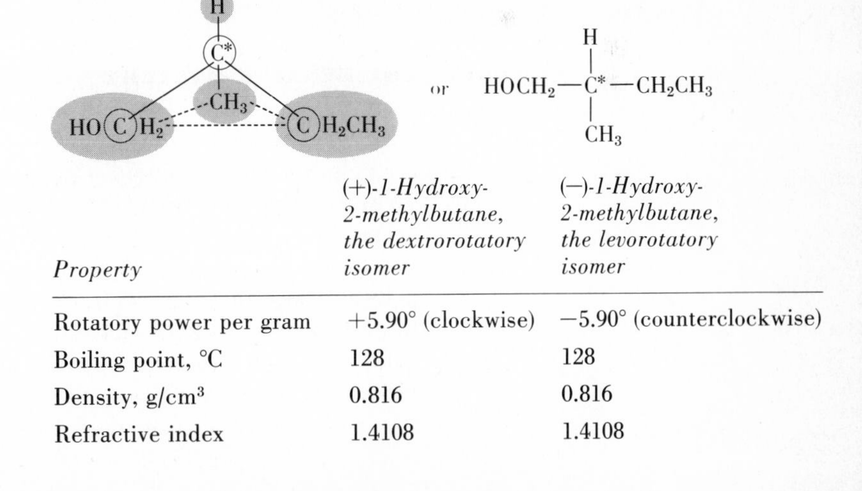

Property	*(+)-1-Hydroxy-2-methylbutane, the dextrorotatory isomer*	*(−)-1-Hydroxy-2-methylbutane, the levorotatory isomer*
Rotatory power per gram	+5.90° (clockwise)	−5.90° (counterclockwise)
Boiling point, °C	128	128
Density, g/cm³	0.816	0.816
Refractive index	1.4108	1.4108

FIGURE 15.25

Alpha amino acids.

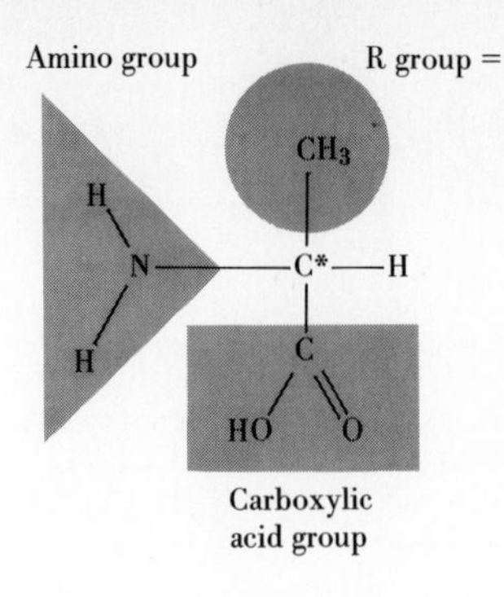

(*a*) *Alanine (Al–uh–neen)*

(*b*) *Valine (Val–een)*

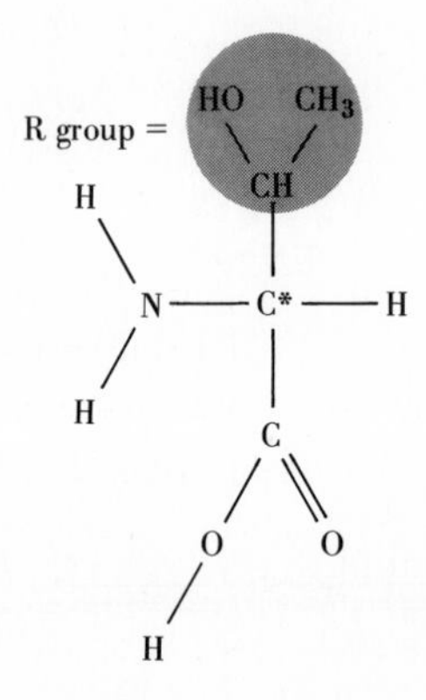

(*c*) *Threonine (Three–o–neen)*

(*d*) *Cysteine (Sis–tuh–een)*

15.10 THE BENZENE MYSTERY

Benzene, one of the most important compounds in all of chemistry, is a colorless, volatile liquid which was discovered in pyrolized whale oil by Michael Faraday in 1825. As chemists began to investigate the properties of this compound, they became more and more puzzled. Its elemental analysis showed that it is composed of only carbon and hydrogen and that it is 93.6 percent carbon, giving it an empirical formula of CH. At 25°C and 97.9 Torr 1 l of benzene vapor weighs 0.0419 g, giving a molecular weight of 78 g; therefore, the molecular formula of benzene is C_6H_6. Such a low hydrogen-to-carbon ratio in this compound suggests that benzene is highly unsaturated, and some chemists suggested structures like

```
    H
    |
    C              H
   / \\           /
  H   C—C≡C—C
      |       \\
      H        C—H
               |
               H
```

A molecule with such a structure would be expected to be very reactive. It would undergo rapid addition reactions. Surprisingly, benzene was found to be relatively inert, like the alkanes, and like these saturated compounds it undergoes slow substitution reactions:

$$C_6H_6(l) + Cl_2(g) \rightarrow C_6H_5Cl(l) + HCl(g)$$

Benzene

$$C_2H_6(g) + Cl_2(g) \rightarrow C_2H_5Cl(l) + HCl(g)$$

Ethane

Cyclic structures were also considered. The ratio of hydrogen to carbon in these compounds is lower than in the straight-chain hydrocarbons because two of the valence bonds are needed to cyclize the compound. The hydrogen-to-carbon ratio in cyclopentane is 2:1, just as in pent*ene*, even though there is no unsaturation in cyclopentane (see Fig. 15.26). Cyclization reduces the hydrogen-to-carbon ratio without introducing double or triple bonds, and the resulting cycloalkanes would be expected to undergo substitution reactions; but the hydrogen-to-carbon ratio in the cycloalkanes is still not as low as it is in benzene. Benzene cannot be a cycloalkane.

The nineteenth-century chemists had a benzene riddle: What

FIGURE 15.26

Comparison of normal and cycloalkanes.

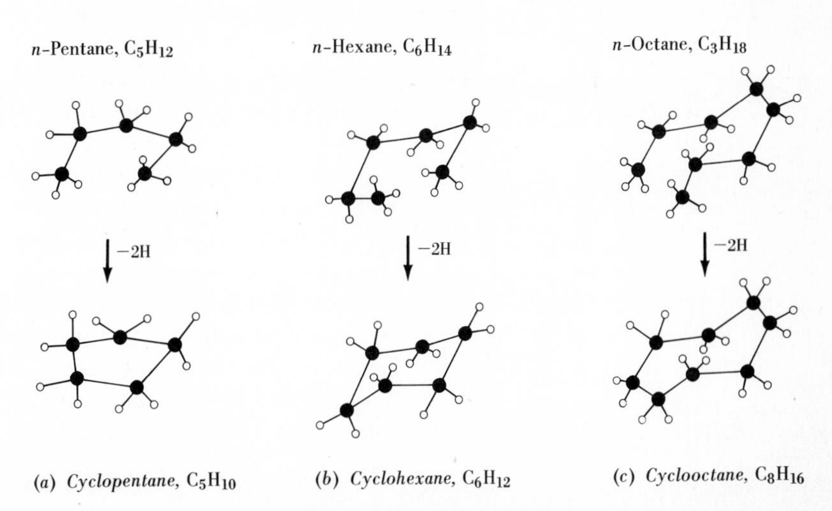

Friedrich August Kekulé von Stradonitz. (Courtesy of Burndy Library, Norwalk, Connecticut.)

looks like (has the formula of) an unsaturated compound and acts like a saturated compound? The riddle was solved in a dream. Friedrich August Kekulé von Stradonitz, the most inventive structural chemist of the nineteenth century, made this report in 1865:

> I was sitting, writing at my text-book; but the work did not progress; my thoughts were elsewhere. I turned my chair to the fire and dozed. Again the atoms were gambolling before my eyes. This time the smaller groups kept modestly in the background. My mental eye, rendered more acute by repeated visions of the kind, could now distinguish larger structures, of manifold conformation: long rows, sometimes more closely fitted together; all twining and twisting in snake-like motion. But look! What was that? One of the snakes seized hold of its own tail, and the form whirled mockingly before my eyes. As if by a flash of lightning I awoke; and this time also I spent the rest of the night in working out the consequences of the hypothesis.
>
> Let us learn to dream, gentlemen, then perhaps we shall find the truth . . . but let us beware of publishing our dreams before they have been put to the proof by the waking understanding.†

In developing his structural formula for benzene Kekulé applied the principle of constant valence, which he found so fruitful when applied to other carbon compounds. Again he insisted that the valence of carbon must always be 4 (see page 93). His early representation of the benzene structure looked like a link of sausages. This is shown in Fig. 15.27*a*, where a carbon atom is represented as a "sausage" with four bulges on it; each bulge serves as a valence site. A double bond between two carbon atoms involves the mutual contact of two bulges (valence sites) on each carbon atom; a single bond involves one bulge on each carbon atom. Each dot in the formula represents a point where carbon can bond another atom or radical. There are six such points, and all benzene derivatives have a general formula of C_6A_6.

A representation of Kekulé's sausage formula using modern symbols is shown in Fig. 15.27*b*, from which it is easy to see how the idea of cyclical benzene may have developed in his mind. Further consideration of the problem led to an evolution of the structural representations of benzene. Kekulé next proposed the awkward cyclical structure with alternating double and single bonds, shown in Fig. 15.27*c*, but eventually the hexagonal representation of benzene shown in Fig. 15.27*d* became popular and is still used today. It is appropriately referred to as the *Kekulé structure*.

Although the Kekulé structure (Fig. 15.27*d*) accounts for many of the chemical and physical properties of benzene, it really does not completely explain the benzene riddle. In fact, it poses more questions than answers. If benzene has a hexagonal structure with alternate single and double bonds between carbon atoms, we would ex-

† Quoted from R. Japp, Kekulé Memorial Lecture, *J. Chem. Soc.*, **73**:100 (1898).

FIGURE 15.27

Kekulé structures for benzene.

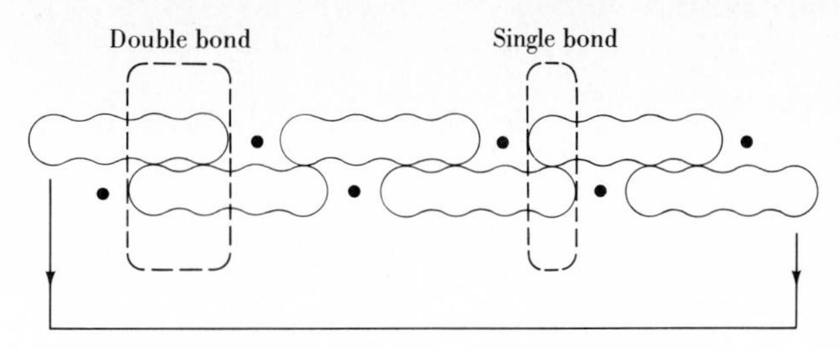

(a) The Kekulé sausage formula for benzene (1866)

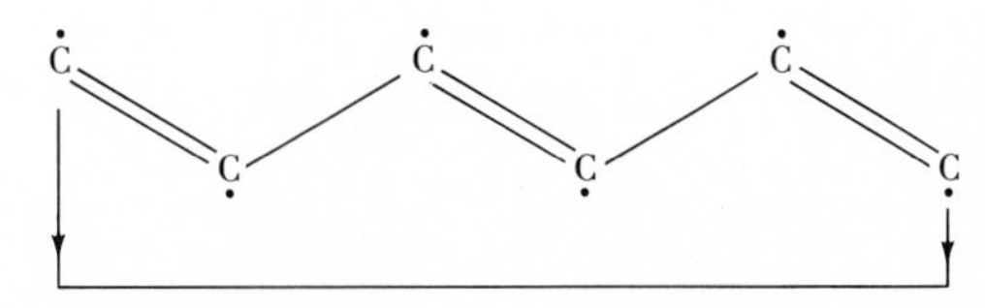

(b) The Kekulé sausage formula using modern symbols (compare with (a), above)

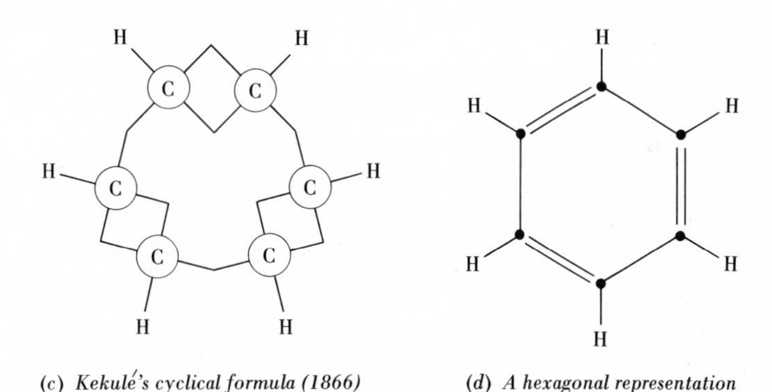

(c) Kekulé's cyclical formula (1866)

(d) A hexagonal representation of benzene

pect certain peculiarities in its derivatives. For example, benzene can react with chlorine to produce substitution products with from one to six chlorine atoms, C_6H_5Cl, $C_6H_4Cl_2$, etc. One of the disubstituted derivatives is called orthodichlorobenzene. The prefix *ortho-* means that two adjacent carbon atoms of the benzene ring carry substituents. Using the Kekulé structure, we can draw two different structures for orthodichlorobenzene. Figure 15.28*a* has the chlorine atoms substituted on carbon atoms joined by a double bond, and Fig. 15.28*b* has chlorine atoms on carbon atoms joined by a single bond. Even the organic chemists of the 1860s and 1870s knew that only one orthodichlorobenzene compound can be prepared. It was clear that every position on the benzene nucleus is chemically equivalent, so that the two structures shown in Fig. 15.28 are different representations of the same molecule. Many suggestions were

FIGURE 15.28

Can there be two forms of orthodichlorobenzene?

(a) (b)

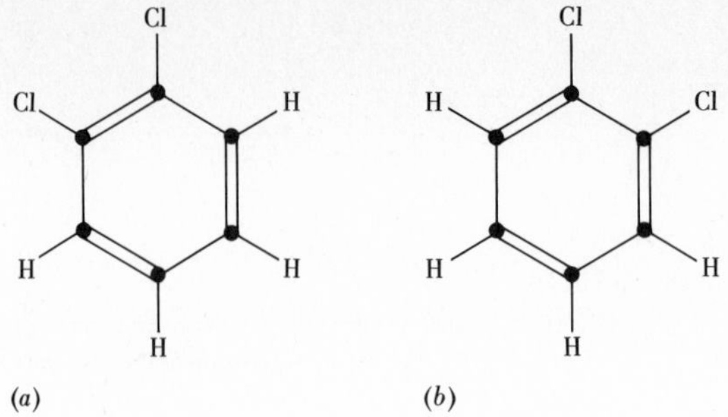

made to explain the fact that only one ortho-disubstituted compound can be prepared. Kekulé proposed a rapid oscillation between two structures so that two adjacent carbon atoms are bonded by a single bond for an instant and then by a double bond as shown in Fig. 15.29. This would account for the chemical equivalence of all sites on the benzene nucleus and the fact that only a single orthodichlorobenzene compound can be prepared.

When twentieth-century physicists and physical chemists had learned how to measure the distance between atoms within a molecule, they pointed out another weakness in the Kekulé structure: all six carbon-carbon bond lengths in benzene are equal (1.39 Å). From studies of other molecules single bonds would be expected to be longer than double bonds. The C—C single-bond length in the alkanes is always 1.54 Å, whereas the C=C double-bond length the alkenes is always 1.34 Å. If the Kekulé structure were a good representation of the benzene molecule, there should be three short bonds between the three pairs of carbon atoms joined by double bonds and three longer bonds between the three pairs joined by single bonds. This is not true.

A weakness more obvious to organic chemists is based on the fact, already mentioned, that benzene is relatively inert and undergoes substitution. It does not behave as an unsaturated compound

FIGURE 15.29

Kekulé proposed rapid oscillation.

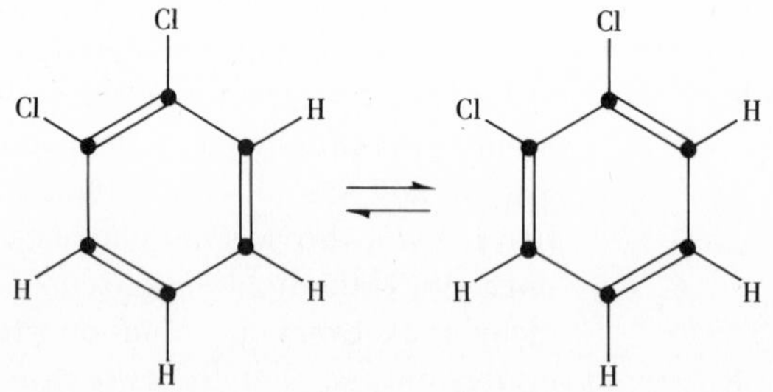

should, and the Kekulé structure is no help here. A satisfactory answer to this perplexity was not available until the development of modern valence theory based on quantum mechanics.

15.11 MODERN VALENCE THEORY AND BENZENE

Before we consider the electronic structure of benzene, let us review what is known about the geometry of the benzene molecule. We have learned by the modern methods of molecular-structure determination that the structure proposed by Kekulé, as far as geometry is concerned, is very close to the truth. The benzene molecule is symmetrical. The six carbon atoms may be considered to reside at the vertices of a regular, planar hexagon. The six hydrogen atoms would then be located in the same plane, on rays emanating from the center of the hexagon and passing through the carbon atoms; the hydrogen atoms could be considered to form a regular hexagon of their own. As shown in Fig. 15.30, all carbon-carbon bond lengths are 1.39 Å, and all carbon-hydrogen bond lengths are 1.19 Å. Each carbon atom is bonded to three atoms in the same plane, and all bond angles are exactly 120°.

FIGURE 15.30

The benzene structure.

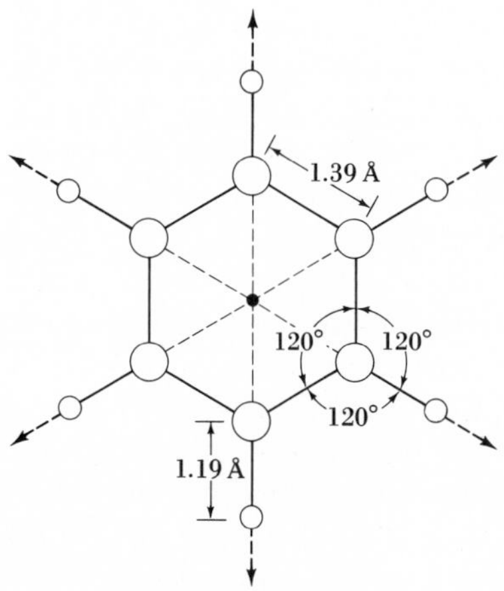

A Lewis structure which satisfies the octet rule can easily be drawn for benzene in terms of the known structure of the molecule. Figure 15.31 shows the Lewis electron-dot structure equivalent to a Kekule structure. In the explanation of the chemical and physical properties of benzene a single Lewis structure is no more adequate

FIGURE 15.31

Lewis structure of benzene.

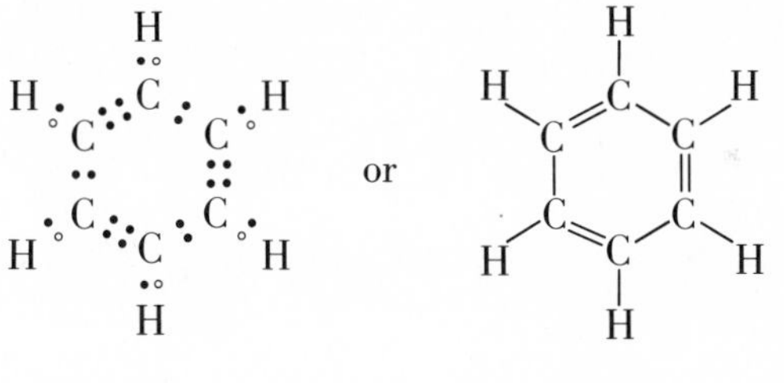

than a single Kekulé structure because it too has alternate single and double bonds in the ring.

Therefore, chemists began writing resonance structures with three pairs of delocalized electrons. Two resonance structures are shown in Fig. 15.32. The resonance structures have no independent existence; the best representation of the benzene molecule is a hybrid of these two structures. (A double-headed arrow is the ordinary symbol for resonance.)

FIGURE 15.32

Resonance in benzene.

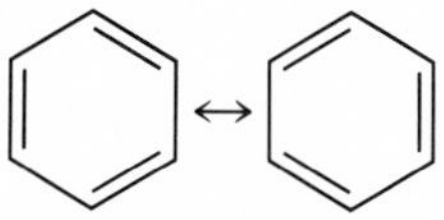

Benzene as a hybrid of resonance structures

Resonance implies that certain valence electrons are delocalized; i.e., certain valence electrons are not restricted to the region between a particular pair of bonded atoms. There are three pairs of delocalized valence electrons in benzene.

On page 146 we indicated that the bonding in the benzene molecule can be explained in terms of the valence-bond theory and hybridization. The geometrical structure of benzene immediately suggests that carbon is involved in sp^2, or trigonal, hybridization (Sec. 5.33).

The atomic number of carbon is 6; therefore, its ground-state electronic configuration is $1s^2$, $2s^2$, $2p_x{}^1$, $2p_y{}^1$, $2p_z{}^0$. In order to exhibit trigonal bonding, hybridization is required. We shall indicate this by a box diagram like the ones used before (see Fig. 15.33). The carbon atoms in benzene are in the same hybrid form as the carbon atoms of ethylene and other alkenes, but, as we shall see, the bonding is quite different.

The three sp^2 hybrid orbitals are used to form three σ bonds between a particular carbon atom and three other atoms; specifically, each carbon atom is bonded to two other carbon atoms and one hydrogen atom via the three sp^2 orbitals. As previously discussed, when sp^2 hybridization occurs, all bonded atoms are coplanar and all bond angles are 120°. This is exactly what is required to produce the known structure of benzene. Figure 15.34*a* and *b* shows a segment of the benzene molecule to emphasize that each carbon atom is bonded to three other atoms in the same plane. Figure 15.34*c* shows all the σ bonds in the benzene molecule.

The σ bonding does not finish the bonding story in benzene. In fact, the most fascinating aspect of bonding in this molecule and all its derivatives is the presence of the unhybridized $2p_z$ orbital on

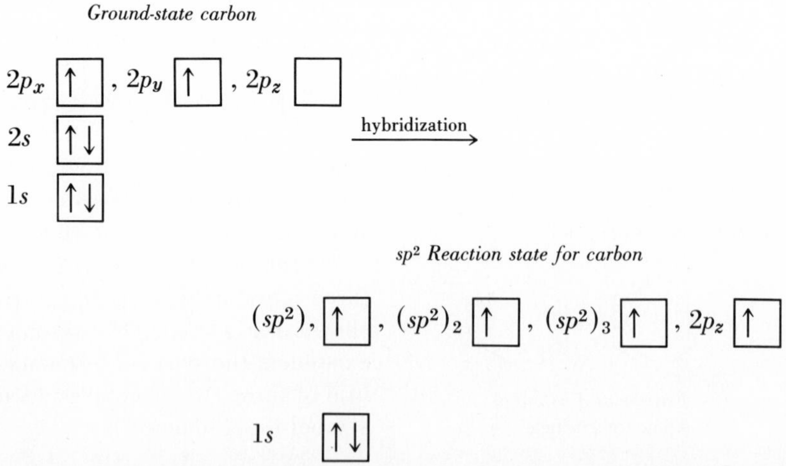

FIGURE 15.33

Hybridization produces an sp^2 state for carbon.

Ground-state carbon

$2p_x$ [↑], $2p_y$ [↑], $2p_z$ [ ]

$2s$ [↑↓]

$1s$ [↑↓]

$\xrightarrow{\text{hybridization}}$

sp^2 Reaction state for carbon

$(sp^2)_1$ [↑], $(sp^2)_2$ [↑], $(sp^2)_3$ [↑], $2p_z$ [↑]

$1s$ [↑↓]

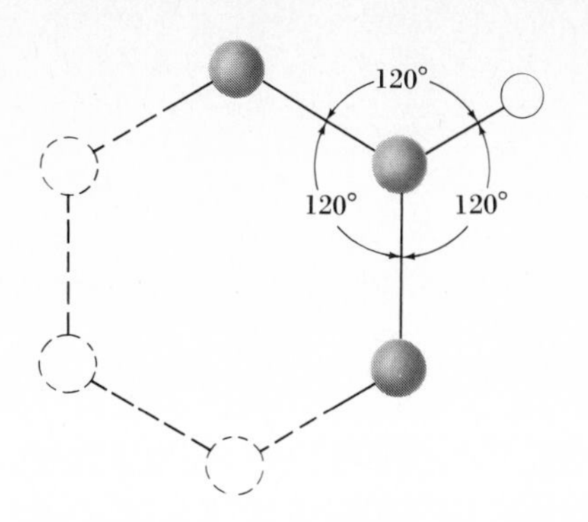

(a) *Three atoms are bonded to each carbon atom*

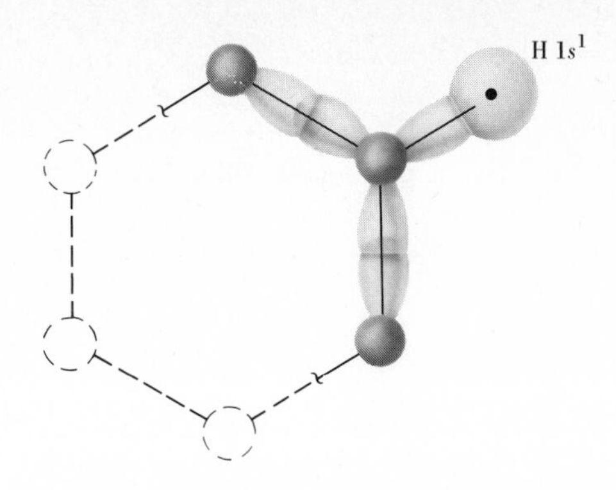

(b) *The bonding results from overlap of the three sp^2 orbitals of a particular carbon with the sp^2 orbitals of the two adjacent carbon atoms and the 1s orbital of hydrogen*

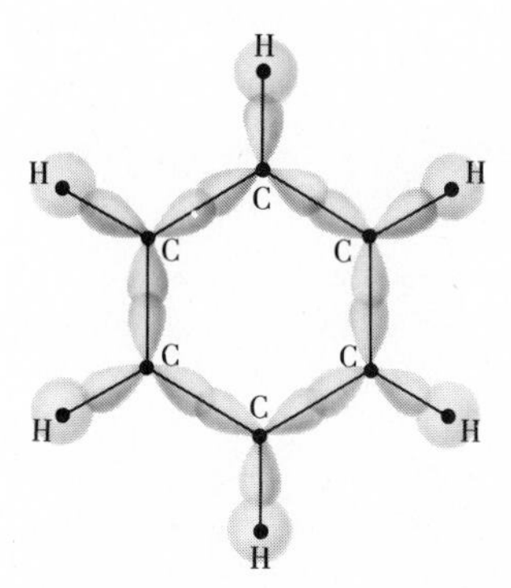

(c) *The sigma–bond skeleton in benzene*

FIGURE 15.34

The bonding and structure of benzene.

each of the carbon atoms. Figure 15.35*a* shows that the axes of the unhybridized $2p_z$ orbitals are perpendicular to the plane of the molecule. According to the valence-bond theory, a pair of half-filled parallel p atomic orbitals on adjacent atoms can form π bonds. But there are six half-filled orbitals, and they can form three π bonds in two different ways, as shown in Fig. 15.35*b*. This is the valence-bond equivalent of the Kekulé and Lewis structures for benzene. The concept of resonance with delocalized electrons is again necessary to explain that each C—C bond is equivalent and that the benzene ring is perfectly symmetrical.

The molecular-orbital theory of benzene is more satisfactory. According to it, the six p atomic orbitals produce six π molecular orbitals. The three molecular orbitals of lowest energy are bonding orbitals, and the next three are antibonding. The six electrons which originally occupied the $2p_z$ atomic orbitals occupy the three lowest-energy bonding orbitals of benzene. Figure 15.35*c* shows the lowest-energy π molecular orbital formed from in-phase interaction of the

FIGURE 15.35

π bonding in benzene.

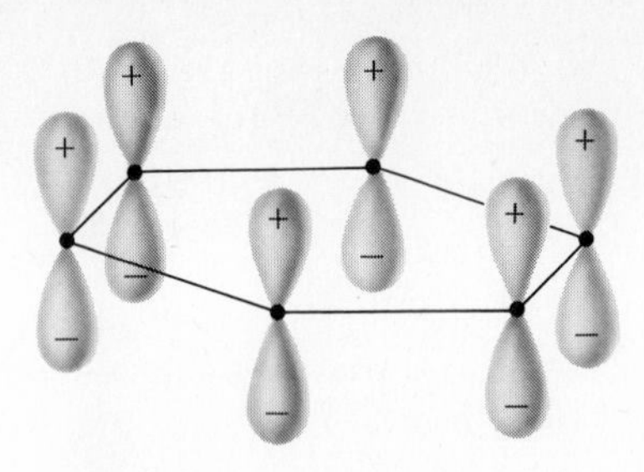

(a) The six $2p_z$ atomic orbitals.

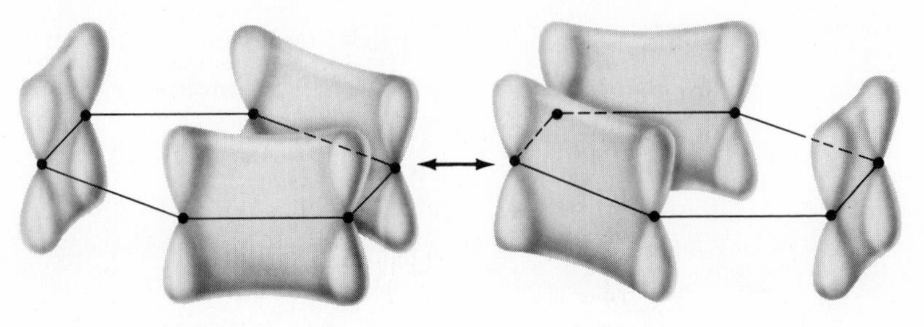

(b) Valence-bond structures for benzene. There are three π bonds in each representation.

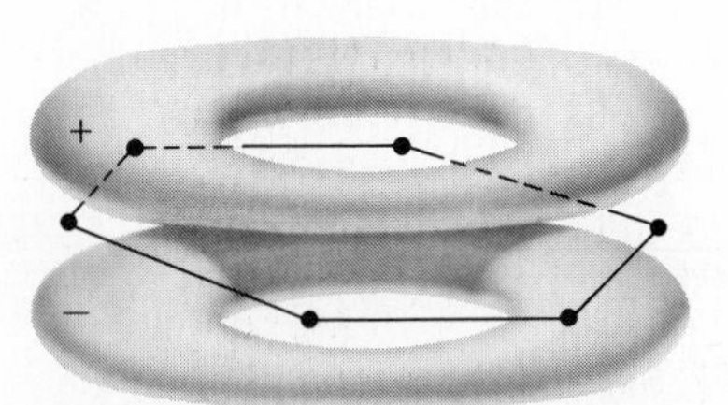

(c) The lowest energy π molecular orbital which bonds all the carbon atoms.

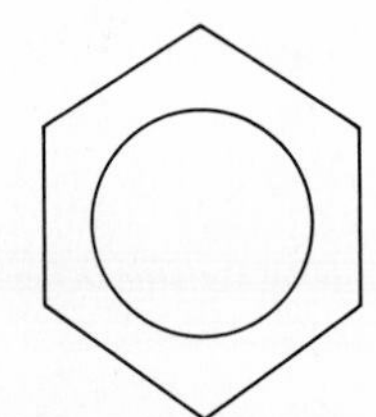

(d) Symbol for benzene which emphasizes the delocalization of valence electrons.

six individual $2p_z$ orbitals shown in Fig. 15.35*a*. Two electrons will occupy this molecular orbital, and the remaining four will go into the next two higher molecular orbitals, which are also bonding orbitals. The advantage of the molecular-orbital theory is that it shows correctly that the bonds between all C—C pairs are equivalent. The molecular orbital encompasses the whole ring system symmetrically. Each pair of adjacent carbon atoms is bonded by a σ bond and the symmetrical all-encompassing π bond. Instead of representing the benzene structure as a ring system of alternating single and double bonds, chemists now prefer the circle in a hexagon, which emphasizes the bonding symmetry of the molecule (Fig. 15.35*d*).

15.12 DERIVATIVES OF BENZENE AND THEIR NOMENCLATURE

Chemists have learned to prepare many useful derivatives of benzene such as *aspirin*, *styrene* (used in the manufacture of synthetic rubber and plastics), *mothballs*, *TNT*, and *aniline* (used in the dye and photographic industries). You should be familiar with the nomenclature of the benzene derivatives and the structure of some of the important derivatives.

Table 15.4 shows the structure and nomenclature of several of the more commonly encountered derivatives. The first, commonly called *toluene*, is simply benzene with one of the hydrogen atoms replaced by a methyl radical, and a nonchemist would naturally call the structure methylbenzene. However, the acceptable name is not related to its structure; it is always called toluene. Note that the acceptable names of the hydroxy and amino derivatives of benzene also have names which are not related to the structure. *Phenol* is the name commonly used for hydroxybenzene and *aniline* is the name used for aminobenzene.

When there is more than one substituent on the benzene nucleus

TABLE 15.4: SOME BENZENE DERIVATIVES

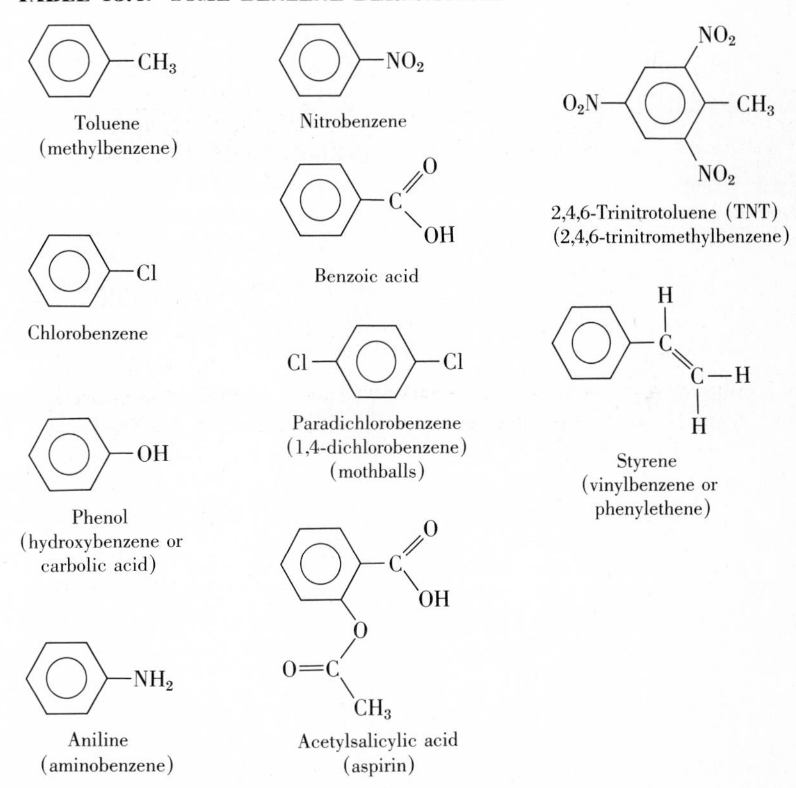

(*a*) *Orthodichlorobenzene, 1,2-dichlorobenzene*

(*b*) *Metadichlorobenzene, 1,3-dichlorobenzene*

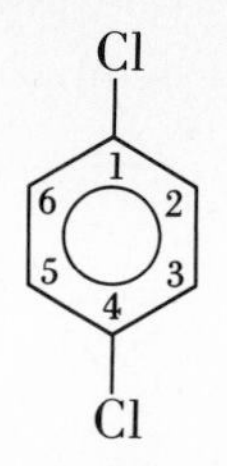

(*c*) *Paradichlorobenzene, 1,4-dichlorobenzene*

FIGURE 15.36

The dichlorobenzenes.

the positions of the substituents must be specified. For example, there are three possible dichlorobenzenes. After the first chlorine atom is substituted on the ring, there are three possible positions for the second atom, i.e., on an adjacent carbon atom, two carbon atoms away, or three carbon atoms away. Substitution occurring on an adjacent carbon atom gives the *ortho* derivative (Fig. 15.36*a*). The *meta* and *para* derivatives are shown in Figs. 15.36*b* and *c*.

The ring-numbering system used in naming the benzene derivatives is a more systematic and useful method. In Fig. 15.36*a* the carbon atom at the top is arbitrarily assigned the number 1; all other positions are numbered clockwise. (We choose clockwise numbering because this will assign the smallest numbers to the carbon atoms carrying substituents.) In orthodichlorobenzene, substitution occurs at carbon atoms 1 and 2, and so the name of this structure is also 1,2-dichlorobenzene. Paradichlorobenzene (Fig. 15.36*c*) is also called 1,4-dichlorobenzene. For complicated benzene derivatives it is imperative to use the nomenclature based on the ring-numbering system. The ortho-meta-para nomenclature can be used only for disubstituted benzene derivatives. Examples of trisubstituted benzene derivatives are given in Fig. 15.37. Benzene derivatives have been prepared in which substituents occur at four, five, and even all six positions.

The radical formed from the benzene when one hydrogen atom is removed is called the *phenyl* group, C_6H_5—. It is analogous to methyl, ethyl, etc., and can be a substituent. It would be permissible (though unconventional) to refer to chlorobenzene as *phenyl* chlo-

FIGURE 15.37

Trisubstituted benzene derivatives.

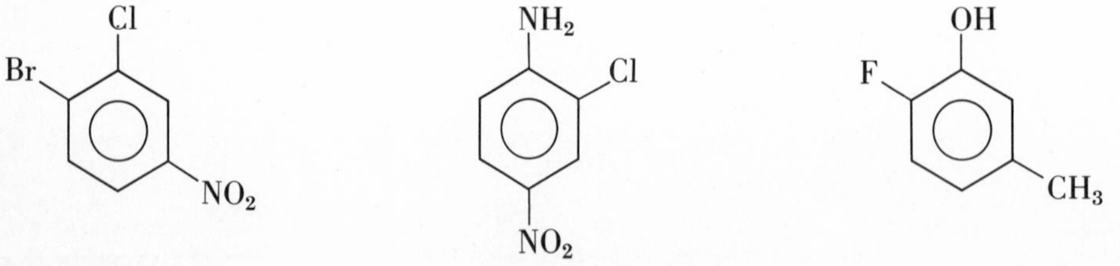

(*a*) *2-Bromo-5-nitrochlorobenzene, or 3-chloro-4-bromonitrobenzene*

(*b*) *2-Chloro-4-nitroaniline, or 2-amino-5-nitrochlorobenzene*

(*c*) *2-Fluoro-5-methylphenol*

ride; analine could be called phenylamine, and phenol could be called phenyl hydroxide. Table 15.4 shows that styrene is also named phenylethene.

15.13 FUNCTIONAL GROUPS

The reactions of the hydrocarbons (alkanes, alkenes, and benzene) are rather limited, but when certain substituents, called *functional groups*, are grafted onto these hydrocarbon backbones, a variety of very important reactions can occur. We shall discuss only those functional groups which prepare us for a study of the important synthetic polymers of the chemical industry (rubber and plastics) and the giant molecules of the living cell (proteins, DNA, and RNA).

Alcohols

The alcohols are compounds in which the functional group is the hydroxy radical, —OH. Methyl alcohol, or methanol, CH_3—OH, can be prepared by the destructive distillation of wood and hence has been called *wood alcohol*. It is a deadly poison which claims victims every year who mistake it for grain alcohol. Ethyl alcohol, or ethanol, CH_3CH_2—OH, forms from the fermentation of sugars and starch by yeast. Since the starch is usually derived from various grains, the product is called *grain alcohol*. Ethanol and other alcohols such as *n*-propanol, isopropanol, and *n*-butanol can also be prepared by the hydrolysis of alkenes in the presence of acid:

$$CH_2{=}CH_2 + HOH \xrightarrow{H_2SO_4} \underset{\text{Ethanol}}{CH_3{-}CH_2OH}$$

$$CH_3{-}CH{=}CH_2 + HOH \xrightarrow{H_2SO_4} \underset{\text{Isopropanol}}{CH_3{-}\underset{\displaystyle OH}{\underset{|}{CH}}{-}CH_3}$$

The suffix *-ol* is used in the systematic nomenclature to indicate the presence of the hydroxy group. Since CH_3—CH_2—OH has two carbon atoms, it is an ethane derivative. To name the corresponding alcohol drop the final *-e* of the alkane and add the suffix *-ol*; the systematic name for CH_3—CH_2—OH is *ethanol*. The position of the hydroxy group is indicated by numbering the carbon chain in higher-molecular-weight alcohols:

$$CH_3{-}\underset{\displaystyle OH}{\underset{|}{CH}}{-}CH_3 \qquad CH_3{-}CH_2{-}\underset{\displaystyle OH}{\underset{|}{CH_2}} \qquad CH_3{-}CH_2{-}\underset{\displaystyle OH}{\underset{|}{CH}}{-}CH_3$$

Isopropyl alcohol (isopropanol or 2-propanol) — Normal propyl alcohol (1-propanol) — 2-Butanol

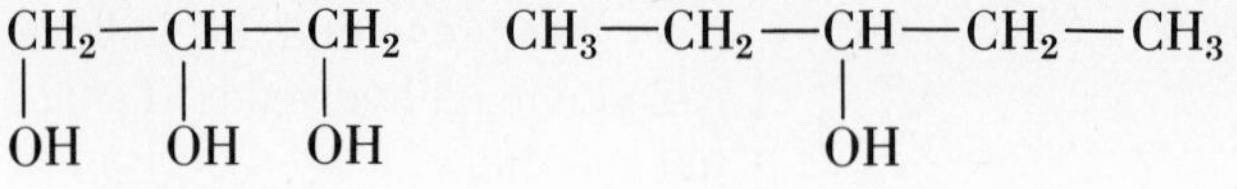

$$\underset{\text{OH}}{\underset{|}{CH_2}}-\underset{\text{OH}}{\underset{|}{CH}}-\underset{\text{OH}}{\underset{|}{CH_2}} \qquad CH_3-CH_2-\underset{\text{OH}}{\underset{|}{CH}}-CH_2-CH_3$$

Glycerol (1,2,3-propan*triol*) — 3-Pentanol

The alcohols are similar to water in their structure and properties, and it is fruitful to consider them as derivatives of water. The water molecule is bent; there are two nonbonding pairs of electrons in the valence energy level of oxygen, and hydrogen bonding is very pronounced. Figure 15.38 shows that the simple alcohols are bent at the H—O—C bonds and that there are two nonbonding pairs of electrons on the oxygen atoms of the alcohols. Alcohols exhibit hydrogen bonding and undergo reactions reminiscent of the reactions of water.

Amines

The most common organic and biochemical bases possess the *amine* or *amino* group, $—NH_2$, or a derivative of it like the *methylamino* group, $—N\begin{smallmatrix} \diagup H \\ \diagdown CH_3 \end{smallmatrix}$. For this reason such compounds are often referred to as *nitrogenous* bases.

Methylamine is produced industrially by treating methanol with ammonia at high temperatures:

$$CH_3OH + NH_3 \xrightarrow[Al_2O_3 \text{ cat.}]{450°C} CH_3NH_2$$

Methylamine (aminomethane)

Higher-molecular-weight amines can be prepared from the corresponding halides by treating them with ammonia and a strong base:

$$CH_3CH_2CH_2Br \xrightarrow[(2)\ NaOH]{(1)\ NH_3} CH_3CH_2CH_2NH_2$$

1-Bromopropane — *n*-Propylamine (1-aminopropane)

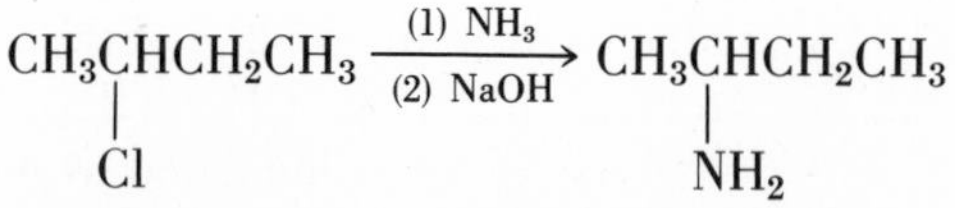

$$CH_3\underset{\text{Cl}}{\underset{|}{C}}HCH_2CH_3 \xrightarrow[(2)\ NaOH]{(1)\ NH_3} CH_3\underset{NH_2}{\underset{|}{C}}HCH_2CH_3$$

2-Chlorobutane — 2-Aminobutane

The systematic nomenclature of the amines is straightforward and is illustrated in the equation above for the production of 2-aminobutane. First, find the longest carbon chain carrying the amino functional group. Number the chain in the direction that gives the amino-bearing carbon atom the smaller number. In this example

FIGURE 15.38

Comparison of the structures of water and alcohol.

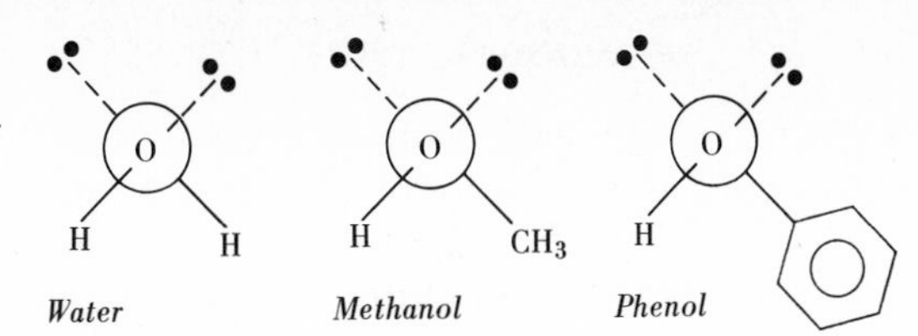

since the chain contains four carbon atoms, the compound is a butane derivative; the amino group is on carbon number 2.

The amines can be considered to be derivatives of ammonia, NH_3, in which one, two, or three of the hydrogen atoms have been replaced by organic radicals. In this light it should not be surprising that the amines are fairly strong bases and react with acids to form ammoniumlike salts:

$$NH_3 + HCl \rightarrow NH_4^+Cl^-$$

Ammonia — Ammonium chloride

$$CH_3NH_2 + HCl \rightarrow CH_3NH_3^+Cl^-$$

Methylamine — Methylammonium chloride

$$(CH_3)_2NH + HCl \rightarrow (CH_3)_2NH_2^+Cl^-$$

Dimethylamine — Dimethylammonium chloride

$$C_6H_5\text{—}NH_2 + HCl \rightarrow C_6H_5\text{—}NH_3^+Cl^-$$

Phenylamine (aniline) — Phenylammonium chloride (anilinium chloride)

The amines preserve the pyramidal structure of the ammonia, as shown in Fig. 15.39, where ammonia, methylamine, and aniline are compared. The presence of the nonbonding pair of electrons on the nitrogen atom confers upon ammonia and the amines the property of being a base. This was discussed in Chap. 13.

FIGURE 15.39

A comparison of the structures of ammonia and the amines.

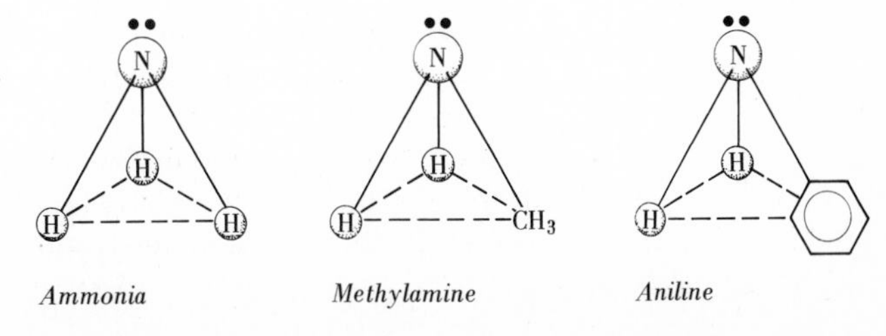

Carboxylic acids

The most common organic acids contain the carboxylic acid group, which is shown for acetic acid, benzoic acid, and the amino acid alanine in Fig. 15.40. Since the polar O—H bond of the carboxylic acid group hydrolyzes, the group acts as a proton donor and produces acidic solutions:

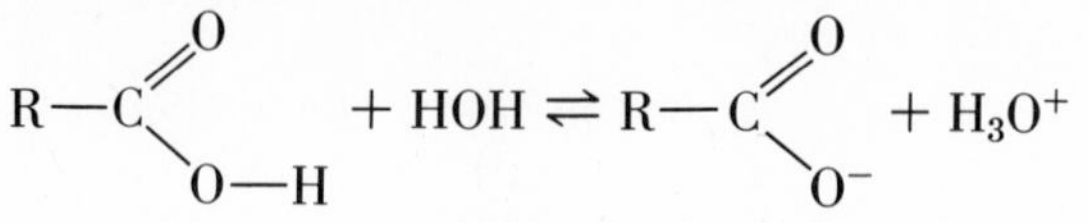

$$R{-}C(=O){-}O{-}H + HOH \rightleftharpoons R{-}C(=O){-}O^- + H_3O^+$$

In Chap. 13, the ionization constant K_i for acetic acid is given as 1.8×10^{-5}. Other carboxylic acids have K_i's ranging from 10^{-4} to 10^{-6}, so that compounds are considered to be moderate to weak acids.

In the carboxylic acid group the carbon atom is bonded to three other atoms which are coplanar with it, and all the bond angles, C—C—O, C—C=O, and O=C—O, are approximately 120° (see Fig. 15.40*d*). Therefore, the structure of the carboxylic acid group can be explained in terms of sp^2, or trigonal, hybridization.

Like inorganic acids the carboxylic acids react with bases to form salts:

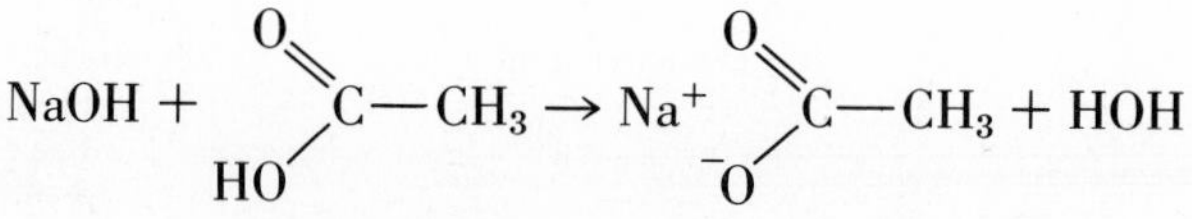

$$NaOH + HO{-}C(=O){-}CH_3 \rightarrow Na^+ \; {}^-O{-}C(=O){-}CH_3 + HOH$$

Acetic acid — Sodium acetate

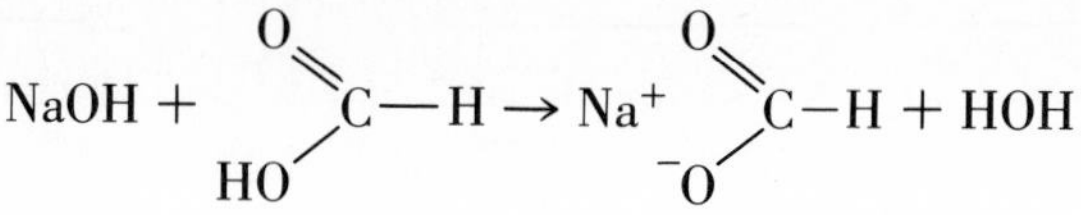

$$NaOH + HO{-}C(=O){-}H \rightarrow Na^+ \; {}^-O{-}C(=O){-}H + HOH$$

Formic acid — Sodium formate

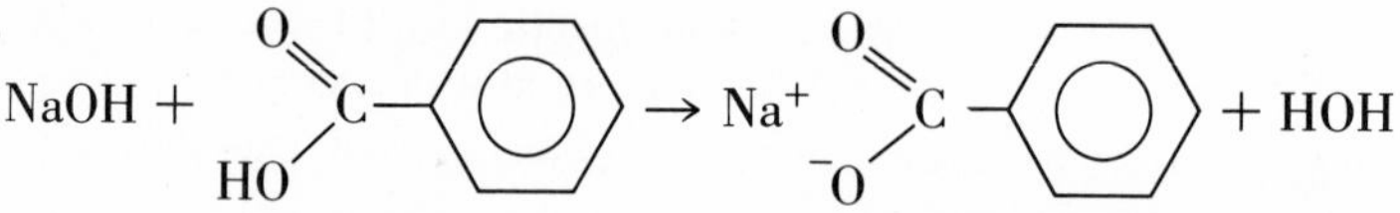

$$NaOH + HO{-}C(=O){-}C_6H_5 \rightarrow Na^+ \; {}^-O{-}C(=O){-}C_6H_5 + HOH$$

Benzoic acid — Sodium benzoate

Many of the common lower-molecular-weight acids have trivial names, which are not derived from the corresponding alkane. These include formic acid, $HCOOH$; acetic acid, CH_3COOH; and butyric acid, $CH_3(CH_2)_2COOH$. The systematic name for a given acid is derived from the corresponding alkane. For example, acetic acid contains two carbon atoms; the two-carbon alkane is ethane. To

FIGURE 15.40

The carboxylic acid function.

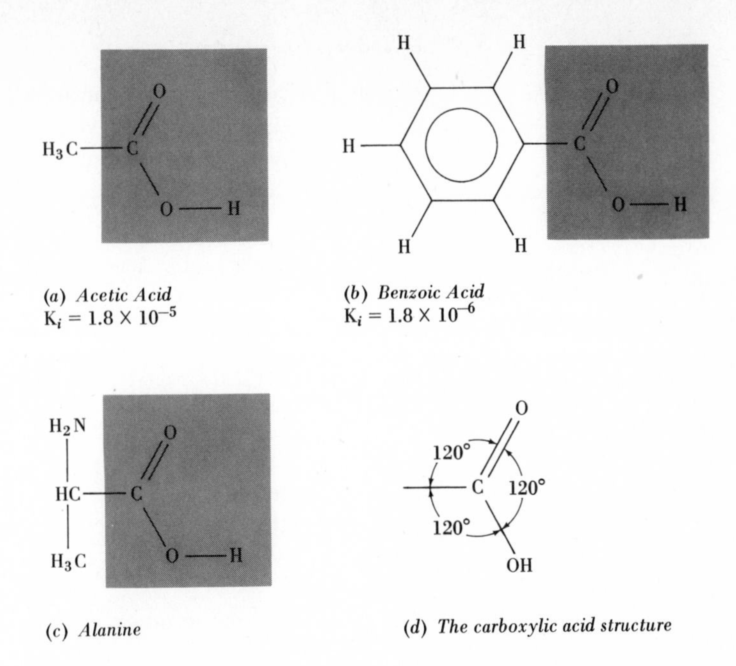

(*a*) *Acetic Acid* $K_i = 1.8 \times 10^{-5}$

(*b*) *Benzoic Acid* $K_i = 1.8 \times 10^{-6}$

(*c*) *Alanine*

(*d*) *The carboxylic acid structure*

develop the systematic name of the carboxylic acid drop the final *-e* of the alkane and add the suffix *-oic* and the word *acid*. Thus, the systematic name for acetic acid is ethan*oic acid*. Table 15.5 gives common and systematic names for some familiar acids. In this table the carboxylic acid group is written —COOH; you will often see it written —CO_2H.

As might be guessed from the equations given for the reactions of carboxylic acids with bases, to name the salts of these acids we replace the suffix *-ic* by *-ate*. A salt of acet*ic* acid is an acet*ate* and a salt of benzo*ic* acid is benzo*ate*.

TABLE 15.5: NAMES OF FAMILIAR CARBOXYLIC ACIDS

Compound	*Number of carbon atoms in chain*	*Common name*	*Systematic name*
HCOOH	1	Formic acid	Methanoic acid
CH_3COOH	2	Acetic acid	Ethanoic acid
CH_3CH_2COOH	3	Propionic acid	Propanoic acid
$CH_3CH_2CH_2COOH$	4	Butyric acid	Butanoic acid
—COOH	. . .	Benzoic acid	Benzoic acid

Esters

The acids undergo very important reactions with the alcohols; the product of such a reaction is called an *ester*. The product of the reaction between ethyl alcohol and acetic acid is the ester ethyl acetate:

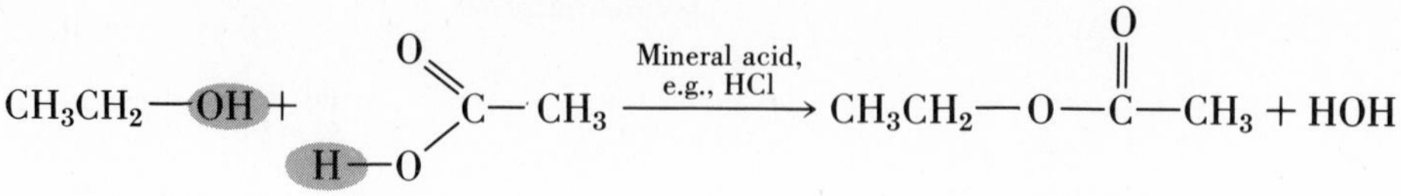

$$CH_3CH_2{-}OH + H{-}O{-}C({=}O){-}CH_3 \xrightarrow[\text{e.g., HCl}]{\text{Mineral acid,}} CH_3CH_2{-}O{-}\overset{\overset{\displaystyle O}{\|}}{C}{-}CH_3 + HOH$$

Ethyl alcohol (ethanol) — Acetic acid (ethanoic acid) — Ethyl acetate, an ester (ethyl ethanoate)

The formation of esters is catalyzed by mineral acids like HCl and H_2SO_4. Water is one of the products. The reaction should not be confused with acid-base reactions; the ester is not a salt. The ester group can be given the general symbol $R{-}O{-}\overset{\overset{\displaystyle O}{\|}}{C}{-}R'$, where R and R′ are organic radicals derived from the alcohol and from the acid, respectively. Some examples are given in Table 15.6.

Amides

Another important group of compounds, the amides, are derived from the reaction of the carboxylic acids with ammonia and the amines. The first product formed when acetic acid reacts with ammonia is a salt, ammonium acetate:

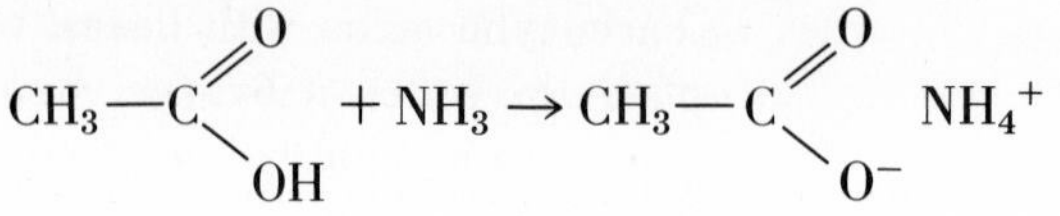

$$CH_3{-}C({=}O){-}OH + NH_3 \rightarrow CH_3{-}C({=}O){-}O^- \quad NH_4^+$$

Ammonium acetate

When ammonium acetate is heated, the *amide* forms, along with one molecule of water:

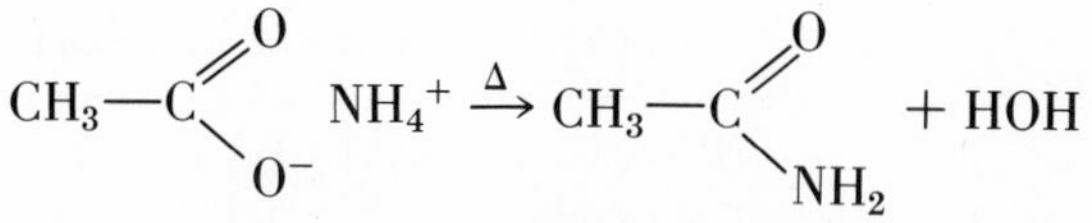

$$CH_3{-}C({=}O){-}O^- \quad NH_4^+ \xrightarrow{\Delta} CH_3{-}C({=}O){-}NH_2 + HOH$$

Acetamide

The amide is named by dropping the suffix *-ic* or *-oic* of the acid name and adding *-amide*. Thus the amides derived from propanoic and butanoic acids are named propan*amide* and butan*amide*, respectively.

When an amine, an ammonia derivative, reacts with a carboxylic

TABLE 15.6: EXAMPLES OF ESTERS

Formula	Name	Derived from: Alcohol	Derived from: Acid
$CH_3CH_2CH_2{-}O{-}\overset{O}{\overset{\parallel}{C}}{-}CH_3$	*n*-Propyl acetate	*n*-Propanol	Acetic acid
$CH_3{-}O{-}\overset{O}{\overset{\parallel}{C}}{-}CH_2CH_2CH_3$	Methyl butanoate	Methanol	Butanoic acid
$CH_3CH_2{-}O{-}\overset{O}{\overset{\parallel}{C}}{-}C_6H_5$	Ethyl benzoate	Ethanol	Benzoic acid

acid, a substituted amide forms. Methylamine and acetic acid give rise to *N*-methylacetamide. The *N*-methyl of the name means that the nitrogen atom carries the methyl substituent.

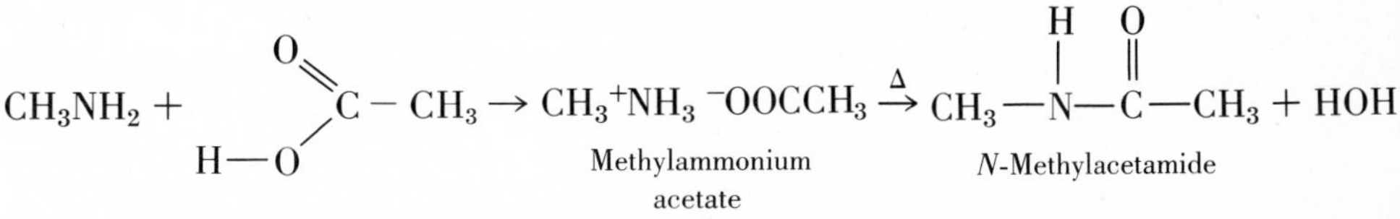

$$CH_3NH_2 + HO{-}\overset{O}{\overset{\parallel}{C}}{-}CH_3 \rightarrow \underset{\text{Methylammonium acetate}}{CH_3{}^+NH_3\ {}^-OOCCH_3} \xrightarrow{\Delta} \underset{N\text{-Methylacetamide}}{CH_3{-}\overset{H}{\overset{|}{N}}{-}\overset{O}{\overset{\parallel}{C}}{-}CH_3} + HOH$$

The general formula for an amide is $R{-}\overset{H}{\overset{|}{N}}{-}\overset{O}{\overset{\parallel}{C}}{-}R'$, where R and R′ can be hydrogen or organic radicals.

Alpha-amino acids and peptides

The α-amino acids, which are the building blocks of proteins, carry both the amino group and the carboxylic acid group. The amino group is bonded to a carbon atom adjacent to the carboxylic group. This carbon atom is called the α-carbon atom and the structure is therefore referred to as an α-amino acid. The amino group of one molecule can react with the carboxylic acid group of another molecule to form the amide linkage:

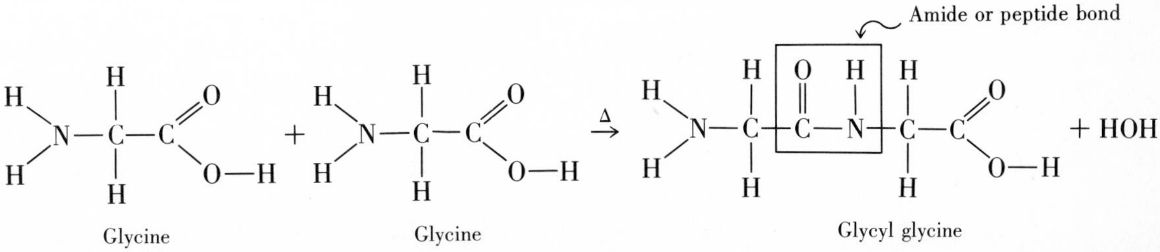

$$\underset{\text{Glycine}}{H_2N{-}CH_2{-}\overset{O}{\overset{\parallel}{C}}{-}O{-}H} + \underset{\text{Glycine}}{H_2N{-}CH_2{-}\overset{O}{\overset{\parallel}{C}}{-}O{-}H} \xrightarrow{\Delta} \underset{\text{Glycyl glycine}}{H_2N{-}CH_2{-}\boxed{\overset{O}{\overset{\parallel}{C}}{-}\overset{H}{\overset{|}{N}}}{-}CH_2{-}\overset{O}{\overset{\parallel}{C}}{-}O{-}H} + HOH$$

Amide or peptide bond

When α-amino acids form an amide, the compound is called a *peptide*. The structure glycyl glycine resulting from the reaction of two molecules of glycine, shown above, is called a *dipeptide*. Proteins are polypeptides derived from hundreds of α-amino acids.

There are other functional groups which are of interest to organic chemists, but we shall not study them in any detail. Table 15.7 shows the most important functional groups including some we have not discussed.

Table 15.8 shows functional groups which can be derived from those in Table 15.7. Ethers can be derived by dehydration of alcohols:

$$C_2H_5O\boxed{H + HO}C_2H_5 \xrightarrow[120°C]{H_2SO_4} \underset{\text{Diethyl ether}}{C_2H_5-O-C_2H_5} + H_2O$$

and the disulfides can be derived from thiols by oxidation with elemental iodine:

$$C_2H_5S\boxed{H + I_2 + H}SC_2H_5 \rightarrow \underset{\text{Diethyldisulfide}}{C_2H_5S-SC_2H_5} + 2HI$$

TABLE 15.7: FUNCTIONAL GROUPS

Functional group	*Name*	*Example*	*Name*
$R-OH$	Hydroxy	$CH_3-CH(OH)-CH_3$	2-Propanol
$R-NH_2$	Amino	$CH_3CH_2-NH_2$	Ethylamine
$R-COOH$	Carboxylic acid	CH_3CH_2-COOH	Propanoic acid
$R-C(=O)-H$	Aldehyde	$CH_3-C(=O)-H$	Acetaldehyde
$R-C(=O)-R'$	Ketone	$CH_3-C(=O)-CH_3$	Acetone
$R-SH$	Thiol	CH_3-SH	Thiomethane (methylmercaptan)
$R-NO_2$	Nitro	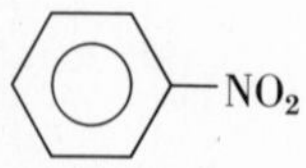 $C_6H_5-NO_2$	Nitrobenzene

TABLE 15.8: DERIVED FUNCTIONAL GROUPS

Derived functional group	*Name*	*Example*	*Name*
$R{-}O{-}\overset{O}{\overset{\|\|}{C}}{-}R'$	Ester	$CH_3{-}O{-}\overset{O}{\overset{\|\|}{C}}{-}C_6H_5$	Methyl benzoate
$R{-}\underset{}{\overset{H}{\overset{\|}{N}}}{-}\overset{O}{\overset{\|\|}{C}}{-}R'$	Amide	$(CH_3)_2CH{-}\overset{H}{\overset{\|}{N}}{-}\overset{O}{\overset{\|\|}{C}}{-}C_2H_5$	*N*-Isopropylpropanamide
$R{-}O{-}R'$	Ether	$C_2H_5{-}O{-}C_2H_5$	Diethyl ether
$R{-}S{-}S{-}R'$	Disulfide	$C_2H_5{-}S{-}S{-}C_2H_5$	Diethyl disulfide

We shall see why functional groups are important when we study industrial polymers in the next section and biochemistry in Chap. 16.

15.14 INDUSTRIAL POLYMERS

The industrial production of polymers is second only to the production of steel as a measure of the vitality of an economic system. In 1970 the worldwide production of synthetic polymers, plastics, and rubbers was estimated at 5 million tons worth 1.5 billion dollars. These synthetics are used for every imaginable product: tires, textiles, toys, paints, auto bodies (Chevrolet's Corvette), windows, coatings, bags, pipes (for plumbing to replace copper and iron), football helmets, and so on. New polymers are being developed and tested every day, and new applications are constantly being discovered.

From the applications of polymers it is obvious that they must be strong, stable, and chemically inert. However, depending on the use, we may want other properties to vary. For example, to make a window we want a polymer that is light- and heat-stable and chemically inert; we also want it to be colorless, rigid, and shatterproof. To make a basketball we want a polymer which is stable and chemically inert but also resilient, elastic, and resistant to abrasion. Polymer chemists have been able to relate the chemical structure of polymer molecules to their physical properties and hence can tailor their polymers to fit the application. Polymer chemists have found the true philosopher's stone. In this section we shall discuss some of the important industrial polymers.

Addition polymers

In Sec. 15.8 we learned that the characteristic reaction of the alkenes is addition. The carbon atoms of the double bond are bonded to only three other atoms; halogens, hydrohalides, and hydrogen molecules can add across the double bond to produce saturated compounds. Of great importance is the fact that alkene molecules can add to themselves under the proper conditions; at high pressures and temperatures and in the presence of oxygen, ethylene undergoes *addition polymerization*. In this reaction ethylene molecules add to themselves and produce long-chain alkanes which are plastics:

$$H_2C{=}CH_2 + H_2C{=}CH_2 + H_2C{=}CH_2 + H_2C{=}CH_2 + \cdots \xrightarrow[O_2]{\text{high temp, pressure}} {-}CH_2{-}CH_2{-}CH_2{-}CH_2{-}CH_2{-}CH_2{-}CH_2{-}CH_2{-}$$

or $n\text{CH}_2{=}\text{CH}_2 \rightarrow {+}\text{CH}_2{-}\text{CH}_2{+}_n$

Polyethylene

This product is called polyethylene. In this reaction ethylene is called the *monomer* (single unit), and the polyethylene is called the *polymer* (many units). Since polyethylene is tough, insoluble in many common liquids, and resistant to chemical attack, it has replaced glass, steel, copper, and aluminum in many applications.

The physical properties of polyethylene and other polymers are related to the chain length or average molecular weight. Depending on the application, the chemical industry produces polyethylenes having average molecular weights between 20,000 to 100,000 g/mol. Note that we speak of the *average* molecular weight of a synthetic polymer. Unlike reactions of small molecules, polymerization reactions give products with a range of molecular weights; polymer chemists have learned how to control the average molecular weight and how to produce a polymer product in which the molecular weights of individual molecules are not very different from the average.

The formation of a polymer is initiated by a free radical or an ion that can add to a monomer and activate it. Such a species is called an *initiator*. If the concentration of initiators is high, many polymer chains will begin to grow simultaneously. If the concentration of initiators is low, only a few chains will begin to grow. All things being equal, when the initiator concentration is high, many short, or low-molecular-weight, chains will form, and when the initiator concentration is low, only a few long, or high-molecular-weight, chains will form. Hence, the major control over chain length is the initiator-to-monomer ratio in the polymerization reaction system.

EXAMPLE 15.1 Calculate the average molecular weight of polyethylene formed (*a*) when the initiator-monomer mole ratio is 5:100 and (*b*) when it is 5:750.

Solution: (*a*) The five initiator molecules will cause five polymer chains to start. On the average each chain will contain 20 monomer residues. Each residue, $—C_2H_4—$, has a weight of 28 g. Therefore, the average molecular weight is

$$20 \text{ residues/chain} \times 28 \text{ g/residue} = 560 \text{ g/chain} = \text{average mol wt}$$

(*b*) By the same logic there will be an average of 150 residues per chain and the average molecular weight will be

$$150 \text{ residues/chain} \times 28 \text{ g/residue} = 4200 \text{ g/chain} = \text{average mol wt}$$

Many other addition polymers can be considered as polyethylene derivatives, e.g., polypropylene, polyvinyl chloride, polystyrene, Orlon, Saran, and Teflon. The reactions for the formation of some of these plastics are shown, and their special properties and uses are given in Table 15.9.

TABLE 15.9: ADDITION POLYMERS

	Monomer			
Polymer	*Formula*	*Name*	*Useful properties*	*Applications*
Polyethylene	$CH_2{=}CH_2$	Ethylene or ethene	Tough, lightweight	Bottles, plumbing
Polypropylene	$CH_2{=}CH{-}CH_3$	Propylene or propene	More heat resistant than polyethylene	Laboratory utensils, (flasks, etc.)
Polyvinyl chloride (PVC)	$CH_2{=}CHCl$	Vinyl chloride or chloroethene	Tough, nonabrasive, easily dyed	Floor and wall covering, leatherette
Polystyrene	$CH_2{=}CH{-}C_6H_5$	Styrene or phenylethene	Clear, rigid	Windows
Orlon	$CH_2{=}CHCN$	Acrylonitrile or cyanoethene	Strong, easily dyed, can be blown into a fiber	Acrylon textile
Teflon	$CF_2{=}CF_2$	Tetrafluoroethene	Heat-resistant, nonsticking	Liner for cooking utensils
Saran	$CH_2{=}CCl_2$	1,1-Dichloroethene	Clear, strong; as thin film adheres to itself	Food wrapping

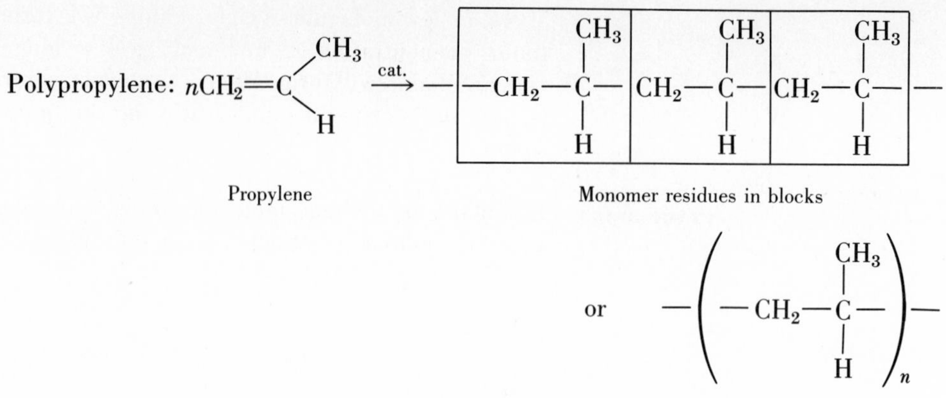

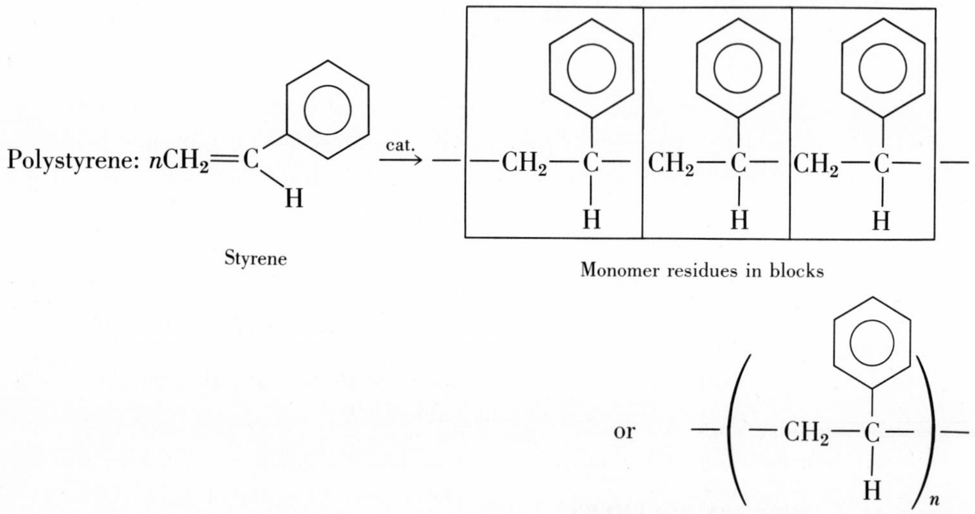

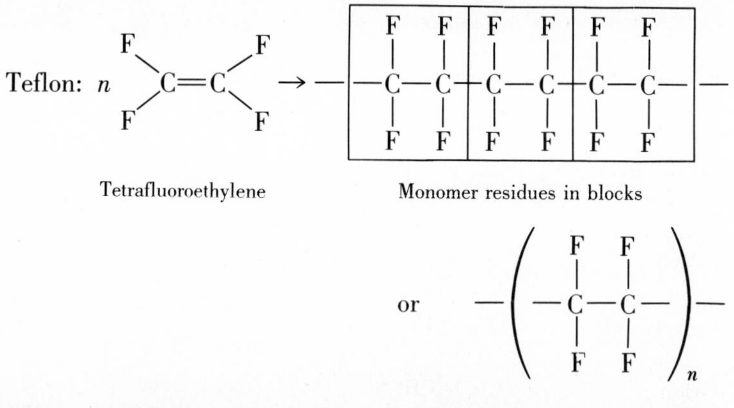

Polypropylene: $nCH_2{=}C(CH_3)H \xrightarrow{\text{cat.}} -\left|-CH_2-\underset{H}{\overset{CH_3}{C}}-\right|CH_2-\underset{H}{\overset{CH_3}{C}}-\left|CH_2-\underset{H}{\overset{CH_3}{C}}-\right|-$

Propylene

Monomer residues in blocks

or $-\left(-CH_2-\underset{H}{\overset{CH_3}{C}}-\right)_n-$

Polystyrene: $nCH_2{=}C(C_6H_5)H \xrightarrow{\text{cat.}} -\left|-CH_2-\underset{H}{\overset{C_6H_5}{C}}-\right|CH_2-\underset{H}{\overset{C_6H_5}{C}}-\left|CH_2-\underset{H}{\overset{C_6H_5}{C}}-\right|-$

Styrene

Monomer residues in blocks

or $-\left(-CH_2-\underset{H}{\overset{C_6H_5}{C}}-\right)_n-$

Teflon: $n\ F_2C{=}CF_2 \rightarrow -\left|-\underset{F}{\overset{F}{C}}-\underset{F}{\overset{F}{C}}-\right|-\underset{F}{\overset{F}{C}}-\underset{F}{\overset{F}{C}}-\left|-\underset{F}{\overset{F}{C}}-\underset{F}{\overset{F}{C}}-\right|-$

Tetrafluoroethylene

Monomer residues in blocks

or $-\left(-\underset{F}{\overset{F}{C}}-\underset{F}{\overset{F}{C}}-\right)_n-$

Butadiene addition polymers: 1,4 addition

An organic molecule with two double bonds is called a *diene*. When the double bonds are separated by one carbon-carbon single bond, the addition reactions of the unsaturated molecule are very peculiar. The simplest case to consider is 1,3-butadiene:

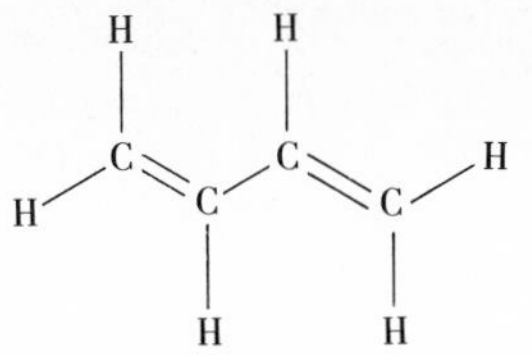

$$H_2C{=}CH{-}CH{=}CH_2 \text{ (structural formula with all H atoms shown)}$$

or

$$H_2C{=}CH{-}CH{=}CH_2$$

When we add only one molecule of a halogen to butadiene, we might expect it to add across a double bond:

$$H_2C{=}CH{-}CH{=}CH_2 + Cl_2 \rightarrow H_2ClC{-}CHCl{-}CH{=}CH_2$$

This addition reaction does occur, but it is not the main addition reaction of butadiene; a 1,4 addition predominates:

$$\overset{1}{H_2C}{=}\overset{2}{CH}{-}\overset{3}{CH}{=}\overset{4}{CH_2} + Cl_2 \rightarrow \overset{1}{H_2ClC}{-}\overset{2}{CH}{=}\overset{3}{CH}{-}\overset{4}{CH_2Cl}$$

In 1,4 addition the chlorine atoms add to the terminal (1 and 4) carbon atoms and the double bond shifts to the middle. Butadiene undergoes 1,4 addition with other halogens, hydrohalides, and hydrogen.

Like ethylene and its derivatives, 1,3-butadiene has been used to prepare polymers. In this case, however, an unsaturated polymer chain results because the addition occurs via the 1,4 route:

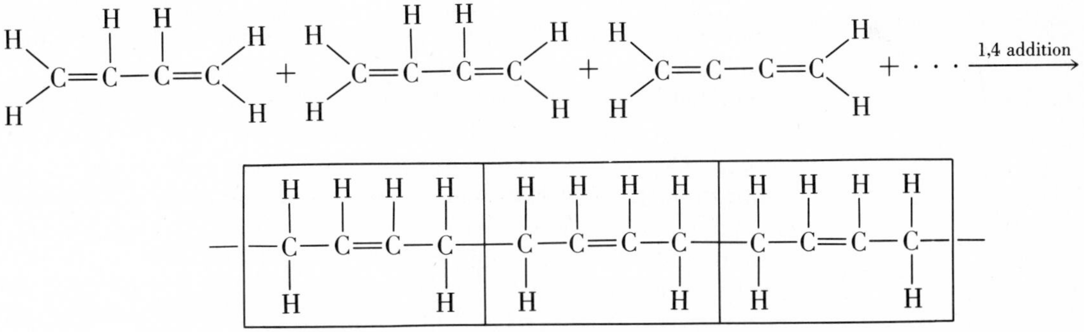

$$H_2C{=}CH{-}CH{=}CH_2 + H_2C{=}CH{-}CH{=}CH_2 + H_2C{=}CH{-}CH{=}CH_2 + \cdots \xrightarrow{\text{1,4 addition}}$$

$$-\boxed{-CH_2{-}CH{=}CH{-}CH_2-}\boxed{-CH_2{-}CH{=}CH{-}CH_2-}\boxed{-CH_2{-}CH{=}CH{-}CH_2-}-$$

Monomer residues of butadiene in rectangles

or

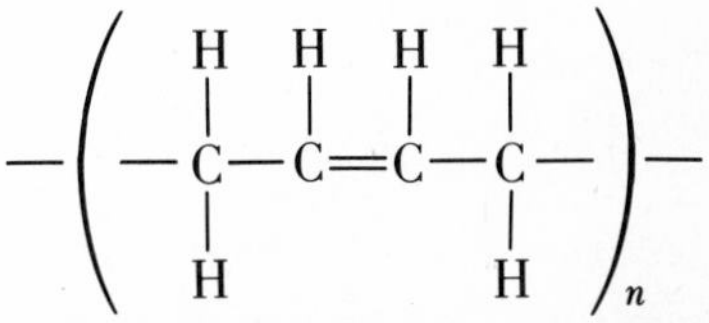

$$-\left(-CH_2{-}CH{=}CH{-}CH_2-\right)_n-$$

Polybutadiene

Polybutadiene is a rubber; it is soft, has low strength, and becomes tacky when it is warmed. It is not very useful without further treatment.

Synthetic polybutadiene is similar in physical properties and structure to natural rubber (latex). Natural rubber is a polymer of isoprene or 2-methylbutadiene, $CH_2{=}C(CH_3){-}CH{=}CH_2$. In natural latex the isoprene units are added to each other so that the methyl group and the hydrogen atom are on the same side of the double bonds in the resulting polymer:

$$H_2C{=}C(CH_3){-}CH{=}CH_2 \xrightarrow{\text{polymerization}} {-}CH_2{-}C(CH_3){=}CH{-}CH_2{-}CH_2{-}C(CH_3){=}CH{-}CH_2{-}CH_2{-}C(CH_3){=}CH{-}CH_2{-}$$

Natural rubber
(cis form)

Certain trees produce *gutta-percha*, in which the methyl groups and hydrogen atoms are on opposite sides of the double bonds:

$${-}H_2C{-}C(CH_3){=}CH{-}CH_2{-}CH_2{-}C(CH_3){=}CH{-}CH_2{-}CH_2{-}C(CH_3){=}CH{-}CH_2{-}$$

Gutta-percha
(trans form)

In comparing natural rubber and gutta-percha we see the importance of geometrical isomerism. The polymers are chemically identical, yet they have very different properties. Natural rubber has many desirable characteristics and is very valuable, whereas gutta-percha, which cannot be vulcanized, has only limited use.

Polybutadiene and natural rubber latex, poly(*cis*-2-methylbutadiene) are not useful without further treatment. Every American schoolchild knows the story of Charles Goodyear's discovery of *vulcanization*. Rubber had been known for centuries, and many attempts had been made to use it, especially for rainwear and boots. Unfortunately, natural rubber becomes brittle in winter cold and sticky in summer heat. Goodyear knew that adding sulfur to rubber reduced its stickiness. He found that heating a mixture of sulfur and rubber not only eliminated the stickiness but produced a material which was resilient and strong in all weather. He named the process *vulcanization* after Vulcan, the god of fire.

Vulcanization of rubber by sulfur and heat is believed to result from the cross-linking of the polymer strands by sulfur atoms:

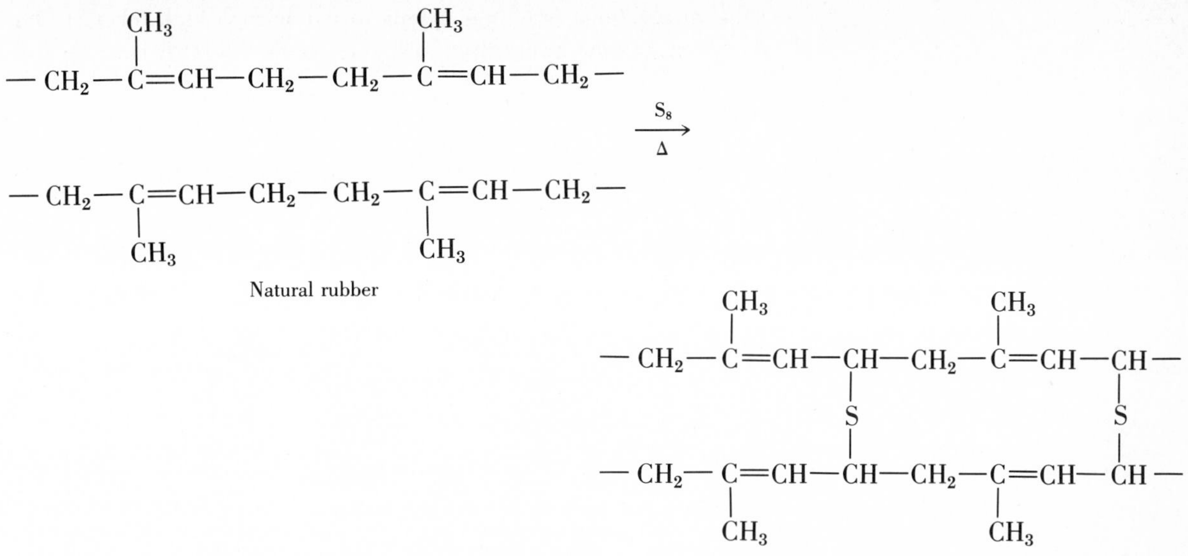

$$
\begin{array}{c}
\begin{array}{l}
-CH_2-\underset{}{\overset{CH_3}{\overset{|}{C}}}=CH-CH_2-CH_2-\overset{CH_3}{\overset{|}{C}}=CH-CH_2- \\
-CH_2-\underset{CH_3}{\underset{|}{C}}=CH-CH_2-CH_2-\underset{CH_3}{\underset{|}{C}}=CH-CH_2-
\end{array}
\xrightarrow[\Delta]{S_8}
\end{array}
$$

Natural rubber

$$
\begin{array}{l}
-CH_2-\overset{CH_3}{\overset{|}{C}}=CH-\underset{\underset{S}{|}}{CH}-CH_2-\overset{CH_3}{\overset{|}{C}}=CH-\underset{\underset{S}{|}}{CH}- \\
\qquad\qquad\qquad\quad\; | \qquad\qquad\qquad\qquad\quad\; | \\
-CH_2-\underset{CH_3}{\underset{|}{C}}=CH-CH-CH_2-\underset{CH_3}{\underset{|}{C}}=CH-CH-
\end{array}
$$

Vulcanized rubber

The hydrogen atoms on carbon atoms next to the double bonds are activated, and during vulcanization they are replaced when the sulfur bridge forms.

Nature uses only isoprene to form rubber, but chemists have learned how to polymerize other butadiene derivatives into synthetic rubbers which in some ways are superior to natural rubber. The first synthetic rubber ever produced industrially was Du Pont's *polychloroprene*, prepared from a butadiene derivative called chloroprene or 2-chlorobutadiene:

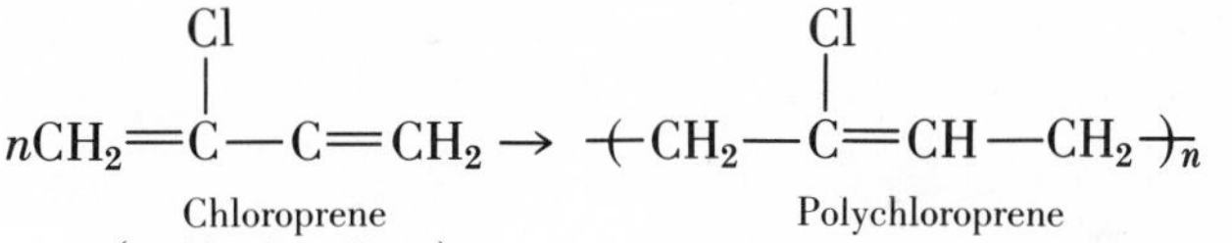

$$
nCH_2=\overset{Cl}{\overset{|}{C}}-C=CH_2 \rightarrow +\!\!\left(CH_2-\overset{Cl}{\overset{|}{C}}=CH-CH_2\right)\!\!+_n
$$

Chloroprene (2-chlorobutadiene) — Polychloroprene

Polychloroprene rubber is more expensive than natural rubber and not as strong. It is not suitable for tires or any application where strength and resistance to abrasion are critical. However, because it is much less soluble in oils and gasoline, hoses for carrying and pumping these liquids are manufactured from polychloroprene.

Condensation polymers

Examination of the equations for the addition reactions which produce the polymers of ethylene, butadiene, and their derivatives will show that the only product is the polymer. Nylon and Dacron represent different kinds of polymers. Nylon can form when a *dibasic* organic acid and a *diamine* react. The polymer is held together by amide linkages and is called a *polyamide*.

One of the most common nylons is nylon 66, which forms when 1,6-diaminohexane and adipic acid are copolymerized. Each of the monomers has 6 carbon atoms, hence, the name nylon 66:

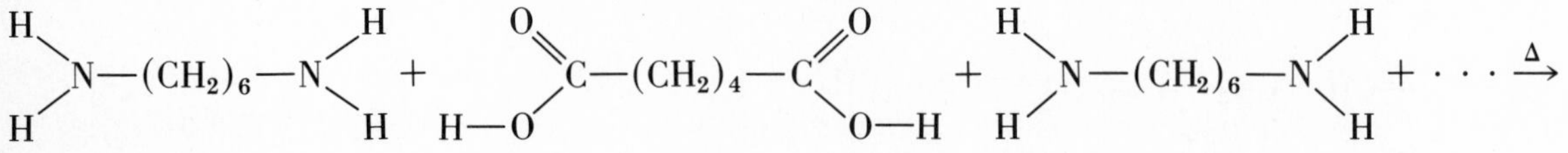

$$H_2N-(CH_2)_6-NH_2 + HO-C(=O)-(CH_2)_4-C(=O)-OH + H_2N-(CH_2)_6-NH_2 + \cdots \xrightarrow{\Delta}$$

1,6-Diammohexane | Adipic acid | 1,6-Diaminohexane

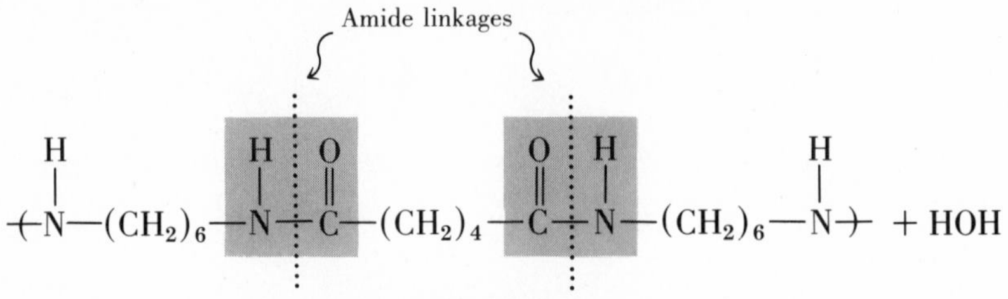

Amide linkages

$$\text{-}\!\!\left[N(H)-(CH_2)_6-N(H)-C(=O)-(CH_2)_4-C(=O)-N(H)-(CH_2)_6-N(H)\right]\!\!\text{-} + HOH$$

Nylon 66, a polyamide

Note that the synthesis of nylon 66 produces water along with the polymer. Chemists say that water is "split out" during the reaction. Reactions which involve the splitting out of water or some other small molecule are called *condensation* reactions. Therefore, nylon is a *condensation polymer.* When adipic acid is used to form nylon, the reaction is slow, and so a derivative of adipic acid, adipyl chloride, is actually used. The —OH group of the carboxylic acid function is replaced by chlorine in this compound:

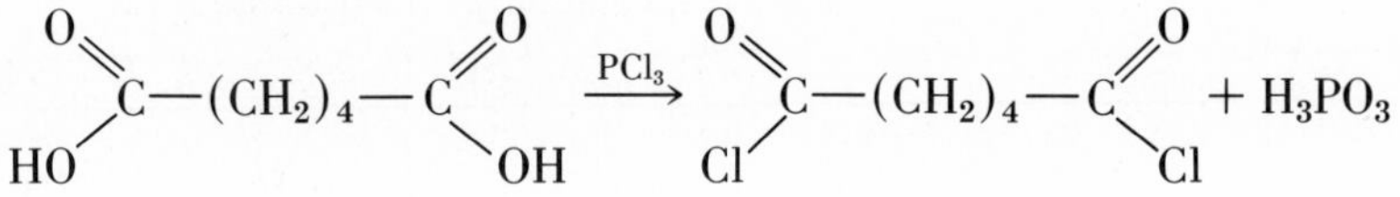

$$HO-C(=O)-(CH_2)_4-C(=O)-OH \xrightarrow{PCl_3} Cl-C(=O)-(CH_2)_4-C(=O)-Cl + H_3PO_3$$

Adipic acid | Adipyl chloride

Adipyl chloride reacts very rapidly with 1,6-diaminohexane to form nylon 66. In this case HCl is split out during the condensation.

Nylon 6, another common nylon is prepared from a single 6-carbon atom compound which contains both the amino group and the carboxylic acid function. The compound is an amino acid, named 6-aminohexanoic acid:

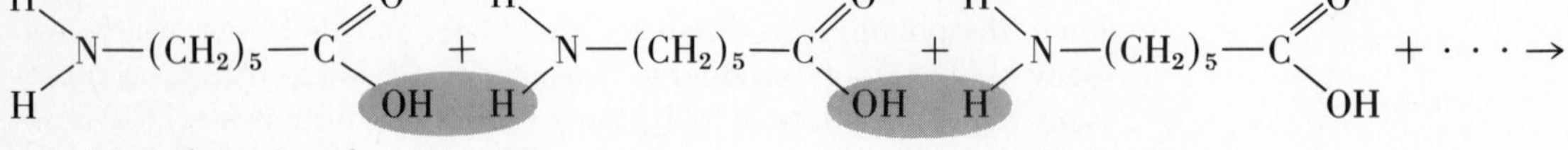

$$H_2N-(CH_2)_5-C(=O)-OH + H_2N-(CH_2)_5-C(=O)-OH + H_2N-(CH_2)_5-C(=O)-OH + \cdots \rightarrow$$

6-Aminohexanoic acid

Amide linkages

$$-(-\overset{\displaystyle H}{\overset{|}{N}}-(CH_2)_5-\overset{\displaystyle O}{\overset{\|}{C}}-\overset{\displaystyle H}{\overset{|}{N}}-(CH_2)_5-\overset{\displaystyle O}{\overset{\|}{C}}-\overset{\displaystyle H}{\overset{|}{N}}-(CH_2)_5-\overset{\displaystyle O}{\overset{\|}{C}}-)- + HOH$$

Nylon 6

The preparation of nylon 6 resembles the synthesis of proteins. The living cell uses α-amino acids to form proteins which are polyamides; 6-aminohexanoic acid is an ϵ-amino acid.

Dacron is a condensation polymer formed from the reaction of a dibasic acid, terephthalic acid, and a *diol,* i.e., a dialcohol, ethylene glycol:

$$HOCH_2CH_2OH + HO-\overset{\displaystyle O}{\overset{\|}{C}}-C_6H_4-\overset{\displaystyle O}{\overset{\|}{C}}-O-H + HOCH_2CH_2OH + HO-\overset{\displaystyle O}{\overset{\|}{C}}-C_6H_4-\overset{\displaystyle O}{\overset{\|}{C}}-OH + \cdots \longrightarrow$$

Ethylene glycol (1,2-ethanediol)

Terephthalic acid

$$+O-CH_2-CH_2-\underbrace{O-\overset{\displaystyle O}{\overset{\|}{C}}}-C_6H_4-\underbrace{\overset{\displaystyle O}{\overset{\|}{C}}-O}-CH_2-CH_2-\underbrace{O-\overset{\displaystyle O}{\overset{\|}{C}}}-C_6H_4-\overset{\displaystyle O}{\overset{\|}{C}}-O+ \; + HOH$$

Ester linkages

Dacron forms when the acid and alcohol form *ester linkages,* and Dacron is a *polyester.* Again, the reaction is much faster if the acid is converted to the corresponding chloride, $C_6H_4(COCl)_2$.

SUMMARY

A little over a century ago chemists believed that organic substances could be synthesized only by living organisms. After Wohler showed that organic substances can be prepared in test tubes, the science of organic chemistry advanced rapidly. Our style of living has been greatly influenced by the work of organic chemists, who have produced synthetic medicines, fabrics, building materials, pigments, furniture, tires, upholstery, lacquers, paints, toys, dishes, shoes, flooring, wall covering, and a host of other things. Organic chemistry is the basis of the chemical and pharmaceutical industries and a foundation stone of our economy. Support your local organic chemist.

QUESTIONS AND PROBLEMS

15.1 (*a*) Why are the alkanes called paraffins? (*b*) Why are they said to be saturated?

15.2 Write structures for all the (*a*) pentane and (*b*) heptane isomers. (*c*) Give the systematic name for each structure.

15.3 How does the valence theory explain the basic tetrahedral structure of the alkanes?

15.4 Draw structures for the alkanes with the formulas C_6H_{12} and C_8H_{16}.

15.5 Are straight-chain hydrocarbons actually straight chains of carbon atoms? What does this term mean?

15.6 Name the following compounds:
(*a*) $CH_3—CH(C_2H_5)—CH_2—CH(CH_3)_2$, (*b*) $CH_3—(CH_2)_6—CH_3$,
(*c*) $CH_3(CH_2)_4CH(CH(CH_3)_2)CH_3$, (*d*) $CH_3—C—(CH_3)_3$.

15.7 How does the valence-bond theory explain the bonding in ethylene and other alkene molecules? Use sketches.

15.8 Draw structures for all the alkene isomers of C_5H_{10} and name them.

15.9 Name the following structures: (*a*) $CH_2=CH—CH_2—CH(CH_3)_2$,
(*b*) $CH_2=CH—CH(CH_3)—CH(C_2H_5)—(CH_2)_2CH_3$,
(*c*) $CH_3—CH=CH—CH=CH_2$, (*d*) $CH_3—CHCl—CH=CH_2$.

15.10 Explain the structure of acetylene and the alkynes in terms of valence-bond theory. Use sketches.

15.11 Name the following structures: (*a*) $CH_3—C\equiv C—CH_3$,
(*b*) $HC\equiv C—(CH_2)_3—CH_3$,
(*c*) $CH_3—CH_2—C(CH_3)(C_2H_5)—C\equiv C—CH—(CH_3)_2$.

15.12 Draw all the possible hexyne isomers and name them.

15.13 What is the basic difference between the reactions of the alkanes and the unsaturated hydrocarbons? Demonstrate the difference by example.

15.14 (*a*) How would you prepare monochloroethane and hexachloroethane from ethane? (*b*) How would you prepare the same compounds from ethylene? Show equations for all chemical changes.

15.15 An organic liquid is either *n*-octane or 2-hexene. What tests could you perform to decide which it is? Give equations for any chemical tests.

15.16 A gaseous hydrocarbon is found to be 82.7 percent carbon by weight. The density of the gas at 25°C and 755 Torr is 2.36 g/l. Is the compound an alkane or an unsaturated hydrocarbon?

15.17 What is the importance of cracking in the petroleum industry?

15.18 Give the products and balance the equations for the following reactions:

(*a*) $CH_3—CH{=}CH_2 + 1\ \text{mol}\ Cl_2 \rightarrow$ (*b*) $CH_3—CH{=}CH_2 + \text{excess}\ Cl_2 \rightarrow$
(*c*) $CH_3—CH{=}CH_2 + O_2 \rightarrow$ (*d*) $CH_3—CH{=}CH_2 + H_2 \xrightarrow{\text{cat.}}$
(*e*) $CH_3—CH{=}CH_2 + HCl \rightarrow$ two products†
(*f*) $H—C{\equiv}C—H + 1\ \text{mol}\ Br_2 \rightarrow$ (*g*) $H—C{\equiv}C—H + 2\ \text{mol}\ Br_2 \rightarrow$
(*h*) $H—C{\equiv}C—CH_3 + \text{excess}\ Br_2 \rightarrow$ (*i*) $H—C{\equiv}C—CH_3 + 2HBr \rightarrow$

15.19 What are the important commercial applications of hydrogenation?

15.20 (*a*) How would you prepare vinyl chloride from acetylene? (*b*) What is the systematic name for vinyl chloride?

15.21 (*a*) Why do olefinic compounds exhibit geometric isomerism? (*b*) Do geometric isomers have different chemical and physical properties? (*c*) Give at least two examples to support your answer to part (*b*).

15.22 Give the name and make a sketch of all the structures and the molecules consistent with the formulas listed below. Note that in several instances molecular, geometric, and/or optical isomerism is possible: (*a*) C_3H_8, (*b*) C_3H_6, (*c*) C_3H_4, (*d*) C_5H_{12}, (*e*) C_5H_{10}, (*f*) C_5H_8, (*g*) CH(OH)ClBr, (*h*) $CH_3CHOHC_2H_5$, (*i*) 2-hydroxypentane, (*j*) CH_3CHNH_2COOH (alanine), (*k*) $C_2H_2Br_2$, (*l*) $C_3H_4Cl_2$. *Note:* Consider cyclic structures.

15.23 Make a sketch of 3-chloro-2-pentene and indicate which atoms of the molecule, if any, are coplanar. Does this structure exhibit cis-trans isomerism?

15.24 Why was benzene so puzzling to organic chemists of the nineteenth century?

15.25 (*a*) Sketch the Lewis structure and the valence-bond structure for benzene. (*b*) Why is the molecular-orbital theory or valence-bond theory superior to the Lewis theory in explaining the structure of benzene?

15.26 Name the following compounds:

(*a*) CH_3 (*b*) CH_3 (*c*) CH_3, Cl (*d*)  Cl, NO_2

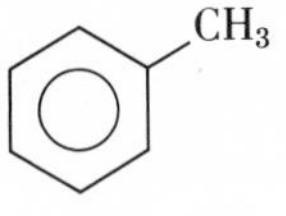

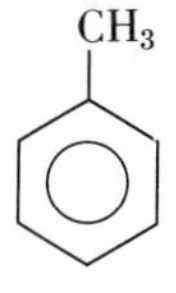

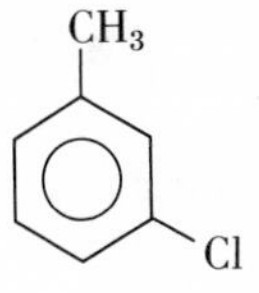

(*e*) 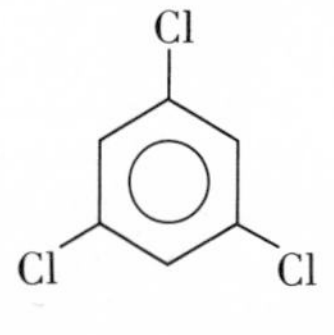 Cl, Cl, Cl (*f*) NH_2, H_2N, NH_2 (*g*) NO_2, OH (*h*) NO_2, OH

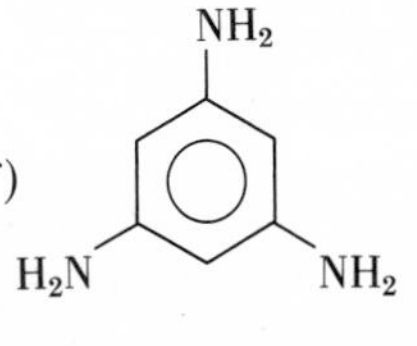

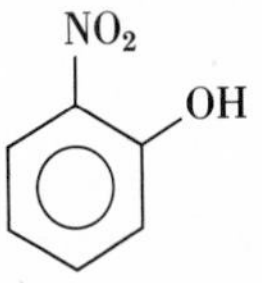

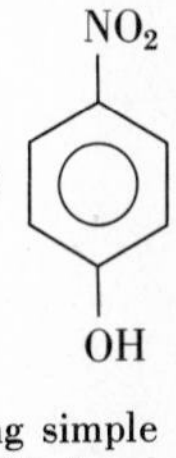

† Actually, according to Markownikoff's rule, in addition reactions involving simple gases like HCl, HBr, etc., the hydrogen adds to the carbon atom of the double bond which is already bonded to more hydrogen, for example, $CHCl{=}CH_2 + HCl \rightarrow CHCl_2—CH_3 + \text{trace}\ CH_2Cl—CH_2Cl$.

15.27 Phenol, C_6H_5OH, exhibits intermolecular hydrogen bonding. (*a*) Make a sketch to explain this. (*b*) Structure (*h*) in Prob. 15.26 exhibits intermolecular hydrogen bonding but (*g*) does not. Explain.

15.28 There are three isomers of dichlorobenzene. Upon chlorination of dichlorobenzene, trichlorobenzene is produced:

$$C_6H_4Cl_2 + Cl_2 \rightarrow C_6H_3Cl_3 + HCl$$

What is the structure of the dichlorobenzene isomer that will yield two different trichlorobenzene isomers?

15.29 From the following information deduce the exact structure of the underlined compounds:

(*a*) $\underline{C_6H_4(NO_2)(NH_2)} + Cl_2 \rightarrow$ two isomers of $C_6H_3(NO_2)(NH_2)Cl + HCl$
(*b*) $\underline{C_6H_4(NO_2)_2 + Cl_2} \rightarrow$ single product, with formula $C_6H_3(NO_2)(NH_2)Cl + HCl$
(*c*) $\underline{C_6H_4(CH_3)_2 + Br_2} \rightarrow$ three isomers of $C_6H_3(CH_3)_2Br + HBr$

15.30 Name the following compounds: (*a*) $CH_3—CHOH—CH_3$, (*b*) $CH_3—CHCH_3—CHOH—C_2H_5$, (*c*) $CH_2OH—CHC_2H_5—CHOH—C_2H_5$.

15.31 Give the structures of all of the alcohols having the formula $C_4H_{10}O$.

15.32 Give the structures of the following compounds: (*a*) 1-butanol, (*b*) 1-hydroxybutane, (*c*) 4-hydroxybutene-1, (*d*) 3-hexanol, (*e*) 2,4-dihydroxyheptane, (*f*) heptan-2,4-diol.

15.33 Show how the alcohols and phenols are similar in structure to water and how they can participate in hydrogen bonding.

15.34 Complete the following equations; name the organic products.

(*a*) Acetic acid + isopropyl alcohol $\rightarrow$ (*b*) Acetic acid + isopropylamine $\rightarrow$
(*c*) Hexanoic acid + 1-butanol $\rightarrow$ (*d*) Propanoic acid + dimethylamine $\rightarrow$
(*e*) Benzoic acid + ethylamine $\rightarrow$ (*g*) 2-Bromohexane $\xrightarrow[(2)\ NaOH]{(1)\ NH_3}$
(*f*) Propanoic acid + dimethylamine $\xrightarrow{\Delta}$
(*i*) Benzoic acid + NaOH $\rightarrow$ (*h*) 3-Methylaniline + HCl $\rightarrow$

15.35 Show the pyramidal structures of ammonia and trimethylamine.

15.36 Name the following compounds: (*a*) HCOOH,
(*b*) $CH_3(CH_2)_5COOH$,
(*c*) $CH_3(CH_2)_3CHCH_3COOH$, (*d*) Cl—C_6H_4—COOH
(*e*) $CH_3CH_2CHClCH_2COOH$.

15.37 Complete the following equations and name the organic products:

(*a*) Benzoic acid + ethyl alcohol $\xrightarrow{HCl}$ (*b*) Butanoic acid + methanol $\xrightarrow{HCl}$
(*c*) Propanoic acid + ammonia $\rightarrow$ (*d*) Acetic acid + *n*-hexylamine $\rightarrow$
(*e*) Ethanoic acid + 1-aminohexane $\xrightarrow{\Delta}$ (*f*) Hexanoic acid + NaOH $\rightarrow$

15.38 Name the following compounds: (*a*) $CH_3(CH_2)_2\overset{\overset{O}{\|}}{C}—OCH(CH_3)_2$, (*b*) $CH_3CH_2\overset{\overset{O}{\|}}{C}—NH_2$, (*c*) $CH_3\overset{\overset{O}{\|}}{C}—O(CH_2)_3CH_3$, (*d*) $CH_3\overset{\overset{H}{|}}{N}—\overset{\overset{O}{\|}}{C}—CH_3$,

(*e*) $(CH_3)_2N—\overset{\overset{O}{\|}}{C}—H$, (*f*) $CH_3(CH_2)_3—O—\overset{\overset{O}{\|}}{C}—(CH_2)_4CH_3$.

15.39 Write an equation for the preparation of all the compounds given in Prob. 15.38.

15.40 (*a*) Write the equation for the reaction of two molecules of the amino acid alanine, $H_2N—CH(CH_3)—COOH$, to yield a dipeptide, (*b*) Show that it is possible to obtain two different dipeptides from the reaction of glycine, $H_2N—CH_2—COOH$, and alanine.

15.41 Show the molecular structures for the following compounds: (*a*) dimethyl ether, (*b*) methylethyl ether, (*c*) diisopropyl ether, (*d*) methyl-*n*-propyl disulfide, (*e*) butanol, (*f*) methyl ethyl ketone. (See Tables 15.7 and 15.8.)

15.42 Distinguish between and give an example of an addition polymer and a condensation polymer.

15.43 Give the equation for the polymerization of (*a*) $CH_2{=}CHCN$, (*b*) $CH_2{=}CCl_2$, and (*c*) $CH_2{=}CHCH_3$; (*d*) give the common names of the polymers resulting from these polymerizations.

15.44 How may the average length of a polymer chain be controlled?

15.45 (*a*) Give an equation for the formation of nylon 44 and of nylon 4. (*b*) In what way is nylon similar to protein? Is nylon an addition polymer?

15.46 (*a*) Give the equation for the formation of a commercial polyester. Is a polyester a condensation polymer?

15.47 Show the equations for the polymerization of 1,3-butadiene, isoprene, chloroprene, and 2-bromobutadiene-1,3.

15.48 What is the chemical role of sulfur in the vulcanization of latex?

15.49 Why is adipyl chloride used instead of adipic acid in the manufacture of nylon and Dacron?

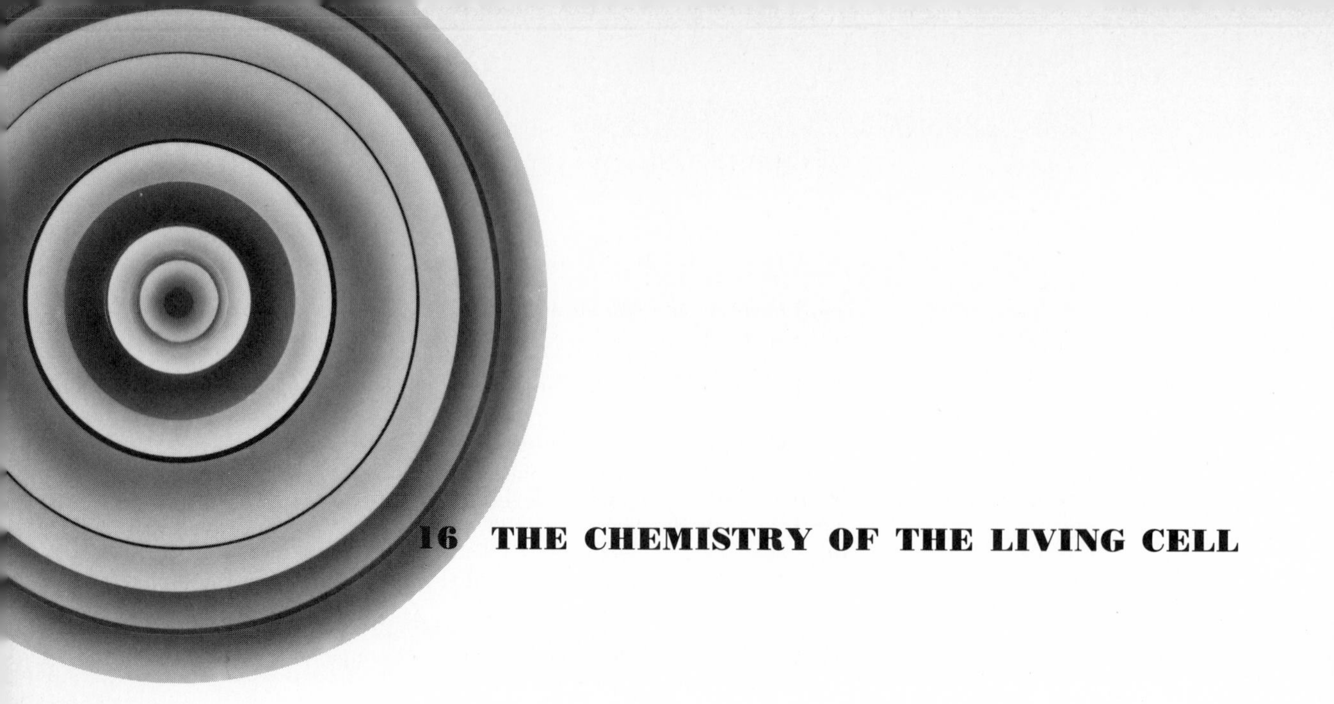

16 THE CHEMISTRY OF THE LIVING CELL

16.1 INTRODUCTION

All living things, including microscopic bacteria and algae, are composed of units, or building blocks, called *cells*. The cellular structure of any organism, plant or animal, is easily discernible even under a low-power microscope. The size varies considerably, but, generally speaking, a living cell is about 0.001 cm in breadth. Biologists have shown that there is a basic similarity between these cells regardless of their source. The single cell of a bacterium has points of similarity with cell from the tissue of a mammal; the cell from the liver of an animal has the same general features as a cell from its heart. There are similarities in the appearance, chemical constituents, and activities of all cells. The study of the chemistry, structure, and reactions of the molecules peculiar to the living cell, called *molecular biology*, is the topic of this chapter.

16.2 THE GEOGRAPHY OF THE CELL

Just as a historian uses a map of a region to show where his story is set, the biochemist uses a "map" of a cell to show where the story of molecular biology is set. Structures within the cell which have special roles are called *organelles*. There are many different kinds, as can be seen from the drawings of generalized animal and plant cells in Fig. 16.1*a* and *b*. Each organelle has an important, specific function in the cell, but we shall be concerned only with the *nucleus*, *mitochondria*,† *chloroplasts*, and *ribosomes* (Fig. 16.1*a* and *b*).

The cell can be compared to a completely automated factory under computerlike control at the nucleus. The nucleus, or command station, contains the information for the development and operation of the cell. This information is stored chemically in long molecular strands called deoxyribonucleic acid (DNA).

To carry out the directions of the nucleus the cell requires energy, which it derives in the same way that a gasoline engine derives its energy: oxidation of a suitable fuel. A gasoline engine causes gasoline to be oxidized internally by very rapid combustion with oxygen. The energy which is generated must be used immediately to drive the pistons down and turn the engine. It is impossible to use more than a small percentage of the total energy under these conditions.

The oxidation of fuel in a cell occurs within the mitochondria, which are more efficient than gasoline engines because mitochondrial oxidations are carried out in small, controlled steps. The chemical potential energy derived from the fuel is trapped in the so-called higher-energy adenosine triphosphate (ATP) molecules. These molecules are relatively small and can diffuse rapidly throughout the cell, delivering energy to sites where it is required for cellular processes. The structure and reactions of ATP will be discussed in more detail in Sec. 16.8.

Chloroplasts are organelles which occur in green plant cells. Their green color is due to the presence of the *chlorophyll* molecule. Chlorophyll and other pigments in the chloroplasts absorb light energy from the sun and use it to produce ATP, glucose, and other carbohydrates along with molecular oxygen.

The nucleus, mitochondria, and chloroplasts are visible under a good light microscope. The ribosomes, which are only ~230 Å in diameter, are so small that they were discovered only after the electron microscope was developed. They appear as tiny spheres in Fig. 16.1*a* and *b*. The nucleus sends directions for protein synthesis to the ribosomes in the form of a special molecule, called messenger RNA (mRNA). The mRNA then causes a specific protein molecule

† Mitochondrion is singular; mitochondria is plural.

FIGURE 16.1

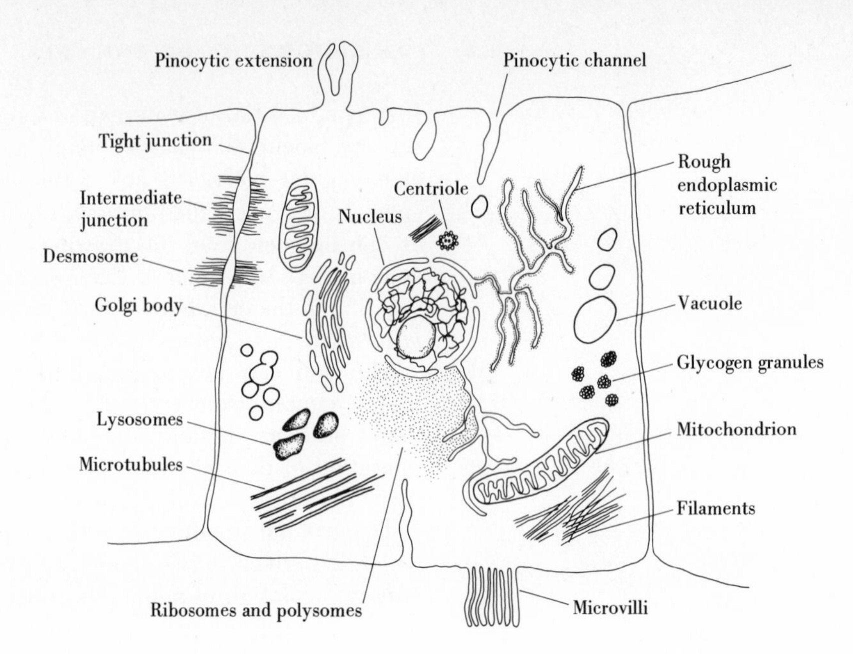

(*a*) *Generalized animal cell*

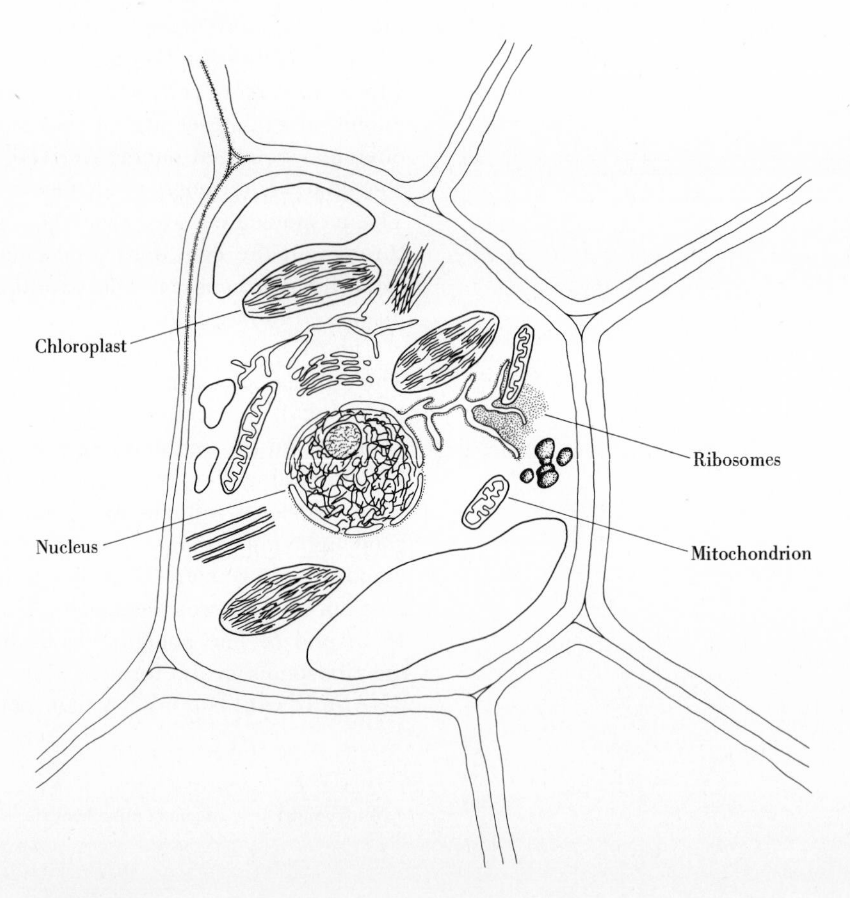

(*b*) *Plant cell*

to be synthesized from the pool of amino acids present in the cell cytoplasm.

We have now discussed the general functions of the most prominent organelles. In the rest of the chapter we shall give some details of the structure and reactions of the molecules involved.

16.3 THE STRUCTURE OF DEOXYRIBONUCLEIC ACID (DNA)

The information required for the development of organisms is stored within the *chromosomes* of the nucleus. The chromosomes are visible under an ordinary light microscope. Each chromosome contains a large number of *genes*, or *hereditary units*, which determine the characteristics of the organism. The number of chromosomes in a nucleus varies with the species: man has 23; the fruit fly, 4; corn, 10; and the mosquito, 3.

When the cell reaches maturity, the nucleus divides and two new cells form. Each daughter nucleus receives the same number of chromosomes the parent cell had. Chromosomes are duplicated before cell division occurs, so that each new cell has a full complement of chromosomes for proper functioning.

On a molecular level each chromosome corresponds to a single giant molecule of *d*eoxyribo*n*ucleic *a*cid (DNA). The weights of these giant molecules are of the order of 1 *billion* or more.

Under a microscope we can see the duplicated chromosomes divide equally, as a cell divides. If chromosomes are giant DNA molecules, the DNA molecule must also be capable of duplicating itself and dividing.

What is the structure of a molecule which can duplicate itself, govern heredity, and contain enough information to regulate all cellular activities? The answer to this question was supplied in 1953 by James D. Watson and Francis H. C. Crick (Nobel laureates, 1962). The DNA molecule is a double helix of two giant molecular strands. Each strand is a chain in which sugar molecules (deoxyribose) and phosphate radicals form alternating links:

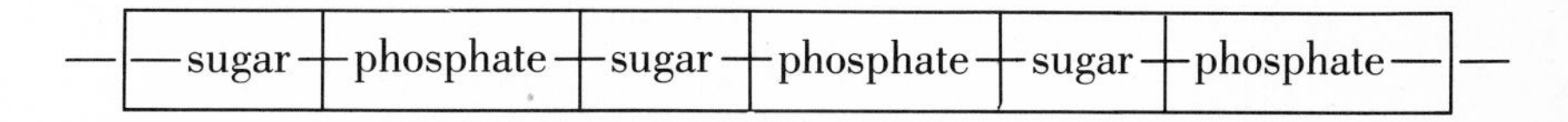

The chain can be considered to be a magnetic tape upon which is recorded the information for the development of the cell and the organism. The information is chemically stored by molecules called nitrogen bases which are bonded to the sugar residues of the sugar-phosphate chain. The nitrogen bases form an alphabet of a molecular language, and the order in which they appear on the chain makes up a molecular message. There are four nitrogen bases; two are purines, which are bicyclic molecules:

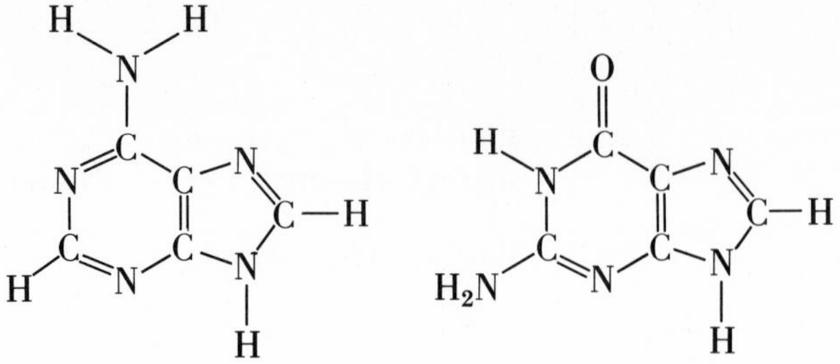

Adenine (A)

Guanine (G)

Purines

and two are pyrimidines, which are monocyclic:

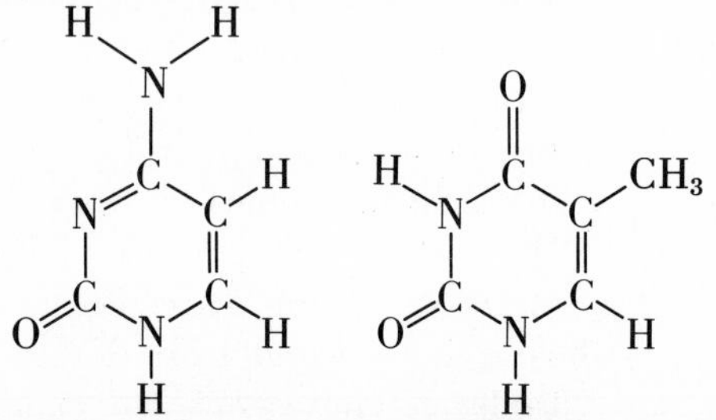

Cytosine (C)

Thymine (T)

Pyrimidines

Using the symbols A, G, C, and T for the nitrogen bases, we can indicate their position on the sugar-phosphate chain and complete the representation of a portion of a single strand of DNA:

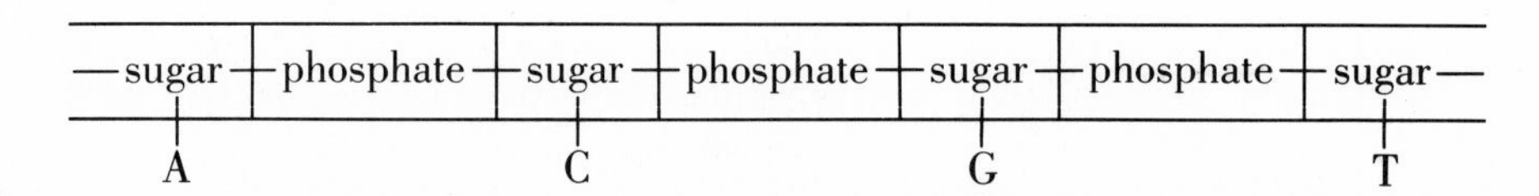

The full chemical details of this portion of the single-strand DNA are given in Fig. 16.2. Note that the sugar-phosphate backbone involves bonds between phosphate and the 3 and 5 positions of the cyclic sugar structure, deoxyribose. The nitrogen base is bonded to carbon number 1 of deoxyribose; the bond is always to a nitrogen atom of the base.

The DNA double strand forms when the bases on two adjacent single strands form hydrogen bonds. The bases must be arranged in a particular way, as Watson and Crick learned when they began

FIGURE 16.2

Structure of DNA. (From R. B. Brown, "General Biology." Copyright 1970 by McGraw-Hill, Inc., Used by permission.)

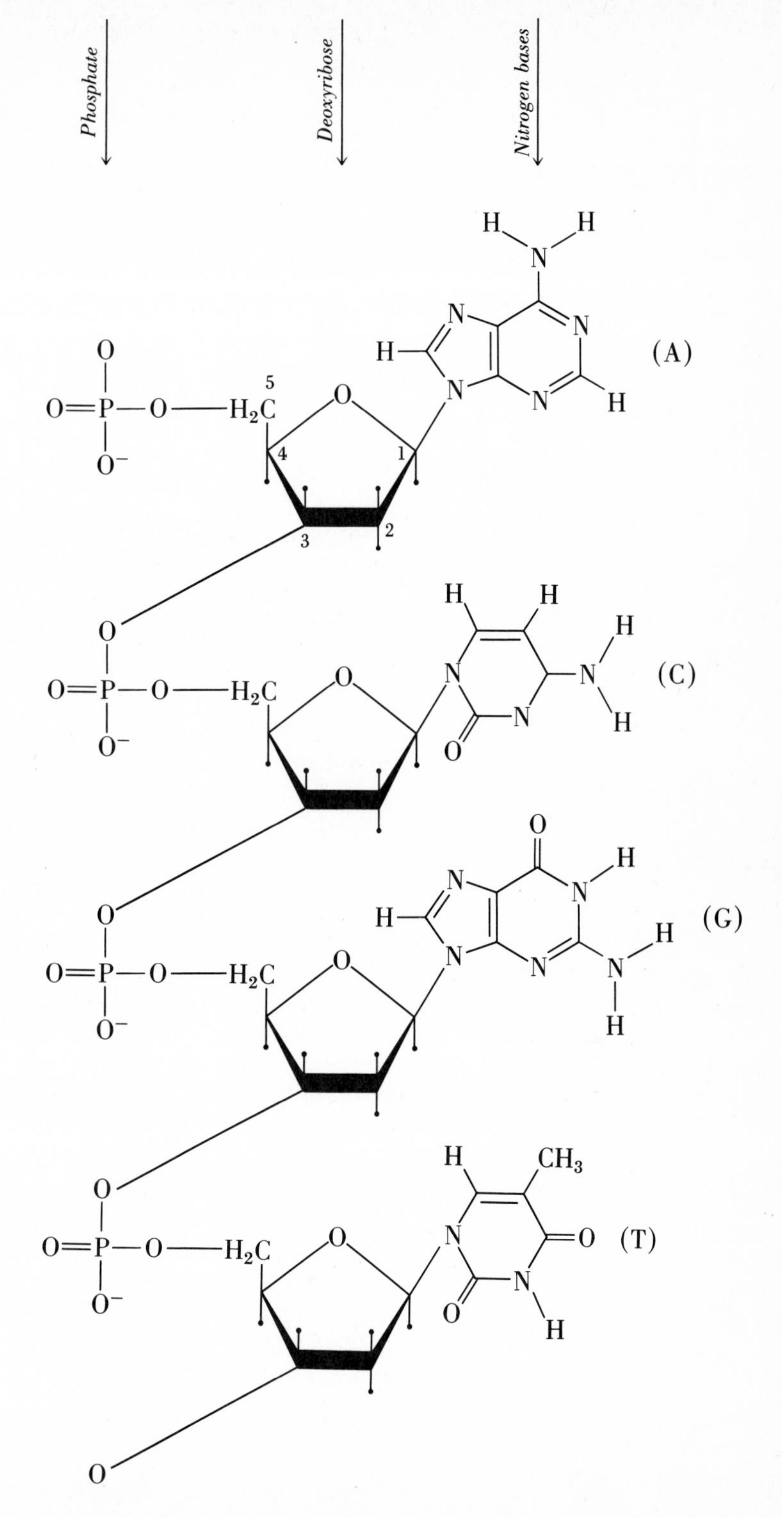

making three-dimensional atomic models for DNA. In the words of Crick, "we found that we could not arrange the bases any way we pleased; the four bases would fit into the structure only in certain pairs. In any pair there must always be one big one (purine) and one little one (pyrimidine)." More specifically Watson and Crick concluded that cytosine, C, forms hydrogen bonds with guanine, G, of the complementary strand, and adenine, A, always forms hydrogen bonds with thymine, T:

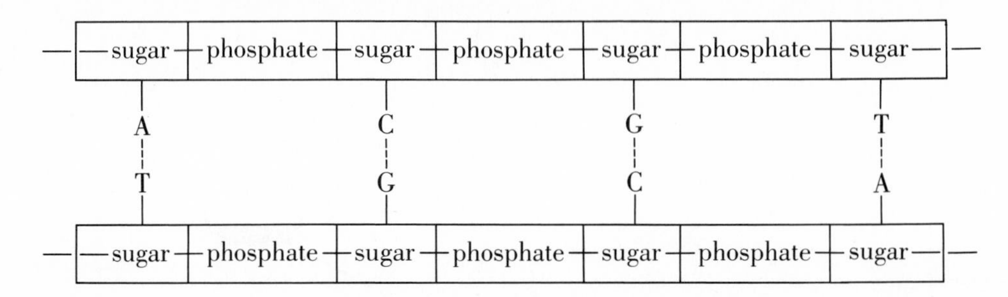

Since it is understood that the backbone is a chain of alternating sugar and phosphate residues, the DNA single strands above can be represented as A C G T and T G C A. The double strand would be

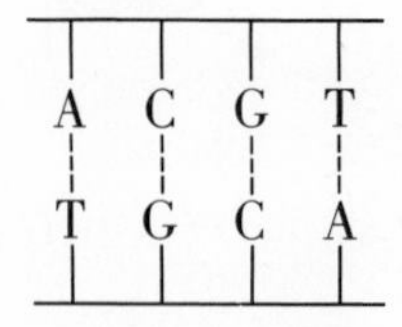

Since adenine, A, and thymine, T, form a hydrogen-bonded pair, they are said to be *complementary-base* pairs. Cytosine, C, and guanine, G, are also complementary-base pairs. Figure 16.3*a* and *b* gives fuller details of the hydrogen bonding in a portion of a DNA double strand.

Watson and Crick proposed that each strand of the DNA structure is in the form of a helix and referred to the DNA molecule itself as a *double helix*. The helix undergoes one complete clockwise twist with every 10 base pairs, which amounts to 10.6° of twist per angstrom along the helix. The helical configuration of the Watson-Crick model explained the regularity of the x-ray diffraction pattern, which indicated a 34-Å repeat distance in the crystal; this corresponds to 10.6° of twist per angstrom along the helix.

The pairing of guanine to cytosine and of adenine to thymine explained data on the base ratios obtained by Erwin Chargaff of

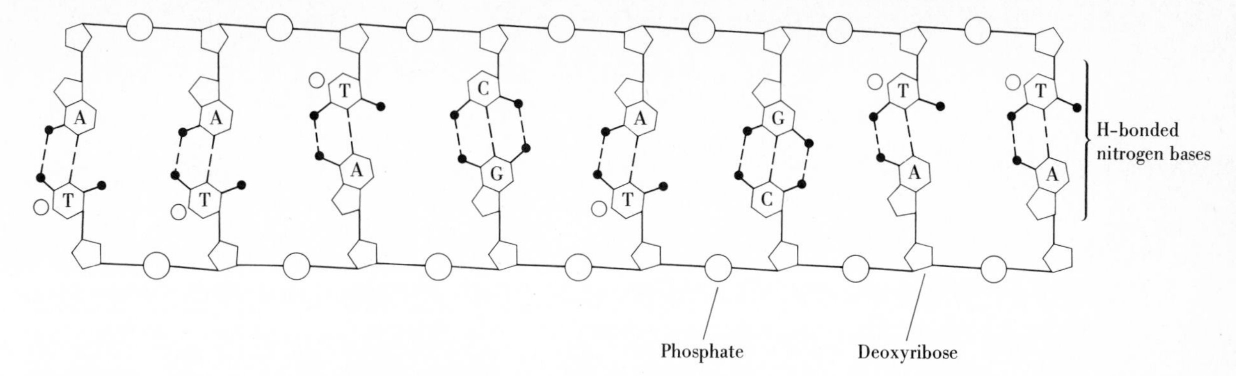

(*a*) *The DNA double strand forms by hydrogen bonding*

Adenine (A) | Thymine (T) | Guanine (G) | Cytosine (C)

Deoxyribose

(*b*) *The details of the hydrogen bonding between complementary pairs of nitrogen bases*

FIGURE 16.3

TABLE 16.1: NITROGEN-BASE CONTENT OF DNA FROM DIFFERENT ORGANISMS†

Species	*Tissue source*	A	G	C	T	$\frac{A}{T}$	$\frac{G}{C}$
Calf	Thymus	29.0	21.2	21.2	28.5	1.01	1.00
Crab	All tissue	47.3	2.7	2.7	47.3	1.00	1.00
Algae (*Euglena*)	Chloroplast	38.2	12.3	11.3	38.1	1.00	1.09
Virus (coliphage X174)	Replicative form	26.3	22.3	22.3	26.4	1.00	1.00

† Data are given in mole percent.

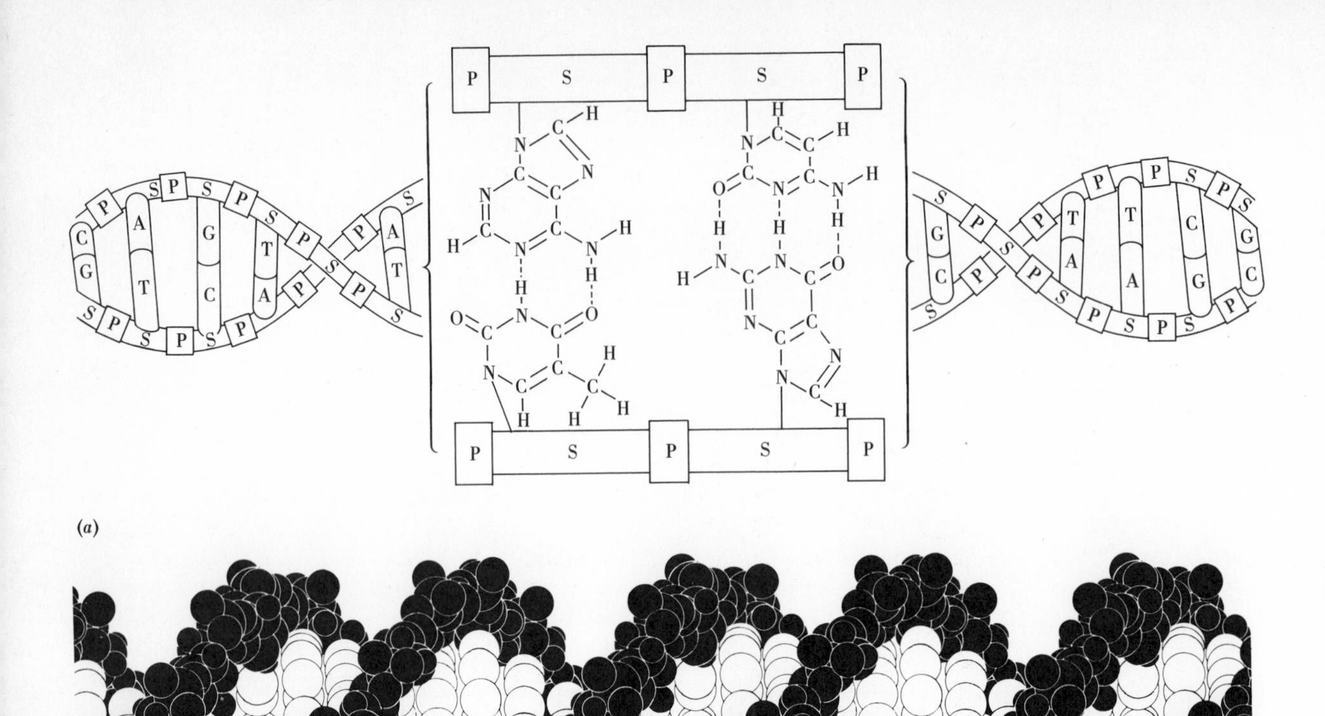

(*a*)

(*b*) Gene (DNA)

FIGURE 16.4

Two representations of the DNA double helix. [Upper representation from Chemistry, G. R. Choppin and B. Jaffe; Courtesy, General Learning Corporation (1970). Lower representation from Charles Yanofsky, Gene Structure and Protein Structure. Copyright © 1967 by Scientific American, Inc. All rights reserved.]

Columbia University. When he disassembled the DNA molecule from different organisms (calf, crab, algae, and virus) and determined the nitrogen-base content of the residue, he found that the number of moles of guanine and cytosine are always equal (G/C = 1) and that the number of moles of adenine and thymine are always equal (A/T = 1). Table 16.1 summarizes the important features of Chargaff's work.

The Watson-Crick model has G-C and A-T as complementary-base pairs and is therefore consistent with Chargaff's results. Representations of the DNA double helix are shown in Fig. 16.4.

16.4 REPLICATION OF DNA

With the double-helix model Watson and Crick were able to explain how the nucleus maintains its control of cell growth and division. During ordinary cell division, called *mitosis*, two new cells result from a single parent, and each daughter cell has the same number of chromosomes as the parent. Hence, if DNA is the molecular stuff of chromosomes, it must be able to reproduce itself accurately. Watson and Crick hypothesized that the DNA double helix can unwind and separate into single strands. As the unwinding occurs, the single strands act as *templates* for the synthesis of new complementary strands (Fig. 16.5). As a base becomes exposed when the double strand unwinds and separates, a new complementary base is hydrogen-bonded to it. When the parent DNA double helix has completed its unwinding, two new DNA double-stranded molecules have been formed. One strand in each new double strand is inherited from the parent DNA, and one strand is new. The two new DNA doubly stranded molecules are identical to each other and to their parent. This ensures that the hereditary information of the chromosomes will be accurately transmitted to new cells during cell division. The process by which new DNA is formed is called *replication*.

FIGURE 16.5

Replication of DNA.

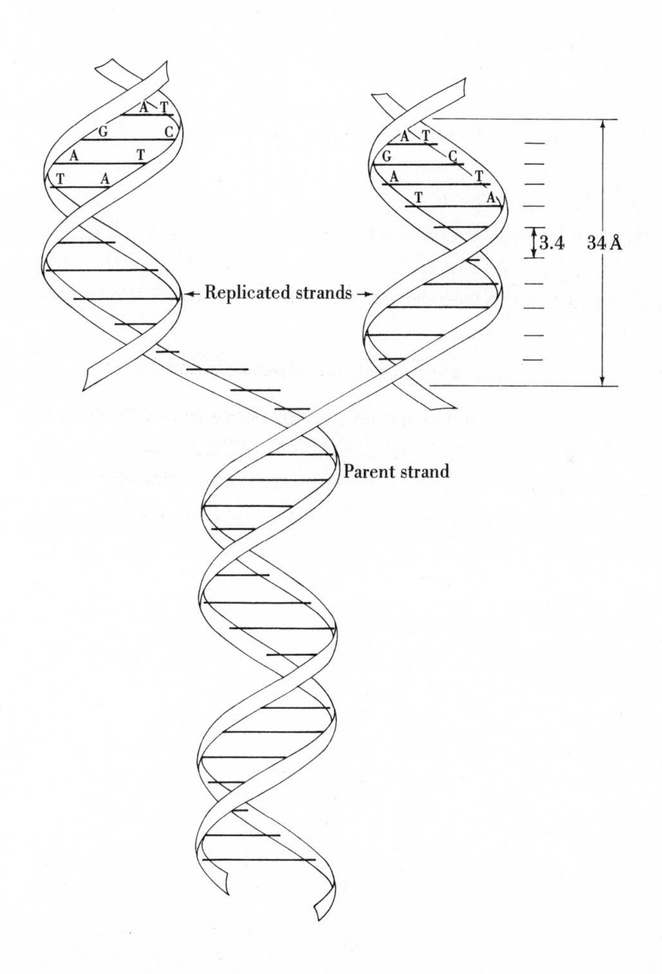

TABLE 16.2: α-AMINO ACIDS

General formula $H_2N-\underset{H}{\overset{R}{C}}-C(=O)-O-H$

Name	*Symbol*	R *group*
Glycine	Gly	$-H$
Alanine	Ala	$-CH_3$
Valine	Val	$-CH(CH_3)_2$
Leucine	Leu	$-CH_2-CH(CH_3)_2$
Isoleucine	Ile	$-CH(CH_3)-CH_2-CH_3$
Proline	Pro	H_2C-CH_2 / H_2C $CHCOOH$ † (ring closed through N–H)
Phenylalanine	Phe	$-CH_2-C_6H_5$ (benzene ring)
Tyrosine	Tyr	$-CH_2-C_6H_4-OH$ (benzene ring)
Cysteine	CysH	$-CH_2-SH$
Cystine	Cys—S—S—Cys	$HOOC\underset{NH_2}{C}HCH_2SSCH_2\underset{NH_2}{C}HCOOH$ †
Serine	Ser	$-CH_2-OH$
Threonine	Thr	$-CH(CH_3)-OH$
Histidine	His	$H-C$ (ring: $N-CH$, N–H) $C-CH_2\underset{NH_2}{C}HCOOH$ †
Aspartic acid	Asp	$-CH_2-C(=O)-O-H$
Asparagine	Asp(NH₂)	$-CH_2-C(=O)-NH_2$

TABLE 16.2: α-AMINO ACIDS (*Continued*)

Name	*Symbol*	R *group*
Glutamic acid	Glu	$-CH_2-CH_2-C(=O)OH$
Glutamine	$Glu(NH_2)$	$-CH_2-CH_2-C(=O)NH_2$
Lysine	Lys	$-CH_2-CH_2-CH_2-CH_2-NH_2$
Arginine	Arg	$-CH_2-CH_2-CH_2-NH-C(=NH)NH_2$
Tryptophan	Try	benzene ring fused to ring with $-C(-CH_2-)=CH-N(H)-$
Methionine	Met	$-CH_2-CH_2-S-CH_3$

† Complete structure.

The power of the model proposed by Watson and Crick for the structure and replication of DNA was immediately recognized. The experiments of the last 15 years have verified their original hypotheses. In Sec. 16.6 we shall see how DNA controls cellular activities by sending information out into the cell for protein synthesis.

16.5 PROTEINS AND PROTEIN STRUCTURE

Proteins occupy a central position in the study of molecular biology. They are very important as structural and chemical agents in the living cell. Among the proteins are *enzymes*, which catalyze a multitude of cellular reactions; some *hormones*, which regulate metabolic processes; *antibodies*, which counteract agents dangerous to the cell; *contractile tissue*, such as *myosin* of muscles; *cell membrane*, which governs the flow of materials in and out of the cell; and *respiratory agents*, like hemoglobin. Although the functions of the various proteins are diverse, on a molecular level they are all polyamide polymers of about 20 different α-amino acids, the structure and names of which are given in Table 16.2. The general formula for the α-amino acids is $H_2N-C^*HR-COOH$. Since they all contain an asymmetric carbon atom, all the α-amino acids used by nature to form proteins are optically active except the simplest one, glycine, $H_2N—CH_2—COOH$, and it is interesting that only one of the two possible optical isomers is produced and used in protein synthesis. This one is called the left-handed or *levo* form. The right-handed or *dextro* form of the amino acids is not used for amino acid synthesis.

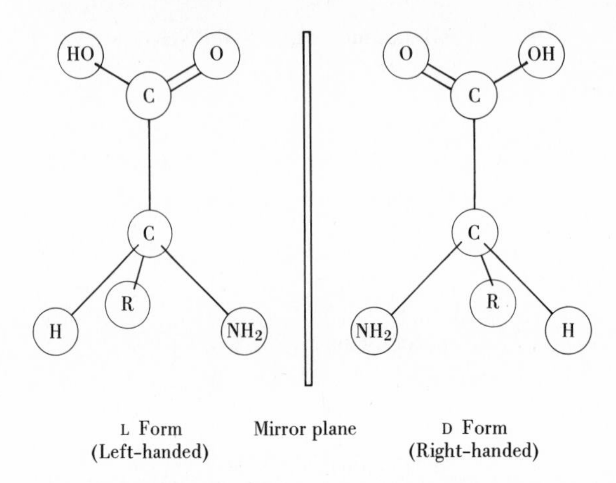

L Form (Left-handed) | Mirror plane | D Form (Right-handed)

This apparent accident of nature which causes living organisms to be "left-handed" has led to much speculation by scientists and science-fiction writers. It is obvious after a study of protein structure and enzyme reactions that there is an advantage for all organisms to use only one form of the amino acid optical isomers, just as it would be easier to manufacture certain tools if all people were either right-handed or left-handed. But the advantage of our levo world over a hypothetical dextro world is not obvious, and if extraterrestrial life is discovered, perhaps it will be dextro.

Under ordinary biological conditions, α-amino acids exist as *zwitterions;* the proton of the carboxylic acid site is donated to the amino group, and the molecule is converted into a double ion:

Amino acid ⇌ *Zwitterion*

The general structure of an α-amino acid and its zwitterion show the asymmetry of the central carbon atom.

The proteins are synthesized from the α-amino acids by a condensation reaction leading to an *amide:*

$$H_3N^+—CHR—COO^- + H_3N^+—CHR—COO^- \rightarrow$$

Peptide bond

$$H_3N^+—CHR—\overset{\overset{\displaystyle O}{\|}}{C}—\overset{\overset{\displaystyle H}{|}}{N}—CHR—COO^- + H_2O \qquad (16.1)$$

This is called a *peptide bond*. The reaction can continue, as amino acids may add to either end of the *di*peptide shown in Eq. (16.1). The proteins are formed by the condensation of hundreds of α-amino acids via peptide linkages. Proteins are high-molecular-weight polypeptides.

The amino acid sequence in a protein is referred to as its *primary* structure. It is a relatively easy task to determine the amino acid content of a protein since a protein can be hydrolyzed to yield its individual amino acids. For example, consider a small protein containing only 10 amino acid residues, a *deca*peptide; hydrolysis would yield 10 amino acids per molecule. The amino acids could be separated and identified, and this would tell us the number of each kind of amino acid present in the decapeptide:

$$H_3N^+—\underset{\underset{\displaystyle R'}{|}}{\overset{\overset{\displaystyle H}{|}}{C}}—\overset{\overset{\displaystyle O}{\|}}{C}—\left(—\overset{\overset{\displaystyle H}{|}}{N}—\underset{\underset{\displaystyle R_{int}}{|}}{\overset{\overset{\displaystyle H}{|}}{C}}—\overset{\overset{\displaystyle O}{\|}}{C}—\right)_8—\overset{\overset{\displaystyle H}{|}}{N}—\underset{\underset{\displaystyle R''}{|}}{\overset{\overset{\displaystyle H}{|}}{C}}—COO^- \xrightarrow[+HOH]{\text{mineral acid}}$$

$$\text{Gly} + 2\text{Ala} + 2\text{Leu} + 3\text{Phe} + \text{Tyr} + \text{Ser} \qquad (16.2)$$

10 amino acid molecules per
molecule of decapeptide

The decapeptide of Eq. (16.2) is given a general formula. The symbol R' represents the organic radical bonded to the amino acid residue at the *amino end* of the decapeptide; R'' is the radical bonded to the amino acid residue at the *carboxylic acid end*, and R_{int} represents the radicals bonded to the eight interior amino acid residues.

At this point of the analysis since we know the amino acid *content* of the decapeptide, we have taken a step toward the determination of the primary structure. To determine the *sequence* of the amino acids biochemists have resorted to all kinds of chemical trickery. It is possible to treat polypeptides with special reagents which react only with the amino end, H_3^+N—, or only with the carboxylic end, —COO^-, of the protein. These reagents are not removed by hydrolysis and therefore can be useful as tags for determining the amino residues at the ends of the polypeptide chain. Let Q_A symbolize a reagent which bonds to the amino end and Q_C a reagent which bonds to the carboxylic end. We would react the decapeptide with Q_A and hydrolyze the peptide. This would give us 10 amino acids, but only one will carry the Q_A tag. This will be the one at the amino end of the original decapeptide. After we perform analogous reactions with Q_C, both terminal amino acids will be known. Let us see what might happen to our hypothetical decapeptide:

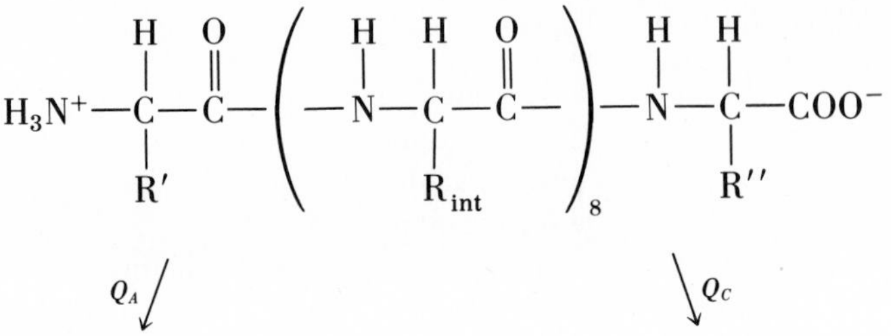

$$H_3N^+\text{—}\underset{\underset{R'}{|}}{\overset{\overset{H}{|}}{C}}\text{—}\overset{\overset{O}{\|}}{C}\text{—}\left(\text{—}\overset{\overset{H}{|}}{N}\text{—}\underset{\underset{R_{int}}{|}}{\overset{\overset{H}{|}}{C}}\text{—}\overset{\overset{O}{\|}}{C}\text{—}\right)_8\text{—}\overset{\overset{H}{|}}{N}\text{—}\underset{\underset{R''}{|}}{\overset{\overset{H}{|}}{C}}\text{—}COO^-$$

$\swarrow Q_A$ $\qquad\qquad$ $Q_C \searrow$

Q_A—decapeptide (amino end tagged) $\xrightarrow{\text{hydrolysis}}$ Gly + Ala + $\underline{\text{Ala—}Q_A}$ + 2Leu + 3Phe + Tyr + Ser

Decapeptide—Q_C (carboxylic end tagged) $\xrightarrow{\text{hydrolysis}}$ Gly + 2Ala + Leu + $\underline{\text{Leu—}Q_C}$ + 3Phe + Tyr + Ser $\qquad$ (16.3)

Equation (16.3) indicates that after treatment with Q_A followed by hydrolysis one of the alanine residues is found to be bonded to Q_A; therefore, alanine is the amino acid residue present at the amino end of the decapeptide. Likewise, treatment with Q_C followed by hydrolysis shows that leucine is present at the carboxylic end. The decapeptide structure becomes more definite:

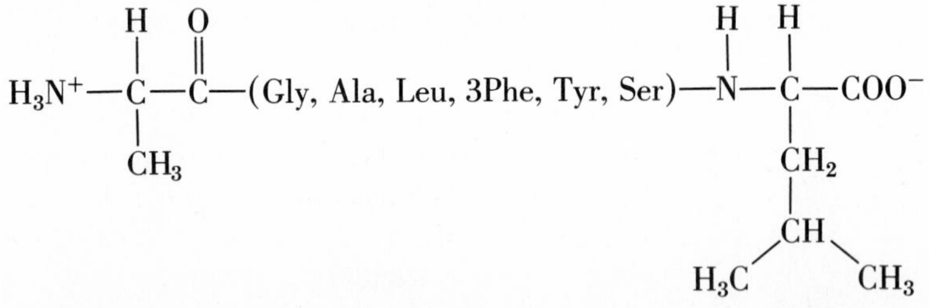

$$H_3N^+\text{—}\underset{\underset{CH_3}{|}}{\overset{\overset{H}{|}}{C}}\text{—}\overset{\overset{O}{\|}}{C}\text{—(Gly, Ala, Leu, 3Phe, Tyr, Ser)—}\overset{\overset{H}{|}}{N}\text{—}\underset{\underset{\underset{\underset{H_3C\ \ \ CH_3}{CH}}{|}}{\underset{|}{CH_2}}}{\overset{\overset{H}{|}}{C}}\text{—}COO^-$$

Alanine $\qquad\qquad$ Leucine

By convention the internal residues are written in parentheses to show that their order is unknown.

Biochemists also use enzymes to hydrolyze protein chains. In this case hydrolysis may be very specific and occur only at particular peptide linkages. For instance, an enzyme attacks the peptide linkage that forms when the amino end of serine is bonded to the carboxylic end of leucine:

$$\underset{\text{Leu-Ser dipeptide}}{H_3N^+—\underset{\underset{\underset{H_3C\quad CH_3}{CH}}{|}}{\underset{CH_2}{\overset{H}{C}}}—\overset{O}{\overset{\|}{C}}—\overset{H}{N}—\underset{\underset{OH}{|}}{\underset{CH_2}{\overset{H}{C}}}—COO^-} \xrightarrow[\text{hydrolysis}]{\text{enzyme}} \underset{\text{Leucine}}{H_3N^+—\underset{\underset{\underset{H_3C\quad CH_3}{CH}}{|}}{\underset{CH_2}{\overset{H}{C}}}—C\begin{matrix}{}^{\nearrow O}\\ {}_{\searrow O^-}\end{matrix}} + \underset{\text{Serine}}{H_3N^+—\underset{\underset{OH}{|}}{\underset{CH_2}{\overset{H}{C}}}—COO^-}$$

Treatment of our decapeptide with this enzyme would cause hydrolysis at any Leu—Ser bond that might be present. The fragments of the enzymic hydrolysis could then be completely analyzed and the structure of the decapeptide would continue to clarify. Suppose that this enzyme caused hydrolysis yielding a tetrapeptide and a hexapeptide from the original decapeptide:

Decapeptide + Leu—Ser hydrolyzing enzyme → fragments A and B

$$\underset{\text{Tetrapeptide}}{\text{Fragment A}} \xrightarrow[\text{HOH}]{\text{mineral acid}} 2\text{Ala} + \text{Gly} + \text{Leu}$$

$$\underset{\text{Hexapeptide}}{\text{Fragment B}} \xrightarrow[\text{HOH}]{\text{mineral acid}} 3\text{Phe} + \text{Ser} + \text{Tyr} + \text{Leu}$$

Alanine shows up only in fragment A, which also contains the leucine. Since we know alanine was originally present at the amino end of the decapeptide, the tetrapeptide, fragment A must be

H_3N-Ala-(Ala, Gly)-Leu-COO^-

The serine and leucine residues would be at the amino and carboxylic ends of the hexapeptide, fragment B:

$H_3{}^+N$-Ser-(3Phe, Tyr)-Lew-COO^-

In the original decapeptide fragments A and B were joined by a Leu—Ser bond, and therefore the structure of the decapeptide must be

$$\underbrace{H_3N^+—\text{Ala}—(\text{Ala, Gly})—\text{Leu}}_{\text{Fragment A}}—\underbrace{\text{Ser}—(3\text{Phe, Tyr})—\text{Leu}—COO^-}_{\text{Fragment B}}$$

With the use of other bond-specific enzymes different fragments can be generated and analyzed after complete hydrolysis and separation.

Linus Pauling, Nobel laureate 1954 and 1963, with his model of the protein alpha helix. (From The Double Helix, Atheneum Publishers, New York. Copyright © 1968 by James D Watson.)

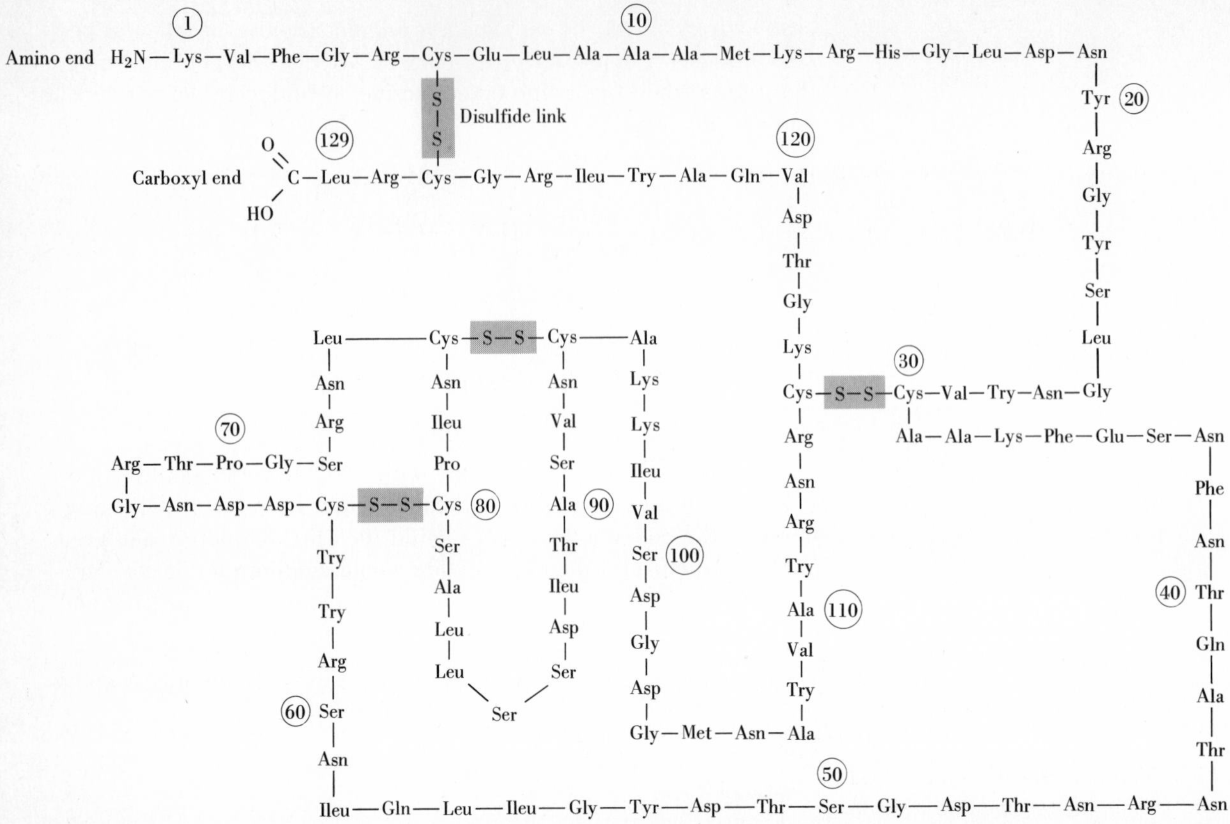

Amino end H_2N—Lys—Val—Phe—Gly—Arg—Cys—Glu—Leu—Ala—Ala—Ala—Met—Lys—Arg—His—Gly—Leu—Asp—Asn (1) (10)

Tyr (20) — Arg — Gly — Tyr — Ser — Leu — Gly

S—S Disulfide link

Carboxyl end HO—C(=O)—Leu—Arg—Cys—Gly—Arg—Ileu—Try—Ala—Gln—Val (129) (120)

Val — Asp — Thr — Gly — Lys — Cys — Arg — Asn — Arg — Try — Ala (110) — Val — Try — Ala

Cys—S—S—Cys—Val—Try—Asn—Gly (30)

Ala—Ala—Lys—Phe—Glu—Ser—Asn

Asn — Phe — Asn — Thr (40) — Gln — Ala — Thr — Asn

Leu———Cys—S—S—Cys———Ala

Leu — Asn — Arg — Ser

Arg—Thr—Pro—Gly—Ser (70)

Gly—Asn—Asp—Asp—Cys—S—S—Cys (80)

Cys — Try — Try — Arg — Ser (60) — Asn — Ileu

Cys — Asn — Ileu — Pro — Cys — Ser — Ala — Leu — Leu — Ser — Ser — Asp — Ileu — Thr — Ala (90) — Ser — Val — Asn — Cys

Ala — Lys — Lys — Ileu — Val — Ser (100) — Asp — Gly — Asp — Gly—Met—Asn—Ala

Ileu—Gln—Leu—Ileu—Gly—Tyr—Asp—Thr—Ser—Gly—Asp—Thr—Asn—Arg—Asn (50)

FIGURE 16.6

The primary structure of lysozyme. Note how the disulfide linkages cause a folding of the structure.

By piecing the results together it is possible to deduce the entire sequence of a polypeptide. Enough additional information is given in Prob. 16.10 to determine the complete amino acid sequence of the decapeptide discussed here.

One can imagine that the complete analysis of an ordinary protein which contains hundreds of amino acid residues requires years of work. Frederick Sanger (1918–) and his colleagues at Cambridge University worked 10 years to establish the primary structure for the relatively small protein insulin, which has only 51 amino acid residues. The primary structure of the bactericidal enzyme lysozyme, composed of 129 amino acid residues (Fig. 16.6) was determined by the joint efforts of Pierre Jolles, of the University of Paris, and Robert E. Canfield, of Columbia University.

The primary structure of a protein is not the only structural feature of importance when its biological function is considered. Hydrogen bonding and van der Waals forces cause the long protein molecules to twist and fold into shapes that are important to their biochemical roles. The conformation, or shape, a polypeptide takes because of hydrogen bonding is called the *secondary* structure. In 1951 Linus Pauling and Robert Corey showed, through x-ray studies, that certain proteins have a helical structure. The polypep-

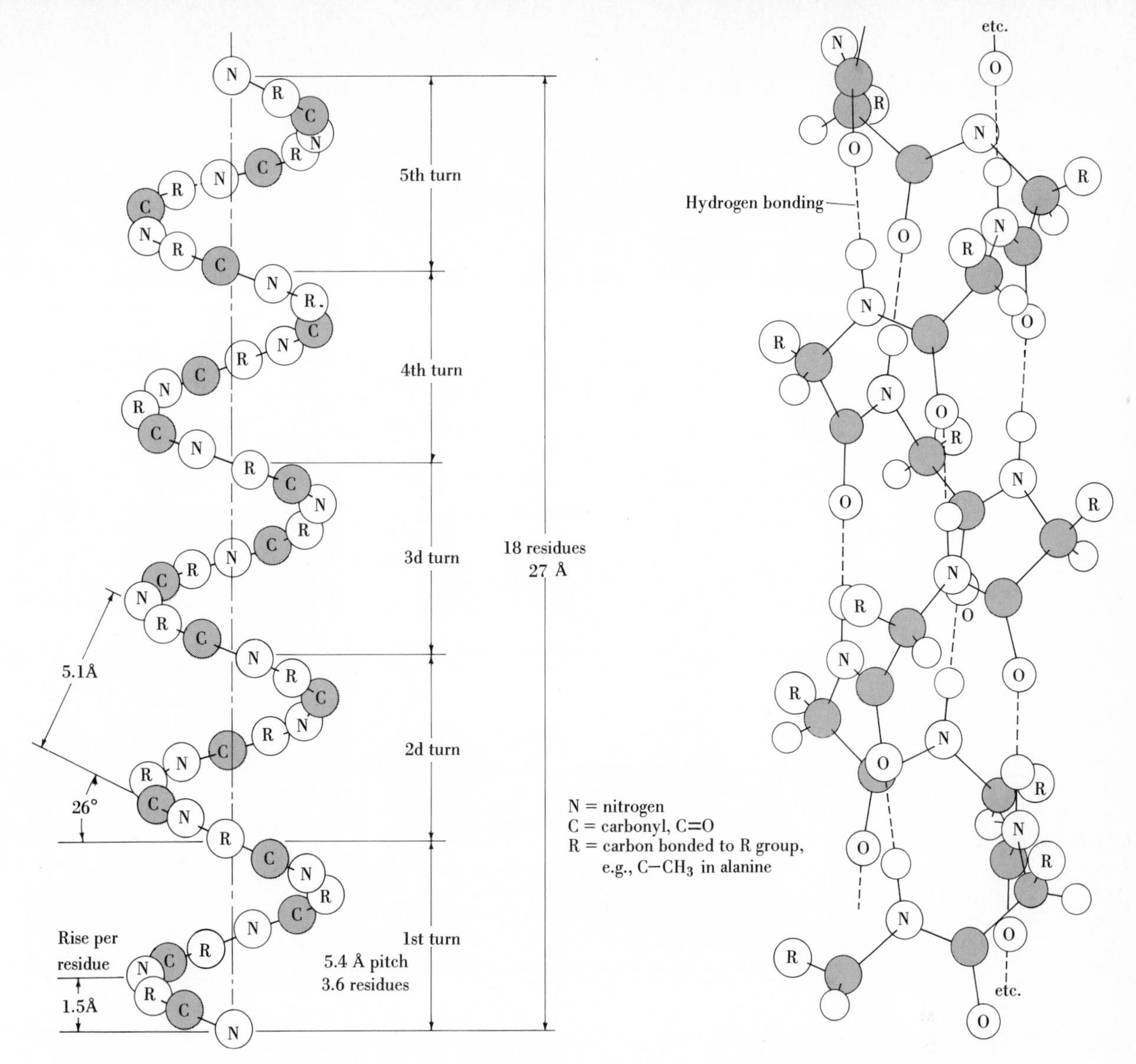

(*a*) *Representation of a polypeptide chain as an α-helical configuration.*

(*b*) *Stabilization of an α-helical configuration by hydrogen bonding to yield a secondary structure. All the shaded balls represent carbon atoms.*

FIGURE 16.7

The alpha helix. [*From L. Pauling and R. B. Corey, Proc. Int. Wool Textile Res. Conf.,* **B249** (*1955*), *as redrawn by C. B. Anfinsen, The Molecular Basis of Evolution, John Wiley & Sons, New York, 1959.*]

tide chain has the form of a single helix which makes a full clockwise turn for every 3.6 amino acid residues. This structure, referred to as an *alpha helix,* is the principal secondary structure of the protein. The details of a segment of a helix are shown in Fig. 16.7*a*. The hydrogen bonding between carbonyl and amino groups, which are adjacent to each other in the helical structure, stabilizes the helix as shown in Fig. 16.7*b*.

Because there are unsatisfied van der Waals forces even in the helical secondary structure, it may undergo folding; the resulting

FIGURE 16.8

The tertiary structure of myoglobin. The folded tertiary structure is imposed over the helical secondary structure. The primary structure is not shown in detail.

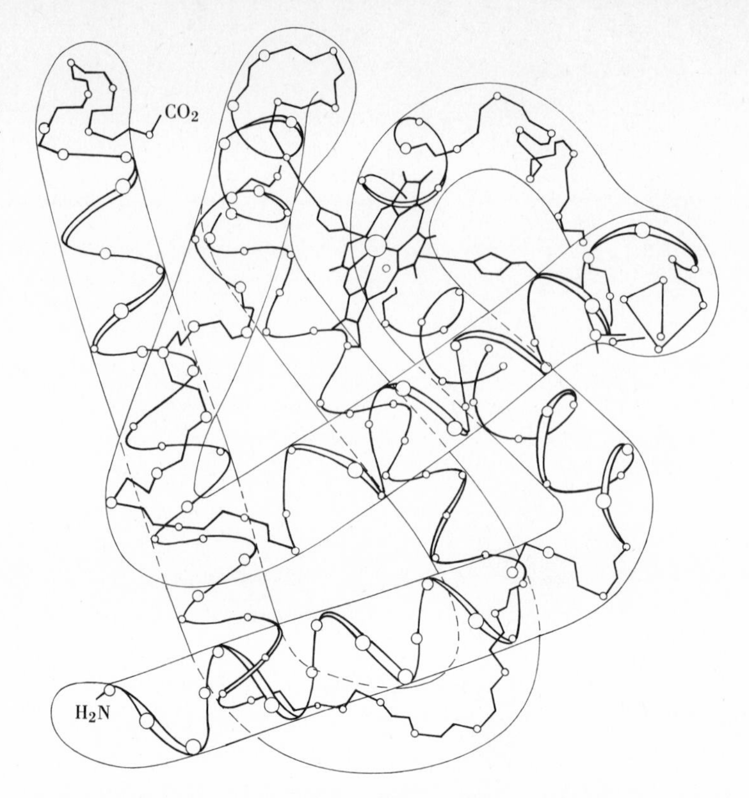

shape is called the *tertiary structure*. Then the primary structure results from covalent bonding, the secondary structure results from hydrogen bonding, and the tertiary structure results from the weak van der Waals forces. The tertiary structure of myoglobin, a muscle protein involved in respiration, is shown in Fig. 16.8. Note that the folding is superimposed on the helical secondary structure.

Some proteins in their tertiary structure form aggregates of two, four, or six units. Hemoglobin is an aggregate of four helical protein molecules which individually resemble myoglobin. The aggregate, which also owes its existence to van der Waals forces, is called the *quaternary structure*.

16.6 MESSENGER RNA

The properties and function of a particular protein depend ultimately on its primary structure. Under biological conditions a protein spontaneously goes into helical or other secondary forms, folds, and aggregates, and the result of these events depends only on the amino acid sequence. More succinctly, the primary structure determines the secondary, tertiary, and quaternary structures of a protein. These structural details are crucial to the biological role of a protein, so that faithful reproduction of the amino acid sequence is vital to a cell. As we shall see, the production of the protein is under the indirect control of the genes.

Each gene contains the information for the complete sequence of a single protein. On a molecular level this means that the DNA in the nucleus is made of many information segments; each segment is identified as a gene, and a gene controls the amino acid sequence of a single protein.

As we mentioned in Sec. 16.3 the nitrogen bases bonded to the sugar-phosphate chain of the DNA molecule are the actual information carriers. Their order on the DNA molecule determines the order of amino acids in the protein molecule. But the DNA is in the nucleus, and proteins are synthesized on the ribosomes outside the nucleus. How does the nucleus transmit information to the site of synthesis?

The information in a gene is a molecular message which must be sent out to the ribosomes so that the correct protein can be synthesized. In order to accomplish this, the gene serves as a *template* for the synthesis of an mRNA molecule. RNA stands for *r*ibo*n*ucleic

FIGURE 16.9

The structural features of mRNA.

Ribose

Uracil, U

Thymine has a methyl group in this position

Deoxyribose has H (not OH) in this position

(a) Ribose and uracil structures are similar to deoxyribose and thymine, respectively.

R = ribose
P = phosphate

(b) The mRNA single strand. Uracil replaces thymine and ribose replaces deoxyribose.

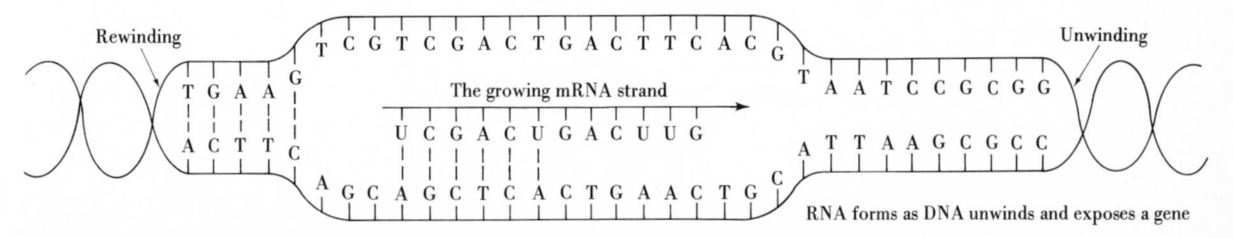

(c) mRNA is synthesized as unwinding proceeds. The mRNA contains the information equivalent to a single gene.

*a*cid. It has a structure similar to DNA but contains the nitrogen base uracil instead of thymine; its backbone is built from alternating phosphate and *ribose* units (instead of the alternating phosphate and deoxyribose units in DNA); and in its biological role it exists as a single strand. Uracil is a pyrimidine and therefore has the same shape and size as thymine. Ribose, like deoxyribose, is a cyclic five-carbon sugar, but it has one oxygen atom more than deoxyribose. The general structures of these sugars are very similar. Uracil and ribose structures are shown in Fig. 16.9*a*, and a segment of an mRNA molecule is shown in Fig. 16.9*b*.

During mRNA synthesis the DNA double helix unwinds and the gene or segment being copied becomes exposed. The actual synthesis of mRNA is analogous to the replication synthesis of DNA. A DNA segment corresponding to a gene is exposed and acts as a template for the synthesis of mRNA (Fig. 16.9*c*). Since mRNA carries only the information of a single gene, it is much shorter than a DNA strand, but nevertheless mRNA molecules are giants by normal standards; their molecular weights are on the order of 10^6. The base sequence of mRNA is determined by the base sequence of the gene on the DNA. The order of bases in the mRNA strand is complementary to the gene, so that the chromosomal information is preserved during mRNA synthesis. After mRNA is synthesized, it is transported out of the nucleus and becomes attached to the ribosomes, where protein synthesis can begin.

16.7 THE CODE AND PROTEIN SYNTHESIS

The order of the bases on the mRNA molecule determines the amino acid sequence in the protein molecule synthesized at the ribosomes. Since there are only four different bases on the mRNA strand and more than 20 different amino acids in an ordinary protein, there cannot be a one-to-one correspondence between the nitrogen bases and the amino acids. The American physicist George Gamow (1904–1968) guessed that the amino acid sequence is determined by a *triplet code* on the mRNA molecule: a group of three adjacent nitrogen bases represent a code or "word" signifying a single amino acid. Since there are four different bases and a word is made up of three bases, there can be 4^3, or 64, different words representing particular amino acids. This is more than enough words for the synthesis of a single protein. A doublet code of the four nitrogen bases would give only 4^2, or 16 different words, which is insufficient for the synthesis of a protein consisting of approximately twenty different amino acid residues.

Molecular biologists have deciphered the triplet code of the mRNA molecule. The correspondence between the nitrogen base

TABLE 16.3: GENETIC CODE†,‡

Codon	Amino acid	Codon	Amino acid	Codon	Amino acid	Codon	Amino acid
UUU	Phe	UCU	Ser	UAU	Tyr	UGU	Cys
UUC		UCC		UAC		UGC	
UUA	Leu	UCA		UAA	Chain termination	UGA	?
UUG		UCG		UAG		UGG	Try
CUU	Leu	CCU	Pro	CAU	His	CGU	Arg
CUC		CCC		CAC		CGC	
CUA		CCA		CAA	Glu(NH_2)	CGA	
CUG		CCG		CAG		CGG	
AUU	Ile	ACU	Thr	AAU	Asp(NH_2)	AGU	Ser
AUC		ACC		AAC		AGC	
AUA	Met	ACA		AAA	Lys	AGA	Arg
AUG		ACG		AAG		AGG	
GUU	Val	GCU	Ala	GAU	Asp	GGU	Gly
GUC		GCC		GAC		GGC	
GUA		GCA		GAA	Glu	GGA	
GUG		GCG		GAG		GGG	

† See Table 16.2 for amino acid abbreviations.

FIGURE 16.10

A pictorial summary of protein synthesis.

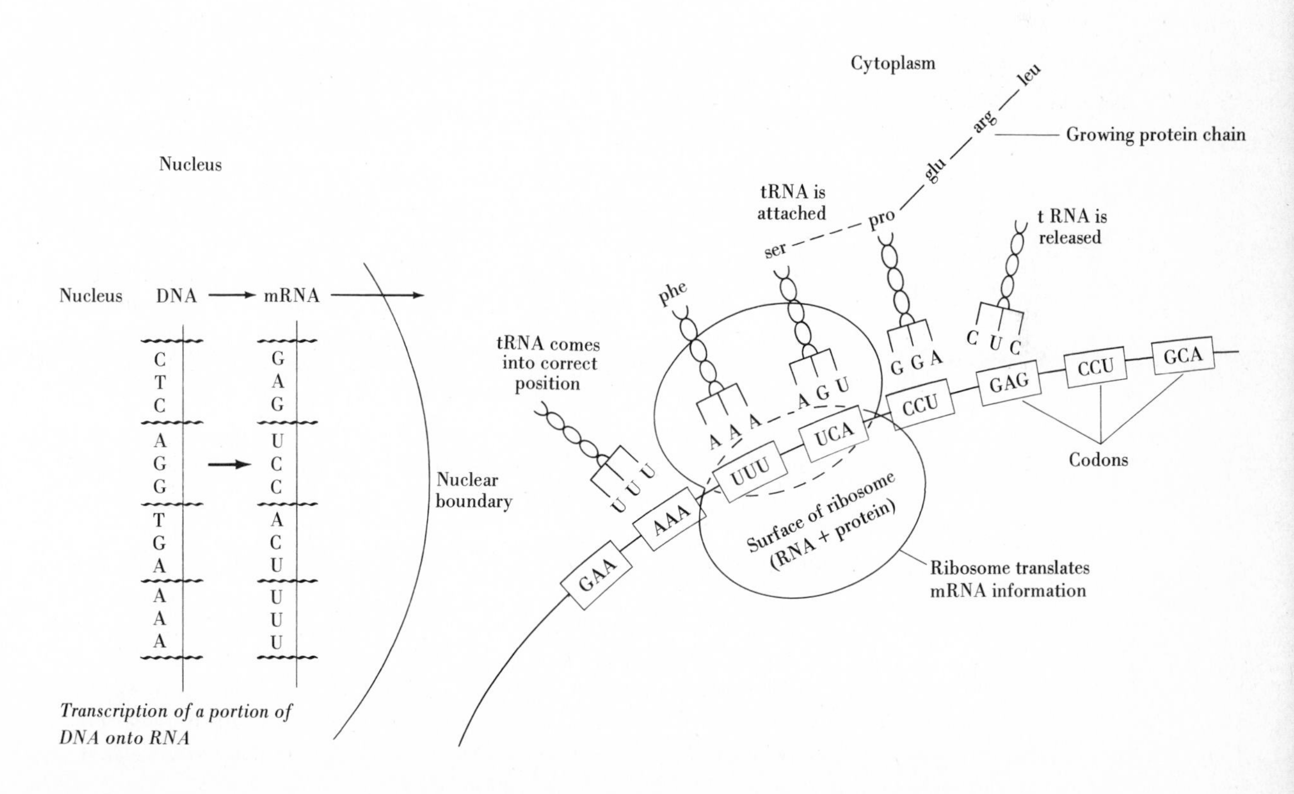

triplets and the amino acids is given in Table 16.3. Each triplet is referred to as a *codon*. Note that there is a high degree of redundancy and that some of the codons appear to signify the end of a gene or the termination of protein synthesis.

When the mRNA strand becomes bound to the ribosomes, the amino acids are brought to it by a much smaller RNA molecule called transfer RNA (tRNA). Each tRNA has a triplet of bases which is complementary to an amino acid codon on mRNA. For instance, one of the mRNA codons for valine is GUU (guanine, uracil, uracil); valine will be brought to this mRNA site by a tRNA which has the *anticodon* CAA (cytosine, adenine, adenine). The anticodon of tRNA is complementary to the codon of mRNA.

The whole process of protein synthesis is summarized in Fig. 16.10, which shows that mRNA is synthesized in the nucleus using DNA as a template. It is then transferred out of the nucleus and becomes attached to the ribosomes. As the ribosomes move along the mRNA chain, they "read out" the amino acid sequence for the synthesis of a complete protein. The tRNA molecules bring the amino acids to the ribosomes as they move along the mRNA strand, and the amino acids are knit into the growing protein chain. After the tRNA has discharged its amino acid passenger, it moves out into the cytoplasm, finds another amino acid, and returns to the ribosome surface.

16.8 ATP

The replication of DNA, the copying of RNA, the synthesis of proteins, the contraction of muscle cells, the luminescence of fireflies, and a multitude of other cellular activities are based on energy-requiring, or *endergonic*, chemical reactions. These energy-requiring reactions can take place because they are coupled to other energy-producing, or *exergonic*, reactions. The chief energy-producing reaction in the cell is the hydrolysis of adenosine triphosphate (ATP) (Fig. 16.11). ATP is formed from adenine, ribose, and three phosphate residues. The aggregate of adenine and ribose without the phosphate residues is called *adenosine*. When ATP hydrolyzes, it yields adenosine diphosphate (ADP) and H_3PO_4; at the same time there is a liberation of 10 kcal/mol of chemical potential energy. In biological reactions the hydrolysis of ATP is intimately coupled with endergonic reactions, so that the energy of hydrolysis (10 kcal/mol) can be used as a driving force. As an example, consider the energy-producing cellular oxidation of glucose. In order to prepare glucose for oxidation (in the cell) it is first converted to glucose 6-phosphate:

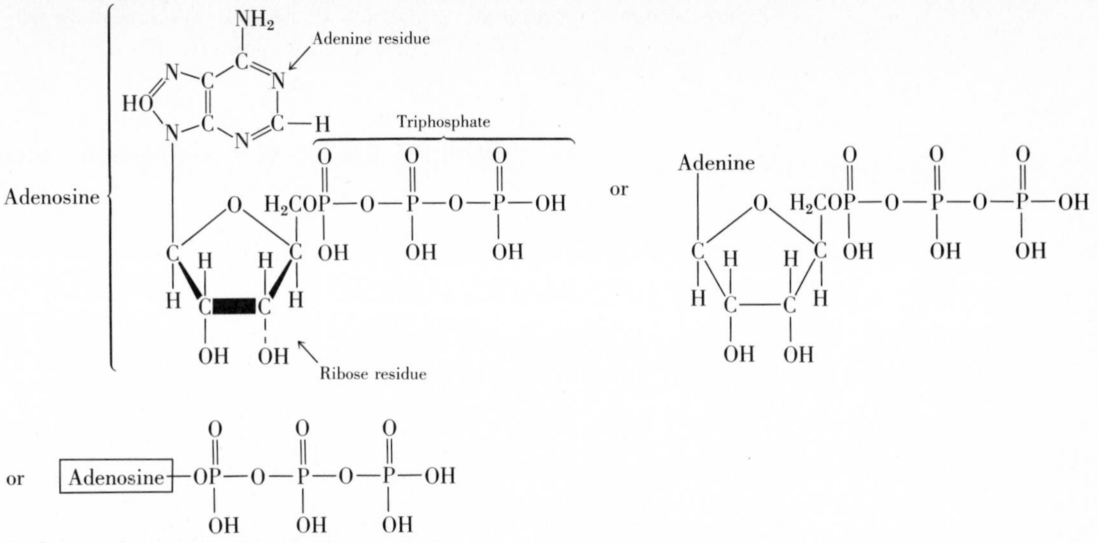

Adenosine — Adenine residue — Triphosphate — Ribose residue

$$\text{Adenosine}\ \text{(adenine–ribose)}-\text{H}_2\text{COP(=O)(OH)}-\text{O}-\text{P(=O)(OH)}-\text{O}-\text{P(=O)(OH)}-\text{OH}$$

or

$$\text{Adenine}-\text{(ribose)}-\text{H}_2\text{COP(=O)(OH)}-\text{O}-\text{P(=O)(OH)}-\text{O}-\text{P(=O)(OH)}-\text{OH}$$

or

$$\boxed{\text{Adenosine}}-\text{OP(=O)(OH)}-\text{O}-\text{P(=O)(OH)}-\text{O}-\text{P(=O)(OH)}-\text{OH}$$

(a) Different representations of adenosine triphosphate, ATP

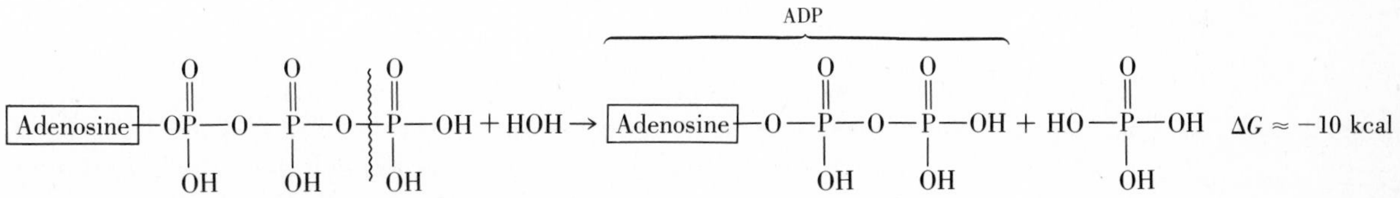

$$\boxed{\text{Adenosine}}-\text{OP(=O)(OH)}-\text{O}-\text{P(=O)(OH)}-\text{O}-\text{P(=O)(OH)}-\text{OH} + \text{HOH} \rightarrow \underbrace{\boxed{\text{Adenosine}}-\text{O}-\text{P(=O)(OH)}-\text{O}-\text{P(=O)(OH)}-\text{OH}}_{\text{ADP}} + \text{HO}-\text{P(=O)(OH)}-\text{OH} \quad \Delta G \approx -10\ \text{kcal}$$

(b) The hydrolysis of ATP yields adenosine diphosphate, ADP, and 10 kcal of energy

FIGURE 16.11

The structure and hydrolysis of ATP.

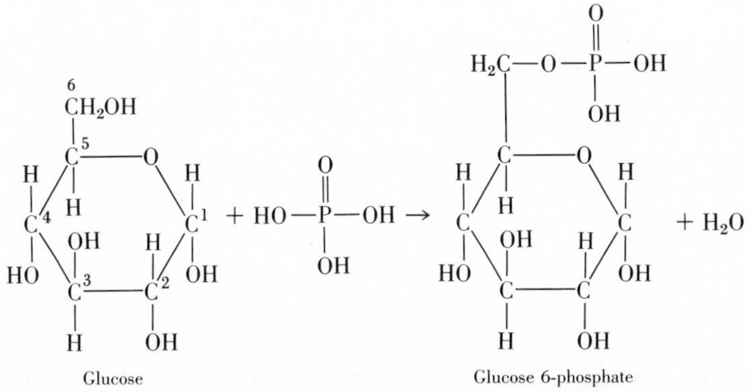

$$\underset{\text{Glucose}}{\text{C}_6\text{H}_{12}\text{O}_6\ (\text{C}^6\text{H}_2\text{OH})} + \text{HO}-\text{P(=O)(OH)}-\text{OH} \rightarrow \underset{\text{Glucose 6-phosphate}}{\text{H}_2\text{C}-\text{O}-\text{P(=O)(OH)}-\text{OH}\ (\text{on C}^6)} + \text{H}_2\text{O} \qquad \Delta G \approx +3\ \text{kcal; unfavorable} \tag{16.4}$$

This reaction is endergonic and takes place only because it is coupled to the hydrolysis of ATP, which is exergonic:

$$\text{ATP} + \text{H}_2\text{O} \rightarrow \text{ADP} + \text{HO}-\overset{\overset{\displaystyle \text{O}}{\|}}{\underset{\underset{\displaystyle \text{OH}}{|}}{\text{P}}}-\text{OH} \qquad \Delta G \approx -10 \text{ kcal} \qquad (16.5)$$

The net reaction for the phosphorylation of glucose is the sum of Eqs. (16.4) and (16.5):

$$\text{Glucose} + \text{ATP} + \text{H}_2\text{O} \rightarrow \text{glucose 6-phosphate} + \text{ADP} \quad \Delta G \approx -7 \text{ kcal}$$

The chemical potential-energy change for the net reaction is −7 kcal, and therefore the coupled process is possible.

The hydrolysis of ATP is the most important energy source for biological processes. The coupling of ATP hydrolysis with endergonic reactions occurs in all organisms and all cells. Biochemists indicate coupled reactions by a curved arrow joining the straight arrow in the equation for a chemical reaction:

$$\text{A} \xrightarrow{\quad\quad\quad} \text{B}$$
$$\text{ATP} \curvearrowright \text{ADP}$$

The chloroplasts and the mitochondria work in concert to supply the cell with ATP. The chloroplasts possess light-absorbing molecules to trap solar energy, which in turn is used to produce carbohydrates such as glucose and starch. In the cell (but chiefly in the mitochondria) these products undergo controlled oxidation, liberating the solar energy which was trapped in photosynthesis. More than 50 percent of the energy produced by the mitochondrial oxidation of carbohydrates is recaptured, as ADP is converted into ATP by concerted reactions occurring on the interior wall of the mitochondria:

$$\underset{\text{Glucose}}{C_6H_{12}O_6} + 6O_2 \xrightarrow{\quad\quad\quad} 6CO_2 + 6H_2O$$

$$\text{ADP} + \text{HO}-\overset{\overset{\displaystyle \text{O}}{\|}}{\underset{\underset{\displaystyle \text{OH}}{|}}{\text{P}}}-\text{OH} \curvearrowright \text{ATP} + \text{H}_2\text{O}$$

16.9 PHOTOSYNTHESIS

Photosynthesis represents a biological process which has evolved over eons to harness the energy of the sun. Most biologists believe that it is very unlikely that life could have come into existence in the absence of the radiation we call visible light. This radiation has just the right energy to excite molecules and make them more reactive, break chemical bonds, and to cause endergonic (energy-requiring) reactions to occur. In the absence of visible light the large ordered organic molecules which are the key characteristics of biological systems could not have formed.

FIGURE 16.12

The light-gathering pigments of the chloroplasts.

(*a*) *Chlorophyll a (the circled* CH_3 *group is replaced by* CHO *group in chlorophyll b). All corners except those marked* N *are occupied by carbon atoms.*

(*b*) *Porphin structure consisting of four pyrrole rings (all corners except those marked* N *are occupied by carbon atoms).*

(*c*) *Structure of β carotene. (All corners on the rings at the two ends are occupied by carbon atoms.)*

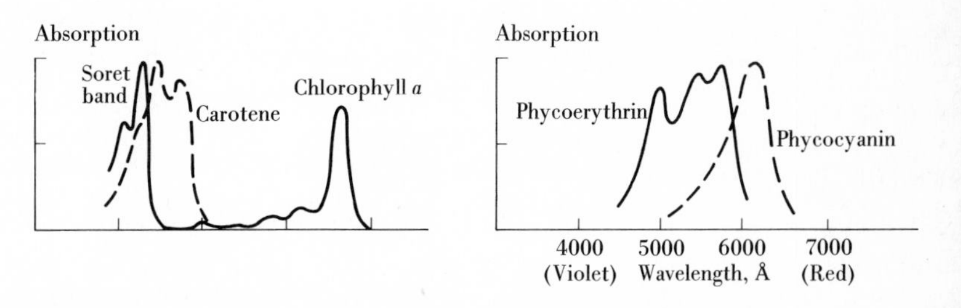

(*d*) *Absorption spectra of three types of chloroplast pigments. (1) chlorophylls. carotenoids; (2) phycoerythrins and phycocyanins.*

The chief pigment molecules of the chloroplasts are the *chlorophylls*. These molecules act like powerful antennas as they strongly absorb the blue and red radiation from the extremes of the visible spectrum. The structure of the most common chlorophyll molecule is shown in Fig. 16.12*a*. All chlorophyll molecules are porphyrins. The simplest porphyrin is porphin (Fig. 16.12*b*). Porphyrins contain the residues of four pyrrole molecules bridged together by a single carbon atom. The porphyrin structure is completely conjugated; i.e., it has alternating single and double bonds. Large conjugated molecules invariably absorb visible light, and so it is not surprising that nature has chosen the chlorophylls, which are magnesium porphyrins, as the chief light-gathering agents for photosynthesis.

In addition to the chlorophylls, the chloroplasts contain auxiliary pigments, which absorb the green light near the middle of the visible spectrum. Carotene is one of the chief auxiliary pigments; its structure and absorption spectrum are shown in Fig. 16.12*c* and *d*. Note that it is also a large conjugated molecule. The absorption spectra of two other auxiliary pigments, phycoerythrin and phycocyanin, are also shown. Taken together, chlorophyll, carotene, phycoerythrin, and phycocyanin absorb light from the entire visible spectrum. The auxiliary pigments transfer light energy to the chlorophylls, which set the photosynthetic apparatus to work.

The first steps in photosynthesis involve the oxidation of water. The solar energy absorbed by the pigments of the chloroplasts is used to strip the water molecule of its hydrogen atoms, producing molecular oxygen. After this occurs, the hydrogen atoms are transferred to a carbon dioxide molecule, whereupon CO_2 is reduced to carbohydrate. These are the two main steps in the process.

The total photosynthesis of carbohydrates and oxygen is a complicated, multistep process which is not completely understood, but the overall reaction is relatively simple:

$$6CO_2^{\ddagger} + 12H_2O^* + \text{light energy} \xrightarrow{\text{chlorophyll}} C_6H_{12}O_6^{\ddagger} + 6O_2^* + 6H_2O^{\ddagger} \qquad (16.6)$$

This equation is arbitrarily written for the formation of 1 mol of glucose. Note also that the oxygen atoms in CO_2 and H_2O are designated differently so that the equation will indicate the known fact that the molecular oxygen is derived from the water:

$$\text{Light energy} + 12H_2O^* \rightarrow 24\ H + 6O_2^*$$

Subsequently, CO_2 is converted into carbohydrate and water when the CO_2, in a second light-induced reaction, accepts the hydrogen atoms which were stripped from the water molecule:

$$\text{Light energy} + 6CO_2^{\ddagger} + 24H \rightarrow C_6H_{12}O_6^{\ddagger} + 6H_2O^{\ddagger}$$

Plant cells store the carbohydrate and use it to synthesize ATP when cellular activities call for an energy supply. When animals ingest plants, they also use the plant-produced carbohydrate as a source of ATP. All life is sustained by the photosynthetic organisms. Any threats to these organisms can cause serious chemical and thermal imbalances in the biosphere and ultimately lead to a barren earth.

16.10 GLYCOLYSIS AND RESPIRATION

The carbohydrate is stored in the cells until energy is required. Then by a degradative process, which occurs chiefly on the mitochondria, it supplies the energy for the synthesis of ATP. The relatively small ATP molecules race to points in the cell where energy is required. Carbohydrate is the energy reserve bank of the living cell which releases ATP as its "energy currency."

There are two major stages of carbohydrate degradation: *glycolysis* and *respiration.* Glycolysis is *anaerobic* because it does not require oxygen; respiration is aerobic. Glycolysis, also called *fermentation,* occurs in the cytoplasm outside the mitochondria. In this process glucose, a six-carbon sugar, is broken down into two three-carbon molecules of pyruvic acid, $CH_3—CO—COOH$. At the same time there is a net production of two molecules of ATP by coupled reactions:

$$\text{Glycolysis: } \underset{\text{Glucose}}{C_6H_{12}O_6} \xrightarrow[\;2ADP + 2P_i \;\curvearrowright\; 2ATP + 2H_2O\;]{} \underset{\text{Pyruvic acid}}{2CH_3—\overset{\overset{\displaystyle O}{\|}}{C}—C\!\!\begin{array}{l}\diagup\!\!\!\!= O\\ \diagdown O—H\end{array}} \qquad (16.7)$$

This equation, which is not balanced, merely summarizes the important aspects of glycolysis; the symbol P_i represents phosphoric acid, or *inorganic phosphate* as the biochemists say. Glycolysis is very complicated and is known to consist of at least 11 different enzyme-catalyzed steps. The first two are endergonic and require an *input* of two ATP molecules; four ATP molecules are produced in the last few steps so that there is a net production of two ATP molecules.

During rapid exercise muscle cells obtain energy by glycolysis. Many marine microorganisms, the deadly *Clostridium botulinum,* and common yeast derive all their ATP in this way.

Everyone knows that breweries and bakeries take advantage of the fermentation of yeast. When mixed in bread dough or sugar-containing fruit juices the yeast multiplies. Fermentation produces pyruvic acid, which then decomposes to produce ethyl alcohol and CO_2. The characteristic odor of a bakery is partially due to the presence of ethyl alcohol in the unbaked dough.

In aerobic cells the product of glycolysis, pyruvic acid, moves into the mitochondria, where respiration occurs. Respiration consists of a cyclic degradation of pyruvic acid which produces CO_2 followed by a series of oxidation-reduction reactions in which molecular oxygen is finally reduced to water. As a result of glycolysis and respiration 40 mol of ATP are produced for every mole of glucose used.

The cyclic part of pyruvic acid degradation, which occurs during respiration, is called the *Krebs cycle* in honor of the Nobel laureate Sir Hans Krebs, who discovered it in the 1930s. In this cycle the hydrogen atoms of pyruvic acid are stripped off and transferred to coenzyme† molecules, which are thereby reduced; the three carbon atoms of pyruvic acid are transformed into CO_2 molecules and are transported out of the cell. (The Krebs cycle is very adaptable; it can accept fats and even proteins as substrates, but it uses carbohydrate whenever it is available.) The reduced coenzymes generated in the Krebs cycle set into motion a chain of enzyme-catalyzed oxidation-reduction reactions which result in the transfer of hydrogen atoms to molecular oxygen and water is formed. This reduction of oxygen is mediated by the *cytochrome* enzymes, which actually make up the interior membrane of the mitochondria. The active site of the cytochromes is an iron porphyrin called *heme*, which is bonded to the protein chain of cytochrome molecule (see Fig. 16.13). The heme undergoes reversible oxidation and reduction during respiration since the iron atom can exist in two oxidation states:

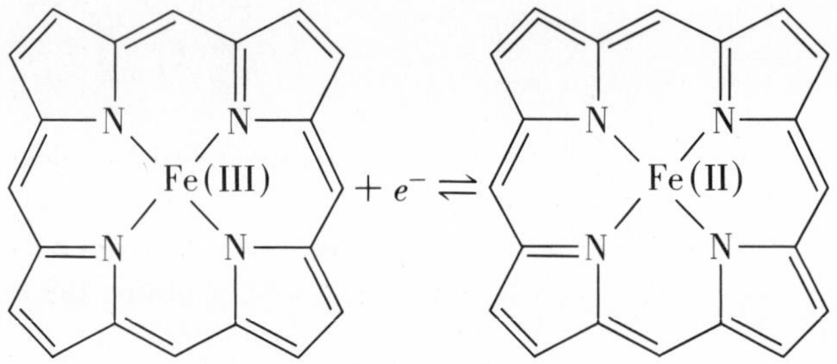

$$\text{Fe(III)} + e^- \rightleftharpoons \text{Fe(II)}$$

Involved in the respiration process are at least four cytochromes, each with a slightly different oxidation potential. The electrons are passed from one to the other in a series of exergonic reactions. The chemical potential energy generated by these reactions is used to produce ATP.

The cytochromes can be considered as a bucket brigade for car-

† A coenzyme is usually a relatively small molecule which activates an enzyme.

FIGURE 16.13

The heme is bonded to the protein at four points. Two cysteine residues form covalent bonds to the porphyrin, and two hystidine residues are coordinated to the iron above and below the porphyrin plane.

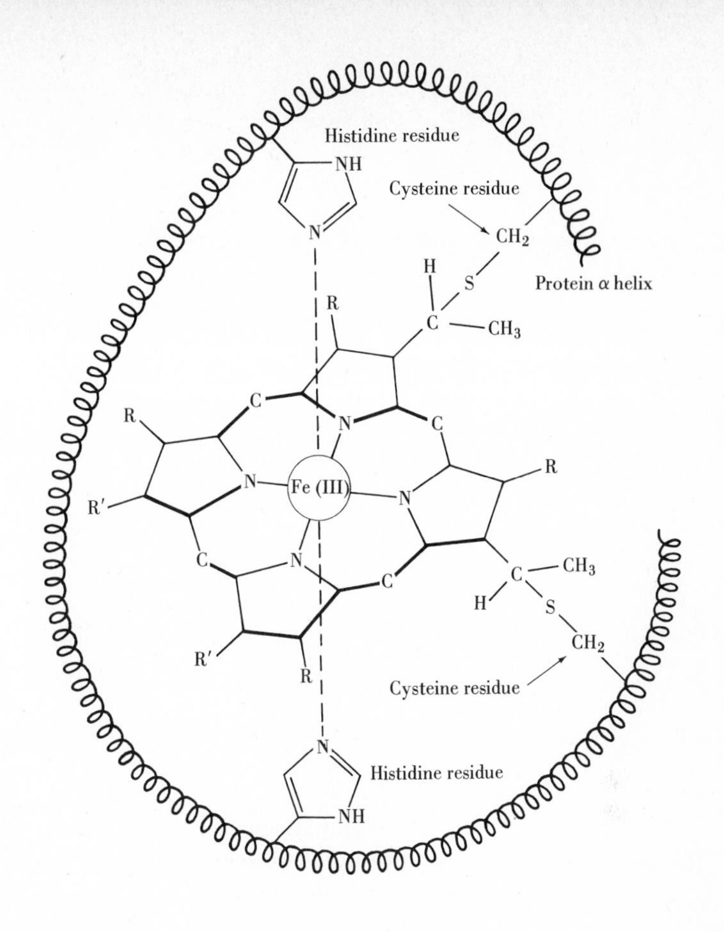

$R = -CH_3; R' = -CH_2CH_2COOH$

rying electrons. The cytochromes are arranged in the interior of the mitochrondria in order of decreasing oxidation potential. Electrons supplied at one end by reduced coenzymes produced in the Krebs cycle are passed from one cytochrome to the next and finally to an oxygen molecule, which is reduced to water:

$$\underset{(2\ \text{mol})}{\text{[porphyrin]Fe(II)}} + \tfrac{1}{2}O_2 + 2H^+ \rightarrow \underset{(2\ \text{mol})}{\text{[porphyrin]Fe(III)}} + H_2O$$

FIGURE 16.14

The cyclic character of photosynthesis plus respiration.

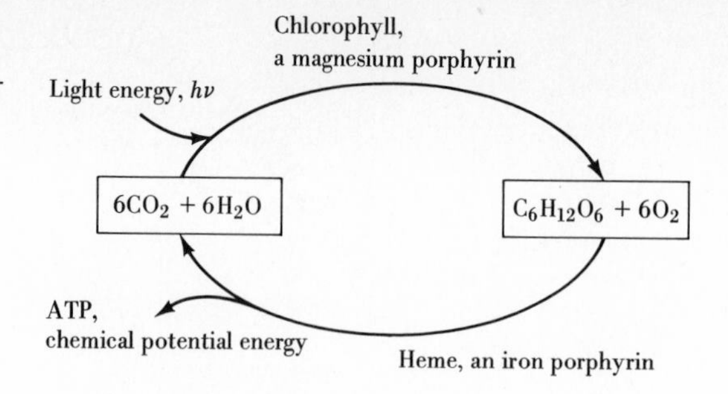

The cellular oxidation of glucose to CO_2 and H_2O during glycolysis and respiration occurs in discrete, controlled chemical steps which are coupled with ATP-forming reactions. In this way the chemical potential energy is made available in small, easy-to-handle packages. In the rapid oxidation which would occur by burning glucose in air the available energy would be generated so rapidly that it would be difficult to capture the energy efficiently. Furthermore, if glucose oxidation occurred rapidly in a living cell, the heat evolved could cause chemical damage. In the glycolysis-respiration process over 50 percent of the chemical potential energy generated by the oxidation of glucose is used in the concurrent formation of ATP. Our best machines have efficiencies which are only one-half that of a living cell.

Comparison of photosynthesis and respiration shows that nature preserves a poetic symmetry. In the initial steps of photosynthesis, in which glucose and O_2 are produced, a porphyrin molecule, chlorophyll, assists the removal of hydrogen atoms from water; in the final steps of respiration, in which glucose is degraded, a porphyrin molecule, heme, presides over the attachment of the hydrogen atoms to oxygen to form water. Another iron porphyrin protein is extremely important in respiration. *Hemoglobin*, a heme-protein molecule present in the red blood cells, carries molecular oxygen from the lungs to cells where respiration occurs.

The overall view of photosynthesis and respiration is summarized in Fig. 16.14, which demonstrates the cyclic nature of the two processes and emphasizes that the net result is the production of ATP from light energy.

16.11 THE STRUCTURE OF FATS AND CARBOHYDRATES

We have dealt with some of the most important cellular processes and have mentioned proteins, starch (carbohydrate), and fats. These are the chemical components which, along with water, are the most abundant substances in the cell. The structure of the proteins was discussed in Sec. 16.5. Fats, which are produced when cellular intake is high, are esters of glycerol, $CH_2OH—CHOH—CH_2OH$, and long-chain carboxylic acids, called *fatty acids*. Stearin, a typical fat, is a derivative of glycerol and a straight-chain 18-carbon acid, stearic acid, $CH_3(CH_2)_{16}COOH$:

$$
\begin{array}{lcl}
3\ CH_3(CH_2)_{16}—C(=O)—OH & + \quad \begin{array}{l} HO—CH_2 \\ HO—CH \\ HO—CH_2 \end{array} \rightarrow & \begin{array}{l} CH_3(CH_2)_{16}—C(=O)—OCH_2 \\ CH_3(CH_2)_{16}—C(=O)—O—CH \\ CH_3(CH_2)_{16}—C(=O)—O—CH_2 \end{array} + 3H_2O \\
\text{Stearic acid} & \text{Glycerol} & \text{Stearin}
\end{array}
$$

Stearic acid is a *saturated* acid because there are no carbon-carbon double bonds in the molecule; the fat derived from stearic acid, stearin, is a *saturated fat*.

Many other fatty acids form fats with glycerol. All naturally occurring fatty acids have an even number of carbon atoms, but many natural fatty acids are unsaturated. Typical of this group is oleic acid, which was discussed on page 473. Oleic acid is an 18-carbon, straight-chain acid with a double bond between the ninth and tenth carbon atoms of the chain: $CH_3—(CH_2)_7—CH{=}CH—CH_2)_7—COOH$. Other fatty acids, like linoleic acids, contain more than one double bond in the carbon chain. These acids form the *polyunsaturated fats* which are of great concern to diet watchers.

Fats are produced by the cell when food intake is more than required for the normal functions of an organism. It is kept as a reserve and used at times of fasting. Unfortunately, in many modern societies there is no time of fasting and energy-consuming physical activity has been reduced by mechanization. As a result, obesity, an inordinate storage of fat, is a common problem in adults past the age of twenty-five years.

We have also spoken of glucose, sugars, and starches. These are the substances produced during photosynthesis. Glucose is a sugar molecule containing six carbon atoms. It normally exists as a cyclic structure; there are five carbon atoms and one oxygen atom in a ring:

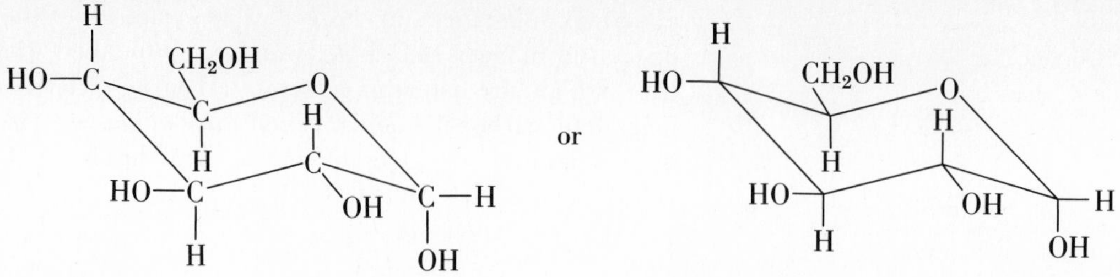

Glucose, $C_6H_{12}O_6$; the structure is tetrahedral around each carbon atom

There are many other simple sugars, but we shall not discuss them.

Glucose can polymerize to form *polysaccharides*. When two glucose molecules join together, a disaccharide forms; three give a trisaccharide; and so on. The starches are huge polysaccharides and, like the fats, are stored by the cell as an energy reserve.

Cellulose, another polysaccharide, is the structural fiber of plant cells, forming the cell walls and other parts where strength and chemical inertness are required. Wood, cotton, linen, straw, corncobs, and many other materials are rich in cellulose. Cellulose and starch differ in the nature of the linkage between the simple sugars which make them up. Figure 16.15*a* and *b* shows that the linkage between the simple sugar residues in starch is slightly different from those in cellulose. In cellulose adjacent glucose residues are rotated 180° with respect to each other, whereas in the starch structure each residue has the same relative orientation. Because of the specificity of the enzymes involved in metabolism, this difference makes cellulose indigestible by most organisms because the ordinary intestinal enzymes cannot break cellulose down into simple sugars. The bacteria which reside in the intestinal tract of the termite convert the cellulose into glucose and simple amino acids, so that the termite can thrive on a diet of wood. The implications of this conversion are quite obvious. Many investigators are studying bacteria which can convert cellulose into simple sugars and amino acids. Imagine how much food could be provided by a single redwood tree.

QUESTIONS AND PROBLEMS

16.1 What important chemical activities occur (*a*) in the nucleus, (*b*) at the ribosomes, (*c*) in the chloroplasts, and (*d*) in the mitochondria?

16.2 A transducer is a device which converts energy from one form into another form. Which of the organelles discussed in this chapter may be considered transducers?

16.3 Chargaff found that the numbers of adenine and thymine residues and the numbers of cytosine and guanine residues in DNA are equal. How does the Watson-Crick model account for this fact?

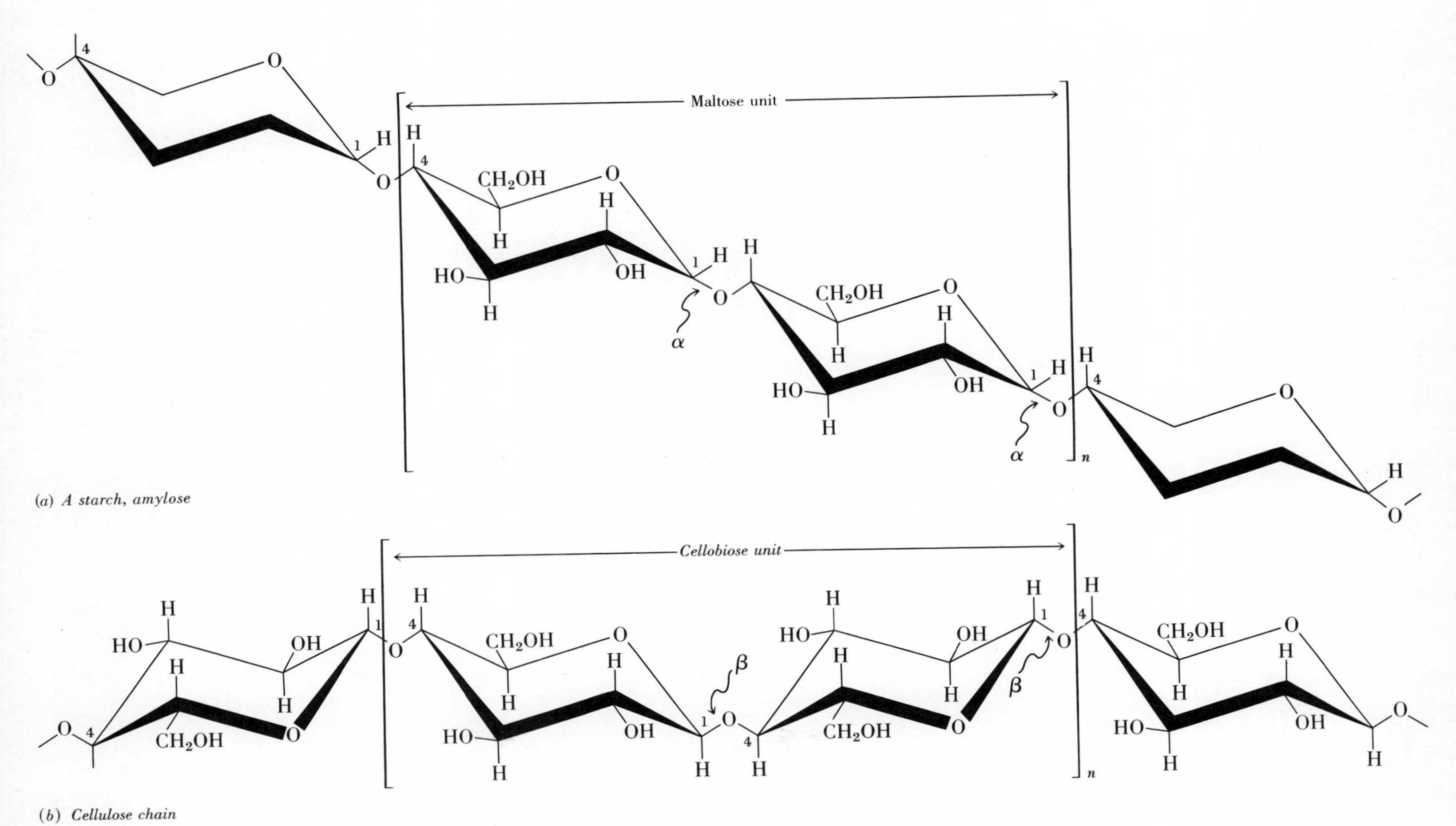

FIGURE 16.15

Polysaccharides, a starch and cellulose.

16.4 Make a sketch of a DNA segment in which the base order is cytosine, guanine, guanine, thymine, guanine, and adenine. Make a sketch of the complementary strand for this segment.

16.5 X-ray studies show that there is a 34-Å repeat distance in the DNA molecule. Show that this amounts to 10.6° of twist per angstrom if DNA is helical.

16.6 How does the Watson-Crick DNA molecule account for ordinary cell division (mitosis) in which two daughter cells, containing the same genetic information as the parent, are produced?

16.7 (*a*) What chemical linkage is the basis of protein structure? (*b*) Describe what is meant by the primary, secondary, tertiary, and quaternary structure of a protein.

16.8 Using Table 16.2, show the details of the primary structure for the polypeptide H_3^+N—Cys-Tyr-Ala-Ala-Phe-Pro-Arg-Met-COO^-.

16.9 (*a*) Using Table 16.2 and Fig. 16.6, show the details of the primary structure of lysozyme from residue 10 to residue 15. Show the different R groups of the amino acid residues. (*b*) Show by a sketch of the atoms involved how residue 30 is bonded to residue 115 in the lysozyme primary structure. What other residues are linked in the same way?

16.10 In Sec. 16.5 the determination of the primary structure of a decapeptide was discussed, and enough information was presented for positions of four of the ten amino acids to be determined. The following information, combined with that in the text, allows the complete amino acid sequence to be determined. An enzyme E2 is known to hydrolyze the Tyr-Phe (not Phe-Tyr) peptide linkage. When the decapeptide is treated with E2, two fragments, C, a tripeptide, and D, a heptapeptide, are produced. Complete hydrolysis of C yields two Phe and one Leu. Complete hydrolysis of D yields two Ala and one each of Gly, Leu, Ser, Phe, and Tyr. After being tagged with Q_A a rapid, nonenzymic hydrolysis (in weak acid) of the decapeptide produced Ala, a dipeptide E, and a heptapeptide, F. Upon hydrolysis E yielded Gly and Ala-Q_A and F yielded two Leu, one Ser, three Phe, and one Tyr. What is the complete primary structure of the decapeptide? Using Table 16.2, show the complete primary structure of the polypeptide.

16.11 Discuss the roles of DNA, mRNA, and tRNA in the synthesis of proteins.

16.12 Give the polypeptides that correspond to the mRNA segments UUUUUC, CCUCAAAGC, and GGUGAUACUCUC.

16.13 (*a*) Give the polypeptides that correspond to the base order GCCACGACCCGG and TCCTTACGC on DNA. (*b*) What mRNA base orders are complementary to these DNA segments? (*c*) What are the anticodons of the corresponding tRNA molecules for each segment?

16.14 What are the chief chemical differences between DNA and RNA?

16.15 (*a*) Give the structure for ATP. (*b*) How does it enable endergonic biochemical reactions to occur?

16.16 What important chemical changes occur during (*a*) glycolysis and (*b*) respiration?

16.17 The chemical potential energy of formation ($\Delta G^\circ_{f,298}$) for $CO_2(g)$, $H_2O(l)$ and glucose, $C_6H_{12}O_6(c)$ are −94.3, −56.7, and −306 kcal, respectively. (*a*) Calculate the chemical-potential-energy change for the combustion of 1 mol of glucose. (*b*) The cell produces 40 mol of ATP per mole of glucose metabolized. Each mole of ATP can deliver 10 kcal of chemical potential energy by hydrolysis when the cell requires energy. What is the efficiency of the cellular oxidation of glucose; i.e., what fraction of the chemical potential energy from the oxidation of 1 mol of glucose is available from the hydrolysis of 40 mol of ATP?

16.18 (*a*) What are the two important chemical steps in photosynthesis? (*b*) What molecules are most significant in photosynthesis?

16.19 (*a*) What is the net result of the cycle which includes the two "opposing" processes, photosynthesis and glycolysis-respiration? (*b*) What similar types of molecules are involved in each process?

16.20 Discuss an experiment by which we could prove or disprove that H_2O is the source of the oxygen gas formed in photosynthesis: $H_2O + CO_2 \rightarrow (CH_2O) + O_2$.

16.21 Draw a structure for the fat formed from glycerol and (*a*) linoleic acid, $CH_3(CH_2)_4{-}CH{=}CH{-}CH_2{-}CH{=}CH{-}(CH_2)_7COOH$, (*b*) palmitic acid, $CH_3(CH_2)_{14}COOH$, and (*c*) lauric acid, $CH_3(CH_2)_{10}COOH$.

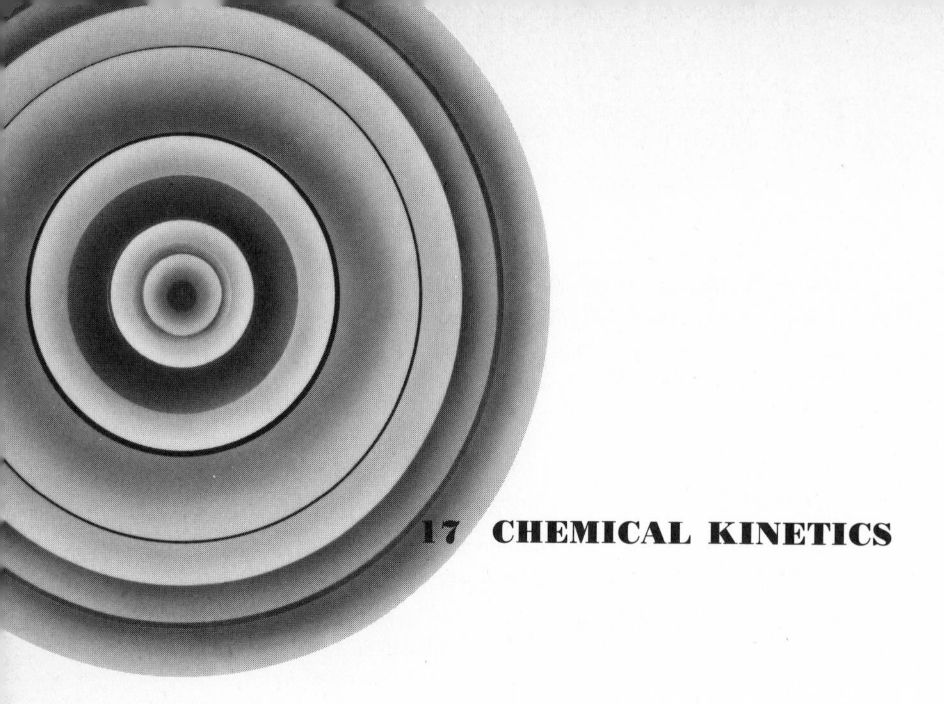

17 CHEMICAL KINETICS

17.1 INTRODUCTION

Many chemical reactions are of industrial importance because they yield products which are important to our economy. Chemists and chemical engineers study these reactions to determine what conditions affect the yield and speed of the reaction. Learning more about why and how chemical reactions occur is the first step in developing new and better ways to manipulate chemistry for our own needs.

In Chap. 11 we examined the effect of temperature, pressure, concentration, and catalysts on the *yield* of an equilibrium reaction. In this chapter we shall be concerned about how these same factors affect the *speed*, or *rate*, of a chemical reaction. The series of steps by which reactants are converted into products is called the *reaction mechanism*. Once a mechanism is known, conditions for a particular reaction can be optimized so that a greater yield of product is achieved in a shorter time. The study of reaction mechanisms has led to the development of several theories of chemical reactions. Hence, the subject is important from both a practical and theoretical point of view.

So far we have used chemical equations to describe the overall change which occurs during a chemical reaction. For instance, when N_2 and H_2 react to produce ammonia, we write

$$N_2(g) + 3H_2(g) \rightarrow 2NH_3(g)$$

But is this reaction slow or fast? Does it occur in one step or in several steps? We cannot answer these questions from the stoichiometric equation alone: it simply states that hydrogen and nitrogen react in a ratio of 1 mol to 3 mol and that 2 mol of ammonia are formed for every 3 mol of hydrogen. The stoichiometric equation does *not* give any details of the chemical transformation. It would be incorrect to interpret this equation to mean that three molecules of H_2 and one molecule of N_2 come together at some point and form two molecules of ammonia. In fact, H_2 and N_2 undergo many rapid, intermediate transformations before ammonia is formed. The stoichiometric equation is always much simpler than the reaction mechanism. For the formation of water the stoichiometry is given by a simple, familiar equation:

$$2H_2(g) + O_2(g) \rightarrow 2H_2O(l)$$

But H_2 and O_2 undergo many rapid, intermediate transformations before water is formed. The mechanism of the reaction between hydrogen and oxygen is very complex, even though its stoichiometry is rather simple. Some of the steps as described by mechanistic chemical equations are believed to be

$$H_2 + O_2 \rightarrow HO_2 + H$$
$$O_2 + H \rightarrow OH + O$$
$$OH + H_2 \rightarrow H_2O + H$$
$$H_2 + HO_2 \rightarrow OH + H_2O$$

As can be seen, many unusual species, such as HO_2, OH, O, and H, are formed during the transformation. These intermediate species, or *intermediates*, are so reactive that they disappear almost as soon as they form, and their concentration is always so low that they cannot be detected by ordinary means.

A mechanistic set of equations indicates what we believe happens at the molecular level. They represent the detailed molecular transformations thought to occur when reactants are converted into products. The stoichiometry of a chemical reaction simply identifies the reactants and products and gives the weight ratios involved in the overall chemical change. It is a *macroscopic* description of the chemical transformation; it does not depend on the concept of atoms or molecules.†

† The stoichiometric equation can be written in terms of masses instead of moles; it does not really depend on the mole concept:

$2H_2 + O_2 \rightarrow 2H_2O$ or 4 g hydrogen + 32 g oxygen $\rightarrow$ 36 g water

We can learn much about the mechanism of a chemical reaction by studying how conditions affect the yield of product and the rate of reaction. In this chapter we shall be chiefly concerned with effects of concentration and temperature changes on the rate of reaction. We shall attempt to translate the results of rate studies into a description of the mechanism for a chemical reaction.

17.2 THE REACTION RATE AND REACTION MECHANISMS

The term *chemical kinetics* refers to the measurement of reaction rates *and* their interpretation in terms of mechanisms. The rate of a reaction may be defined as the rate of appearance of a product or the rate of disappearance of a reactant, but, as we shall see, these two are not necessarily equal. If we choose to define the rate of a reaction as the rate of appearance of a product, we must measure the concentration of the product at timed intervals. The rate of a reaction will be equal to the increase in the concentration of the product per unit time. Consider a reaction with simple stoichiometry $A \rightarrow B$, like the rearrangement of *N*-chloroacetanilide

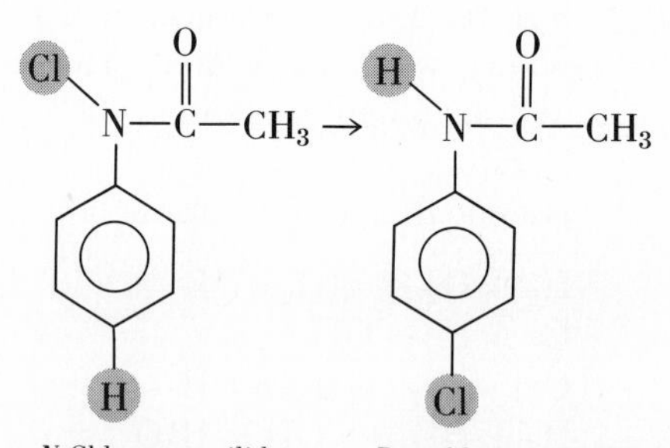

We could define the rate of the reaction as the rate of appearance of the product, B, parachloroacetanilide:

Rate of reaction = rate of appearance of product

Suppose that the concentration of the product is $[B]_1$ at time t_1 and $[B]_2$ at time t_2. The increase in the concentration of B per unit time is

$$\frac{[B]_2 - [B]_1}{t_2 - t_1} = \frac{\Delta[B]}{\Delta t}$$

which is, by definition, the rate of the reaction:

$$\text{Rate} = \frac{\Delta[B]}{\Delta t} \tag{17.1}$$

If we had chosen to define the rate of the reaction as the decrease in the reactant concentration per unit time, we would write

$$\text{Rate} = -\frac{\Delta[\text{A}]}{\Delta t} \tag{17.2}$$

The negative sign is introduced so that the rate will be a positive quantity; $\Delta[\text{A}]$, which equals $[\text{A}]_2 - [\text{A}]_1$, must be negative if A is the reactant. The rate always has units of moles per liter per unit time, e.g., moles per liter-second, moles per liter-minute.

We can make similar definitions for reactions with more complicated stoichiometries. For example, in the decomposition of hydrogen peroxide

$$2\text{H}_2\text{O}_2(l) \rightarrow 2\text{H}_2\text{O}(l) + \text{O}_2(g)$$

the rate could be defined as the rate of disappearance of hydrogen peroxide, H_2O_2, or the rate of appearance of either H_2O or O_2. It is important to make an explicit definition of the rate before any analysis is begun. In this case the rate of disappearance of $H_2O_2(l)$ would be equal to the rate of formation of water:

Rate of disappearance of H_2O_2 = rate of formation of H_2O

or

$$-\frac{\Delta[\text{H}_2\text{O}_2]}{\Delta t} = \frac{\Delta[\text{H}_2\text{O}]}{\Delta t} \tag{17.3}$$

Also, because 1 mol of O_2 is produced by the decomposition of 2 mol of H_2O_2, the rate of formation of O_2 will be one-half the rate of disappearance of H_2O_2:

$$\frac{\Delta[\text{O}_2]}{\Delta t} = -\frac{1}{2}\frac{\Delta[\text{H}_2\text{O}_2]}{\Delta t} \tag{17.4}$$

Equations (17.3) and (17.4) will hold only if the conversion of H_2O_2 into products does not involve intermediates which persist. If H_2O_2 forms a series of intermediates which last a long time before they are converted into O_2 and H_2O, the rate of disappearance of H_2O_2 will not equal the rate of appearance of H_2O. The rate of appearance of product is not necessarily equal or proportional to the rate of disappearance of reactant; if there are two or more reactants, their rates of disappearance are not necessarily equal or proportional; if there are two or more products, their rates of appearance are not necessarily equal or proportional. The relationship between the different rates must be determined experimentally and cannot be deduced from the stoichiometric equation. If we detect a difference between the experimental rates of appearance and disappearance, we have obtained a valuable clue to the mechanism.

Consider the reaction between $CuCl_2$ and porphin:

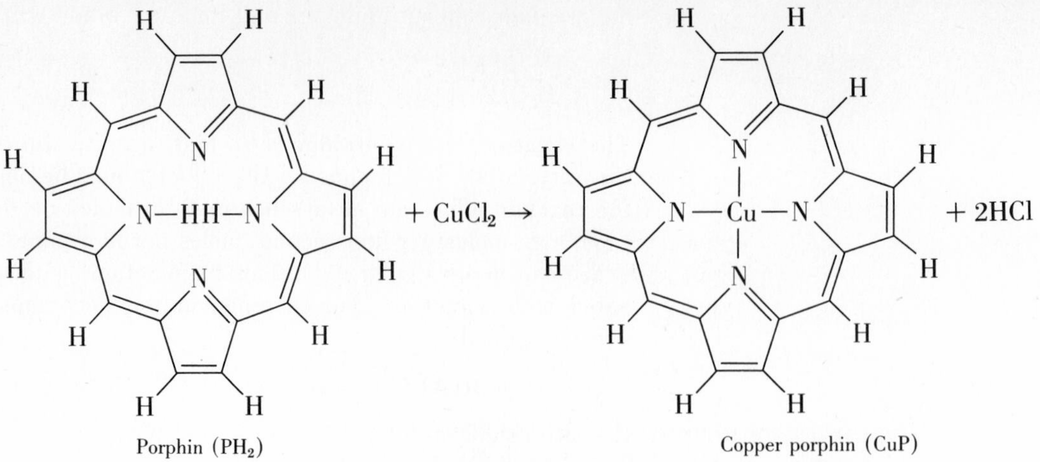

$$\text{Porphin } (PH_2) + CuCl_2 \rightarrow \text{Copper porphin } (CuP) + 2HCl$$

Porphin (PH_2)

Copper porphin (CuP)

The Cu(II) ion replaces two protons in the center of the porphin molecule, and copper porphin forms. (Since only the two protons in the center of the porphin molecule are replaceable, it is customary to symbolize porphin as PH_2 and copper porphin as CuP.) During this reaction, the concentrations of Cu^{2+} and PH_2 decrease rapidly and at the same rate; the CuP concentration increases at a rate equal and opposite to the rate of disappearance of either reactant; and the total concentration of the porphyrins, PH_2 and CuP, remains constant (see Fig. 17.1). We conclude from these observations that the reaction is relatively simple on the molecular level. The Cu^{2+} ion and the porphin molecule come together directly and form some intermediate aggregate which "immediately" produces CuP.

$$Cu^{2+} + H_2P \rightarrow \overset{\text{Intermediate}}{[Cu^{2+}\text{---}H_2P]} \rightarrow CuP + 2H^+$$

The product molecule, CuP, is formed directly from the intermediate aggregate, and thus CuP and $2H^+$ form at equivalent rates.

We must devise different mechanisms when the rates of appearance and disappearance are not equal.

EXAMPLE 17.1 In a reaction for which the stoichiometry is $A + B \rightarrow C$, the rate of disappearance of A and B are equal, but in the beginning of the reaction these rates are greater than the rate of appearance of C. Present a mechanism consistent with these observations.

Solution: Since A and B disappear at the same rate, we must guess that they react with each other directly, otherwise their *equal rates* of disappearance would be unrelated or accidental, which is not very probable since they are reactants. In the first step one molecule each of reactants A and B come together to form an intermediate:

FIGURE 17.1

The results of an experiment in which the $CuCl_2$ *concentration is initially twice the concentration of* PH_2. *The disappearance rates of* $CuCl_2$ *and* PH_2 *are equal. The appearance rate of* CuP *is equal but opposite to the disappearance rate of* $CuCl_2$ *and* PH_2.

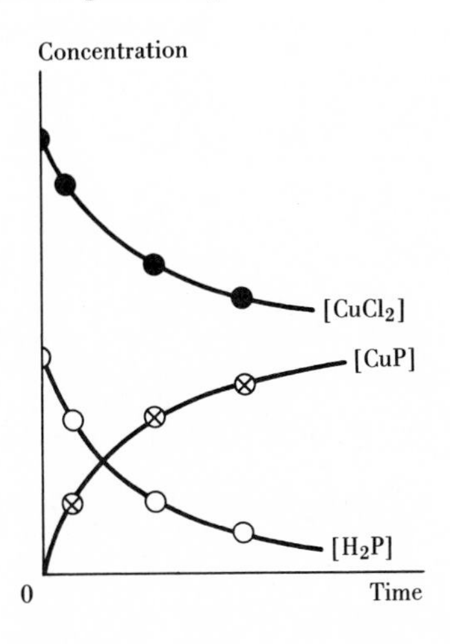

Step 1: $A + B \rightarrow [A \cdots B]$

From the stoichiometry ($A + B \rightarrow C$) we know that the intermediate $[A \cdots B]$ must eventually yield product C, but the rate at which C is formed from $[A \cdots B]$ must be much lower than the rate at which $[A \cdots B]$ is formed from the reactant molecules.

Step 2: $[A \text{---} B] \xrightarrow[\text{slow}]{\text{very}} C$

The interpretation of this reaction may or may not be correct; other mechanisms can be devised. However, the one given is the simplest way of explaining the characteristics of the reaction and can easily be tested. Since it is hypothesized that the intermediate $[A \cdots B]$ reacts only slowly, its concentration should become high and we should be able to devise an experiment to detect it. If it is not detectable, we must revise our mechanism. As we learn more about the nature of a reaction, the mechanism becomes more detailed.

The above examples were presented to demonstrate how the study of reaction rates permits the chemist to gain some understanding of the mechanism. The rates of chemical reactions are greatly affected by changes in the concentrations of the reactants, changes in temperature, and the introduction of catalysts; and again, these effects give us more facts which can be used to develop a detailed mechanism.

17.3 EFFECT OF CONCENTRATION ON THE RATE OF A REACTION

Wilhelmy's early work

The first quantitative studies of reaction rates were reported by the German physicist Ludwig Wilhelmy (1812–1864) in 1850. He found that when sucrose is dissolved in water, it is converted into a different form of sugar, called *invert sugar*. Today we know that sucrose is a disaccharide which hydrolyzes in acidified aqueous solution to produce two monosaccharides, glucose and fructose:

$$\underset{\text{Sucrose}}{H_2O + C_{12}H_{22}O_{11}} \xrightarrow[\text{acid}]{\text{mineral}} \underset{\text{Glucose}\dagger}{C_6H_{12}O_6} + \underset{\text{Fructose}\dagger}{C_6H_{12}O_6} \qquad (17.5)$$

The acid is indicated over the arrow because it is a catalyst not a reactant.

† Fructose and glucose are isomeric monosaccharides. See p. 546.

Wilhelmy found that the rate of disappearance of sucrose S is directly proportional to its concentration:

Rate of disappearance of sucrose $\propto$ [S]

or, since a proportionality may be converted into an equation by introducing a proportionality constant:

$$\text{Rate} = -\frac{\Delta[\text{S}]}{\Delta t} = k[\text{S}] \qquad (17.6)$$

The proportionality constant k is called a *rate constant;* the greater the value of the rate constant k, the greater the rate of the reaction.

FIGURE 17.2

Plots of data for the hydrolysis of sucrose at 35°C in a solution in which $[\text{HCl}]_0 = 0.099$.

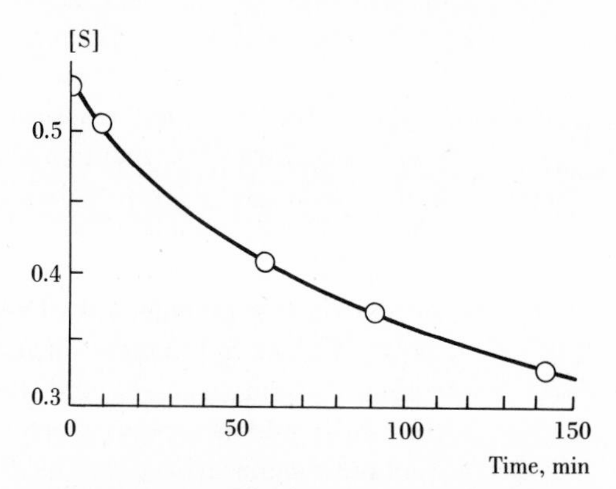

(*a*) *A plot of the concentration of sucrose vs. time is not linear*

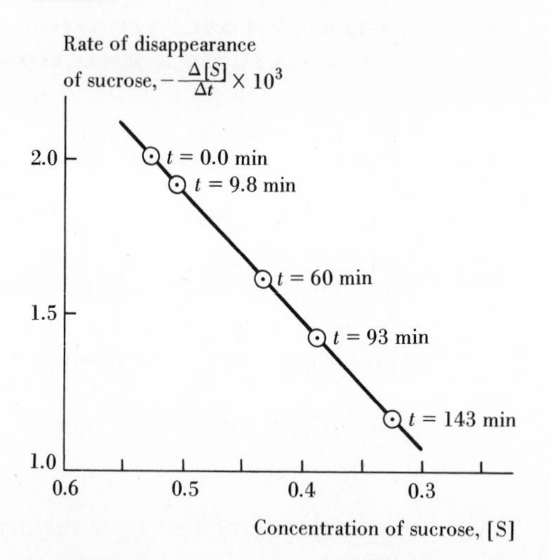

(*b*) *A plot of the rate vs. sucrose concentration is linear and the slope equals the rate constant*

Figure 17.2*a* and *b* shows plots of the sucrose concentration versus time and a plot of the rate of the disappearance of sucrose $\Delta[S]/\Delta t$ versus its own concentration. Figure 17.2*b* shows that the rate decreases in direct proportion to the decrease in the concentration of sucrose, in keeping with Eq. (17.6). Also, in view of Eq. (17.6) the value of the rate constant is just equal to the slope of the plot shown in Fig. 17.2*b*, and hence k turns out to be approximately 0.004 min^{-1}. This means that in a solution which contains 1 mol of sucrose per liter, 0.004 mol of sucrose is hydrolyzed per minute; in a solution containing 0.5 mol of sucrose per liter, 0.002 mol of sucrose is hydrolyzed per minute. Since the rate of hydrolysis changes with concentration, it is more appropriate to write the equation in differential form:

$$\text{Rate} = \frac{d[S]}{dt} = -k[S]$$

We shall occasionally use the differential notation. Students unfamiliar with the calculus should simply interpret the quantity $d[S]/dt$ to mean the instantaneous rate, or the value of $\Delta[S]/\Delta t$ at a particular instant.

The law of mass action and rate expressions

The systematic approach to finding a rate law for a given reaction involves the law of mass action, which states that the rate of a reaction is proportional to the concentrations of the reactants. We have already used this law in discussing chemical equilibrium. If we apply this law to the hydrolysis of sucrose, we write

$$\text{Rate} = -k[S]^x[H_2O]^y$$

because sucrose and water are reactants and we expect the rate of the reaction to depend on their concentrations. Wilhelmy found that $x = 1$ under the conditions of his experiments; he did not determine y.

If we apply the law of mass action to reactions in general, we can write general rate expressions. For example, if the stoichiometry is

$$aA + bB + cC + \cdots \rightarrow \text{products}$$

by the law of mass action the rate expression for this reaction is

$$\text{Rate} \propto \text{concentrations of reactions}$$

or

$$\frac{\Delta[A]}{\Delta[t]} = -k[A]^x[B]^y[C]^z \cdots$$

where A, B, and C are reactants *and* catalysts and x, y, and z are exponents of the concentration terms. The chemical kineticist must determine k and the values of the exponents $x, y, z, \ldots$. These exponents may be zero, positive or negative integers, or fractions. The sum of the exponents of the concentration terms in the experimental rate equation is called the *order* of the reaction. When the sum equals 1, the reaction is first order; when the sum equals 2, the reaction is second order, and so on. Wilhelmy showed that the acid hydrolysis of sucrose is *first order.*

17.4 FIRST-ORDER REACTIONS

A reaction is first order if it follows the experimental rate law of the form

$$\text{Rate} = \pm k_1[\text{R}]^1 \qquad \text{or} \qquad \frac{\text{rate}}{[\text{R}]} = \pm k_1 = \text{const}$$

In this equation [R] is the concentration of the reactant, and rate means either the rate of disappearance of reactant or the rate of appearance of product. In differential form the rate law for the disappearance of reactant is

$$\text{Rate of disappearance of reactant} = \frac{d[\text{R}]}{dt} = -k_1[\text{R}] \tag{17.7}$$

and for the appearance of products P

$$\text{Rate of appearance product} = \frac{d[\text{P}]}{dt} = +k_1[\text{R}]$$

By the methods of calculus, Eq. (17.7) can be converted into an equivalent logarithmic form:

$$2.303 \log \frac{[\text{R}]}{[\text{R}]_0} = -k_1(t - t_0) \tag{17.8}$$

where $[\text{R}]_0$ = concentration of reactant at time t_0
$[\text{R}]$ = concentration at some later time t

The constant k_1 is called the first-order rate constant, and, as can be determined from the unit analysis of Eq. (17.8), it has the units of reciprocal time, for example, s^{-1}, min^{-1}, h^{-1}. The magnitude of the rate constant is a direct measure of the speed of a reaction. When we wish to compare reactions from the point of view of chemical kinetics, we use the rate constant as an index.

We can use Eq. (17.8) to determine the concentration of the reactant or product at any time during the reaction.

EXAMPLE 17.2 At the beginning of an experiment at 35°C the concentration of sucrose is 0.530 mol/l. It is hydrolyzed at 35°C in a solution which contains 0.099 mol/l of HCl. The first-order rate constant under these conditions is 0.004 min^{-1}. Determine the concentration of the reactant sucrose and the product fructose at the end of 60 min.

Solution: Write down Eq. (17.8)

$$2.303 \log \frac{[\text{R}]}{[\text{R}]_0} = -k_1(t - t_0)$$

Let t_0 equal the time at the beginning of the experiment:

$$t_0 = 0.0 \text{ min} \qquad \text{and} \qquad [\text{R}]_0 = 0.530\ M$$

We wish to know [R], which is the concentration of sucrose at the end of 60 min:

$$t = 60 \text{ min} \qquad \text{and} \qquad [\text{R}] = x$$

Substitution into Eq. (17.8) yields

$$2.303 \log \frac{x}{0.530} = -0.004 \text{ min}^{-1}(60 \text{ min} - 0 \text{ min})$$

$$\log x = -0.380 = 0.620 - 1$$

$$x = 10^{0.620} \times 10^{-1} = [\text{R}] = 0.417 \text{ mol/l}$$

The concentration of sucrose is 0.407 M after 60 min. By examining the stoichiometry of the reaction [Eq. (17.5)] we can calculate the concentration of the product fructose:

$$\underset{\text{Sucrose}}{C_{12}H_{22}O_{11}} + H_2O \xrightarrow{H^+} \underset{\text{Fructose}}{C_6H_{12}O_6} + \underset{\text{Glucose}}{C_6H_{12}O_6}$$

The equation shows that 1 mol of fructose forms for each mole of sucrose which hydrolyzes. Since the initial concentration of sucrose was 0.530 mol/l and the concentration at the end of 60 min was 0.417 mol/l, the amount which hydrolyzed was 0.113 mol/l. Hence, the concentration of fructose (and of glucose) is 0.113 mol/l:

Concentration of fructose = decrease in concentration of sucrose

or

$$[\text{F}] = \Delta[\text{S}] = 0.530\ M - 0.417\ M = 0.113\ M$$

The half-life $t_{1/2}$ for first-order reactions

Suppose that we wish to calculate the time necessary for a reactant to be completely consumed in a first-order reaction; i.e., we wish to calculate t when $[\text{R}] = 0$. Substitution into Eq. (17.8) yields

$$\log \frac{[\text{R}]_0}{0} = \frac{k_1(t - t_0)}{2.303}$$

an equation which could be satisfied only if $t=\infty$. Therefore, since in theory an infinite time is required for the completion of any first-order reaction, this time is not a useful characteristic of a first-order reaction. However, the time required for a definite fraction of the reactant to disappear varies from reaction to reaction. Chemists often determine the half-life $t_{1/2}$, which is the time required to reduce the concentration of a reactant by one-half. Thus, when reactant concentration drops from $[R]_0$ to $\frac{1}{2}[R]_0$, the time elapsed, $t-t_0$, is equal to $t_{1/2}$. By Eq. (17.8), $[R]=\frac{1}{2}[R]_0$ when $t-t_0=t_{1/2}$; substitution gives

$$2.303 \log \frac{\frac{1}{2}[R]_0}{[R]} = -k_1 t_{1/2}$$

This reduces to

$$2.303 \log \tfrac{1}{2} = -k_1 t_{1/2} \qquad \text{or} \qquad 2.303 \log 2 = +k_1 t_{1/2}$$

and since 2.303 log 2 equals 0.693, we have

$$t_{1/2} = \frac{0.693}{k_1} \tag{17.9}$$

which shows that a simple relationship exists between the first-order rate constant k_1 and the half-life $t_{1/2}$. Note that $t_{1/2}$ is independent of concentration for the first-order reaction. This means that the concentration of the reactant will decrease by a factor of $\frac{1}{2}$ during each half-life period, regardless of the concentration at the beginning of the period.

EXAMPLE 17.3 Using the data in Example 17.2, evaluate $t_{1/2}$ for the acid hydrolysis of sucrose at 35°C.

Solution: Use Eq. (17.9):

$$t_{1/2} = \frac{0.693}{0.004\ \text{min}^{-1}} = 173\ \text{min}$$

EXAMPLE 17.4 What is the concentration of sucrose after hydrolysis at 35°C (*a*) for a period of one half-life and (*b*) for a period of five half-lives if the initial concentration is 1.00 M?

Solution: (*a*) After one half-life the concentration is reduced by one-half.

$$\tfrac{1}{2}(1.00\ M) = 0.500\ M$$

(*b*) In each half-life period the concentration is reduced by one-half. At the end of the first half-life it is 0.500 M. At the end of the second half-life it is reduced by one-half again:

$\frac{1}{2}(0.500\ M) = 0.25\ M$ in two half-lives

or

$\frac{1}{2}(\frac{1}{2} \times 1.00\ M) = (\frac{1}{2})^2(1.00\ M)$

At the end of the third half-life the concentration will be $(\frac{1}{2})^3(1.00\ M)$. In n half-life periods the concentration is

$$(\tfrac{1}{2})^n C_0$$

where C_0 is the initial concentration. At the end of five half-lives $n = 5$ and

$$(\tfrac{1}{2})^n C_0 = (\tfrac{1}{2})^5(1.00\ M) = \tfrac{1}{32}(1.00\ M) = 0.0313\ M$$

EXAMPLE 17.5 A reaction of the stoichiometric form $A \rightarrow B$ is found to be first order. The rates of appearance and disappearance are equal, and the rate constant is $1.20 \times 10^{-3}\ s^{-1}$. In a particular reaction, the initial concentration of A is 0.450 mol/l, and the initial concentration of B is zero. Find (*a*) the concentrations of A and B at the end of 1 h and (*b*) the half-life for the reaction.

Solution: (*a*) Using Eq. (17.8), we have

$$\log \frac{[R]}{[R]_0} = -\frac{k_1(t - t_0)}{2.303}$$

Substitute $t = 1$ h or 3600 s, $t_0 = 0$ s, $[R]_0 = [A]_0 = 0.450$ mol/l, $[R] = [A]$, and solve for $[A]$:

$$\log \frac{[A]}{0.450} = -\frac{(1.20 \times 10^{-3}\ s^{-1})(3600\ s - 0\ s)}{2.303}$$

$$\log [A] - \log 0.450 = -1.88$$
$$\log [A] = -1.88 + \log (4.50 \times 10^{-1}) = -2.23 = 0.77 - 3$$
$$[A] = 5.89 \times 10^{-3}\ M$$

Since 1 mol of A produces 1 mol of B, the concentration of B must be equal to the decrease in concentration of A:

$$[A]_0 - [A] = 0.450 - 0.00589 \approx 0.444 = [B]$$

(*b*) From Eq. (17.8)

$$t_{1/2} = \frac{0.693}{1.20 \times 10^{-3}\ s^{-1}} = 578\ s$$

In every 578-s period one-half of the reactant disappears. At the end of the first 578 s, one-half of A has reacted, or $\frac{1}{2}(0.45)$ mol in each liter. The concentration of A is then 0.225 mol/l. At the end of the next 578-s period another one-half has reacted, leaving $\frac{1}{2}(0.225)$, or 0.113 mol/l, and so on.

Hydrogen peroxide is decomposed by thiosulfate ion in slightly acidic solution. The stoichiometric equation is

$$H_2O_2 + \underset{\text{Thiosulfate ion}}{2S_2O_3^{2-}} + 2H^+ \rightarrow 2H_2O + \underset{\text{Tetrathionate ion}}{S_4O_6^{2-}}$$

The experimental rate law for the disappearance of $S_2O_3^{2-}$ at 25°C is

$$\text{Rate} = -k_2[H_2O_2][S_2O_3^{2-}] \qquad k_2 = 8.3 \times 10^{-3} \text{ l/mol-s}$$

By our definition the reaction is second order since the sum of the exponents of the concentration terms equals 2. The order has no obvious relationship to the stoichiometric coefficients. To generalize, if the stoichiometric equation is

$$a\text{A} + b\text{B} + c\text{C} + \cdots \rightarrow \text{products}$$

the rate of disappearance of A is

$$\frac{d[\text{A}]}{dt} = -k_n[\text{A}]^x[\text{B}]^y[\text{C}]^z \cdots$$

If the reaction is second order,

$$x + y + z + \cdots = 2$$

The two most common rate expressions for second-order reactions are

$$\text{Rate} = \frac{d[\text{A}]}{dt} = -k_2[\text{A}][\text{B}] \qquad (17.10a)$$

and

$$\text{Rate} = \frac{d[\text{A}]}{dt} = -k_2[\text{A}]^2 \qquad (17.10b)$$

A reaction which obeys the rate law given by Eq. (17.10*a*) is said to be first order in each reactant, A and B, and second order overall. A reaction which obeys the rate law given by Eq. (17.10*b*) is said to be second order in one reactant and second order overall. Both these equations can be integrated by the methods of the calculus to yield more useful forms, but we shall be satisfied with them as they are.

The second-order equations indicate that doubling the concentration of reactants will quadruple the rate. This is a characteristic of a second-order reaction and is useful in experimental work aimed at discovering the order of a reaction. Examination of Eq. (17.7) or (17.8) will reveal that doubling the concentration of the reactant in a first-order reaction simply doubles the rate of the reaction.

The rate constant for a second-order reaction will be symbolized by k_2. An analysis of Eq. (17.10a) or (17.10b) shows that k_2 has the units of liters per mole-second, i.e., reciprocal concentration multiplied by reciprocal time, 1/(concentration × time). Substitution into Eq. (17.10a) demonstrates this:

$$\frac{d[\mathrm{A}]}{dt} = -k_2[\mathrm{A}][\mathrm{B}]$$

$$\frac{\mathrm{mol/l}}{\mathrm{s}} = -k_2 \ \mathrm{mol/l} \times \mathrm{mol/l}$$

Therefore, in order for the right side to have the same units as the left side of the equation k_2 must be in liters per mole-second.

The half-life concept for a second-order reaction is meaningful only if the concentrations in the rate expression are initially equal. Under these conditions it can be shown that the half-life of a second-order reaction is inversely proportional to the concentration of the reactants:

$$t_{1/2} = \frac{1}{k_2 C} \tag{17.11}$$

Therefore, we have another feature which distinguishes second-order from first-order reactions: the half-life of a first-order reaction is independent of concentration [Eq. (17.9)], while the half-life of a second-order reaction is inversely proportional to concentration.

Third-order reactions

Third-order reactions are rare. A few gaseous reactions involving nitric oxide, NO, seem to follow third-order kinetics, among them are:

$$2\mathrm{NO}(g) + \mathrm{Br_2}(g) \rightarrow 2\mathrm{NOBr}(g) \qquad \frac{d[\mathrm{NO}]}{dt} = -k_3[\mathrm{NO}]^2[\mathrm{Br_2}]$$

$$2\mathrm{NO}(g) + \mathrm{O_2}(g) \rightarrow 2\mathrm{NO_2}(g) \qquad \frac{d[\mathrm{NO}]}{dt} = -k_3[\mathrm{NO}]^2[\mathrm{O_2}]$$

Third-order reactions in solution have been reported, but some chemical kineticists believe that they may actually be second order.

17.6 MOLECULARITY

The rate law for a chemical reaction is something we can measure; we say that it is an *experimental observable*. Because we have a firm belief in the molecular nature of matter, we always examine our results for such implications. The experimental rate law leads us to the hypothetical mechanism. Each chemical step of a mechanism involves a certain number of molecules. The number of reactant molecules involved in a chemical step is called the *molecularity*. The thermal decomposition of acetaldehyde, $CH_3{-}CHO$, leads to the production of methane, ethane, and CO. The experimental observations can be explained in terms of a three-step mechanism involving reactive methyl free radicals, $CH_3^{\cdot}$, as intermediates:

Step 1: $CH_3CHO \rightarrow CH_3^{\cdot} + CHO$
Step 2: $CH_3CHO + CH_3^{\cdot} \rightarrow CH_4 + CO + CH_3^{\cdot}$
Step 3: $CH_3^{\cdot} + CH_3^{\cdot} \rightarrow C_2H_6$

Step 1 involves one reactant molecule and the molecularity is 1, or step 1 is said to be *uni*molecular; steps 2 and 3 are *bi*molecular.

The reaction between H_2 and I_2 has been studied very carefully by many investigators. Its stoichiometry and rate law are simple:

$$H_2(g) + I_2(g) \xrightarrow{300°C} 2HI(g) \qquad \frac{d[I_2]}{dt} = -k_2[H_2][I_2]$$

Until recently it was believed that this reaction proceeded by a mechanism involving a bimolecular step producing an intermediate:

$H_2 + I_2 \rightarrow$ H------------H
⋮ ⋮
I-------------I

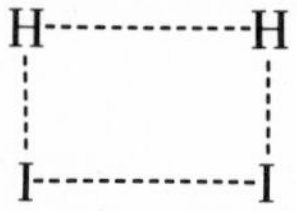

Intermediate

followed by the decomposition of the intermediate in a unimolecular step:

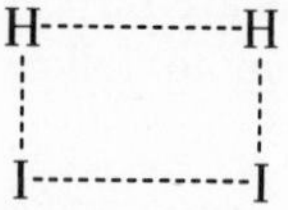

H------------H
⋮ ⋮ $\rightarrow 2HI$
I-------------I

More recent data show that the reaction probably proceeds by a different two-step mechanism. The first is a unimolecular decomposition of I_2 into atoms:

$I_2(g) \rightarrow 2I(g)$

followed by a termolecular step which produces product:

$2I(g) + H_2(g) \rightarrow 2HI(g)$

Both mechanisms are consistent with the stoichiometry and the rate law for the reaction. Only the recent development of kinetic techniques allowing us to follow extremely rapid chemical transformations makes it possible to get a better glimpse of the hydrogen-iodine mechanism; these techniques have proved the older mechanism incorrect.

We have seen that a step in a mechanism may involve one, two, or three molecules. A step involving only one molecule is said to be *unimolecular*. The simultaneous action of two and three molecules would result in *bimolecular* and *termolecular* steps, respectively. Since the probability of a termolecular collision is quite low, termolecular steps in a mechanism are very slow steps. Any steps involving the collision of more than three molecules are so highly improbable that they are hardly ever considered. Since a net chemical transformation may consist of a series of steps which are of different molecularities and speeds, the order may be nonintegral. The thermal decomposition of acetaldehyde is $\frac{3}{2}$ order; the gaseous reaction involving NO and H_2 is $\frac{5}{2}$ order:

$$2NO + 2H_2 \rightarrow N_2 + 2H_2O \qquad \frac{d[NO]}{dt} = -k[NO][H_2]^{3/2}$$

In other cases the experimental rate law is extremely complicated, and the concept of order is meaningless. For example, the reaction between H_2 and Br_2, which has the same stoichiometry as the reaction between hydrogen and iodine, is very complex, as can be seen from the rate law

$$H_2 + Br_2 \rightarrow 2HBr \qquad \frac{d[HBr]}{dt} = \frac{k'[H_2][Br_2]^{1/2}}{k'' + [HBr][Br_2]^{-1}}$$

Nevertheless, when an accurate mathematical expression of the rate law is discovered, the chemical kineticist has a powerful aid in deducing the reaction mechanism. The reaction mechanism, which is a microscopic model for the chemical reaction, must be consistent with the experimental rate law and the stoichiometry of the reaction.

17.7 ZERO-ORDER REACTIONS AND CATALYSIS

The law of mass action states that the rate of a reaction is directly proportional to the concentration of the reactants. In view of this very reasonable principle it may seem surprising that some reactions have rates that appear to be independent of the concentration of reactants. Again, for the general reaction

$$aA + bB + cC + \cdots \rightarrow \text{products}$$

the general rate law would have the form

$$\text{Rate} = \frac{d[A]}{dt} = -k_n[A]^x[B]^y[C]^z \cdots$$

$$n = x + y + z + \cdots = \text{order}$$

In many reactions $n = 0$; that is, the reaction is *zero* order, and the rate law has the form

$$\text{Rate} = \frac{d[A]}{dt} = -k_0[A]^0[B]^0[C]^0 \cdots = -k_0 \qquad (17.12)$$

The zero exponent of each concentration term indicates that the rate is independent of the concentration. The reaction proceeds at a constant rate.

Most of the zero-order reactions which have been studied involve a catalyst, i.e., a chemical agent which changes the speed of a chemical reaction but does not undergo any *net* chemical change (see page 325). Almost all reactions in living organisms are catalyzed by protein catalysts called *enzymes*.

Since the catalyst undergoes no net chemical change, it does not appear in the stoichiometric equation. For example, NH_3 is rapidly converted into its elements on the surface of tungsten metal, the rate of this reaction being independent of the concentration or pressure of the gaseous NH_3 above the metal.

$$NH_3(g) \xrightarrow{W} \tfrac{1}{2}N_2(g) + \tfrac{3}{2}H_2(g) \qquad \frac{d[NH_3]}{dt} = -k_0[NH_3]^0 = -k_0$$

However, an increase in the surface area of the tungsten causes a proportional increase in the reaction rate. Obviously, there must be some interaction between the metal surface and the reactant. We presume that one step of the mechanism involves bond formation between an ammonia molecule and a tungsten atom at the surface (Fig. 17.3). Furthermore, since H_2 and N_2 are formed rapidly in the presence of tungsten, there must be a rapid severing of N—H bonds after the NH_3 molecule is bonded at the tungsten surface. Perhaps the bond which forms between nitrogen and tungsten causes the N—H bonds of ammonia to lengthen and weaken. In any event a rapid severing of N—H bonds must occur, and H—H and N—N bonds must form.

On the basis of the picture shown in Fig. 17.3 we can explain why the rate of decomposition of ammonia on the tungsten surface appears to be independent of the ammonia concentration. Suppose that all the sites on the catalytic surface are occupied by ammonia at ordinary pressures. Then the catalyst is working at maximum speed; all tungsten atoms which can form bonds with NH_3 are doing so. An increase in NH_3 pressure or concentration will therefore not cause an increase in the reaction rate since the reaction must in-

FIGURE 17.3

The catalytic surface.

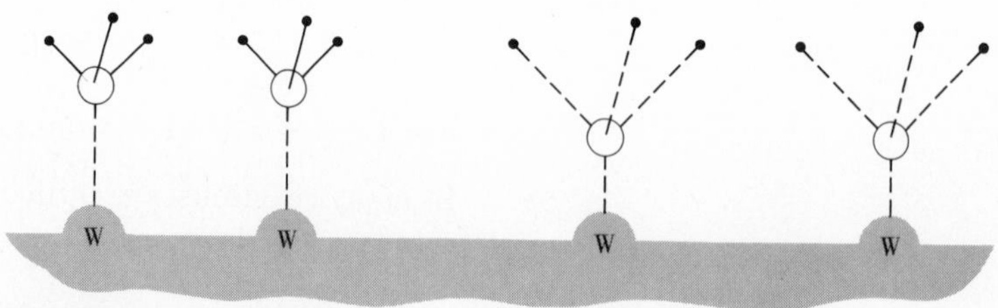

(a) The NH_3 molecule bonds to W atoms at the surface

(b) As the N—W bonds shorten and strengthen the N—H bonds lengthen and weaken

FIGURE 17.4

Pressure effects in presence of a catalytic surface.

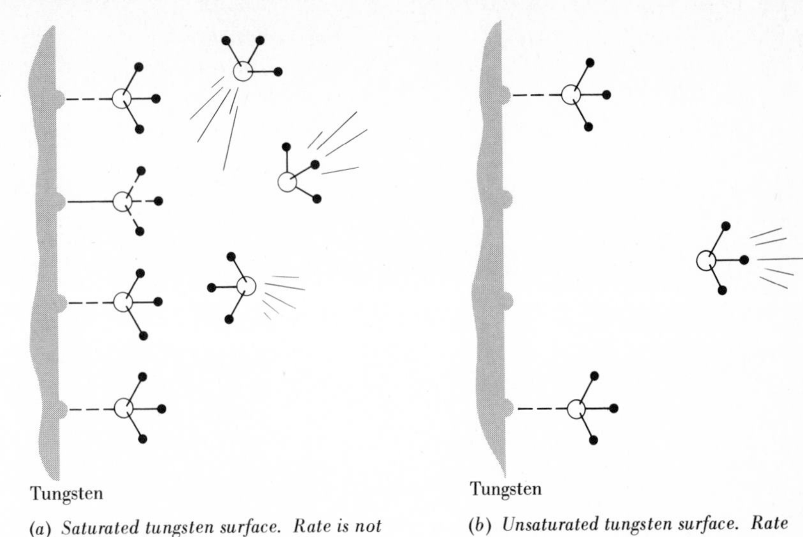

(a) Saturated tungsten surface. Rate is not increased when pressure increases

(b) Unsaturated tungsten surface. Rate is increased when pressure increases

volve the surface (see Fig. 17.4). We can represent this hypothesis by the following mechanism:

Step 1: $NH_3(g)$ + W surface → H_2—N---W surface (three H atoms bonded to N by solid lines)

Step 2: H_3N---W surface (solid N—H bonds) → H_3N---W surface (dashed N---H bonds)

Step 3: H_3N---W surface (dashed N---H bonds) → $\frac{3}{2}H_2(g) + \frac{1}{2}N_2(g)$ + W surface

Step 1 represents the formation of a bond between an ammonia molecule and a tungsten atom. Step 2 represents the lengthening and weakening of the N—H bonds; this is followed in step 3 by bond severing and product formation.

To test our mechanism we can determine the rate of NH_3 decomposition at very low pressures where the catalytic surface is not saturated. Under these conditions an increase in pressure should cause an increase in rate. This is exactly what is observed. At low NH_3 concentration the rate of decomposition of NH_3 on tungsten surfaces is proportional to the concentration of NH_3 and the surface area of the tungsten catalyst S_W:

$$\text{Rate} = k[NH_3]S_W$$

We can substitute pressures for concentrations of gases since the pressure of a gas is proportional to its concentration. For an ideal gas $pV = nRT$; the quantity n/V equals concentration in moles per liter, and $p = cRT$. Therefore, we can rewrite the rate equation for the NH_3 decomposition as

$$\text{Rate} = k[NH_3]S_W = k\frac{P_{NH_3}}{RT}S_W = \frac{k}{RT}P_{NH_3}S_W$$

$$= k'P_{NH_3}S_W \qquad \text{where } k' = \frac{k}{RT}$$

In a particular experiment the surface area of the tungsten S_W does not vary; hence, $kS_W = \text{const}$, and the rate is dependent only on the pressure of ammonia:

$$\text{Rate} = \text{const} \times P_{NH_3}$$

As the pressure of ammonia increases, more and more of the catalytic positions of the tungsten surface are taken up. A point is reached where all catalytic positions are occupied, and a further increase in pressure will have no effect on the reaction rate. At this point the reaction is zero order:

$$\text{Rate at low pressure} = \text{const} \times P^1_{NH_3}$$

$$\text{Rate at high pressure} = \text{const} \times P^0_{NH_3} = \text{const}$$

Zero-order kinetics is observed when a reaction occurs at an appreciable rate only in the presence of a catalyst. In order to be effective the catalyst must form a temporary bond with the reactant. When the catalyst is completely tied up, the reaction rate will not be affected by an increase in the concentration of the reactant.

Since the tungsten-catalyzed decomposition of NH_3 occurs at the surface of the catalyst, this is an example of surface, or *heterogeneous*, catalysis. When the catalyst is dissolved in the reaction phase, it is *homogeneous* catalysis. An important example of homogeneous catalysis involves the catalytic action of nitric oxide, NO, on the oxidation of sulfur dioxide, a step in the manufacture of sulfuric acid. The direct oxygen oxidation of SO_2 is very slow:

$$SO_2(g) + \tfrac{1}{2}O_2(g) \xrightarrow{\text{slow}} SO_3(l)$$

In the presence of a trace of nitric oxide, NO, the reaction proceeds rapidly. The mechanism of the reaction in the presence of this catalyst involves a rapid oxidation of NO:

$$NO(g) + \tfrac{1}{2}O_2(g) \rightarrow NO_2(g)$$

The nitrogen dioxide which forms quickly oxidizes the SO_2, and the catalyst is regenerated:

$$NO_2(g) + SO_2(g) \rightarrow NO(g) + SO_3(l)$$

Both these steps are much faster than the reaction between SO_2 and O_2, so that the reaction rate, $d[SO_3]/dt$, is enhanced.

An examination of the mechanisms for the decomposition of ammonia and the oxidation of SO_2 shows that the catalyst is intimately involved in the overall chemical transformation and undergoes chemical changes, *but* it does not undergo a *net* chemical change. Since it is continually regenerated, a catalyst can be very effective even when it is present at a very low concentration.

Many industrial chemical processes employ homogeneous and heterogeneous catalysts. It is economically feasible to use rather expensive materials, like platinum, as *catalysts;* the initial investment is justified because the product is obtained at an appreciable reduction of labor, energy, and time, i.e., at less cost.

EXAMPLE 17.6 From the data in Table 17.1, determine whether the reaction is first or zero order. Calculate the rate constant.

TABLE 17.1: DATA FOR A → B

Time, min	[B]
0	0
5.00	0.0160
7.00	0.0224
11.00	0.0352

Solution: The data show that the rate of formation of B, $d[B]/dt$, or $\Delta[B]/\Delta t$, is constant. Hence, the reaction is zero order.

Proof 1: In the time interval 0 to 5 min:

$$\frac{\Delta[B]}{\Delta t} = \frac{0.0160 - 0}{5 - 0} = 0.0032 \text{ mol/l-min}$$

In the interval 5 to 7 min:

$$\frac{\Delta[B]}{\Delta t} = \frac{0.0224 - 0.0160}{7.00 - 5.00} = \frac{0.0064}{2} = 0.0032 \text{ mol/l-min}$$

In the interval 7 to 11 min:

$$\frac{\Delta[B]}{\Delta t} = \frac{0.0352 - 0.0224}{11.00 - 7.00} = \frac{0.0128}{4} = 0.0032 \text{ mol/l-min}$$

Proof 2: If the reaction is zero order, the equation $[B] = k_0 t$ governs the time course of the reaction. At 5 min, we have

$$0.0160 \text{ mol/l} = k_0 (5 \text{ min})$$

or

$$k_0 = \frac{0.0160 \text{ mol/l}}{5 \text{ min}} = 0.0032 \text{ mol/l-min}$$

If this is truly a zero-order reaction, we should obtain the same value for k_0 no matter which data are used. Testing the 11-min data gives

$$0.0352 \text{ mol} = k_0 (11 \text{ min})$$

$$k_0 = 0.0032 \text{ mol/l-min}$$

Since the same value of k_0 is obtained, we have proved that the reaction is zero order. If we had tested the data with the first-order equation, $-2.303 \log [A]/[A]_0 = k_1 t$, we would have obtained a different value of k_1 for each set of data. Hence, the reaction is zero order, and the rate constant is 0.0032 mol/l-min.

EXAMPLE 17.7 From the data in Table 17.2 determine whether the reaction is first zero order and calculate the rate constant.

TABLE 17.2: DATA FOR A → B

Time, min	[A]	[B]
0.00	0.0100	0.0000
9.82	0.009646	0.000354
59.60	0.008036	0.001964

Solution: The rate of formation of B is not constant. In the interval from 0 to 9.82 min

$$\frac{\Delta B}{\Delta t} = \frac{0.000354 \text{ mol/l}}{9.82 - 0 \text{ min}} = 0.0000360 \text{ mol/l-min}$$

In the interval from 9.82 to 59.60 min

$$\frac{\Delta B}{\Delta t} = \frac{0.001964 - 0.000354 \text{ mol/l}}{59.60 - 9.82 \text{ min}}$$

$$= \frac{0.00161 \text{ mol/l}}{49.78 \text{ min}} = 0.0000323 \text{ mol/l-min}$$

Since the rate is not constant, the reaction is not zero order. (We could have shown that the rate of disappearance of A is not constant, which also signifies that the reaction is not zero order.) By using the first-order equation, it will be possible to prove that the reaction is first order because the same value of k_1 will be obtained from any set of data. For first-order reactions

$$2.303 \log \frac{[A]}{[A]_0} = -k_1(t - t_0) \qquad \text{or} \qquad k_1 = \frac{2.303}{t - t_0} \log \frac{[A]_0}{[A]}$$

At the end of 9.82 min

$$k_1 = \frac{2.303}{9.82 - 0 \text{ min}} \log \frac{0.0100}{0.009646} = 0.0367 \text{ min}^{-1}$$

In the interval from 9.82 to 59.60 min

$$k_1 = \frac{2.303}{59.60 - 9.82 \text{ min}} \log \frac{0.009646}{0.008036} = 0.00367 \text{ min}^{-1}$$

Note that $[A]_0$ and t_0 correspond to the concentration and time at the beginning of the time interval. The reaction is first order because the rate constant k_1 has the same value using either set of data. (You should calculate k_1 using the data in the interval from 0 to 59.60 min.)

EXAMPLE 17.8 Determine the order with respect to each reactant and the overall order of the reaction A + B → products, from the data in Table 17.3.

Solution: By doubling the initial concentration of B and holding

TABLE 17.3: DATA FOR A + B → PRODUCTS

Experiment	$[A]_0$	$[B]_0$	*Initial rate*, mol/l-min
1	0.0100	0.0100	0.000352
2	0.0100	0.0200	0.001408
3	0.0050	0.0200	0.000704

$[A]_0$ constant (experiments 1 and 2) the rate was quadrupled, which means that the reaction is second order with respect to B. By reducing the initial concentration of A by a factor of $\frac{1}{2}$ (experiments 2 and 3), the rate is decreased by a factor of $\frac{1}{2}$. This indicates that the reaction is first order in A. The rate expression is

$$\text{Rate} = k_3[A]^1[B]^2 = k_3[A][B]^2$$

and therefore the *reaction is third order* overall.

17.8 THE EFFECT OF TEMPERATURE ON REACTION RATE

The rate of a chemical reaction is dramatically affected by an increase in temperature. Benzaldehyde and pyrrole react to produce a porphyrin:

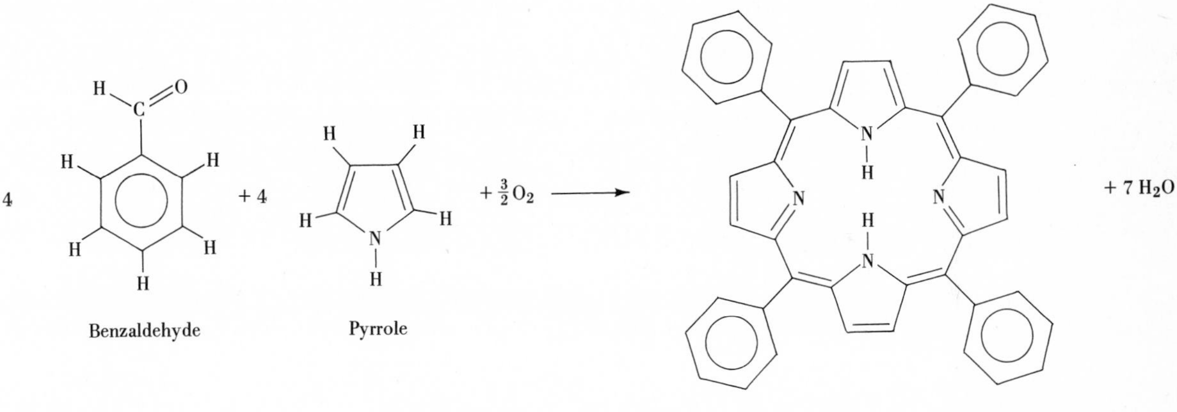

$$4\ \text{Benzaldehyde} + 4\ \text{Pyrrole} + \tfrac{3}{2}O_2 \longrightarrow \text{Tetraphenylporphin} + 7\,H_2O$$

Benzaldehyde

Pyrrole

Tetraphenylporphin

This reaction requires more than 3 weeks for completion at room temperature, whereas at 140°C it is complete in 30 min. There are many systems in which no measurable reaction occurs at room temperature. The Haber process for the industrial production of NH_3 from N_2 and H_2 is an example. At 25°C a mixture of H_2 and N_2 is very stable, but above 400°C the reaction occurs rapidly. Biochemical reactions which are catalyzed by enzymes are very slow below body temperature. Certain bacteria show no growth at all at 20°C, while at 37°C their growth and division, which are chemical processes, occur at incredible rates. A single bacterium, like *Clostridium botulinum*, which causes a fatal disease of the nervous system, can give rise to 5 billion progeny in about 36 h.

Table 17.4 gives the values of the second-order rate constants at several temperatures for the appearance of HI by the reaction of H_2 and I_2. Note that in going from 550 to 630 K the rate constant increases by a factor of about 68; in going from 700 to 800 K it increases by a factor of almost 100. The effect of temperature on the reaction rate is quite variable, but it is always appreciable, and it cannot be neglected even for small temperature variations. As a rough rule the rate constant doubles for every 10 K rise in temperature. Referring to Table 17.4, we would guess that the rate constant at 640 K would be twice the value tabulated for 630 K, that is, 13.60×10^{-3} l/mol-s.

In 1889 Arrhenius discovered that in many cases the rate constant k is related to the absolute temperature T by a relatively simple equation:

$$\ln k = -\frac{B}{T} + \ln A \tag{17.13}$$

or

$$\log k = -\frac{B}{2.303T} + \log A \tag{17.14}$$

The logarithmic equation (17.13) can also be expressed in exponential form by taking the antilog of each side of the equation:

$$k = Ae^{-B/T} \tag{17.15}$$

where e = base of natural logarithms
A = const

Equations (17.13) to (17.15) are equivalent forms describing the effect of the temperature on the rate constant.

TABLE 17.4: Reaction: $H_{2(g)} + I_{2(g)} \rightarrow 2HI_g$
Rate = $k_2[H_2][I_2]$

T, K	$\frac{1}{T} \times 10^3$	k_2, l/mol-s	$\log k_2$
550	1.81	1.14×10^{-4}	−3.944
630	1.59	6.80×10^{-3}	−2.169
700	1.43	1.68×10^{-1}	−0.776
800	1.25	12.5	1.096

The formal equation for a straight line is

$$y = mx + b \tag{17.16}$$

where y = dependent variable
x = independent variable
m = slope
b = y-axis intercept

If we plot y versus x, we obtain a straight line with a slope equal to m and an intercept equal to b. If we consider $\log k$ to be the dependent variable in Eq. (17.14) and $1/T$ to be the independent variable, this equation is also in the form of an equation for a straight line:

$$\log k = -\frac{B}{2.303}\frac{1}{T} + \log A$$

A plot of $\log k$ versus $1/T$ should produce a straight line with a slope equal to $-B/2.303$ and an intercept equal to $\log A$ if the Arrhenius equation holds. The data from Table 17.4 for the HI reaction are plotted according to Eq. (17.14) in Fig. 17.5, and, as can be seen, a straight line is obtained.

The applicability of the Arrhenius equation permits us to make good estimates of rate constants at different temperatures. With the determination of the rate constant at two or more temperatures we can evaluate B and $\log A$ in Eq. (17.14). Substituting the values of B and $\log A$ obtained from analysis of the HI data, we obtain

$$\log k_2 = -\frac{20{,}300}{2.303T} + 11.86 \tag{17.17}$$

($B = 20{,}300$ and $\log A = 11.86$)

Equation (17.17) permits us to calculate the rate constant for the HI reaction at any temperature. This is one of the important uses of the Arrhenius equations. We can deduce equations of the form of Eq. (17.17) for any reaction for which we have obtained the rate constants at two temperatures.

EXAMPLE 17.9 Using Eq. (17.17), calculate the rate constant for the HI reaction at 750 K.

Solution

$$\log k_2 = -\frac{20{,}300}{2.303T} + 11.86 = -\frac{20{,}300}{2.303 \times 750} + 11.86$$

$$= -11.74 + 11.86 = +0.12$$

$$k_2 = 1.32 \text{ l/mol-s}$$

The calculated value seems reasonable in view of the data given in

Table 17.4. We expect the rate constant at 750 K to lie somewhere between the values given for 700 and 800 K, and it does.

We can also estimate k_2 from the plot shown in Fig. 17.5, where $\log k_2$ is plotted versus the reciprocal temperature. The reciprocal of 750 K is 1.333×10^{-3}. The value of $\log k$ on the plot corresponding to the reciprocal of 750 K is about 0.13:

$$\log k_2 = 0.13$$

$$k_2 = 10^{0.13} = 1.35 \text{ l/mol-s}$$

EXAMPLE 17.10 The decomposition of nitrogen pentoxide, given by the stoichiometric equation

$$N_2O_5(g) \rightarrow N_2O_4(g) + \tfrac{1}{2}O_2(g)$$

is a first-order reaction. The rate of disappearance of N_2O_5 is directly proportional to its concentration:

$$\text{Rate of disappearance} = k_1[N_2O_5]$$

The rate constant is $1.72 \times 10^{-5}\ s^{-1}$ at 25°C and $2.40 \times 10^{-3}\ s^{-1}$ at 65°C. Using the Arrhenius equation, calculate the rate constant at 45°C. (The experimental value at this temperature is $2.49 \times 10^{-4}\ s^{-1}$.)

There are two ways to solve this problem: (1) we can derive an equation like Eq. (17.17) and determine the rate constant at 45°C by substitution, or (2) we can plot $\log k_1$ versus $1/T$ and find k_1 at 45°C by graphical interpolation. We shall solve the problem both ways.

Solution 1: Substitute both sets of data into Eq. (17.14):

$$\text{At 25°C: } \log (1.72 \times 10^{-5}) = -\frac{B}{2.303}\frac{1}{298} + \log A$$

FIGURE 17.5

Plot of the data for the HI *reaction according to the Arrhenius equation (17.14) (see Example 17.13)*

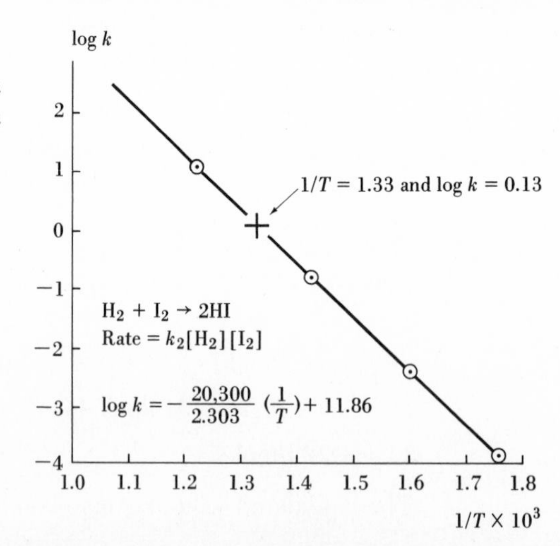

$$\text{At } 65°\text{C: } \log (2.40 \times 10^{-3}) = -\frac{B}{2.303} \frac{1}{338} + \log A$$

The simultaneous solution of these equations yields

$B = 1.24 \times 10^4$ and $\log A = 13.36$

Equation (17.14) can be written specifically for the N_2O_5 reaction as

$$\log k_1 = -\frac{1.24 \times 10^4}{2.303T} + 13.36$$

At 45°C, $T = 318$ K and

$$\log k_1 = -\frac{1.24 \times 10^4}{2.303 \times 318} + 13.36 = -3.57 = 0.43 - 4$$

$$k_1 = 10^{0.43} \times 10^{-4} = 2.7 \times 10^{-4}\ \text{s}^{-1}$$

The value of k_1 obtained from the Arrhenius equation is in good agreement with the experimental value.

We should add one note of caution. Although it may seem that we should be able to obtain the value of k at *any* temperature by this method, it turns out that log k is only an approximately linear function of the reciprocal temperature $1/T$. Therefore, we cannot hope to obtain good values of k if the constants A and B have been evaluated at temperatures very different from the temperature of interest. For example, from the data given in Example 17.10 we should not expect to obtain a good value of the rate constant at 200°C. In general, we can rely on *interpolated* values of k, but we should be suspicious of any value obtained by an *extrapolation* of more than ±25°C outside the temperature range for which we have experimental rate constants. Since B and log A were obtained from data between 25 and 65°C, we can be confident in any calculated rate constant within this interval. In addition, extrapolated estimates of k down to 0°C and up to 90°C will probably be of the correct order of magnitude.

Solution 2: We can also determine k_1 at 45°C from a plot of the data. When log k_1 is plotted versus $1/T$, a straight line should result and by graphical interpolation we can obtain log k when $1/T$ equals $\frac{1}{318}$. The experimental values of k_1 at 298 and 338 K are given in Table 17.5. Plotting log k_1 versus $1/T$, we obtain the graph in Fig. 17.6. We are interested in the value of k_1 at 45°C, or 318 K. At this

TABLE 17.5: EFFECT OF TEMPERATURE

T, K	k_1	$\log k_1$	$\frac{1}{T}$
298 (25°C)	1.72×10^{-5}	−4.763	3.36×10^{-3}
338 (65°C)	2.40×10^{-3}	−2.619	2.96×10^{-3}

FIGURE 17.6

Graphical interpolation: find the value of log k *corresponding to* $1/T = 1/318 = 3.14 \times 10^{-3}$.

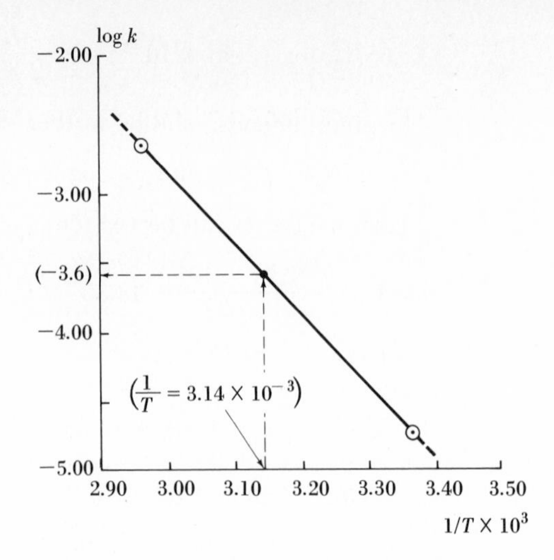

temperature $1/T = 3.14 \times 10^{-3}$. The corresponding value of $\log k_1$ is -3.6.

$$\log k_1 = -3.6 = 0.4 - 4$$

$$k_1 = 10^{0.4} \times 10^{-4} = 2.51 \times 10^{-4}\ \text{s}^{-1}$$

Again our result is in good agreement with experimental value for the rate constant at 45°C.

17.9 THE COLLISION THEORY OF REACTION RATES

In order for a reaction to occur the reactant molecules must collide or come in contact with each other; but in terms of our modern picture of atoms and molecules it is difficult to define what is meant by *contact*. Remember that the electrons in molecular orbitals are not at fixed distances from the nuclei of the atoms in the molecule. How close do molecules have to come before they can be considered to be in contact? The simplest way of dealing with this problem is to consider molecules as hard spheres with definite radii. Then two molecules are in collision when their centers are separated by a distance equal to the sum of the radii of the two molecules. At any greater distance the molecules are not in collision. Assuming that molecules must collide in order to react, we can make an estimate of the reaction rate for gaseous reactions because we can calculate the number of bimolecular collisions which occur in a gaseous sample. If we guess that every bimolecular collision leads to reaction, we can estimate the time required for the completion of a reaction and show that this guess is absurd. In 1 l of an ordinary gas, like O_2, at 0°C and 1 atm there are approximately 10^{31} collisions per second, and since two molecules are involved in each collision, there are 2×10^{31} molecules colliding. The number of molecules in 1 l of gas at 0°C and 1 atm is about 3×10^{22}. Then, on the average, each molecule will collide $\sim 10^9$ (1 billion) times each second, and therefore we estimate that all reactions involving O_2 and similar simple molecules would be over in a very small fraction of a second. This is not observed; in fact, it is an extremely poor estimate. We must conclude that not every collision can lead to chemical reaction since so many observed gaseous reactions require hours or even days for completion. Only a small fraction of the collisions leads to reaction.

Arrhenius suggested that a collision can lead to a chemical transformation only when the colliding molecules have a high relative velocity. This suggestion seems reasonable since molecules with high velocity or high kinetic energy can undergo forceful collisions, causing bonds to break and chemical transformations to occur. Then, according to Arrhenius, a molecule must have a certain minimum of kinetic energy in order to undergo reaction upon collision. He called this minimum the *activation energy*, and we symbolize it E_a. Molecules with a kinetic energy of E_a or more are called *activated* molecules.

The kinetic theory tells us how the velocities and energies are apportioned to the molecules of a gas and allows us to calculate the fraction of activated molecules. The details of the theory are highly mathematical, but the results are easy to understand and use. In principle, the kinetic energies of molecules can vary from zero to infinity. However, there will be very few molecules with very high or very low energies. Most of the molecules will be in a smaller energy range. Also, we shall find many molecules with the same energy. This situation is analogous to the test-grade distribution in a large lecture section. There are very few students with zeros or hundreds; approximately 90 percent of the grades will lie within the relatively narrow range of 60 to 85. In addition, many students will have the same grade; e.g., there will be several 70s, 71s, 72s. Figure 17.7 shows a typical plot of the grade results in the examination of a large group. Note that the maximum, at $G = 72$, means that more students obtained this grade than any other grade; 72 is called the *most probable grade*. If a 90 is the minimum for an A, the total number of A's will be proportional to the shaded area to the right of $G = 90$. The fraction of A grades will equal the shaded area divided by the total area under the curve.

If we plot the number of molecules N possessing a particular kinetic energy versus the kinetic energy (KE), we obtain the bell-shaped curve of Fig. 17.8. The curve is not perfectly symmetrical because our theory allows the maximum energy to be infinite while the minimum energy is zero. The maximum in the curve represents the most probable value of the energy. The average energy E_{av} is slightly greater because the curve is not symmetrical about the max-

FIGURE 17.7

A grade distribution.

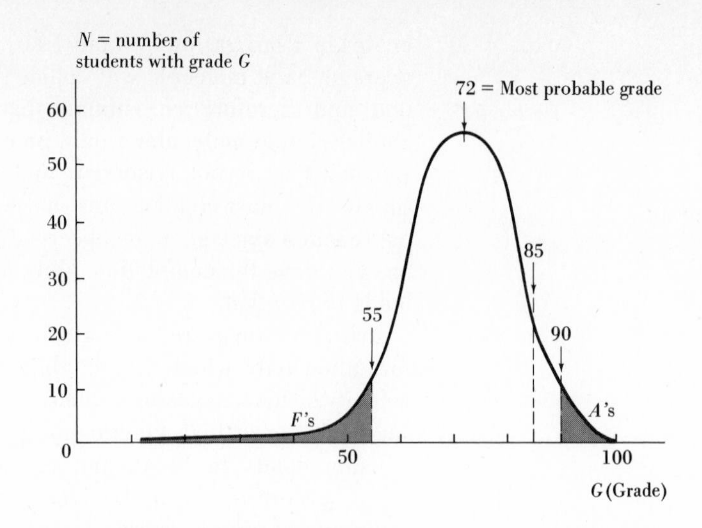

imum but is displaced to higher energies. In the plot, E_a represents the activation energy for a particular reaction. Then all the molecules which have this energy or more are accounted for by the shaded area to the right of E_a in Fig. 17.8. If we divide the shaded area by the total area under the curve, we obtain the fraction of molecules activated. We do not have to make a plot like the one in Fig. 17.8 to determine this fraction; using statistical theory, we can calculate it very easily. It is just $e^{-E_a/RT}$:

$$\text{Fraction of activated molecules} = e^{-E_a/RT} \tag{17.18}$$

where e = base of natural logarithms
R = gas-law constant
T = absolute temperature

FIGURE 17.8

An isothermal energy distribution.

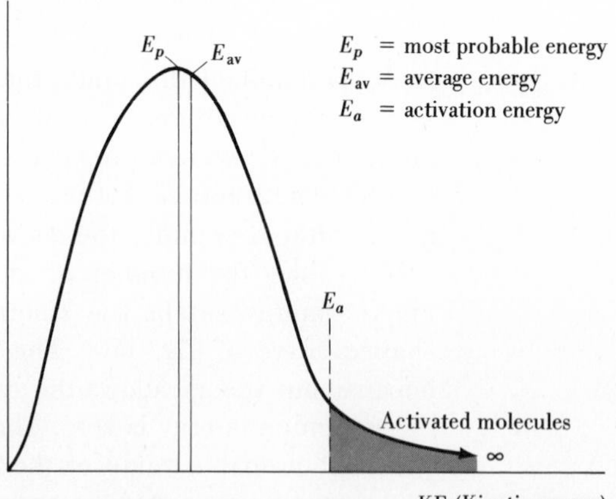

The exponential term $e^{-E/RT}$ is called the *Boltzmann factor*. In general, the fraction of molecules with energies lying between the energy E and infinity is given by $e^{-E/RT}$. For instance, the number of molecules which possess a minimum energy equal to the average energy, E_{av}, is just $e^{-E_{av}/RT}$. This fraction is proportional to the area under the curve to the right of the E_{av} in Fig. 17.8. The fraction of molecules with a minimum of 5000 cal/mol at 25°C is

$$e^{-(5000\,\text{cal/mol})/(1.987\,\text{cal/mol-K}\times 298\,\text{K})} = e^{-8.44} = 2.15 \times 10^{-4}$$

or only about 0.000215 of the molecules.

If we raise the temperature of our gaseous sample, we increase the fraction of activated molecules. This is obvious from Fig. 17.9, which shows the energy distribution at two different temperatures. At the higher temperature T_2 the curve is broader, and the maximum is shifted to higher energy; more important, the number of activated molecules, which is proportional to the shaded area, has increased. An examination of Eq. (17.18) also shows that the fraction of activated molecules will increase as the temperature increases. The following examples will demonstrate the concepts we have just discussed.

EXAMPLE 17.11 Calculate the fraction of molecules which are activated (*a*) at 27 and at 127°C for a reaction in which the activation energy is 6000 cal/mol and (*b*) at 27°C if the activation energy is 12,000 cal/mol.

Solution: (*a*) We use Eq. (17.18):

$$\text{Fraction of activated molecules} = e^{-E_a/RT}$$

$$E_a = 6000\ \text{cal/mol} \qquad R = 1.987\ \text{cal/mol-K} \qquad T = 300\ \text{K}$$

$$\text{Fraction} = e^{-6000/(1.987\times 300)}$$

$$\approx e^{-10} = 0.000045 \qquad \text{from mathematical tables}$$

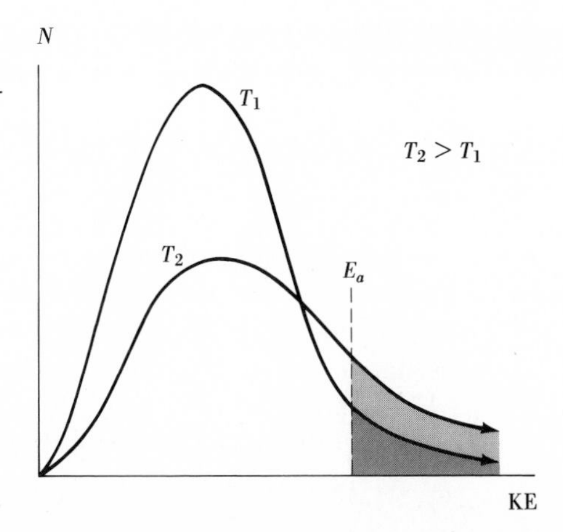

FIGURE 17.9

The effect of temperature on the energy distribution.

Our result shows that at 27°C, only 45 in every 1,000,000 molecules are activated. Let us see what a 100-deg rise in temperature will do to the fraction of activated molecules:

$$127°\text{C} = 400 \text{ K}$$

$$\text{Fraction} = e^{-6000/(1.987\times400)} = e^{-7.5} = 0.00053$$

Therefore, at 400 K, 53 molecules in every 100,000 are activated; this is a twelvefold increase over the number at 300 K, and we can expect a reaction at 400 K to be much faster.

(*b*) Again we use Eq. (17.18) and examine the effect of an increased activation energy:

$$\text{Fraction} = e^{-E/RT} = e^{-12{,}000/(1.987\times300)} \approx e^{-20} = 21 \times 10^{10}$$

Only 21 molecules in every 10 billion are activated. Doubling the activation energy has decreased the number of activated molecules by a factor of 1/20,000 at 27°C. The higher the activation energy the smaller the fraction of activated molecules and the slower the reaction. This can be seen by referring to Fig. 17.8 or 17.9; by shifting E_a to higher values a smaller fraction of molecules will lie in the energy range between E_a and infinity.

Derivation of the Arrhenius equation

Since the rate constant k is a measure of the reaction velocity, it should be proportional to the fraction of molecules which can react. This is the fraction which has a minimum energy E_a and is given by $e^{-E_a/RT}$. We can express this idea mathematically as a proportionality:

$$k \propto e^{-E_a/RT}$$

Introducing a proportionality constant A, we write

$$k = Ae^{-E_a/RT} \tag{17.19}$$

and the equivalent forms

$$\ln k = -\frac{E_a}{RT} + \ln A \tag{17.20}$$

or

$$\log k = -\frac{E_a}{2.303RT} + \log A \tag{17.21}$$

These equations are different forms of the Arrhenius equation. Equation (17.21) has the same form as the experimental equation (17.14), and a comparison of these two equations shows that the constant B in Eq. (17.14) is just equal to the activation energy divided by R, the gas-law constant:

$$B = \frac{E_a}{R} \tag{17.22}$$

Since we have already learned how to calculate B, the Arrhenius equation permits us to determine the activation energy for a reaction. A plot of log k versus $1/T$ gives a straight line with a slope of $-B/2.303$ or $-E_a/2.303R$.

EXAMPLE 17.12 Calculate the activation energy for the N_2O_5 decomposition from the information given in Example 17.10.

Solution: In Example 17.10 we obtain a value of $B = 1.24 \times 10^4$ K. Equation (17.22) relates B and E_a:

$$B = \frac{E_a}{R} \qquad R = 1.987 \text{ cal/mol-K}$$

or

$$E_a = BR = (1.24 \times 10^4 \text{ K})\ (1.987 \text{ cal/mol-K}) = 24{,}600 \text{ cal/mol}$$

EXAMPLE 17.13 Calculate the activation energy for the reaction $H_2 + I_2 \rightarrow 2HI$.

Solution: Figure 17.5 shows a plot of log k versus $1/T$. According to Eq. (17.21), the slope of this plot should be $-E_a/2.303R$. From the figure the slope has a value of -8.84×10^3 K. Hence

$$\text{Slope} = -\frac{E_a}{2.303R}$$

or

$$\begin{aligned} E_a &= -(\text{slope})2.303R \\ &= -(-8.84 \times 10^3 \text{ K})2.303(1.987 \text{ cal/mol-K}) \\ &= 40{,}400 \text{ cal/mol} \end{aligned}$$

By using kinetic theory we have been able to give a molecular basis to the Arrhenius equations. The constant B is related to the activation energy. The constant A is called the *frequency factor*, and in the early development of chemical kinetics it was believed to be exactly equal to the number of collisions occurring per cubic centimeter per second. This has proved incorrect, but it is now believed that A depends on the frequency of collision *and* the orientation of the molecules as they collide.

Catalysis and the Arrhenius equation

A catalyst acts by changing the mechanism of a chemical reaction. In the presence of a catalyst a reaction can occur by a faster series of steps, and the rate constant for the catalyzed reaction is higher. Examination of the Arrhenius equation shows that the presence of a catalyst must reduce the activation energy: at a fixed temperature A, R, and T in the Arrhenius equation ($\log k = -E_a/2.303RT + \log A$) are constants. The increase in the rate constant can be explained only if the catalyst decreases the activation energy E_a. Then from the point of view of the Arrhenius equation the effect of a catalyst is to reduce the activation energy, increasing both the fraction of activated molecules and the reaction rate. This is understandable in terms of Fig. 17.8: as E_a decreases, it will move to the left in the plot and the shaded area representing the activated molecules will grow larger.

In summary, the Arrhenius equation explains how temperature and catalysts affect the rate of a reaction. An increase in temperature actually changes the way the energy is apportioned or distributed to the molecules and increases the number of molecules which have energies greater than E_a (see Fig. 17.9). The presence of a catalyst does not change the way the energy is apportioned to the molecules; it has an accelerating effect because it reduces the activation energy. An increase of temperature or the addition of a catalyst increases the Boltzmann factor, i.e., the fraction of activated molecules.

17.10 TRANSITION-STATE THEORY

The collision theory depicts a reaction that occurs as a result of an abrupt collision between reactant molecules which are hard spheres. This picture is oversimplified because molecules are not hard spheres. The transition-state theory modifies the picture to give a more realistic description of what happens. A "soft" cloud of valence electrons envelops each molecule. As they begin to approach each other, their valence-electron clouds are mutually affected at relatively large separations and consequently bond lengths and the general shape of the molecules begin to change. There is no sharp point at which a collision occurs. The transition-state theory pictures a continuous transformation as reactant molecules approach each other, form an aggregate, and are finally converted into products. The reactant molecules must take up a definite configuration in order for a reaction to occur, and therefore the orientation of the molecules as they approach each other is important. This definite configuration is called the *transition state* or *activated complex*. When the activated complex forms, it may be transformed into products or revert to reactants.

Let us consider a hypothetical sequence for the reaction between O_2 and N_2 to produce NO. As shown in Fig. 17.10*a* the equilibrium bond distances for the N_2 and O_2 molecules are 1.10 and 1.21 A, respectively. As the molecules approach each other, these distances

FIGURE 17.10

A chemical transformation according to the transition-state theory.

(a) *Highly separated molecules*

(b) *N_2 and O_2 begin to approach; N—N and O—O bonds begin to lengthen*

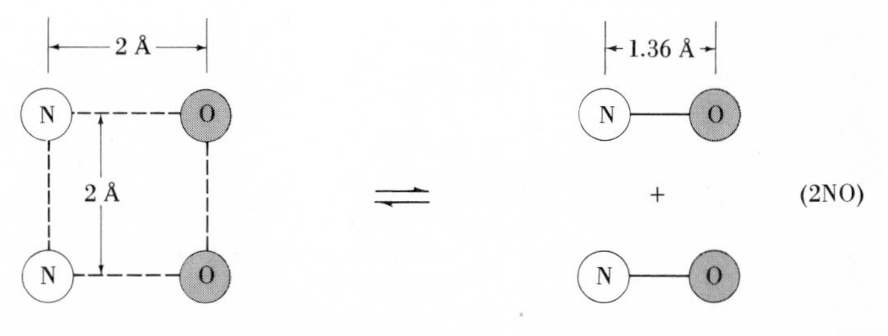

(c) *A transition state has formed in which all bonds are slightly longer than covalent bonds*

(d) *The N—O bonds shorten and products form*

grow longer; and the distances between the nonbonded atoms grow shorter. A transition-state configuration is achieved in which all bond lengths are slightly greater than the equilibrium covalent-bond length (Fig. 17.11*c*). At this point the product, NO, may form, or the transition state may revert to the reactants, N_2 and O_2.

The reactants must overcome the strong repulsive forces of their electron clouds as they approach each other. Therefore, only reactant molecules with relatively high kinetic energy can come close enough together to form the transition state. Also, the energy of the transition state is much higher than the average energy of the reactants. When the transition state disintegrates (into products or reactants), the energy decreases. The course of a chemical reaction can be viewed in terms of the energy changes which occur as the reactant molecules approach. Figure 17.11 shows a plot of the energy versus the configuration of the molecules as they are transformed from reactants to products. This plot is called a *reaction profile*. Initially, the average energy of the reactants is represented on the plot as U_i. Only molecules with higher than the average energy can approach close enough to form the transition state. As they approach, the N—N and O—O bonds lengthen and the energy of nitrogen-oxygen aggregate increases (see Fig. 17.11*b*). Finally the transition state forms, represented at the maximum in the reaction profile. If the N—O bonds continue to grow shorter, the transition state is transformed into products; otherwise it reverts to reactants.

FIGURE 17.11

The reaction profile for the formation of NO. The letters correspond to the same configurations in Fig. 17.10.

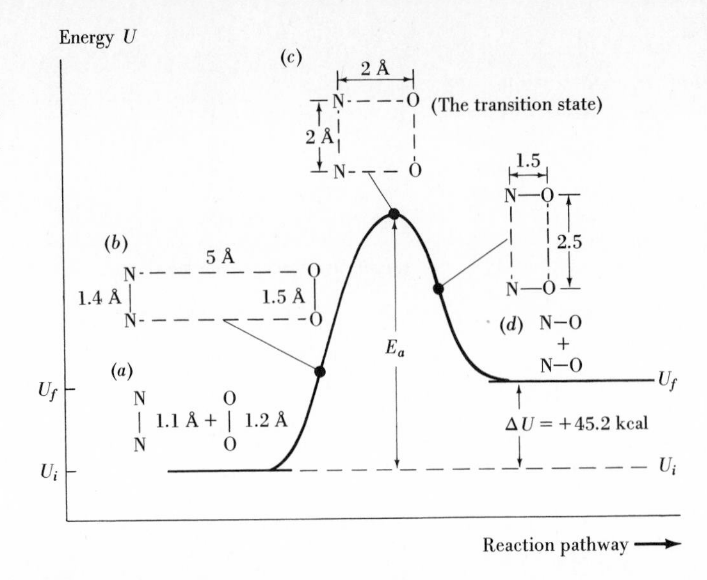

The difference between the average energy of the reactants and the energy of the transition state is equal to the activation energy E_a. In the transition-state theory the activation energy is viewed as a barrier over which reactants must pass in order to become products.

The final energy of the products is U_f. The difference between the energy of the products and the energy of the reactants is the thermodynamic energy change ΔU which occurs during the chemical reaction. In the specific case under consideration the products have a higher energy than the reactants, so that ΔU is positive and the reaction is endothermic.

Figure 17.12 shows the reaction profile for an exothermic reaction, both with and without a catalyst. Note that the final energy of the products is less than the initial energy of the reactants, and ΔU, the thermodynamic energy change, is negative. The activation-energy barrier for the catalyzed path E_a is lower than the activation energy for the uncatalyzed path, so that the catalyzed reaction is faster. The catalyst has no effect on ΔU; in the view of the transition-state theory the catalyst speeds the reaction to its completion by lowering the energy barrier over which the reactants must pass. The catalyst has no effect on ΔU because it does not change the nature or the concentration of the final product. It causes the reaction to proceed more rapidly to the same final state that would be achieved in the absence of a catalyst, but by a different mechanism and by way of a lower-energy transition state. In terms of the transition-state theory a catalyst speeds up a reaction by lowering the energy barrier over which reactants must pass. This effect is shown in the reaction profiles of Fig. 17.12, where the maximum in the catalyzed path is much lower than the maximum in the uncatalyzed path.

FIGURE 17.12

The reaction profile for an exothermic reaction with and without catalysis.

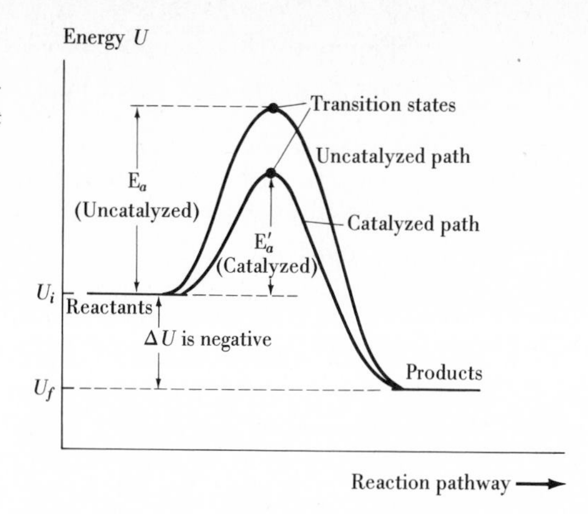

The rate-controlling step

Many reactions proceed by multistep mechanisms. Each step involves its own transition state so that the reaction profile would exhibit many maxima, one for each transition state. Let us consider one which involves two steps:

$$\text{Step 1: A + B} \xrightarrow{\text{via } T_1} P_1$$

$$\text{Step 2: } P_1 \xrightarrow{\text{via } T_2} P_2$$

$$\text{Overall reaction: A + B} \rightarrow P_2$$

The reactants, A and B, proceed to form the product of the first step P_1 by way of the transition state T_1. Then, in the second step, P_1 is converted into the final product P_2 by way of the transition state T_2.

The reaction profile for the two-step process is shown in Fig. 17.13; there are two maxima, corresponding to the two transition states. In this case the activation energy for the second step is greater than the activation energy for the first step. Hence, the energy barrier for the second step is greater, and this step will be slower than the first step. The slowest step in a reaction mechanism is called the *rate-controlling step;* the reaction can never proceed any faster than the rate-controlling step. This is true for any process, not just a chemical process. For instance, suppose you did a time study on hamburger preparation at a neighborhood drive-in restaurant; you might obtain the following results for the different steps leading to a cooked hamburger:

FIGURE 17.13

Reaction profile for a two-step reaction.

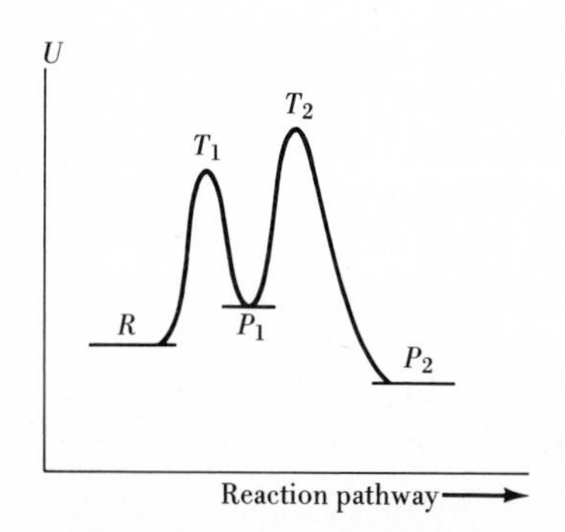

1 Cutting the rolls, 10 per minute

2 Grinding the meat, enough for 15 hamburgers per minute

3 Shaping the hamburger patties, 5 per minute

4 Grilling the hamburgers, 2 per minute

The slowest, or rate-controlling, step is step 4, grilling the hamburgers. In order to increase the rate of preparing hamburgers something must be done to increase the rate of step 4. Perhaps a larger and hotter grill could be used. If the restaurant owner hired more help to cut rolls or grind meat, these items would just pile up because the bottleneck or rate-controlling step is grilling. Likewise, in the chemical reaction with the reaction profile depicted in Fig. 17.13, the overall rate can be increased only if the rate-controlling step is speeded up. Finding a catalyst which would reduce the activation-energy barrier of the second step would increase the overall rate of the reaction; reducing the activation energy of the first step would not affect the overall rate.

17.11 ENZYME CATALYSIS

Chemical kinetics has been used very successfully in developing rapid, economical methods of manufacture for many valuable materials. The attention of many kineticists has recently turned to the enzyme-catalyzed reactions which occur in living cells, one of the most exciting frontiers of modern chemical research. In Chap. 16 we learned that enzymes are giant polypeptides with molecular weights of 20,000 or more. Each peptide residue carries an organic substituent, or R group (see Table 16.2). When the enzyme folds into its tertiary structure, the polar R groups protrude into the aqueous medium of the cell. It is believed that the catalytic effect of an enzyme is due to the arrangement of these R groups on a small region of the surface of the folded enzyme. The substrate, which is the small molecule upon which the enzyme acts, is held at the surface of the enzyme by hydrogen bonds to the polar R groups.

Figure 17.14*a* represents a hypothetical sequence of the events during the enzymatic transfer of a phosphate group from ATP to glucose (this initiates glycolysis). The substrate, glucose, and the coenzyme, ATP, are attracted to the enzyme because the glucose and ATP molecules are so shaped that they fit the R-group pattern on the active site. The rapid transfer of phosphate can occur because the R-group pattern ensures a favorable orientation of the substrate and coenzyme.

There are other hypotheses concerning the action of enzymes. The popular lock-and-key model (Fig. 17.14*b*) is self-explanatory. No single model for enzyme action is completely satisfactory, but so many scientists are at work in this field, our understanding is improving rapidly.

FIGURE 17.14

Possible modes of enzyme action. (a) Molecules are bonded long enough for the reaction to occur only if the sites on the active area of the enzyme match the substrate. [After D. H. Andrews, Introductory Physical Chemistry, McGraw-Hill Book Co., New York, 1970. (b) After J. D. Watson, The Molecular Biology of the Gene, W. A. Benjamin, Inc., New York, 1965.]

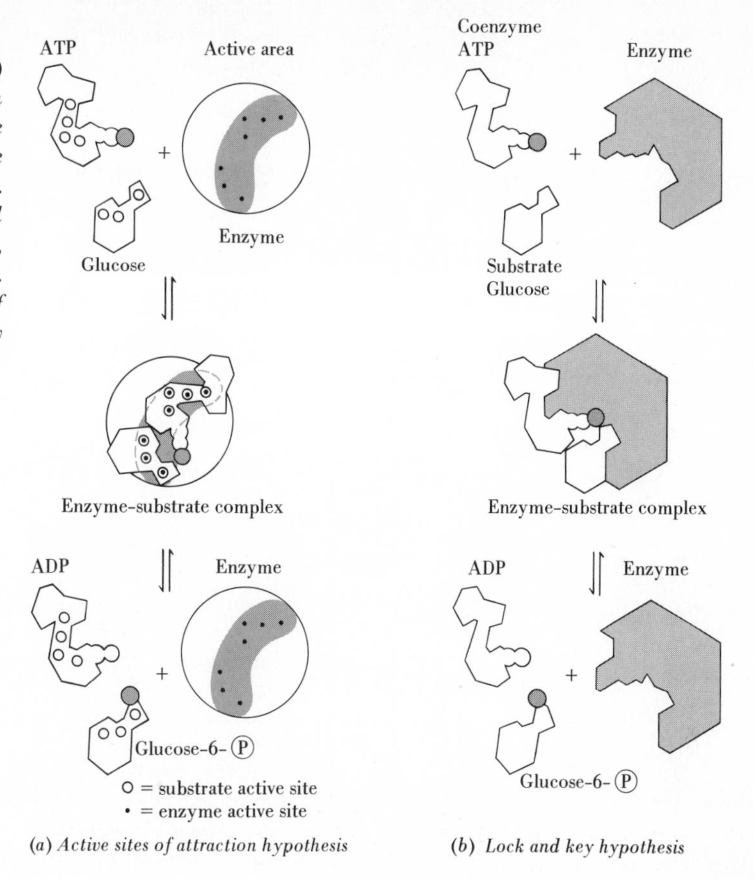

(a) *Active sites of attraction hypothesis*

(b) *Lock and key hypothesis*

QUESTIONS AND PROBLEMS

17.1 List the factors which influence the speed of a reaction.

17.2 (*a*) What is the difference between the stoichiometry and the mechanism of a chemical transformation? (*b*) Explain why, as we study a chemical reaction, its mechanism is modified and becomes more detailed but its stoichiometry remains the same.

17.3 Give three mathematical expressions for the rate of the reaction $H_2 + \frac{1}{2}O_2 \rightarrow H_2O$ in terms of the three different species involved [see Eqs. (17.1) and (17.2)].

17.4 In the reaction $A + B \rightarrow C$ the initial rates of disappearance of A and of B are equal, but these rates do not equal the rate of appearance of C. The rate at which C appears is much slower than the rate at which A or B disappears. Present a set of chemical steps, i.e., a mechanism to explain these observations.

17.5 In the reaction $A + B \rightarrow C$, the initial rate of disappearance of A is greater than the rate of disappearance of B. Furthermore, the rate of appearance of C is lower than the rate of disappearance of either A or B.

Finally, the reaction does not go to completion. Present a mechanism to explain these observations.

TABLE P17.6:

Exp.	$[A]_0$	$[B]_0$	*Initial rate,* mol/min
1	0.100	0.0500	0.00134
2	0.100	0.100	0.00268
3	0.200	0.100	0.00268

17.6 (*a*) From the data given in Table P17.6, determine the order of the reaction $A + B \rightarrow$ products with respect to A, to B, and the overall reaction. (*b*) Write the differential rate expression for the above reaction.

17.7 (*a*) From the data given in Table P17.7, determine the order of the reaction $A + B \rightarrow$ products with respect to A, to B, and the overall reaction. (*b*) Write the differential rate expression for the reaction.

17.8 Explain by example: the law of mass action allows us to write a mathematical relationship between the reaction rate and the concentration of reactants.

17.9 Distinguish between order and molecularity.

TABLE P17.7:

Exp.	$[A]_0$	$[B]_0$	*Initial rate,* mol/min
1	0.150	0.300	0.000280
2	0.300	0.600	0.00158
3	0.300	0.150	0.0000989

17.10 The first-order rate constant for the hydrolysis of methyl acetate at 25°C is 1.26×10^{-4} s^{-1}. (*a*) Calculate $t_{1/2}$. (*b*) If we start a reaction with methyl acetate concentration of 0.500 mol/l, how long will it take to reduce the concentration to 0.0625 *M*?

17.11 The reaction $A \rightarrow B$ is first order, and the rate constant is 3.30×10^{-5} s^{-1}. (*a*) In a particular experiment if the initial concentration of A is 0.300 mol/l and the initial concentration of B is zero, what are the concentrations of A and of B at the end of 200 min? (*b*) Theoretically, how long will it take for all of A to disappear?

17.12 The decomposition of N_2O_5 into NO_2 and NO_3 is first order at 55°C ($N_2O_5 \rightarrow NO_2 + NO_3$). The rate constant is 1.50×10^{-3} s^{-1}. If the concentrations of N_2O_5, NO_2, and NO_3 are 0.500, 0.150, and 0.100 *M*, respectively, what are the concentrations of all species after 10.0 min?

17.13 The acid hydrolysis of sucrose in an aqueous solution of HCl follows first-order kinetics. This means that the rate of the reaction

$$\underset{\text{Sucrose}}{C_{12}H_{22}O_{11}} + H_2O \xrightarrow{H^+} \underset{\text{Fructose}}{C_6H_{12}O_6} + \underset{\text{Glucose}}{C_6H_{12}O_6}$$

depends only on the concentration of sucrose. The rate constant for the reaction is 6.34×10^{-5} s. (*a*) Calculate the half-life for this reaction. (*b*) How much time will be required for the hydrolysis of 65 percent of the sucrose in a 1.00 *M* sucrose solution? (*c*) Repeat part (*b*) for a 2.00 *M* solution.

17.14 Write a rate expression for the disappearance of species A in the reaction $A + B + C \rightarrow$ products as if the reaction were (*a*) zero order, (*b*) first order in all reactants, (*c*) first order in A and B and zero order in C, and (*d*) second order but independent of B.

17.15 The half-life of a certain species in a chemical reaction is determined in a series of experiments. A plot of the half-life versus the reciprocal of the initial concentration gives a straight line with a negative slope. (*a*) What is the order of the reaction? (*b*) If the plot of the half-life had given a straight horizontal line, what would the order be?

TABLE P17.20:

Time, min	[A]
0.00	0.100
5.00	0.085
8.00	0.076
11.00	0.067
15.00	0.055

TABLE P17.21:

Time, min	[*N-chloroacetanilide*]
0.00	0.0245
15.0	0.0181
30.0	0.0133
45.0	0.0097

TABLE P17.26:

t, min	P_{HI}, atm
0.00	0.00
3.00	0.015
5.00	0.025
8.60	0.043

TABLE P17.28:

t, °C	k_1, s^{-1}
0	2.5×10^{-5}
20	4.75×10^{-4}
40	5.77×10^{-3}
60	5.49×10^{-2}

17.16 (*a*) Distinguish a catalyst from an ordinary reactant. (*b*) In the broadest sense of the term is a catalyst a reactant? Explain.

17.17 Why do the rates of catalyzed reactions often *appear* to be independent of the reactant concentration?

17.18 Explain why increasing the H_2O_2 concentration in a solution containing the enzyme peroxidase does not increase the rate of O_2 evolution.

17.19 (*a*) Distinguish between a heterogeneous and homogeneous catalyst. (*b*) Give an example of each. (*c*) Is an enzyme a homogeneous or heterogeneous catalyst? Explain.

17.20 The reaction A → B may be zero or first order. (*a*) Determine the order and the rate constant from the data in Table P17.20. (*b*) If the reaction is zero order, make a plot of [A] versus time; if it is first order, make a plot of log [A] versus time.

17.21 The isomerization of *N*-chloroacetanilide to parachloroacetanilide in the presence of hydrochloric acid may be first or zero order. (*a*) From the data in Table P17.21, determine the order and the rate constant for the reaction. (*b*) If the reaction is zero order, make a plot of the concentration of *N*-chloroacetanilide versus time; if it is first order, make a plot of the log of the *N*-chloroacetanilide concentration versus time.

17.22 Explain why an increase in temperature increases the rate of a chemical reaction.

17.23 (*a*) Using the data in Table 17.4, decide whether our rule of thumb that the rate of a chemical reaction doubles for a temperature of 10°C is fairly accurate or not. (*b*) Are the estimates in this case too high or too low?

17.24 The rate constant for the decomposition of N_2O_5 in carbon tetrachloride is $6.14 \times 10^{-4}\ s^{-1}$ at 45°C. Estimate the rate constant at 55°C and then estimate the activation energy. See Prob. 17.23.

17.25 The rate constants for the reaction A → B are $1.06 \times 10^{-5}\ s^{-1}$ at 0°C and 2.92×10^{-3} at 45°C. (*a*) Calculate the activation energy for this reaction. (*b*) Calculate the rate constant at 25°C.

17.26 (*a*) Using the data in Table P17.26 for the reaction

$$C_2H_5I \xrightarrow{Pt} C_2H_4 + HI$$

determine whether the reaction is first or zero order and determine the rate constant at 25°C. (*b*) The energy of activation is 23,100 cal. Determine the rate constant at 50°C.

17.27 (*a*) If the rate constant for a reaction doubles in going from 25 to 35°C, what is the activation energy of the reaction? (*b*) Would the activation energy of a reaction whose rate doubles in going from 125 to 135°C be the same as in part (*a*)?

17.28 Acetonedicarboxylic acid decomposes in aqueous solution by first-order kinetics:

Rate of disappearance $= -k_1$[acetonedicarboxylic acid]

(*a*) Make a plot of log k_1 versus $1/T$ from rate constants obtained at various temperatures (Table P17.28). (*b*) Determine the value of E_a and log A. (*c*) Express the rate constant as a function of temperature by an equation of the form $k_1 = Ae^{-E_a/RT}$.

17.29 In terms of the collision theory why does an increase in temperature increase the reaction rate? Include some sketches to make your explanation clear.

17.30 In terms of the transition-state theory why does an increase in temperature increase the reaction rate? Sketches would be helpful.

17.31 Explain why the transition-state theory is considered more realistic than the collision theory.

18 THE PRINCIPAL ELEMENTS OF THE REGULAR FAMILIES

18.1 INTRODUCTION

The regular families, i.e., those designated as the A families or groups of the periodic table, are referred to as regular because their properties show more regular trends within families and periods than the transition elements of the B families.

We have discussed many of the properties of the regular elements, but in this chapter we give a more systematic treatment of their chemistry, paying special attention to the elements which are prominent in technology. It will be very helpful to refer to the periodic table frequently to determine the relative positions of the families and the individual elements. In this way it will be easier to understand and predict the properties of the elements.

18.2 THE ALKALI METALS, GROUP IA

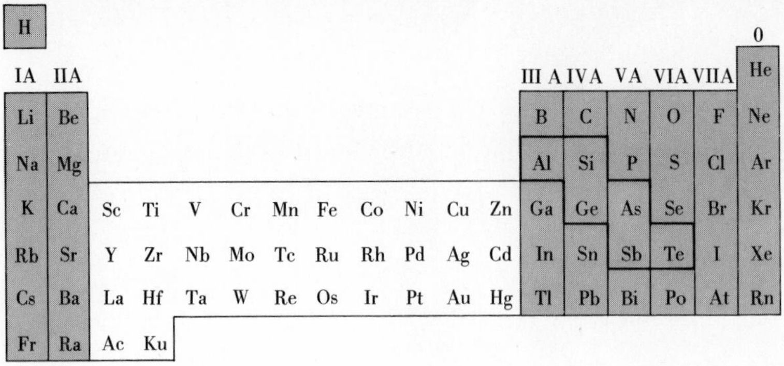

IA	IIA											III A	IVA	VA	VIA	VIIA	0
H																	He
Li	Be											B	C	N	O	F	Ne
Na	Mg											Al	Si	P	S	Cl	Ar
K	Ca	Sc	Ti	V	Cr	Mn	Fe	Co	Ni	Cu	Zn	Ga	Ge	As	Se	Br	Kr
Rb	Sr	Y	Zr	Nb	Mo	Tc	Ru	Rh	Pd	Ag	Cd	In	Sn	Sb	Te	I	Xe
Cs	Ba	La	Hf	Ta	W	Re	Os	Ir	Pt	Au	Hg	Tl	Pb	Bi	Po	At	Rn
Fr	Ra	Ac	Ku														

The elements of group IA are the most active metals known. This means that they are very powerful reducing agents, and many common substances react with them. Their compounds are ionic, and they exhibit a +1 oxidation state exclusively. The high reactivity of these metals can be understood by examination of Table 18.1, which reveals that in the +1 oxidation state they have the electronic configuration of the very inert noble gases. The inertness of the +1 state often means that the chemical behavior of an alkali-metal salt depends on the nature of the anion: Na_2CO_3 is basic because CO_3^{2-} is a Brönsted-Lowry base; $KMnO_4$ and $K_2Cr_2O_7$ are oxidizing agents because of the presence of MnO_4^- and $Cr_2O_7^{2-}$; NaCN is poisonous because of the cyanide ion. The great stability and chemical inertness of the +1 cations means that the alkali metals occur in nature only as compounds which cannot be reduced to the metals without a great expenditure of energy.

One of the most common ways of producing alkali metals is by

TABLE 18.1: SOME PROPERTIES OF THE ALKALI METALS

				Radius, Å					
Element	*at wt*	*Abundance in earth's crust,* wt %	*Electronic configuration*	*In metallic crystal*	*Ionic*	*Ionization potential,* kcal	*Density,* g/cm³	*mp,* °C	*bp,* °C
Lithium, $_3$Li	6.939	0.0065	$1s^2, 2s^1$ or [He] $2s^1$	1.55	0.78	124	0.53	186	1336
Sodium, $_{11}$Na	22.990	2.83	[Ne] $3s^1$	1.90	0.98	118	0.97	97.5	880
Potassium, $_{19}$K	39.102	2.59	[Ar] $4s^1$	2.35	1.33	99.7	0.86	62.3	760
Rubidium, $_{37}$Rb	85.87	0.029	[Kr] $5s^1$	2.48	1.49	94.0	1.53	38.5	700
Cesium, $_{55}$Cs	132.91	0.0007	[Xe] $6s^1$	2.67	1.65	89.5	1.90	28.5	670
Francium, $_{87}$Fr	(223)	. . .	[Rn] $7s^1$	?	?	?	?	?	?

FIGURE 18.1

The Downs cell.

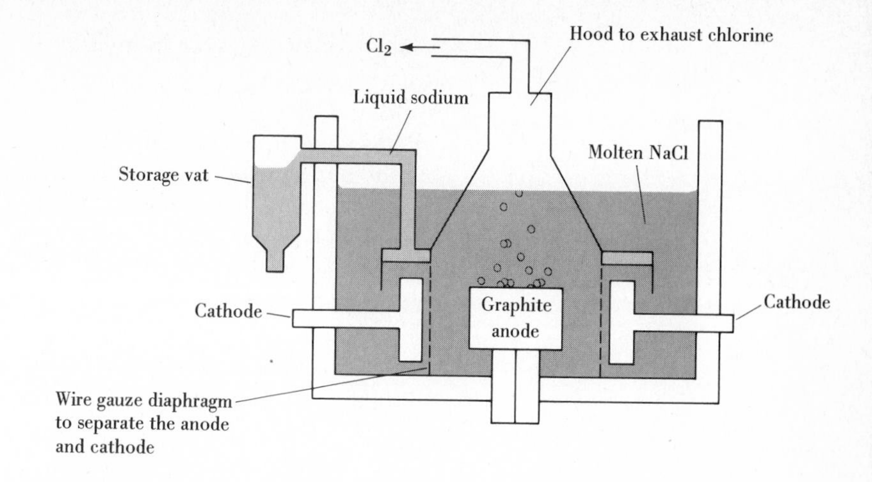

electrolysis; e.g., sodium is produced industrially by the electrolysis of molten NaCl in a *Downs cell* (Fig. 18.1). Chlorine gas is also a product:

$$NaCl(l) \rightarrow Na(l) + \tfrac{1}{2}Cl_2(g)$$

In the early nineteenth century Sir Humphry Davy electrolyzed salts of sodium and potassium and became the first man to see the very reactive silvery, soft metals. Lithium, rubidium, and cesium have also been obtained by electrolysis, but francium, which is radioactive, was discovered as a product of the radioactive decay of actinium, ${}^{227}_{89}Ac$:

$${}^{227}_{89}Ac \rightarrow {}^{4}_{2}He + {}^{223}_{87}Fr$$

Since only trace amounts are produced in this manner, all we know about francium is its atomic number and approximate atomic weight, that its chemical reactions are very similar to cesium, and that it is radioactive, having a half-life of about 2 min.

Table 18.1 shows that sodium and potassium are common elements in the earth's crust but the other elements of group IA are quite rare. Sodium is ubiquitous, and can be detected in any substance unless great care is taken to remove it.

Chemical reactions of the metals

All the alkali metals react with molecular oxygen. Lithium gives the ordinary oxide, in which the oxidation state of oxygen is −2:

$$2Li(s) + \tfrac{1}{2}O_2(g) \rightarrow Li_2O(s)$$

Sodium forms a peroxide, O_2^{2-}, in excess oxygen:

$$2Na(s) + O_2(g) \rightarrow Na_2O_2(s)$$

The more active metals, potassium, rubidium, and cesium, form superoxides, O_2^-; for example,

$$K(s) + O_2(g) \rightarrow KO_2(s)$$

Potassium superoxide produces molecular oxygen when it is treated with CO_2. We take advantage of this reaction to replenish air with oxygen on submarines and in high-altitude aircraft. The exhaled CO_2 is passed over KO_2, liberating molecular oxygen:

$$2KO_2(s) + CO_2 \xrightarrow[(CuCl_2)]{\text{cat.}} K_2CO_3(s) + \tfrac{3}{2}O_2(g)$$

The alkali metals react violently with the halogens and combine directly with sulfur and phosphorus:

$$2Na(s) + Cl_2(g) \rightarrow 2NaCl(s) \qquad \text{explosion}$$

$$16K(s) + S_8(s) \rightarrow \underset{\substack{\text{Potassium}\\ \text{sulfide}}}{8K_2S(s)}$$

$$12K(s) + P_4(s) \rightarrow \underset{\substack{\text{Potassium}\\ \text{phosphide}}}{4K_3P(s)}$$

The alkali metals react with hydrogen to form hydrides, LiH, NaH, KH, etc. For example,

$$2Na(s) + H_2(g) \rightarrow 2NaH(s)$$

When sodium hydride is fused and electrolyzed, H_2 gas is produced at the anode, proving that hydrogen must be in a negative oxidation state in the hydrides:

$$2H^- \rightarrow H_2 + 2e^- \qquad \text{anode reaction}$$

The reaction of the group IA metals with water proceeds explosively. There is a generation of H_2 gas and a great deal of heat, which causes the H_2 to ignite explosively in air:

$$2K(s) + 2HOH(l) \rightarrow 2KOH(aq) + H_2(g) + \text{heat}$$

and

$$H_2(g) + \tfrac{1}{2}O_2(g)(\text{air}) \rightarrow H_2O(g) \qquad \text{explosion}$$

Obviously, these metals must be kept away from water and humidity. They are ordinarily stored under liquid hydrocarbons such as kerosene.

The alkali metals dissolve in pure liquid ammonia to form a stable solution which is blue when dilute and resembles a liquid bronze when concentrated. The bronze solutions have all the properties of liquid metals. These solutions have received a lot of attention because it is postulated that the valence electrons escape from the

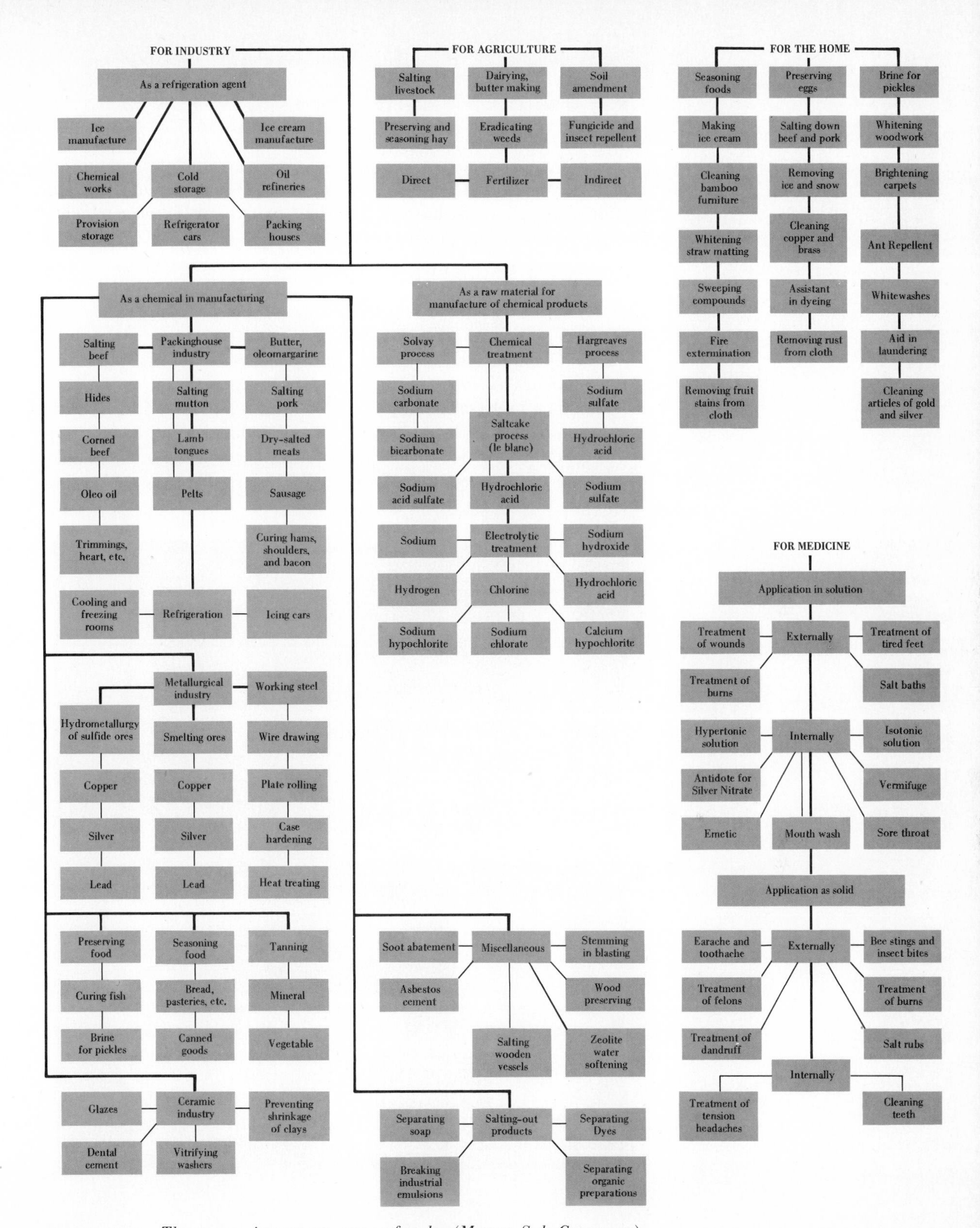

FIGURE 18.2 *The many important uses of salt.* (*Morton Salt Company.*)

metal atoms and exist in the solvent as *solvated electrons:*

$$\mathrm{K^0 + \mathit{x}NH_3(\mathit{l}) \rightarrow \underset{\text{Solvated electron}}{\mathit{e}^-(NH_3)} + \underset{\text{Dissolved in } NH_3(\mathit{l})}{K^+(amm)}}$$

Metal surfaces and certain salts catalyze the reaction of the alkali metals and ammonia in these solutions to give a reaction analogous to that of the alkali metals and water:

$$\mathrm{K + NH_3 \rightarrow \underset{\text{Potassium amide}}{KNH_2} + \tfrac{1}{2}H_2(\mathit{g})}$$

The amide ion, NH_2^-, in liquid ammonia is analogous to the hydroxide ion, OH^-, in water.

Common table salt is one of our most valuable minerals. Its presence is vital to all living organisms and indispensable to industry. The amount used for flavoring represents only a trivial fraction of the total NaCl consumed. As Fig. 18.2 shows, NaCl is a chemical

FIGURE 18.3

An underground salt mine at Grand Saline, Texas. (Morton Salt Company.)

cornucopia. In the United States alone, 20 million tons is mined annually (Fig. 18.3). It is used to synthesize hydrogen, chlorine, sodium hydroxide, sodium metal, and sodium carbonate, all essential to a large cross section of industries. Metallic sodium is used to prepare tetraethyllead, the antiknock ingredient of gasoline. Sodium carbonate is the world's most widely used industrial base, 8 million tons being consumed annually by technology in the United States alone. Much of the Na_2CO_3 is mined, but most of it is manufactured by the Solvay process.

The Solvay process is an ideal industrial process in that all starting materials (limestone, salt, water, and ammonia) are inexpensive and all by-products are either recycled or have commercial value. The limestone, $CaCO_3$, is thermally decomposed to produce lime, CaO, and carbon dioxide:

$$\text{Step 1: } CaCO_3(s) \xrightarrow{\Delta} CaO(s) + CO_2(g)$$

The CO_2 gas is then passed into a *carbonating tower* (Fig. 18.4), where it is mixed with an aqueous spray of concentrated NaCl and NH_3. The solution reaching the bottom of the carbonating tower contains several ions which form from the salt, ammonia, and carbon dioxide:

$$\text{Step 2: } \underbrace{Na^+(aq) + Cl^-(aq) + NH_3(aq)}_{\text{spray}} + CO_2(g) + H_2O(l) \rightarrow$$

$$Na^+(aq) + Cl^-(aq) + NH_4^+(aq) + HCO_3^-(aq)$$

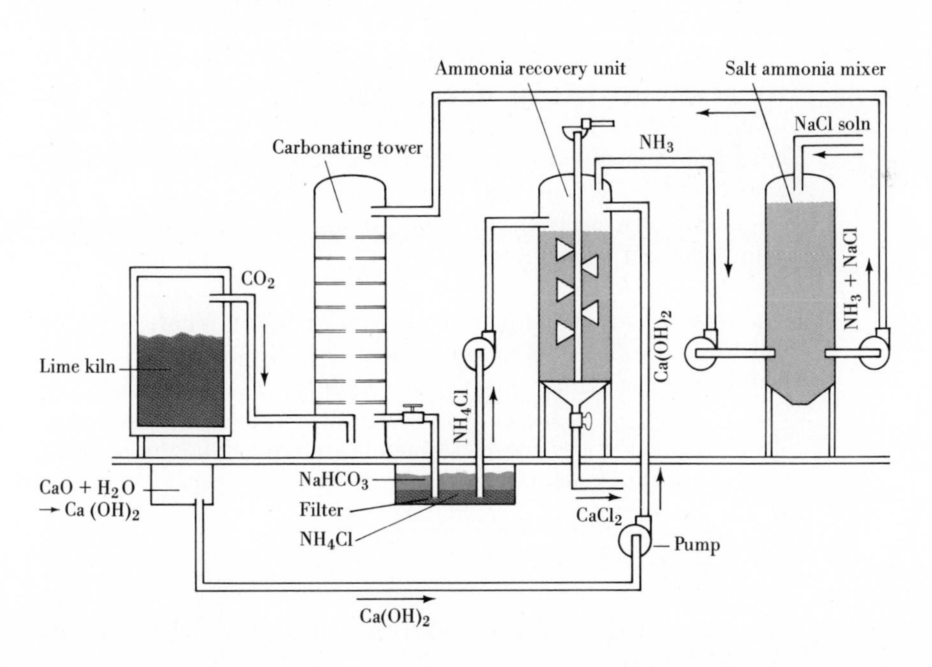

FIGURE 18.4

The Solvay process. Carbon dioxide forms in the kiln and passes into the carbonating tower, where it is mixed with NH_3, NaCl, *and water. The desired sodium bicarbonate is produced along with* NH_4Cl.

Of the four different salts which can form, $NaCl$, $NaHCO_3$, NH_4Cl, and NH_4HCO_3, the least soluble is sodium bicarbonate, $NaHCO_3$, which precipitates from the concentrated solution, leaving an aqueous solution of NH_4Cl. In the next step of the Solvay process the NH_4Cl solution is treated with the $Ca(OH)_2$ formed by dissolving the lime (obtained in step 1) in water:

$$CaO(s) + H_2O(l) \rightarrow Ca(OH)_2(aq)$$

The $Ca(OH)_2$, a strong base, liberates $NH_3(g)$ from the NH_4Cl solution:

$$\text{Step 3: } 2NH_4Cl(aq) + Ca(OH)_2(aq) \rightarrow 2NH_3(g) + CaCl_2(aq) + 2H_2O(l)$$

The NH_3 is recycled, and the $CaCl_2$ is dried and sold as a dehydrating agent widely used in industry and research.

The sodium bicarbonate is sold directly or converted into sodium carbonate:

$$2NaHCO_3(s) \xrightarrow{\Delta} Na_2CO_3(s) + CO_2(g) + H_2O(g)$$

Sodium carbonate is second only to lime, CaO, in importance as a base; it owes its basic character to the CO_3^{2-} ion, which is a Brönsted-Lowry base:

$$CO_3^{2-}(aq) + H_3O^+(aq) \rightleftharpoons HCO_3^-(aq) + H_2O(l)$$

Potassium and lithium have commercial value; rubidium and cesium, on the other hand, are not used in large amounts.

18.3 THE ALKALINE-EARTH METALS, GROUP IIA

The alkaline earths are harder, denser, and have higher melting points than the alkali metals. Since each alkaline-earth atom contributes two valence electrons, the atoms in the metallic crystal are more strongly bonded.

In general, the chemical reactions of the alkaline earths are similar to those of the alkali metals but less vigorous. Whereas sodium tarnishes immediately in air and reacts violently with cold water, its alkaline-earth neighbor magnesium undergoes a very slight tarnishing in air and an imperceptible reaction with cold water.

The reactions of the group IIA metals are typified by the reactions of calcium with elements:

$$3Ca(s) + N_2(g) \rightarrow Ca_3N_2(s)$$
$$3Ca(s) + P_4(s) \rightarrow Ca_3P_4(s)$$
$$Ca(s) + \tfrac{1}{2}O_2(g) \rightarrow CaO(s)$$

$$Ca(s) + \frac{1}{8}S_8(s) \rightarrow CaS(s)$$
$$Ca(s) + Cl_2(g) \rightarrow CaCl_2(s)$$

and with simple compounds:

$$Ca(s) + 2H_2O(l) \rightarrow Ca(OH)_2(aq) + H_2(g)$$
$$Ca(s) + 2HCl(aq) \rightarrow CaCl_2(aq) + H_2(g$$
$$Ca(s) + 2NH_3(l) \rightarrow Ca(NH_2)_2(\text{amm}) + H_2(g)$$

Although the oxides, hydroxides, and carbonates of the alkaline-earth metals are strong bases, they are weaker and less soluble than the corresponding alkali-metal compounds. Generally speaking, the alkaline-earth salts are less soluble than the corresponding alkali-metal salts.

The difference between the chemistry of the group IA and group IIA metals can be explained in terms of kernel charge and electronic configuration. The two *s* valence electrons of the alkaline earths are attracted by a kernel charge of +2, while the single *s* electron of the alkali metals is attracted by a kernel charge of +1. Hence, the alkaline earths do not as easily give up their valence electrons and are poorer reducing agents. The differences in the other physical properties of these two groups (Tables 18.1 and 18.2) can also be explained in these terms.

Calcium carbonate, lime, and portland cement

Calcium is the most abundant element of group IIA and the sixth most abundant element in the earth's crust. As calcium carbonate it is found widely distributed in four different forms. As *limestone*, $CaCO_3$, it is the major component of sedimentary-rock formations. Under the great pressures generated by volcanic upheavals limestone is converted into *marble*, which is dense and hard and can be polished. The mineral *calcite* is virtually pure $CaCO_3$ existing as a hard, transparent, crystal; *chalk* is a soft white form of $CaCO_3$.

The chief use of $CaCO_3$ is in the production of lime, CaO. Approximately 20 million tons of lime is produced annually in the United States by *calcination* in giant kilns:

$$CaCO_3(s) \xrightarrow{\Delta} CaO(s) + CO_2(g)$$

Most of the lime produced is used in the construction industry as the basic ingredient of plaster, mortar, and concrete. Lime also finds important applications as a flux to remove silicates from ores in metallurgical processes and to produce bleach for the textile industry and the home.

All the alkaline-earth metals can be produced by electrolysis of their molten chlorides. The metals can also be produced by reduction of their compounds with magnesium or aluminum metal

TABLE 18.2: SOME PROPERTIES OF THE ALKALI EARTHS

Element	at wt	Abundance in earth's crust, wt %	Electronic configuration	Radius, Å		Ionization potential, kcal		Oxidation potential, V	Density g/cm³	mp, °C	bp, °C
				In metallic crystal	Ionic	$M^0(g) \rightarrow M^+(g) + e^-$	$M^+(g) \rightarrow M^{2+}(g) + e^-$	$(M^0(g) \rightleftharpoons M^{2+}(g) + 2e^-)$			
Beryllium, $_4Be$	9.013	0.0002	[He] $2s^2$	1.12	0.31	215	420	+1.70	1.86	1350	1500
Magnesium, $_{12}Mg$	24.32	2.09	[Ne] $3s^2$	1.60	0.65	176	347	+2.37	1.75	651	1110
Calcium, $_{20}Ca$	40.08	3.63	[Ar] $4s^2$	1.97	0.99	141	274	+2.87	1.55	810	1170
Strontium, $_{38}Sr$	87.63	0.022	[Kr] $5s^2$	2.15	1.13	131	253	+2.89	2.6	800	1150
Barium, $_{56}Ba$	137.36	0.1	[Xe] $6s^2$	2.22	1.35	120	229	+2.90	3.61	850	1140
Radium, $_{88}Ra$	(226)	1.3×10^{-10}	[Rn] $7s^2$	~2.5	~1.5	122	233	+2.92	~4.5	960	?

which is usually more economical:

$$BeCl_2(l) + Mg^0(l) \xrightarrow{\Delta} Be^0(l) + MgCl_2(l)$$

and

$$\tfrac{3}{2}CaCl_2(l) + Al^0(l) \xrightarrow{\Delta} \tfrac{3}{2}Ca^0(l) + AlCl_3(g)$$

Only magnesium has real industrial importance as a metal, 100,000 tons being produced annually in the United States. It is used in alloys and as a reducing agent in metallurgical processes.

Beryllium finds its major use as an alloying agent. Strontium and barium have very limited uses. Radium is employed in cancer therapy because it is radioactive and can be used to control tumors; actually radon gas, the disintegration product of radium decay, is ordinarily used in the treatment. Radon is the heaviest member of the inert- or noble-gas family (group O), and it is radioactive.

18.4 THE ANOMALOUS SECOND PERIOD

As expected, the metallic character of the elements in groups IA and IIA increases as we descend in the group. The most metallic elements are cesium, francium, and barium; compounds of these heavier elements are highly ionic. The light elements, lithium and beryllium, have a less pronounced metallic nature, and their compounds are somewhat covalent. Lithium compounds are predominantly ionic and beryllium compounds predominantly covalent. The molten halides of beryllium, like $BeCl_2$, are poor conductors of electricity; the hydroxide of beryllium may act as either an acid or a base; i.e., it is *amphoteric:*

Reaction with acid: $Be(OH)_2(s) + 2H_3O^+(aq) \rightarrow Be^{2+}(aq) + 4HOH(l)$
Reaction with base: $Be(OH)_2(s) + 2OH^-(aq) \rightarrow Be(OH)_4^{2-}$

Beryllium is the least typical element of the very active metals, and lithium is also anomalous. All the elements of the second period differ somewhat from the heavier members of their respective families.

Lithium does not react as vigorously as the other alkali metals and resembles magnesium in its physical and chemical properties. Like magnesium, it is light and hard and reacts only slowly with cold water. The carbonates and nitrates of all the alkali metals are thermally stable except for Li_2CO_3 and $LiNO_3$, which decompose like the magnesium salts:

$$Li_2CO_3(s) \xrightarrow{\Delta} Li_2O(s) + CO_2(g)$$

$$MgCO_3(s) \xrightarrow{\Delta} MgO(s) + CO_2(g)$$

and

$$LiNO_3(s) \xrightarrow{\Delta} \frac{1}{2}Li_2O(s) + NO_2(g) + \frac{1}{4}O_2(g)$$

$$Mg(NO_3)_2(s) \xrightarrow{\Delta} MgO(s) + 2NO_2(g) + \frac{1}{2}O_2(g)$$

Beryllium forms covalent compounds and an amphoteric hydroxide; as we shall see, in its chemical reactions it resembles aluminum (group IIIA) more than any member of its own family. Boron, the first member of group IIIA, resembles silicon (group IVA). From our study of organic chemistry we already know that carbon, the next member of the second period, is unusual. Not only is it not like other members of group IVA, it is not like any other element. It is the unique chain former.

The remaining elements of the second period—nitrogen, oxygen, and fluorine—are extremely electronegative; their compounds are highly polar, and their hydrides are involved in hydrogen bonding. Hence, the whole second period is an unusual set of elements.

We have implicitly alluded to a *diagonal* relationship exhibited by the first three members of the second period: lithium resembles magnesium, beryllium resembles aluminum, and boron resembles silicon. The diagonal relationship in this portion of the periodic table

```
Li  Be  B  C
  ↘   ↘  ↘
Na  Mg  Al  Si
```

occurs because the ions of each pair have a similar charge-to-radius ratio.

18.5 THE BORON-ALUMINUM FAMILY, GROUP IIIA

Elemental boron forms a very hard, covalent crystal with high electric resistance. Its oxides and hydroxides are acidic, and its compounds are covalent. Boron is clearly a nonmetal, but the remaining elements of the family are metalloids, i.e., intermediate between metals and nonmetals. They resemble nonmetals in that their oxides and hydroxides are amphoteric and their compounds covalent, but they are like true metals in that they are relatively soft, low-melting, malleable, silvery elements with a high electric conductance.

The chemical properties of the group IIIA elements can be explained in terms of the high charge and small kernel radius of these atoms. (see Table 18.3). A very high energy is required to remove three electrons from small kernels, and as a result the group IIIA elements form compounds by sharing electrons; i.e., they form covalent compounds.

The reactions of the boron-aluminum family are typified by the

TABLE 18.3: SOME PROPERTIES OF THE GROUP IIIA ELEMENTS

Element	*at wt*	*Abundance in earth's crust,* wt %	*Electronic configuration*	*Radius,* Å		*Third ionization potential,* kcal $M^{2+}(g) \rightarrow M^{3+}(g) + e^-$	*Oxidation potential,* V $M^0(s) \rightleftharpoons M^{3+}(aq) + 3e^-$	*Density,* g/cm³	*mp,* °C	*bp,* °C
				Ionic	*Covalent*					
Boron, $_5B$	10.81	0.0002	[He] $2s^2$, $2p^1$	0.21†	0.82	870	. . .	2.33	2300	. . .
Aluminum, $_{13}Al$	26.98	8.13	[Ne] $3s^2$, $3p^1$	0.52	1.20	660	1.66	2.70	660	2270
Gallium, $_{31}Ga$	69.72	0.0016	[Ar] $3d^{10}$, $4s^2$, $4p^1$	0.62	1.26	710	0.52	5.93	30	2070
Indium, $_{49}In$	114.82	0.000011	[Kr] $4d^{10}$, $5s^2$, $5p^1$	0.78	1.45	645	0.34	7.29	155	1450
Thallium, $_{81}Tl$	204.37	0.00013	[Xe] $4f^{14}$, $5d^{10}$, $6s^2$, $6p^1$	0.95	1.50	690	0.33‡	11.8	303.5	1457

† Because the simple ion is very small and highly charged, these elements form covalent compounds. The 3+ ion does not exist under ordinary chemical conditions.

‡ For $Tl^0(s) \rightleftharpoons Tl^+(aq) + e^-$.

reactions of aluminum:

$$2Al + 3X_2 \rightarrow 2AlX_3 \qquad X_2 = Cl_2, Br_2, I_2$$

$$4Al + 3O_2 \rightarrow \underset{\text{Alumina}}{2Al_2O_3} \qquad \Delta H = -399 \text{ kcal/mol}$$

$$2Al + \tfrac{3}{8}S_8 \rightarrow Al_2S_3(s)$$

$$Al + \tfrac{1}{2}N_2 \xrightarrow{\Delta} AlN(s)$$

$$Al + \tfrac{1}{4}P_4 \xrightarrow{\Delta} AlP(s)$$

$$Al + 3HCl(aq) \rightarrow AlCl_3(aq) + \tfrac{3}{2}H_2(g) \qquad \text{(not boron)}$$

$$Al(s) + NaOH(aq) + H_2O(l) \rightarrow NaAlO_2(aq) + \tfrac{3}{2}H_2(g) \qquad \text{(not boron)}$$

Some details of these reactions and exceptions to them should be noted. Aluminum metal is fairly stable in the ordinary atmosphere thanks to the formation of a tenacious oxide coating which retards further oxidation. However, when it is heated in oxygen, it ignites, liberating an extraordinary amount of light and heat (aluminum is used to make flashbulbs). When a mixture of powdered aluminum and iron(III) oxide, called Thermit, is ignited, the aluminum reduces the iron and the heat liberated by the reaction causes the iron to melt:

$$2Al(s) + Fe_2O_3(s) \rightarrow Al_2O_3(s) + 2Fe(l)$$

Thermit is used for on-the-spot welding. Other metals can be produced in an analogous manner.

The chlorides and bromides of aluminum and gallium exist as dimers, for example, Al_2Cl_6. In the iodide of thallium, TlI_3, the iodine is probably present as the triiodide ion, I_3^-, rather than as three individual iodide ions, which means that thallium is in the +1 oxidation state, $Tl^+I_3^-$. When thallium reacts with oxygen or sulfur at high temperature, it exhibits its +1 oxidation state and forms Tl_2O and Tl_2S, compounds more ionic than Tl_2O_3 and Tl_2S_3, in which thallium exhibits a +3 oxidation state. Thallium(I) hydroxide, TlOH, is a strong, nonamphoteric base. It should also be noted that elemental boron does not react with acids or bases. Broadly speaking, boron and thallium show the greatest deviations from the normal behavior of the group IIIA elements. Boron is clearly a nonmetal, and thallium in its +1 oxidation state behaves like a true metal by forming ionic compounds and a basic hydroxide.

Boron

As can be seen from Table 18.3, boron is a rare element. It is found in Death Valley and the Mojave Desert of California, where it occurs as borax, $Na_2B_4O_7 \cdot 10H_2O$, sodium tetraborate decahydrate, a valuable mineral, and in parts of Italy, where it occurs as orthoboric acid, H_3BO_3. It is used in the production of flameproof glass, glazes for china, and enamel. Elemental boron can be prepared by the electrolysis of the melted oxide but has found no extensive uses.

The hydrides of boron, the *boranes*, ranging from B_2H_6 to $B_{10}H_{14}$, have attracted interest for both practical and theoretical reasons. They all exhibit a positive enthalpy of formation, which means that they are unstable with respect to their elements. They have high heats of combustion and ignite explosively in moist air. Their reactivity makes them potential rocket fuels, but they have not yet been found suitable for this purpose.

All attempts to synthesize BH_3 have yielded diborane, B_2H_6, a structure analogous to ethane, C_2H_6. Although the bonding in ethane is easily explained, since it has 14 valence electrons to form seven covalent bonds, bonding in the diborane molecule, with 12 valence electrons, is more difficult to explain. The structure of the diborane molecule (Fig. 18.5) gives us a clue. Six of the atoms, the two boron atoms and four of the hydrogen atoms, are coplanar. The other two hydrogen atoms are positioned symmetrically above and below the plane, forming a *hydrogen bridge* between the two boron atoms. Each boron atom is bonded tetrahedrally to four hydrogen atoms, but the four bonds are not identical. The B—H bond length of the bridge is 1.33 Å, and the B—H bond length to the corner atoms is only 1.19 Å. The shorter bond is an ordinary covalent bond, but the longer bond, which involves the bridge hydrogen atom, is probably similar to a hydrogen bond (page 110). Thus, the electron deficiency of diborane is remedied by an extraordinary hydrogen bridge.

The higher-molecular-weight boranes are not polymers of BH_3. As can be seen (Table 18.4), the boron-hydrogen ratio varies considerably. There are two pentaboranes; one has nine hydrogen atoms (pentaborane 9) and one has eleven (pentaborane 11). The structure of the higher boranes is more complex than that of diborane, but all exhibit hydrogen bridging.

Boron forms some unusual nitrogen compounds which are reminiscent of graphite, diamond, and benzene (see Fig. 18.6). When boron

FIGURE 18.5

The structure of diborane, B_2H_6.

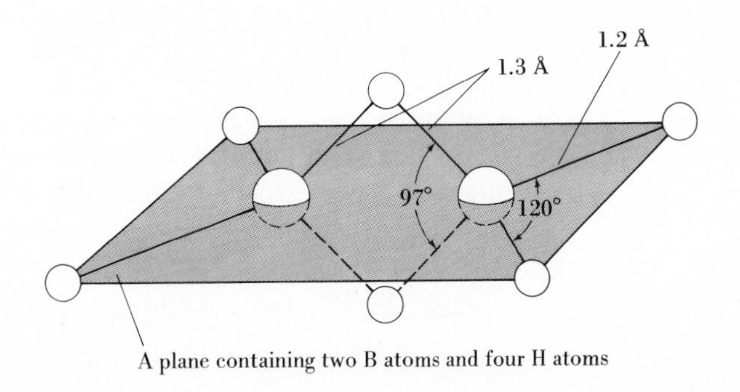

TABLE 18.4: SOME BORON HYDRIDES

General formula	*Compound*	*Formula*	*bp, °C*
B_nH_{n+4}	Diborane	B_2H_6	−93
	Pentaborane 9	B_5H_9	58
	Decaborane 14	$B_{10}H_{14}$	213
B_nH_{n+6}	Tetraborane	B_4H_{10}	16
	Pentaborane 11	B_5H_{11}	65
	Decaborane 16	$B_{10}H_{16}$	

is heated in the presence of NH_3 or molecular nitrogen, a white solid nitride forms. Boron nitride has an empirical formula of BN and the crystalline structure of graphite. At high pressures boron nitride is converted into *borazon*, which is very hard and has the structure of diamond.

When diborane is treated with ammonia, a boron-nitrogen analog of benzene, called borazole, $B_3N_3H_6$, is formed. Borazole is also referred to as *inorganic benzene*.

The similarities of these materials to the carbon compounds is excellent verification of the modern electronic theory of valence. Boron and nitrogen are adjacent to carbon in the second period, and all three atoms use the same atomic orbitals in forming molecular orbitals. In addition the B—N bond is isoelectronic with the C—C bond. Boron has three valence electrons, and nitrogen has five; each carbon atom has four. The similarities are therefore due to the fact that the same kinds of molecular orbitals are formed and have identical occupancies in both kinds of compounds.

The most common hydroxide of boron, $B(OH)_3$, orthoboric acid, is usually written H_3BO_3. It is a weak acid ($K_i = 7.3 \times 10^{-10}$) and unlike the hydroxides of the other group IIIA elements it is not amphoteric. It can be converted into borax by treatment with Na_2CO_3:

$$4H_3BO_3(s) + Na_2CO_3(s) + 4H_2O(l) \rightarrow Na_2B_4O_7 \cdot 10H_2O(s) + CO_2(g)$$

This is a common industrial method for preparing borax.

Aluminum, aluminum, everywhere

Aluminum was discovered in 1825, and until almost the end of the nineteenth century it was an expensive curiosity, used primarily in jewelry. This may seem surprising because (Table 18.3) it is not a rare element. In fact, aluminum is the most abundant metal in the earth's crust, occurring as a component of many rocks, clays, and soils in silicate forms such as feldspar, $KAlSi_3O_8$; kaolin, $H_4Al_2Si_2O_9$; and bauxite, a hydrated alumina, $Al_2O_3 \cdot xH_2O$, the ore from which metallic aluminum is obtained commercially.

FIGURE 18.6

The boron-nitrogen compounds resemble carbon structures.

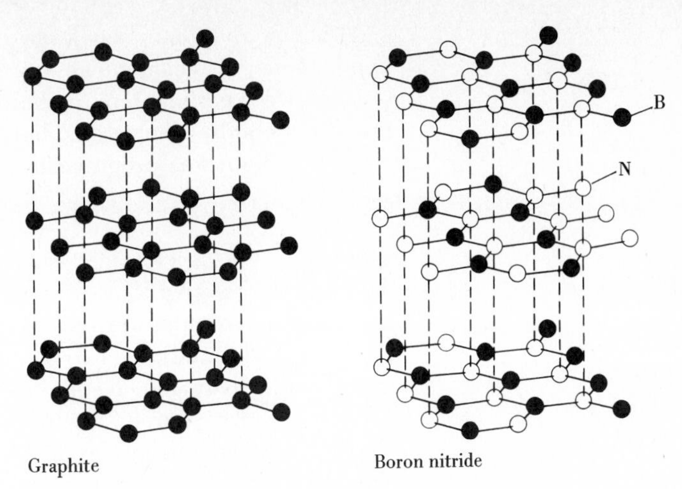

(*a*) *Comparison of graphite and boron nitride*

B

N

Diamond

Borazon

(*b*) *Comparison of diamond and borazon*

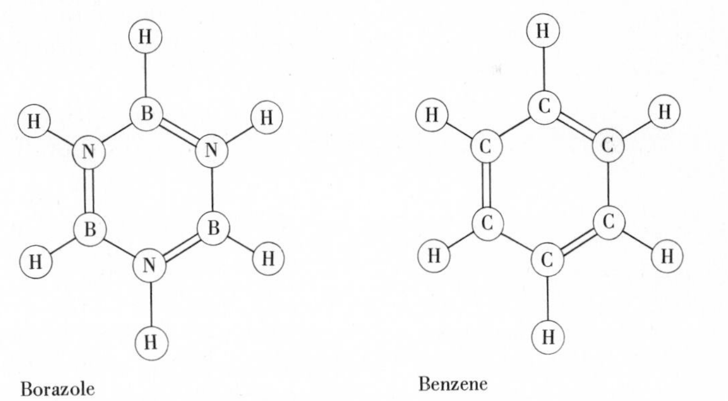

(*c*) *Comparison of borazole and benzene*

Although aluminum was known to have ideal properties for many applications, no economical way could be found to win it from its ores. Many experimenters, including Sir Humphry Davy, tried to obtain the metal by electrolysis, but all attempts were fruitless. Before 1886 aluminum was prepared by reduction with sodium metal:

$$6Na^0(l) + Al_2O_3(s) \rightarrow 2Al(s) + 3Na_2O(s)$$

This kept the price high since sodium is both expensive and dangerously reactive.

Alumina, the chief ore, is an amphoteric oxide, dissolving in an acidic or basic medium. The electrolysis of aqueous solutions, however, yields hydrogen gas at the cathode and oxygen gas at the anode. In other words, the water, not the dissolved alumina, is electrolyzed. Obviously, to obtain aluminum pure molten Al_2O_3 should be electrolyzed, but alumina melts at 2050°C; it would not be practical or economical to work at such a high temperature. Not only would a lot of fuel be required, but feasible materials for the equipment in such a process would be limited and expensive. (Steel and iron melt well below 2000°C.)

In 1886 two young men, Paul Heroult, of France, and Charles Martin Hall, of the United States, independently hit upon the idea of dissolving the alumina in a lower-melting aluminum ore, cryolite Na_3AlF_6. Cryolite melts near 1000°C and is highly ionic; molten solutions of Al_2O_3 in Na_3AlF_6 are easily electrolyzed, producing pure aluminum.

The essential features of the Hall-Heroult process are depicted in Fig. 18.7. A cast-iron tank lined with graphite serves as the cathode. Large graphite anodes dip into the molten Al_2O_3-Na_3AlF_6 solution. As the metallic aluminum forms at the cathode, it falls to the bottom and is drained off.

With the introduction of the Hall-Heroult process aluminum metal became more than just a curiosity, and its price tumbled from $12 to 25 cents per pound.

Aluminum has many attractive features; it is cheap, abundant, corrosion-resistant, and decorative; it has a low density and a high

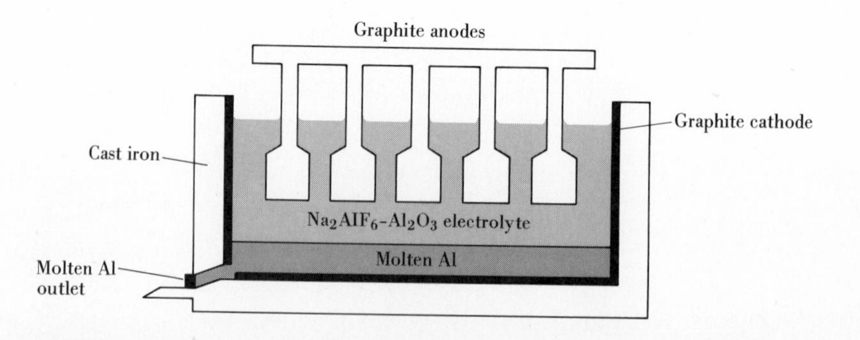

FIGURE 18.7

The Hall-Heroult process.

electric conductivity. Alloyed with small amounts of copper and other metals, it has a tensile strength comparable with that of steel. Naturally it has replaced steel, copper, and wood in many applications from electric cable to flip-top cans. Three million tons are produced in the United States each year. Since aluminum is used by every major industry, its production rate is as an indicator of economic health.

18.6 THE CARBON-SILICON FAMILY, GROUP IVA

The study of carbon, i.e., organic chemistry, has contributed more to our understanding of chemical bonding than that of all the other elements combined. Half the chemists in the world are organic chemists, devoting their professional lives to the synthesis and examination of carbon compounds. The variety and multitude of carbon compounds is due to a unique combination of three properties of the element: (1) carbon can catenate without limit, (2) it forms multiple bonds, and (3) the hydrides of carbon (the hydrocarbons) are very stable and fairly inert. Thus we find single molecules with chains containing thousands of carbon atoms. We see double and triple bonds: C=C, —C≡C—, >C=O, >C=N—, C≡N. The variety of the groups which can be grafted onto the stable hydrocarbon skeleton results in compounds with an enormous range of chemical and physical properties.

Although carbon is not the most plentiful element, it occurs in substantial quantities in both the free and combined forms. Carbon in the form of graphite has been used as a pigment and fuel for at least 15,000 years. In fact, the word *graphite* is derived from the Greek word meaning "to write." The United States has ample resources of graphite and anthracite and bituminous coal. Anthracite is approximately 90 percent carbon, while the softer bituminous coal is 60 to 70 percent carbon. Of course, limited amounts of elemental carbon are found in nature as the valuable diamond. Limestone, $CaCO_3$, petroleum, and natural gas are the chief sources of combined carbon.

Silicon, the second most common element in the earth's crust, occurs only in the chemically combined form in nature. The bulk of the rocky materials of the earth's crust are silicates of iron, magnesium, calcium, and aluminum. Just as carbon is the key element of the organic, living world, silicon is the key element of the inorganic, inanimate world.

Germanium, the rarest of the group IVA elements, is obtained from the ore *germanite*, a mixed sulfide of iron, copper, and germanium, which occurs in significant amounts only in South Africa.

It is also obtained from the residues of zinc- and lead-ore refining. The chief use of germanium is the manufacture of semiconductors for solid-state diodes and transistors, which to a great extent have replaced vacuum tubes in electronic circuitry (page 209.)

Tin and lead have been known and used for thousands of years. These elements are mentioned in the Bible and in the writings of Homer and Pliny. The wealthy Romans used cooking utensils made of lead because it sweetened the food.† Actually, most lead compounds are poisonous, and there is a theory that these utensils caused wholesale poisoning of the Roman leadership, contributing significantly to the fall of the Roman Empire. Tin was used in bronzes by the Persians as early as 3000 B.C. It has been found in the form of tools, utensils, weapons, ornaments, and chariot fittings. Tin is now obtained from the mineral cassiterite, SnO_2, found in Bolivia, and lead is obtained from galena, PbS, which occurs in the western United States.

Chemical behavior

The group IVA elements show a fully developed gradation from nonmetallic to metallic behavior. That is obvious from examination of the properties listed in Table 18.5. Carbon and silicon are definitely nonmetals, both in chemical and physical properties; germanium is a metalloid; and tin and lead are metallic. As expected, the electric resistivity is very high for carbon and silicon, very low for tin and lead, and intermediate for germanium. The gradation is nicely demonstrated by the chemical reactions of the elements. Carbon, silicon, and germanium do not react with dilute nonoxidizing acids, but, like true metals tin and lead, displace hydrogen:

$$Sn(s) + 2HCl(aq) \rightarrow SnCl_2(aq) + H_2(g)$$
$$Pb(s) + H_2SO_4(aq) \rightarrow PbSO_4(aq) + H_2(g)$$

Note that the two metallic elements of group IVA commonly exhibit a stable, lower +2 oxidation state as well as the +4 state. Concentrated nitric acid oxidizes carbon and yields molecular carbon dioxide, whereas tin and lead give ionic salts with the same reagent:

$$C(s) + 4HNO_3(aq) \xrightarrow{\Delta} CO_2(g) + 4NO_2(g) + 2H_2O(l)$$
$$Pb(s) + 4HNO_3(aq) \rightarrow Pb(NO_3)_2(aq) + 2NO_2(g) + 2H_2O$$

In excess oxygen all these elements combine to give the dioxide except for lead, which forms the lower oxide, consistent with its more metallic nature:

$$X + O_2(g) \xrightarrow{\Delta} XO_2 \qquad X = C, Si, Ge, Sn$$
$$Pb(s) + \tfrac{1}{2}O_2(g) \xrightarrow{\Delta} PbO(s)$$

† Lead acetate, $Pb(C_2H_3O_2)_2$, is called *sugar of lead* because it is sweet.

TABLE 18.5: SOME PROPERTIES OF THE GROUP IVA ELEMENTS

Element	*at wt*	*Abundance in earth's crust,* wt %	*Electronic configuration*	*Covalent radius,* Å	*Fourth ionization potential, kcal* $M^{3+}(g) \rightarrow M^{4}(g) + e^-$	*Oxidation potential,* V $M^0 \rightleftharpoons M^{2+}(aq) + 2e^-$	*Density,* g/cm³	*mp,* °C	*bp,* °C
Carbon, $_6C$	12.0111	0.03	[He] $2s^2, 2p^2$	0.77	1487	. . .	2.26,† 3.51‡	3570‡	4827‡
Silicon, $_{14}Si$	28.086	27.7	[Ne] $3s^2, 3p^2$	1.17	1041	. . .	2.32	1420	2355
Germanium, $_{32}Ge$	72.59	0.0007	[Ar] $3d^{10}, 4s^2, 4p^2$	1.22	1030	. . .	5.35	959	2700
Tin, $_{50}Sn$	118.69	0.004	[Kr] $4d^{10}, 5s^2, 5p^2$	1.41	939	0.140	7.28	232	2360
Lead, $_{82}Pb$	207.19	0.0015	[Xe] $4f^{14}, 5d^{10}, 6s^2, 6p^2$	1.54	976	0.126	11.3	327	1755

† Graphite.
‡ Diamond.

Carbon dioxide and silicon dioxide are acidic, whereas all the other oxides are amphoteric.

Similar behavior is observed when the group IVA elements react with the halogens, which, like O_2, are fairly strong oxidizing agents.

$$X(s) + 2Cl_2(g) \rightarrow XCl_4 \qquad X = Si, Ge, Sn$$

Carbon does not combine directly with any halogen except fluorine, which produces CF_4. Lead again demonstrates its more metallic nature by forming ionic dihalides, for example, $PbCl_2$, except when it reacts with fluorine and forms the more covalent lead tetrafluoride.

Hydrides

It was natural to hope that the chemistry of the heavier elements of group IVA would prove similar to carbon chemistry. Many chemists have dreamed of synthesizing silicon, germanium, and tin analogs of the carbon compounds, the alkanes, alkenes, benzene derivatives, ethers, carboxylic acids, and others, but now these dreams appear unattainable. The ability of the group IVA elements to form covalently bonded chains (to catenate) falls off quickly with increasing atomic weight. There appears to be no limit on the number of carbon atoms which can catenate, but the highest hydride of silicon, or *silane*, which has been characterized is Si_6H_{14}. The highest *germane* is Ge_3H_8, whereas tin and lead form only the simplest hydrides, SnH_4 and PbH_4, called *stannane* and *plumbane*, respectively.

The ability to form double and triple bonds seems limited to carbon atoms. No analogs of ethylene or acetylene have been prepared. Unlike the alkanes, the hydrides of the other elements become more unstable as the atomic weight of the element increases. Stannane decomposes at 150°C and plumbane at 0°C. In addition, all the hydrides ignite immediately in air and hydrolyze rapidly in the presence of moisture. It appears that the only similarity between the stable inert alkanes and the hydrides of the other group IVA elements is in their general formula, M_nH_{2n+2}.

Carbon

Much of the chemistry of this vital element was discussed in Chaps. 15 and 16. The structures of the two important crystalline forms were shown in Figs. 7.15 and 7.16. In this section we limit our attention to inorganic and industrial aspects of carbon chemistry. Table 18.5 shows that the density of diamond is much higher than that of graphite. Therefore, a simple application of Le Châtelier's principle suggests that it should be possible to convert graphite into diamond by use of high pressures:

$$\text{Graphite} \rightleftharpoons \text{diamond}$$

Density, g/cm_3: 2.26 3.51

In 1893, Henri Moisson implemented this suggestion with an ingenious experiment, mixing small chunks of pure graphite in molten iron and rapidly quenching the whole mass by dropping it in cold water. When he dissolved the iron in acid, he obtained tiny diamonds; the contraction of the cooling iron about the trapped graphite generated pressures sufficient for the conversion. Diamonds are manufactured industrially in an apparatus which can generate over 10,000 atm of pressure. These industrial diamonds are used chiefly for cutting and drilling tools.

Sixty million tons of carbon in the form of *coke* are produced annually from the destructive distillation of bituminous coal. This process involves roasting the coal in the absence of air. As a result, volatile components are vaporized, and a residue of coke is left. The volatile components, valuable in themselves, consist of ammonia, fuel gas, and coal tar, from which pigments, plastics, drugs, and explosives are manufactured. Most of the coke is used as a reducing agent in metallurgy, more than 80 percent of it in the production of pig iron.

Since coke is also used as a fuel to supply the energy to turn the giant turbines of our electric power plants, we see why carbon is indispensable to modern society. We use it to obtain steel and electricity, two of our most important commodities.

Carbon dioxide, carbonic acid, and the carbonates

Carbon dioxide is present in the atmosphere to the extent of only 0.03 percent by volume, but it is far from being a negligible constituent. Carbon dioxide and water are the raw materials for photosynthesis, which occurs in the chloroplasts of green plants and produces carbohydrates, the basic nutrient for all living organisms. Until this century the rate of removal of CO_2 from the air by photosynthesis was equal to the rate at which it was produced by glycolysis and respiration in the cells of living organisms and by the combustion of carbonaceous compounds. A steady state existed, in which the CO_2 concentration in the air remained constant. With the burgeoning and spread of heavy industry, air transport, and the automobile the concentration of CO_2 is increasing, but it has not yet approached a dangerous level.

Carbon dioxide is a colorless, odorless gas approximately 1.5 times denser than air. It is relatively inert at ordinary temperatures, but at elevated temperatures it may act as an oxidizing agent. *After* magnesium is ignited, it will continue to burn in a pure CO_2 atmosphere:

$$CO_2(g) + 2Mg^0(s) \rightarrow C^0(s) + 2MgO(s)$$

Carbon dioxide dissolves to some extent in water and forms an acidic solution which can be attributed to the formation of carbonic acid:

$$CO_2(g) + H_2O(l) \rightleftharpoons H_2CO_3(aq)$$

Greater amounts of CO_2 can be made to dissolve in water by increasing the pressure of the gas, and this is the basis for the commercial preparation of carbonated water, or "soda." Since its critical temperature is 31°C, CO_2 can be liquefied at room temperature. When the liquid is forced through a nozzle, it vaporizes immediately, and, since the process is so rapid, the required-heat of vaporization is supplied by the carbon dioxide itself. As a result of the rapid cooling the vapor condenses into a solid, Dry Ice. Under 1 atm the temperature of Dry Ice is about −78°C so that it is much colder than ordinary ice. The fact that it will sublime rather than melt gives it another advantage as a refrigerant.

Carbonic acid is a weak acid which ionizes in two steps to produce bicarbonate and carbonate ions:

$$\text{Step 1: } H_2CO_3(aq) + H_2O(l) \rightleftharpoons \underset{\text{Bicarbonate ion}}{HCO_3^-(aq)} + H_3O^+(aq) \qquad K_i = 3.5 \times 10^{-7}$$

$$\text{Step 2: } HCO_3^-(aq) + H_2O(l) \rightleftharpoons \underset{\text{Carbonate ion}}{CO_3^{2-}(aq)} + H_3O^+(aq) \qquad K_i = 4.4 \times 10^{-11}$$

Salts of the bicarbonate and carbonate are important. Only the alkali-metal carbonates are soluble. The Solvay process is used in the production of $NaHCO_3$ and $NaCO_3$. Limestone is another important industrial carbonate. The carbonates of sodium and calcium, which we have already discussed, are bases and indeed are the most important basic chemicals of industry.

Sodium bicarbonate, $NaHCO_3$, is a component of baking powder, the function of which is to liberate bubbles of CO_2 in a quick-bread dough or cake batter. When $NaHCO_3$ is mixed with water and acidified, CO_2 is generated:

$$NaHCO_3(s) + H_3O^+(aq) \rightarrow Na^+(aq) + 2H_2O(l) + CO_2(g)$$

In baking, a nonpoisonous acid like tartaric acid is used; mixed with the $NaHCO_3$, it forms baking powder.

Silicon

The second member of group IVA is significantly different from carbon. If one of the key characteristics of carbon chemistry is catenation, then one of the key characteristics of silicon chemistry must be *con*catenation: most silicon compounds are oxycompounds, the basic structure of which is a chain of alternating silicon and oxygen atoms.

Silicon is a hard, brittle, gray solid with a hint of a metallic luster. It crystallizes in a diamond structure and has a fairly high electric resistance. Like carbon, elemental silicon is not very reactive at room temperature and is a nonmetal.

Silicon can be prepared by reducing silicon dioxide (silica) with carbon or an active metal like sodium or magnesium. The industrial process involves heating SiO_2 with coke in an electric furnace:

$$SiO_2(s) + C(s) \xrightarrow{\Delta} Si(s) + CO_2(g)$$

Since the major consumption of silicon is in the manufacture of ferrosilicon steels, most silicon is prepared by the reduction of a combination of silica and iron ores.

In order to produce pure silicon a careful control of the ratio of silica to coke is required. Too much coke causes the formation of silicon carbide, SiC (Carborundum), a very hard substance used as an abrasive, and too little coke leaves the product silicon contaminated with unreduced silica. Even under carefully controlled conditions the product is slightly contaminated, and silicon to be used for semiconductor devices, where ultrapurity is required, must be *zone-refined*. In zone refining a melted zone is passed very slowly through the sample to be purified (Fig. 18.8). The impurities tend to accumulate in the melted zone, and after several passes 99.999999 percent pure silicon can be prepared. Aside from its uses in steel and semiconductor production, elemental silicon has no major applications.

There is probably no greater difference between the oxides of two nonmetals of the same family than that between CO_2 and SiO_2. Carbon dioxide melts at −57°C; silica melts above 1700°C. Whereas solid carbon dioxide (Dry Ice) is composed of discrete linear CO_2 molecules held together by weak van der Waals forces (Fig. 18.9*a*), the SiO_2 unit in silica is covalently bonded to four other such units in a tetrahedral array. Each silicon atom is located at the center of a regular tetrahedron with an oxygen atom at each vertex (see Fig. 18.9*b*). The most common natural form of silica is quartz, in which endless SiO_2 tetrahedra are bonded in a regular three-dimensional framework. Silica forms a covalent crystal with a diamond structure, whereas solid carbon dioxide forms a van der Waals crystal with weak intermolecular forces.

FIGURE 18.8

In zone refining the impurities tend to accumulate in the melted zone as it moves up (or down) through the rod.

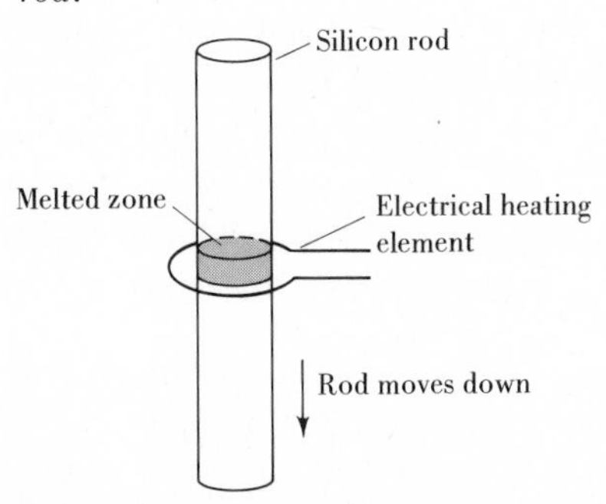

The inability of silicon to form double bonds explains the sharp difference between the two dioxides. Thus silicon must bond to four oxygen atoms to achieve the rare-gas configuration, while carbon, in forming double bonds with two oxygen atoms, can form discrete linear molecules, O═C═O.

Silica is acidic and does not react with acids except HF, with which it produces gaseous SiF_4:

$$SiO_2(s) + 4HF(aq) \rightarrow SiF_4(g) + 2H_2O(l)$$

The HF reaction is the basis for etching glass. SiO_2 reacts with

FIGURE 18.9

Comparison of the crystal structures of CO_2 and SiO_2.

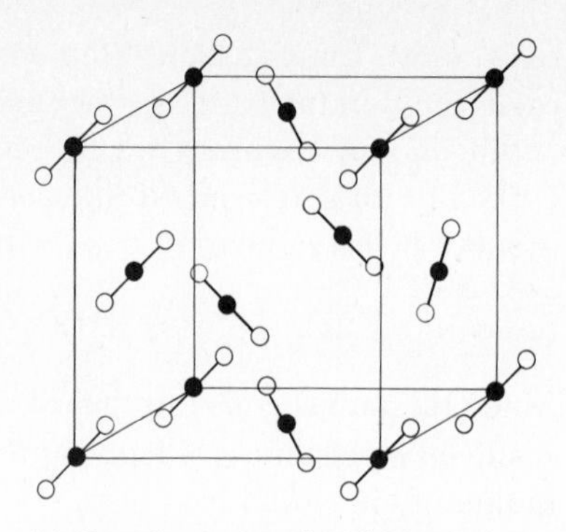

(a) Crystal structure of solid CO_2, dry ice. Carbon dioxide forms a face-centered cubic crystal consisting of discreet molecular units. The carbon atoms of the CO_2 molecules are at the corners and the faces of the unit cell.

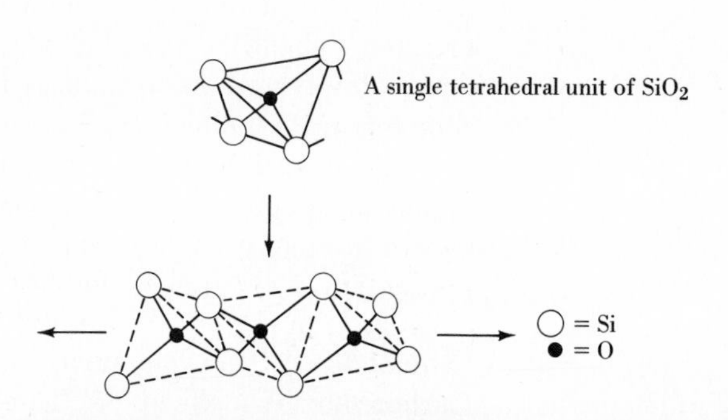

(b) Silica is a crystal of condensed tetrahedra.

strong bases to yield silicates:

$$SiO_2(s) + 4NaOH(aq) \rightarrow Na_4SiO_4(aq) + 2H_2O(l)$$

Most rocky materials in the earth's crust are silicates of common metals. Fundamental to all these minerals is the silicate ion, SiO_4^{2-}, or a polysilicate anion. The silicate ions are tetrahedral units which can bind to themselves in a variety of ways to give rise to the multitude of mineral structures that occur in nature. Some of the minerals have definite, discrete anions, while others have giant silicate chains of indefinite length. When the polysilicates form parallel chains, the result is a fibrous material like asbestos; giant layers give rise to soft minerals, like mica. If a three-dimensional polysilicate forms, the result is a very hard mineral like quartz (see Fig. 18.10).

The best-known use of silica is in the manufacture of glass. Soft glass, used for ordinary windows, is produced by fusing silica with Na_2CO_3 and $CaCO_3$; the result is a mixture of sodium and calcium silicates. Soft glass is attacked by alkali and because it has a high

FIGURE 18.10

Structure of silicate and polysilicate anions.

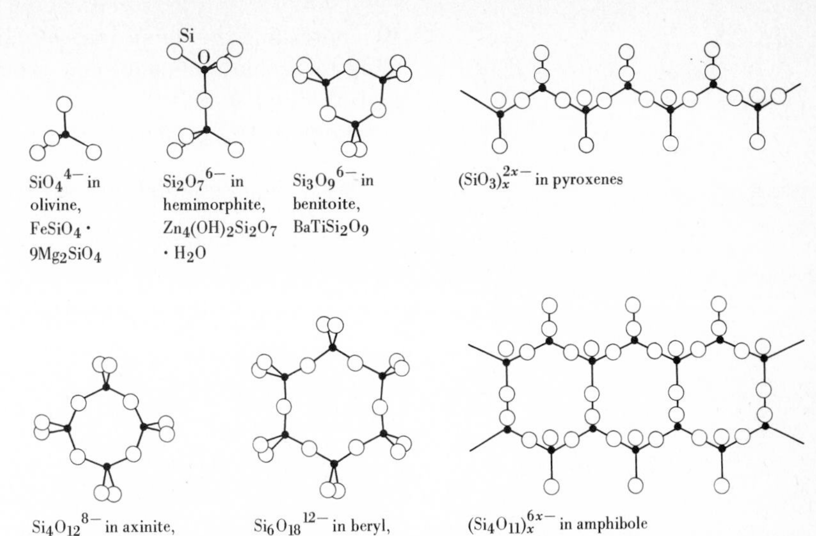

(*a*) *Discrete silicate anions found in some silicate minerals*

(*b*) *Giant silicate anion chains found in some silicate minerals*

coefficient of expansion, it shatters under thermal shock and cannot be used for baking or in industrial and laboratory procedures. For these reasons Pyrex and Kimax, made with SiO_2, Al_2O_3, and B_2O_3, have found wide use.

Zeolite is a giant polysilicate complex in which some of the silicon (+4) has been replaced by aluminum (+3), so that there is a negative charge on the molecular framework. This unbalance is offset by cations held within the framework by ionic forces. If the zeolite is flushed with a brine solution, the offsetting cations can be replaced by sodium ions; i.e., there is an *ion exchange*. The ion-exchange capability of the zeolites makes them valuable in water treatment. Water is *hard* when it contains a high concentration of calcium and magnesium salts because these ions form an insoluble scum with soap. When hard water is passed through a bed of sodium zeolite, the calcium and magnesium ions are replaced by sodium ions and the result is soft water. Sodium salts of the fatty acids present in soap are water-soluble, and a scum does not form. (See page 728.)

The high chemical and thermal stability of the Si—O—Si bond of quartz and the other silicates is the basis for the desirable properties of the *silicones*, or *organic* polysilicates. The giant chain of alternating Si and O atoms carries two methyl groups on each silicon atom:

$$\cdots\!-\!\underset{}{Si(CH_3)_2}\!-\!O\!-\!Si(CH_3)_2\!-\!O\!-\!Si(CH_3)_2\!-\!O\!-\!Si(CH_3)_2\!-\!O\!-\!\cdots$$

By controlling the chain length and extent of cross-linking it is possible to prepare silicone oils, plastics, and rubbers with extremely high thermal stability.

Silicones are prepared by condensing dihydroxydimethylsilane:

$$HO-\underset{CH_3}{\overset{CH_3}{\underset{|}{\overset{|}{Si}}}}-OH + HO-\underset{CH_3}{\overset{CH_3}{\underset{|}{\overset{|}{Si}}}}-OH \rightarrow HO-\underset{CH_3}{\overset{CH_3}{\underset{|}{\overset{|}{Si}}}}-O-\underset{CH_3}{\overset{CH_3}{\underset{|}{\overset{|}{Si}}}}-OH + H_2O$$

and

$$HO-\underset{CH_3}{\overset{CH_3}{\underset{|}{\overset{|}{Si}}}}-O-\underset{CH_3}{\overset{CH_3}{\underset{|}{\overset{|}{Si}}}}-OH + n\text{Si}(CH_3)_2(OH)_2 \rightarrow HO-\underset{CH_3}{\overset{CH_3}{\underset{|}{\overset{|}{Si}}}}-O-\left(-\underset{CH_3}{\overset{CH_3}{\underset{|}{\overset{|}{SiO}}}}-\right)_n-\underset{CH_3}{\overset{CH_3}{\underset{|}{\overset{|}{Si}}}}-OH$$

Silicone

By including a small amount of trihydroxymethylsilane it is possible to prepare a cross-linked rigid polymer.

18.7 THE NITROGEN-PHOSPHORUS FAMILY, GROUP VA

As we move across the periodic table in our review of the regular elements, we see an increasing number of nonmetals within the families. All the elements of group IA are metals; beryllium of group IIA shows some nonmetallic character, which is developed further in groups IIIA and IVA (boron; carbon, silicon, and germanium). And in group VA the behavior of the elements in predominantly nonmetallic. Even though the last member of group VA is considered a metal, in the +5 oxidation state even bismuth compounds behave like compounds of a nonmetal. Nitrogen and phosphorus are nonmetals, arsenic and antimony are metalloids, and bismuth is a metal. This is consistent with the fact that the first four elements can form molecular solids; we find N_2, P_4, As_4, and Sb_4, but bismuth exhibits a metallic solid only.

The elements of group VA are not extremely abundant (Table 18.6). Thanks to its accessibility as 80 percent of the earth's atmosphere, nitrogen is not considered rare. Phosphorus, the most abundant element of group VA, is distributed widely through the earth's crust, occurring mainly as rock phosphate, $Ca_3(PO_4)_2$, and apatite, $CaF_2 \cdot Ca_3(PO_4)_2$. The chief ores of the other elements of this group are sulfides, for example, As_2S_3, Sb_2S_3, and Bi_2S_3. However, like nitrogen, arsenic and bismuth have been found in the free elemental state.

TABLE 18.6: SOME PROPERTIES OF THE GROUP VA ELEMENTS

Element	*at wt*	*Abundance in earth's crust*, wt %	*Electronic configuration*	*Covalent radius*, Å	*First ionization potential*, kcal $M^0(g) \rightarrow M^+(g) + e^-$	*mp*, °C	*bp*, °C	*Density*, g/cm³ (*for solids*)
Nitrogen, $_7N$	14.0067	0.0046	[He] $2s^2$, $2p^3$	0.74	334	−210	−196	
Phosphorus, $_{15}P$	30.9738	0.118	[Ne] $3s^2$, $3p^3$	1.10	243	44.1	280	1.82
Arsenic, $_{33}As$	74.9216	0.0005	[Ar] $3d^{10}$, $4s^2$, $4p^3$	1.21	226	615	814	5.73
Antimony, $_{51}Sb$	121.75	0.0001	[Kr] $4d^{10}$, $5s^2$, $5p^3$	1.41	199	630	1380	6.68
Bismuth, $_{83}Bi$	208.980	0.00002	[Xe] $4f^{14}$, $5d^{10}$, $6s^2$, $6p^3$	1.52	168	271	1470	9.80

TABLE 18.7: COMPOUNDS OF NITROGEN IN DIFFERENT OXIDATION STATES

Species	*Oxidation state*
NH_3	−3
N_2H_4	−2
N_2H_2	−1
N_2	0
N_2O	+1
NO	+2
HNO_2	+3
NO_2	+4
HNO_3	+5

TABLE 18.8: THE TRIHYDRIDES OF THE GROUP VA ELEMENTS

Formula	*Name*	*bp*, °C
NH_3	Ammonia	−33
PH_3	Phosphine	−88
AsH_3	Arsine	−55
SbH_3	Stibine	−18
BiH_3	Bismuthine	+22

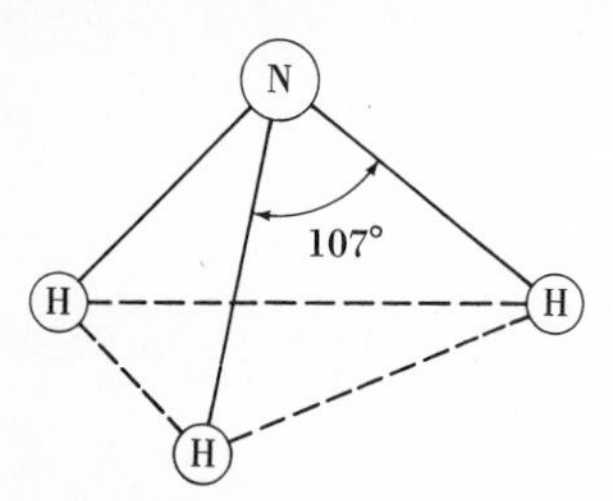

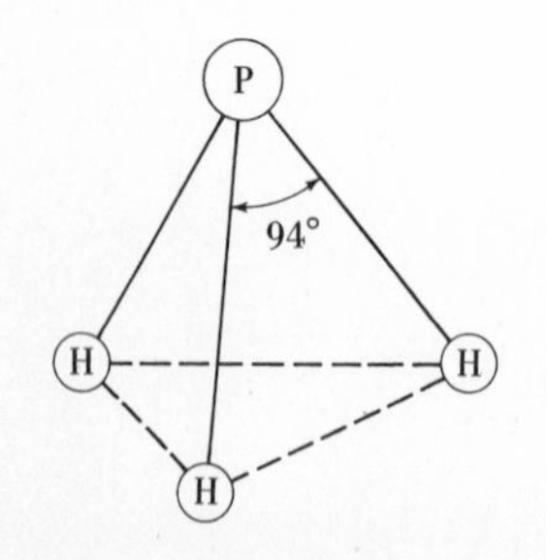

Since the number of oxidation states exhibited by the family is quite large, the chemistry is varied and complex. Nitrogen exhibits every oxidation state from −3 to +5 (Table 18.7). The most common oxidation states of the family are +3 and +5, which correspond to the characteristic electronic configuration of the valence shell: ns^2, np^3. The +5 oxides of the elements are acidic, and, as expected, the acidity of this oxidation state decreases from nitrogen to bismuth. The +3 oxides of nitrogen, phosphorus, and arsenic are weak acids, whereas the +3 oxide of bismuth is basic, showing that bismuth exhibits a metallic nature in its +3 state and a nonmetallic nature in its +5 state.

Each of these elements forms a pyramidal trihydride molecule. Nitrogen trihydride, i.e., ammonia, appears to be anomalous because of hydrogen bonding (Table 18.8). Because of its relatively high boiling point and heat of vaporization it has been widely used as a refrigerant. Ammonia is a relatively strong base, acting as a proton acceptor in aqueous media:

$$NH_3(aq) + H_2O(l) \rightleftharpoons NH_4^+(aq) + OH^-(aq)$$

Phosphine reacts to produce the *phosphonium* ion, PH_4^+

$$PH_3(g) + HCl(g) \rightarrow PH_4Cl(s)$$

but *not* in aqueous solutions. Phosphine is a much weaker base than water, so that the PH_4^+ ion is a strong acid, donating H^+ to H_2O:

$$PH_4Cl(aq) + HOH(l) \rightarrow PH_3(g) + H_3O^+(aq) + Cl^-(aq)$$

The remaining trihydrides are nonpolar and almost water-insoluble. They are very weak bases and thermally unstable. Very gentle heating of arsine, AsH_3, and stibine, SbH_3, causes them to decompose into their elements.

Only two elements show catenation in the hydrides. Nitrogen forms hydrazine, $H_2N—NH_2$, and phosphorus forms diphosphine, $H_2P—PH_2$. Both these compounds are reactive and unstable, so that among the group VA elements there are no series corresponding to the alkanes or even the boranes and silanes.

One of the chief differences between nitrogen and phosphorus chemistry lies in the ability of phosphorus to expand its valence shell to 10 or even 12 electrons so that it can form five or six bonds to other elements. For instance, PCl_3 can react with Cl_2 to produce the pentavalent chloride:

$$PCl_3(g) + Cl_2(g) \rightarrow PCl_5(g)$$

In addition, there are solids in which the PCl_6^- ion exists. There are other species in which phosphorus exhibits a valence of 5 or 6. The expanded valence shell of phosphorus is explained by the accessibility of vacant *d* orbitals in the electronic structure of the atom.

The valence electrons of phosphorus are in the third energy level, where there are vacant d orbitals: P $1s^2$, $2s^2$, $2p^6$, $3s^2$, $3p^3$, $3d^0$. A participation of the $3d$ orbitals would allow phosphorus to expand its valence shell and become pentavalent or hexavalent.

Nitrogen fixation

Elemental nitrogen as the diatomic molecule, N_2, is the major component of the air we breathe. Bonding in the N_2 molecule was discussed on page 129. The presence of the triple bond makes the species very stable and chemically inert. We use millions of tons of ammonia, nitrates, and nitric acid each year in the manufacture of explosives, fertilizers, plastics, and other products. The amount of nitrogen in compound form, i.e., the amount of *fixed* nitrogen present in nature, cannot keep up with the demands of agriculture and industry. The conversion of molecular nitrogen to more useful forms has always been a major concern of chemists and biochemists.

In the most important industrial process for fixing nitrogen, the Haber process, N_2 and H_2 react at very high temperatures and pressures to produce ammonia, which can then be converted into other nitrogen compounds. In living organisms nitrogen is required primarily in the synthesis of proteins, but, in general, molecular nitrogen cannot be used for this purpose. Certain soil bacteria are able to fix nitrogen when they attack *leguminous* plants and form nodules in their roots (Fig. 18.11). The nitrogen of the atmosphere is

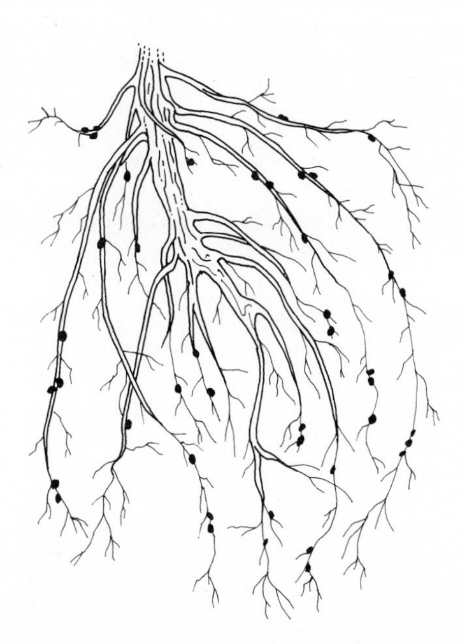

FIGURE 18.11

Nodules containing nitrogen-fixing bacteria on roots of a soybean plant.

first converted into ammonia, which then is used to form the plant protein. The bacterial fixation of nitrogen has always intrigued chemists and biochemists because it occurs under such mild conditions. Great savings would be realized if we could learn to imitate the soil bacteria, and major research is directed toward this goal.

A schematic of the Haber process is shown in Fig. 18.12. The gases in a hydrogen-nitrogen mole ratio of 3:1 are passed into a compressor, where their pressure is increased to about 200 atm. Next, the gases are heated to 500°C and passed over a catalyst of alumina and K_2O embedded in a porous iron support; the conversion to NH_3 is then about 20 percent. Because it has a much higher boiling point than either N_2 or H_2, ammonia can be separated from the reaction mixture by liquefaction. The unreacted gases are recycled, and the liquid ammonia is tapped and stored in steel cylinders.

Over 10 million tons of ammonia are produced by the Haber process each year. The predominant use of ammonia is in the preparation of ammonium salts (nitrate, sulfate, and phosphate) used as fertilizers. Ammonium chloride is used in the manufacture of dry cells. Half the ammonia produced each year is converted into nitric acid, to make ammonium nitrate and explosives.

Nitric acid is produced commercially by the Ostwald process, which involves the catalytic oxidation of ammonia. A 10 percent mixture of ammonia in air is heated to 800°C at a pressure of 10 atm and passed over a platinum catalyst. The product of the first step is nitric oxide:

$$2NH_3(g) + \tfrac{5}{2}O_2(g)(\text{air}) \xrightarrow{\text{Pt}} 2NO(g) + 3H_2O(g)$$

The nitric oxide is then oxidized to nitrogen dioxide, NO_2, which is dissolved in water, whereupon HNO_3 forms:

$$3NO_2(g) + H_2O(l) \rightarrow 2HNO_3(l) + NO(g)$$

The by-product of the last reaction, NO, is recycled to form more NO_2.

Nitric acid is a strong acid, being 100 percent ionized in aqueous solutions. It is also a powerful oxidizing agent and reacts with

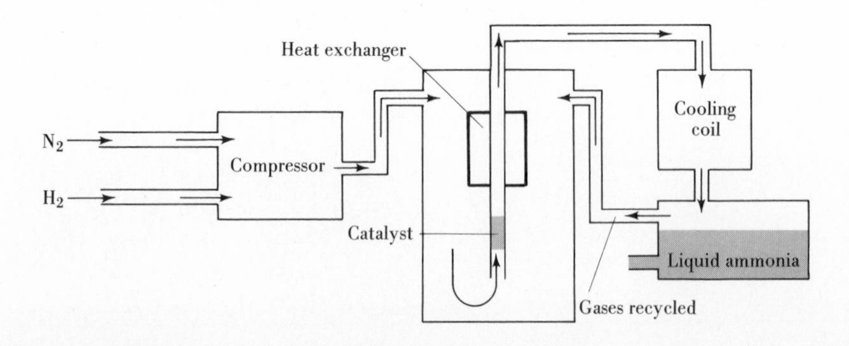

FIGURE 18.12

The Haber process.

metals that do not displace hydrogen from acid solutions. In addition to its use as a nitrate source for fertilizer, HNO_3 is used as a nitrating agent in the preparation of pigments.

Phosphorus

Elemental phosphorus exists in 11 allotropic forms, but only the white, red, and black forms are common. White phosphorus was first obtained by alchemists from urine. They marveled at this substance which glows in the dark and ignites spontaneously in air. Its preparation was a closely guarded secret until the substance aroused the curiosity of Robert Boyle, who published a method for its preparation and named it *phosphorus* (from the Greek meaning "to bear light").

Phosphorus in the form of phosphate is extremely important in living organisms. We have already learned that DNA, the genetic molecule in the nucleus of a cell, is formed from a giant phosphate-sugar chain, and phosphate has the same significance in the structure of the various RNAs. Adenosine triphosphate (ATP) is involved in the most important energy-transfer process in the living cell. Phosphate is also important as a structural material in bones and teeth.

White phosphorus is obtained commercially by reducing rock phosphate with coke in the presence of sand in an electric furnace (1500°C) (see Fig. 18.13). The reactions involve a conversion of rock phosphate to slag, $CaSiO_3$, which is tapped off at the bottom of the furnace, and gaseous P_4O_{10}, which rises through the coke and is reduced to phosphorus:

$$\underset{\text{Rock phosphate}}{2Ca_3(PO_4)_2(s)} + \underset{\text{Sand}}{6SiO_2(s)} \xrightarrow{\Delta} \underset{\text{Slag}}{6CaSiO_3(l)} + P_4O_{10}(g)$$

$$P_4O_{10}(g) + 10C(s) \xrightarrow{\Delta} \underset{\substack{\text{White}\\ \text{phosphorus}}}{P_4(g)} + 10CO(g)$$

White phosphorus and carbon monoxide pass out of the electric furnace as gases, whereupon the white phosphorus is condensed and stored under water to prevent it from igniting.

It should be noted that white phosphorus is composed of tetratomic molecules, P_4, which have a pyramidal shape. Each phosphorus atom of the P_4 molecule is bonded to three others, so that P_4 forms a perfect tetrahedron (see Fig. 18.14).

The tetrahedral geometry is preserved when P_4 reacts to form phosphorus(III) oxide, P_4O_6, and phosphorus(V) oxide, P_4O_{10}. The oxygen atoms enter sites between the phosphorus atoms to form P_4O_6; four additional oxygen atoms are bonded to the phosphorus atoms in P_4O_{10} (Fig. 18.14).

FIGURE 18.13

The production of phosphorus in the electric furnace.

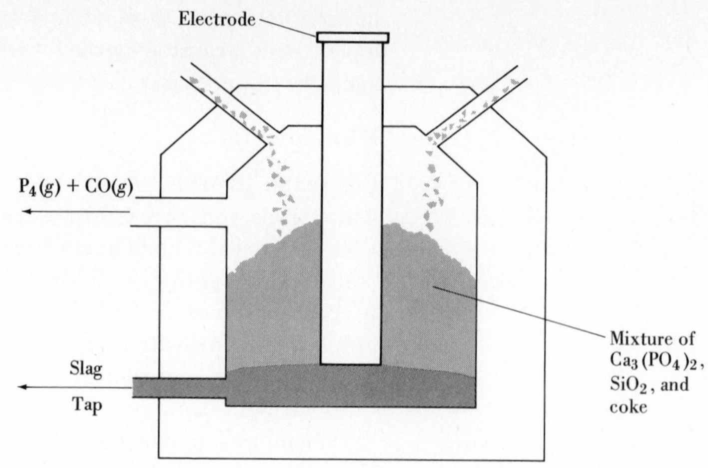

FIGURE 18.14

Phosphorus structures.

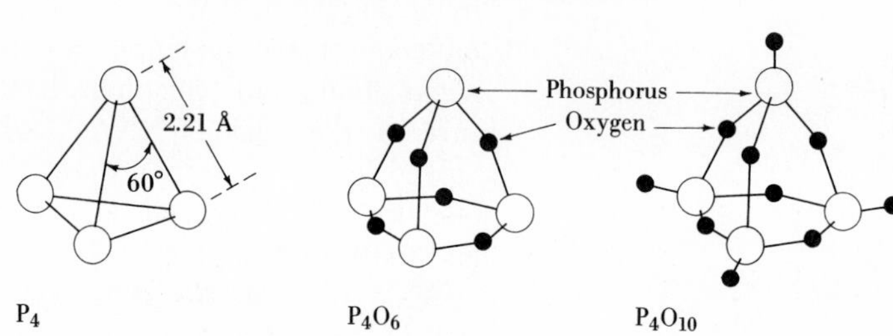

(*a*) *The structures of* P_4, P_4O_6, *and* P_4O_{10}

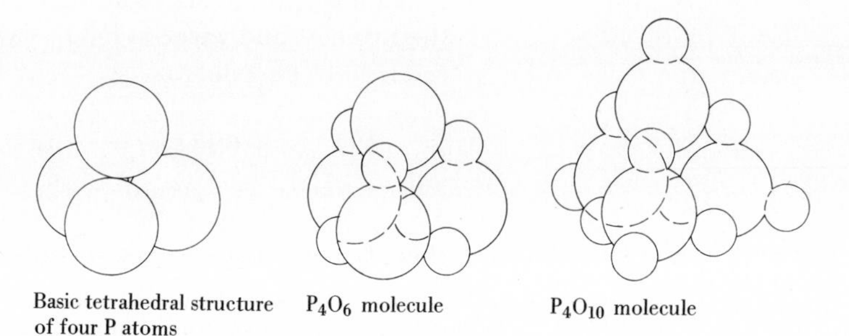

(*b*) *Structure of the* P_4, P_4O_6, *and* P_4O_{10} *molecules using spheres with relative diameters of* P *and* O *atoms*

White P will cause severe, slow healing burns if it touches bare skin

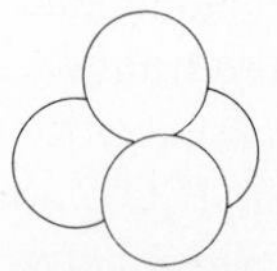

White phosphorus, P_4, a molecular solid

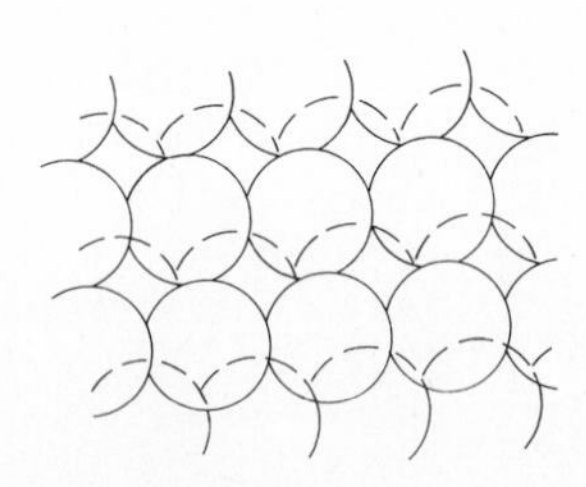

Red phosphorus, P_x, a covalent solid

(c) *Two allotropes of phosphorus*

The oxides of phosphorus are the acid anhydrides of two series of acids. The most important acid derived from P_4O_{10} is phosphoric acid, H_3PO_4, and the most important acid derived from P_4O_6 is phosphorous acid H_3PO_3. The Lewis structures of these two acids (page 425) show that H_3PO_4 is tribasic and H_3PO_3 is dibasic; in H_3PO_3 one of the hydrogen atoms is bonded to the central phosphorus and is not an acidic hydrogen.

Red phosphorus does not ignite spontaneously in air; it has a higher melting point and lower vapor pressure than white phosphorus. That red phosphorus is a covalent solid and white phosphorus is a molecular solid (Fig. 18.14*c*) explains the relative inertness and other physical differences between the two allotropes.

Phosphorus in the form of phosphate is extremely important to the proper growth of plants because it is involved in vital structures and processes of the living cell. It should not be surprising that most of the phosphorus produced is eventually converted into phosphoric acid and its salts for use as a fertilizer. Rock phosphate is converted directly to superphosphate fertilizer by treatment with sulfuric acid:

$$Ca_3(PO_4)_2(s) + 2H_2SO_4(l) \rightarrow \underbrace{2CaSO_4(s) + Ca(H_2PO_4)_2(s)}_{\text{Superphosphate}}$$

18.8 THE OXYGEN-SULFUR FAMILY, GROUP VIA

Oxygen is the most abundant element on the earth's surface. The crust of the earth is almost 50 percent oxygen by weight, the ocean waters are 88 percent, and the atmosphere is 23 percent. The number of oxygen atoms in the earth's crust is greater than that of all the other elements combined. Sulfur is not highly abundant, but it is widely distributed, both as the free element and as metal sulfides and sulfates. The remaining elements of group VIA are very rare (see Table 18.9).

Polonium, the heaviest element, is not only rare but radioactive, so that it has not been studied as extensively as the other elements. Selenium and tellurium are nonmetallic, but since tellurium has a fairly high electric conductivity, it may be considered a metalloid. Selenium exhibits photoconductivity; i.e., when exposed to light its electric resistance decreases sharply and it remains a good conductor as long as light shines on it. Its conductivity decreases as the intensity of the light decreases and becomes very low in the dark. Because of its photoconductivity selenium finds widespread application in light meters, photoelectric eyes, and other electronic devices.

Traces of selenium and tellurium occur in nature as the selenides, Se^{2-}, and tellurides, Te^{2-}, of the heavy elements mixed with the heavy-metal sulfides like Cu_2S, Ag_2S, etc. These metals are recov-

TABLE 18.9: SOME PROPERTIES OF THE OXYGEN-SULFUR FAMILY, GROUP VIA

Element	at wt	Abundance in earth's crust, wt %	Electronic configuration	Radius, Å		Electro-negativity	First ionization potential, kcal $M^0(g) \rightarrow M^+(g) + e^-$	mp, °C	bp, °C	Density, g/cm³ (for solids)
				Covalent	Ionic (X^{2-})					
Oxygen, $_8O$	15.994	46.6	[He] $2s^2, 2p^4$	0.74	1.35	3.5	313	−218	−183	
Sulfur, $_{16}S$	32.064	0.052	[Ne] $3s^2, 3p^4$	1.04	1.84	2.5	249	119	444.6	2.07
Selenium, $_{34}Se$	78.96	9×10^{-6}	[Ar] $3d^{10}, 4s^2, 4p^4$	1.17	1.98	2.4	224	217	688	4.79
Tellurium, $_{52}Te$	127.60	1.8×10^{-7}	[Kr] $4d^{10}, 5s^2, 5p^4$	1.37	. . .	2.1	207	452	1390	6.25
Polonium, $_{84}Po$	(210)	3×10^{-8}	[Xe] $4f^{14}, 5d^{10}, 6s^2, 6p^4$	(1.5)	. . .	. . .	194			

ered from the residues of copper refining. In their elemental states the group VIA elements exhibit several allotropic forms. Gaseous oxygen may exist as O_2 or O_3 (ozone). Rhombic sulfur, the stable crystalline form under ordinary conditions, is composed of puckered S_8 rings (see Fig. 18.15). When the temperature is raised above 96°C and maintained there, rhombic sulfur is transformed into a monoclinic crystalline form, still composed of S_8 rings. When the sulfur is raised to its melting point, 119°C, a nonviscous liquid composed of S_8 molecules is formed. Continued heating of the liquid causes the S_8 rings to split open, and long molecular chains form; the liquid thus becomes very viscous. Continued heating breaks up the long chains, and at the boiling point the vapor is composed of S_2 molecules. Selenium also exhibits a crystalline form composed of Se_8 molecules.

FIGURE 18.15

Crystalline structures and allotropes of oxygen and sulfur.

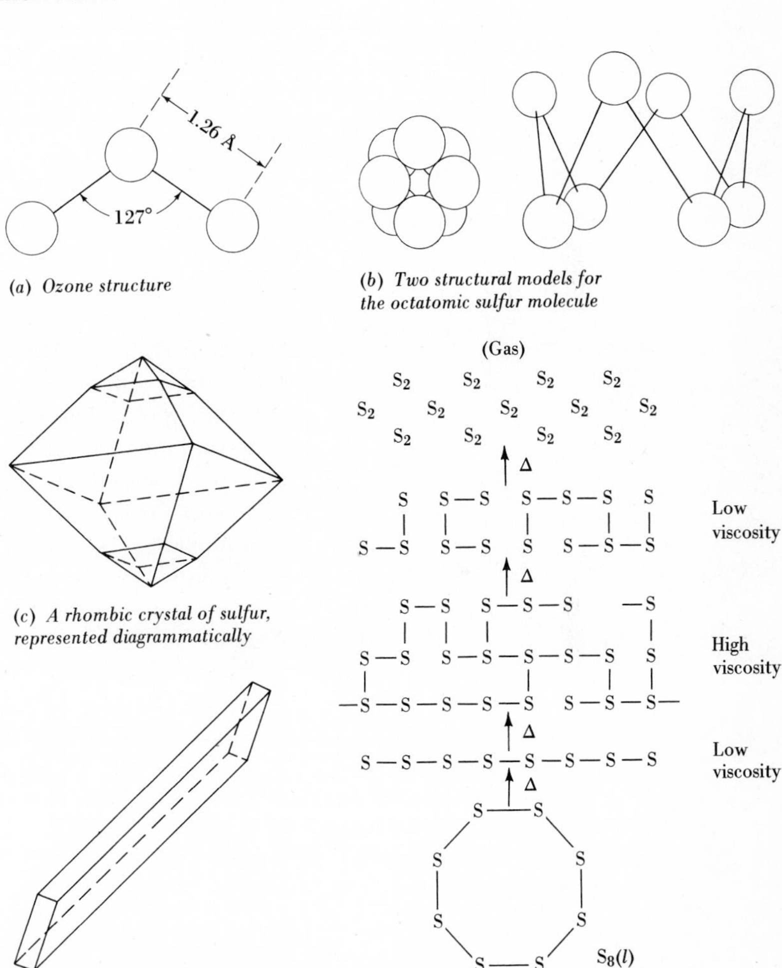

(a) *Ozone structure*

(b) *Two structural models for the octatomic sulfur molecule*

(c) *A rhombic crystal of sulfur, represented diagrammatically*

(d) *A monoclinic crystal of sulfur, represented diagrammatically*

(e) *A model to interpret the physical changes that occur in liquid sulfur*

All the elements form dihydrides, XH_2. The aqueous solutions of the hydrides of sulfur down through tellurium are acidic. The acid strength increases with increasing atomic weight of the group VI element. Oxygen hydride (water) is anomalous in many of its properties because the molecule is polar and is extensively hydrogen-bonded. Its melting point, boiling point, and enthalpies of fusion and vaporization are very high, reflecting strong intermolecular forces. The remaining hydrides exhibit the expected trends in their chemical and physical properties. All are extremely poisonous and foul-smelling. The odor of rotten eggs is due to sulfur hydride, H_2S, more commonly called hydrogen sulfide. As a group, the hydrides of these elements are more stable and inert than the hydrides of any other elements except for the halogen hydrides. The possibility of catenation to form a series of hydrides like the alkanes or silanes is severely limited because of the lower valence of these elements. Hydrogen peroxide, H—O—O—H, is the only well-characterized example.

Oxygen

Many of the major topics covered in this book could be appropriately placed under the heading of "oxygen chemistry," e.g., acid-base theory; water, the alcohols, and hydrogen bonding; combustion and oxidation-reduction reactions; gases; photosynthesis and respiration; and the inorganic oxides, hydroxides, carbonates, silicates, phosphates, and sulfates. Almost all the topics we have discussed have an important, direct relationship to oxygen or its most important compound, water. Therefore, in this section we shall be satisfied with discussing a few of its unusual compounds.

As Table 18.10 shows, air is about 20 percent by volume oxygen. The major component is nitrogen. There are small amounts of argon and carbon dioxide with traces of the other noble gases, ozone, and hydrogen. By liquefaction and fractional distillation of air it is possible to separate it into its different gaseous components (see Fig. 18.16). This is the chief commercial method for the preparation of oxygen, nitrogen, and the noble gases. A purer sample of oxygen can be prepared by the electrolysis of aqueous electrolyte solutions, but this method is more expensive. The relatively constant concentration of O_2 in our atmosphere has been maintained by a balance of photosynthesis, which produces O_2, and respiration, combustion, and other forms of oxidation, which consume it. Any change in the environment which alters the rate of any of these processes could have very serious consequences on higher organisms which depend on O_2 for the mitochondrial production of ATP.

The trace of ozone, O_3, in the atmosphere is due to a conversion of O_2 to O_3 promoted by ultraviolet radiation and electric storms. It is a powerful oxidizing agent, which can be prepared chemically by

TABLE 18.10: AVERAGE COMPOSITION OF DRY AIR

Gas	*Percent* *By volume*	*By weight*
Nitrogen	78.09	75.51
Oxygen	20.95	23.15
Argon	0.93	1.28
Carbon dioxide	0.03	0.046
Neon	0.0018	
Krypton†, Helium, Xenon	0.0025	
Radon	‡	
Hydrogen	5×10^{-5}	
Ozone, O_3	1×10^{-6}	

† The noble gases are given in order of decreasing concentration.
‡ Radon, the radioactive noble gas, is by far the least abundant gas in this table.

reacting water with fluorine:

$$3F_2 + 3H_2O \rightarrow O_3 + 6HF$$

The structure of O_3 is shown in Fig. 18.15.

Hydrogen peroxide, H_2O_2, another unusual compound, has the same molecular weight as H_2S but melts at $-0.9°C$ and boils at 150°C. The dielectric constant is very high, 93 at 25°C, compared with 78 for water. It is more polar and more extensively hydrogen-bonded than water. Hydrogen peroxide would be an excellent sol-

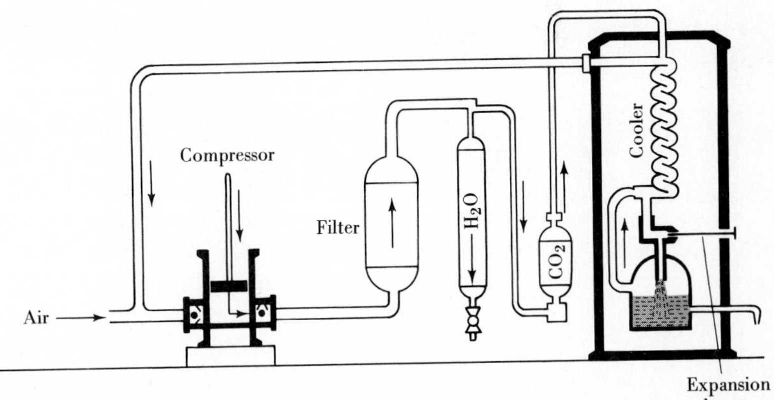

FIGURE 18.16

Liquefaction of air. Air is compressed and filtered. H_2O and CO_2 can be removed as liquids since their critical temperatures are quite low. The remaining gases are passed into a cooler and then allowed to expand, whereupon the air is cooled further. The cooled air is used to reduce the temperature of incoming air. Finally, the temperature is lowered sufficiently so that liquefaction results.

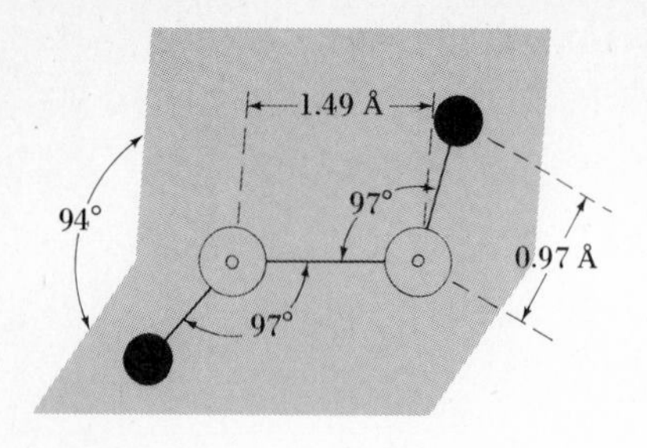

vent for ionic and polar materials but is not used for this purpose because H_2O_2 is such a powerful oxidant that it is dangerous.

The oxidation state of oxygen in H_2O_2 is -1; oxygen compounds in which the oxidation state is -2 are much more stable, and hence H_2O_2 has high oxidizing power. This compound may also act as a reducing agent, in which case molecular oxygen forms; it appears that oxygen is more stable in an oxidation state of 0 or -2 than in a state of -1.

The peroxide linkage involves an O—O single bond; it is present in the ionic peroxides of the alkali metals and in many covalent substances like peroxysulfuric acid, $HOSO_2—O—O—SO_2OH$, and peroxyacetic acid, $CH_3—CO—O—OH$.

Hydrogen peroxide can be prepared by treating barium peroxide with sulfuric acid:

$$BaO_2(s) + H_2SO_4(aq) \rightarrow BaSO_4(s) + H_2O_2(aq)$$

It is produced commercially by hydrolyzing peroxysulfuric acid, which is obtained from the electrolysis of pure H_2SO_4:

$$2H_2SO_4(l) \rightarrow \underset{\text{Peroxysulfuric acid}}{HO—SO_2—O—O—SO_2—OH(l)} + H_2(g)$$

$$\underset{\text{Peroxysulfuric acid}}{H_2S_2O_8(l)} + 2H_2O(l) \rightarrow 2H_2SO_4(aq) + H_2O_2(aq)$$

Hydrogen peroxide finds application as a decolorizing or bleaching agent because many pigmented substances lose their color when they are oxidized. It is commonly used to bleach hair and natural textile fibers such as cotton and wool. As a 90% solution it is an extremely powerful and dangerous oxidizing agent which is used in combination with hydrazine, a strong reducing agent, as a rocket fuel.

Sulfur

Sulfur—along with salt, NaCl, coal, and limestone—is considered one of the four basic raw materials of the chemical industry. Actually, water and air should be included, but since they are virtually free materials they are usually omitted. The value of these "free" materials has come into sharper focus because of the worldwide concern over the abuse and pollution of our natural environment.

The huge deposits of elemental sulfur discovered in Louisiana, Texas, and Mexico during the search for petroleum in the second half of the nineteenth century are believed to have been formed by the bacterial decomposition of $CaSO_4$. They lie from 200 to 2000 feet below the earth's surface under muck and quicksand, so that the sulfur cannot be reached by conventional mining techniques.

These deposits were inaccessible until the 1890s, when Herman Frasch (1851–1914) invented an ingenious process for obtaining the sulfur by melting it underground and forcing the liquid to the surface. In the Frasch process three coaxial pipes are lowered into a deep shaft drilled into the deposit (Fig. 18.17). Superheated water is forced through the outer jacket formed by the pipes; hot air is forced through the central pipe. The superheated water melts the sulfur. The hot air produces a sulfur froth, which rises in the inner jacket. Thanks to this process, the world production of sulfur increased from 7000 tons in 1901 to 220,000 tons in 1905. The present annual production rate is about 1500 million tons.

Sulfur combines directly with all metals except gold and platinum to form sulfides, in which the oxidation state of sulfur is -2. It combines with many nonmetals and forms compounds in which the oxidation state of sulfur is usually $+4$ or $+6$; it does exhibit states of -1 and $+2$, but more rarely.

FIGURE 18.17

The Frasch process for obtaining sulfur.

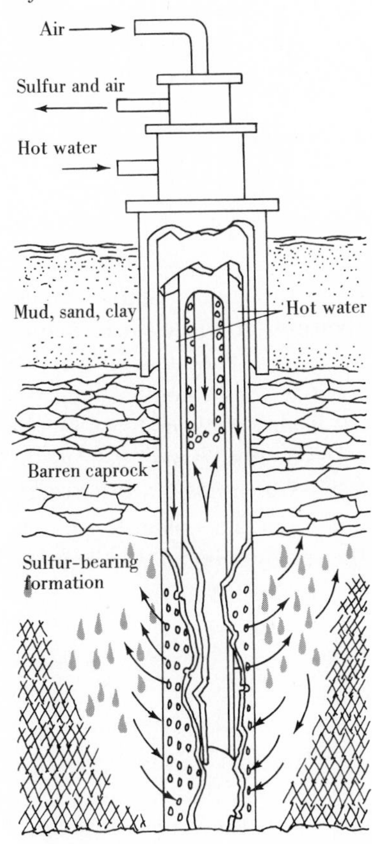

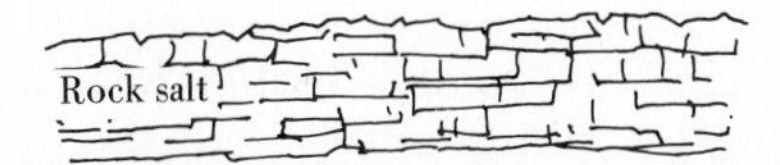

Sulfuric acid

Sulfuric acid is employed in the manufacture of almost all industrial products, over 28 million tons being produced annually in the United States alone. Its wide applicability is due to its cheapness and chemical versatility. It is an oxidizing agent, a dehydrating agent, and a sulfonating agent. It acts as a catalyst in many reactions and is a strong electrolyte. Each of these properties is industrially important.

Almost all sulfuric acid is manufactured by the *contact* process; small amounts are produced by an older method called the *lead-chamber* process. Both methods are designed to achieve a catalytic oxidation of SO_2. In the contact process SO_2 is oxidized when it makes contact with the surface of a solid catalyst; the lead-chamber process uses a homogeneous catalyst (page 568).

In the contact process sulfur is burned in excess air to produce SO_2 (Fig. 18.18), an exothermic reaction requiring no catalyst. The SO_2 gas is filtered to remove dust particles and scrubbed to dissolve out chemical impurities which would reduce the lifetime of the catalyst. The purified SO_2 is heated and passed over the catalyst with air, whereupon the exothermic oxidation of SO_2 occurs:

$$SO_2(g) + \tfrac{1}{2}O_2(g) \xrightarrow{\text{cat.}} SO_3(g) \qquad \Delta G^\circ = -16.7$$

Note that the oxidation of SO_2 to SO_3 is thermodynamically favorable, but in the absence of a catalyst the reaction is imperceptible.

The sulfur trioxide is the acid anhydride of sulfuric acid, but SO_3 does not dissolve readily in water and a fine mist forms when SO_3 and H_2O are mixed together. In the contact process the SO_3 gas is dissolved in pure sulfuric acid forming pyrosulfuric acid, $H_2S_2O_7$,

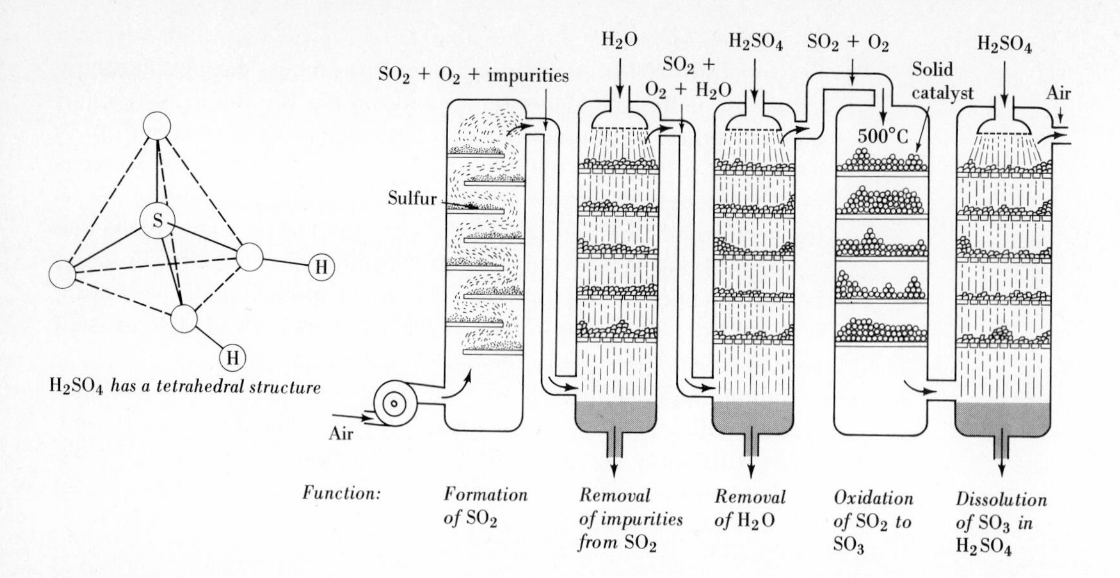

FIGURE 18.18

The contact process for manufacture of H_2SO_4.

which is then converted to sulfuric acid by dilution with water:

$$SO_3(g) + H_2SO_4(l) \rightarrow \underset{\text{Pyrosulfuric acid}}{HO{-}SO_2{-}O{-}SO_2{-}OH(l)}$$

and

$$H_2S_2O_7(l) + H_2O(l) \rightarrow H_2SO_4(l)$$

Although most of the SO_2 produced in the United States is used to manufacture sulfuric acid, great quantities are used for bleaching wood pulp (for paper), straw, and other fibers. The bleaching action of SO_2 is due to its reducing power; it is much gentler than hydrogen peroxide, an oxidizing bleach. It is also a raw material for the production of sodium thiosulfate, $Na_2S_2O_3$, the "hypo" used to fix photographic films. In the fixing process the thiosulfate ion complexes the nonsensitized silver ions in the film. The silver-thiosulfate complex dissolves leaving behind the photoreduced silver metal which forms the image.

18.9 THE HALOGENS, GROUP VIIA

The halogens are widely distributed through the earth's crust and ocean waters. They are very reactive nonmetals and therefore are never found in the free state. Except for fluorine, all the halogens are obtained from seawater or other brines, where they occur as soluble halide salts. The proportion of iodides in seawater is very low, but since iodides are concentrated by growing seaweed, this plant is used as the direct source of the element. The chief sources of fluorine are feldspar, CaF_2, and cryolite, Na_3AlF_6.

Preparation of the halogens, X_2

Elemental chlorine, Cl_2, is obtained by the electrolysis of brine or molten NaCl (page 360) and in turn is used to free bromine and iodine from the bromides and iodides:

$$\tfrac{1}{2}Cl_2(g) + NaBr(aq) \rightarrow \tfrac{1}{2}Br_2(l) + NaCl(aq)$$
$$\tfrac{1}{2}Cl_2(g) + NaI(aq) \rightarrow \tfrac{1}{2}I_2(s) + NaCl(aq)$$

The displacement of Br_2 or I_2 by chlorine is an irreversible oxidation-reduction reaction which demonstrates a useful property of the halogen series. A lighter halogen can displace a heavier one from the halide salt; bromine, Br_2, displaces I_2 from an iodide, but Br_2 cannot displace Cl_2 from a chloride.

Fluorine cannot be used to displace Cl_2 from brine because fluorine reacts violently with the water in the brine and produces ozone, O_3, and oxygen, O_2:

$$F_2(g) + H_2O(l) \rightarrow 2HF(aq) + \tfrac{1}{2}O_2(g)$$

As Table 18.11 shows, fluorine is the most electronegative element in the periodic table. It therefore should not be surprising that it is a very potent oxidizing agent. Fluorine does liberate chlorine from anhydrous NaCl.

Elemental fluorine is prepared by the electrochemical decomposition of molten potassium hydrogen fluoride, KHF_2 (anion: $F—H—F^-$). The electrolysis is carried out in a copper vessel equipped with graphite electrodes. Initially, the F_2 generated oxidizes the copper surface, but a protective film of CuF_2 forms and resists further attack. Fluorine is collected at the anode and hydrogen at the cathode, so that the process may be thought of as an electrolysis of HF dissolved in anhydrous KF (see Fig. 18.19).

The ease with which the halides are oxidized to the free halogens is indicated by the standard oxidation potentials:

$$2F^-(aq) \rightarrow F_2(g) + 2e^- \qquad \mathscr{E}^0 = -2.80\ V$$
$$2Cl^-(aq) \rightarrow Cl_2(g) + 2e^- \qquad \mathscr{E}^0 = -1.36\ V$$
$$2Br^-(aq) \rightarrow Br_2(l) + 2e^- \qquad \mathscr{E}^0 = -1.06\ V$$
$$2I^-(aq) \rightarrow I_2(s) + 2e^- \qquad \mathscr{E}^0 = -0.54\ V$$

The oxidation potentials increase from F^- to I^-, which means that I^- is the most easily oxidized halide. There are no known chemical reagents which can oxidize F^- to F_2, but there are many relatively weak oxidants which can oxidize I^- to I_2.

Chemical properties

Some of the important properties of the halogens are presented in Table 18.11. Note that the configuration in the valence level is always ns^2, np^5 so that the rare-gas configuration is most easily

TABLE 18.11: SOME PROPERTIES OF THE HALOGENS

Element	at wt	Abundance in earth's crust, wt %	Electronic configuration	Radius, Å		Electro-negativity	First ionization potential, kcal $X^0(g) \rightarrow X^+(g) + e^-$	mp, °C	bp, °C	State under ordinary conditions
				Covalent	Ionic (X^-)					
Fluorine, $_9F$	18.9984	0.070	[He] $2s^2, 2p^5$	0.72	1.34	4.0	401	−223	−188	Pale yellow gas
Chlorine, $_{17}Cl$	35.453	0.15	[Ne] $3s^2, 3p^5$	0.99	1.80	3.0	299	−102	−34.6	Greenish-yellow gas
Bromine, $_{35}Br$	79.909	0.000162	[Ar] $3d^{10}, 4s^2, 4p^5$	1.14	1.90	2.8	272	−7.0	58	Red liquid
Iodine, $_{53}I$	126.9044	0.00003	[Kr] $4d^{10}, 5s^2\ 5p^5$	1.33	2.23	2.5	240	†		Violet solid
Astatine, $_{85}At$	(211)	‡	[Xe] $4f^{14}, 5d^{10}, 6s^2, 6p^5$							

† Sublimes.

‡ Artificial element.

FIGURE 18.19

Electrochemical production of F_2.

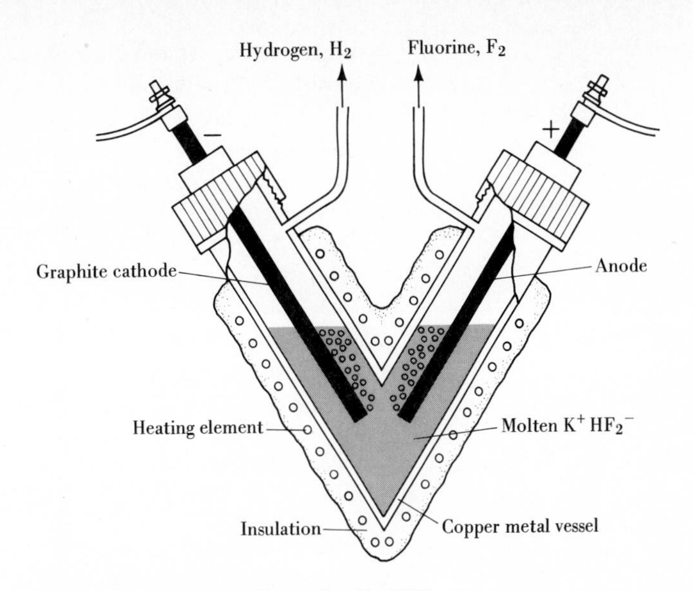

Electrolysis of potassium hydrogen fluoride, KHF_2.

achieved by adding an electron to form the halide ion, X^-. In the elemental state the halogens exist in form of diatomic molecules, F_2, Cl_2, etc. Astatine, the heaviest element of the group, is a synthetic element first prepared in 1940. It has no appreciable distribution in nature because all its isotopes are radioactive.

The halogens, as a group, are the most electronegative elements, reacting with all the elements of the periodic table except the noble metals like platinum and noble gases. Fluorine is so powerful an oxidizing agent that it can form compounds with xenon and krypton. The halogens form ionic compounds with many of the metals. In fact, the name "halogen" is derived from the Greek meaning "salt former."

The halogen oxides and hydroxides are acidic, perchloric acid, ClO_3OH or $HClO_4$, being the strongest common acid. The halogen hydrides are all stable covalent compounds which are gaseous at room temperature. (The halogen hydrides are most often referred to as hydrogen halides because hydrogen is more electropositive than any halogen and it is considered to be in the +1 oxidation state. The term "hydride" usually refers to hydrogen in the −1 state.)

Dissolved in water, the hydrogen halides act as strong proton donors and produce an acidic solution:

$$HCl(g) + H_2O(l) \rightarrow H_3O^+(aq) + Cl^-(aq)$$

Hydrogen chloride　　　　Hydrochloric acid

A solution of hydrogen chloride in water is called hydrochloric acid. The other hydrogen halides form analogous aqueous solutions, producing hydrofluoric, hydrobromic, and hydriodic acids. Only HF is considered to be a weak acid in water ($K_i = 3.35 \times 10^{-4}$); all the other hydrogen halides are completely ionized. The acid strength increases from HF to HI.

The hydrogen halides can be prepared by direct combination of the elements, but for fluorine and chlorine the reactions are so violent that they are dangerous. Since bromine and iodine react slowly with hydrogen, the hydrogen halides are usually prepared by treating the metal halides with a nonvolatile mineral acid; the hydrogen halides escape as gases:

$$CaF_2(s) + H_2SO_4(l) \rightarrow CaSO_4(aq) + 2HF(g)$$
$$2NaCl(s) + H_2SO_4(l) \rightarrow Na_2SO_4(aq) + 2HCl(g)$$

Sulfuric acid cannot be used to prepare HBr or HI from the corresponding salt because it oxidizes them, producing Br_2 and I_2, respectively. Instead phosphoric acid, which is a much weaker oxidizing agent than H_2SO_4, may be used:

$$NaBr(s) + H_3PO_4(l) \rightarrow NaH_2PO_4(aq) + HBr(g)$$

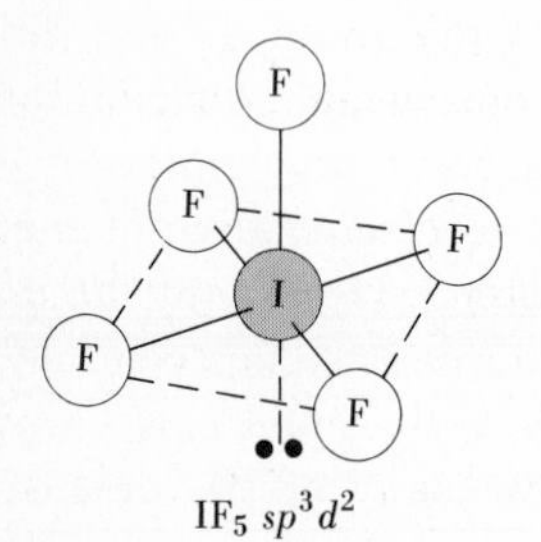

IF_5 sp^3d^2

Each of the halogens except fluorine forms a series of oxyacids like HClO, $HClO_2$, $HClO_3$, and $HClO_4$ (page 426), and corresponding salts. In addition, the halogens can form compounds among themselves. These *interhalogens* are composed of an atom of a halogen combined with an odd number of atoms of a lighter halogen, for example, IF_5, IF_7, ICl, BrF_5, BrCl, and ClF_3.

The structure of the interhalogen compounds can be explained in terms of *d*-orbital participation in the bonding. Iodine pentafluoride involves a sp^3d^2 hybridization producing six identical bonding orbitals oriented toward the vertices of a regular octahedron. The IF_5 molecule is in the shape of a pyramid with a square base, but if the lone pair is considered, the octahedral geometry of IF_5 is apparent. In IF_7 iodine forms seven covalent bonds by an sp^3d^3 hybridization, and the molecule has the shape of a pentagonal bipyramid, which is expected for the hybridization. Note that iodine can expand its valence shell to accommodate up to 14 electrons.

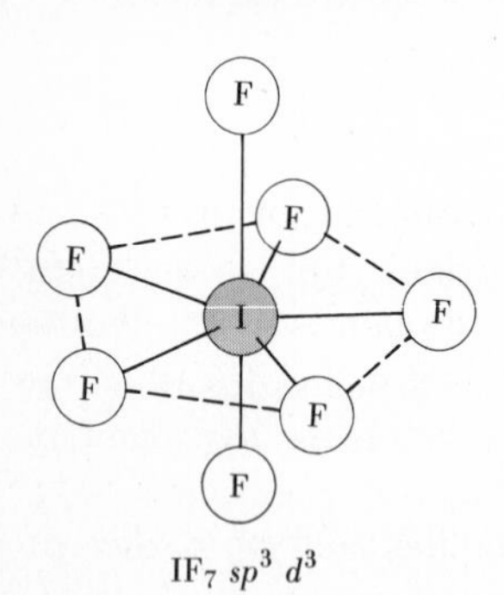

IF_7 sp^3 d^3

Fluorine

Fluorine, maintaining the pattern of the second-row elements, exhibits a behavior unlike that of any other halogen. The anomalous effects are due to its small radius and electronic configuration, which make fluorine much more electronegative than any other element and unable to expand its valence shell. It exhibits only the −1 oxidation state in its compounds, and no oxyacids or oxy salts have been prepared. It is such a powerful oxidizing agent that storage is a serious problem. Fluorine hydride is the only halogen hydride which exhibits hydrogen bonding; aggregates of up to six HF molecules are present in the pure liquid, and liquid HF shows the anomalous physical properties associated with strong intermolecular forces.

Uses of the halogens

One of the important uses of fluorine is in the manufacture of the Freon refrigerants. It is also used to produce fluorinated polymers like Teflon and has found application as a tooth-decay preventative in the form of NaF and SnF_2.

The chief use of chlorine is as an oxidizing bleach, either as free chlorine or as an unstable compound like $CaOCl_2$ (bleaching powder). Laundry bleach is produced by passing chlorine through a solution of sodium hydroxide:

$$Cl_2(g) + 2NaOH(aq) \rightarrow \underset{\text{Sodium hypochlorite}}{NaOCl(aq)} + NaCl(aq) + H_2O(l)$$

Chlorine is used as a bactericide in water purification. All these uses depend on its oxidizing ability.

Most of the bromine produced is used to prepare 1,2-dibromoethane, more commonly called ethylene bromide, component of the antiknock mixture used in gasolines. The key antiknock compound, tetraethyllead, keeps the gasoline from igniting prematurely in the cylinder. In the absence of ethylene bromide the tetraethyllead would produce PbO, a solid that would deposit in the engine. When ethylene bromide is present, the lead is converted into volatile $PbBr_2$, which can be exhausted (into the atmosphere):

$$Pb(C_2H_5)_4(g) + C_2H_4Br_2(g) + 16O_2(g) \rightarrow PbBr_2(g) + 10CO_2(g) + 12H_2O(g)$$

The chief light-sensitive component of photographic films is silver bromide; when it is sensitized by light, it can be reduced to elemental silver, which appears black. The amount reduced depends on the intensity of the light, so that shades from light to dark can form an image on the film.

Iodine is important to health. In places where it does not naturally occur, it must be added to the diet to prevent malfunction of the thyroid gland, which produces a hormone controlling metabolism and growth. For this reason most table salt is iodized, i.e., has a small amount of NaI added to it. Elemental iodine dissolved in alcohol, called *tincture of iodine*, is used as an antiseptic for minor wounds.

18.10 THE NOBLE GASES, GROUP 0

The noble gases form the most inert family of elements. Only the three heavier elements form compounds and then only in combination with very electronegative elements. Argon, the most abundant of the noble gases, is present in air to the extent of 1 percent by volume. It was first detected by William Ramsay when a gaseous atmospheric N_2 was passed over hot magnesium. The residue, which was inert and unlike any other known gases, was identified as a new element, argon, which means "lazy one." A more careful analysis of the inert-gas residue shows that it is a mixture of similar gases. Argon is predominant, but there are traces of two lighter gases, helium and neon, and two heavier ones, xenon and krypton. The first member of the family, helium, was known to be present in the chromosphere of the sun before it was discovered on earth. It emits a characteristic spectrum unlike that of any other element and excited helium atoms in the sun's chromosphere were identified by spectroscopic examination. Before the end of the nineteenth century Ramsay had found helium on our planet and is therefore responsible for discovering all the noble gases except radon, a radioactive noble gas produced by the disintegration of radium.

The important properties of these elements are shown in Table 18.12. They have very low melting and boiling points; no other substances have lower enthalpies of fusion or vaporization. All properties show that very little energy is required to overcome the attractive forces between the molecules.

Since early attempts to prepare compounds of the noble gases met with failure, they are also referred to as the inert gases. The electronic configuration of the noble gases gives the clue to their inertness. The valence shell has an ns^2, np^6 structure. The valence shell of the atoms is filled so that there is only the slightest tendency

TABLE 18.12: SOME PROPERTIES OF THE NOBLE GASES

Element	*at wt*	*mp*, °C	*bp*, °C	*Enthalpy*, kcal/mol		*First ionization potential*, kcal/mol	*Relative concentration in air*, Ar = *100*
				Fusion	*Vaporization*	$X(g) \rightarrow X^+(g) + e^-$	
Helium, $_2He$	4.003	−271.9†	−268.9	. . .	0.02	566	0.055
Neon, $_{10}Ne$	20.184	−248.7	−245.9	0.08	0.45	496	0.16
Argon, $_{18}Ar$	39.946	−189.2	−186	0.27	1.51	362	100
Krypton, $_{36}Kr$	83.80	−15.7	−153	0.36	2.30	322	5×10^{-4}
Xenon, $_{54}Xe$	131.31	−112	−107	0.49	3.3	279	6×10^{-5}
Radon, $_{86}Rn$	(222)	−71	−62	0.8	3.9	247	10^{-13}

† Helium has a triple point at −271.9°C and 26 atm.

to form chemical bonds with other atoms. No compounds of helium, neon, and argon are known. The heavier elements combine directly with elemental fluorine. For instance, xenon can form XeF_2, XeF_4, and XeF_6. The product can be controlled by varying ratio of reactants. Krypton and radon also react to form fluorides.

Table 18.12 shows that the ionization potentials for the three heavier elements are comparable to the more reactive elements, and so it is not surprising that fluorine, the most electronegative element, is able to oxidize the heavier noble gases.

The first noble-gas compound was prepared by Neil Bartlett in 1962 by the reaction of Xe with platinum(VI) fluoride, PtF_6, an extremely powerful oxidizing agent; PtF_6 is capable of converting molecular oxygen to O_2^+:

$$PtF_6(s) + O_2(g) \rightarrow O_2^+PtF_6^-(s)$$

Since the ionization potentials of the O_2 molecule and xenon atom are approximately equal, Bartlett reasoned correctly that it might be possible to prepare $XePtF_6$. Since 1962 many reactions of xenon and krypton have been reported. Some of these compounds and their properties are presented in Table 18.13.

The uses of the noble gases depend chiefly on their chemical inertness. Helium has replaced hydrogen for inflating airships and observation balloons. The *Hindenberg*, the last dirigible filled with hydrogen, exploded over Lakehurst, New Jersey, in 1937 after making an Atlantic crossing. Helium is also used to attain temperatures close to the absolute of zero. Liquid helium has a boiling point of -268.9°C, or about 4.3 K, but it has been used to achieve temperatures as low as 0.001 K.

Argon is used as a gaseous blanket during the arc welding of the reactive metals like magnesium to prevent the metals from igniting

TABLE 18.13: SOME OF THE COMPOUNDS OF XENON AND KRYPTON

Compound	*Preparative reaction*	*Oxidation state*	*mp,* °C	*Enthalpy of formation,* kcal/mol
XeF_2	$Xe + F_2$	+2	140	−40
XeF_4	$Xe + 2F_2$	+4	114	−70
XeF_6	$Xe + 3F_2$	+6	48	−100
$XeOF_4$	$XeF_6 + 3H_2O$	+6	−28	
XeO_3	$XeF_6 + 4H_2O$	+6	†	96
$CsXeF_7$	$CeF + XeF_6$	+6		
KrF_2	$Kr + F_2$	+2	‡	

† Explodes.

‡ Decomposes at room temperature.

and forming oxides or nitrides. Some food packaging is carried out under an argon blanket to retard spoiling. Argon is also used in incandescent light bulbs to prevent the tungsten filament from subliming. Other gases are used to make lighting displays ("neon" signs) since they emit a colored glow when a high-voltage current passes.

18.11 HYDROGEN

In 1815 an English chemist, William Prout (1785–1850), observed that the atomic weights of most of the elements are very close to whole-number multiples of the atomic weight of hydrogen. Prout felt that hydrogen might be the simple chemical principle from which all other elements are formed. This hypothesis led to a careful examination of the atomic weights. An examination of the accepted atomic weights of the elements shows that there are instances where the value is really not a whole-number multiple of the atomic weight of hydrogen. Perhaps if chemists had realized that atoms of the same element can have different masses, Prout's hypothesis would have gained more adherents, but the existence of isotopes was not known until the twentieth century. All isotopes do have masses which are very nearly whole-number multiples of the mass of a hydrogen atom, and the mathematical probability of such a circumstance is so low that we could never explain it on the basis of pure coincidence. It must be considered further.

The most abundant element in the universe is hydrogen. It is present in interstellar space, and young stars are nearly 100 percent hydrogen. Our own sun is approximately 99 percent hydrogen by weight. One widely accepted theory for the creation of the elements holds that stars are formed by the condensation of an interstellar gas composed mostly of hydrogen. Gravitational forces cause a tremendous compression of the gas, and incredible temperatures and pressures are reached. At a temperature of 10^6 K fusion of hydrogen begins, and the heavier elements are formed. The chief nuclear process occurring on the surface of the sun is the conversion of hydrogen into helium, but isotopes as massive as iron are also forming. The very heavy elements form later in the evolutionary cycle of a star. So we see that Prout's hypothesis is alive today in astrophysical theory. However, whereas Prout considered heavier elements to be formed from hydrogen by some vague kind of polymerization, the modern theory proposes that these heavier elements form by nuclear fusion, a process requiring tremendous energies.

Hydrogen also holds a special niche in the periodic table. With only one electron in its valence shell, it might be expected to be an alkali metal. However, if it loses its single electron, it does not achieve a rare-gas configuration. It has a smaller radius and a much higher ionization potential than any alkali metal. The hydrogen nucleus, which is a single proton, has a very strong hold on the lone valence electron. In its +1 oxidation state hydrogen shares electrons to achieve the rare-gas configuration of helium. Hence, in contrast to the ionic compounds of the alkali metals, the +1 compounds of hydrogen are covalent. Even in its −1 oxidation state hydrogen achieves the electronic configuration of helium, for example, Na^+H^-. These chemical properties suggest that hydrogen has the characteristics of the halogen family, but there are a number of differences here also, the most obvious being the difference between the reactivities of the hydrides and the halides. For example, hydrogen and chlorine form −1 ions when treated with sodium; however, NaH reacts with water to produce H_2 gas, whereas NaCl dissolves without gas evolution. Hydrogen is usually placed in group IA of the periodic table, sometimes in both groups IA and VIIA, and sometimes in neither.

Very pure hydrogen can be prepared by the electrolysis of many electrolyte solutions, most commonly dilute H_2SO_4 solutions. This method is too expensive for the production of large amounts of hydrogen. Most commercial hydrogen is produced by the reduction of steam with coke in the presence of a catalyst:

$$2H_2O(g) + C(s) \xrightarrow[\text{cat.}]{\Delta} 2H_2(g) + CO_2(g)$$

Since CO_2 is an acidic gas, the products can be separated by bubbling the mixture through a basic solution. The CO_2 reacts to form a nonvolatile salt, leaving only H_2 in the gas phase.

Small amounts of hydrogen can be prepared quickly by the action of mineral acids on the active metals. Tin, zinc, iron, and many other common elements are strong enough reductants to release H_2 from acid solution.

We have already discussed the reactions of hydrogen with many other elements and compounds and have noted the general and special properties of the products. A reconsideration of the material in previous sections will lead you to conclude that hydrogen is extremely important in industry. It is used in the metallurgical reduction of many oxides like Cu_2O, ZnO, WO_3, and Fe_2O_2. Large quantities of the gas are used in the catalytic reduction of vegetable oils to produce solid fats (Spry, Crisco, etc.). Perhaps the most important commercial use of hydrogen is in fixing nitrogen by the Haber process. Smaller but appreciable amounts are used as a heating fuel and in welding.

QUESTIONS AND PROBLEMS

18.1 (*a*) Contrast the chemical properties of metals, metalloids, and nonmetals. (*b*) Contrast the physical properties of metals and nonmetals.

18.2 What chemical and physical properties are useful in distinguishing metals from nonmetals? Give examples.

18.3 Write chemical equations for the reactions which would occur when (*a*) potassium is dropped into water, (*b*) sodium hydride is dropped into water, (*c*) brine is electrolyzed, (*d*) molten table salt is electrolyzed.

18.4 Show that the bicarbonate ion is amphoteric.

18.5 By comparing Tables 18.1 and 18.2 explain why the alkaline-earth metals are less reactive than the alkali metals.

18.6 Give the equation for the reaction of strontium with (*a*) nitrogen, (*b*) sulfur, (*c*) bromine, (*d*) water, and (*e*) ammonia.

18.7 (*a*) List some chemical and physical properties of the second-period elements which seem to be anomalous in comparison with other elements of the same group. (*b*) Explain why these elements are anomalous.

18.8 What is the diagonal relationship, and what is its basis?

18.9 Why is it impossible to obtain aluminum by electrolysis of one of its salts in aqueous solution?

18.10 In terms of valence theory explain why borane, BH_3, dimerizes.

18.11 Give the structure for borazole. Would you expect this molecule to show the inertness exhibited by benzene? Explain.

18.12 What three structural metals are most important in our economy? List them in order of tons produced per year.

18.13 What is the chief difference between carbon and other elements of group IVA?

18.14 How would you prove or disprove that diamond is elemental carbon?

18.15 What are the chief uses of silicon and germanium?

18.16 Why does lead exhibit a +2 oxidation state rather than a +4 oxidation state in its reactions with Cl_2, Br_2, and I_2?

18.17 Coke is used to reduce silica, SiO_2, to elemental silicon. Would silicon be capable of reducing CO_2 to carbon? Explain.

18.18 Silicon carbide is a very hard substance. What is its crystalline form?

18.19 On the basis of what we learned about freezing-point depression (page 259), explain why the impurities would be expected to accumulate in the melted zone during zone refining.

18.20 Explain the bonding in the linear CO_2 molecule, $O{=}C{=}O$, in terms of the valence-bond theory.

18.21 Silicones are formed by the condensation polymerization of dihydroxydimethylsilane. Show how the inclusion of a small amount of trihydroxymethylsilane results in a cross-linked polymer.

18.22 What were the sources of fixed nitrogen for agriculture before the use of synthetic chemical fertilizers?

18.23 What is the most important commercial method of nitrogen fixation? Why are chemists searching for different methods?

18.24 (*a*) Compare the hydrides of families IA, IVA, and VIA with respect to chemical reactivity, acidity (or basicity), and stability. (*b*) Which of the elements in these families show catenated hydrides? Give examples.

18.25 In terms of the state of aggregation explain the difference between the properties of white and red phosphorus.

18.26 Write the chemical equation for the reactions between (*a*) phosphoric acid and excess NaOH and (*b*) phosphorous acid and excess NaOH.

18.27 Selenium is an intrinsic semiconductor. Explain what this means in terms of the discussion in Sec. 7.9.

18.28 Why are no values shown for the *ionic* radii (X^{2-}) for tellurium and polonium in Table 18.9?

18.29 Oxygen hydride is a most unusual compound. If it did not occur in nature, we certainly would synthesize it. Make a list of some of its unusual properties.

18.30 Why don't the group VIA elements form catenated hydrides?

18.31 What volume of dry air would be required to produce (*a*) 1 l of argon and (*b*) 1 l of neon?

18.32 What processes are responsible for the constant concentration of O_2 in the atmosphere?

18.33 (*a*) Give the Lewis electron-dot formula for ozone. (*b*) Explain its structure in terms of valence-bond theory.

18.34 (*a*) In what properties is H_2O_2 similar to water? (*b*) What is the chief dissimilarity?

18.35 From the structure of H_2O shown on page 630 do you think that there is appreciable sp^3 hybridization on the oxygen atom? Explain.

18.36 Write equations demonstrating that H_2O_2 can act as reductant or as an oxidant.

18.37 List the most important inorganic raw materials of industry and indicate the major uses of each.

18.38 (*a*) What are the main difficulties in converting sulfur to sulfuric acid by the contact process? (*b*) Why must the SO_2 be physically and chemically pure before it is oxidized?

18.39 Write the equation which explains the action of hypo on the unreduced silver halides in a photographic emulsion.

18.40 What chemical property of SO_2 makes it very useful in industry?

18.41 In view of the data in Table 18.11 explain why a lighter halogen can displace a heavier halogen from a halide salt but never vice versa.

18.42 Give physical and chemical evidence demonstrating that fluorine is the most electronegative element.

18.43 Why must *anhydrous* molten KHF_2 be used to obtain F_2 by electrolysis?

18.44 Instead of giving a melting point for I_2 Table 18.11 shows that it sublimes. Does this mean that liquid I_2 cannot exist?

18.45 If HCl, HBr, and HI are completely ionized in water, how can we say that HI is a stronger acid?

18.46 Show the reaction of H_2SO_4 with (*a*) CaF_2, (*b*) $MgCl_2$ (*c*) NaBr, and (*d*) KI.

18.47 Bromine trichloride, $BrCl_3$, involves sp^3d hybridization; five equivalent orbitals are produced. Predict its structure from the electron-repulsion theory (Sec. 5.11).

18.48 What are the chief uses of each halogen?

18.49 How does the electronic structure of the noble-gas atoms explain their inertness?

18.50 Why is it reasonable that fluorine reacts with the heavier halogens?

18.51 Are all the noble gases rare?

18.52 (*a*) What is Prout's hypothesis? (*b*) It is acceptable in modern science? Explain.

18.53 What chemical property of hydrogen makes it useful in industry? Explain.

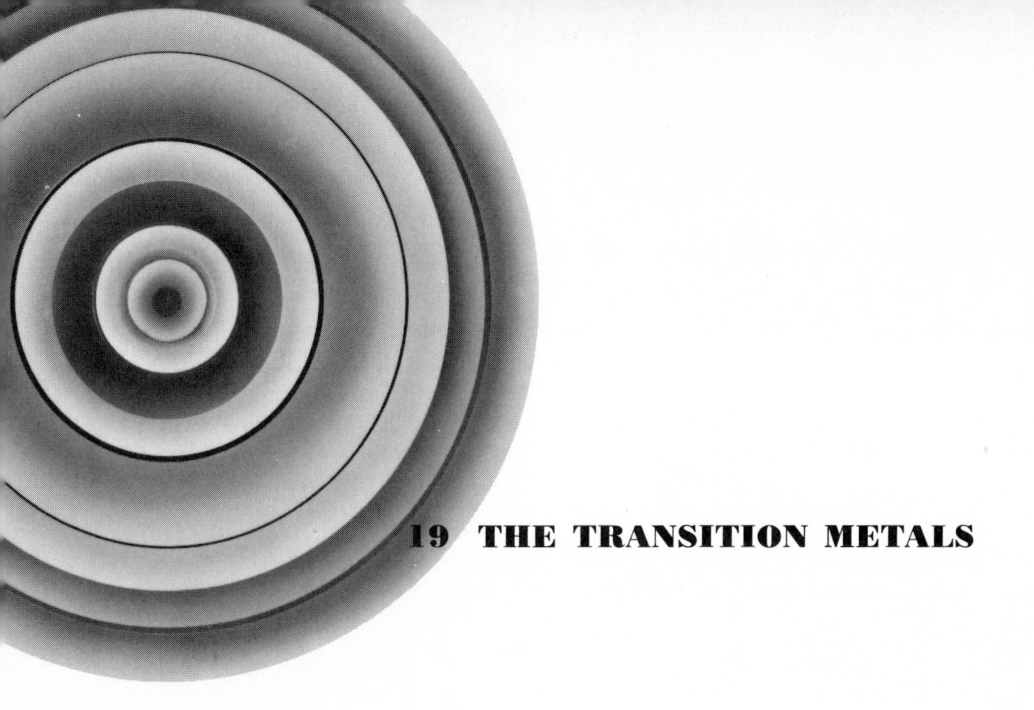

19 THE TRANSITION METALS

19.1 GENERAL PROPERTIES

In this chapter we consider the transition-metal elements, i.e., members of the *B* families of the periodic table. If we follow the periodic classification of the elements according to atomic number, the first transition metal is scandium, ${}_{21}Sc$, the third member of period 4. The next nine elements of the period ending with zinc, ${}_{30}Zn$, are also transition metals. There are 10 transition elements in periods 5 and 6 also. All these elements are hard and lustrous, exhibit high electric and thermal conduction, and have high melting and boiling points; they possess the *physical* properties generally associated with metals. In terms of chemical properties they are significantly different from the regular metals.

The regular metals, which are in families IA, IIA, and IIIA, are good reducing agents; i.e., they give up electrons easily and are oxidized in chemical reactions:

$$Mg(s) + H_2SO_4(aq) \rightarrow MgSO_4(aq) + H_2(g)$$
$$2Na(s) + 2H_2O(l) \rightarrow 2NaOH(aq) + H_2(g)$$
$$4Al(s) + 3O_2(g) \rightarrow 2Al_2O_3(s)$$

In contrast, the transition metals of families IB, IIB, and IIIB vary considerably in their reactivity.

Family IB

The IB family, consisting of copper, silver, and gold, is characterized by chemical inertness. These elements are therefore used for coins and jewelry and referred to as the *coinage* metals. Gold and silver are found as free elements in nature, and copper is easily obtained by reduction of its ores. The IB elements exhibit the +1 oxidation state, but in all other respects they differ from the IA elements, like sodium and potassium, which are extremely reactive reducing agents. This is emphasized in Table 19.1, which shows that the oxidation potentials of the coinage metals have negative values, in contrast to the IA metals, which have high positive values. The alkali metals react violently with water, but gold is not affected by water, while copper and silver corrode slowly in water if oxygen is also present.

Family IIB

The chemistry of the IIB family is relatively simple because of the limited number of oxidation states exhibited by these elements. Mercury can exist in the +2 oxidation state as the monatomic ion Hg^{2+} and in the +1 oxidation state as the diatomic ion Hg_2^{2+}. Zinc and cadmium are much more reactive than mercury, as can be seen from the oxidation potentials in Table 19.1, and exhibit only the +2 oxidation state.

Mercury compounds can be reduced to elemental mercury by

TABLE 19.1: THE TRANSITION ELEMENTS AND THEIR OXIDATION POTENTIALS†

						VIIIB				
Period	IIIB	IVB	VB	VIB	VIIB	Fe	Co	Ni	IB	IIB
4	$_{21}$Sc Very high	$_{22}$Ti High	$_{23}$V 1.5	$_{24}$Cr 0.557	$_{25}$Mn 1.029	$_{26}$Fe 0.409	$_{27}$Co 0.23	$_{28}$Ni 0.28	$_{29}$Cu −0.340	$_{30}$Zn 0.763
5	$_{39}$Y Very high	Zr	Nb	Mo	Te	Ru	Rh	$_{46}$Pd −0.7	$_{47}$Ag† −0.799	$_{48}$Cd 0.403
6	$_{57}$La† +2.4	Hf	Ta	W	Re	Os	Ir	$_{78}$Pt −1.2	$_{79}$Au† −1.42	$_{80}$Hg −0.851
7	$_{89}$Ac‡									

† Except for La, Ag, and Au the oxidation potentials are given for $M^0 \rightarrow M^{2+} + 2e^-$. For Au and La the potentials are for $M^0 \rightarrow M^{3+} + 3e^-$, and for Ag it is for $Ag^0 \rightarrow Ag^+ + e^-$.

‡ No other transition elements of period 7 are known.

thermal decomposition even in the presence of air or oxygen. The chief ore of mercury is mercuric sulfide (cinnabar), HgS, which is converted to mercury by roasting in air:

$$HgS(s) + O_2(g) \rightarrow Hg(l) + SO_2(g)$$

Mercury is very valuable because it is the only inert liquid metal. It is used as an electric switch in many thermostating devices. As it warms it expands and closes a circuit which causes a furnace or heater to shut off. As it cools down it contracts and opens a circuit and the heater goes on again.

Zinc and cadmium ores, ZnS and CdS, are found mixed in nature. As might be expected from their high oxidation potentials of these metals (Table 19.1), they are not converted to the elements when they are roasted:

$$2ZnS(s) + 3O_2(g) \xrightarrow{\text{roasting}} 2ZnO(s) + 2SO_2(g)$$

The oxide obtained in the roasting process is thermally stable and must be subjected to the action of a good reducing agent to obtain the free metal:

$$ZnO(s) + C(\text{coke}) \rightarrow Zn(g) + CO(g)\dagger$$

Zinc and cadmium are both used as protective coatings on many items fabricated from steel, such as wire cable, garbage cans, jar tops, and hub caps.

Family IIIB

The IIIB elements, scandium, yttrium, lanthanum, and actinium, are the most reactive transition metals. They have oxidation potentials as high as the alkali metals and react quite vigorously with water. They exhibit only the +3 oxidation state, and their compounds show some resemblance to the compounds of aluminum in family IIIA.

Family IVB

Titanium, zirconium, and hafnium, group IVB, are more abundant in nature than lead or copper, but "freeing" them from their ores requires very strong reducing agents, and as a result these metals are relatively expensive. For instance, the commercial preparation of titanium metal involves the reduction of $TiCl_4$ with one of the alkali or alkaline-earth metals, which are expensive reducing agents:

$$TiCl_4(l) + 4Na_0(s) \xrightarrow{\Delta} Ti_0(s) + 4NaCl(s)$$

† Note that the gases SO_2 and CO produced in the metallurgical processes would have an extremely deleterious effect if they were allowed to escape into the environment.

Taking advantage of titanium's high strength and melting point, aeronautical engineers have used the metal in the manufacture of jet engines. As TiO_2 it is a superior white pigment used in paints, plastics, and rubber products.

Titanium exhibits +2, +3, and +4 oxidation states, but only in the +4 state is it stable in the presence of air or oxygen. The atomic radii of zirconium and hafnium are equal and have almost identical chemical properties, so that it is extremely difficult to separate these elements. Both exhibit a +4 oxidation state; zirconium also exhibits the +3 state, which is easily oxidized to +4.

Family VB

All elements of family VB—vanadium, niobium, and tantalum—exhibit the +5 oxidation state. Vanadium exhibits three other oxidation states, so that we have compounds like $VOCl_3$, $VOCl_2$, VOCl, and VSO_4 in which the oxidation states are +5, +4, +3, and +2, respectively. Niobium exhibits only the +3 and +5 states, whereas tantalum shows only the +5 state.

The members of group VB have an unexpected inertness in view of their high oxidation potentials. Vanadium, which has a standard oxidation potential twice that of zinc (Table 19.1), is not attacked by strong mineral acids or the strong oxidizing agents chlorine and bromine. One of the obvious applications of these metals is in the manufacture of corrosion-resistant containers. Tantalum is used to handle reactive chemicals; it is also used as a surgical wire when long implantation is necessary and in dental instruments. Vanadium is added to steel to decrease corrosion and increase strength.

Family VIB

The VIB elements, chromium, molybdenum, and tungsten, are well known for their high resistance to corrosion. Considering the fact that the oxidation potential of chromium is higher than that of very reactive cadmium, it is surprising that chromium is not easily oxidized. Careful examination shows that the chromium surface becomes coated with an oxide film which protects it from further oxidation. All three metals are very hard and have high melting points and are added to steel to improve its characteristics. Steel containing 10 percent or more chromium is called *stainless steel*. Tungsten is used as the filament in the common incandescent light bulb.

All these metals can be obtained from their ores by a process called *aluminothermy*. A mixture of the powdered ore and powdered aluminum is heated until the mixture ignites, whereupon the aluminum reduces the ore to the free metal:

$$WO_3(s) + 2Al^0(s) \rightarrow W(s) + Al_2O_3$$

Tungsten is also produced by carbon reduction of WO_3 and molybdenum by hydrogen reduction of MoO_3:

$$WO_3(s) + 3C(\text{coke}) \xrightarrow{\Delta} W(s) + 3CO(g)$$

$$MoO_3(s) + 3H_2(g) \xrightarrow{\Delta} Mo(s) + 3H_2O(g)$$

All elements of the VIB family exhibit all oxidation states from +2 to +6. The +6 state is the most common for compounds of molybdenum and tungsten, but chromium also forms many compounds in the +3 state. Chromium compounds in the +6 state, like potassium dichromate, $K_2Cr_2O_7$, potassium chromate $KCrO_4$, and chromic acid, H_2CrO_4, are excellent oxidizing agents. The oxidation potential for $Cr_2O_7^{2-} \mid Cr^{3+}$ in acidic solution is +1.33 V:

$$Cr_2O_7^{2-} + 14H^+ + 6e^- \rightarrow 2Cr^{3+} + 7H_2O \qquad \mathscr{E} = +1.33 \text{ V}$$

These compounds are commonly used for controlled oxidation of organic substances:

$$K_2Cr_2O_7 + H_2SO_4 + C_6H_5\text{—}CH_3 \rightarrow C_6H_5\text{—}COOH + Cr_2(SO_4)_3$$

Toluene Benzoic acid

$$H_2CrO_4 + C_6H_5\text{—}CH_3 \rightarrow CO_2 + H_2O + Cr_3^+$$

$(CrO_3 + H_2SO_4)$ Toluene

Chromic acid is prepared by dissolving CrO_3 in concentrated H_2SO_4; it is used as a cleaning solution in the laboratory for cleaning glass apparatus.

Family VIIB

The only common element of the VIIB family is manganese. Technetium is radioactive with a short half-life and is not found in nature. As its name suggests, it is a synthetic element produced in uranium fission;† in fact it was the first artificial element ever produced. Only manganese, which is the thirteenth most abundant element in the earth's crust, has been studied in any detail. It commonly forms compounds with oxidation states of +2, +4, and +7, but it actually exhibits all states between +2 and +7.

Manganese occurs in nature chiefly as the ore pyrolusite, which is mostly MnO_2. It can be reduced by aluminothermy, but it is ordinarily reduced along with iron ore since most manganese metal is used in the manufacture of an abrasion-resistant steel.

† Atomic fission, the "splitting" of the nucleus, is discussed in Chap. 20.

$$3MnO_2(s) + 4Al^0(s) \rightarrow 3Mn^0(s) + 2Al_2O_3(s)$$

$$\underbrace{MnO_2(s) + Fe_2O_3(s)}_{\text{Iron ore}} + 5C(\text{coke}) \xrightarrow{\Delta} \underbrace{Mn^0 + 2Fe^0}_{\text{Liquid alloy}} + 5CO(g)$$

Manganese in the +7 oxidation state, for example, $KMnO_4$, is a strong oxidizing agent used in the controlled oxidation of many kinds of organic materials. For instance, xylene can be converted to terephthalic acid (a monomer for the manufacture of Dacron) by oxidation with $KMnO_4$:

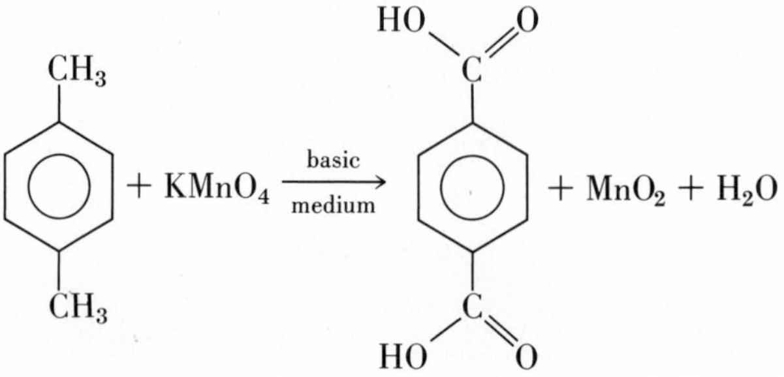

$$C_6H_4(CH_3)_2 + KMnO_4 \xrightarrow[\text{medium}]{\text{basic}} C_6H_4(COOH)_2 + MnO_2 + H_2O$$

Paraxylene (1,4-dimethylbenzene) — Terephthalic acid

Family VIIIB

Family VIIIB is made up of nine elements, usually categorized as the iron group, consisting of the fourth-period elements iron, cobalt, and nickel, and the platinum group, consisting of the fifth- and sixth-period elements ruthenium, rhodium, palladium, osmium, iridium, and platinum. The platinum-group metals are noted for their chemical inertness and are used for jewelry and laboratory apparatus when corrosion is a problem. The iron-group metals are extremely important in our economy, the amount of iron or steel produced per year being an indicator of our economic growth.

As can be seen by examination of Table 19.2, the iron-group metals have similar atomic and ionic radii. It should not be surprising that iron, cobalt, and nickel also have similar chemical properties even though they are members of a horizontal family. Iron commonly exhibits a +3 oxidation state. Although it is not dif-

TABLE 19.2: PROPERTIES OF THE IRON GROUP

			Radius, Å	
Element	*Atomic configuration*	*at wt*	*Atomic*	*Ionic* (M^{2+})
Iron, $_{26}Fe$	$3d^6, 4s^2$	55.85	1.16	0.76
Cobalt, $_{27}Co$	$3d^7, 4s^2$	58.93	1.16	0.74
Nickel, $_{28}Ni$	$3d^8, 4s^2$	58.71	1.15	0.72

ficult to prepare compounds in which iron is in the +2 state, they must be treated with care to prevent air oxidation to the +3 state. Cobalt and nickel also show the +2 and +3 states in their compounds, but for these elements the +2 state is more stable. In fact Co^{3+} is as strong an oxidizing agent as ozone and can oxidize water:

$$2Co^{3+}(aq) + H_2O(l) \rightarrow 2Co^{2+}(aq) + \tfrac{1}{2}O_2(g) + 2H^+(aq)$$

The electronic configuration of the ground-state atoms shows that when iron loses three electrons, it has a half-filled set of *d* orbitals, whereas Co and Ni must lose four and five electrons to attain the stable half-filled configuration.

There are several important ores of iron, cobalt, and nickel. Cobalt is found combined with arsenic as in smaltite, $CoAs_2$, and cobaltite, CoAsS. These ores are roasted to yield CoO, which can be reduced to produce the metal:

$$CoAs_2(s) \xrightarrow[\text{air } (O_2)]{\Delta} CoO(s) + As_2O_5(s)$$

$$CoO(s) + H_2(g)^{\dagger} \xrightarrow{\Delta} Co^0(s) + H_2O(g)$$

Nickel ores are also treated in a similar manner, but the purification involves an unusual step. After the ore has been converted into the metal, many impurities are present; the crude metal is melted and CO gas is bubbled through it. A volatile substance called nickel carbonyl, $Ni(CO)_4$, having a boiling point of 43°C, is formed and escapes as a gas. When nickel carbonyl is heated to high temperature, it decomposes into pure nickel and CO gas.

The reduction and purification of iron is extremely important because this metal is a key raw material of our economic system. The metallurgy of iron involves reduction of the ores to a crude form of the metal called *pig iron*, purifying the metal, and alloying it with other elements to produce steel.

There are large deposits of iron-rich ores, *hematite*, Fe_2O_3, and *magnetite*, Fe_3O_4, in Minnesota surrounded by large quantities of a lower-grade ore called *taconite*. The industrial reduction of these ores has been studied in great detail. The actual reactions are quite complicated, but the basic processes are quite simple. The ore is crushed and mixed with limestone, $CaCO_3$, and coke and heated in a blast furnace. The blast-furnace operation is *continuous*. The furnace is charged at the top with the ore-limestone-coke mixture, and at the same time preheated air is blasted in from the bottom (Fig. 19.1). The oxidation of the coke results in the formation of carbon monoxide, CO, and high temperatures. The limestone dissociates into CaO and carbon dioxide, CO_2. As the charge descends into the furnace, carbon monoxide reduces the ore stepwise to metallic iron. Near the top of the furnace, where the temperature is close to

† Or other reducing agents.

FIGURE 19.1

The blast furnace.

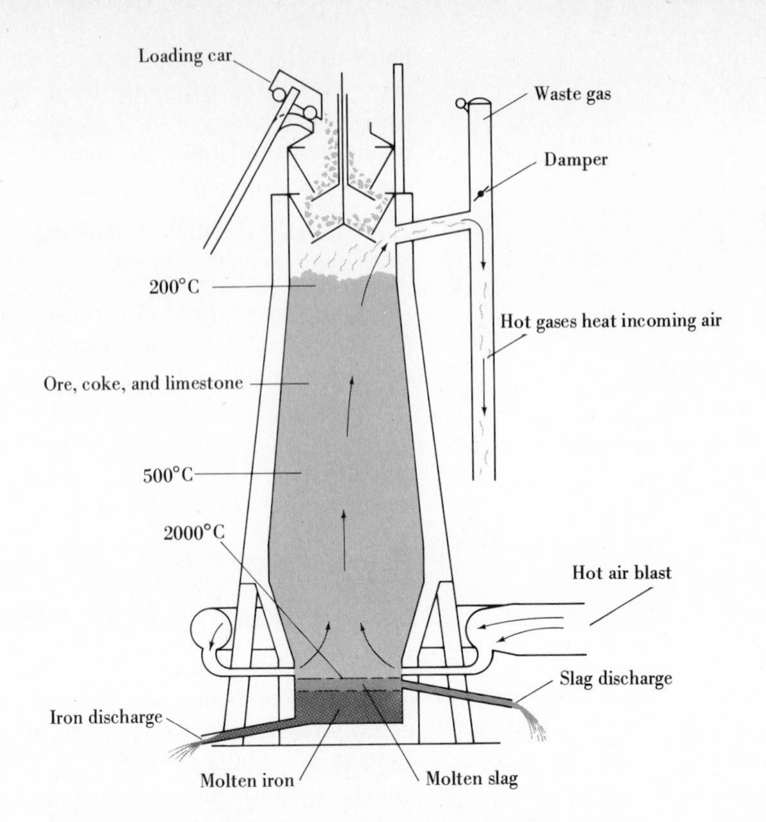

200°C, the ore undergoes the first step, a reduction:

$$3Fe_2O_3(s) + CO \rightarrow 2Fe_3O_4(s) + CO_2$$

As the charge descends into the hotter regions of the blast furnace, it is reduced further to FeO and finally to Fe^0. At the same time the ore impurities, such as silica and alumina, react with the CaO formed from the limestone:

$$CaCO_3(s) \xrightarrow{\Delta} CaO(s) + CO_2(g)$$

$$CaO(s) + SiO_2(s) \rightarrow \underset{\text{Calcium silicate}}{CaSiO_3(s)}$$

$$CaO(s) + Al_2O_3(s) \rightarrow \underset{\text{Calcium aluminate}}{Ca(AlO_2)_2(s)}$$

Toward the bottom of the blast furnace the temperature is near 2000°C. The new iron and the calcium salts, or *slag*, are liquids at this temperature. The slag floats on the liquid metal and is continuously drained off while the metal is tapped from below (Fig. 19.1). A modern blast furnace is capable of producing 3000 tons of iron daily.

Steelmaking

The product of the blast furnace is called pig iron, and it is not extremely pure. It contains up to 4 percent carbon and an equal amount of other impurities such as phosphorus, aluminum, and silicon. The excess carbon, present as Fe_3C, makes the metal brittle, and pig iron is not very useful without further treatment. Several industrial processes are used to purify pig iron and convert it into steel; all convert the impurities into oxides which escape as gas or react to form an insoluble slag. The chief impurity, carbon, is converted into CO_2. After the iron is purified, controlled amounts of carbon and metals are added to make a steel with the desired set of properties.

19.2 THE ELECTRONIC STRUCTURE OF TRANSITION-METAL ATOMS

It is instructive to examine the electronic configurations of the transition metals shown in Table 19.3 for the first transition series. The valence-electron configuration of first transition series is representative of those in the fifth and sixth periods. Scandium, the first member of the series, represented as $1s^2, 2s^2, 2p^6, 3s^2, 3p^6, 4s^2, 3d^1$, has an argon-core configuration plus three valence electrons; two are in the 4*s* orbital, and one is in a 3*d* orbital. Note how the *d*-orbital occupancy changes with increasing atomic number. The or-

TABLE 19.3: THE FIRST TRANSITION SERIES

		Configuration of valence electrons								
Element	*Family*	*3d*					*4s*	*Atomic radius*	*First ionization potential*, kcal	*Oxidation states*
$_{21}Sc$	IIIB	↑					↑↓	1.44 Å	151	3
$_{22}Ti$	IVB	↑	↑				↑↓	1.32	158	2†, 3, 4
$_{23}V$	VB	↑	↑	↑			↑↓	1.22	155	2†, 3, 4†, 5
$_{24}Cr$	VIB	↑	↑	↑	↑	↑	↑	1.17	156	2, 3, 6
$_{25}Mn$	VIIB	↑	↑	↑	↑	↑	↑↓	1.17	171	2, 3†, 4, 5†, 6†, 7
$_{26}Fe$	Iron	↑↓	↑	↑	↑	↑	↑↓	1.16	182	2, 3, 4, 6†
$_{27}Co$	triad	↑↓	↑↓	↑	↑	↑	↑↓	1.16	181	2, 3, 4†
$_{28}Ni$	VIII	↑↓	↑↓	↑↓	↑	↑	↑↓	1.15	176	2, 3, 4
$_{29}Cu$	IB	↑↓	↑↓	↑↓	↑↓	↑↓	↑	1.17	178	1, 2
$_{30}Zn$	IIB	↑↓	↑↓	↑↓	↑↓	↑↓	↑↓	1.25	217	2

† Uncommon oxidation states.

bitals are represented by boxes and the electrons by arrows to indicate the spin direction. As we pointed out in Chap. 4, the filling scheme in the *d* orbitals is regular except for chromium and copper. In these cases we see the tendency toward either half-filled or completely filled orbitals as the 4*s* orbital becomes half occupied. The important distinction between the regular metals and the transition metals lies in the *d* orbitals. In the transition metals electrons are being added to the *d* orbitals as the atomic number increases. The *d*-orbital electrons are valence electrons along with the outer *s*-orbital electrons.

A chemical characteristic of transition metals is the multiplicity of oxidation states. Most exhibit more than two, whereas the regular metals usually have a single oxidation state or two at the most, for example, Na(+1), Ba(+2), Al(+3), Pb(+2 and +4). Table 19.3 shows that the number of different oxidation states increases from one for scandium (+3) to six for manganese (+2, +3, +4, +5, +6, +7) and then decreases to one again for zinc (+2). The multiplicity of oxidation states is due to the fact that the 3*d* and 4*s* electrons are at approximately the same energy level and many of the electrons can be involved in bonding. The decrease in the number of oxidation states after manganese can be explained in terms of the relative constancy of the atomic radius. As electrons are added to the *d* orbitals, the radius changes only slightly. Since the nuclear charge increases, the electrons experience a greater attraction by the nucleus.

The atomic radii shown in Table 19.3 for the transition metals should be compared with the radii given in Table 4.9 for the third period, which includes regular elements only. From such a comparison it is obvious that the radius change from scandium to zinc is relatively small. The elements from $_{24}Cr$ to $_{29}Cu$ have identical atomic radii. Therefore, the ionization potentials given in Table 19.3 primarily reflect the effect of the increasing nuclear charge. The general increase in the energy necessary to remove an electron must be due to the greater attraction of the nucleus. As a result there is a general decrease in reactivity of the elements across a transition-metal series.

19.3 THE LANTHANIDES AND ACTINIDES

Lanthanum, $_{57}La$, is the first member of the third transition-metal series. It is in family IIIB with scandium, $_{21}Sc$; yttrium, $_{39}Y$; and actinium, $_{89}Ac$. Between lanthanum and hafnium, $_{72}Hf$, the second member of the third transition series, are 14 elements, called *lanthanides* or *rare earths;* they are members of an *inner* transition series. The electronic configuration of the lanthanides is shown in Table 4.5. Note that the seven $4f$ orbitals of these elements are filling up as the atomic number increases. The lanthanide elements all exhibit the +3 oxidation state and may be placed in family IIIB. The +2 and +4 states are also exhibited.

The decrease in atomic radius over the 14 lanthanide elements is only about 0.2 Å. This is much smaller than the decrease between adjacent regular elements $_{55}Cs$ (2.6 Å) and $_{56}Ba$ (2.17 Å); it is approximately the same as the change between adjacent transition elements $_{39}Y$ (1.80 Å) and $_{40}Zr$ (1.57 Å). Because of the very small *lanthanide contraction*, these elements have almost identical atomic radii and consequently very similar chemical and physical properties. Since the lanthanides are found together in nature, it is very difficult to separate and purify them; separation depends on some chemical or physical dissimilarity.

Actinium, $_{89}Ac$, is the first and only known member of the fourth transition series. (Period 7 of the periodic table is incomplete.) Like lanthanum, which is directly above it in the periodic table, actinium is followed by 14 elements of an inner transition series, called *actinides*. The actinides exhibit more varied oxidation states (from +2 to +6) than the lanthanides. All the actinides are radioactive, and those with atomic numbers greater than 92, i.e., above uranium, $_{92}U$, decay so rapidly that none can be found in nature. The heavier elements have been synthesized by fusion of smaller nuclei in high-energy accelerators. Because of the scarcity of the *transuranium* elements their chemical and physical behavior has not been well studied, but we assume that their properties are very much like those of the corresponding lanthanides.

19.4 WERNER AND A VALENCE THEORY FOR INORGANIC COMPOUNDS

In Chaps. 5 and 15 we showed how the structural theory of organic chemistry based on the assumption that there is a single, constant valence for each element led to a spectacular advance of the science (even though the assumption was incorrect). While this was true for organic chemistry, the same assumption impeded progress in inorganic chemistry. When the principle of constant valence was applied here, the structural formulas derived proved incorrect. For instance, iron and chlorine form two compounds, ferrous chloride and ferric chloride, with the empirical formulas $FeCl_2$ and $FeCl_3$, respectively. Assuming that iron has a constant valence of 3, i.e., iron always forms three bonds, we could write the following structural formulas:

```
Cl        Cl
  \      /
   Fe—Fe          Fe2Cl4
  /      \
Cl        Cl
```

Ferrous chloride, mol wt 253.6 g

and

```
     Cl
     |
     Fe          FeCl3
    /  \
  Cl    Cl
```

Ferric chloride, mol wt 162.3 g

This would mean a higher molecular weight for ferrous chloride than for ferric chloride. Experiments proved that the necessary conclusion is incorrect.

For the stannous and stannic chlorides the following constant valence formulas were proposed:

```
Cl        Cl
  \      /
   Sn=Sn
  /      \
Cl        Cl
```

Stannous chloride,
empirical formula, $SnCl_2$

and

```
     Cl
     |
     Sn
    / | \
  Cl  |  Cl
     Cl
```

Stannic chloride,
empirical formula, $SnCl_4$

These tin compounds are obviously analogous to carbon compounds. Both tin and carbon are in the same chemical family and would be expected to show similarities. Again, as with the iron chlorides, these structures proved incorrect since experiments showed that stannic chloride has a greater molecular weight than stannous chloride.

In many other instances the application of the principle of constant valence gave rise to incorrect molecular formulas, but for the transition-metal complexes the situation was becoming absurd. For example, cobaltic chloride, $CoCl_3$, appeared to form compounds in which ammonia was chemically incorporated:

$$CoCl_3 + NH_3 \rightarrow CoCl_3 \cdot xNH_3$$

The compounds were called ammoniates. The highest ammoniate was the hexaammoniate, $CoCl_3 \cdot 6NH_3$. Chemists assumed that the cobalt atom was directly bonded to three chlorine atoms in $CoCl_3$ and $CoCl_3 \cdot 6NH_3$, and thus a constant valence of 3 for Co is preserved:

```
     Cl              Cl
     |               |
     Co              Co      · 6NH3
    /  \            /  \
  Cl    Cl        Cl    Cl
```

$CoCl_3$ $\qquad$ $CoCl_3 \cdot 6NH_3$

They were puzzled about the arrangement of the six ammonia molecules around the $CoCl_3$ unit.

Chemists were not surprised that heating the hexaammoniate

yields the pentaammoniate and NH_3 gas:

$$CoCl_3{\cdot}6NH_3(s) \xrightarrow{\Delta} CoCl_3{\cdot}5NH_3(s) + NH_3(g)$$

What surprised them was the difference between the reactions of the hexa- and pentaammoniate with $AgNO_3$. When $AgNO_3$ is added to a solution of the hexaammoniate, all the chloride precipitates:

$$CoCl_3{\cdot}6NH_3(aq) + 3AgNO_3(aq) \rightarrow Co(NO_3)_3{\cdot}6NH_3(aq) + 3AgCl(s)$$

but when $AgNO_3$ is added to the pentaammoniate only two-thirds of the chloride precipitates:

$$CoCl_3{\cdot}5NH_3(aq) + 2AgNO_3(aq) \rightarrow Co(NO_3)_2Cl{\cdot}5NH_3 + 2AgCl$$

In terms of the accepted structures of the two ammoniates it was not possible to explain the difference in the $AgNO_3$ reaction. Similar observations were made for many other transition-metal compounds like $CuCl_2{\cdot}4NH_3$ and $CrBr_3{\cdot}6H_2O$. It was especially difficult to explain the properties of the so-called "double salts" like $CrCl_3{\cdot}3KCl$ and $Co(NO_3)_2{\cdot}4KNO_3$.

In 1893 the Swiss chemist Alfred Werner (1886–1919) attacked the structural problem of inorganic compounds from a fresh point of view. After studying the known properties of the ammoniates of several transition-metal salts, he discarded the principle of constant valence. From carefully tabulated data on known ammonia complexes Werner found that he could explain all the properties of these compounds if he assumed the metal atom to be at the center of a large complex ion. For example, consider the four different ammoniated cobalt nitrites shown in Table 19.4. Werner proposed that each central cobalt ion is directly bonded to a total of *six* other molecules or ions. In the Werner formulas of the complex salts the square brackets include all the molecules and ions bonded to the central Co^{3+}. In the first compound, the hexaammoniated salt, there are six ammonia molecules bonded to central Co^{3+}; since NH_3 is neutral, the charge on the complex ion is +3. In the next compound there are five NH_3 molecules and a single NO_2^- ion bonded to central Co^{3+}; the net charge on the complex ion is +2. As we inspect

TABLE 19.4: WERNER FORMULAS FOR COMPLEX SALTS

Compound	*Werner formula*	*Charge on complex ion*	*Molar electric conductance, arbitrary units*
$Co(NO_2)_3{\cdot}6NH_3$	$[Co(NH_3)_6](NO_2)_3$	+3	440
$Co(NO_2)_3{\cdot}5NH_3$	$[Co(NH_3)_5NO_2](NO_2)_2$	+2	250
$Co(NO_2)_3{\cdot}4NH_3$	$[Co(NH_3)_4(NO_2)_2]NO_2$	+1	90
$Co(NO_2)_3{\cdot}3NH_3$	$[Co(NH_3)_3(NO_2)_3]$	0	0

Werner's formulas we see that NO_2^- ions can replace the NH_3 molecules within the complex ion, and as this occurs, the charge on the resulting aggregates is reduced stepwise from +3 to 0. The ammonia molecules and the NO_2^- ions which are bound to the central metal ion are called *ligands;* the ligands are said to be in the *coordination sphere* of the central ion. Werner indicated the coordination sphere by using square brackets to enclose the species within it. The number of ligands in the coordination sphere is called the *coordination number* of the central ion.

Werner proposed that each metal ion has a *primary* and *secondary* valence. The secondary valence is equal to the coordination number and can be satisfied by ions or neutral molecules. The primary valence can be satisfied by anions only. Hence, in the four ammoniated cobalt nitrites there are always three NO_2^- ions present to satisfy the primary valence of Co^{3+}, but these anions may be inside or outside the coordination sphere. The coordination sphere of Co^{3+}, indicated by square brackets, always contains a total of six ligands (ions and/or molecules).

The electric conductance of solutions of the cobalt nitrites (Table 19.4) is understandable in terms of the Werner formulas. The ability of an electrolyte to conduct current is directly proportional to the charge on the ions and the number of ions present in a solution; any decrease is reflected by a decrease in electric conductance. When NO_2^- replaces NH_3 within the coordination sphere, the charge on the complex ion and the number of anions decrease because the ligands are covalently bonded to the central metal ion. Hence, the molar conductance decreases.

Werner also showed that "double salts" like $Co(NO_2)_3 \cdot KNO_2 \cdot 2NH_3$, $Co(NO_2)_3 \cdot 2KNO_2 \cdot NH_3$, and $Co(NO_2)_3 \cdot 3KNO_2$ are members of the cobalt nitrite series of compounds. If we consider the last compound in Table 19.2, we see that the replacement of one more NH_3 molecule by a NO_2^- ion would cause the neutral complex to become an anion:

$$[Co^{3+}(NH_3)_3(NO_2^-)_3]^0 \rightarrow [Co^{3+}(NH_3)_2(NO_2^-)_4]$$

Werner concluded that the double salt $Co(NO_2)_3 \cdot KNO_2 \cdot 2NH_3$ is better represented as $K^+[Co(NH_3)_2(NO_2)_4]^-$. He included seven compounds in the series of ammoniated cobalt nitrites, making up three different classes:

1. The coordination sphere is a cation:

$$[Co(NH_3)_6]^{3+}(NO_2)_3,\ [Co(NH_3)_5NO_2]^{2+}(NO_2)_2,\ [Co(NH_3)_4(NO_2)_2]^+NO_2$$

2. The coordination sphere is neutral:

$$[Co(NH_3)_3(NO_2)_3]^0$$

3. The coordination sphere is an anion:

$K[Co(NH_3)_2(NO_2)_4]^-$, $K_2[Co(NH_3)(NO_2)_5]^{2-}$, $K_3[Co(NO_2)_6]^{3-}$

All these compounds have a coordination sphere containing six ligands about a central Co^{3+} ion.

As we go from $K[Co(NH_3)_2(NO_2)_4]$ to $K_3[Co(NO_2)_6]$, the number of cations and the charge on the complex anion increase. The electric conductance should increase, and in fact it does. The conductance was a key property used by Werner in deducing the structure of the complex salts of the transition metals. Figure 19.2 is Werner's graph of the conductances of the ammoniated cobalt nitrites.

Returning to the hexa- and pentaammoniated cobalt chlorides, their behavior toward $AgNO_3$ is easily explained when the equations for the reactions are written with Werner formulas:

$$[Co(NH_3)_6]Cl_3(aq) + 3AgNO_3(aq) \rightarrow [Co(NH_3)_6](NO_3)_3(aq) + 3AgCl(s)$$

$$[Co(NH_3)_5Cl^*]Cl_2(aq) + 2Ag(NO_3)(aq) \rightarrow [Co(NH_3)_5Cl^*](NO_3)_2(aq) + 2AgCl(s)$$

When the hexaammoniate is treated with $AgNO_3$, all the chloride precipitates immediately since it is all outside the coordination sphere and ionic. In the pentaammoniate one of the chloride ions is in the coordination sphere covalently bonded to cobalt and only two-thirds of the chloride in the pentaammoniate can precipitate immediately.

Werner extended his theory to include all the transition metals in various complexes such as the hydrates, $[Cr(H_2O)_6]^{3+}$, cyanides, $[Fe(CN)_6]^{3-}$, and many others in which the coordination number

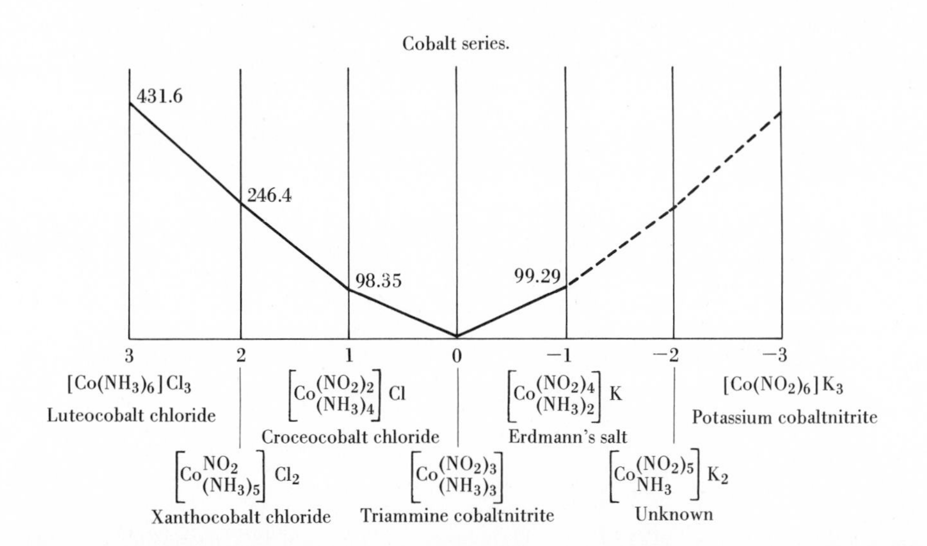

FIGURE 19.2

Conductivity of complex cobalt compounds. [From A. Werner, New Ideas on Inorganic Chemistry, (1911).]

differed from 6. He showed that most metals have coordination numbers of 4 or 6 but there are examples of 2 and 8 and, much less frequently, 3, 5, and 7.

19.5 A SYSTEMATIC NOMENCLATURE FOR COMPLEX IONS

It is important to understand the meaning of several terms commonly used in discussing coordination compounds. The coordination sphere is the unit including the central metal ion and all the ligands bonded to it. The formula for the coordination sphere is usually bracketed: $[Ag(NH_3)_2]^+$, $[Cu(NH_3)_4]^{2+}$. Within the coordination sphere the atom directly attached to the metal is called the *donor atom*. In $[Ag(NH_3)_2]^+$ the donor atom is N; in $[Fe(H_2O)_6]^{3+}$ the donor atom is 0. The donor atom has at least one unbonded pair of electrons. The Lewis structures for NH_3 and H_2O show this. Other ligands like NO_2^-, Cl^-, F^-, CN^-, and CO offer at least one unbonded electron pair. The atoms directly bonded to the central metal are called donor atoms because they donate electron pairs to the metal ion. The central metal is sometimes called the *acceptor atom*.

FIGURE 19.3

The structure of $[Co(H_2N—CH_2—CH_2NH_2)]^{3+}$.

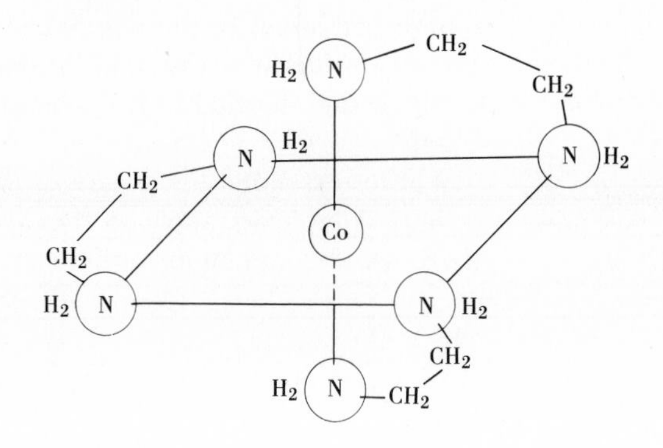

or

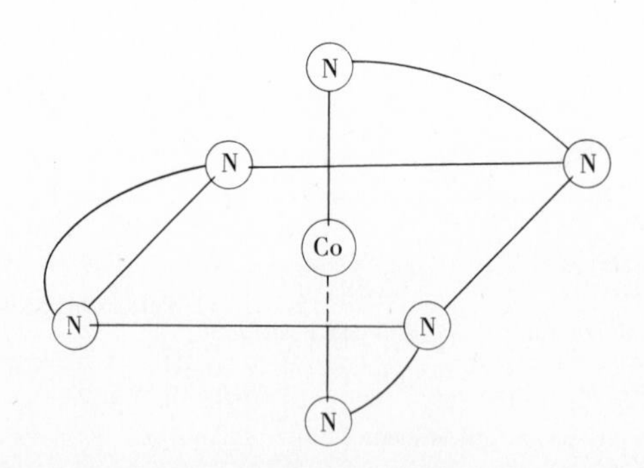

The *coordination number* is the number of donor atoms bonded to the central metal ion. The coordination number in $[Cu(NH_3)_4]^{2+}$ is 4, and in $[Fe(CN)_6]^{3-}$ it is 6. Here the coordination number is also equal to the number of ligands, but this is not always the case. Some ligands, such as ethylenediamine, $H_2N—CH_2—CH_2—NH_2$, furnish two donor atoms per molecule. Thus in the complex ion $[Co(H_2N—CH_2—CH_2NH_2)_3]^{3+}$ there are three molecules of ethylenediamine, but there are six donor atoms and the coordination number is 6, as shown in Fig. 19.3. When one ligand molecule bonds to the central metal through more than one donor atom, the ligand molecule is called a *chelating agent* and the resulting complex is referred to as a *metal chelate*. Chelating agents are very important chemical reagents in the laboratory and in living systems. The compound ethylenediamine tetraacetic acid (EDTA) is used by analytical chemists for the quantitative determination of many different metals.

In basic solution EDTA exists as the ethylenediamine tetraacetate ion, and in this form it possesses six possible donor sites, four oxygen and two nitrogen atoms:

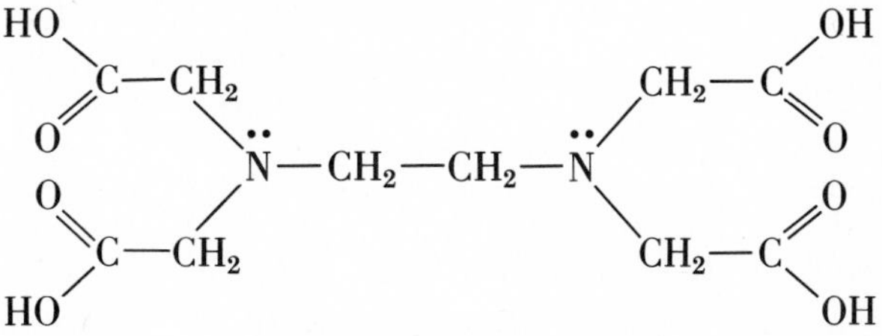

```
HO                                    OH
  \                                  /
   C—CH2                     CH2—C
  //     \                   /      \\
 O        N—CH2—CH2—N                O
 O        /   ••     ••  \           O
  \\     /                \         //
   C—CH2                     CH2—C
  /                                  \
HO                                    OH
```

Ethylenediamine tetraacetic acid

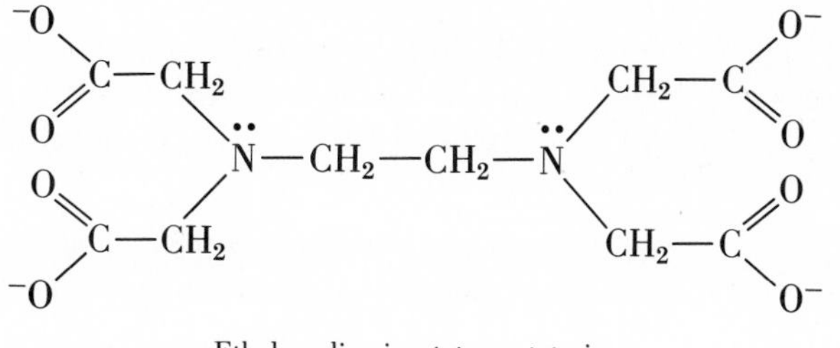

```
⁻O                                    O⁻
  \                                  /
   C—CH2                     CH2—C
  //     \                   /      \\
 O        N—CH2—CH2—N                O
 O        /   ••     ••  \           O
  \\     /                \         //
   C—CH2                     CH2—C
  /                                  \
⁻O                                    O⁻
```

Ethylenediamine tetraacetate ion

EDTA forms colored chelates with many metal ions. The color of the chelate serves as a means of identifying the metal ion, and the intensity of the color serves as a means of determining the concentration of the ion. A chelating agent like ethylenediamine with two donor atoms is called a *bidentate* chelating agent; when there are three donor atoms in a single chelating molecule, it is called a *tridentate*. Heme, which is the active site of the hemoglobin molecule, is a complex of Fe(II) and a *quadridentate* chelating agent with four nitrogen donor atoms (Fig. 19.4). In the heme structure,

FIGURE 19.4

Heme, a quadridentate chelate of Fe(II).

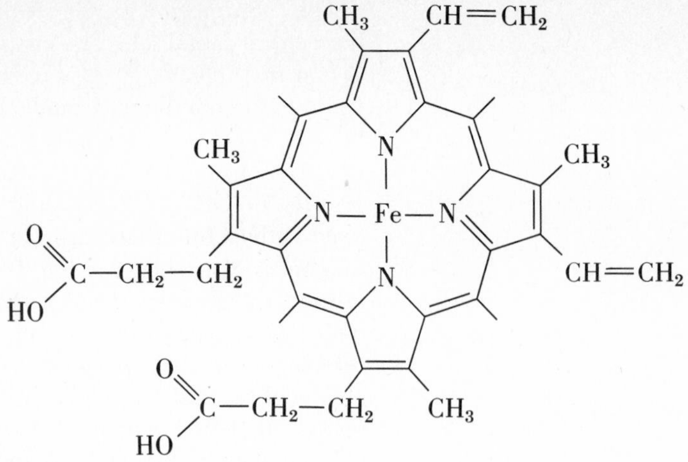

the Fe atom in the +2 oxidation state is at the center of a planar, organic structure called a *porphyrin*. The inner four nitrogen atoms of the porphyrin act as donor atoms. EDTA, which has six donor atoms, is called a *sexadentate*.

The nomenclature of the coordination compounds is patterned after the system originally suggested by Werner. The following rules will be useful:

1. The names used for the common ligands are fluoro- (F^-), chloro- (Cl^-), bromo- (Br^-), iodo- (I^-), cyano- (CN^-), nitro- (NO_2^-), nitrito- (ONO^-), nitrato- (ONO_2^-), hydroxo- (OH^-), oxo- (O^{2-}), amido- (NH_2^-), oxalato- ($O_4C_2^{2-}$), ammine- (NH_3), and aquo- (OH_2). The formulas for the ligands are given with the donor atom as the first symbol.

2. In naming the complex ion, whether it is a cation, an anion, or a neutral molecule, the ligands are named first and then the central metal: $[Fe(NH_3)_6]^{3+}$ is hexaammineiron(III) ion.

3. In cases where there are different ligands in the coordination sphere, it is conventional to name the anionic ligands first and the neutral (molecule) ligands second; $[FeCl_2(H_2O)_4]^+$ is called the dichlorotetraaquoiron(III) ion.

4. The prefixes bis- and tris- are used to denote the presence of two or three chelating groups in the coordination sphere. The ethylenediamine complex of copper, $[Cu(NH_2CH_2CH_2NH_2)_2]^{2+}$ or $[Cu(en)_2]^{2+}$, is called bis(ethylenediamine)copper(II) ion.

5. When the coordination sphere is a cation or a neutral molecule, the ordinary name of the central metal is used followed by a Roman numeral in parentheses to designate its oxidation state: $[Cu(OH_2)_4]^{2+}$ is tetraaquocopper(II) ion. When the coordination sphere is an anion, the ending *-ate* is added to the English or Latin

TABLE 19.5: NOMENCLATURE FOR WERNER COMPLEXES

Chemical formula	*Name*
Complex cations:	
$[Co(NH_3)_6]Cl_3$	Hexaamminecobalt(III) chloride
$[Pt(NH_3)_4(ONO_2)_2]^{2+}$	Dinitratotetrammineplatinum(IV) ion
$[Ag(NH_3)_2]^+$	Diamminesilver(II) ion
$[Cr(H_2O)_4Cl_2]ClO_4$	Dichlorotetraaquochromium(III) perchlorate
$[Co(en)_3]_2(SO_4)_3$	Tris(ethylenediammine)cobalt(III) sulfate
Neutral molecules:	
$[Pt(NH_3)_2(OH)_4]$	Tetrahydroxodiammineplatinum(IV)
$[Co(NH_3)_3(NO_2)_3]$	Trinitrotriamminecobalt(III)
Complex anions:	
$K_3[Co(NO_2)_6]$	Potassium hexanitrocobalt*ate*(III)
$[PtI_6]^{2-}$	Hexaiodoplatin*ate*(IV) ion
$Na_2[SnCl_6]$†	Sodium hexachlorostann*ate*(IV)

† Tin, Sn, is not a transition metal, but the rules of nomenclature apply to such complexes.

stem for the name of the central metal followed by the Roman numeral in parentheses: $[Cu(CN)_4]^{2-}$ is tetracyanocuprate(II) ion.

6. In naming a compound the name of the cation is given first and then the name of the anion, in accord with the usual nomenclature rules: $[Co(NH_3)_6]Cl_3$ is hexaamminecobalt(III) chloride, and $K_3[Co(ONO)_6]$ is potassium hexanitritocobaltate(III).

Table 19.5 gives examples of the application of these rules.

19.6 STEREOCHEMISTRY OF COMPLEX IONS

Another important conclusion Werner reached is that the coordination sphere of the transition-metal complexes is also associated with a definite geometry. Metals with a secondary valence or coordination number of 6 always form complex ions with an octahedral geometry. When the coordination number is 4, there are two common geometries for the complex ion, tetrahedral and square planar (tetragonal). For coordination numbers of 2 and 8 the geometry of the ions are linear and dodecahedral, respectively (see Fig. 19.5).

The central-metal theory led Werner to infer the possibility of geometrical isomers, complex ions with the same molecular formulas but with different structures and properties. He could show two different geometric structures for the complex ion $[Co(NH_3)_4Cl_2]^+$. To understand this consider Fig. 19.6, which shows a regular octahedron. The six vertices are equivalent. A 90°

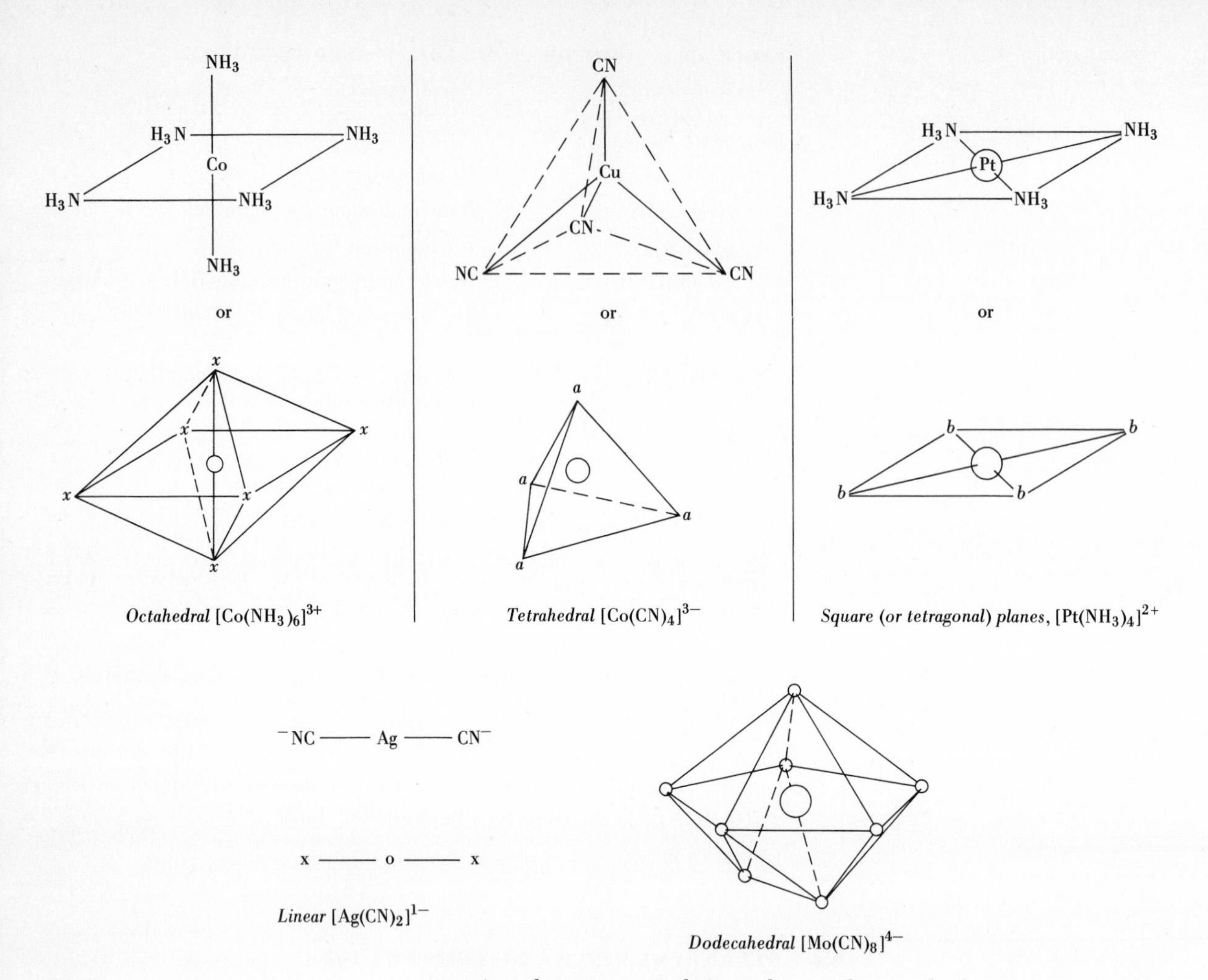

FIGURE 19.5

Shapes of complex ions.

rotation about any axis brings the regular octahedron into an equivalent configuration. A 90° rotation about the 3-5 axis interchanges vertices 1, 6, 2, and 4. The principal cross sections of a regular octahedron, for example, 1, 6, 2, 4; 3, 4, 5, 6, are squares. Any particular vertex is *adjacent* to four others. For example, vertex 5 is adjacent to vertices 1, 6, 2, and 4; it is opposite vertex 3. In chemical language we say vertex 5 is cis to 1, 6, 2, and 4 and trans to 3. A complex with the general formula MX_4Y_2 can therefore have two structural representations. The two Y ligands can be placed at adjacent vertices or at opposite vertices. (There are no other possibilities.) When the Y ligands are at adjacent vertices, the metal complex is said to be in the cis configuration; when the Y ligands are at opposite vertices, the metal complex is in the trans configuration.

Structures *a* and *b* in Fig. 19.7 are the cis and trans isomers of a complex octahedral ion with the general formula MX_4Y_2. The cis and trans forms represent different chemical entities with different chemical and physical properties. Structures *c* and *d* are the cis and

FIGURE 19.6

Rotation of the regular octahedron by 90° about any axis brings it into an equivalent configuration.

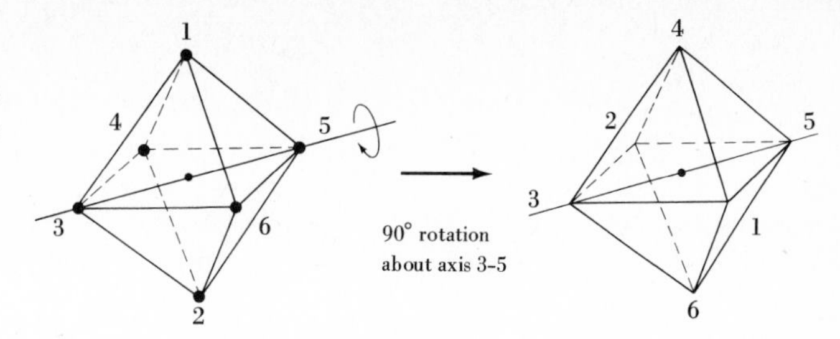

trans forms postulated by Werner for $[Co(NH_3)_4Cl_2]^+$, one violet and the other green.

A complex with the general formula $[M(X_2Y_2Z_2)]^{\pm\nu}$, like $[Cr(Br_2(NH_3)_2(H_2O)_2)]^+$, can exhibit several geometric isomers. Each ligand may be trans to one like itself or cis to one like itself, as shown in Fig. 19.8*a* and *b*. The ammonia molecules can be trans to each other while the Br and H_2O ligands are cis to each other, as shown in Fig. 19.8*c*. Similarly, the Br (or H_2O) ligands may be trans to each other while the other pairs are cis (Fig. 19.8*d* and *e*). Hence, there are five geometric isomers of $[Cr(Br_2(NH_3)_2(H_2O)_2)]^+$.

Octahedral complexes which result from the coordination of two or three bidentate molecules can also give rise to *optical* isomers,

FIGURE 19.7

Geometric isomerism.

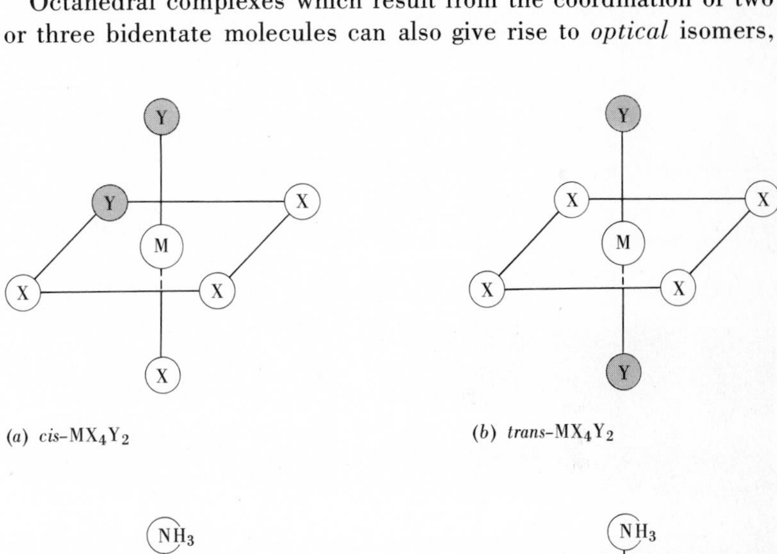

(*a*) *cis*-MX_4Y_2

(*b*) *trans*-MX_4Y_2

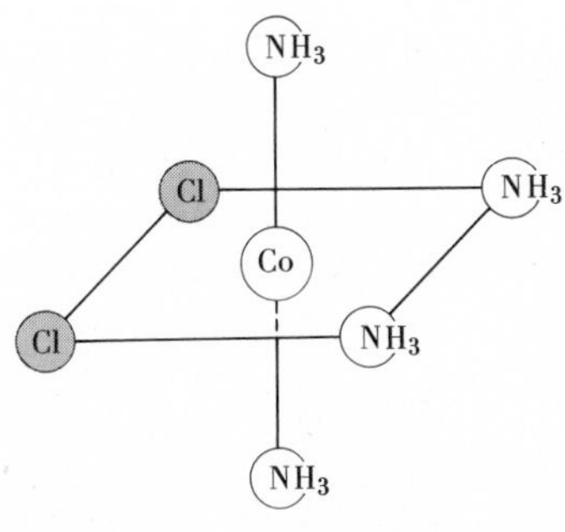

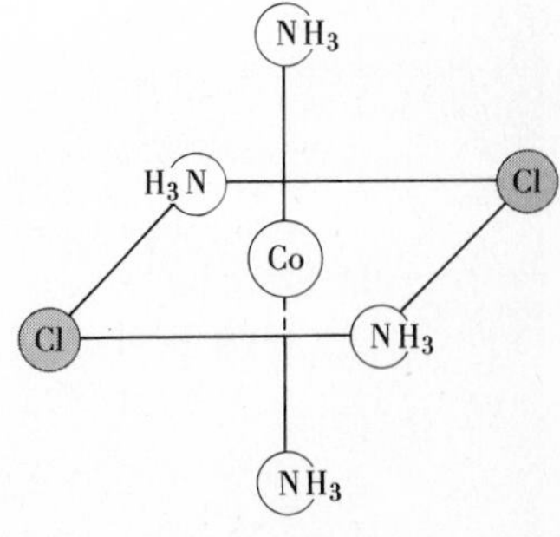

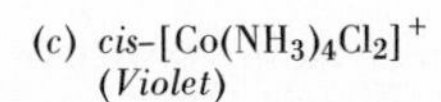

(*c*) *cis*-$[Co(NH_3)_4Cl_2]^+$
(*Violet*)

(*d*) *trans*-$[Co(NH_3)_4Cl_2]^+$
(*Green*)

FIGURE 19.8

The geometric isomers of $[Cr(Br_2(NH_3)_2(H_2O)_2)]^+$.

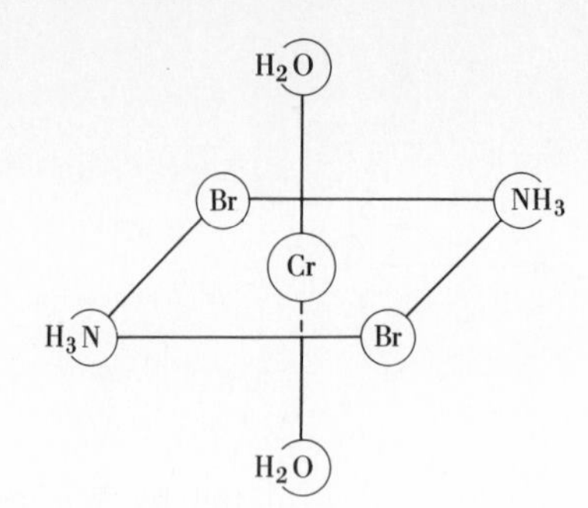

(*a*) *All trans*

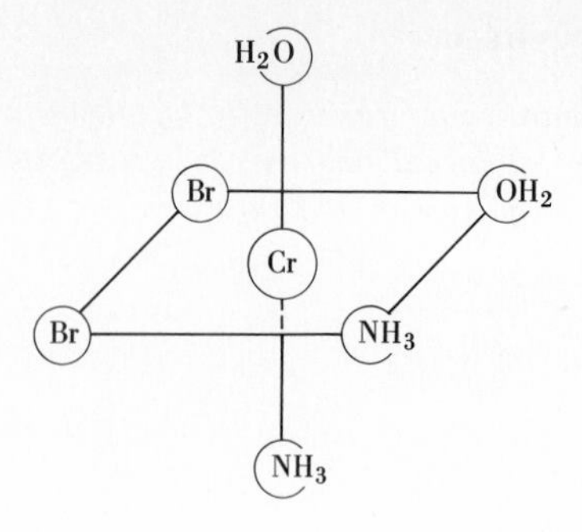

(*b*) *All cis*

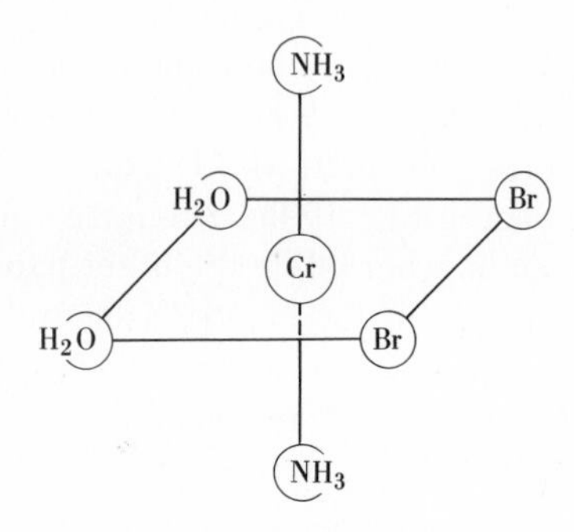

(*c*) *Trans* NH_3; *cis* H_2O *and* Br

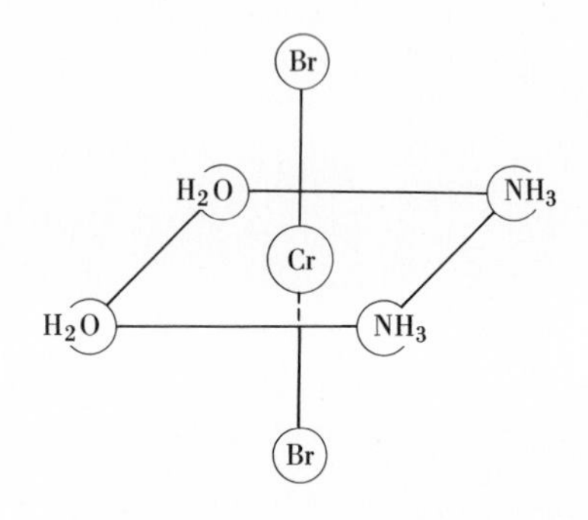

(*d*) *Trans* Br; *cis* H_2O *and* NH_3

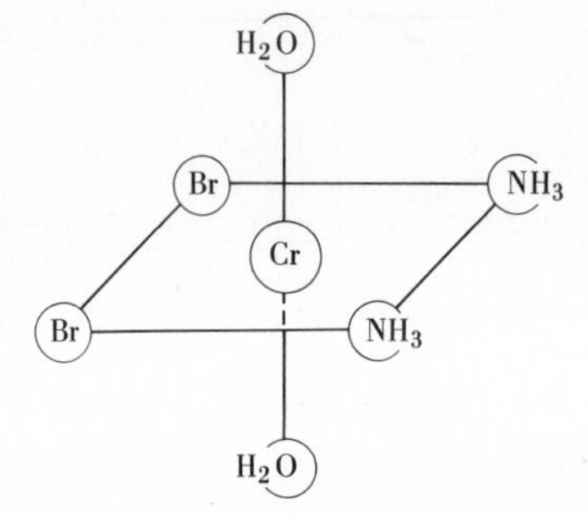

(*e*) *Trans* H_2O; *cis* Br *and* NH_3

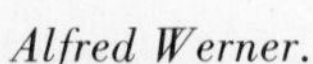

Alfred Werner.

structures which are nonsuperimposable mirror images. Figure 19.9 shows three isomers of dichlorobis(ethylenediamine) cobalt(III) ions. In one case two chlorine atoms are trans to each other and in the other two the chlorine atoms are cis to each other. The two cis forms are nonsuperimposable mirror-image optical isomers; one isomer rotates the plane of polarized light clockwise and the other counterclockwise. Whenever a complex ion has the formula $[M(X_2(BD)_2)]$ or $[M(XY(BD)_2)]$, where BD represents a bidentate, there will be three isomers, two of which will be optically active forms

FIGURE 19.9

There are three isomers of $[Co(Cl_2en_2)]^{3+}$*; two are optical isomers.*

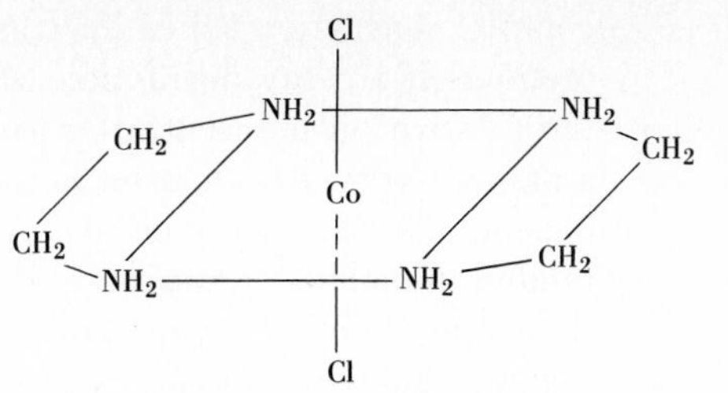

(*a*) *The trans dichloro isomer*

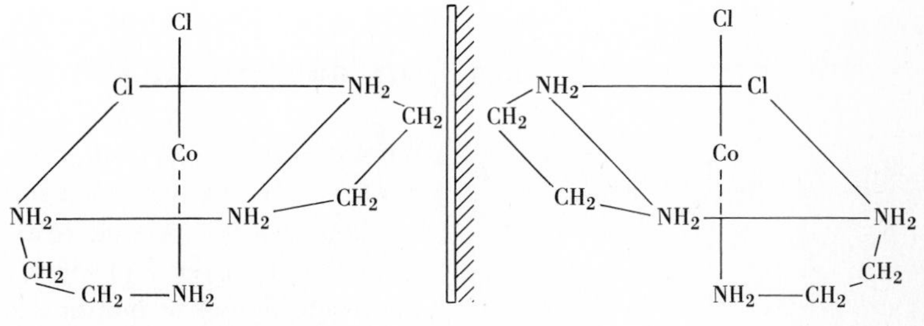

(*b*) *The cis dichloro optical isomers*

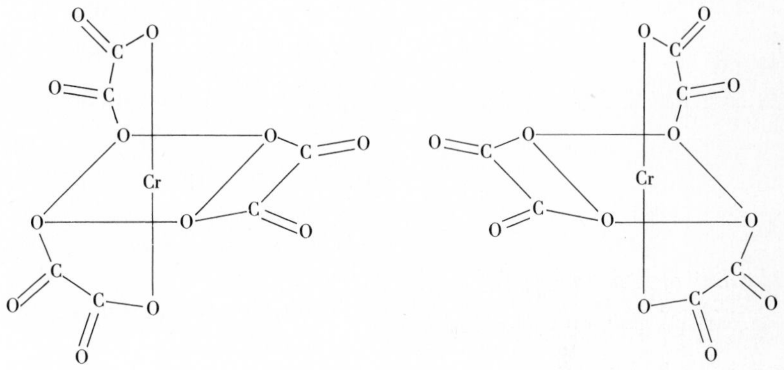

(*c*)

with two X ligands or an X and a Y ligand in cis positions. Werner actually obtained the optical isomers of chloroamminebis-(ethylenediamine) cobalt (III) compounds, $[Co(ClNH_3en_2)]^{2+}$. In this case the optical isomers have Cl and NH_3 in cis positions.

The trisbidentate complexes like trisoxalatochromate(III) ion also exhibit optical isomerism as shown in Fig. 19.9*c*. The trisoxalato isomers are nonsuperimposable mirror images. Any octahedral complex with the general formula $[M(BD)_3]$ may exhibit optical isomerism.

By a careful examination of the chemistry and physical properties of complex compounds Werner correctly deduced the structure of complex ions. He reached his conclusions before the development of methods like x-ray diffraction, which give direct information about the relative positions of atoms in a molecule. Since 1915 we have amassed structural data on complex compounds to verify the conclusions Werner reached by analyzing properties only indirectly related to molecular structure. The Werner theory accomplished for inorganic chemistry what the Kekulé theory had accomplished for organic chemistry 40 years before. Werner's laboratory in Zurich attracted many students, and in 1913 he received the Nobel prize.

19.7 THE BONDING THEORY OF COMPLEX IONS

When Werner proposed his central-metal hypothesis in 1893, the electron had not yet been identified, and, of course, the theory of bonding was not related to the electronic structure of the atom. It was not until after Werner received his Nobel prize in 1913 that the electronic theory of bonding attracted many adherents.

In the 1920s the English chemist Nevil Vincent Sidgwick (1873–1952) proposed the coordinate covalent bond chiefly to bring the Werner formalism into the realm of the Lewis-Langmuir valence theory. He recognized that the donor atoms of the ligands or chelating agents had at least one unshared pair of electrons in their valence shells:

$\ddot{N}H_3$ Ammonia

$H_2\ddot{\underset{..}{O}}$ Water

$:\ddot{\underset{..}{Br}}:^-$ Bromide

$:C{\equiv}N:^-$ Cyanide

$H_2\ddot{N}—CH_2—CH_2—\ddot{N}H_2$

Ethylenediamine

He suggested that the donor atom uses an unshared pair to supply both electrons required for the formation of the bond with the central metal. The resulting bond, called a *dative* or *coordinate covalent bond,* was discussed in Chap. 5.

For $[PtCl_4]^{2-}$ ion Sidgwick suggested a structure like the one shown in Fig. 19.10*a*, where the arrow indicates that the Cl^- is contributing both electrons for each electron-pair bond between Pt and Cl. For octahedral complexes like $[Fe(H_2O)_6]^{3+}$ and $[Fe(CN)_6]^{3-}$ Sidgwick proposed structures in which the donor atoms of the six ligands furnish electron pairs for the six metal-ligand bonds, as shown in Fig. 19.10*b* and *c*. These structures satisfy the rule of two (page 97) which Lewis considered to be the most important part of

FIGURE 19.10

Sidgwick structures for the complex ions.

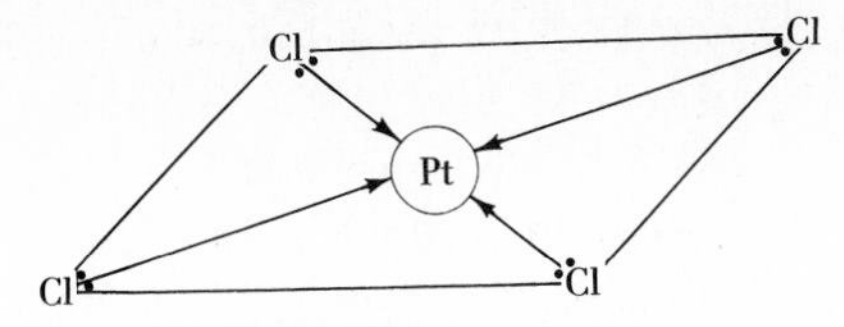

(*a*) *Square planar* $PtCl_4^{2-}$; *four coordinate covalent bonds*

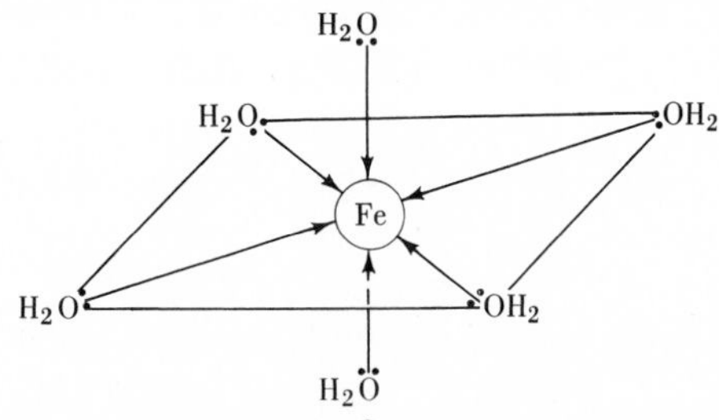

(*b*) *Octahedral* $[Fe(H_2O)_6]^{3+}$; *six coordinate covalent bonds*

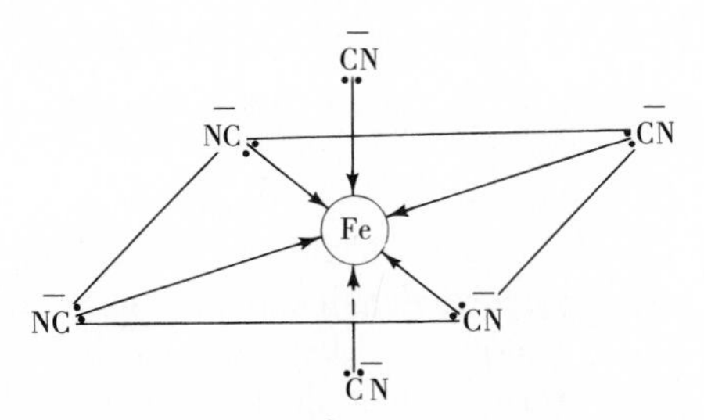

(*c*) *Octahedral* $[Fe(CN)_6]^{3-}$; *six coordinate covalent bonds*

valence theory; but some features of the structures proposed by Sidgwick were unacceptable. To say that the ligands contribute all the electron pairs implies a total charge on the central ion that would be unrealistically high. More obvious is the objection based on the comparative values of the paramagnetic susceptibilities of ions such as $[Fe(H_2O)_6]^{3+}$ and $[Fe(CN)_6]^{3-}$. According to Sidgwick, the ligands contribute *paired* electrons to the complex, and hence the number of unpaired electrons should be equal to the number in the simple Fe^{3+} ion. This would be the same in any Fe^{3+} complex ion, but magnetic-susceptibility studies show that $[Fe(H_2O)_6]^{3+}$ has five unpaired electrons and $[Fe(CN)_6]^{3-}$ has only a single unpaired electron.

Further weaknesses in the Sidgwick formulation become obvious when we consider some nickel complexes. The nickel ion, Ni^{2+}, forms complexes with NH_3 and CN^-: $[Ni(NH_3)_4]^{2+}$ and

$[Ni(CN)_4]^{2-}$, respectively tetrahedral and square planar. In addition, $[Ni(NH_3)_4]^{2+}$ is paramagnetic, due to the presence of two unpaired electrons, whereas $[Ni(CN)_4]^{2-}$ is diamagnetic, with no unpaired electrons. The difference in the structural and magnetic properties of these two complex ions cannot be understood in terms of the Sidgwick coordinate-covalent-bond idea.

19.8 APPLICATION OF VALENCE-BOND AND HYBRIDIZATION THEORY

The vast majority of the transition-metal complexes exhibit one of three structures: tetrahedral, square planar, or octahedral. The tetrahedral and square planar structures have four ligands and are said to be *four-coordinated.* The octahedral species are six-coordinated. The valence-bond theory enables us to explain the structural characteristics and magnetic properties of these complexes in terms of the electronic configurations of the central metal. The electronic configurations for the transition metals of the fourth period are presented in Table 4.4. The spatial properties of the d orbitals are important to an understanding of the geometry of the complex ions, and representations of the $3d_{x^2-y^2}$ and the $3d_{z^2}$ orbitals are shown in Fig. 19.11. Note that the $3d_{x^2-y^2}$ orbital has its highest electron density along the x and y axes; the $3d_{z^2}$ orbital has its highest density along the z axis. When we consider square planar complexes like the tetrachloroplatinate(IV) ion, $[PtCl_4]^{2-}$, we picture the ligands as lying along the x and y axes. Hence, the ligands interact strongly with the $3d_{x^2-y^2}$ orbital. When we consider an octahedral complex like the hexacyanoferrate(III) ion, $[Fe(CN)_6]^{3-}$, we picture the ligands as lying along the x, y, and z axes so that the ligands interact strongly with both $3d_{x^2-y^2}$ and $3d_{z^2}$.

Let us consider the two complexes $[Ni(NH_3)_4]^{2+}$ and $[Ni(CN)_4]^{2-}$ to see how the valence-bond theory accounts for the structural and

FIGURE 19.11

The $3d_{x^2-y^2}$ and $3d_{z^2}$ orbitals.

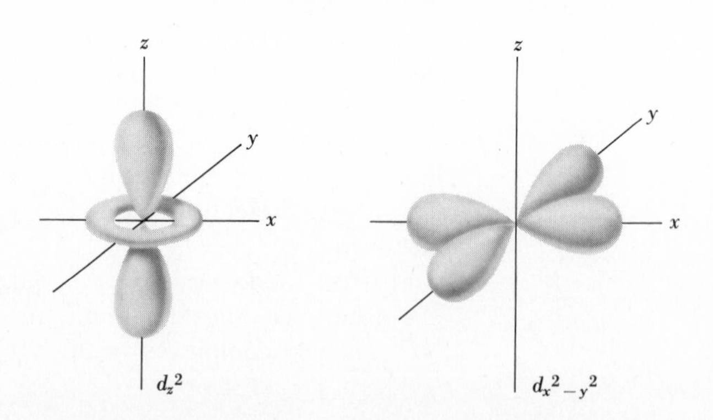

magnetic differences in these species. From Table 4.4 we see that the electronic configuration of the nickel atom, Ni^0, can be represented as an argon core with 10 additional electrons in the $3d$ and $4s$ orbitals:

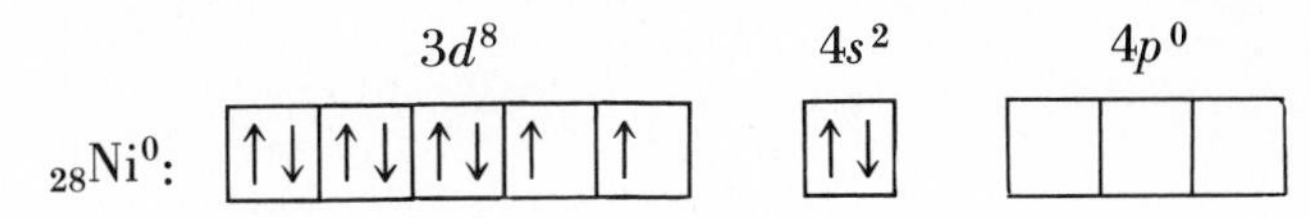

When nickel goes to the +2 state, two electrons are removed from the $4s$ orbital:

$3d^8$ $4s^0$ $4p^0$

$_{28}Ni^{2+}$: ↑↓ ↑↓ ↑↓ ↑ ↑

By losing its two $4s$ electrons in going from Ni^0 to Ni^{2+}, the $4s$ orbital is vacant along with the initially vacant three $4p$ orbitals.

The structure of $[Ni(NH_3)_4]^{2+}$ is tetrahedral. We know from our previous study of bonding theory that this can be explained in terms of sp^3 hybridization. In this case the $4s$ orbital and the three $4p$ orbitals hybridize to produce four equivalent sp^3 orbitals with tetrahedral orientation. The resulting hybridized orbitals accept electron pairs from four ligands. We can represent $[Ni(NH_3)_4]^{2+}$, where the electron pairs in the sp^3 orbitals are donated by the nitrogen atoms of the ammonia ligands NH_3, by the electron-box method:

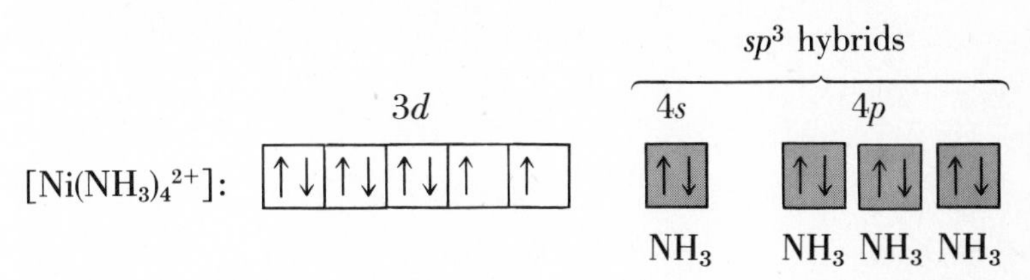

The valence-bond description is very similar to the Sidgwick coordinate-covalent-bond idea, but there is one important difference: we have explained the geometry of the ion, and we can decide whether it is paramagnetic or not. An examination of the electron-box description of $[Ni(NH_3)_4]^{2+}$ reveals the presence of two unpaired electrons in the d orbitals. Hence, this complex should be paramagnetic. Magnetic-susceptibility experiments have verified this prediction.

The valence-bond approach seems fruitful; let us now consider $[Ni(CN)_4]^{2-}$, which has a square planar or tetragonal† planar structure. In this case the hybridization would not be sp^3 because such hybridization results in a tetrahedral geometry. It has been determined that hybridization of the $3d_{x^2-y^2}$ with the $4s$, $4p_x$, and $4p_y$ orbitals produces equivalent orbitals with electron densities lying

† A tetragon is a four-sided structure; it may be rectangular or square.

along the x and y axes. Since the axes of all four of the unhybridized atomic orbitals initially lie in the same plane (Fig. 19.12), it should not be surprising that the resultant hybridized orbitals also lie in the same plane. Note that the $3d_{x^2-y^2}$ orbital has its highest electron density along the x and y axes, and it seems natural that this orbital would be involved in the formation of a square planar complex. We refer to this kind of hybridization as dsp^2; four atomic orbitals (a d orbital, an s orbital, and two p orbitals) produce four identical hybrid orbitals.

This electronic configuration of the uncomplexed ion, Ni^{2+}, shows that all the $3d$ orbitals are at least half occupied. In order to have a completely vacant $3d$ orbital so that it can accept an electron pair from a donor atom we imagine that the two unpaired electrons of uncomplexed Ni^{2+} are forced to pair up, giving

	3d					4s	4p		
$_{28}Ni^{2+}$:	↑↓	↑↓	↑↓	↑↓					

This frees one of the d orbitals. Then the electronic configuration of $[Ni(CN)_4]^{2-}$ with a donation of four electron pairs from the CN^- ligands can be represented as

	3d					4s	4p		
$[Ni(CN)_4]_2^-$:	↑↓	↑↓	↑↓	↑↓	↑↓	↑↓	↑↓	↑↓	
					CN^-	CN^-	CN^-	CN^-	
					dsp^2				

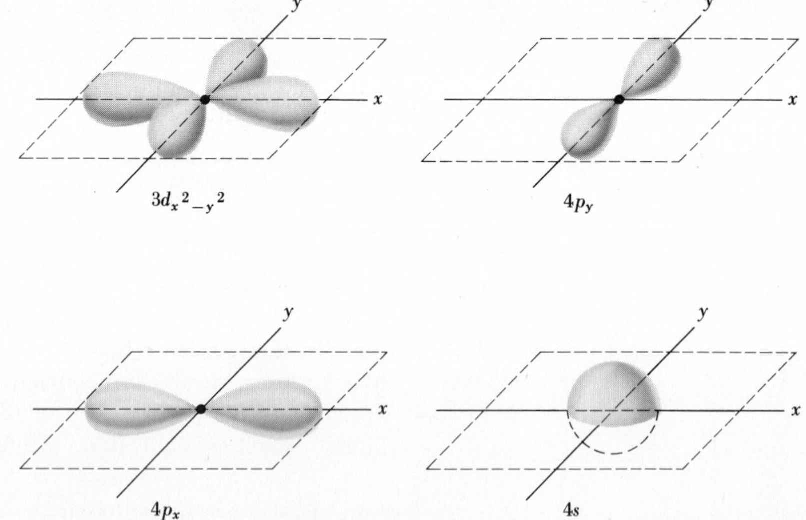

FIGURE 19.12

The axes of the $3d_{x^2-y^2}$, $4p_y$, $4p_x$, and 4s orbitals are coplanar.

Not only does this configuration account for the square planar structure of $[Ni(CN)_4]^{2-}$, but it predicts correctly that the complex is diamagnetic; no unpaired electrons are present.

Let us now consider complex ions of octahedral symmetry in terms of the valence-bond theory. In these cases six vacant atomic orbitals are required to accommodate the six electron pairs from the donor atoms of the ligands. The strongest bonds which can be formed are those permitting maximum overlap between metal and ligand orbitals. The orbitals $3d_{x^2-y^2}$ and $3d_{z^2}$ have their highest electron densities along the x, y, and z axes. Since the ligands bond the central metal atom along these axes, the $3d_{x^2-y^2}$ and $3d_2$ orbitals will be very important in the octahedral complexes (see Fig. 19.13).

In addition the three $4p$ orbitals will be very important in the octahedral complexes since the $4p$ orbitals also have their highest densities along the x, y, and z axes. This gives us a total of five orbitals, but six identical orbitals are required. We use the nondirectional $4s$ orbital to complete the set. Therefore, we say that bonding in an octahedral complex is a result of the hybridization of two d or-

FIGURE 19.13

The $3d_{x^2-y^2}$ and $3d_{z^2}$ are very important in the explanation of the bonding in octahedral complexes.

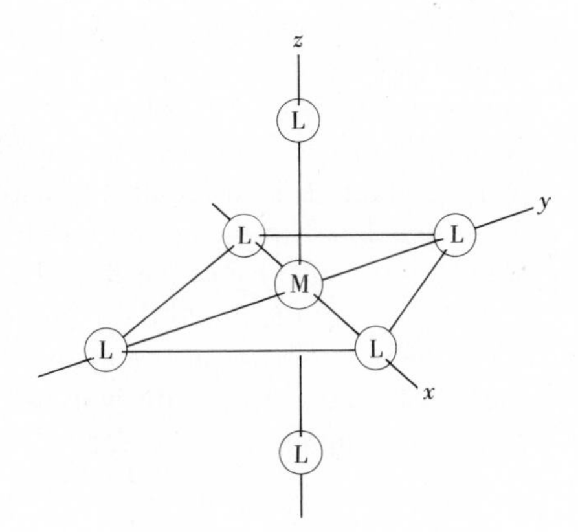

(*a*) *The ligands lie along the x, y, and z axes*

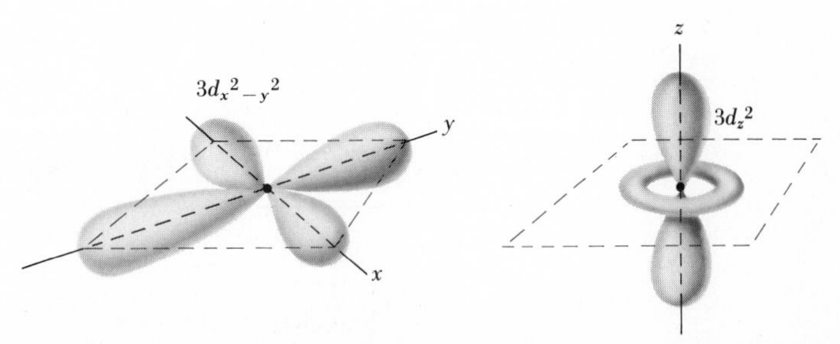

(*b*) *The $3d_{x^2-y^2}$ and $3d_{z^2}$ orbitals have their highest densities along the bonding axes.*

bitals, one *s* orbital, and three *p* orbitals. Hybrids of six such orbitals are designated as d^2sp^3 when the *d* orbitals are in a lower energy level than the *s* or *p* orbitals and sp^3d^2 when the *d* orbitals are in a higher energy level. For example, in the fourth period octahedral hybridization can involve either of the following sets of six orbitals:

$$3d_{x^2-y^2},\ 3d_{z^2},\ 4s,\ 4p_x,\ 4p_y,\ 4p_z \rightarrow d^2sp^3$$

or

$$4s,\ 4p_x,\ 4p_y,\ 4p_z,\ 4d_{x^2-y^2},\ 4d_{z^2} \rightarrow sp^3d^2$$

In Table 4.4 the electron configuration of the $_{26}Fe^0$ atom is given:

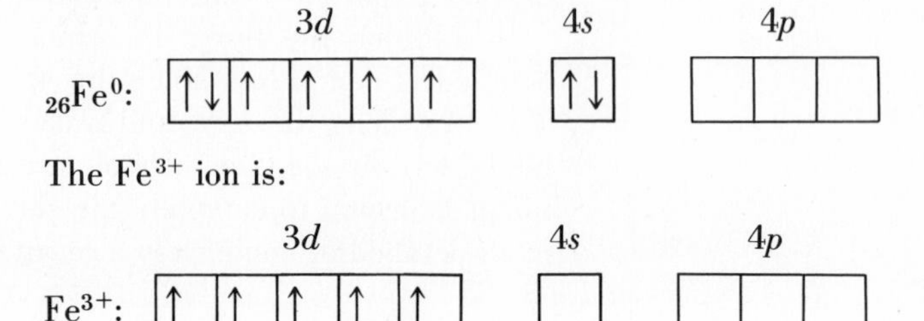

The hexacyanoferrate(III) ion, $[Fe(CN)_6]^{3-}$, has a magnetic susceptibility corresponding to one unpaired electron, whereas the hexammineiron(III) ion, $[Fe(NH_3)_6]^{3+}$, has a magnetic susceptibility corresponding to five unpaired electrons. The hexacyanoferrate(III) ion is very stable; it is difficult to displace a cyanide group from the complex, but the hexammineiron(III) ion is unstable; NH_3 can be displaced easily by other ligands. We can explain the structural, magnetic, and chemical data for these two ions if we say the d^2sp^3 hybridization occurs in the cyano derivative and sp^3d^2 hybridization occurs in the ammine derivative. The electron-box descriptions of the two complex ions are given:

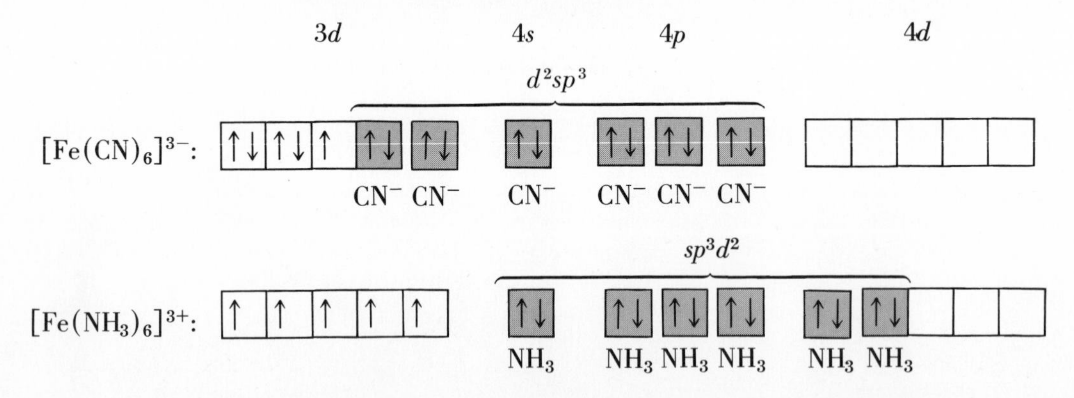

In $[Fe(CN)_6]^{3-}$ the electrons in the 3*d* orbitals are forced to pair; energy is required for this process since electrons generally prefer to avoid each other. This pairing-up energy is more than recovered by the formation of very strong bonds between Fe^{3+} and CN^-. Note that the d^2sp^3 description of the cyano complex shows one unpaired electron, in agreement with experiment. In the hexammine complex the hybridization involves the $4d$ orbitals which have larger average radii than the $3d$ orbitals. The bond between the central iron atom and the NH_3 ligands is not as strong as the Fe—CN bond, and the cyano complex of Fe(III) is more stable than the ammine complex. In the valence-bond structure for the hexammine complex there are five unpaired electrons, in agreement with the results of magnetic-susceptibility experiments.

The $[Fe(CN)_6]^{3-}$ ion, to which we ascribe d^2sp^3 hybridization, is called an *inner-orbital* complex since the closer (inner) $3d$ orbitals are used in the bonding; $[Fe(NH_3)_6]^{3+}$, to which we ascribe sp^3d^2 hybridization, is called an *outer-orbital* complex, since the $4d$ orbitals are used.

We see that the valence-bond theory and hybridization allows us to account for the chemical, magnetic, and structural features of the complex ions.

EXAMPLE 19.1 Cobalt in the oxidation state +3 forms hexammine- and hexanitro complexes which are octahedral. Account for the fact that the nitro complex is more stable than the ammine complex and that the nitro complex is diamagnetic whereas the ammine complex is paramagnetic.

Solution: Write the electronic configuration for Co^0 and the uncomplexed Co^{3+}:

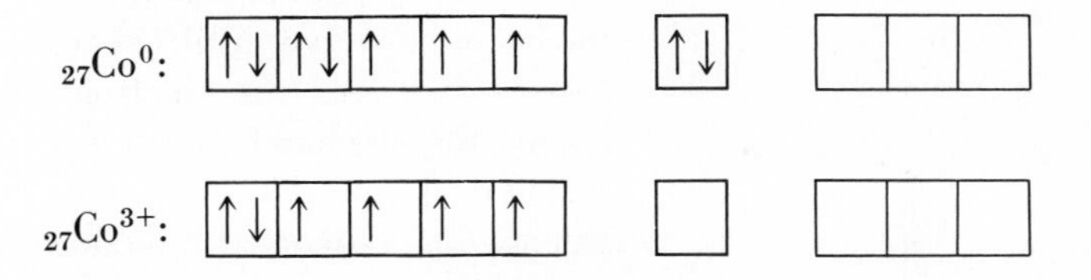

Both complexes are octahedral; hence, we should consider either d^2sp^3 or sp^3d^2. Since $[Co(NO_2)_6]^{3-}$ is diamagnetic, the unpaired electrons of the $3d$ orbitals of the uncomplexed Co^{3+} ion must become paired when this complex ion forms and we have an inner-orbital complex with d^2sp^3 hybridization:

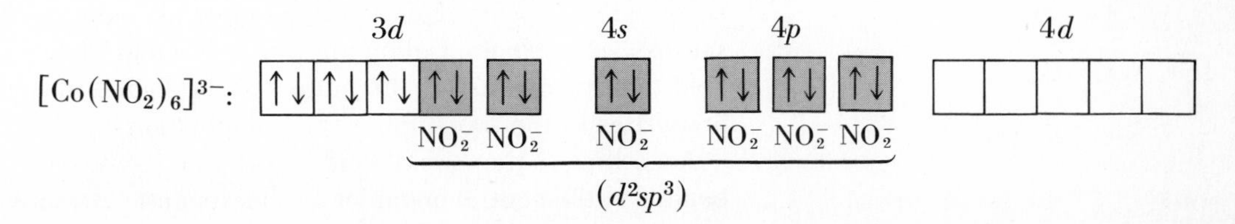

Since $[Co(NH_3)_6]^{3+}$ is octahedral and paramagnetic, we would postulate an sp^3d^2 outer-orbital complex. The $3d$ orbitals of the uncomplexed Co^{3+} ion would be undisturbed, and this ion would have four unpaired electrons:

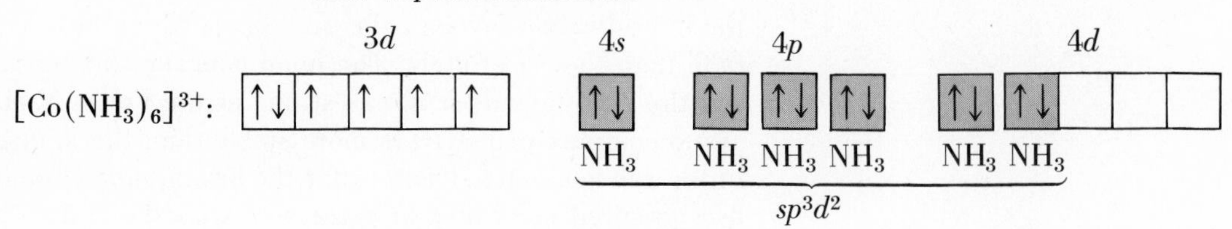

In agreement with experiment, the outer sp^3d^2 complex formed by the ammine is less stable than the inner d^2sp^3 complex formed by the nitro complex.

When Ag(I) complexes, it forms a two-coordinated linear species like $[Ag(NH_3)_2]^+$ and $[Ag(CN)_2]^-$, where linear means that the central metal and the donor atoms of the ligand are collinear

$H_3N —(Ag^+)— NH_3$ $NC^- —(Ag^+)— CN^-$

We learned in Chap. 5 that linear triatomic molecules, like $MgCl_2$ and BeF_2, can be explained in terms of *sp* hybridization. This kind of hybridization also explains the structure of linear two-coordinated complex ions (see Prob. 19.24).

In a sense the valence-bond-theory approach to the transition-metal complex ions is like the Sidgwick theory. In both cases the nature of the ligand is ignored except for the requirement that it have an unshared pair of electrons in the valence shell. Valence-bond theory is more powerful because it presents us with a simple relationship between the electronic configuration of the central metal ion, on the one hand, and the geometric structure and chemical and magnetic properties of its complexes, on the other. Nevertheless, valence-bond theory is a qualitative, arbitrary explanation of experimental data.

19.9 THE CRYSTAL-FIELD THEORY

The crystal-field theory accounts for the chemical, magnetic, and structural properties of the coordination compounds as well as the valence-bond theory does but in addition accounts for the colors, or *spectral* properties. Unlike the valence-bond theory, the crystal-field theory emphasizes the nature of the ligand in determining the properties of the transition-metal complex ions.

According to the crystal-field theory, the ligands are attracted and bonded to the central metal ion by electrostatic attractive forces. The

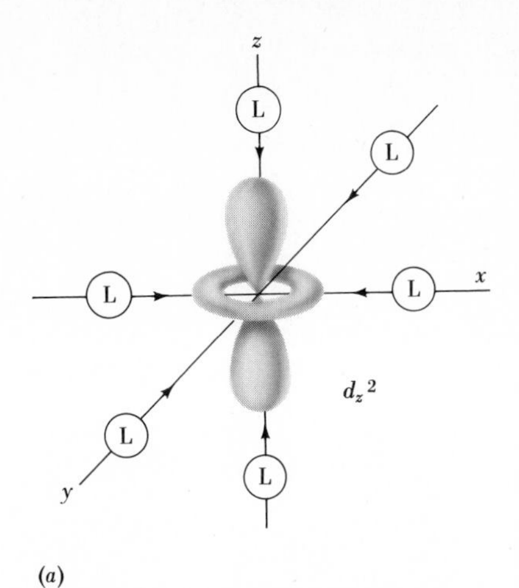

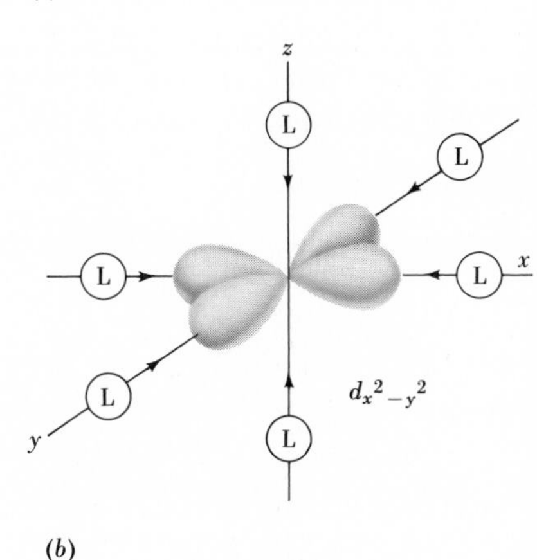

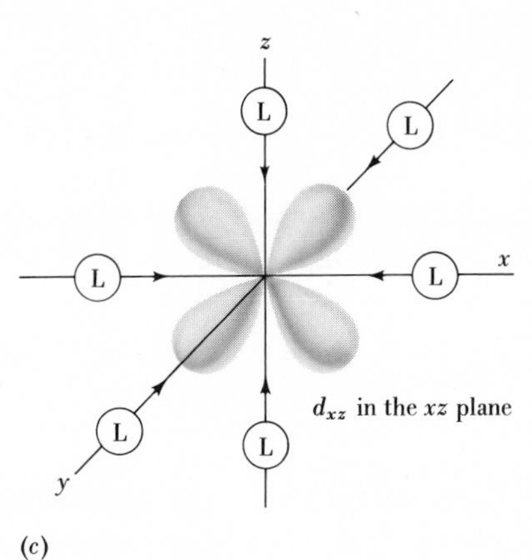

FIGURE 19.14

The ligands have a greater affect on the d_{z^2} and $d_{x^2-y^2}$ orbitals than on the other three d orbitals.

central point of the theory is the electric effect of the ligands as they approach five d orbitals of the transition-metal ion. The ligands are either negative ions or dipoles. The electrons in the d orbitals feel the repulsion due to the negative electric fields of the ligands. The energy of the d orbitals is raised, *but* not all five d orbitals are raised by an equal amount. To understand why, look at Fig. 19.14, which shows an octahedral complex. The ligands approach along the x, y, and z axes and have their greatest effect on the $d_{x^2-y^2}$ and d_{z^2} orbitals because these orbitals have their highest electron densities along these axes. The d_{xy}, d_{xz}, and d_{yz} orbitals have their highest densities midway between the axes and do not feel the full electrostatic effect of the approaching ligands. The electric field of the ligands *splits* the d orbitals, i.e., the d orbitals of the complex ion no longer have the same energy. In the uncomplexed ion the five d orbitals are of equal energy i.e., they are degenerate. The ligands remove this fivefold degeneracy, splitting the five d orbitals into two groups. The $d_{x^2-y^2}$ and d_{z^2} form one group, and the d_{xy}, d_{xz}, and d_{yz} form a lower-energy group. The difference in energy between these two groups is the *crystal-field splitting energy*, written CFSE or Δ. Figure 19.15a shows the degenerate set of five d orbitals. Figure 19.15b shows the effect of the ligands on these orbitals. All the d orbitals are raised in energy in the formation of the octahedral complex, but the $d_{x^2-y^2}$ and d_{z^2} orbitals are at higher energy than the other three. Instead of a single set of five degenerate orbitals we have two sets. The lower-energy set consists of three orbitals of equal energy and is triply degenerate; this set is called the t_{2g} set because t is the symbol indicating triple degeneracy.† The higher-energy set is doubly degenerate and is called the e_g set because e is the symbol indicating double degeneracy. The difference in energy between the t_{2g} set and the e_g set is the crystal-field-splitting energy Δ.

We have considered how the six ligands of an octahedral complex interact with the five d orbitals of the central metal. In species where the coordination number and geometry are different the splittings are different. In tetrahedral complexes the splitting pattern is reversed; the triply degenerate set d_{xy}, d_{xz}, and d_{yz} will be higher in energy than the doubly degenerate set $d_{x^2-y^2}$ and d_{z^2}. This splitting can be understood from Fig. 19.16 and is left as an exercise (see Prob. 19.27). The splitting in the square planar complexes is more complicated and will not be considered.

We have not discussed the effect the ligands have on the 4s and 4p orbitals. In the octahedral complexes these orbitals are also raised in energy because of the electrostatic fields of the ligands, but there is only one s orbital, and hence in this case there can be

† These symbols come from group-theory applications to molecular structure; the subscript g means that the orbitals have a center of inversional symmetry. We shall not need to know the full significance of these symbols.

FIGURE 19.15

All the d orbitals are raised in energy by the presence of the ligands but not by equal amounts. This results in splitting.

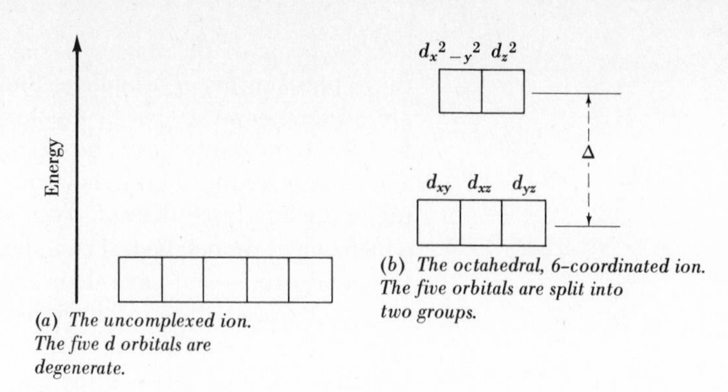

(a) The uncomplexed ion. The five d orbitals are degenerate.

(b) The octahedral, 6-coordinated ion. The five orbitals are split into two groups.

no splitting. The three triply degenerate *p* orbitals, because they lie along the *x*, *y*, and *z* axes, are all affected in the same way and will be raised in energy by equal amounts (in octahedral complexes); hence, they maintain their triple degeneracy. In conclusion the important effect of the ligand field of the octahedral complex is the splitting of the *d* orbitals.

FIGURE 19.16

The tetrahedral complex splits the d orbitals differently. The black dots represent the ligands. Note that the $d_{x^2-y^2}$ and d_{z^2} orbitals are oriented "between" the ligands.

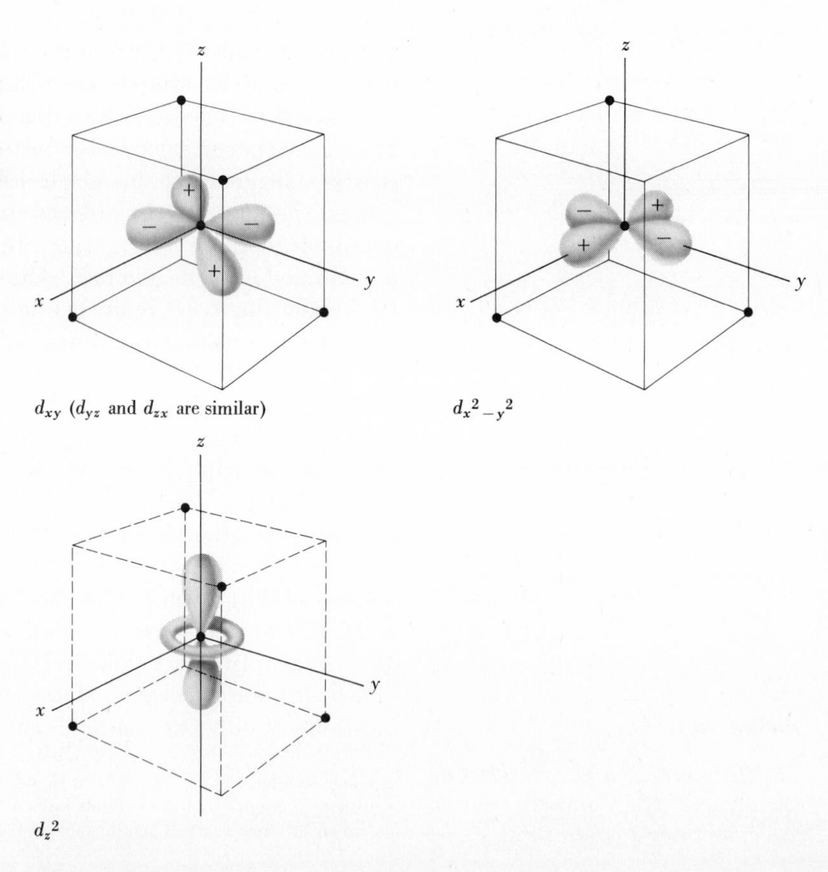

19.10 FACTORS INFLUENCING CRYSTAL-FIELD SPLITTING

Since the crystal-field splitting Δ is the result of electrostatic repulsion between the negative ligands and the electrons in the metal-ion d orbitals, any factors which influence this repulsion will influence Δ. One obvious factor is the negative charge on the ligand. When the ligand is an anion, the value of Δ will increase as the *charge density*, or charge-to-radius ratio, increases. Hence, with a given central metal ion, the value of Δ increases as we go from Br^-, to Cl^-, to F^- because the charge density increases in this order. In addition, we would expect doubly charged ligand anions to cause a greater splitting than singly charged anions. When the ligands are polar but neutral molecules, any factor which makes the molecule more polar will result in a high negative charge in a portion of the molecule. When such highly polar molecules are ligands, they can cause an appreciable splitting of the d orbitals. If we place the more common ligands in order of increasing Δ, we obtain

$$I^- < Br^- < Cl^- < SCN^{-\dagger} < F^- < OH^- < H_2O < SCN^{-\ddagger} < NH_3 < SO_3^{2-} < NO_2^- < CN^-$$

This arrangement is called the *spectrochemical* series because, as we shall see, Δ values are obtained from spectral studies of the complex ions. From this series we conclude, for example, that the crystal-field splitting energy is greater in $[Co(CN)_6]^{3-}$ than in $[Co(H_2O)_6]^{3+}$.

The central metal ion also has an important effect on the Δ values. As the charge on the positive central ion increases, the negative ligands are brought closer to the d orbitals and the repulsive force between the negative ligands and the d-orbital electrons increases. Hence, the Δ values generally increase as we increase the charge on the central metal. The series for metal ions analogous to the spectrochemical series for the ligands is

$$Mn^{2+} < Ni^{2+} < Co^{2+} < Fe^{2+} < V^{2+} < Fe^{3+} < Cr^{3+} < V^{3+} < Co^{3+} < Mn^{4+}$$

From this series we would conclude that the Δ value for $[Fe(H_2O)_6]^{3+}$ is greater than that for $[Fe(H_2O)_6]^{2+}$ and that the Δ value for $[Co(CN)_6]^{3-}$ is greater than that for $[Cr(CN)_6]^{3-}$.

19.11 CRYSTAL-FIELD REPRESENTATIONS OF COMPLEX IONS

To understand how the crystal-field theory explains the magnetic and spectral properties of the complex ions we must consider the effect the splitting of the d orbitals has on the electronic configuration

† Donor atom is S.
‡ Donor atom is N.

of the central metal ion. This is best done by specific examples. In an uncomplexed Fe^{3+} ion the electronic configuration is

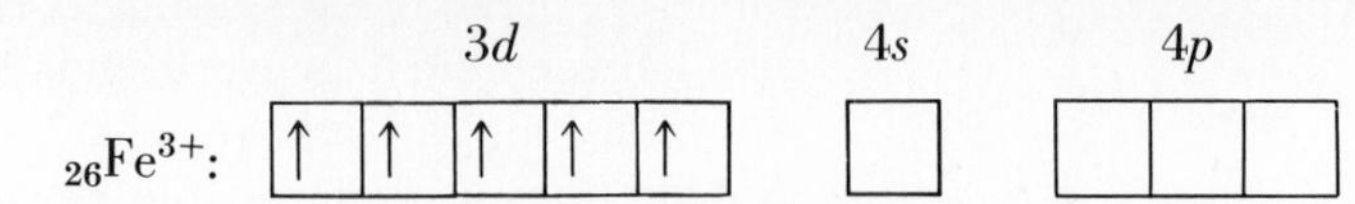

A shorthand notation of the electronic configuration would be Fe^{3+} d^5, where it is understood that all electrons are unpaired, in conformity with Hund's rule. When the six ligands of the octahedral complex surround the Fe^{3+} ion, the d orbitals split into the t_{2g} and e_g sets. In species like $[FeBr_6]^{3-}$ and $[Fe(H_2O)_6]^{3+}$ the splitting is relatively small, whereas in $[Fe(CN)_6]^{3-}$ the splitting is large. Figure 19.17 shows the different effects. The weak field created about the Fe^{3+} ion by the six H_2O ligands splits the d orbitals slightly, while the strong field of the CN^- ions causes a large splitting. The difference in the splitting means that there is a difference in the occupancy of the d orbitals in $[Fe(H_2O)_6]^{3+}$ and $[Fe(CN)_6]^{3-}$: electrons repel each other and tend to occupy separate orbitals; i.e., they tend to remain unpaired. In the uncomplexed Fe^{3+} ion the five d orbitals are degenerate, and there is one electron in each. When Fe^{3+} forms an octahedral complex, the five d orbitals split; they are no longer at the same energy. If the splitting between the t_{2g} and e_g is very large, the three t_{2g} orbitals fill up before any electrons go into the two e_g orbitals. This is what occurs in the $[Fe(CN)_6]^{3-}$ complex (see Fig. 19.17). In $[Fe(H_2O)_6]^{3+}$ the splitting is small, and the electrons remain unpaired.

FIGURE 19.17

Weak-field and strong-field complexes of Fe(III).

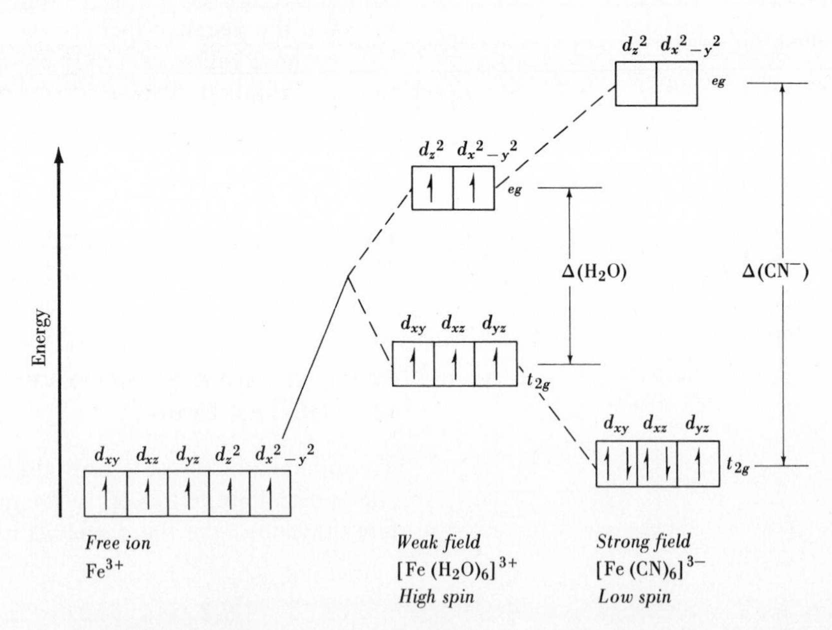

Energy is required to force electrons to occupy the same orbital; i.e., energy is required to cause electrons to pair up. Pairing in the lower t_{2g} orbitals occurs when the crystal-field splitting energy Δ is greater than the electron-repulsion energy or the electron-pairing energy P. Then it is the quantity $\Delta - P$ which controls the electron distribution in the d orbitals of the complex ions. In Fig. 19.18 the arrows representing the electron spins have a length equal to the electron-pairing energy P. When Δ is less than P, that is, when $\Delta - P$ is negative, no pairing occurs. When $\Delta - P$ is positive, pairing in the t_{2g} orbitals can occur before the e_g orbitals are occupied. In Fig. 19.18 we see the electronic configuration for a weak-field d^5 and a strong-field d^5 complex ion.

Gouy balance measurements (page 115) have shown that $[Fe(CN)_6]^{3-}$ and $[Fe(H_2O)_6]^{3+}$ are paramagnetic, indicating unpaired electrons in both species. From these measurements it can be determined that there is a single unpaired electron in the hexacyano complex and five unpaired electrons in the hexaaquo complex, in agreement with the prediction of the crystal-field theory (Figs. 19.17 and 19.18).

The crystal-field theory accounts for the differences in the magnetic properties of the complex ions by the mechanism of crystal-field splitting. This should be contrasted with the valence-bond theory, which uses inner and outer complexes to account for the differences in magnetic properties. (Review page 674 for the valence-bond descriptions of $[Fe(CN)_6]^{3-}$ and $[Fe(NH_3)_6]^{3+}$.)

Figures 19.17 and 19.18 show the weak- and strong-field configurations of a d^5 transition-metal ion, Fe^{3+}. These configurations can also be indicated in shorthand as $(t_{2g})^3, (e_g)^2$ and $(t_{2g})^5, (e_g)^0$. There is no difference in the weak- and strong-field configurations of the d^1,

FIGURE 19.18

*Pairing up occurs in the strong-field complex because the energy requirement for pairing is lower than the energy requirement for elevation to the higher e_g orbitals. [After S. A. Mayper, J. Chem. Educ., **47**: 786 (1970).]*

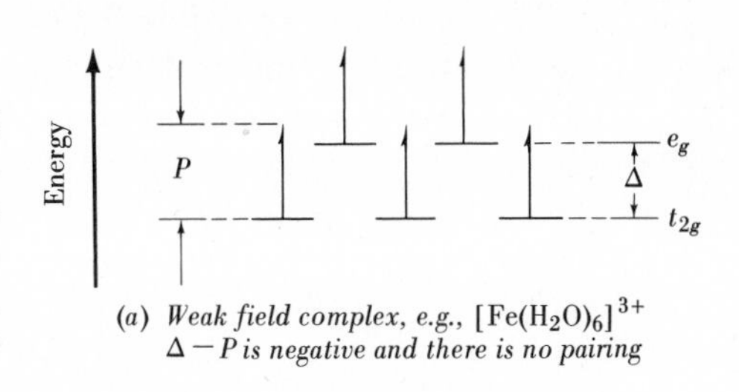

(a) *Weak field complex, e.g.,* $[Fe(H_2O)_6]^{3+}$
$\Delta - P$ is negative and there is no pairing

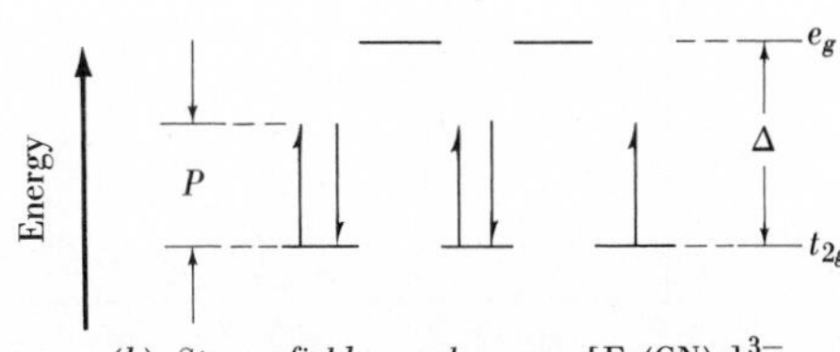

(b) *Strong field complex, e.g.,* $[Fe(CN)_6]^{3-}$
$\Delta - P$ is positive and pairing occurs

d^2, d^3, d^8, d^9, and d^{10} ions (see Prob. 19.30). The ions which have 4 to 7 electrons in the d orbitals show different strong- and weak-field configurations. For instance, a d^4 ion, like Fe^{2+} or Co^{3+}, may have the following electronic configurations:

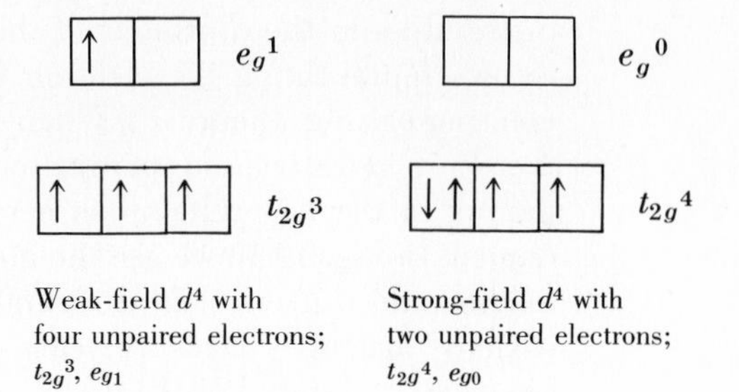

The shorthand descriptions for the weak- and strong-field d^4 complexes are $(t_{2g})^3$, $(e_g)^1$ and $(t_{2g})^4$, e_g^0, respectively.

19.12 THE PROPERTY OF COLOR

The color of a substance is due to the color, or wavelength, of light it *reflects*. A red substance appears red because it reflects red light and absorbs blue or green light. If a substance which looks red in sunlight is irradiated with pure green light, it will appear black, since no light will be reflected. A black body is one that absorbs all the visible radiation falling on it, and a white body is one that reflects all the visible radiation falling on it. Chlorophyll a, the important pigment of photosynthesis, absorbs light at the extremes of the visible spectrum (Fig. 19.19). Since these are the blue and red regions, chlorophyll a is green because it reflects the green, middle portion.

The absorption of visible light is attributed to transitions involving the electrons in the molecules and atoms which make up a substance. We discussed the absorption and emission of light by hy-

FIGURE 19.19

Chlorophyll absorption spectrum.

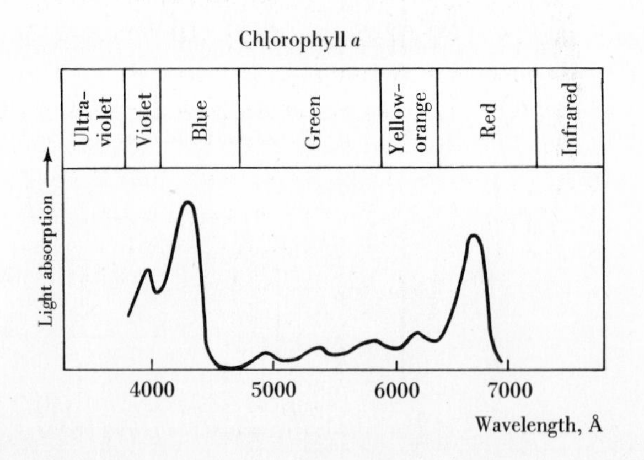

drogen atoms in Chap. 3. When the hydrogen atom absorbs a photon of visible light, its single electron is promoted to a higher energy state, or orbital. Likewise, when other atoms and molecules absorb photons, the electrons are promoted to higher-energy states. In transition-metal complexes, which are generally colored, we attribute the absorption of visible light to transitions involving the *d*-orbital electrons. In the octahedral complexes electrons in the t_{2g} orbitals are promoted to the e_g orbitals. One piece of evidence supporting this idea is the observation that transition-metal ions with a d^0 or d^{10} configuration form colorless complexes. For example, the electronic configuration for the +1 oxidation state of the atoms of the IB family (copper, silver, and gold) is d^{10}, and many +1 compounds, for example, CuCN and AgCl, are colorless. In their +2 and +3 oxidation states the ions form colored compounds; for example, $Cu(CN)_2$ is green, AgF_2 is brown, and $Au(NO_3)_3$ is yellow. Titanium, $_{22}Ti$, has two common oxidation states, +3 and +4. The +3 state is a d^1 configuration, and the +3 compounds of titanium are colored; $TiCl_3$ is violet, and TiF_3 is red. The +4 compounds of titanium, like $TiCl_4$, TiF_4, and TiO_2, which have a d^0 configuration are colorless. Many other examples indicate that the criterion for color or visible-light absorption appears to be the presence of partially filled *d* orbitals. Since the absorption of visible light is due to the promotion of electrons from the t_{2g} to the e_g orbitals, there must be some electrons in the t_{2g} orbitals and vacancies in the e_g orbitals. In a d^0 ion there are no electrons in the t_{2g} orbitals, and in a d^{10} ion there are no vacancies in the e_g orbitals. Ions with d^0 or d^{10} configurations are colorless.

The crystal-field theory with its *d*-orbital splitting caused by the electrostatic repulsion of the ligands gives us a mechanism that explains the absorption of visible light. The difference in the energy between the t_{2g} and e_g corresponds to the energy of the photons of visible or near-visible radiation. An octahedral transition-metal complex can absorb this radiation, whereupon electrons are promoted from the t_{2g} to the higher e_g orbitals. In Fig. 19.20, showing the light-absorption curves for four cobalt complex ions, the wavelength is plotted across the bottom and the corresponding energy is plotted across the top. Note that the wavelengths increase from left to right while the corresponding energies increase from right to left because the energy of a photon is inversely proportional to the wavelength (see Planck's equation, Sec. 3.5).

When hydrogen absorbs visible radiation, its single electron is promoted to a higher *atomic* orbital. We also believe that when molecules and complex ions absorb light (and infrared and ultraviolet radiation), the electrons are promoted to higher-energy *molecular orbitals.* Let us interpret Fig. 19.20 in terms of electronic transitions. $[Co(CN)_6]^{3-}$ has an absorption peak at 2970 Å, which corresponds

FIGURE 19.20

Light-absorption curves for CoX_6 *perchlorates.*

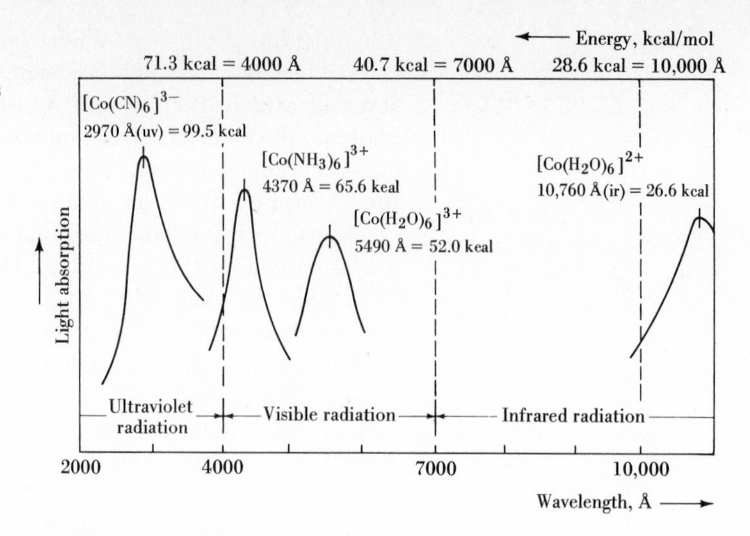

to an energy of 99.5 kcal.† This means that 99.5 kcal is required to promote a ground-state electron to a higher-energy state in 1 mol of $[Co(CN)_6]^{3-}$ ions. In terms of the crystal-field theory the electron is promoted from a t_{2g} orbital to an e_g orbital, and therefore the difference in energy, or the splitting of the t_{2g} and e_g orbitals, is 99.5 kcal/mol, which is quite high. In the ground state the three t_{2g} orbitals will fill before any electrons go into the e_g orbitals, and $[Co(CN)_6]^{3-}$ has the electronic configuration $(t_{2g})^6$, $e_g{}^0$. There are six electrons in the three t_{2g} orbitals, and $[Co(CN)_6]^{3-}$ is diamagnetic:

$$\_\_\ \_\_\ \cdots\ (e_g)^0$$

$$\uparrow\ \Delta = 99.5 \text{ kcal}$$

$$\underline{\uparrow\downarrow}\ \underline{\uparrow\downarrow}\ \underline{\uparrow\downarrow}\ \downarrow$$

$$[Co(CN)_6]^{3-},\ (t_{2g})^6,\ (e_g)^0$$

The change in electronic configuration which occurs when ultraviolet radiation at 2970 Å is absorbed can be written

$$(t_{2g})^6,\ (e_g)^0 \xrightarrow[\lambda = 2970 \text{ Å}]{\text{absorption}} (t_{2g})^5,\ (e_g)^1$$

A single electron is promoted from the t_{2g} orbitals to an e_g orbital.

The hexaaquocobalt(III) ion, $[Co(H_2O)_6]^{3+}$, has an absorption maximum in the green region of the spectrum at 5490 Å, corresponding to an energy of 52.0 kcal/per mol. The splitting of the t_{2g}

† One mole of photons with a wavelength of 2970 Å can deliver 99.5 kcal of energy.

and e_g orbitals is relatively small, and the electron repulsion can prevent the electrons from pairing up. This is shown in the following diagram, where the arrow lengths represent the electron-repulsion energy P. P is greater than Δ.

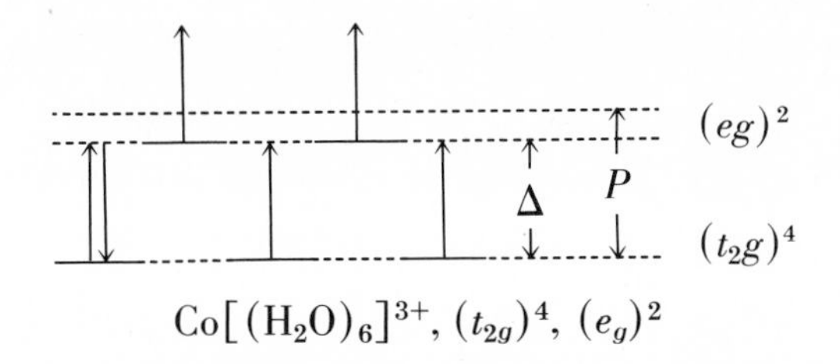

$Co[(H_2O)_6]^{3+}$, $(t_{2g})^4$, $(e_g)^2$

$[Co(H_2O)_6]^{3+}$ is paramagnetic, having four unpaired electrons. The transition occurring when this hexaaquocobalt(III) complex absorbs green light can be represented as

$$(t_{2g})^4(e_g)^2 \xrightarrow[\lambda = 5490\ \text{Å}]{\text{absorption}} (t_{2g})^3(e_g)^3$$

The hexaamminecobalt(III) ion has an absorption maximum between the hexacyano and hexaaquo derivatives, which means that the splitting of the five d orbitals caused by the ligand, NH_3, is intermediate between that caused by CN^- and H_2O.

Note also (Fig. 19.20) that the hexaaquocobalt(II) ion has a relatively low energy-absorption maximum in the infrared at 10,760 Å. This demonstrates that the splitting is proportional to the charge of the central cation. With a lower charge the central metal ion cannot bring the ligands in as close to itself as the Co(III) ion can. The repulsion between the electrons in the d orbitals and the electrons of the ligands is lower, and the splitting is smaller.

19.13 CRITIQUE OF CRYSTAL-FIELD THEORY

We see that the crystal-field theory accounts nicely for the structure and the magnetic and spectral properties of the complex ions. Unfortunately, the hypothesis of purely electrostatic bonding between the ligand and the metal ion, which is the central idea of this theory, is unacceptable. Most ligands behave as if they were covalently bonded to the central metal ion. Electrostatically or ionically bonded substances like $AgNO_3$ and KCl undergo rapid exchange reactions in solution:

$$Ag^+NO_3^-(aq) + K^+Cl^-(aq) \xrightarrow{\text{fast}} K^+NO_3^-(aq) + AgCl(s)$$

We know that analogous reactions involving transition-metal complexes are slow:

$$Ag^+NO_3^-(aq) + [FeCl_6]^{3-}(aq) \xrightarrow{\text{slow}} [FeCl_5NO_3]^{3-}(aq) + AgCl(s)$$

In the language of the Lewis-Langmuir theory this means that the ligand behaves as if it were involved in an electron-pair bond with the central metal ion. In this respect the crystal-field theory is unrealistic. A good bonding theory for the transition-metal complexes must account for the structure, the physical properties, and the chemical behavior. The most promising theory for the transitional-metal complex ions, the ligand-field theory, accounts for all properties in a semiquantitative way, and it also accounts nicely for the chemical behavior of these substances.

19.14 THE LIGAND-FIELD THEORY

The ligand-field theory is a special case of the molecular-orbital theory discussed in Chap. 5. Since in its full, rigorous application this theory requires the use of sophisticated mathematics and quantum mechanics, we shall simply give a qualitative picture of the method applied to octahedral ions.

To explain the bonding according to molecular-orbital theory, we consider the orbitals of the central metal ion which have the proper geometry to make strong bonds by overlap with the ligand orbitals. In the transition-metal ions there are nine atomic orbitals which can be considered as the basis for bond formation; we might say there are nine valence orbitals. For example, in $_{26}Fe$ the five $3d$ orbitals, one $4s$ orbital, and three $4p$ orbitals are of approximately equal energy, and all are involved in bonding. These atomic orbitals will overlap with ligand orbitals to form molecular orbitals for the entire complex. In octahedral complexes there is at least one molecular (or atomic) orbital from each of the six ligands. Hence, we have a total of 15 orbitals from which to construct the molecular orbitals for transition-metal complexes. Examination of Fig. 19.21 shows that the d_{z^2} and $d_{x^2-y^2}$ orbitals are favorably oriented for overlap with the ligand molecular orbitals in an octahedral complex since the six ammonia ligands lie along the x, y, and z axes. The $4p_x$, $4p_y$, and $4p_z$, which lie along the x, y, and z axes, can also overlap nicely with the ligand orbitals, and therefore these orbitals are strongly involved in the bonding of an octahedral complex. The nondirectional character of the $4s$ orbital allows it to participate in the bonding, also. The remaining $3d$ orbitals ($3d_{xy}$, $3d_{xz}$, and $3d_{yz}$) are directed between the six ligands and lack the proper orientation for strong overlap with the ligand orbitals. The six metal orbitals, $3d_{x^2-y^2}$, $3d_{z^2}$, $4s$, $4p_x$, $4p_y$, and $4p_z$, and the six ligand orbitals produce twelve σ molecular orbitals; six are bonding orbitals, and six are higher-energy anti-

FIGURE 19.21

Nonbonding and bonding interactions of the $3d_{xy}$ and p orbitals.

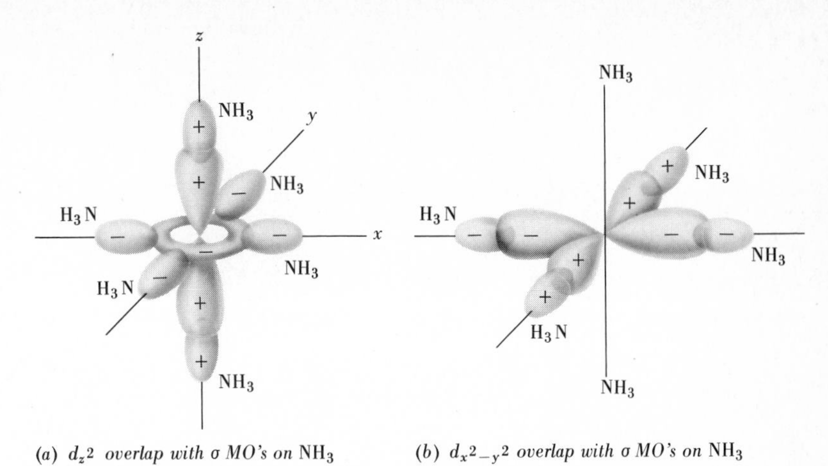

(a) d_{z^2} overlap with σ MO's on NH_3 (b) $d_{x^2-y^2}$ overlap with σ MO's on NH_3

bonding orbitals (σ^*). Recall from Figure 5.13 that all σ orbitals give electron clouds which are symmetrical with respect to rotation about the bond axis; a bonding σ orbital "piles up" electron density between the bonded atoms, while an antibonding σ^* orbital has low electron density and a node in the electron cloud between the two atoms.

As mentioned above and shown in Figs. 19.14 and 3.13, the three metal orbitals, $3d_{xy}$, $3d_{xz}$, and $3d_{yz}$, are directed in between the six ligands of an octahedral complex and cannot form strong σ bonds. The reason for this is demonstrated in Fig. 19.22, where the p_x orbital of a ligand (Cl^-) overlaps equally with positive and negative regions of the metal-ion $3d_{xy}$ orbital. There is no net bonding in this configuration. Therefore, the three p orbitals of the central metal are referred to as the nonbonding orbitals to signify that they are not involved in the σ bonding, but it would be a serious error to suppose that these orbitals are not used in bonding at all. They can be involved in the formation of molecular π orbitals. [Recall that π orbitals may be bonding (π) or antibonding (π^*) and that they are antisymmetric with respect to rotation about the bond axis (page 123).] A π bond formed from overlap of the p_y orbital of a ligand and the $3d_{xy}$ orbital of a central metal ion is represented in Fig. 19.22*b*. The $3d_{xz}$ and $3d_{yz}$ orbitals can also form π bonds with appropriate ligands. Hence, some authors refer to the d_{xy}, d_{xz}, and d_{yz} as the π orbitals, π_{xy}, π_{xz}, and π_{yz}.

The twelve σ orbitals are split into the lower-energy bonding orbitals and the higher-energy antibonding orbitals. The nonbonding or π-bonding orbitals, π_{xy}, π_{xz}, and π_{yz}, are of intermediate energy. A qualitative energy-level diagram is shown in Fig. 19.23*a*. Note that the six bonding orbitals are not all equal energy. In an octahedral complex the six σ bonding orbitals will be filled because in such a complex there are at least six electron pairs which come from the

FIGURE 19.22

Nonbonding and bonding interactions of $3d_{xy}$ with a p orbital.

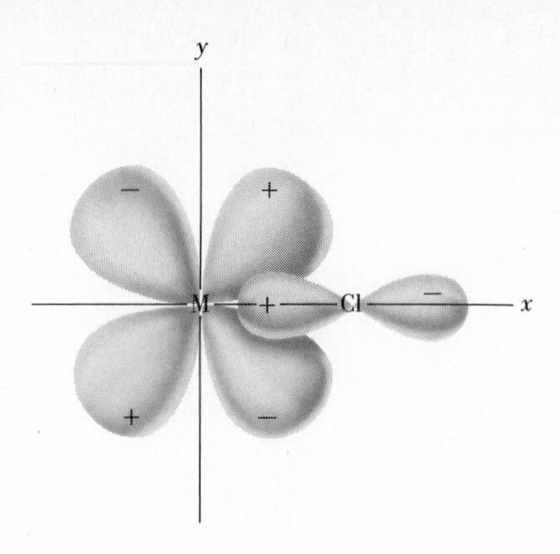

(a) No σ bond can form by overlap of a $3d_{xy}$ and a ligand p_x orbital which is parallel to the bond axis. The positive lobe of the ligand p orbital overlaps the negative and positive lobes of the metal d orbital equally and there is no net bonding.

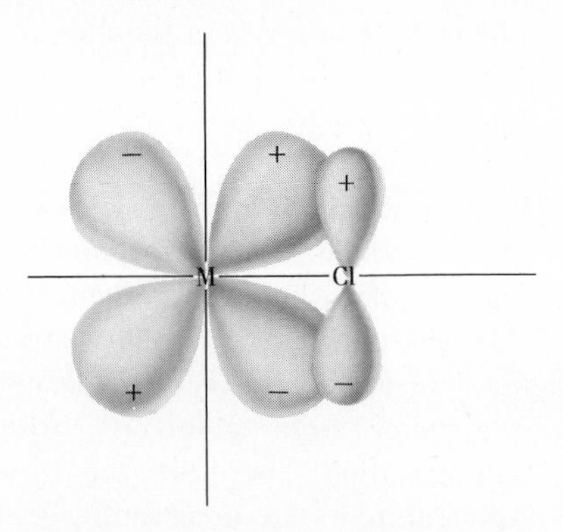

(b) A π bond can form by overlap with a ligand p orbital which is perpendicular to the bond axis.

six ligands. Additional electrons must go into the nonbonding and antibonding orbitals. The lowest-energy antibonding σ orbitals are derived from the $3d_{x^2-y^2}$ and $3d_{z^2}$ of the metal atomic orbitals and hence are referred to as $\sigma^*_{x^2-y^2}$ and $\sigma^*_{z^2}$; this is a doubly degenerate pair, designated e_g, which means doubly degenerate. The difference in energy between the triply degenerate π_{xy}, π_{xz}, π_{yz}, on the one hand, and the double degenerate $\sigma^*_{x^2-y^2}$ and $\sigma^*_{z^2}$, on the other, is called the *ligand-field splitting energy* Δ_0 in analogy to the crystal-field splitting energy Δ.

Figure 19.23*b* and *c* shows the orbital occupancies for high- and low-spin complexes, $[Fe(H_2O)_6]^{3+}$ and $[Fe(CN)_6]^{3-}$. The ligand-field splitting is greater in the cyano complex; this produces the difference in orbital occupancy and the consequent differences in the

FIGURE 19.23

Energy-level diagrams for octahedral complexes according to the ligand-field theory.

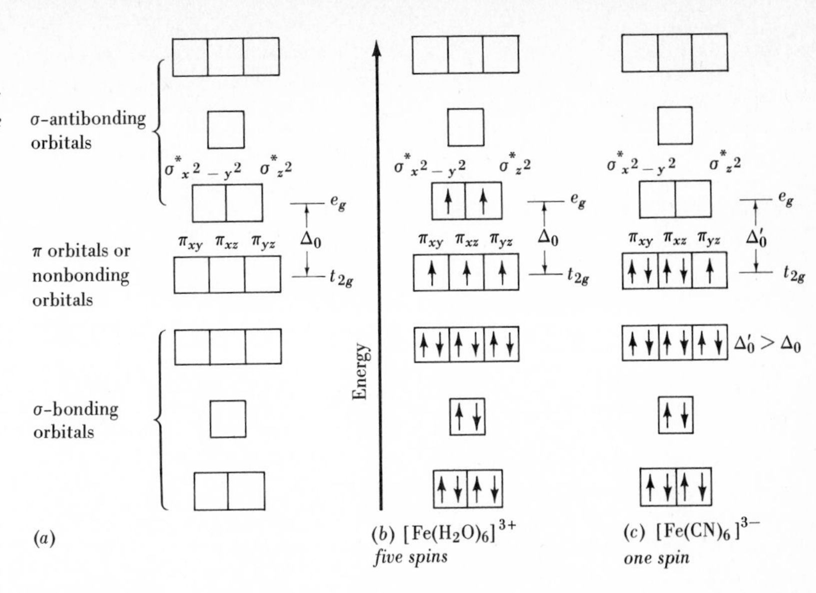

properties of the two complexes. The similarity of the results of the crystal-field and ligand-field theories is obvious. Both approaches produce an orbital splitting, Δ or Δ_0, the magnitude of which permits us to explain the magnetic and spectral properties of the metal-ion complexes.

The crystal-field descriptions of $[Fe(H_2O)_6]^{3+}$ and $[Fe(CN)_6]^{3-}$ were presented on page 680. A comparison of Figs. 19.17 and 19.23 shows the difference in the two approaches. In the crystal-field theory the ligand–metal-ion bonding is simply dismissed as being due to electrostatic attraction. The ligand-field theory generates molecular orbitals from the available metal-atom and ligand orbitals and gives a satisfactory explanation of the bonding within the complex. The ligand-field theory, a subfield of molecular-orbital theory, is the most promising approach to understanding the behavior and properties of the complexes of the transition-metal ions.

QUESTIONS AND PROBLEMS

19.1 What are the chief differences between the regular metals and the transition metals?

19.2 (*a*) What are the chief reactions which occur during the reduction of iron ore in a blast furnace? (*b*) What is the difference between iron and steel?

19.3 Using Table 19.3 as a guide, (*a*) write the electronic configurations for uncomplexed ions of Fe(II), Fe(III), Ti(III), and Ti(IV). (*b*) How many unpaired electrons are present in each of these ions?

19.4 Define or explain what is meant by ligand, donor atom, coordination sphere, coordination number, chelate, polydentate, inner-orbital complex, outer-orbital complex, and paramagnetism.

19.5 (*a*) What is the lanthanide contraction, and what is its effect on the properties of the lanthanide elements? (*b*) What effect, if any, does the lanthanide contraction have on elements other than lanthanides?

19.6 In terms of electronic configuration why do the lanthanides and actinides have similar chemical properties?

19.7 In keeping with the principle of constant valence for the hexaammoniate of cobaltic chloride, $CoCl_3 \cdot 6NH_3$, S. M. Jorgensen of Copenhagen (1837–1914) proposed the chain structure (*a*). Why did he not suggest structure (*b*)? (*Hint:* Consider the reaction of $AgNO_3$ with the hexa- and pentaammoniates [see J. B. Kaufman, *J. Chem. Educ.* **36**: 521 (1959)].

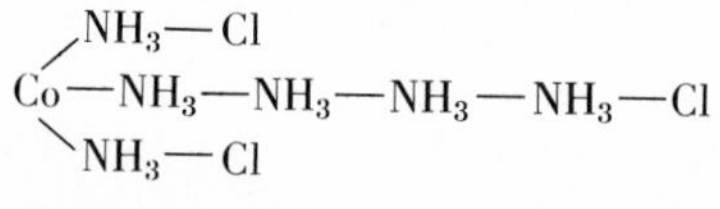

```
   ╱NH3—Cl                                   ╱NH3—NH3—Cl
Co—NH3—NH3—NH3—NH3—Cl                     Co—NH3—NH3—Cl
   ╲NH3—Cl                                   ╲NH3—NH3—Cl
```

(*a*) 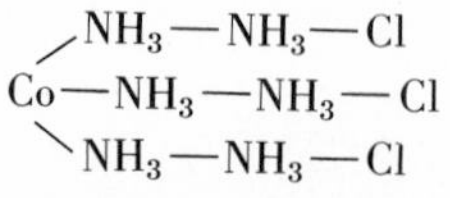 (*b*)

19.8 What did Werner mean by the terms primary valence and secondary valence?

19.9 (*a*) Explain how Werner used the molar conductance of the transition-metal compounds to obtain the structures of the complex ions. (*b*) Assuming octahedral symmetry for the complex ions of the "double salts" $CrCl_3 \cdot NaCl \cdot 2H_2O$ and $CrCl_3 \cdot 2NaCl \cdot H_2O$, give the Werner structure of each and tell which would have the higher molar conductance.

19.10 Before Werner it was difficult to explain why all the chloride in hexaammoniated $CoCl_3$ precipitates when $AgNO_3$ is added while only two-thirds of the chloride in the pentaammoniated $CoCl_3$ precipitates. Explain the difference in the reactions in terms of Werner structures.

19.11 Show the cis and trans isomers of (*a*) $[Cr(H_2O)_4Br_2]^+$, (*b*) $[Cr(H_2O)_4BrCl]^+$, (*c*) $[Fe(NO_2)_4F_2]^{3-}$, (*d*) $[Co(H_2NCH_2CH_2NH_2)_2Br_2]^+$.

19.12 For octahedral complex ions of the form, $[MX_2Y_2Z_2]$, where X, Y, and Z represent ligands, it is possible to have optical isomers. These are nonsuperimposable mirror-image structures. Sketch optical isomers of $[Cr(NH_3)_2(H_2O)_2Br_2]^+$ and $[Co(H_2NCH_2CH_2NH_2)_3]^{3+}$. (Be sure that the mirror images are nonsuperimposable. Superimposable mirror images represent the same structure.)

19.13 (*a*) Make sketches of the tetrahedral complexes $[VCl_4]^0$ and $[CoCl_4]^{2-}$ and the square planar complex $[PdCl]^{2-}$. (*b*) Account for the bonding in the first two structures by using valence theory.

19.14 Two compounds have the formula $K_2(PtCl_2Br_2)$. This is proof that the complex ion $[PtCl_2Br_2]^{2-}$ is square planar and not tetrahedral. Explain.

19.15 How many different forms of the octahedral complex ion $[FeCl_3Br_3]^{4-}$ can exist? Draw them.

19.16 Draw all the isomers of $[Cr(NO_2)_2(H_2O)_2(CN)_2]^-$. Consider geometrical and optical isomers.

19.17 Show all the isomers of (*a*) $[Co(en)_3]^{3+}$ and (*b*) $[CoI_2(en)_2]^+$. Encircle any optical isomer pairs.

19.18 (*a*) Show the *optical* isomers of the chloroamminebis-(ethylenediamine) cobalt(III) ion. (*b*) Are there any other isomers?

19.19 Give the correct names for the following complex ions and compounds:

(*a*) $[Ti(H_2O)_6]^{3+}$ (*b*) $[Ti(H_2O)_3F_3]^0$
(*c*) $[Rh(NO_2)_3]^{3+}$ (*d*) $[Ir(H_2NCH_2CH_2NH_2)_3]Cl_3$
(*e*) $K_3[Ni(H_2O)(NH_2)_5]$ (*f*) $[Fe(NO_2)_4F_2]^{4-}$
(*g*) $[Cu(NH_3)_4]SO_4$

Which of these complexes can exhibit cis-trans isomerism?

19.20 Give the Werner formulas for the following complex ions and compounds: (*a*) hexaaquovanadium(IV) ion; (*b*) bis(ethylenediamine) cobalt(II) bromide: (*c*) potassium tetracyanoaurate(II); (*d*) *cis*-dihydroxobis(ethylenediamine) chromium(III) chloride.

19.21 With exception of $_{25}Mn$ the elements from $_{21}Sc$ to $_{28}Ni$ commonly exhibit the +3 oxidation state (see Table 19.3). Use the electronic configurations of the atoms of these elements to explain why.

19.22 The ion $[Cu(CN)_4)]^{3-}$ is tetrahedral, and $[Ni(CN)_4]^{2-}$ is square planar. (*a*) Discuss the bonding in these ions in terms of the valence-bond theory. (*b*) Is $[Cu(CN)_4]^{3-}$ paramagnetic? (*c*) Is $[Ni(CN)_4]^{2-}$ paramagnetic? (*d*) How many unpaired electrons are there in each ion?

19.23 (*a*) What is the hybridization state of the inner-orbital octahedral complex? (*b*) Explain why an inner-orbital octahedral complex ion would be less paramagnetic than an outer-orbital complex.

19.24 (*a*) Show that the bonding in linear $[Ag(NH_3)_2]^+$ can be explained in terms of *sp* hybridization. (*b*) Would $[Ag(NH_3)_2]^+$ be paramagnetic?

19.25 How does the crystal-field theory account for the color of the complex ions?

19.26 Upon what factors does the crystal-field splitting energy depend?

19.27 (*a*) For the *tetrahedral* complexes the crystal-field theory predicts that the d_{xy}, d_{xz}, and d_{yz} orbitals will be higher in energy than the d_{z^2} and $d_{x^2-y^2}$. Using Fig. 19.16, explain why. (*b*) Give the crystal-field description for the tetrahedral ions $[FeCl_4]^-$ and $[Cu(NH_3)_4]^{2+}$.

19.28 (*a*) What is the relationship between the electric field of the ligand and Δ? (*b*) What is the relationship between the oxidation state of the central metal and Δ?

19.29 (*a*) In terms of crystal-field theory, show the electronic configurations of the complex ions $[Co(H_2O)_6]^{3+}$ and $[Co(CN)_6]^{3-}$. (*b*) Show the electronic configurations of these species according to valence-bond theory.

19.30 Show that there is no difference in the weak- and strong-field configurations of the d^1, d^2, d^3, d^8, d^9, and d^{10} ions.

19.31 Compare the valence-bond, crystal-field, and ligand-field theories. Point out the weaknesses and strengths of each.

19.32 (*a*) Show the electronic configurations of $[Co(H_2O)_6^-]^{3+}$. $[Co(H_2O)_6]^{2+}$, and $[Cr(CN)_6]^{3-}$ according to the ligand-field theory. (*b*) Put the ions in order of increasing paramagnetism.

20 NUCLEAR CHEMISTRY

20.1 THE DISCOVERY OF RADIOACTIVITY

Experiments with cathode-ray tubes led to significant discoveries in physics and chemistry, among them the identification of the electron, the discovery of x-rays and isotopes, and, indirectly, the discovery of radioactivity. We have already discussed Thomson's work with cathode tubes, which led him to conclude that the cathode ray is a stream of negatively charged particles, or electrons (Chap. 3). At the same time, in Germany, Wilhelm Konrad Röntgen (1845–1923) discovered that a very powerful radiation (x-ray) emanates from the spot where the cathode ray strikes the glass envelope of the tube. X-rays penetrate substances which are constituted from the lighter elements; they are absorbed by bones, heavy metals, and the compounds of heavy metals. Almost immediately after Röntgen's discovery x-rays became a diagnostic tool in medicine. Further sophistication led to their use in the study of crystal and molecular structure (Chap. 7).

Along with the x-radiation, visible light is also produced by the

FIGURE 20.1

Becquerel investigated the possibility that visible fluorescence was always accompanied by invisible x-rays.

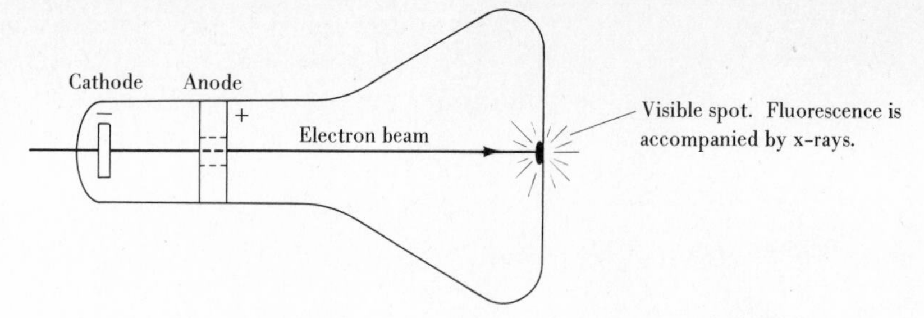

(a) Cathode beam generates visible light and x-rays when it strikes the glass tube

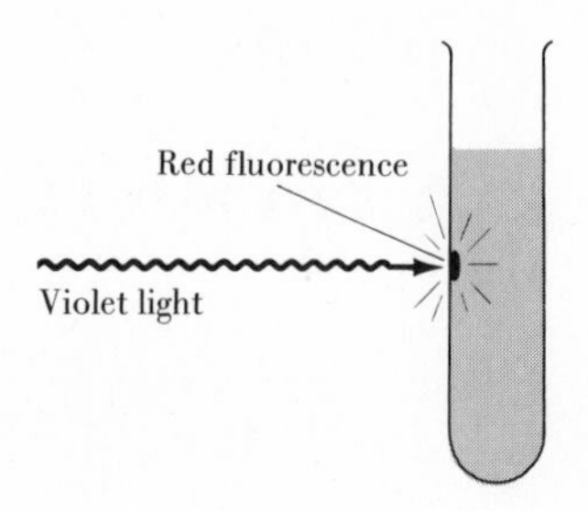

(b) A chlorophyll solution emits a red fluorescence

electron beam of the cathode ray. The electron beam causes a spot to light up at the end of the tube; this is called *fluorescence* (see Fig. 1*a*). Because fluorescence and x-radiation were produced simultaneously, the French physicist Antoine Henri Becquerel (1852–1908) set out to investigate the possible relationship and in so doing took his first step toward a Nobel prize.

Many materials were known to fluoresce when irradiated with ordinary light or ultraviolet radiation, which is the "black light" of the psychedelic world. When chlorophyll, the green pigment found in plant leaves, is irradiated with blue or violet light, it fluoresces brilliant red (Fig. 1*b*). Becquerel wanted to know whether materials which fluoresce also produce x-radiation. His experiments were quite simple. Knowing that x-rays penetrate black paper whereas the ordinary visible light produced by fluorescence does not, he wrapped a photographic plate in heavy, black paper, placed a sample of fluorescent material upon it, and set the whole package in bright sunlight. The sunlight caused the material to fluoresce. If x-rays were also generated, they would pass through the paper and darken the plate. Becquerel made an extensive study of fluorescent materials, like ZnS, CdS, and $BaPt(CN)_4$, and finally discovered that compounds of uranium, e.g., potassium uranyl sulfate, $K_2UO(SO_4)_2$, cause a darkening of the photographic plate. But the big surprise came when he found that the radiation responsible for the darkening was in no way related to fluorescence: he found that uranium compounds darken a wrapped photographic plate even in a closed drawer, i.e., in the dark.

It was soon discovered that compounds of thorium also darken photographic plates. By 1912, 30 elements which spontaneously emit a powerful radiation were known; they are called *radioactive*.

The discoveries of Becquerel intrigued Marie Sklodowska Curie (1867–1934), who came from Poland to study physics at the University of Paris, where she met and married Pierre Curie. The Curies collaborated in studying radioactive substances and learned that the radiation they produce is not affected by chemical combination, physical state, or temperature. They found that purified uranium oxide is *not* as radioactive as the unpurified ore, pitchblende, which is only 80 percent uranium oxide. This was a very important observation since it led to the inescapable conclusion that there are elements in pitchblende besides uranium that are radioactive. From the residue remaining after the uranium had been extracted from the pitchblende the Curies separated a new radioactive element having chemical properties similar to bismuth; this was named *polonium*, Po, in honor of Madame Curie's native land. Another radioactive element in the residue seemed to have chemical properties similar to the alkaline-earth barium. The Curies obtained a crude sample of the chloride of this new element. Gram for gram it was much more radioactive than uranium, and since it even glowed in the dark, it was appropriately named radium, Ra. To obtain a larger sample of radium so they could study its properties the Curies began work with several tons of pitchblende residue, in which radium occurred in infinitesimal amounts. After 3 years of tedious extractions and purifications they produced 0.1 g of pure radium chloride. Since it was chemically similar to barium, they assumed that radium was a heavy member of the alkaline-earth family with an oxidation number of +2. From the percent composition of its chloride they determined that the atomic weight is 225 (see Prob. 20.4).

For their discoveries Becquerel and Pierre and Marie Curie shared the Nobel prize in physics in 1903.

20.2 α, β, AND γ RADIATION

The most fruitful studies on the nature of the radiation produced by radioactive substances were conducted in Ernest Rutherford's laboratory at Cambridge between 1900 and 1920. Rutherford and his coworkers showed that these substances produce three different kinds of radiation: alpha (α), beta (β), and gamma (γ). Whereas α radiation is easily absorbed by metal foils, β radiation is approximately 100 times more penetrating. α and β radiation are deflected in mag-

Lord Rutherford. (Courtesy of the Edgar Fahs Smith Collection, Van Pelt Library, University of Pennsylvania, Philadelphia.)

netic and electric fields, but γ radiation is not. The deflection in these fields indicates that α radiation is a stream of positively charged particles and β radiation is a stream of less massive, negatively charged particles. Rutherford guessed correctly that the γ ray, which is even more penetrating than the β ray, is similar to an x-ray. He proved his guess by showing that γ rays undergo crystal diffraction and determined that the wavelength corresponds to a short x-ray. γ rays travel at the speed of light and have a zero rest mass. Figure 20.2 demonstrates the distinguishing features of α, β, and γ radiation. Note that the α ray, because it is made up of more massive particles than the β ray, undergoes a much smaller deflection as it passes through the magnetic field; the fact that the α and β rays are deflected in different directions proves that they are oppositely charged. Becquerel observed that the β ray has all the characteristics of a cathode ray and concluded that the β ray is a high-speed beam of electrons.

Rutherford measured the charge-to-mass ratio of α particles and concluded that they might be *hydrogen molecule-ions*, that is, H_2^+, or helium ions, He^{2+}. Both particles have the same charge-to-mass ratio, but Rutherford favored the He^{2+} ions because helium gas is always associated with uranium and thorium minerals. In a beau-

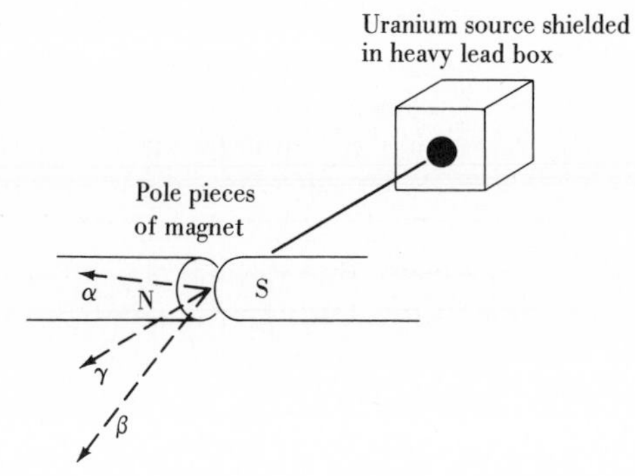

(a) As the radiation passes through a magnetic field the α ray is deflected up and β ray is deflected down. The γ ray is undeflected.

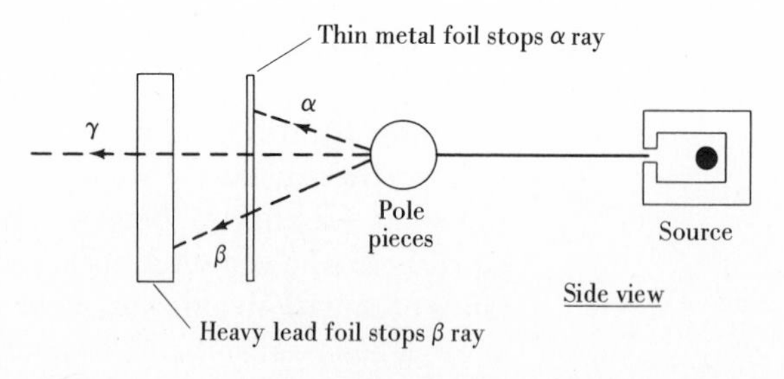

(b) The γ ray is highly penetrating. α rays are most easily stopped.

FIGURE 20.2

The distinguishing features of α, β, and γ rays.

FIGURE 20.3

α particles are identified as He^{2+} *ions.*

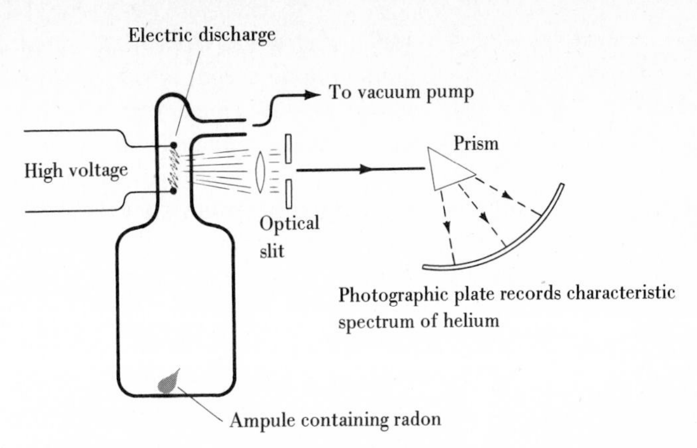

tifully simple experiment Rutherford and one of his students, Thomas D. Royds, proved that α particles are helium ions. They sealed a sample of radon gas, an α emitter, in a small ampule having very thin glass walls. The ampule was placed in a heavy-walled glass tube, from which the air was pumped out (Fig. 20.3). After several days an electric discharge in the tube caused the emission of spectral lines characteristic of helium. The α particles had penetrated the thin walls of the small ampule and, accumulating as helium atoms in the glass tube, they had reached a detectable concentration after several days.

20.3 DETECTION OF RADIATION

The three emanations produced in radioactive decay have such high energy that they cause the ionization of molecules and atoms upon collision. This is the basis of many instruments used to detect α, β,

FIGURE 20.4

Geiger counter.

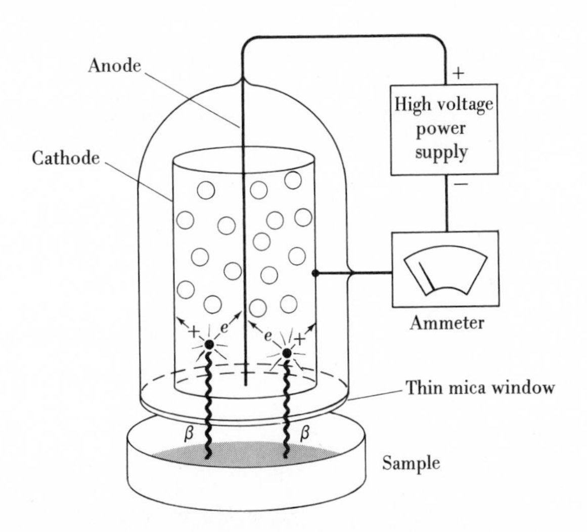

and γ radiation. One of the most popular detectors is the *Geiger* or *Geiger-Müller counter* (Fig. 20.4). Radioactive material is placed under the thin mica window of the tube, which is filled with argon gas. A high voltage is maintained between the central anode wire and the coaxial cylindrical cathode. When, for instance, β particles pass into the tube, they cause the ionization of the argon atoms; argon cations and electrons are produced. The cations move to the cathode and the electrons move to the anode, causing the ammeter to indicate a current surge. Another form of detector, the scintillation counter, is based on the fact that certain materials light up, fluoresce, or *scintillate* when α, β, and γ rays impinge upon them. In this counter the scintillations are detected by a photocell, the number of scintillations per minute being proportional to the radioactivity of the sample.

20.4 THE RATE OF RADIOACTIVE DECAY

As we shall see later in this chapter, the nuclear makeup of a radioactive element changes when it emits α or β radiation. The number of neutrons and protons changes, and the element is said to undergo *radioactive disintegration* or *decay*. The rate of decay decreases as the radioactive element disintegrates. The rate of decay of a sample of any radioactive element is directly proportional to the mass of the sample. We can state this law as a simple mathematical expression

$$\text{Rate of decay} = -km \tag{20.1}$$

where k = proportionality constant or rate constant
m = mass

The rate constant k has units of reciprocal time and has different values for different radioactive elements. Equation (20.1) can be expressed as a differential equation:

$$\text{Rate} = \frac{dm}{dt} = -km \tag{20.2}$$

or

$$\frac{dm}{m} = -k\,dt \tag{20.3}$$

Integration of Eq. (20.3) yields

$$\ln m = -kt + I \tag{20.4}$$

The integration constant I in Eq. (20.4) is evaluated in terms of the mass by designating the mass of the sample at the beginning of an experiment as m_0; at $t = 0$, $m = m_0$. Substitution into Eq. (20.4) shows that the integration constant is just $\ln m_0$:

$$\ln m_0 = 0 + I \qquad \text{or} \qquad \ln m_0 = I$$

Finally, we have the useful form of the radioactive-decay law:

$$\ln \frac{m_0}{m} = kt \tag{20.5}$$

where m_0 = initial mass of radioactive element when $t = 0$
m = mass at end of time t

It should be obvious that the Eq. (20.5) is identical with the rate expression for a first-order reaction, where the reaction rate is directly proportional to the concentration of a single reactant (see Chap. 17).

The rate constant k, which appears in Eqs. (20.1) to (20.5), is a characteristic property of a particular radioactive element. The greater the value of k, the greater the rate of the decay. Radiochemists characterize a radioactive element by its half-life $t_{1/2}$, that is, the time required for one-half of a sample of a radioactive element to decay.

By using Eq. (20.5) we can show that k and $t_{1/2}$ are very simply related; in fact, they represent equivalent information. If m_0 is the initial mass of a sample, its mass at the end of a half-life period, i.e., when $t = t_{1/2}$, will be $\frac{1}{2}m_0$. Substitution of $m = \frac{1}{2}m_0$ at $t = t_{1/2}$ in Eq. (20.5) gives

$$\ln \frac{m_0}{\frac{1}{2}m_0} = kt_{1/2}$$

or

$$t_{1/2} = \frac{\ln 2}{k} = \frac{0.693}{k} \tag{20.6}$$

This equation shows that if k is known, $t_{1/2}$ can be calculated, and vice versa. It also shows that the half-life is independent of the initial mass of a radioactive element.

EXAMPLE 20.1 What percentage of the original mass of a radioactive element remains after four half-lives have elapsed?

Solution: Let m_0 be the initial mass. At the end of the first half-life $\frac{1}{2}m_0$ has disintegrated and $\frac{1}{2}m_0$ remains; at the end of the second half-life $\frac{1}{2}(\frac{1}{2}m_0)$, or $(\frac{1}{2})^2m_0$, remains. The amount remaining after four half-lives is $(\frac{1}{2})^4m_0 = \frac{1}{16}m_0 = 0.0625m_0$, or 6.25 percent of the original mass of the element.

It should be obvious from the solution of Example 20.1 that the amount of a radioactive element which remains after n half-lives is $(\frac{1}{2})^n m_0$.

EXAMPLE 20.2 Radon has a half-life of 3.83 d. Starting with an initial sample of 100 g, what weight remains after 10 d?

Solution: We must determine the value of m at $t = 10$ d in Eq. (20.5):

$$\ln \frac{m_0}{m} = kt$$

$m_0 = 100$ g, $t = 10$ d and, from Eq. (20.6), $k = 0.693/t_{1/2} = 0.693/3.83 \text{ d} = 0.181 \text{ d}^{-1}$. Substitution of these values into Eq. (20.5) gives

$$\ln \frac{100}{m} = (0.181 \text{ d}^{-1})(10 \text{ d})$$

or

$$\log m = \log 100 - \frac{1.81}{2.303} = 1.214$$

$$m = 16.4 \text{ g}$$

The answer, 16.4 g, seems reasonable since 10 d is about two and one-half half-life periods. In two half-life periods, 7.66 d, the sample would have been reduced to $(\frac{1}{2})^2 m_0$, or 25 g, and in three half-life periods, 11.49 d, the sample would have been reduced to 12.5 g.

The disintegration of certain elements is the basis for estimating the age of archeological findings, fossils, minerals, meteorites, moon rocks, and the earth itself. Potassium 40 and uranium 238 are often used for meteorites and minerals; carbon 14 is used to date archeological findings and fossils.

Scientists believe that the level of ^{14}C is maintained constant in the atmosphere. Cosmic rays from the sun indirectly cause the transmutation of ^{14}N, producing ^{14}C; this transmutation is balanced by the decay of ^{14}C, which has a half-life of 5700 yr. In the form of carbon dioxide, the ^{14}C is incorporated into carbohydrates during photosynthesis. Since the carbohydrates are primary nutrients, all living organisms maintain a constant ^{14}C level. Upon death an organism ceases to ingest nutrients, and its ^{14}C level is reduced at a rate predicted by Eq. (20.5). The carbon taken from a living organism will produce 15 β disintegrations (dis) per minute per gram. Carbon taken from an organism which has been dead for a long time will give less than this. The number of disintegrations per minute is in direct proportion to the mass of ^{14}C present.

EXAMPLE 20.3 Carbon prepared from a cypress beam found in the tomb of an ancient Egyptian pharaoh showed 8 dis/min-g. Estimate the time of death of the pharaoh.

Solution: We must solve for t in Eq. (20.5):

$$\ln \frac{m_0}{m} = kt$$

From Eq. (20.6) we can obtain the value of k from the half-life:

$$k = \frac{0.693}{t_{1/2}} = \frac{0.693}{5700 \text{ yr}} = 1.22 \times 10^{-4} \text{ yr}^{-1}$$

We assume that the rate was 15 dis/min-g when the tomb was built. Then

$$m \propto 8 \text{ dis/min-g}$$
$$m_0 \propto 15 \text{ dis/min-g}$$

and

$$k = 1.22 \times 10^{-4} \text{ yr}^{-1}$$

Substitution into Eq. (20.5) gives

$$\ln \tfrac{15}{8} = (1.22 \times 10^{-4} \text{ yr}^{-1})t$$

or

$$2.30 \log \tfrac{15}{8} = (1.22 \times 10^{-4} \text{ yr}^{-1})t$$

or

$$t = 5150 \text{ yr}$$

The time elapsed since the death of the cypress beam is about 5150 yr which means the tomb was built around 3170 B.C. The pharaoh died not before 3170 B.C.

20.5 RADIOACTIVE-DECAY SERIES

When compounds of uranium and thorium are carefully purified, their radioactivity is greatly diminished; at the same time the radioactivity of the separated impurities is quite high. On standing, the purified uranium compounds slowly regain their original activity while the impurities lose theirs (see Fig. 20.5). These observations led to the conclusion that the decay of uranium and thorium must lead to the production of other radioactive elements which decay relatively rapidly to stable isotopes.

A careful analysis of uranium minerals showed that several radioactive elements are admixed with the uranium. They are produced in a series of steps, starting with the disintegration of $^{238}_{92}U$ nuclei and ending with the production of a stable, nonradioactive isotope of lead, $^{206}_{82}Pb$.

In the $^{238}_{92}U$ decay series the first step involves the loss of an α particle. This means that the uranium isotope loses two protons and two neutrons from the nucleus:

$$^{238}_{92}U \xrightarrow[4.9 \times 10^9 \text{ yr}]{t_{1/2} =} {}^{234}_{90}X + {}^{4}_{2}He$$

The new element, X, has a *mass number* of 234 and an atomic number of 90. The mass number is equal to the number of protons and neutrons in the nucleus; it is approximately equal to the atomic weight of the isotope. The new element, X, is identified by its atomic number and therefore must be thorium; $^{234}_{90}X = {}^{234}_{90}Th$. Thorium 234 is radioactive and decays fairly rapidly by β-ray emission, which is always accompanied by an increase in atomic number but no change in mass number. β particles are formed when a neutron in the nucleus is transformed into a proton and an electron; the electron is ejected as a β particle. The equation for the β decay of ^{234}Th is

FIGURE 20.5

The time dependence of the activity of purified uranium and separated impurities.

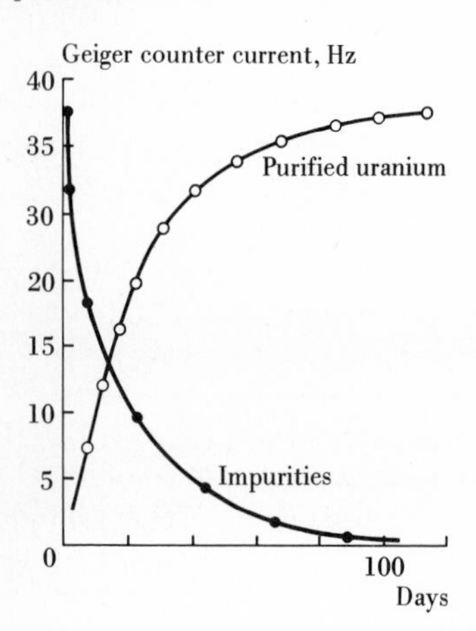

$$^{234}_{90}Th \xrightarrow[24.5 \text{ d}]{t_{1/2} =} {}^{234}_{91}X + {}^{0}_{-1}e$$

Since the element produced has an atomic number of 91, it must be protactinium, $^{234}_{91}Pa$; it is also radioactive, and the decay chain continues. In a radioactive decay chain the loss of an α particle results in a decrease of 2 units in atomic number and 4 units in mass number; the loss of a β particle results in an increase of 1 unit in atomic number and no change in mass number.

The radioactive-decay series beginning with ^{238}U is shown in Table 20.1. Each step involves a nuclear transformation by ejection of an α or β particle. The process continues until the stable isotope $^{206}_{82}Pb$ (lead 206) forms. The isotopes appearing in the series are listed in the first column in the order they are generated. The second column indicates the type of emission which accompanies the disintegration of the nucleus. The third and fourth columns show the change in the nuclear composition, and the fifth column gives the half-life of the isotope. As an example of the interpretation of Table 20.1 the data for ^{214}Bi indicate that this isotope decays by β emission and forms ^{214}Po. There is no change in the mass number, but since a neutron is converted to a proton, there is an increase of atomic number by 1 unit. The half-life for the decay of a ^{214}Bi sample is 19.7 min.

Table 20.1 is simplified because it omits alternate decay paths. For instance, a small fraction of the ^{218}Po decays by β emission, forming the astatine isotope $^{218}_{85}At$. Astatine 218 decomposes by α emission, and ^{214}Bi appears, which can disintegrate in the mode in-

TABLE 20.1: ^{238}U RADIOACTIVE-DECAY SERIES

Isotope	*Decay mode*	*Change in mass number*	*Change in atomic number*	*Half-life*
$^{238}_{92}U$	α	−4	−2	4.9×10^9 yr
$^{234}_{90}Th$	β	0	+1	24.5 d
$^{234}_{91}Pa$	β	0	+1	1.14 min
$^{234}_{92}U$	α	−4	−2	2.3×10^5 yr
$^{230}_{90}Th$	α	−4	−2	8.3×10^4 yr
$^{226}_{88}Ra$	α	−4	−2	1,590 yr
$^{222}_{86}Rn$	α	−4	−2	3.83 d
$^{218}_{84}Po$	α	−4	−2	3.05 min
$^{214}_{82}Pb$	β	0	+1	26.8 min
$^{214}_{83}Bi$	β	0	+1	19.7 min
$^{214}_{84}Po$	α	−4	−2	1.5×10^{-4} s
$^{210}_{82}Pb$	β	0	+1	22 yr
$^{210}_{83}Bi$	β	0	+1	5 d
$^{210}_{84}Po$	α	−4	−2	140 d
$^{206}_{82}Pb$	Stable			

dicated by Table 20.1. There are other alternate paths in the ^{238}U decay series which are not indicated in the table.

The ^{238}U decay series is one basis for estimating the age of the earth. The method can best be demonstrated by example.

EXAMPLE 20.4 If a mineral sample assumed to be of the same age as the earth contains 1.00 g of ^{238}U per 0.838 g of ^{206}Pb, what is the age of the earth?

Solution: Calculate the weight of ^{238}U necessary to produce 0.838 g of ^{206}Pb. It must be realized that, starting with 1 mol of ^{238}U, 1 mol of ^{206}Pb is eventually produced, meaning that 238 g of ^{238}U is required to produce 206 g of ^{206}Pb. This fact is used to calculate the weight of ^{238}U necessary to produce 0.838 g of ^{206}Pb:

$$0.838 \text{ g } ^{206}\text{Pb} \times \frac{238 \text{ g } ^{238}\text{U}}{206 \text{ g } ^{206}\text{Pb}} = 0.968 \text{ g } ^{238}\text{U}$$

Therefore, 0.968 g ^{238}U is required to produce the 0.838 g ^{206}Pb. Calculate the initial mass m_0 of ^{238}U in the sample.

$$m_0 = (\text{mass now present}) + (\text{mass which decayed})$$
$$m_0 = m + 0.968 \text{ g} = 1.00 + 0.968 = 1.97 \text{ g } ^{238}\text{U}$$

Calculate k for ^{238}U:

$$k = \frac{0.693}{t_{1/2}} = \frac{0.693}{4.5 \times 10^9 \text{ yr}} = 0.154 \times 10^{-9} \text{ yr}^{-1}$$

Calculate the age of the earth using Eq. (20.5):

$$\ln \frac{m_0}{m} = kt \qquad \text{or} \qquad t = \frac{2.303}{k} \log \frac{m_0}{m}$$

$$t = \frac{2.303}{0.154 \times 10^{-9}\ \text{yr}^{-1}} \log \frac{1.97}{1.00}$$

$$= 4.38 \times 10^9\ \text{yr} = 4.4\ \text{billion yr}$$

20.6 TRANSMUTATION OF THE ELEMENTS

The alchemist's search for the philosopher's stone ended in the early twentieth century, when chemists and physicists learned the secret of transmutation. Rutherford observed that the bombardment of nitrogen gas with α particles produces ^{17}O:

$$^4_2\text{He} + ^{14}_7\text{N} \rightarrow ^1_1\text{H} + ^{17}_8\text{O} \tag{20.7}$$

This was the first induced transmutation ever reported. An extensive study of the effect of α bombardment showed that transmutation of all the light elements from boron to potassium (except carbon and oxygen) can be effected. This type of transmutation is indicated in shorthand as (α,p), which means that a target, in this case nitrogen nuclei, is subjected to α radiation and protons are produced. The transmutation of ^{14}N given in Eq. (20.7) can be indicated as

$$^{14}_7\text{N}(\alpha,p)^{17}_8\text{O} \tag{20.8}$$

In these experiments it was often observed that the ejected protons have a higher kinetic energy than the α particles used as projectiles.

In 1934 Irène Joliot-Curie (daughter of Pierre and Marie Curie) and her husband, Frédéric, observed the first induced transmutation leading to a radioactive element. They found that ^{27}Al is transformed by α radiation into ^{30}Si by an (α,p) reaction:

$$^{27}_{13}\text{Al} + ^4_2\text{He} \rightarrow ^{30}_{14}\text{Si} + ^1_1\text{H}$$

and into ^{30}P by an (α,n) reaction:

$$^{27}_{13}\text{Al} + ^4_2\text{He} \rightarrow ^{30}_{15}\text{P}^* + ^1_0n$$

The Joliots were surprised to find that ^{30}P is radioactive and decays by *positron* emission:

$$^{30}_{15}\text{P}^* \rightarrow ^{30}_{14}\text{Si} + ^{\ 0}_{+1}e$$

A positron is a particle with the mass of an electron; its charge is equal but opposite. This is the first reported instance of induced radioactivity, but hundreds more were to follow.

Because the nucleus of a heavy element has a high positive charge, it exerts a strong electrostatic force and α particles are prevented from making a close approach. It became obvious that particles with greater kinetic energies than α particles from natural sources would be required to induce the transmutation of the heavier elements. In the early 1930s a proton accelerator developed in Rutherford's laboratory was capable of producing hydrogen ions with kinetic energies up to 3 million electron volts (MeV).† The α particles produced by natural radioactive decay have kinetic energies up to 2000 eV. Accelerated protons with their high kinetic energies and lower positive charges are much more effective nuclear projectiles.

One of the first experiments performed with the accelerated protons was the bombardment of ^{7}Li, which yielded two α particles for each ^{7}Li atom bombarded:

$$^{1}_{1}H + ^{7}_{3}Li \rightarrow 2\ ^{4}_{2}He$$

The helium ions produced had a much greater energy than the proton projectiles, an enhancement exactly accounted for by the decrease in mass which accompanies the process:

Reaction:	$^{1}_{1}H$	+	$^{7}_{3}Li$	$\rightarrow$	$2\ ^{4}_{2}He$
Exact mass, g:	1.00788		7.015956		2(4.00260)

$$\Delta m = \text{change in mass} = 2(4.00260) - (1.00788 + 7.015956)$$
$$= -0.01864\ \text{g}$$

By Einstein's formula, $\Delta E = c^2\ \Delta m$, the energy equivalent of 0.01864 g is 2.8×10^{-5} erg per atom of lithium bombarded or 1.4×10^{-5} erg per α particle formed. The experimental value of the kinetic energy of the α particle was 1.38×10^{-5} erg.

Other workers developed different kinds of particle accelerators, the names of which are familiar to anyone who reads the newspapers. They include the *cyclotron*, the *synchrotron*, the *betatron*, the *Van de Graaff generator*, and *linear* accelerators.

Figure 20.6 demonstrates the principle of the linear accelerator. A positive ion or (electron) is admitted to a long, evacuated chamber containing a series of tubular electrodes. Their polarity is rapidly alternated by a high-frequency oscillator. The ion enters the first tube and drifts through it while the tube is negative. Just as it reaches the end, the polarity of the electrodes is reversed; the first tube becomes positive, and the second becomes negative. The ion is thereby accelerated toward the second tube while it is in the gap. This

† This corresponds to a proton velocity of 2.4×10^{9} cm/s, which is 8 percent of the speed of light. See Appendix Table B-3 for definition of electron volt.

FIGURE 20.6

The principle of the linear accelerator.

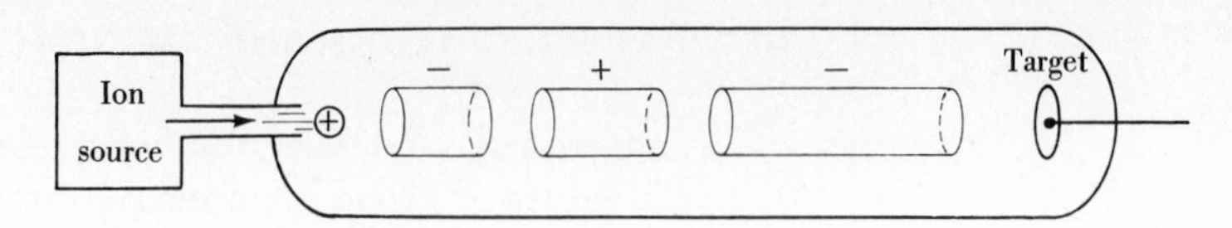

(*a*) *Ion enters accelerator*

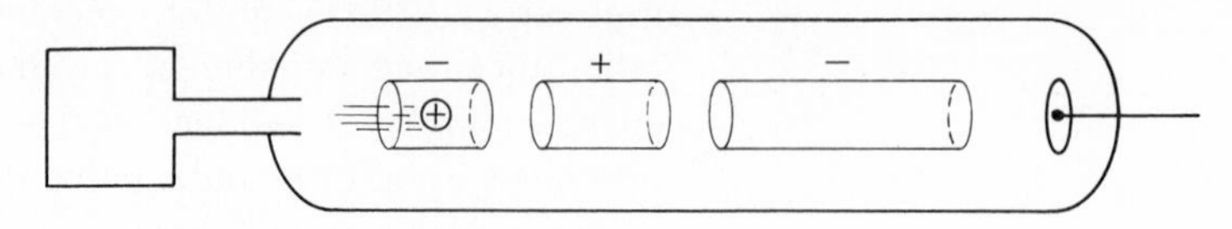

(*b*) *Ion is accelerated by first tubular electrode*

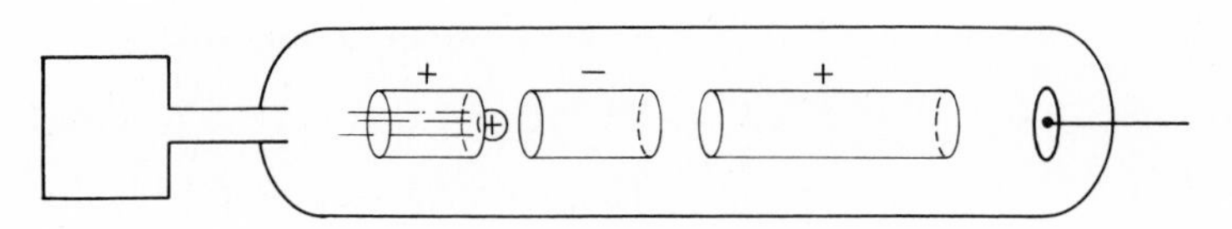

(*c*) *Electrode polarity reverses as ion leaves first tube and is accelerated for second time*

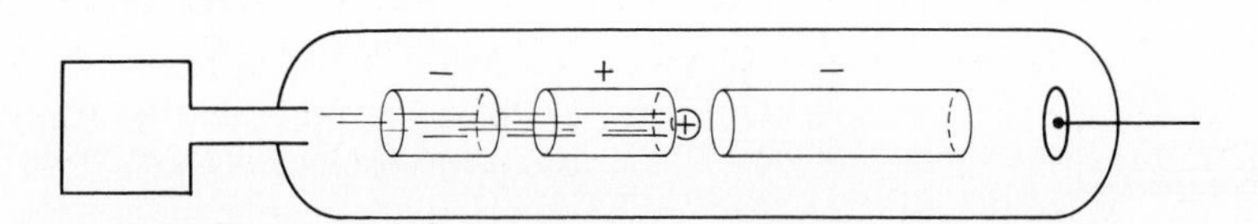

(*d*) *Electrode polarity reverses again and ion is accelerated for third time*

process is repeated each time the ion reaches the end of a tubular electrode. At each gap the ion is accelerated so that it travels at higher speeds in consecutive tubes. Note that the tubular electrodes are progressively longer, allowing the ion to spend the same time in each tube and reach the end just as the polarities are being reversed.

The final velocity of the particle depends on the number of times it is accelerated, i.e., the number of gaps. This number can be increased by increasing the frequency of the reversal of electrode polarity and by lengthening the chamber. Both these approaches have been used. The linear accelerator designed for Stanford University in California is 2 mi long and capable of producing electrons with energies of 100 billion electron volts (BeV).

In cyclotrons charged particles are accelerated by alternating electrode polarities, as in linear accelerators, but in addition a strong magnetic field is imposed on the particles, causing them to travel in circular paths (see Fig. 20.7). A positive ion is released at

FIGURE 20.7

Cyclotron.

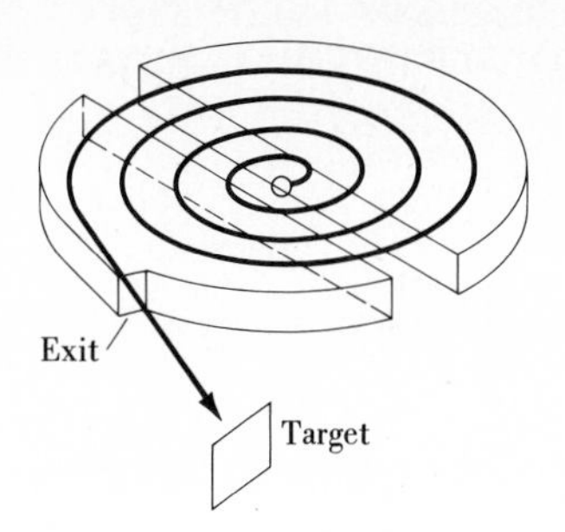

(*a*) *The path of an ion in a cyclotron*

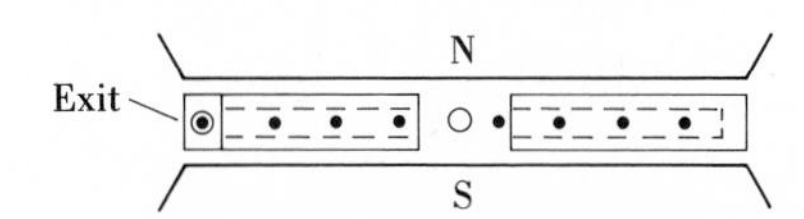

(*b*) *Side view showing magnetic pole pieces*

the center of two hollow semicircular electrodes, called *dees* because of their shape. The ion is accelerated toward the negative dee. At the same time it undergoes circular motion parallel to the plane of the dee-shaped electrodes. As it approaches the gap between the electrodes, the polarity reverses and the ion is accelerated across the gap into the other dee. This process is repeated, the velocity increasing each time the ion crosses the gap. As the velocity increases, the particle moves in progressively larger circular paths. When it is close enough to the wall, it is deflected into an exit slit, onto a target.

In a linear accelerator the time required for an ion to move through any tubular electrode is constant because the tube lengths increase in direct proportion to the particle velocity. In a cyclotron the time required for an ion to move through the semicircular path of a dee is constant because the circumference of the path increases in direct proportion to the particle velocity. We see that the linear accelerator and the cyclotron are very similar in principle. The cyclotron requires much less space because the particle path is a spiral.

One limit on the velocity a particle can reach in a cyclotron is due to a relativistic increase in mass. According to Einstein's theory of relativity, the mass of a particle increases as its velocity increases:

$$m = m_0 \frac{1}{(1 - v^2/c^2)^{1/2}} \tag{20.9}$$

As can be seen from Eq. (20.9), when the velocity v is near the velocity of light c, the mass of a particle m is appreciably greater than its rest mass m_0. Because of this relativistic change in mass, the time

TABLE 20.2: SYNTHESIS OF THE TRANSURANIUM ELEMENTS

Element synthesized	Reaction at no	Target	Projectile	Products	Notation
Neptunium (planet Neptune)	93	$^{238}_{92}U$	$^{2}_{1}H$	$^{238}_{93}Np + 2\ ^{1}_{0}n$	$^{238}_{92}U(d,2n)^{238}_{93}Np$
Plutonium (planet Pluto)	94	$^{238}_{92}U$	$^{4}_{2}He$	$^{240}_{94}Pu + 2\ ^{1}_{0}n$	$^{238}_{92}U(\alpha,2n)^{240}_{94}Pu$
Americium (the Americas)	95	$^{239}_{94}Pu$	$^{4}_{2}He$	$^{241}_{95}Am + ^{1}_{0}n$	$^{239}_{94}Pu(\alpha,n)^{241}_{95}Am$
Curium (P. and M. Curie)	96	$^{239}_{94}Pu$	$^{4}_{2}He$	$^{240}_{96}Cm + 3\ ^{1}_{0}n$	$^{239}_{94}Pu(\alpha,3n)^{240}_{96}Cm$
Berkelium (University of California, Berkeley)	97	$^{244}_{96}Cm$	$^{4}_{2}He$	$^{245}_{97}Bk + ^{1}_{1}H + 2\ ^{1}_{0}n$	$^{244}_{96}Cm(\alpha,p,2n)^{245}_{97}Bk$
Californium (California)	98	$^{238}_{92}U$	$^{12}_{6}C$	$^{245}_{98}Cf + 5\ ^{1}_{0}n$	$^{238}_{92}U(^{12}C,5n)^{245}_{98}Cf$
Einsteinium (A. Einstein)	99	$^{238}_{92}U$	$^{14}_{7}N$	$^{247}_{99}Es + 5\ ^{1}_{0}n$	$^{238}_{92}U(^{14}N,5n)^{247}_{99}Es$
Fermium (E. Fermi)	100	$^{238}_{92}U$	$^{16}_{8}O$	$^{250}_{100}Fm + 4\ ^{1}_{0}n$	$^{238}_{92}U(^{16}O,4n)^{250}_{100}Fm$
Mendelevium (D. Mendeleev)	101	$^{253}_{99}Es$	$^{4}_{2}He$	$^{256}_{101}Md + ^{1}_{0}n$	$^{253}_{99}Es(\alpha,n)^{256}_{101}Md$
Nobelium (A. Nobel)	102	$^{246}_{96}Cm$	$^{13}_{6}C$	$^{251}_{102}No + 8\ ^{1}_{0}n$	$^{246}_{96}Cm(^{13}C,8n)^{251}_{102}No$
Lawrencium (E. O. Lawrence)	103	$^{252}_{98}Cf$	$^{10}_{5}B$	$^{257}_{103}Lr + 5\ ^{1}_{0}n$	$^{252}_{98}Cf(^{10}B,5n)^{257}_{103}Lr$
Khurchatovium	104	$^{242}_{94}Pu$	$^{22}_{10}Ne$	$^{260}_{104}eHf + 4\ ^{1}_{0}n$	$^{242}_{94}Pu(^{22}Ne,4n)^{260}_{104}eHf$

required for an ion to complete a semicircular path changes, destroying the synchronization of the cyclotron. The ions do not arrive at the gap just as the dee polarity is reversed; instead of being accelerated to higher velocities, the ions are slowed down. This problem is overcome by the *synchrocyclotron*, in which the frequency of polarity reversal is decreased gradually to match the relativistic increase in particle mass, and by the *synchrotron*, in which the strength of the external magnets is increased steadily to match the increase in mass.

With the development of high-energy accelerators it became possible for nuclear scientists to experiment with heavier projectiles along with the protons and helium ions. Ions of boron, carbon, nitrogen, and neon have been used effectively in the synthesis of isotopes and elements which do not occur in nature. Approximately 1400 isotopes have been identified, of which only 280 occur naturally.

The other 1100 have been produced by artificial transmutations of various kinds.

No elements with an atomic number greater than 92 (uranium) are found in nature at a measurable concentration because their half-lives are so short. Various isotopes of the elements 93 to 104, the transuranium elements, have been synthesized with the use of accelerators. The synthesis reactions are given in Table 20.2. Elements 90 (thorium) through 103 (lawrencium) constitute the second series of rare earths and are called the *actinides*. Khurchatovium, element 104, is a member of family IVB, along with the stable elements titanium, zirconium, and hafnium.

20.8 TRANSMUTATION BY NEUTRON CAPTURE

It is not possible to increase the kinetic energy of neutrons in an accelerator because neutrons are uncharged particles. Nevertheless, neutrons are highly effective projectiles for nuclear transmutations. The fact that these particles are uncharged allows them to approach a nucleus without experiencing electrostatic repulsion, and they do not need the high energy required for ion-induced transmutation. Slow, or *thermal*, neutrons, generated by passing neutrons through heavy water or graphite, can probably be absorbed by all isotopes. Neutrons have been extremely useful in the preparation of radioactive isotopes of the lighter elements. Among the more important isotopes prepared by neutron capture are carbon 14, ${}^{14}_{6}C$; tritium, ${}^{3}_{1}H$; and cobalt 60, ${}^{60}_{27}Co$. These three isotopes are radioactive, which is what makes them useful. The reactions for their syntheses are

$${}^{14}_{7}N + {}^{1}_{0}n \rightarrow {}^{14}_{6}C^* + {}^{1}_{1}H$$
$${}^{6}_{3}Li + {}^{1}_{0}n \rightarrow {}^{3}_{1}H^* + {}^{4}_{2}He$$
$${}^{59}_{27}Co + {}^{1}_{0}n \rightarrow {}^{60}_{27}Co^*$$

Many radioactive elements have been synthesized in nuclear reactors which, as we shall see, are rich sources of neutrons.

20.9 NUCLEAR FISSION

Before 1930 the known natural and artificial transmutations involved a relatively small change in mass number and atomic number. The known neutron-induced transmutations gave an element with an atomic number 1 or 2 units higher than that of the target. Enrico Fermi (1901–1954) and his associates reported that neutron bombardment of uranium resulted in at least three different isotopes which decay by β emission. To explain these observations they proposed that a neutron is captured by ^{238}U, producing ^{239}U, which then decays, by ejection of β particles, to element 93 and then to element 94:

$$^{1}_{0}n + {}^{238}_{92}U \rightarrow {}^{239}_{92}U \rightarrow {}^{239}_{93}X + {}^{0}_{-1}e$$

$$\hookrightarrow {}^{239}_{94}X + {}^{0}_{-1}e$$

But a chemical analysis of the products formed by the neutron bombardment failed to give any evidence to support the synthesis of these new elements. Nevertheless, almost all nuclear scientists believed that the new elements had been formed in such small amounts that they could not be detected chemically. They were blinded by the incorrect notion that neutron bombardment could cause transmutations involving only small changes in atomic number and mass number. The careful chemical analysis of the products by Otto Hahn and his associates in Germany gave astonishing results. They showed that neutron bombardment of ^{238}U yielded barium, lanthanum, and cerium, elements with atomic masses near 140. They were so surprised that they were reluctant to publish their findings, realizing that their observations meant neutron bombardment could split the atom. Carrying this hypothesis one step further, they reasoned that if uranium splits or undergoes fission yielding an isotope with a mass near 140, another isotope with a mass near 98 should also be produced. A reexamination of the fission products proved the presence of strontium (88), rubidium (85), and krypton (84). The results indicate that an atom of ^{238}U splits into two smaller atoms, one with mass near 140 and the other with a mass in the range 85 to 95. The mass numbers of the two smaller atoms total to less than 238 because two or three neutrons are also emitted during fission.

Continued studies showed that ^{238}U undergoes fission only when bombarded with fast neutrons while ^{235}U, which is present in natural uranium at only 0.7 percent, is fissioned with slow, or *thermal*, neutrons. A typical splitting of a ^{235}U atom involves the capture of a neutron followed by the formation of barium 144, krypton 89, and three neutrons:

$$^{235}_{92}U + {}^{1}_{0}n \rightarrow {}^{236}_{92}U \rightarrow {}^{144}_{56}Ba + {}^{89}_{36}Kr + 3\,{}^{1}_{0}n \qquad (20.10)$$

This is only one of the many ways that ^{235}U fission proceeds. In fact, over 100 isotopes of 40 different elements have been identified as products of neutron-induced fission of ^{235}U.

The fission of uranium results in the liberation of an enormous amount of energy. Most of the energy of the fission reaction is taken up as kinetic energy by the fragments of fission and is quickly dissipated as heat.

EXAMPLE 20.5 (*a*) Using Einstein's equation for the mass equivalent of energy, $\Delta E = c^2\ \Delta m$, calculate the energy produced when 1 mol of ^{235}U fissions according to Eq. (20.10). (*b*) If the burning of coal produces $\sim 3 \times 10^6$ cal/lb, how many tons of coal are required to produce the same amount of energy as the fission of 1 mol of ^{235}U?

Solution: (*a*) The exact masses of ^{144}Ba and ^{89}Kr have not been determined, but they must be near 143.91 and 88.91, respectively. The exact mass of a neutron is 1.008665; we shall use 1.009 for this calculation. The exact mass of ^{235}U is 235.043915; we shall use 235.04. The difference in mass between products and reactants is

$$\begin{aligned}\Delta m &= (m_{^{144}Ba} + m_{^{89}Kr} + 3m_n) - (m_{^{235}U} + m_n)\\ &= m_{^{144}Ba} + m_{^{89}Kr} + 2m_n - m_{^{235}U}\\ &= 143.91 + 88.91 + 2(1.009) - 235.04 = -0.20\end{aligned}$$

When 1 mol of ^{235}U fissions by Eq. (20.10), there is a conversion of about 0.20 g of mass into energy. Einstein's equation gives the energy equivalent:

$$\Delta E = c^2\ \Delta m = (2.9979 \times 10^{10}\ \text{cm/s})^2 \times 0.20\ \text{g} = 1.8 \times 10^{20}\ \text{ergs}$$

Or since 8.134×10^7 ergs = 1.9872 cal,

$$\Delta E = 1.8 \times 10^{20}\ \text{erg} \times \frac{1.9872\ \text{cal}}{8.314 \times 10^7\ \text{erg}} = 4.3 \times 10^{12}\ \text{cal}$$

(*b*) When 1 mol of ^{235}U fissions into ^{144}Ba and ^{89}Kr, 4.3×10^{12} cal of energy is liberated. We can convert the energy liberated by ^{235}U fission into its equivalent in tons of coal since 1 lb of coal $\approx 3 \times 10^6$ cal.

$$\frac{4.3 \times 10^{12}\ \text{cal}}{1\ \text{mol}\ ^{235}\text{U}}\ \frac{1\ \text{lb coal}}{3.0 \times 10^6\ \text{cal}}\ \frac{1\ \text{ton coal}}{2000\ \text{lb coal}} = 717\ \text{tons coal}$$

The fission of about $\frac{1}{2}$ lb of ^{235}U produces the same amount of energy as the burning of 717 tons of coal.

The extremely high energy yield of uranium fission has industrial implications, as almost everyone knows. It is the rapidity with which this energy can be released that makes the fission process the basis for very powerful weapons. Equation (20.10) indicates that *one* neutron splits the ^{235}U nucleus and *three* neutrons are produced. These three neutrons may be captured by neighboring ^{235}U nuclei which undergo fission and produce nine more neutrons. As fission proceeds, an avalanche of neutrons is created which can result in a very rapid release of a tremendous amount of energy, an atomic explosion.

In a very small sample of ^{235}U most of the neutrons generated by fission escape without striking another ^{235}U nucleus. In larger samples the neutrons must travel long distances through the sample before they can escape, and the probability of their capture is high.

In order for fission to result in an explosion a minimum mass, called the *critical mass,* must be present so that an avalanche of neutrons can be produced. The fission of ^{235}U is a *branching chain reaction* because the products cause the reaction to accelerate.

20.10 NUCLEAR REACTORS

The energy of the nucleus has been put to work to generate power for industry and the home. A nuclear reactor uses ^{235}U or plutonium, $^{239}_{94}Pu$. The ^{235}U is separated from natural uranium ores, but ^{239}Pu must be synthesized since it does not occur in nature. The predominant isotope of natural uranium, ^{238}U, can be transmuted into ^{239}Pu by bombardment with fast neutrons:

$$^{238}_{92}U + ^{1}_{0}n \rightarrow ^{239}_{92}U \rightarrow ^{239}_{93}Np + ^{0}_{-1}e$$

$$^{239}_{93}Np \rightarrow ^{239}_{94}Pu + ^{0}_{-1}e$$

Both ^{235}U and ^{239}Pu are radioactive, but they have long half-lives. Since they undergo branched chain fission when bombarded with slow neutrons, they are excellent fuels for nuclear reactors.

A schematic drawing of a typical fission-fuel nuclear reactor is shown in Fig. 20.8. To increase the probability that the neutrons will be captured they must be slowed down after they are ejected from a nucleus. This is accomplished by mixing graphite with the fuel to reduce neutron velocities. Materials used to slow down neutrons are called *moderators.*

The *power* of the reactor, which is the energy it produces per unit time, depends on the number of neutrons captured. There are materials, like cadmium metal, which can conduct neutrons away and prevent them from initiating nuclear fission in the reactor fuel. The *control rods* of a nuclear reactor are made of cadmium. The power of a reactor is controlled by the depth at which the control rods are set into the fuel.

As fission proceeds in a nuclear reactor, intense heat is generated; a heat exchanger like liquid potassium, liquid sodium, or heavy water, D_2O, carries the heat energy to steam turbines where it can be used to generate electric power.

As we deplete the fossil fuels (coal, natural gas, and petroleum), we shall have to rely more and more on alternate energy resources such as fissionable nuclear fuels; but with the growing world population and the concomitant per capita increase in energy consumption, even fissionable fuels will be exhausted in the foreseeable future. The chief fuel, ^{235}U, is present at only a small percentage in natural uranium. The most promising way to extend our nuclear fuel sup-

FIGURE 20.8

Nuclear reactor.

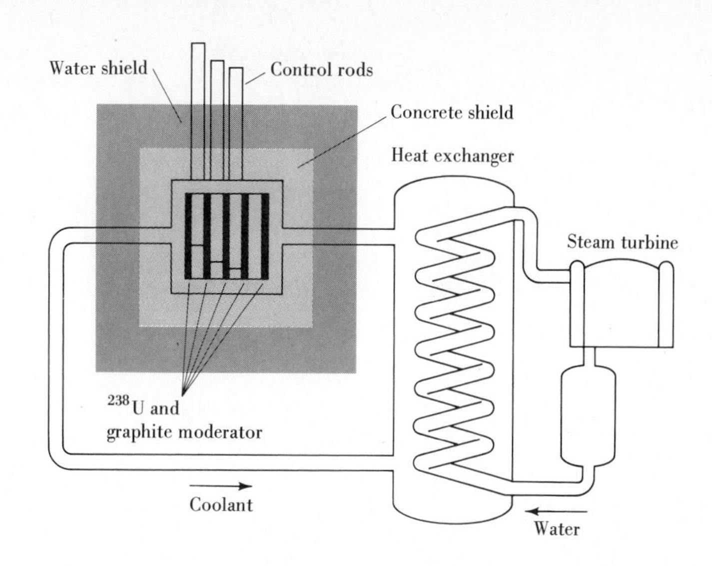

plies seems to be through the development of *fast breeder* reactors. These reactors are called *fast* because they do not employ moderators to slow the neutrons down. The fast neutrons are used to convert ^{238}U into ^{239}Pu; that is, these reactors breed their own fuel. One experimental breeder reactor built by the Atomic Energy Commission contains a sample of ^{235}U about the size of a football at the center. This is surrounded by a blanket of ^{238}U. The fast neutrons generated by ^{235}U fission cause the transmutation of ^{238}U. The object is to achieve a design in which the breeding of ^{239}Pu occurs at about the same rate as the consumption of ^{235}U. Thorium 232 can also be used as a blanket; it is transmuted into fissionable ^{233}U by fast neutrons. By solving the technological problems associated with breeder reactors it will be possible to meet man's energy needs for the next 1000 years. Such reactors are now under construction in Russia and in Western Europe. The United States is involved in an intensive developmental program aimed at designing an economical breeder reactor which will have only minimal deleterious environmental effects.

20.11 NUCLEAR STABILITY AND BINDING ENERGY

The nuclei of the stable, light elements have neutron-proton ratios near unity. As the mass increases, the neutron-proton ratio increases. These facts are shown in Fig. 20.9, a plot of the number of neutrons versus the number of protons for various isotopes. Note that the deviation from a 1:1 ratio becomes appreciable around an atomic number of 20 to 25. The points which lie on the zigzag belt, called the *belt of stability*, represent stable isotopes; the points falling off the belt of stability represent radioactive isotopes.

FIGURE 20.9

Number of neutrons versus number of protons.

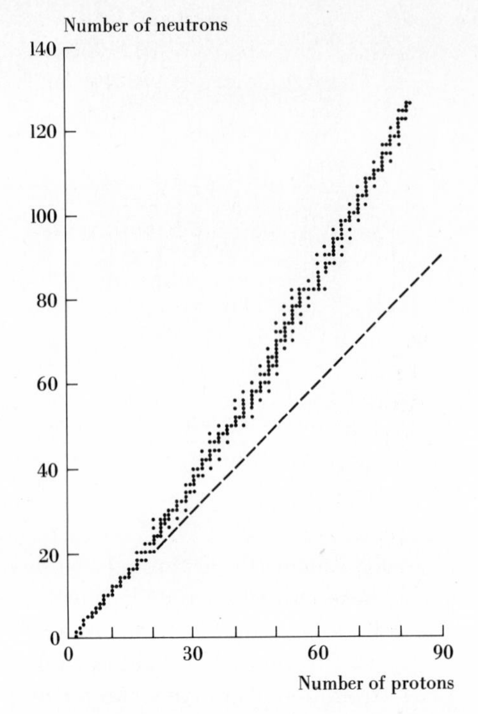

It is believed that neutrons reduce the enormous repulsive forces existing between the protons crowded into a tiny nucleus. Figure 20.9 suggests that more and more neutrons must be added to stabilize the nucleus as the atomic number increases. No nuclei with an atomic number of 84 (polonium) or greater are stable. Apparently, no matter how high the neutron-proton ratio, it is impossible to stabilize nuclei containing 84 or more protons.

The mass of a stable nucleus is less than the sum of the masses of its *nucleons* (protons and neutrons) taken separately. For instance, the mass of a helium nucleus, ${}^4_2He$, is less than the sum of the masses of two protons plus two neutrons. The exact atomic weight of ${}^4_2He$ is 4.002603 g; on the same scale the sum of the weights of two protons and two neutrons is 4.032980 g. There is a difference of 0.030377 g in mass, which is equivalent to 6.5250×10^9 kcal. This means that 6.5250×10^9 kcal would be evolved if two neutrons and two protons were fused into a single helium nucleus. This is called the *nuclear binding energy*. The higher the binding energy the more stable the nucleus.

The binding energy per nucleon is used as a comparative measure of the stability of a particular nucleus. It is calculated by (1) finding the difference between the mass of an isotope and the mass of its nucleons taken independently, (2) converting this difference in mass into an energy equivalent, and (3) dividing the energy equivalent by the total number of nucleons in the nucleus.

EXAMPLE 20.6 Calculate the binding energy per nucleon for the isotope $^{56}_{26}Fe$, which has an exact mass of 55.95286 u.

Solution: There are 26 protons and 30 neutrons in the nucleus of this isotope. The total mass of these nucleons on a molar basis is

$$26m_p + 30m_n = 26(1.007825 \text{ g}) + 30(1.008665 \text{ g}) = 56.46717 \text{ g}$$

The difference between the isotope mass and the total mass of the nucleons is

$$56.46717 \text{ g} - 55.95286 \text{ g} = 0.51431$$

The mass 0.51431 g is the difference in mass between 1 mol of the isotope $^{56}_{26}Fe$ on the one hand and 26 mol of protons and 30 mol of neutrons on the other.

This difference can be converted into its energy equivalent by using Einstein's equation, $\Delta E = c^2 \Delta m$

$$\begin{aligned}\Delta E &= (2.9979 \times 10^{10} \text{ cm/s})^2(0.51431 \text{ g}) \\ &= 4.6223 \times 10^{20} \text{ g-cm}^2/\text{s}^2 = 4.62 \times 10^{20} \text{ ergs}\end{aligned}$$

Since

$$8.314 \times 10^7 \text{ ergs} = 1.987 \text{ cal} = 1.987 \times 10^{-3} \text{ kcal}$$

$$\Delta E = 4.62 \times 10^{20} \text{ ergs/mol} \times \frac{1.987 \times 10^{-3} \text{ kcal}}{8.314 \times 10^7 \text{ ergs}} = 1.10 \times 10^{10} \text{ kcal}$$

ΔE is the binding energy per mole of $^{56}_{26}Fe$.

The binding energy per nucleon is found by dividing the binding energy per mole by the total number of nucleons.

$$\frac{1.10 \times 10^{10} \text{ kcal}}{56 \text{ nucleons}} = 1.97 \times 10^8 \text{ kcal/nucleon}$$

Figure 20.10 is a plot of the binding energy per nucleon versus the mass number of the various isotopes. Note that this plot has a broad maximum near a mass number of 55 units. In terms of nuclear stability this maximum means that the most stable isotopes are those with mass numbers around 55, like iron, cobalt, and nickel. The heaviest isotopes should tend to split up into lighter ones, and the lightest ones should tend to fuse into heavier ones. But just because this tendency exists does not mean that a nuclear reaction will occur spontaneously. Just as a mixture of H_2 and O_2 must be activated before a reaction will occur, the various nuclei must be activated before fission or fusion will occur. The fission of isotopes like ^{235}U and ^{239}Pu is initiated by neutron capture; the binding energy per nucleon of the isotopes which result is greater, and they are more stable.

The atomic bombs detonated over Hiroshima and Nagasaki during World War II are believed to have been fission weapons

FIGURE 20.10

Binding energy per nucleon versus mass number.

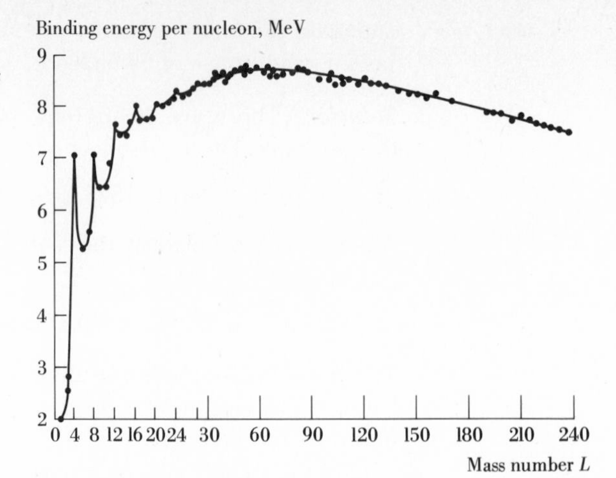

employing ^{235}U. Two near-critical masses were brought together rapidly, probably with the use of a cannon inside the bomb, and a branched-chain fission explosion occurred. Each bomb released an energy equivalent to 20,000 tons of TNT. The temperature on the ground immediately under the point of detonation reached 3000°C. The bombs caused complete destruction within a radius of 0.5 mi and severe damage within a radius of 1.0 mi. Most of the damage was caused by the shock wave and the heat.

Biological damage continues after the explosion of an atom bomb due to the presence of radioactive fission products such as ^{137}Cs, ^{131}I, and ^{90}Sr, which settle to earth and constitute the A-bomb *fallout*. When the dust of a nuclear explosion settles, large amounts of the strontium can be ingested by foraging animals such as dairy cows. Since the isotope is similar to calcium, the ^{90}Sr is concentrated in cows' milk. The ^{90}Sr concentrates (as calcium does) in the bones and teeth of growing babies and children, where it undergoes radioactive decay and produces a powerful radiation which could cause drastic changes in the chromosomes and lead to genetic defects.

20.12 NUCLEAR FUSION

The plot in Fig. 20.10 shows that a conversion of hydrogen into helium is favored since the binding energy per nucleon is much greater for helium. The conversion of hydrogen into helium is a fusion process. It is believed that the energy radiated by the sun and other stars is produced by a conversion of hydrogen into helium in a series of nuclear reactions called the *carbon cycle*. This series occurs in the sun's interior, where the temperature becomes so high ($\sim 2 \times 10^6$ K) that atoms are stripped of electrons.

Carbon cycle:

$$^{12}_{6}C + ^{1}_{1}H \rightarrow ^{13}_{7}N$$
$$^{13}_{7}N \rightarrow ^{13}_{6}C + ^{0}_{+1}e$$
$$^{13}_{6}C + ^{1}_{1}H \rightarrow ^{14}_{7}N$$
$$^{14}_{7}N + ^{1}_{1}H \rightarrow ^{15}_{8}O$$
$$^{15}_{8}O \rightarrow ^{15}_{7}N + ^{0}_{+1}e$$
$$^{15}_{7}N + ^{1}_{1}H \rightarrow ^{12}_{6}C + ^{4}_{2}He$$

The net reaction is the conversion of four hydrogen ions into one helium ion and two positrons:

$$4\ ^{1}_{1}H \rightarrow ^{4}_{2}He + 2\ ^{0}_{+1}e + \text{energy}(\sim 6 \times 10^{8}\ \text{kcal})$$

Because the net reaction involves hydrogen as the only reactant, the process has been called *hydrogen burning*. "Burning" 1 lb of hydrogen would yield the same amount of energy as the conventional burning of 10,000 tons of coal. Years ago, when coal was used for heating, large houses required 10 tons of coal per winter; the heat produced from the nuclear burning of 1 lb of hydrogen would heat one of these houses for 10 centuries.

Nuclear fusion has obvious advantages over fission. The development of a fusion reactor would permanently satisfy man's energy needs since man has an almost inexhaustible supply of fusionable materials at his disposal, the energy yield of fusion is much greater, and there are no radioactive products to contaminate the environment. Unfortunately, extremely high temperatures would be required to initiate fusion. It is estimated that a temperature of 2×10^{6} K would be required to initiate the fusion of hydrogen nuclei. The *uncontrolled* fusion of hydrogen, which is the energy source of the hydrogen bomb, is initiated by a fission explosion of ^{235}U. It has been impossible to construct a reactor in which *controlled* fusion takes place; there are no known materials which can withstand the high temperatures. At 2×10^{6} K matter is in a *plasma* state; atoms are stripped of extranuclear electrons, and only positive ions and electrons exist. There is hope that a hydrogen plasma can be contained by powerful magnetic fields (in a "magnetic bottle") and that controlled nuclear fusion can be initiated.

TABLE 20.3: EXACT REST MASS OF SOME PARTICLES, u

Proton, p	1.00797
Neutron, n	1.008665
α, $^{4}_{2}He$	4.0026
Deuterium, $^{2}_{1}H$	2.01410
Carbon, $^{12}_{6}C$	12.00000
Oxygen, $^{16}_{8}O$	15.99491
Electron, $^{0}_{-1}e$	0.000549

QUESTIONS AND PROBLEMS

20.1 (*a*) Why did Becquerel hypothesize that materials which fluoresce also emit invisible x-rays? (*b*) By what experimental technique did he test his hypothesis? (*c*) Design an experiment using a Geiger-Müller counter in place of a photographic plate to test his hypothesis. (*d*) What modern household items could be used as detectors to test Becquerel's hypothesis?

20.2 The Curies found that the radiation produced by radioactive materials is not affected by chemical combination. What does this mean about the effect of the valence electrons on the energy of the nucleus?

20.3 Why did the Curies conclude that uranium is not the only radioactive material in uranium ore?

20.4 The Curies found that radium has properties similar to barium. Assuming that radium is an alkaline-earth metal, calculate its atomic weight from the following data. When an excess of $AgNO_3$ is added to a solution containing 0.00530 g of radium chloride, a precipitate of AgCl weighing 0.00512 g is formed.

20.5 Describe two experimental methods of distinguishing (*a*) an α ray from a β ray, (*b*) an α ray from a γ ray, and (*c*) a β ray from a γ ray.

20.6 How did Rutherford prove that α rays consist of helium ions?

20.7 (*a*) By what principle does the Geiger-Müller counter detect radiation? (*b*) How does a scintillation counter work? (*c*) Cloud chambers and bubble chambers (not discussed in Chap. 20) are also used to detect high-energy particles and radiation. What principle do they employ?

20.8 What fraction of a radioactive sample will remain after (*a*) one half-life, (*b*) two half-lives, (*c*) four half-lives?

20.9 Sodium 22 is radioactive and has a half-life of 2.60 yr. What fraction of an ^{22}Na sample will remain after (*a*) 2.60 yr, (*b*) 5.20 yr, and (*c*) 10.40 yr?

20.10 At one time a sample of sodium chloride with a trace of ^{22}Na in it gave a reading of 20.0 dis/min on a Geiger-Müller counter. When it was tested again, it gave a reading of 14.3 dis/min. How much time had elapsed between readings? ($t_{1/2} = 2.60$ yr.)

20.11 (*a*) Carbon 14, which is radioactive and decays by β emission, is formed after a nitrogen 14 nucleus captures a neutron. Write the equation for this transformation. (*b*) Write the equation for the β decay of ^{14}C.

20.12 A 0.010-g sample of carbon from a wooden utensil discovered in some diggings in Rome gave 87 dis in 1000 min. Determine the age of the utensil.

20.13 It is implicitly assumed in the solution of Example 20.4 that the isotopes between ^{238}U and ^{206}Pb in the ^{238}U decay series are present in negligible amounts compared to ^{238}U and ^{206}Pb. Why is this justifiable? *Hint:* See half-life data in Table 20.1.

20.14 Write completely balanced equations to represent the following transformations: (*a*) ^{252}Fm emits an α particle, (*b*) ^{115}Cd emits a β particle, (*c*) ^{138}Pr emits a positron, (*d*) ^{214}Bi emits an electron, (*e*) ^{210}Po emits a helium nucleus.

20.15 Complete and balance the following equations

(*a*) $^{238}U + {}^{4}He \rightarrow$ ____ $+ 2\,{}^{1}_{0}n$

(*b*) $^{14}N + {}^{4}He \rightarrow {}^{17}O +$ ____

(*c*) $^{9}Be + {}^{4}He \rightarrow {}^{12}C +$ ____

(*d*) $^{24}Na \rightarrow {}^{24}Mg +$ ____

20.16 How are the linear accelerator and cyclotron (*a*) similar and (*b*) different?

20.17 At what velocity is the mass of a body 25 percent greater than its rest mass?

20.18 Tritium, ^{14}C, and ^{60}Co are useful radioisotopes which decay by β emission. Give equations for (*a*) their syntheses from stable isotopes and (*b*) their disintegration.

20.19 Why did Fermi believe that the neutron bombardment of ^{238}U gives rise to new elements with atomic numbers 93 and 94?

20.20 Give three equations for the possible ways that ^{235}U can fission, using the following isotopes to write the equations: ^{143}Ba, ^{145}Ce, ^{153}Nd, ^{141}Xe, ^{90}Kr, ^{87}Se, ^{79}Ge, and ^{90}Sr.

20.21 Solve Example 20.1 by using Eqs. (20.5) and (20.6).

20.22 One possible mode of neutron induced fission of ^{235}U is given by

$$^{235}_{92}U + ^{1}_{0}n \rightarrow ^{95}_{42}Mo + ^{139}_{57}La + 2n + 7\ ^{0}_{-1}e^{-}$$

If the exact masses of ^{95}Mo, ^{139}La, and ^{235}U are 94.946, 138.955, and 235.101 u, respectively, what is the energy release in kilocalories per mole of ^{235}U?

21 THE ENVIRONMENT

21.1 THE EFFECT OF A GROWING POPULATION

The population of any living species depends on the availability of food and the number of natural predators. As long as these two factors are constant, the death rate will become equal to the birth rate and the population will reach a stable level. The total human population on the earth 10,000 years ago was about 5 million (about equal to the population of greater Philadelphia) and fairly stable. At that time man was a hunter and gatherer of food. He depended entirely on the wild animal and plant populations for his food, and hence, the food supply was fairly constant. Occasionally there may have been an ideal growing season resulting in an abundance of food, but this would have been of little consequence since food-storage methods were extremely primitive. On the other hand, droughts, floods, and other catastrophes resulted in famine, which served as a natural check on population growth.

The discovery of cooking made man's food tastier but, more important, it increased his food supply. Man can derive nutrients from uncooked flesh and certain raw nuts, fruits, and berries. With the development of cooking man was able to derive more nutrients from plant tissue. Cooking breaks down the cellulose of the plant cell wall and releases the nutritious intracellular material which is otherwise available only to herbivores.

Man took his first giant step toward an understanding and manipulation of the environment when he became a farmer approximately 8000 years ago. Plants and animals were domesticated, and the food supply rose sharply. In addition man had learned to cooperate in hunting, and as a result the number of his natural predators diminished significantly.

With an increase in food supply and a decrease in natural predators the human population could not be held in check. It grew along the natural waterways and, like an infection, spread inland where the soil was rich. Everywhere man went other animal species were threatened, and many became extinct. As his understanding of nature grew, his mastery over the environment was enhanced. For his comfort forests and grasslands were sacrificed for buildings and firewood. When his technology became very sophisticated, he invented machines powered by wood and fossil fuels. Technological advances accelerated the farm output, and the population was spurred on, along with the per capita fuel consumption. Large cities sprang up, and the urban air became filled with smoke and poisonous gases. Giant rivers and lakes, reeking with industrial wastes, garbage, and excreta, could no longer support aquatic life. The human species is a more serious threat to the living inhabitants of the earth (including man himself) than all the volcanoes, earthquakes, Ice Age glaciers, floods, and pestilences combined. Fortunately, man is beginning to learn to repair and prevent environmental damage.

TABLE 21.1: WORLD POPULATION GROWTH†

Year, A.D.	*Population, billions*
0	0.25
1650	0.50
1850	1.1
1930	2.0
1975	4.0‡
2000	7.5§
2100	50.0

† H. F. Dorn, *Science*, **125**: 283ff. Copyright 1962 by the American Association for the Advancement of Science.

‡ Good estimate.

§ A guess based on present trends.

Table 21.1 shows how the world's population has grown in the last 2000 years. At the time of Christ there were 250 million human beings scattered over the face of the earth. Today that number lives in Russia alone, and there are almost 4 billion people in all. The earth's human population is doubling every 35 to 40 years; at the present rate of growth the population will be about 7.5 billion in the year 2000 A.D. and 50 billion in the year 2100 A.D. Anyone who has given the slightest consideration to the present plight of our cities understands that such a population would be intolerable and therefore the growth rate must be reduced. This can be done by decreasing the birth rate or increasing the death rate. Which will man choose? If he waits too long he will have no choice.

In poisoning and defacing our natural environment we are wasting our raw materials, fertile soils, and energy resources. Today, in the United States alone, we allow 200 million tons of smoke and poisonous gases to escape into our atmosphere annually. Each year we junk 10 million automobiles and dump 20 million tons of paper, 50 billion cans, 30 billion glass bottles and jars, and 100 million tires on our landscape. Most of our excreta, amounting to 200 million tons per year, along with the poisonous chemical and thermal discharges of our industries, find their way into the nation's waterways. This results in ugliness and pollution, and, furthermore, it wastes our limited raw materials and energy reserves; and as we shall see, our energy reserves have reached a critically low level.

We no longer "have forever" to develop an attack on the poisoning of our environment. Practical solutions have been developed to turn back this trend. We must make ourselves aware of these solutions, convince the general population of the seriousness of the environmental problems, and devise ways to implement our technology to improve the quality of life.

In the following sections we shall investigate the chemical aspects of some of our environmental problems and possible solutions.

21.2 OUR AIR

Table 18.10 shows the gaseous makeup of a healthful sample of air. The air from a city or a heavily industrialized area contains many additional components resulting from the combustion of fuels in power plants, factories, and automobile engines. For the most part, when fossil fuels like natural gas, petroleum, and coal are burned, they produce carbon dioxide and water. The CO_2 concentration of the atmosphere has risen 13 percent in the last 100 years; at the same time an additional 9×10^{11} gal of water has accumulated in our oceans and waterways.

Neither carbon dioxide nor water is considered poisonous, but the continued increase of these substances could cause serious climatological effects. Carbon dioxide in the atmosphere can act like glass in a greenhouse, which transmits sunlight but not infrared radiation. Sunlight passes through the glass of the greenhouse and is absorbed by the plants. It is then partially reemitted as infrared radiation, and in this form it cannot pass through the glass. The trapped infrared radiation causes the temperature of the greenhouse to rise above that of the surroundings. This also explains why a closed automobile heats up in bright sunlight. Carbon dioxide gas also transmits visible light but not infrared radiation. A higher concentration of carbon dioxide in the atmosphere could cause a *greenhouse effect*, resulting in a higher average temperature on the surface of the earth. In turn the giant glaciers and polar ice caps would melt, increasing ocean levels and flooding vast regions of our continents. The higher temperatures would also affect rainfall patterns and general climatological conditions.

Although the greenhouse effect may in the end be the most deleterious result of industrialization, a more imminent danger is due to the side products of fuel combustion. The fossil fuels we use are not pure carbon and hydrocarbon. An appreciable amount of sulfur is present, and on combustion SO_2 and SO_3 are formed, two very irritating substances which can cause serious damage to the mucous membrane of the respiratory tract.

Under ideal conditions, i.e., when there is an excess of O_2, the combustion of fossil fuels yields CO_2 as the only product containing carbon. But under actual conditions the combustion is incomplete:

$$\underset{}{C_2H_6(g)} + \underset{\text{Limited}}{O_2(g)} \rightarrow \underset{\text{Soot}}{C(s)} + CO(g) + H_2O(l) + \underset{\text{Unburned}}{C_2H_6(g)}$$

Particles of carbon, or soot, and carbon monoxide are produced, and an appreciable amount of fuel is unburned. The hot combustion gases carry the soot and fuel into the atmosphere. When coal is used as fuel, there is an appreciable inorganic residue, called *ash*. It too is swept into the atmosphere as a very alkaline dust.

Under ordinary atmospheric conditions gaseous nitrogen is inert and will not combine with oxygen, but at the high temperatures produced by burning fuel, and especially in the internal combustion engines of automobiles, atmospheric nitrogen is converted into a number of oxides, chiefly nitric oxide, NO, which is readily converted into nitrogen dioxide, NO_2, by further reaction with oxygen:

$$N_2(g) + O_2(g) \xrightarrow{\Delta} 2NO(g)$$

$$NO(g) + \tfrac{1}{2}O_2(g) \rightarrow NO_2(g)$$

All the nitrogen oxides are very acidic and poisonous.

Small amounts of any of the gases CO, SO_2, SO_3, and NO_2 or soot particles present no health hazard since atmospheric dispersal is quite efficient. However, when hundreds or thousands of tons are released into a region in a very short time, the natural dispersing capability of the atmosphere is overtaxed and life is endangered. When this situation occurs in valleys where a weather condition called *inversion* can trap the poisonous emissions, a *smog* results.

Smog

Under certain weather conditions the gases and particles produced by the combustion of fuels (*sm*oke) can combine with f*og* and produce an irritating haze, called *smog*, which is extremely hazardous to health. Smog occurs frequently in our major cities, but the first major disaster occurred in the Meuse Valley of Belgium, when a very heavy smog hung over the region for six days in early December 1930. The episode took place in a relatively small basin surrounded by hills. Several industries including lime kilns, zinc-refining plants,

FIGURE 21.1

Industrial smog. (Planned Parenthood–World Population.

steel mills, power plants, glassworks, a coking plant, and a sulfuric acid plant were in operation in the basin, and almost all used high-sulfur soft coal as a source of energy. As the factories and plants were spewing their poisons into the air a temperature *inversion* occurred: the denser, cooler air was trapped in the basin by a warm body of air passing over it. Fog formed at the interface of the two bodies of air, and the sky became dark with soot, fog, and dust particles upon which sulfur and nitrogen oxides had been adsorbed. Thousands of people became sick, reporting chest pains, coughing, shortness of breath, and irritation of the respiratory system and eyes. The majority of the more than 60 deaths attributed to the 6-day smog was among people who had a history of heart or lung ailments.

There have been other severe smog catastrophes, the worst of which occurred in London (1952), when 350 perished. All the serious incidents occurred in low-lying, heavily industrialized areas during a temperature inversion.†

From a chemical point of view there are two important kinds of smog. The one discussed above, *industrial* smog, results from a combination of smoke, poisonous gases, and fog. *Photochemical* smog arises in areas where there is a high density of automobiles, trucks, and buses. It is initiated when a mixture of nitrogen oxides, oxygen, and unburned hydrocarbons is irradiated by sunlight. The brown gas NO_2 absorbs light energy and undergoes a *photochemical* decomposition:

$$NO_2(g) + \text{sunlight} \rightarrow NO(g) + O(g)$$

† Others include Donora, Pa. (1948), 6000 sick and 30 dead; Poza Rica, Mexico (1950), 320 sick and 22 dead.

(*a*) *A clear day in Los Angeles*

(*b*) *Automobile smog trapped by inversion in Los Angeles*

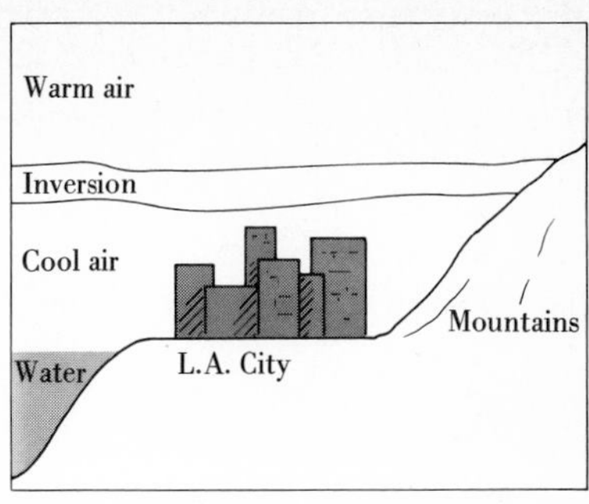

(*c*) *Diagram of inversion trapping air*

FIGURE 21.2

Automobile smog can be trapped by temperature inversions. (*a and b Los Angeles County Air Pollution Control District.*)

The oxygen produced is in atomic form and very reactive. It attacks oxygen molecules, producing ozone:

$$O(g) + O_2(g) \rightarrow O_3(g)$$

It can also react with hydrocarbons, initiating a series of reactions which produce several irritating organic pollutants:

$$O(g) + \text{hydrocarbons} \rightarrow \underset{\text{Aldehyde}}{R{-}C({=}O){-}H} + \underset{\text{Oxyacyl radical}}{R{-}C({=}O){-}O\cdot}$$

The oxyacyl radicals quickly undergo further transformations, and finally peroxyacyl nitrates are formed:

$$R{-}C({=}O){-}O\cdot + O_2 + NO \rightarrow \underset{\text{Peroxyacyl nitrate}}{R{-}C({=}O){-}O{-}O{-}NO_2}$$

Peroxyacyl nitrates, along with aldehydes and ozone, are the major components of the automobile smog occurring in large cities like Los Angeles (Fig. 21.2).

The detailed mechanism for the production of photochemical smog is complex, involving at least 15 interdependent chemical steps, but the essential aspects of the process are easily summarized. Hydrocarbons and nitrogen oxides emitted from gasoline and diesel engines react with oxygen in the presence of sunlight to produce ozone, aldehydes, and peroxyacyl nitrates:

$$\text{Hydrocarbons} + O_2 + NO + NO_2 \xrightarrow{\text{sunlight}} O_3 + R{-}C({=}O){-}H + \text{peroxyacyl nitrate}$$

The assault on our air

In 1971 there were 2.1 million automobiles and 0.7 million trucks in greater Philadelphia. These vehicles consumed 4.5 million gal of gasoline and 85,000 gal of diesel fuel *per day*. In their *daily* emissions were included 1000 tons of unburned hydrocarbons, 300 tons of nitrogen oxides, and 2000 tons of carbon monoxide.

Add to this the barrage from industry—metallic dusts, silicates, fluorides, aromatic compounds, and a whole host of finely divided solids flood the air. Power plants using coal to produce electricity discharged 13 million tons of sulfur into the atmosphere in 1970 in the form of SO_2 and SO_3. This is almost equal to the total amount of sulfur obtained by the Frasch process in that same year.

As the population increases, the amount of discharge into our atmosphere increases. Even with more efficient power plants and internal combustion engines we shall have serious air-pollution problems unless we are able to reduce the absolute amount of fossil fuels we consume.

21.3 OUR WATER

In transforming energy and raw materials into valuable products man has been wasteful and inconsiderate of his surroundings. He has treated his environment as if it were an infinite, self-cleaning dump. Nowhere is this more evident than in the waterways near the big cities. Although the total volume of these vast bodies of water on the earth is an enormous 3.5×10^{20} gal, man has managed to pollute a significant portion of them. The conditions leading to water pollution, like those leading to air pollution, are basically due to the rapid, uncoordinated, and unbridled growth of technology, agriculture, and population.

Lake Erie represents the kind of environmental catastrophe which results when we make no careful studies of water usage and do not make use of available knowledge to prevent serious pollution. Ten million city dwellers and many giant industries draw their water from Lake Erie. Each day 5.5 billion gal is taken from the lake and returned, at appreciably higher temperature, as inadequately treated sewage or solutions containing high concentrations of industrial chemicals. In addition, enormous quantities of phosphates from household detergents have accumulated in the lake. The organic substances in the sewage and industrial wastes serve as nutrients for bacteria. When great quantities of organic substances are released into natural waters, bacteria flourish, rapidly consuming the dissolved oxygen. Further degradation of organic material must occur by *an*aerobic processes, which release incompletely oxidized, foul-smelling gases such as hydrogen sulfide, organic sulfur compounds, and methane. The hydrogen sulfide reacts with many metal ions, producing a dark, insoluble metal sulfide, which floats as a scum on the water.

The condition of Lake Erie is further aggravated by phosphates, which enhance the growth of algae. The algae have clogged the lake and at their death serve as nutrients for bacteria which multiply and deplete the oxygen. The middle 1.6 million acres of Lake Erie are virtually devoid of oxygen except at the surface, where there is direct contact with air. The once-sparkling fresh water now has a murky appearance and the foul odor of organic decay. Where once whitefish, pike, and other valuable fish thrived and supported a large fishing industry, now only the carp survives, . . . the only fish that can exist at the low oxygen level and higher temperatures.

Hard water

The "pure" water which we deem acceptable for most household and industrial uses must be clear, odorless, and free from particles, strong taste, and disease-causing organisms. But our drinking water is not chemically pure. In fact, chemically pure water would probably be unacceptable for drinking because of its flat taste. The crystal-clear water from a mountain stream often has all the desirable features for general human needs. Dissolved in it are atmospheric gases plus traces of NH_3, H_2S, SO_2, and CH_4, many inorganic salts leached from the rocks and soil over which the water has passed, and traces of organic substances. When any of these substances reaches a high concentration, the water becomes unsuitable for human consumption or industrial use and is said to be hard or polluted.

Appreciable concentrations of ions such as Mg^{2+}, Ca^{2+}, Fe^{3+} cause water to be hard. These ions react with soap and form an insoluble solid or scum:

$$\mathrm{M^{2+}}(aq) + \underset{\text{Sodium stearate}}{\mathrm{2Na^+C_{17}H_{35}COO^-}(aq)} \rightarrow \mathrm{2Na^+}(aq) + \underset{\text{Stearate scum}}{\mathrm{M(C_{17}H_{35}COO)_2}(s)}$$

Thus hard water reduces the cleaning power of soap.

Many of the operations in our factories are steam-powered. Water is converted into steam at a central power plant. The steam is sent to various machines in the factory where an energy transfer occurs, and the recondensed water returns to a boiler to be heated again. If hard water containing appreciable amounts of bicarbonate is used in the power plant, the boiler and the pipes become clogged with *boiler scale*, made up of the carbonates of Mg^{2+}, Ca^{2+}, and Fe^{2+}. They precipitate from hard water when it is heated:

$$\mathrm{M^{2+}}(aq) + \mathrm{2HCO_3^-}(aq) \rightarrow \mathrm{MCO_3}(s) + \mathrm{H_2O}(l) + \mathrm{CO_2}(g)$$

FIGURE 21.3

A boiler pipe blocked by scale. (Courtesy of Betz Labs, Inc., Philadelphia, Pa.)

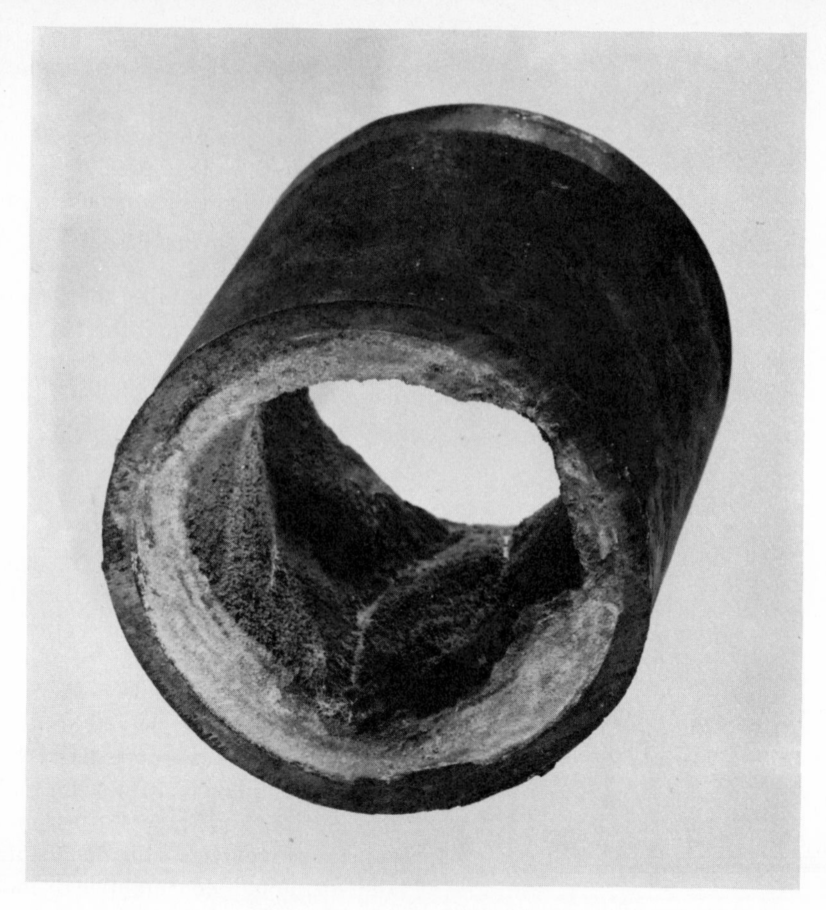

FIGURE 21.4

The negative framework of the zeolite faujasite, showing the large cavities in which cations can be trapped. The open circles represent aluminum silicon atoms; the black circles represent oxygen atoms.

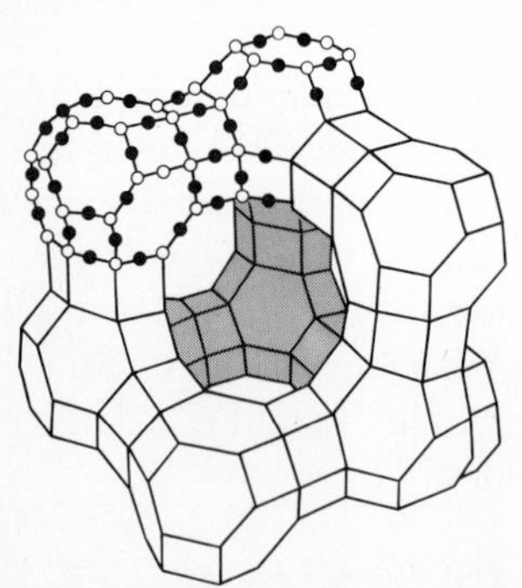

An example of the problems which can develop is shown in Fig. 21.3. This same reaction is responsible for the residue formed in pots and pans used for boiling water in the kitchen.

Hard water can be softened by using complex aluminosilicate minerals called zeolites. The zeolite crystal has an open, negatively charged covalent framework (Fig. 21.4). The negative charge is offset by cations which reside in the cavities of the structure. The sodium ions in the cavities can be replaced by other cations. If hard water is passed through a bed of zeolite, the Ca^{2+}, Mg^{2+}, and Fe^{3+} ions exchange with Na^+ ions, and the effluent is soft water:

$$\text{Zeolite}—\textcircled{Na^+} + Ca^{2+}(aq) \rightarrow \text{zeolite}—\textcircled{Ca^{2+}} + Na^+(aq)$$

In the process the sodium zeolite is converted into a calcium form by *ion exchange*. The presence of sodium ions in the water is tolerable because all common sodium salts are soluble. When the zeolite becomes saturated with divalent ions, it is regenerated by passing a concentrated NaCl solution through the bed.

$$\text{Zeolite}-\!\!\text{(Ca}^{2+}\text{)} + \text{NaCl}(aq) \xrightarrow[\text{exchange}]{\text{ion}} \text{Zeolite}-\!\!\text{(Na}^{+}\text{)} + \text{CaCl}_2(aq)$$

Many synthetic polymers are capable of cation and anion exchange. Usually referred to as *ion-exchange resins,* they are giant organic chains on which anionic or cationic sites are grafted. They can be used to remove essentially all ions from a sample of water; i.e., they can *deionize* water. The ion-exchange resins are used for chemical analysis and chemical synthesis as well as water purification.

21.4 SEWAGE TREATMENT

The city of Philadelphia generates about 500 million gal of sewage per day. The discharge of this much sewage without treatment would hopelessly pollute the Delaware River. Larger downstream communities like Wilmington, Delaware, would find water purification prohibitively expensive.

The water we discharge from our homes and factories eventually reaches lakes, rivers, and wells. What is a sewage dump for one community is often a water resource for another. If we release raw sewage into our environment, we can cause a health hazard for millions of people, endanger fish and wildlife populations, and threaten the whole ecology of a region. Modern technology offers the know-how to convert sewage into water pure enough to drink, but because of the expense no American community can afford to do it.

A modern sewage plant uses the same purification processes that operate on our natural waters: filtration, sedimentation, and bacterial decomposition. Incoming sewage is filtered through a metal screen to remove any large pieces that could damage the pumps. It is then passed into large sedimentation tanks, where the grease and oil collect at the top and the solids settle to the bottom (see Fig. 21.5). The settled solids, called *sludge,* are predominately organic; the sludge is partially decomposed by bacterial action while it is still in the tank. The sedimentation process is referred to as *primary treatment.*

Finer solids which do not settle on their own are made to precipitate by addition of aluminum and iron salts, for example, $Al_2(SO_4)_3$ or $FeCl_3$, which form gelatinous hydrated oxides capable of adsorbing the fine, suspended solids. The effluent liquid from the primary treatment is sprayed over a deep bed of gravel covered with a bacterial slime. As the water trickles through the bed, aerobic

FIGURE 21.5

Sewage treatment.

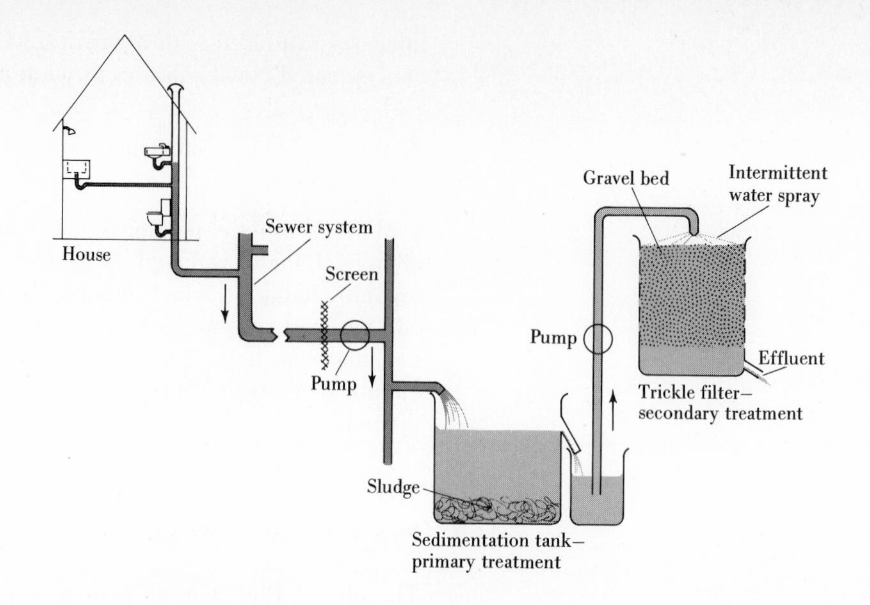

decomposition occurs, transforming the suspended and dissolved organic matter into stable, odorless inorganic substances. The gravel beds are referred to as *trickle filters*, and their action is called *secondary treatment*. The combined action of primary and secondary treatment reduces the suspended solids in sewage by 85 to 90 percent and the organic material by 95 percent. The effluent from the trickle filters is treated with chlorine to kill harmful bacteria before it is released.

The *activated-sludge* method combines primary and secondary treatment into a single step. Air is blown through a sedimentation tank, and then the bacteria in the sludge are capable of decomposing most of the organic substances present. After aeration the remaining solids are allowed to settle, and the effluent is chlorinated and released.

One of the important steps in sewage treatment is the removal of organic substances. As we mentioned above, when high concentrations of organic substances are present in our waters, bacterial action depletes the dissolved oxygen and leads to a Lake Erie condition. A meaningful way of measuring the quality of treated sewage is to determine the amount of oxygen required to assist the biochemical decomposition of the dissolved organic substances. This is called the *biochemical oxygen demand* (BOD). If the water released into streams and lakes has a low BOD, we have a better chance of reviving some of our magnificent rivers. Unfortunately, at present, 25 percent of our municipalities are dumping raw sewage into our streams and 30 percent use only primary treatment. It appears that federal legislation and funding will be required before all communities give adequate treatment to their domestic and industrial sewage.

21.5 THE EFFECT OF AGRICULTURAL CHEMICALS ON THE ENVIRONMENT

Although our giant mechanized farms produce much more food per man-acre than those of any other country, and we are viewed with envy by many foreign nations, the use of insecticides and fertilizers has had a detrimental effect on the environment. Fertilizers high in nitrate and phosphate have seeped into the groundwaters and streams, where they have had a deleterious effect (Sec. 21.4).

Because vast areas are planted with a single crop on many of our large farms, insect pests would multiply out of control were it not for the use of insecticides. With the development of the powerful organic insecticides like DDT in the 1940s it seemed as if many of the agricultural and disease-carrying pests would be permanently exterminated. But within the insect population were individuals immune to the effects of these agents. They survived, and within a few years their vast progeny was threatening the farmlands again. At present, insect pests account for a loss of 30 percent of the farmer's product.

The initial success of DDT made it the most popular insecticide in the world. Unfortunately it is very stable and does not undergo chemical degradation. There is now virtually no spot on the earth where it cannot be detected, at least in trace amounts. It is accumulating in our soil, water, and in the fatty tissue of all living organisms. Apparently, DDT does not present a health hazard to man. Even when the amount present in human fatty tissue is 200 times greater than the national average, there is no noticeable physiological effect; tests on other mammals give similar results, but we do not know what the long-term effects of DDT are. Furthermore, birds, fish, and invertebrates show a high sensitivity to DDT. Because of its accumulation in the environment and the growing insect resistance to it, the use of DDT is controlled by law and severely limited. New organic insecticides which undergo chemical degradation in 2 to 3 days have become more popular. Chemical insecticides will probably never permanently eliminate pests because there will always be some individuals in the vast insect population that are immune. Figure 21.6 shows the structures of some insecticides in wide use today.

The Science Advisory Committee, set up by President Kennedy in 1963, has urged the development of chemicals which will attack specific pests and then decompose into safe residues. It emphasizes

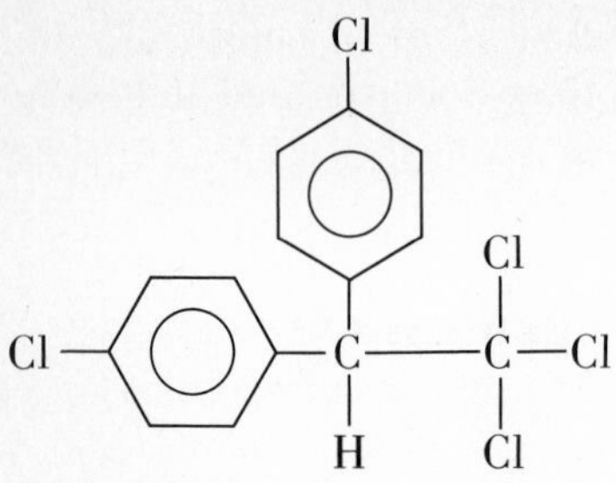

Cl
Cl
Cl—C—C—Cl
H Cl
(a) *Dichlorodiphenyltrichloroethane (DDT)*

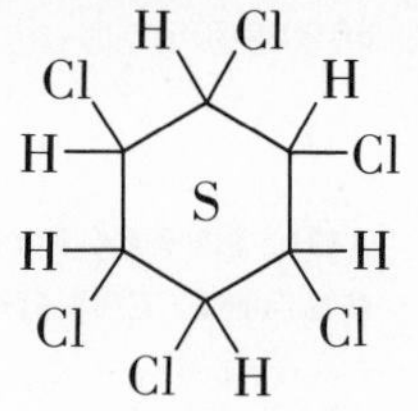

H Cl
Cl H
H— —Cl
S
H— —H
Cl Cl
Cl H
(b) *Lindane (a benzene derivative)*

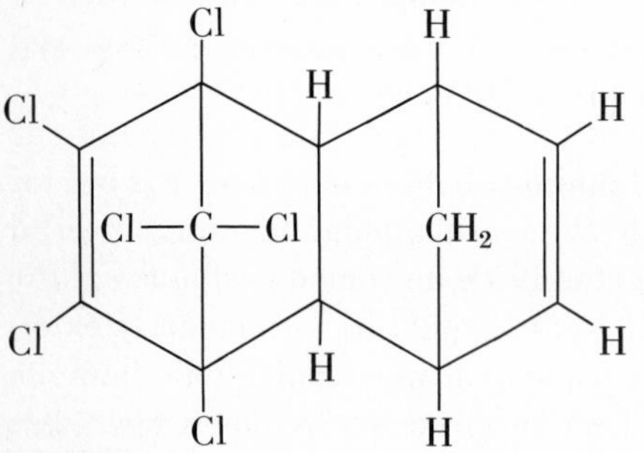

Cl H
Cl H H
Cl—C—Cl CH_2
Cl H H
Cl H
(c) *Aldrin*

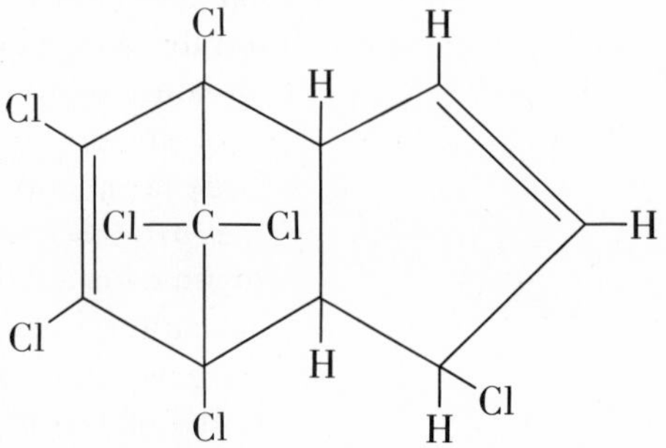

Cl H
Cl H
Cl—C—Cl —H
Cl H
Cl H Cl
(d) *Heptachlor*

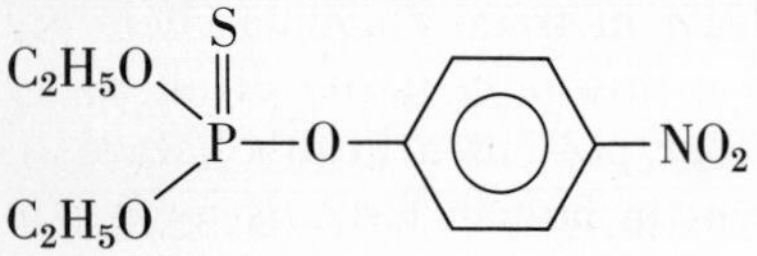

S
C_2H_5O
P—O— —NO_2
C_2H_5O
(e) *Parathion*

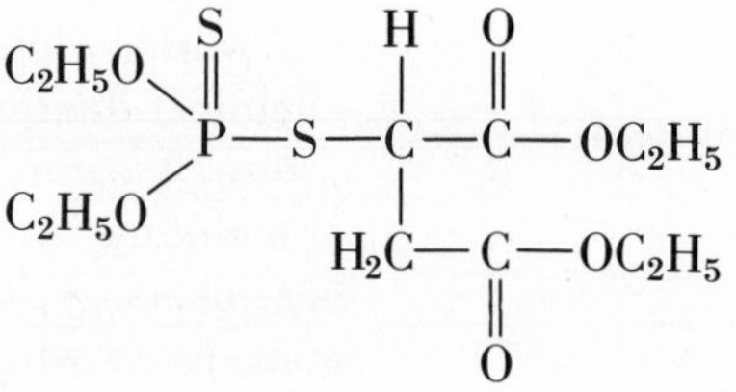

S H O
C_2H_5O
P—S—C—C—OC_2H_5
C_2H_5O
H_2C—C—OC_2H_5
O
(f) *Malathion*

FIGURE 21.6

Molecular structure of some insecticides.

that much more attention should be given to biological methods of fighting pests. In Florida, where screwworm flies had become a menace, sterile males were released to mate with females. Since the female mates only once in a lifetime, her entire semen supply was sterile and the pest population disappeared. A more general biological attack on insects may come from the use of the juvenile hormone found in all growing insect larvae. When the larvae are ready to undergo metamorphosis into adult insects, the hormone is suppressed. If the hormone is administered at this time, normal metamorphosis will not occur and the larvae will not reach the adult reproductive stage. It is unlikely that even a few individuals in the huge insect population could be immune to hormonal treatment because these agents are necessary for normal development. Because of the promise of biological control, international study groups have been formed under the auspices of the United Nations.

21.6 POLLUTION ABATEMENT

An environmental action conference commissioned by the United Nations was convened at Stockholm, Sweden, in the summer of 1972. Represented were 113 nations and 37 international agencies. The conferees attempted to prepare a plan for a concerted effort to improve man's environment. National jealousies and concerns impeded any real progress, but the meeting does represent the first step toward a coordinated attack on spreading pollution. Future meetings are planned, and it may be hoped that beneficial measures will be adopted before our predicament becomes irreversible.

There are many suggestions for restoring some of the beauty to our waters and landscapes. You have probably heard words like "recycling," "biodegradable," and "emission control." These approaches certainly have some beneficial effects, but they are not aimed at the basic problem which is an exploding population coupled with ever increasing amounts of energy consumed per person. Table 21.1 shows how the world's population has grown in the last 2000 years. In the United States the population has doubled in the last 30 years while the energy consumption has quadrupled.

By recycling used materials and producing biodegradable products we can reduce the unsightly trash found throughout our big cities and in nearby streams. But this represents only a superficial attack on our problems. Control of poisonous emissions from power plants, factories, and vehicles gets a little closer to the heart of the matter, but the real solution to the problem depends on using less energy. We can do this only by stabilizing the world's population and manufacturing fewer products per capita. This does not mean that the quality of life or our living standard must be lowered. It means that our products should be of high quality and have high endurance. Stylish, comfortable homes can be constructed to last for centuries, efficient automobiles providing 20 to 30 years' service can be produced today, and household appliances could last for a lifetime. In terms of energy consumption the more permanent products are cheaper. They may require a little more energy input initially, but this is redeemed by their longevity and a healthier environment. If we can reduce our absolute energy requirement, we can reduce the amount of heat, CO_2, and poisonous gases we discharge into our environment and at the same time we shall conserve our limited energy resources.

A serious study of our predicament leads to the hard conclusion that energy conservation is the only real solution to the worldwide environmental pollution problem. A guiding rule to the environmentalist in his professional considerations is: Reduce energy consumption. Obviously, we cannot continue to be a paper-plate, throwaway society, not merely because of the litter which results, but because this manner of living is extravagant in terms of energy.

Of course, if we only manufacture products which will endure, we will require only a very small fraction of our population for the labor force. This emphasizes a dilemma which already exists in the United States. We are capable of producing enough goods to satisfy our needs and our most fantastic desires with only a fraction of the available labor force. This is true in spite of only limited automation, flagrant featherbedding, and built-in obsolescence. To reduce energy consumption we should institute full automation, eliminate featherbedding, and produce only high-endurance goods. The resulting sociological problems are obvious, but they really stem from an outmoded economic system, a system that cannot cope with the blessings of modern technology. In the past and at present, scientific and technological advances have been forced into an established economic framework. Modern economic theories have become virtual religions, insensitive to changes. A good economic system serves its community; it changes, always resting on the technology of the day. In the future it will be required that the economic system be fitted into the technological framework.

QUESTIONS AND PROBLEMS

21.1 Make a list of the three environmental conditions which most annoy you. What would you recommend to alleviate each condition? Do your recommendations eliminate the problem or transfer it to another location or community?

21.2 If the world's population and energy consumption were to stabilize at 1970 levels, would we be able to eliminate serious pollution problems? Explain.

21.3 In 1971 there were 2.1 million automobiles in greater Philadelphia. A 1970 federal standard limits CO emission to 23 g/mi per vehicle. (*a*) In terms of this limit if the average Philadelphian drives 20 mi/d how many tons of CO are dumped into the urban atmosphere daily? (*b*) The combustion of CO yields 26 kcal/mol. How much heat energy is lost by CO emission?

21.4 On the average there are 0.000102 g of particulate matter (particles) per cubic meter of urban air. Philadelphia has an area of 130 mi^2. Calculate the mass of particulate matter within the city's atmosphere up to an altitude of 1.5 mi.

21.5 An appreciable amount of atmospheric SO_2 is generated by the combustion of coal at power plants, steam plants, etc. Suggest a method of extracting SO_2 from stack gases and converting it into H_2SO_4, an extremely important industrial chemical. Use sketches and chemical equations to explain your method.

21.6 Suppose a power plant daily burns 600 tons of coal containing 1.5 percent sulfur and that 90 percent of the sulfur is converted to SO_2. What is the volume of the SO_2 at STP?

21.7 Many environmentalists believe that the introduction of a battery-powered automobile would be more deleterious to the atmosphere than the conventional gasoline-powered automobiles has been. Explain.

21.8 The atmospheric CO_2 concentration has increased 13 percent in the last century. (*a*) What are the possible deleterious effects of this trend? (*b*) What natural processes use up CO_2? (*c*) What experiments, observations, and research would you perform to estimate the amount of CO_2 we could discharge into the atmosphere without changing its present concentration?

21.9 (*a*) What is soot, and why is it harmful? (*b*) How can industry reduce soot emissions? (*c*) Devise an experiment to measure the average weight of soot per unit volume of air. (Use a sketch.)

21.10 Table 18.10 gives the contents of "ordinary" air. How would this differ from city air?

21.11 There are 2.1 million private automobiles in the Philadelphia area. The average annual usage is 10,800 mi. (*a*) Assuming that the average each car travels 12 mi/gal, calculate the tons of gasoline used annually in Philadelphia. (*b*) Assuming that 90 percent of the gasoline (taken as octane, C_8H_{18}, density = 0.7 g/ml) is completely burned, 6 percent is converted to CO and H_2O, and the remainder is exhausted unburned, calculate the weight in tons of CO_2, CO, H_2O, and gasoline emissions.

21.12 (*a*) What are the essential ingredients and conditions for the development of automobile smog? (*b*) What is the difference between industrial smog and automobile smog? (*c*) Would Chicago have a serious industrial smog problem? Why? (*d*) Would you expect London to have a serious automobile smog problem? Why? (*e*) What time of day is automobile smog most likely to develop?

21.13 (*a*) How do organic wastes make our streams and lakes unfit for fish? (*b*) What are the sources and mechanism of phosphate and nitrate water pollution?

21.14 What is BOD and how would you measure it? (Consider combustion with O_2.)

21.15 What are the products of (*a*) aerobic bacterial decomposition of organic matter and (*b*) anaerobic decomposition? (*c*) Why are the anaerobic products undesirable?

21.16 When 100 l of hard water is passed through a zeolite bed, the final solution contains 0.060 g of NaCl per liter. If the initial concentration of NaCl was 0.010 g/l, how many moles of divalent ions (Ca^{2+} and Mg^{2+}) were exchanged?

21.17 How is bicarbonate ion involved in the formation of boiler scale?

21.18 Show how ion-exchange resins can be used to (*a*) remove objectionable cations and anions from water, (*b*) remove all cations and anions, i.e., deionize, water, and (*c*) prepare $NH_4^+SCN^-$ from $NH_4^+Cl^-$.

21.19 Is deionized water equivalent to distilled water? Explain.

21.20 (*a*) What processes are used in sewage treatment? What is the basic effect of (*b*) primary treatment and (*c*) of secondary treatment?

21.21 On the average each person in the world produces 120 gal of sewage and 5 lb of trash per day. (*a*) What is the daily world total for sewage and trash? (*b*) If the present average is stabilized, what would the total be in the year 2000 A.D.? 2100 A.D.? (See Table 21.1.) (*c*) Considering the fact that only 1 percent of the total volume of water on earth is fresh water, do you suppose we shall have any trouble in 21 A.D.? Explain.

21.22 On the average an urban population produces 100 million gal of sewage per 500,000 residents daily. Each million gallons contains 1 ton of suspended solids. How many tons of sewage solids must New York City contend with daily? Annually? How do you suppose these solids are disposed of?

21.23 Philadelphia generates 2.5 million gal of raw sewage per day. A trickle filter can handle 300,000 gal per acre of surface per foot of depth daily. What area would be required to give secondary treatment to all of Philadelphia's sewage if the trickle filters are to be 8 ft deep?

21.24 By examining the structures given in Fig. 21.6 for the most common insecticides, guess which two would undergo chemical degradation most rapidly.

21.25 Why are biological methods of battling insecticide pests preferable to chemical agents? List the dangers which might accompany the use of biological methods.

21.26 Why are recycling and the manufacture of biodegradable products only stopgaps or superficial measures in the war against pollution?

21.27 Assuming that the population of New Jersey is the maximum that could occupy that state under strictly enforced environmental-protection regulations, what is the maximum population the earth could support under the same conditions? Explain all the assumptions you have made in arriving at your estimate.

21.28 Why does the determination of whether an area is overpopulated depend on the per capita energy consumption?

21.29 (*a*) In what broad ways should our industries, businesses, and home activities be changed in order to bring about an improvement in our physical environment? (*b*) What detailed steps can be taken immediately?

Appendix A SCIENTIFIC MEASUREMENTS AND CALCULATIONS

A.1 MEASUREMENTS AND SIGNIFICANT FIGURES

Every physical measurement is accompanied by uncertainty, the degree of which is due primarily to the sensitivity of the measuring device. For instance, the length of a board determined with a meterstick graduated in centimeters is uncertain to the extent of approximately 0.1 cm because the meterstick can be read only to about 0.1 cm.

The board shown in Fig. A-1*a* is 98.3 cm long. Our estimate of the length is probably good to within 0.1 cm. To indicate this we write the length as 98.3 ± 0.1 cm, which means that we are certain of the first two digits and that we are fairly certain of the third digit but the fourth digit is completely uncertain.

If we use a meterstick calibrated in millimeters, we can make a better measurement of the length. Figure A-1*b* shows that the length of the board lies between 983 and 984 mm (or 98.3 cm and 98.4 cm) and is about $\frac{1}{3}$ mm greater than 983 mm. We would indicate the length of the board determined with the second meterstick as

FIGURE A.1

The uncertainty in the length depends on the measuring instrument.

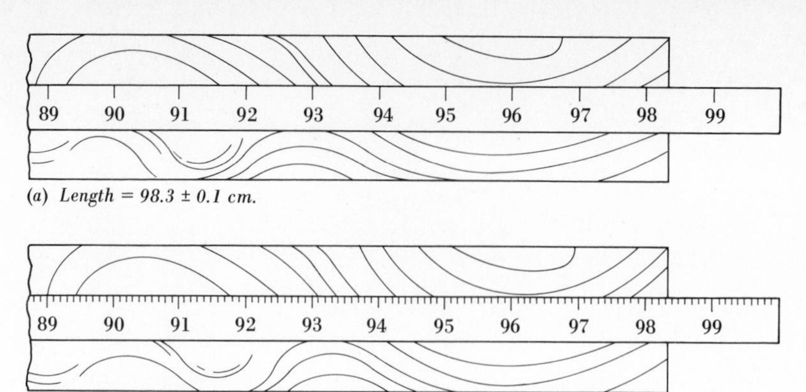

(*a*) *Length* = *98.3 ± 0.1 cm.*

(*b*) *Length* = *983.3 mm or 98.33 cm.*

983.3 ± 0.1 mm or 98.33 ± 0.01 cm. This means that the length of the board lies between 983.2 and 983.4 mm; we have no idea of what the fifth significant figure is.

The Cent-o-gram balances in the laboratory are sensitive to 0.01 g. This means that a mass of 0.01 g will cause the pointer to be deflected by one division on the scale. It is very important to know the sensitivity of a balance. It would be ridiculous to attempt to weigh a 10-mg sample on a Cent-o-gram balance because the uncertainty in the weighing is equal to the weight of the sample. The pointer would hardly deflect. However, for a 1-g sample the Cent-o-gram balance is very good. The uncertainty would be 0.01 g in 1.00 g, or only 1 percent of the mass of the sample.

The sensitivity of a measuring device and the size of the specimen govern the number of significant figures which can be reported for a particular experiment. Using this line of reasoning, we see that the mass of a 1-g sample as measured on a Cent-o-gram balance should be recorded to three significant figures, i.e., as 1.00 ± 0.01 g. The reliability of this observation extends only to 0.01 g.

In any quantitative experiment one must always be concerned with both the precision and the accuracy of measurements. These terms are not synonymous. The *precision* of a measurement is its degree of reproducibility. The *accuracy* of a measurement is its degree of correspondence with a true value.

The following experiment illustrates the difference between precision and accuracy. A sample was carefully weighed five times at 15-min intervals. The following data were obtained: 5.12, 5.11, 5.12, 5.13, and 5.12 g (average 5.12 g). The sample was then compared with primary weight standards at the National Bureau of Standards, and the true weight was found to be 5.24 g. These results indicate that our weighings were precise but inaccurate; the results were reproducible but did not correspond very well to the true weight of the sample.

Addition and subtraction

In adding a series of numbers obtained from physical measurements, the uncertainty in the sum is equal to the uncertainty in the number with the greatest uncertainty.

EXAMPLE A.1 Add 0.923 ± 0.002 cm, 10.1 ± 0.2 cm, and 1.12 ± 0.02 cm.

Solution: Do the arithmetic as if you were working with pure numbers which have no uncertainty:

$$\text{Sum} = 0.923 + 10.1 + 1.12 = 12.143$$

Since the greatest uncertainty in this series of numbers is ± 0.2, the sum, 12.143, must be rounded off to $\frac{1}{10}$ cm:

$$\text{Sum} = 12.1$$

And then we indicate the uncertainty in the sum:

$$\text{Sum} = 12.1 \pm 0.2 \text{ cm}$$

The uncertainty in subtraction is also equal to that in the number with the greater uncertainty.

EXAMPLE A.2 Subtract 0.931 ± 0.003 cm from 10.2 ± 0.1 cm.

Solution:

$$\text{Difference} = 10.2 - 0.931 = 9.269$$

Round off to ± 0.1 cm giving 9.3 cm, and indicate the uncertainty:

$$9.3 \pm 0.1 \text{ cm}$$

Multiplication and division

The simplest rule to follow in multiplication or division is that a product or a quotient has the same uncertainty as the number with the highest uncertainty from which the product or quotient was obtained. This simple rule is not absolutely correct, but it is good enough for most calculations.†

EXAMPLE A.3 Calculate the product of 0.931 ± 0.003 cm and 10.2 ± 0.1 cm.

Solution:

$$\text{Product} = 0.931 \times 10.2 = 9.4962$$

† The exact method of estimating the uncertainty in a product or quotient is described in many handbooks and laboratory textbooks like F. Daniels et al., "Experimental Physical Chemistry," 7th ed., McGraw-Hill, New York, 1970, or F. R. Longo et al., "Experiments in Chemistry," McGraw-Hill, New York, 1968.

Assume the product is uncertain to 0.1 unit and round off:

Product $= 9.5 \pm 0.1 \text{ cm}^2$

It should be noted that we assume that an uncertainty of 0.1 cm in the *length* results in an uncertainty of 0.1 cm^2 in the *area*. This is not absolutely correct. Using the correct methods the uncertainty in the above product is $\pm 0.11 \text{ cm}^2$, which is very close to our assumed uncertainty.

If the uncertainty in both numbers had been ± 0.1, that is, if we were to obtain the product of 10.2 ± 0.1 cm and 0.9 ± 0.1 cm, the uncertainty in the product would be much greater than $\pm 0.1 \text{ cm}^2$. In fact it would be $\pm 1.1 \text{ cm}^2$. So we see that the simple rule for estimating the uncertainty in a product (or quotient) can give poor results.

A.3 QUICK ESTIMATION

In complex multiplication and division problems often all that is required is an estimate rather than an accurate answer. With a little practice complex arithmetic problems can be reduced to simple ones so that answers with one or two significant figures can be obtained quickly, in your head.

EXAMPLE A.4 Estimate the answer to the following arithmetic:

$$\frac{0.0732 \times 10172 \times 0.266}{1724.2 \times 81.2}$$

Solution: Convert all numbers to the power-of-10 form:

$$\frac{\overset{\textcircled{-2}}{7.32} \times \overset{\textcircled{+4}}{1.0172} \times \overset{\textcircled{-1}}{2.66}}{\overset{\textcircled{+3}}{1.7242} \times \overset{\textcircled{+1}}{8.12}} \times 10^{-3}$$

The small encircled whole numbers are the powers of 10 for each number in the original quotient. These result in a -3 as a power of 10 for the quotient. Next convert each number to the nearest whole number:

$$\frac{7.32 \times 1.0172 \times 2.66}{1.7242 \times 8.12} \times 10^{-3} \approx \underbrace{\frac{7 \times 1 \times 3}{2 \times 8} \times 10^{-3}}_{\text{An approximation}}$$

Now, we have a simple quotient which approximates the actual quotient:

$$\frac{7 \times 1 \times 3}{2 \times 8} \times 10^{-3} = \tfrac{21}{16} \times 10^{-3} = 1\tfrac{5}{16} \times 10^{-3} \approx 1.3 \times 10^{-3}$$

The quotient equals $\frac{21}{16}$, which is $1\frac{5}{16}$, or about 1.3. So that the approximate answer is 1.3×10^{-3}. Using a slide rule and three significant figures from the original quotient, we get 1.47×10^{-3}.

The following examples are done by the approximation method:

EXAMPLE A.5

$$\frac{6.02 \times 10^{23} \times 0.115 \times 190.2}{22.4 \times 96{,}500} = \frac{\overset{(+23)}{6.02} \times \overset{(-1)}{1.15} \times \overset{(+2)}{1.902}}{\underset{(+1)}{2.24} \times \underset{(+4)}{9.65}} \times 10^{19}$$

$$\approx \frac{6 \times 1 \times 2}{2 \times 10} \times 10^{19} = \tfrac{12}{20} \times 10^{19} = 0.6 \times 10^{19}$$

The approximate answer is 6×10^{18}. With a slide rule the answer to three significant figures is 6.09×10^{18}.

EXAMPLE A.6

$$\frac{0.509 \times 21{,}100}{0.0038 \times 4350} = \frac{\overset{(-1)}{5.09} \times \overset{(+4)}{2.1100}}{\underset{(-3)}{3.8} \times \underset{(+3)}{4.350}} \times 10^{3}$$

$$\approx \frac{5 \times 2}{4 \times 4} \times 10^{3} = \frac{10}{16} \times 10^{3} = 0.625 \times 10^{3} = 6.25 \times 10^{2}$$

The slide rule gives 6.50×10^{2} to three significant figures.

This method is useful not only in obtaining an approximate solution but in enabling us to decide where the decimal point belongs when we do calculations with a slide rule or other aids which do not fix the decimal point.

A.4 LOGARITHMS

Arithmetic with logarithms is simple if you remember that a logarithm is an exponent of a number to the base 10 or base e. If the logarithm of x is 1.272, the number, x, is $10^{1.272}$. When we write

$\log x = 1.272$

we mean

$x = 10^{1.272}$

If the logarithm of y is 0.673, the number y is $10^{0.672}$.

Since exponents are added when numbers of the same base are multiplied, we have

$$xy = 10^{1.272} \times 10^{0.673} = 10^{1.272+0.673} = 10^{1.945}$$

The log of the product xy is 1.945.

When two numbers expressed in the same base form a quotient, the exponent of the denominator is subtracted from the exponent of the numerator:

$$\frac{x}{y} = \frac{10^{1.272}}{10^{0.673}} = 10^{1.272-0.673} = 10^{0.599}$$

The log of the quotient x/y is 0.599.

If we divide y by x, we obtain a negative logarithm, which means that x/y is less than 1:

$$\frac{y}{x} = \frac{10^{0.673}}{10^{1.272}} = 10^{-0.599}$$

The log of the quotient y/x is -0.599. Only positive logarithms are tabulated in handbooks. In order to find the number corresponding to the logarithm -0.599 we must convert it to a difference containing a whole number:

$$10^{-0.599} = 10^{0.401-1}$$

$$\frac{y}{x} = 10^{-0.599} = 10^{0.401-1} = 10^{0.401} \times 10^{-1}$$

The negative logarithm -0.599 equals $0.401 - 1$, and y/x equals $10^{0.401} \times 10^{-1}$. We need only find the number which corresponds to $10^{0.401}$ since 10^{-1} is one-tenth.

EXAMPLE A.7 Obtain the logarithm of 519.

Solution:

$$\log 519 = \log (5.19 \times 10^2)$$

Since the log of a product equals the sum of the logs of the individual members of the product,

$$\log xyz = \log x + \log y + \log z$$

we have

$$\log (5.19 \times 10^2) = \log 5.19 + \log 10^2$$

But by definition the log of 10^2 is the exponent of 10^2 to the base 10, which is 2. Hence

$$\log 519 = (\log 5.19) + 2$$

The log of 5.19 is found in the tables (inside back cover) to be 0.7152.

$$\log 519 = (\log 5.19) + 2 = 0.7152 + 2 = 2.7152$$

Before taking the log of any number convert it to a power of 10 with a coefficient between 1 and 10. All numbers between 1 and 10 have logs between 0 and 1 (see Table A.1).

EXAMPLE A.8 Find log 0.00371.

Solution:

$$0.00371 = \underset{\text{Coefficient between 1 and 10}}{3.71} \times \underset{\text{Power of 10}}{10^{-3}}$$

$$\log (3.71 \times 10^{-3}) = (\log 3.71) - 3$$

The log of 0.00371 is the log of 3.71 minus 3. The log of 3.71 must lie between 0 and 1 because 3.71 is between 1 and 10. From the log tables, log 3.71 is 0.5694. Hence

$$\log 0.00371 = (\log 3.71) - 3 = 0.5694 - 3 = -2.4306$$

Note that log 0.00371 is negative because $0.00371 < 1$.

EXAMPLE A.9 The pH is defined as $-\log [H_3O^+]$. Calculate $[H_3O^+]$ in a solution which has a pH of 1.56.

Solution:

$$\text{pH} = 1.56$$
$$-\log [H_3O^+] = 1.56$$
$$\log [H_3O^+] = -1.56 = 0.44 - 2$$
$$[H_3O^+] = 10^{0.44} \times 10^{-2}$$

Since the logarithm or exponent of 10 lies between 0 and 1, the corresponding number lies between 1 and 10. From the tables the number corresponding to the logarithm, 0.44, is 2.76. Hence

$$[H_3O^+] = 2.76 \times 10^{-2} = 0.0276$$

TABLE A.1:

Number, N	log *N*
$0.0000001 = 10^{-6}$	−6
$0.00001 = 10^{-5}$	−5
$0.0001 = 10^{-4}$	−4
$0.001 = 10^{-3}$	−3
$0.01 = 10^{-2}$	−2
$0.1 = 10^{-1}$	−1
$1.0 = 10^{0}$	0
$10 = 10^{1}$	1
$100 = 10^{2}$	2
$1{,}000 = 10^{3}$	3
$10{,}000 = 10^{4}$	4
$100{,}000 = 10^{5}$	5
$1{,}000{,}000 = 10^{6}$	6

The limits of the value of a logarithm can be ascertained by examination of Table A.1. A number with a value between 0.000001 (one-millionth) and 0.00001 (one-hundred-thousandth) has a logarithm between −6 and −5. For instance, the log of 0.0000062 is −5.2076.

The logarithm of 215, a number between 100 and 1000, must be between 2 and 3; the logarithm of 0.115, a number between 0.1 and 1 must be between −1 and 0.

The natural logarithms use e (=2.71828) as a base and often use the abbreviation ln instead of log. Since e is smaller than 10, the logarithm of a number to the base e is larger than the logarithm of that same number to the base 10. In fact, the logarithm to the base e is always 2.303 times greater than the logarithm to the base 10:

$$\ln X = 2.303 \log X$$

Table A.2 compares the logarithms of a few numbers in the two bases.

TABLE A.2: COMPARISON OF THE LOGARITHMIC BASES e AND 10

n	$\log n$	$\ln n$	$\frac{\ln n}{\log n}$
2	0.3010	0.6931	2.303
5	0.6990	1.609	2.303
10	1.0	2.303	2.303
25	1.3979	3.219	2.303

Appendix B CONVERSION FACTORS AND PHYSICAL CONSTANTS

TABLE B.1: PREFIXES FOR POWERS OF 10

d	deci	10^{-1}			
c	centi	10^{-2}			
m	milli	10^{-3}	k	kilo	10^{3}
μ	micro	10^{-6}	M	mega	10^{6}
n	nano	10^{-9}	G	giga	10^{9}
p	pico	10^{-12}	T	tera	10^{12}
f	femto	10^{-15}			

TABLE B.2: CONVERSION FACTORS

Length:
- 1 μm (micrometer, formerly micron) = 10^{-6} m
- 1 Å = 10^{-8} cm
- 1 in. = 2.54 cm
- 1 mile = 1.61×10^5 cm = 1.61 km

Mass:
- 1 lb = 453.59 g
- 1 metric ton = 10^6 g
- 1 kg = 1000 g

Energy:
- 1 joule (J) = 10^7 ergs
- 1 cal = 4.184 J = 4.184×10^7 ergs
- 1 l-atm = 24.22 cal

Volume:
- 1 qt = 946.4 cm^3
- 1 in.^{3} = 16.387 cm^3
- 1 l = 1000 ml = 1000 cm^3
- 1 ml = vol of 1 g H_2O at 4°C

TABLE B.3: FUNDAMENTAL CONSTANTS

Symbol	*Name*	*Value*
c	Speed of light‡	2.997925×10^{10} cm/s
h	Planck's constant‡	6.6262×10^{-27} erg-s
k	Boltzmann's constant‡	1.38062×10^{-16} erg/K-molecule
e	Charge on the electron‡	4.80325×10^{-10} esu 1.60219×10^{-19} C
m_e	Rest mass of electron‡	9.1096×10^{-28} g
m_p	Rest mass of proton‡	1.67261×10^{-24} g 1.0072766 u
m_n	Rest mass of neutron‡	1.0086655 u
N_A	Avogadro's number‡	6.022169×10^{23} molecules/mol
R	Gas constant	82.055 cm^3-atm/K-mol 0.082054 l-atm/K-mol 8.3143 J/K-mol 1.9872 defined cal/K-mol
V_0	Standard gas volume‡	2.24136×10^4 cm^3/mol
	Ice point (0°C)	273.15 K
	Triple point of H_2O (base of international temperature scale)	273.16 K
g	Standard gravity	980.655 cm/s^2
$\mathscr{F}$	Faraday constant‡	9.64867×10^4 abs C/equiv
eV	Electron volt‡	1.60210×10^{-12} erg/molecule 2.3060×10^4 cal/mol

‡ Values of the physical constants are those recommended by the Committee on Fundamental Constants of the National Academy of Sciences-National Research Council.

TABLE B.4: EQUILIBRIUM VAPOR PRESSURE OF WATER

Temperature, °C	*Pressure*, Torr
0	4.6
5	6.5
8	8.0
9	8.6
10	9.2
11	9.8
12	10.5
13	11.2
14	11.9
15	12.7
16	13.5
17	14.4
18	15.4
19	16.3
20	17.4
21	18.5
22	19.7
23	20.9
24	22.2
25	23.6
26	25.1
27	26.5
28	28.1
29	29.8
30	31.5
100	760.00

ANSWERS TO SELECTED PROBLEMS

CHAPTER 1

1.4(*a*) Methane 0.333, ethylene 0.167; (*b*) nitrous oxide 0.571, nitric oxide 1.14

1.5(*a*) Since equal volumes contain the same number of molecules: 1 nitrogen molecule +2 hydrogen molecules → 1 hydrazine molecule, (*b*) 4 N atoms and 4 H atoms, (*c*) 7:1, (*d*) yes, (*e*) 14:1

CHAPTER 2

2.1(*a*) $r = 6.37 \times 10^8$ cm; $d = 1.27 \times 10^9$ cm; $c = 4.00 \times 10^9$ cm
$r = 6.37 \times 10^{16}$ Å; $d = 1.27 \times 10^{17}$ Å; $c = 4.00 \times 10^{17}$ Å
(*b*) $A = 5.09 \times 10^{18}\ \text{cm}^2 = 5.09 \times 10^{14}\ \text{m}^2$
(*c*) $V = 1.08 \times 10^{27}\ \text{cm}^3 = 1.08 \times 10^{24}\ \text{l} = 1.08 \times 10^{21}\ \text{m}^3 = 3.81 \times 10^{22}\ \text{ft}^3$
2.2(*a*) 2.04×10^2 kg **2.3** 1.79×10^4 cm/sec **2.4** 21.1°C, 37.0°C
2.7 ^{20}Na: 11*p*, 9*n*, 11*e* **2.8** $\frac{V_A}{V_N} \approx 10^{12}$
2.9(*a*) 4.5×10^{-22} g, (*b*) 1.11×10^{17} atom **2.10**(*a*) 2.04 mol

2.11(*a*) 1.13

2.13(*a*) 1.6×10^{-13} C **2.14**(*b*) 0.730 g atom **2.15**(*a*) 2.01×10^{23} atoms

2.16(*c*) 2 mol **2.17**(*a*) 1.23×10^{23} molecules and 8.75×10^{23} atoms
2.18(*b*) 52.2 C, 13.0 H **2.19** $MgBr_2$ **2.30**(*b*) 34.0 g Cl_2
2.32(*b*) 2.35 g O_2

CHAPTER 3

3.1 $m_e = 8.20 \times 10^{-27}$ g, $m_{nuc} = 3.18 \times 10^{-23}$ g for fluorine
3.3(*a*) 0.862×10^{-19} cal, (*b*) 5.19×10^4 cal
3.5(*a*) 0.25308×10^{-19} cal, (*b*) $\lambda = 18{,}730$ Å, (*c*) no, infrared, (*d*) Paschen
3.15(*a*) 2.645 Å **3.16**(*b*) 75, (*c*) 73.12
3.17(*a*) ψ^2 values: 0.8397, 0.3263, 0.0493, 0.007442, 0.001124, 2.563×10^{-5} 5.846×10^{-7}, 1.333×10^{-8}
3.19(*a*) $4\pi r^2\psi^2$ values: 0.6595, 1.0251, 0.08963, 0.6195, 5.65×10^{-2}, 2.899×10^{-3}, 4.188×10^{-6}
3.20(*a*) ψ values: 0°, 0; 30°, 0.0665, 60°, 0.115; 90°, 0.133; 120°, 0.115; 180°, 0
3.21(*b*) 6493 kg, (*c*) 2.133×10^6 l **3.22**(*b*) 3.34×10^{23} molecules (*c*) 31.08 g

CHAPTER 4

4.7(*a*) ns^2, np^1, (*b*) ns^2, np^2, (*c*) ns^2, np^4, (*d*) ns^2, np^6 **4.13** eTl_2O_3
4.16 3.12 lb **4.17**(*a*) $FeBr_3$ **4.19**(*b*) C_2H_6O, (*c*) C_2H_6O

CHAPTER 5

5.6(*a*) pyramidal, (*b*) angular, (*d*) tetrahedral, (*e*) tetrahedral, (*f*) tetrahedral **5.9** mp ≈ -100°C, bp ≈ -75°C

CHAPTER 6

6.2(*a*) 1.006×10^6 dyn/cm², (*b*) 1028 g/cm², (*c*) 2100 lb/ft², (*d*) 29.72 in. Hg, (*e*) 755 mm Hg = 755 Torr, (*f*) 0.993 atm
6.5 0.667 atm **6.7** 395 ml **6.8** 2.32 l **6.9** 35.3 l
6.10 139.7 ml **6.11**(*a*) $P_{He} = 0.667$ atm, $P_{N_2} = 1.33$ atm
6.13(*c*) 0.0053 mol O_2 **6.14** 0.0933 atm
6.15(*b*) 0.00845 mol N_2, 0.00131 mol benzene **6.16**(*b*) 5.42 atm
6.17(*b*) 2.80 l **6.18**(*c*) 1.27×10^{26} molecules
6.22(*a*) $\bar{v} = 24.66$ km/s; $\overline{v^2} = 1104$ km²/s²; $\bar{v}^2 = 608.1$ km²/s²; $v_{rms} = 33.23$ cm/s
6.24(*a*) $v_{rms,He} = 13.04 \times 10^4$ cm/s, (*b*) 1.35×10^{-21} cal/molecule, 815 cal/mol
6.25 1.0043 **6.26** 38.95 g/mol **6.27** 1.25 g/l

6.29(*a*) $n = 2.46 \times 10^{22}$ molecules
6.31(*a*) 98.4 atm, (*b*) 100.4 atm

CHAPTER 7

7.5(*c*) Cu: 1.27 Å, Rb: 2.43 Å **7.6**(*b*) 19.4 g/cm^3 **7.7** 6.06×10^{23} atoms
7.9(*b*) 4 atoms **7.23**(*b*) 1.529 mol, (*c*) 37.86 l
7.24(*b*) 27,600 g, (*c*) 34,400 g

CHAPTER 8

8.3(*b*) Yes, because resulting vapor pressure of H_2O is less than 26.5 Torr **8.4** 10,099 cal/mol **8.5** 7.05 atm
8.11 12,185 cal/mol

CHAPTER 9

9.6(*a*) 0.8224 *m*, 0.0146X **9.8**(*a*) 10.20 *m*; for solute, $X = 0.1552$; 7.143 M
9.12(*a*) 1.68×10^{-6}, (*b*) 5.111×10^{-4} M, (*c*) 1.144×10^{-5} l (*d*) $k_{H_2} = 1.26 \times 10^{-6}$
9.17 0.658 atm **9.23** 23.4 Torr **9.24** 184.2 g/mol **9.27** 51.8 g/mol
9.29 −10.1°C **9.30** 14.6% **9.31** −9.53°C
9.35(*a*) $C_2H_3O_2$, (*b*) mol wt 118 g, (*c*) $C_4H_6O_4$
9.38 Br 171 Torr, Cl 615.3 Torr
9.43(*a*) 5.12×10^{-8} ml, 4.61×10^5 Å, (*b*) 6.97×10^{-11} mol, 4.20×10^{13} molecules/globule
9.45(*b*) $\pi = 13.2$ Torr

CHAPTER 10

10.4(*a*) 121 cal, (*b*) 1379 cal, (*c*) $mgh = 0.3569$ cal
10.5(*a*) 48.87 l, (*b*) 24.44 l, (*c*) −592 cal, (*d*) 149 K, (*e*) −1508 cal
10.7(*a*) 276 cal **10.8** 22,000 cal **10.9** = 40°C
10.11(*a*) 0.0556, (*b*) 115 g/mol **10.12**(*a*) 1184 cal, (*b*) 0, (*c*) 1930 cal/5 mol, (*d*) 0
10.13(*a*) (1) 185 cal, (*b*) (1) 1300 cal, (*c*) (1) 310 cal, (*d*) (1) 926 cal
10.14(*a*) 753 cal/g, (*b*) 88,100 cal/mol, (*c*) $q = \Delta U$, (*d*) −86,100 cal/mol
10.21(*a*) −66.06 kcal/mol **10.22**(*f*) −191.76 kcal, (*g*) −103.65 kcal
10.23(*f*) −191.64 kcal, (*g*) −93.92 kcal **10.26**(*f*) −1.22 e.u., (*g*) −27.93 e.u.
10.29(*a*) 24.4 cal/mol-K, (*b*) −9.41 cal/mol-K **10.30**(*f*) −191 kcal, (*g*) −95.3 kcal

CHAPTER 11

11.10(*a*) 0.125, (*d*) 0.00287 **11.11**(*b*) 0.625 **11.12**(*b*) 10.0
11.13(*b*) 16.0, (*c*) 16.0 **11.15**(*b*) 2.55
11.17(*a*) 14,076 cal, (*b*) 293 at 45°C **11.19**(*c*) 0.894
11.20(*a*) 2649 cal, (*c*) −4844 and −6023 cal, (*d*) −6561 cal
11.21(*b*) 630.4 cal, (*c*) 1160 cal at 25°C, (*d*) 1.78 cal/mol-K at 25°C

CHAPTER 12

12.6 (Only the coefficients for reactants and products are given.)

(*a*) 1, 4 → 1, 2, 2
(*c*) 1, 8 → $\frac{5}{2}$, 1, 1, 4
(*e*) $\frac{3}{2}$, 2, $\frac{7}{2}$ → 3, 2
(*g*) 1, 14 → 2, 1, 10, 6
(*i*) 2, 5, 3 → 2, 5, 1, 8
(*k*) 1, 14, 3 → 3, 2, 7
(*m*) 1, 3, 2 → 3, 1, 1

12.7(*a*) 63 g, 31.8 g, (*c*) 31.6 g, 80.9 g, (*e*) 21 g, 49.5 g, (*g*) 63 g, 15.9 g
12.9(*a*) 0.793 *N*, (*c*) 2.45 *N*, (*e*) 0.235 *N*
12.10(*b*) 2.24 $\mathscr{F}$
12.11(*b*) 4.03 g, (*c*) 0.21 l, (*d*) 0.0627 *N*
12.14(*c*) 0.0310 l at each anode
12.16 1.02 V
12.18 Eq. (1): (*b*) 0.359 V, (*c*) −8.28 kcal
Eq. (2): (*b*) 1.06 V, (*c*) −24.4 kcal
12.20(*c*) 1.035 V, (*d*) 3.63×10^{17}, (*e*) 0.995 V, (*f*) −22.9 kcal

CHAPTER 13

13.9(*a*) 2.2 to 5.5, (*b*) 6.3 to 8.3, (*c*) 7.6 to 10.1
13.10 0.1115 *N* **13.13** 0.0155 *M*, 6.45×10^{-13} *M*, 1.81
13.15 6.78×10^{-10} **13.16** 2.51×10^{-3} *M*, 2.60, 6.3×10^{-5}
13.17(*a*) 0.424%, 4.24×10^{-3} *M*
13.18 4.8, 9.2 **13.19** 9.68, 4.32 **13.22**(*c*) pH = pK_i, (*d*) pOH = pK_i
13.23 9.60, 9.45 **13.24** 4.05, 9.95, 8.89×10^{-5} *M*, 2.0×10^{-15} *M*
13.26 8.34 **13.27**(*a*) 1.64×10^{-5} **13.28**(*a*) 0.00182 g/l
13.29 AgBr **13.32**(*b*) 11.238 **13.33**(*c*) 10^{-6}, (*e*) 0.16 *M*
13.34 12.1 kcal

CHAPTER 14

14.3 HCl, 412 ku; HAc, 16.3 ku; KCl, 114 ku
14.7 $(0.8 \pm 0.1) \times 10^{-5}$ **14.8** 1.4×10^{-3}
14.12 8.02°, 2.24°, 22.1 Torr **14.14**(*a*) 0.10, (*c*) 1.50, (*e*) 0.50
14.15(*a*) 0.89 **14.16**(*a*) 0.244 atm, (*b*) 0.227 atm

CHAPTER 15

15.2(*a*) 3 isomers, (*b*) 8 isomers
15.4 Both are cyclic
15.6(*a*) 2-Methyl-4-ethylpentane, (*c*) 2,3-dimethyloctane

15.8 6 isomers (includes cis-trans isomers)
15.9(*a*) 4-Methyl-l-pentene, (*c*) 1,3-pentadiene
15.11(*b*) 1-Hexyne **15.12** 7 isomers **15.16** Alkane
15.18(*a*) $CH_3CHClCH_2Cl$, (*c*) CO_2, H_2O, (*e*) $CH_3—CH_2—CH_2Cl$, $CH_3—CHCl—CH_3$, (*g*) $CHBr_2—CHBr_2$, (*i*) $CH_3CBr_2CH_3$
15.20(*b*) Chloroethene
15.26(*c*) Meta or 3-chlorotoluene, (*e*) 1,3,5-trichlorobenzene, (*g*) 2-nitrophenol
15.28 1,2-Dichlorobenzene
15.29(*a*) 4-Nitroaniline, (*b*) 1,4-dinitrobenzene, (*c*) 1,3-dimethylbenzene
15.30(*b*) 2-Methyl-3-pentanol
15.34(*a*) Isopropyl acetate, (*c*) *n*-butyl hexanoate, (*e*) ethylammonium benzoate, (*g*) 2-aminohexane, (*i*) sodium benzoate
15.36(*a*) Formic or methanoic acid, (*c*) 2-methylhexanoic acid, (*e*) 3-chloropentanoic acid
15.38(*a*) Isopropyl butanoate, (*d*) *N*-methylacetamide

CHAPTER 16

16.1(*a*) DNA replication, mRNA synthesis, (*b*) protein synthesis, (*c*) photosynthesis, (*d*) respiration (ATP synthesis)
16.3 A forms hydrogen bonds only to T
16.5 $360°/34\ Å = 10.6°/Å$
16.12 Phe-Phe, Pro-Glu(NH_2)-Ser, Gly-Asp-Thr-Leu
16.15 Uracil and ribose in RNA
16.17(*a*) 600 kcal, (*b*) 67%
16.18(*b*) CO_2, H_2O, chlorophyll

CHAPTER 17

17.3 Rate $= -\Delta[H_2]/\Delta t$, rate $= -\Delta[O_2]/\Delta t$, rate $= +\Delta[H_2O]/\Delta t$
17.4 $A + B \xrightarrow{\text{fast}} \text{intermediate} \xrightarrow{\text{slow}} C$
17.6(*b*) Rate $= k[A]^0[B]^1$
17.10(*a*) 5.49×10^3 s, (*b*) 1.65×10^4 s
17.11(*a*) $[A] = 0.202$, $[B] = 0.098$, (*b*) an infinite time
17.13(*a*) 1.09×10^4 s, (*b*) 16,560
17.14(*a*) $d[A]/dt = -k$, (*b*) $d[A]/dt = -k[A][B][C]$
17.20(*a*) $k_0 = 0.003$ mol/l-min
17.24 Rate approximately doubles every 10°; $k_1 = 12.28 \times 10^{-4}\ s^{-1}$ at 55°C; $E_a = 14{,}400$ cal
17.26(*a*) $k_0 = 0.005$ atm/min, (*b*) $k_0 = 0.102$ atm/min (50°C)
17.28 $E_a = 22.8$ kcal/mol, $\log A = 14.6$

CHAPTER 18

18.2 Electrical conductivity, ionization potentials, electronegativity, reducing or oxidizing power, acidity of hydroxides, etc.

18.4 $HCO_3^- + NaOH \rightarrow Na^+ \; CO_3^{2-} + HOH$

$HCO_3^- + HCl \rightarrow H_2CO_3 + Cl^-$

18.6 (*a*) $3Sr + N_2 \rightarrow Sr_3N_2$

(*c*) $Sr + Br_2 \rightarrow SrBr_2$

(*e*) $Sr + NH_3 \rightarrow Sr(NH_2)_2 + H_2$

18.8 Li → Mg, Be → Al, B → Si; similarity in atomic radius

18.10 The BH_3 molecule is electron deficient

18.12 Iron, aluminum, copper

18.14 Burn a weighed sample and react gas with known volume of standard base. The amount of base neutralized is proportional to the amount of CO_2 formed

18.18 Hint: Si and C are in same family

18.20 C *sp* hybrid

18.26 $H_3PO_4 + xsNaOH \rightarrow Na_3PO_4$

$H_3PO_3 + xsNaOH \rightarrow Na_2HPO_3$

18.31(*b*) 5.56×10^4 l air

18.33(*a*) $H \overset{\circ\circ}{\underset{\circ\circ}{\times O :}} \overset{\circ\circ}{\underset{\circ\circ}{O \times}} H$

18.36 Hint: Peroxide has oxygen in −1 state

18.39 $AgBr(s) + 2S_2O_3^{2-}(aq) \rightarrow Ag(S_2O_3)_2^{3-}(aq) + Br^-(aq)$

18.41 Hint: Compare electronegativities

18.45 See Sec. 13.14

CHAPTER 19

19.3 Fe(II) $3d^6$, $4s^0$; 4 unpaired electrons

Ti(III) $3d^1$, $4s^0$; 1 unpaired electron

19.5(*a*) Properties of lanthanides are similar because radii are similar

19.9(*b*) $Na[CrCl_4(H_2O)_2]$; $Na_2[CrCl_5(H_2O)]$. The second compound has a higher conductance

19.14 Hint: There is only one possible structure of CH_2Cl_2 which is tetrahedral

19.19(*a*) Hexaaquotitanium(III) ion, (*c*) trinitrorhodium(VI) ion, (*e*) potassium pentaamidoaquonickel(II)

19.22 $[Cu(CN)_4]^{3-}$ is sp^3 and diamagnetic, $[Ni(CN)_4]^{2-}$ is dsp^2 and diamagnetic

19.23(*a*) d^2sp^3 from 3*d*, 4*s*, 4*p*, (*b*) pairing up in the 3*d* orbital reduces paramagnetism

19.25 Transitions from t_{2g} to e_g requires energy approximately equal to the energy of a photon of visible light

19.28(*a*) Hint: H_2O is a weak field ligand and CN^- is a strong field ligand

19.32(*a*) For $[Co(H_2O)_6]^{3+}$: the six σ-bonding orbitals and π_{xy} are filled; π_{xz}, π_{yz}, $\sigma^*_{x^2-y^2}$ and $\sigma^*_{z^2}$ are half-filled

CHAPTER 20

20.4 226 g

20.8(*b*) $\frac{1}{4}$

20.10 1.26 yr **20.12** 4460 yr

20.14(*a*) $^{252}_{100}Fm \rightarrow {}^{248}_{98}Cf + {}^{4}_{2}He$

(*c*) $^{138}_{59}Pr \rightarrow {}^{138}_{58}Ce + {}^{0}_{+1}e$

20.17 1.8×10^{10} cm/s [Use Eq. (20.9)]

20.20 $^{235}_{92}U + {}^{1}_{0}n \rightarrow {}^{141}_{54}Xe + {}^{90}_{38}Sr + 5{}^{1}_{0}n$

20.22 4.10×10^{12} cal/mol

CHAPTER 21

21.3(*a*) 1060 ton CO, (*b*) 8.97×10^5 kcal

21.4 8.3×10^7 g = 91.3 ton **21.6** 5.15×10^6 l

21.11(*a*) 5.52×10^6 ton gasoline, (*b*) 1.53×10^7 ton CO_2, 6.65×10^5 ton CO, 7.53×10^6 ton H_2O, and 2.21×10^5 ton gasoline

21.16 0.0428 mol **21.21**(*a*) 4.8×10^{11} gal, 2×10^{10} lb **21.23** 1.04 acres

INDEX

INDEX

LOGARITHMS OF NUMBERS

No.	0	1	2	3	4	5	6	7	8	9
10	0000	0043	0086	0128	0170	0212	0253	0294	0334	0374
11	0414	0453	0492	0531	0569	0607	0645	0682	0719	0755
12	0792	0828	0864	0899	0934	0969	1004	1038	1072	1106
13	1139	1173	1206	1239	1271	1303	1335	1367	1399	1430
14	1461	1492	1523	1553	1584	1614	1644	1673	1703	1732
15	1761	1790	1818	1847	1875	1903	1931	1959	1987	2014
16	2041	2068	2095	2122	2148	2175	2201	2227	2253	2279
17	2304	2330	2355	2380	2405	2430	2455	2480	2504	2529
18	2553	2577	2601	2626	2648	2672	2695	2718	2742	2765
19	2788	2810	2833	2856	2878	2900	2923	2945	2967	2989
20	3010	3032	3054	3075	3096	3118	3139	3160	3181	3201
21	3222	3243	3263	3284	3304	3324	3345	3365	3385	3404
22	3423	3444	3464	3483	3502	3522	3541	3560	3579	3598
23	3617	3636	3655	3674	3692	3711	3729	3747	3766	3784
24	3802	3820	3838	3856	3874	3892	3909	3927	3945	3962
25	3979	3997	4014	4031	4048	4065	4082	4099	4116	4133
26	4150	4166	4183	4200	4216	4232	4249	4265	4281	4298
27	4314	4330	4346	4362	4378	4393	4409	4425	4440	4456
28	4472	4487	4502	4518	4533	4548	4564	4579	4594	4609
29	4624	4639	4654	4669	4683	4698	4713	4728	4742	4757
30	4771	4786	4800	4814	4829	4843	4857	4871	4886	4900
31	4914	4928	4942	4955	4969	4983	4997	5011	5024	5038
32	5051	5065	5079	5092	5105	5119	5132	5145	5159	5172
33	5185	5198	5211	5224	5237	5250	5263	5276	5289	5302
34	5315	5328	5340	5353	5366	5378	5391	5403	5416	5428
35	5441	5453	5465	5478	5490	5502	5514	5527	5539	5551
36	5563	5575	5587	5599	5611	5623	5635	5647	5658	5670
37	5682	5694	5705	5717	5729	5740	5752	5763	5775	5786
38	5798	5809	5821	5832	5843	5855	5866	5877	5888	5899
39	5911	5922	5933	5944	5955	5966	5977	5988	5999	6010
40	6021	6031	6042	6053	6064	6075	6085	6096	6107	6117
41	6128	6138	6149	6160	6170	6180	6191	6201	6212	6222
42	6232	6243	6253	6263	6274	6284	6294	6304	6314	6325
43	6335	6345	6355	6365	6375	6385	6395	6405	6415	6425
44	6435	6444	6454	6464	6474	6484	6493	6503	6513	6522
45	6532	6542	6551	6561	6571	6580	6590	6599	6609	6618
46	6628	6637	6646	6656	6665	6675	6684	6693	6702	6712
47	6721	6730	6739	6749	6758	6767	6776	6785	6794	6803
48	6812	6821	6830	6839	6848	6857	6866	6875	6884	6893
49	6902	6911	6920	6928	6937	6946	6955	6964	6972	6981
50	6990	6998	7007	7016	7024	7033	7042	7050	7059	7067
51	7076	7084	7093	7101	7110	7118	7126	7135	7143	7152
52	7160	7168	7177	7185	7193	7202	7210	7218	7226	7235
53	7243	7251	7259	7267	7275	7284	7292	7300	7308	7316